MACMILLAN
DICTIONARY
OF
CHEMISTRY

MACMILLAN
DICTIONARY
OF
CHEMISTRY

D. B. Hibbert
A. M. James

MACMILLAN
REFERENCE
BOOKS

First published 1987 by
THE MACMILLAN PRESS LTD
London and Basingstoke

Associated companies in Auckland, Delhi, Dublin, Gaborone,
Hamburg, Harare, Hong Kong, Johannesburg, Kuala Lumpur,
Lagos, Manzini, Melbourne, Mexico City, Nairobi, New York,
Singapore, Tokyo.

British Library Cataloguing in Publication Data
Hibbert, D.B.
    Macmillan dictionary of chemistry.
    1. Chemistry—Dictionaries
    I. Title   II. James, A.M.
    540′.3′.21   QD4
    ISBN 0-333-39081-4
    ISBN 0-333-43471-4 Pbk

Typeset by Florencetype Ltd, Kewstoke, Avon
Printed in Great Britain

Contents

Preface

This is a dictionary of chemical terms and concepts and is intended as a reference volume for physical and biological scientists in education, research and industry. The aim is to provide the reader with a brief and lucid definition of each headword, including such relevant information as equations, formulae, preparation and properties of compounds, applications and use. SI units are used throughout, and tables of conversion factors between these and other units are provided.

In contrast to rival works, more balanced consideration is given to topics of a physical chemical nature. To achieve this, within the space available, the extensive catalogue of inorganic and organic compounds has largely been replaced by more general entries. In these the general properties of the compounds within their natural groupings are considered either in the form of tables or reaction schemes. Extensive physical data of compounds are not included since these are available in standard reference books.

It is hoped that this Dictionary, with over 5200 entries, some with illustrations, will be of use to science students, research workers in cognate disciplines and also to lay persons. Further it will be an aid for student revision.

We appreciate the many useful discussions we have had with our colleagues. Our sincere thanks are due to our wives Marion and Mary (Dr. M.P. Lord) for their understanding, patience, help and encouragement during the preparation of the manuscript.

We thank the staff of Macmillan Reference Books for their help and encouragement. We are particularly indebted to Dr. Ann Ralph for her practical help, advice and useful discussions during the production and editing of the manuscript.

DBH
AMJ

Notes on Use

Entries are arranged strictly in alphabetical order (numerals and Greek letters used as prefixes as in the naming of compounds are ignored). The IUPAC convention is used for the systematic nomenclature of inorganic and organic compounds, trivial and other names in current use are fully cross-referenced.

Common symbols and abbreviations that occur throughout the dictionary are used without definition. Lists of these are included at the start of each letter (Greek letters are classified under the appropriate letter (e.g., γ, gamma, will be found under G), where they are either defined or a reference given for further information. Abbreviations for common organic radicals (e.g., Et, Ph) will be found in the appropriate place in the main text.

In the text, words in small capitals indicate a reference to another entry which will provide further information and could be of help; in some instances the reference is in parentheses.

Formulae for simple compounds are given in linear form; structural, conformational and stereochemical formulae are included where appropriate. Wherever possible, compounds have been grouped together (e.g., alkanes, alkenes, alkaloids, proteins, terpenes); there is full cross-referencing between individual compounds and the group. Under the group heading the general methods of preparation, properties, similarities and differences are highlighted. In these instances, only the most important compounds of the group have separate entries (e.g., the first three or four members of a homologous series).

Entries for elements include an outline of their occurrence, isolation, their general physical and chemical properties and a brief account of the chemistry of their inorganic and organic compounds. The general properties of inorganic salts are grouped under such entries as halides, nitrates, sulphates. Only the more important inorganic compounds have a separate entry. Symbols for the elements are given in Table 4, and the full Periodic Table appears on the final page.

Commonly Used Abbreviations

aq	aqueous	$\lg x$	common logarithm of x (to base 10)
bp	boiling point		
ES	electronic structure	$\lvert x \rvert$	absolute value of x
fp	freezing point	$p!$	factorial $p = 1 \times 2 \times 3 \times \dots$ $\times (p-1) \times p$
vp	vapour pressure		
		$e^x, \exp(x)$	exponential of x
		$\propto$	proportional to
		∞	infinity

superscripts

$\ominus$	indicating standard value of a property
0	indicating value of a property at infinite dilution

subscripts

A, B	referring to species A, B
e	value of property at equilibrium
i	referring to a typical species i
P, V, T, S	indicating constant pressure, volume, temperature, entropy
$+, -$	referring to positive, negative ion
$1, 2$	referring to different systems or states of system
g, l, s (or c)	referring to gaseous, liquid, or solid (crystalline) state

mathematical symbols

$\approx$	approximately equal to
$\simeq$	asymptotically equal to
$\sum_i$	sum of i terms
$\prod_i$	product of i terms
$\ln x$	natural logarithm of x

greek alphabet

A	α	alpha
B	β	beta
Γ	γ	gamma
Δ	δ	delta
E	ε	epsilon
Z	ζ	zeta
H	η	eta
Θ	θ	theta
I	ι	iota
K	κ	kappa
Λ	λ	lambda
M	μ	mu
N	ν	nu
Ξ	ξ	xi
O	o	omicron
Π	π	pi
P	ρ	rho
Σ	σ	sigma
T	τ	tau
Y	υ	upsilon
Φ	ϕ	phi
X	χ	chi
Ψ	ψ	psi
Ω	ω	omega

A

A. *See* AMPERE.

Å. *See* ANGSTRÖM.

A. *See* ABSORBANCE, ACTIVITY, ELECTRON AFFINITY, HELMHOLTZ FUNCTION, NUCLEON NUMBER.

A_r. *See* RELATIVE ATOMIC MASS.

a. *See* ACTIVITY.

α. Degree of association or dissociation (*see* DISSOCIATION); POLARIZABILITY.

α_m. *See* SPECIFIC OPTICAL ROTATORY POWER.

$[\alpha]_D^t$. *See* SPECIFIC ROTATION.

α, β, γ, ... group. The carbon chain away from a particular feature in a molecule is labelled α, β, γ, ..., ω (at the end of the chain). For example, the carbon chain in β-hydroxyketones is –C(OH)–C–C(O)–, and in α-chloroacids is –CCl–COOH. A hydrogen atom α to a double bond (C=C, C=O) is usually activated because of the possibilities of conjugation in the cation formed on its removal.

α-particle. *See* ALPHA-PARTICLE.

AAS (atomic absorption spectroscopy). *See* ATOMIC SPECTROSCOPY.

abherent. Substance that reduces or prevents ADHESION between two surfaces. Examples include dry powders (starch, talc), suspensions (bentonite–water) and soap solutions. Fats and oils are used as abherents in the food industry, fluorocarbon resins (e.g., Teflon, *see* POLYTETRAFLUOROETHENE) are used to make 'non-stick' pans.

abietic acid. *See* TERPENES.

***ab initio* calculation.** Solution of the SCHRÖDINGER EQUATION by direct computation which uses no experimentally determined values of integrals.

abrasive. Extremely hard substance used to grind or polish surfaces of metal, glass, wood or plastic (e.g., sand, emery, silicon carbide, alumina, boron carbide, boron nitride, diamond). The smoothness of the finish depends on the hardness (*see* MOH'S SCALE) and particle size of the abrasive and the mechanical force exerted.

absolute rate theory (activated complex theory). Theory of the rates of reactions that assumes an equilibrium between reactants and the ACTIVATED COMPLEX (i.e. $A + B \rightleftharpoons AB^\ddagger \rightarrow$ products). The equilibrium constant is given by statistical thermodynamics (*see* STATISTICAL MECHANICS) using the partition functions of A, B and $AB^\ddagger$, and it is assumed that the activated complex differs from the reactants in only one loose vibrational mode which corresponds to the REACTION COORDINATE. The equation for the second-order RATE CONSTANT (k_2) for $A + B \rightarrow$ products becomes

$$k_2 = \frac{kT}{h} \frac{Q_\ddagger}{Q_A Q_B} \exp\left(- E_0/RT\right)$$

or

$$k_2 = \frac{kT}{h} \exp\left(- \Delta G^\ddagger/RT\right)$$

where Q_A, Q_B are the partition functions of A and B and $Q_\ddagger$ the partition function of the activated complex minus the one vibrational mode. E_0 is the difference in zero point energies between reactants and activated complex and $\Delta G^\ddagger$ is the free energy of activation. *See also* EYRING EQUATION, RRKM THEORY.

absolute temperature. Fundamental TEM-
PERATURE SCALE used in theoretical physics
and chemistry.

absorbance (optical density, A).   Measure
of the amount of radiation absorbed by a
sample

$$A = \lg (I_0/I)$$

where I is the intensity of radiation passed
through the sample and I_0 is the intensity
incident on the sample. *See also* BEER–
LAMBERT LAW, TRANSMITTANCE.

absorption.   Incorporation into the body
of a substance. Thus light may be absorbed
by a compound, or a gas may be absorbed
by a liquid.

absorption coefficient of a gas.   Volume (in
cm^3) of a gas at 0°C and 1 atm pressure
that will dissolve in 1 cm^3 of a liquid; for
water, absorption coefficients are N_2 0.024,
O_2 0.049, C_2H_4 0.25, CO_2 1.71, H_2S 4.7,
SO_2 79.8, HCl 506, NH_3 1300.

abundance.   (1) Relative amount (percentage by weight) of an element or substance in the earth's crust (including the
atmosphere and oceans). The abundance of
the elements is: O 49.2, Si 25.7, Al 7.5, Fe
4.9, Ca 3.4, Na 2.6, K 2.4, Mg 1.9, H 0.9, Ti
0.6, Cl 0.2, P 0.1, Mn 0.1, C 0.09, S 0.05,
Ba 0.05, all others 0.51; of inorganic compounds (excluding water): SiO_2 55, Al_2O_3
15, $CaCO_3$ 8.8, MgO 1.6, Na_2O 1.6, K_2O
1.6. (2) Relative number of atoms of a
particular isotope, known as the natural
abundance (before any enrichment): the
abundance of uranium-235 in natural uranium is 0.71 percent.

Ac.   Actinium (*see* ACTINIDE ELEMENTS).

acac (acetylacetonato group, 2,4-pentane-
dionato group).   *See* DIKETONES.

accelerator.   (1) Machine (e.g., CYCLO-
TRON) for increasing the kinetic energy of
charged particles. It is used for research in
nuclear and particle physics. (2) Substance
used to increase the rate at which thermosetting resins harden or cure by accelerating
the desired cross-linking reaction in polymers. *See also* VULCANIZATION.

acceptor.   Species that acts as a Lewis acid
(*see* LEWIS ACIDS AND BASES) by accepting
an electron pair from a donor (Lewis base)
into a vacant orbital to form a COORDINATE
BOND. *See also* ACIDS AND BASES.

accumulator.   Rechargeable or secondary
BATTERY (e.g., LEAD–ACID BATTERY).

acenaphthene (1,2-dihydroacenaphthylene,
$C_{12}H_{10}$).   POLYCYCLIC   HYDROCARBON
(structure XXXV), occurring in coal tar;
obtained   when   1-ethylnaphthalene   is
passed through a red-hot tube (hydrogen is
liberated). It is a white solid (mp 96°C) that
undergoes substitution most readily in the
5- and 6-positions (e.g., bromination, nitration and sulphonation give the corresponding 5-derivative). Oxidation using CrO_3
gives   naphthalene-1,8-dicarboxylic   acid
and PbO_2 the polycyclic hydrocarbon acenaphthylene (structure XXXIV, $C_{12}H_8$).

acenaphthylene.   *See* POLYCYCLIC HYDRO-
CARBONS (structure XXXIV).

acesulphame K.   *See* SWEETENING AGENTS.

acetaldehyde.   *See* ETHANAL.

acetals.   Diethers formed from ALDE-
HYDES and ALCOHOLS in the presence of, for
example, dry HCl.

$$RCHO + R'OH \rightarrow RCH (OR')_2$$

See also KETALS.

acetamide.   *See* ETHANAMIDE.

acetamido group (acetamino group, acetylamino group).   The group CH_3CONH-.

4-acetamidophenol (paracetamol, 4-CH_3-
$CONHC_6H_4OH$).   White powder (mp
170°C), prepared by reacting 4-amino-
phenol with ethanoic anhydride. It is slightly soluble in cold water, soluble in methanol
and dimethylmethanamide (DMF). Paracetamol is used as an analgesic and antipyretic, and in the manufacture of dyes and
photographic chemicals.

acetamino group.   *See* ACETAMIDO GROUP.

acetanilide ($C_6H_5NHCOCH_3$).   White,
crystalline solid (mp 114°C); the ANILIDE of

aniline, prepared by the ethanoylation of aniline. It is hydrolyzed to ANILINE by dilute acids and alkalis. It is an intermediate in the preparation of derivatives of aniline, where it is required to protect the amino group.

acetates.   *See* ETHANOATES.

acetenyl group.   *See* ETHYNYL GROUP.

acetic acid.   *See* ETHANOIC ACID.

acetic anhydride.   *See* ETHANOIC ANHYDRIDE.

acetic ester.   *See* ETHYL ETHANOATE.

acetins.   ETHANOATES of GLYCEROL. The five possible acetins (1- and 2-mono-, 1,2- and 1,3-di-, and 1,2,3-triethanoate) are prepared as mixtures from glycerol, ethanoic acid and sulphuric acid. They are used commercially as solvents and PLASTICIZERS.

acetoacetic acid (3-oxobutanoic acid).   *See* OXOCARBOXYLIC ACIDS.

acetoacetic ester.   *See* ETHYL 3-OXOBUTANOATE.

acetogenins.   ETHANOIC ACID in the form of acetyl-coenzyme A on condensation with malonyl-coenzyme A gives rise to long, linear β-diketone chains; these undergo cyclic ALDOL CONDENSATIONS for the formation of a wide range of natural products, the acetogenins.

acetoin   (3-hydroxy-2-butanone).   *See* ACYLOINS.

acetol.   *See* HYDROXYPROPANONE.

acetolysis.   Reaction in which the ETHANOYL GROUP (acetyl group) is removed from a molecule. This is usually accomplished by heating with an aqueous or alcoholic solution of a base.

acetone.   *See* PROPANONE.

acetone alcohol.   *See* HYDROXYPROPANONE.

acetonedicarboxylic acid (3-oxoglutaric acid, β-ketoglutaric acid).   *See* 3-OXOPENTANEDIOIC ACID.

acetonitrile.   *See* ETHANENITRILE.

acetonylacetone (2,5-hexanedione).   *See* DIKETONES.

acetonyl group.   The group $CH_3COCH_2–$. *See also* DIKETONES.

acetophenone   (methyl   phenyl   ketone, $C_6H_5COCH_3$).   KETONE (mp 20°C), prepared by the FRIEDEL–CRAFTS REACTION between benzene and ethanoyl chloride. It is manufactured by the catalytic oxidation $(Mn(OOCCH_3)_2)$ of ethylbenzene, and is used as a hypnotic, in perfumes and as a solvent. Acetophenone undergoes the BAEYER–VILLIGER OXIDATION to phenyl ethanoate and reactions typical of an aromatic KETONE.

2-acetoxybenzoic acid.   *See* ASPIRIN.

acetoxy group.   The group $CH_3COO–$. *See also* CARBOXYL GROUP.

acetylacetonato group (acac, 2,4-pentanedionato group).   *See* DIKETONES.

acetylacetone (2,4-pentanedione).   *See* DIKETONES.

acetylamino   group.   *See*   ACETAMIDO GROUP.

acetylation.   *See* ETHANOYLATION.

acetyl   chloride.   *See* ETHANOYL CHLORIDE.

acetylcholine   chloride $((CH_3)_3\overset{+}{N}CH_2CH_2$- $OOCCH_3\,Cl^-)$.   Ethanoyl   (acetyl)   ester of choline (*see* VITAMINS) which acts as a neurotransmitter. It is rapidly hydrolyzed by acetylcholine esterase, and causes dilation of the arteries.

acetyl-coenzyme A.   *See* COENZYMES.

acetylene.   *See* ETHYNE.

acetylene complexes.   *See* CARBON-TO-METAL BOND.

acetylenedicarboxylic acid.   *See* 2-BUTYNEDIOIC ACID.

acetylene dichloride (1,2-dichloroethene). *See* DICHLOROETHENES.

acetylene tetrachloride. *See* 1,1,2,2-TETRACHLOROETHANE.

acetyl group. *See* ETHANOYL GROUP.

acetylides. Salts containing $[C_2]^{2-}$ (*see* CARBIDES) or C_2R^- (R = alkyl or H). Acetylides are named as derivatives of ETHYNE (acetylene) and are prepared by the action of ethyne on electropositive metals or metal compounds, or by the direct reaction of the elements at high temperatures. All acetylides are reactive, especially those of transition elements which explode when dry. They are hydrolyzed by water to ethyne. Calcium carbide (CaC_2) is the most important acetylide. Copper carbide (Cu_2C_2) is used in ETHYNYLATION and high-pressure reactions.

acetylsalicylic acid. *See* ASPIRIN.

acid–base catalysis. Acceleration of reactions in acidic or basic media. The mechanism of catalysis of organic reactions (e.g., the hydrolysis of esters) is via the protonation of a group (in the example, of an OXO GROUP). On a wider scale, LEWIS ACIDS AND BASES by their ability to accept and donate electron pairs can show catalytic activity. In HETEROGENEOUS CATALYSIS, acidic sites on alumina give activity to bifunctional HYDROCRACKING catalysts.

acid–base indicator. *See* INDICATORS.

acid–base titration. *See* TITRATIONS.

acid dissociation constant (K_a). EQUILIBRIUM CONSTANT for the reaction HA $\rightleftharpoons$ $H^+ + A^-$, where HA is an acid. It may be expressed as pK_a (= $-\lg K_a$). For example, ethanoic acid has a pK_a of 4.8. K_a may be determined from acid–base TITRATIONS (pH = pK_a at half the end-point) or from conductivity measurements (*see* OSTWALD'S DILUTION LAW). *See also* BASE DISSOCIATION CONSTANT.

acid dyes. Water-soluble, anionic DYES; usually the sodium salts of carboxylic or sulphonic acids. Metallization (i.e. chela-tion to transition metal ions, especially chromium) improves the fastness of the dye. If the material is pretreated the dye is referred to as a mordant dye (*see* MORDANT).

acid exchange resin. *See* ION EXCHANGE CHROMATOGRAPHY.

acidic oxides. Oxides of non-metals that dissolve in water to give acidic solutions: for example, sulphur(VI) oxide gives sulphuric acid.

acid rain. Form of air pollution caused by the emission of SULPHUR OXIDES (and to a lesser extent NITROGEN OXIDES), which are precipitated in rain as acids. The major sources of sulphur dioxide, which ultimately falls as sulphuric acid, are coal-burning power stations, copper smelters and some brick-making plants.

acids and bases. Compounds defined according to their ability to accept or donate protons (*see* BRØNSTED–LOWRY ACIDS AND BASES) or lone pairs of electrons (*see* LEWIS ACIDS AND BASES). The reaction between an acid and a base is called neutralization, and the product is a salt of the acid. The concentration of acids may be determined by TITRATION. The strength of an acid (or base) is measured by the ACID DISSOCIATION CONSTANT (BASE DISSOCIATION CONSTANT). *See also* pH, SUPERACID.

acinitro. *See* NITROALKANES.

aconitic acid (propene-1,2,3-tricarboxylic acid). *See* 2-HYDROXYPROPANE-1,2,3-TRICARBOXYLIC ACID.

acridine ($C_{13}H_9N$). Nitrogen HETEROCYCLIC COMPOUND (structure XXXVI) that occurs in and is extracted from coal tar; planar, feebly basic molecule (mp 105–10°C). It is steam volatile, slightly soluble in hot water and soluble in ethanol and ether. It dissolves in mineral acids and reacts with alkyl and aryl halides to form quaternary ammonium (acridinium) compounds. It irritates the skin and mucous membrane. Acridine is used in the manufacture of dyes and antibacterial agents.

acriflavine.   Mixture of the hydrochlorides of 3,6-diamino-10-methylacridinium chloride (*see* ACRIDINE) and 3,6-diaminoacridine. It is an orange granular powder, soluble in water and slightly soluble in ethanol. The dark red solution fluoresces on dilution. It is used as an antiseptic with trypanocidal action.

acrolein.   *See* PROPENAL.

acrylaldehyde.   *See* PROPENAL.

acrylamide polymers.   *See* POLYMERS.

acrylate resins.   *See* 2-METHYLPROPENOIC ACID, POLYMERS, PROPENOIC ACID, RESINS.

acrylic acid.   *See* PROPENOIC ACID.

acrylonitrile.   *See* PROPENENITRILE.

acrylonitrile–butadiene rubbers.   *See* POLYMERS, RUBBER.

actinide compounds.   Oxidation state +3 is known for all elements. Later members of the series show the +2 state, and higher oxidation states are shown by earlier members (e.g., uranium(VI)) (*see* ACTINIDE ELEMENTS, Table). In aqueous solution, complexes and polymeric ions are formed. The ability of ions to hydrolyze increases with charge and decreases with ionic radius, thus americum > plutonium > neptunium > uranium and $M^{4+} > MO_2^{2+} > M^{3+} > MO_2^{+}$. Sulphates, halides, perchlorates, nitrates and sulphides are soluble, whereas fluorides and ethanedioates (oxalates) are insoluble in acid. Oxoanions MO_2^{+} and MO_2^{2+} (where M = U, Np, Pu or Am) are found, the latter ion being more stable. Binary compounds with carbon, nitrogen, silicon and sulphur are very stable

actinide elements

Element	Symbol	Z	ES [Rn]+	Common isotopes		Oxidation states[b]
				Mass number	Half-life	
Actinium	Ac	89	$6d^1 7s^2$	227[a]	21.7 yr	3
Thorium	Th	90	$6d^2 7s^2$	232[a]	1.39×10^{10} yr	(3) 4
Protactinium	Pa	91	$5f^2 6d^1 7s^2$	231[a]	3.28×10^5 yr	(3, 4) 5
Uranium	U	92	$5f^3 6d^1 7s^2$	235[a]	7.13×10^8 yr	(3, 4, 5) 6
				238[a]	4.50×10^9 yr	
Neptunium	Np	93	$5f^5 7s^2$	237	2.20×10^6 yr	(3, 4) 5 (6)
Plutonium	Pu	94	$5f^6 7s^2$	238	86.4 yr	(3) 4 (5, 6)
				239	2.44×10^4 yr	
Americium	Am	95	$5f^7 7s^2$	241	458 yr	3 (4, 5, 6)
				243	7951 yr	
Curium	Cm	96	$5f^7 6d^1 7s^2$	242	162.5 d	3 (4)
				244	17.6 d	
Berkelium	Bk	97	$5f^8 6d^1 7s^2$	249	314 d	3 (4)
				247	1×10^4 yr	
Californium	Cf	98	$5f^{10} 7s^2$	252	2.2 yr	(2) 3
Einsteinium	Es	99	$5f^{11} 7s^2$	253	20 d	(2) 3
				254	270 d	
Fermium	Fm	100	$5f^{12} 7s^2$	257	80 d	(2) 3
Mendelevium	Md	101	$5f^{13} 7s^2$	256	1.5 h	2, 3
				258	60 d	
Nobelium	No	102	$5f^{14} 7s^2$	255	3 min	2 (3)
Lawrencium	Lr	103	$5f^{14} 6d^1 7s^2$	260	3 min	3

[a] Naturally occurring.
[b] Values in parentheses are less important oxidation states.

at high temperatures. NON-STOICHIO-
METRIC COMPOUNDS—oxides and hydrides
—are found (e.g., $UO_{2.25}$, $NpH_{2.7}$). *See also*
URANIUM, PLUTONIUM.

actinide elements. TRANSITION ELE-
MENTS—actinium ($Z = 89$) to lawrencium
($Z = 103$) inclusive—having partially filled
$5f$-orbitals (*see* Table). All are radioactive.
Elements after uranium ($Z = 92$) are artificial. The most important elements of this
series are URANIUM and PLUTONIUM, which
are used as energy sources in nuclear power
stations. The properties are similar to those
of the LANTHANIDE ELEMENTS. A steady,
although less marked, reduction in size of
the atoms and ions is seen across the series
(*see* LANTHANIDE CONTRACTION). The
energies of the orbitals $5f$, $6d$, $7s$ and $7p$ are
very close, which allows a wider range of
oxidation states and results in the formation
of complexes. Bonding may be covalent or
ionic. Separation of the actinides from each
other and of the actinides from the lanthanides is by ion exchange, which uses the
ability of the actinides to form chloride
complexes by eluting with strong lithium
chloride solution. Elution with citrate
separates the actinides themselves.

actinium (Ac).   *See* ACTINIDE ELEMENTS.

actinium decay series.   *See* RADIOACTIVE
DECAY SERIES.

actinometry.   Measurement of the amount
of radiation in a chemical compound. *See
also* ISOTOPES, RADIOACTIVITY.

actinon (An).   Obsolete name for $^{219}_{86}Rn$.
See also RADIOACTIVE DECAY SERIES.

activated adsorption.   CHEMICAL ADSORP-
TION on a solid surface that requires
ACTIVATION ENERGY. This may arise from
the transition from a physically adsorbed
molecule to one that is chemisorbed.

activated charcoal.   *See* CARBON.

activated complex.   Molecular complex of
maximum energy formed at the transition
state of a chemical reaction. The activated
complex may be distinguished from an
intermediate by its instability with respect
to the REACTION COORDINATE. A treatment

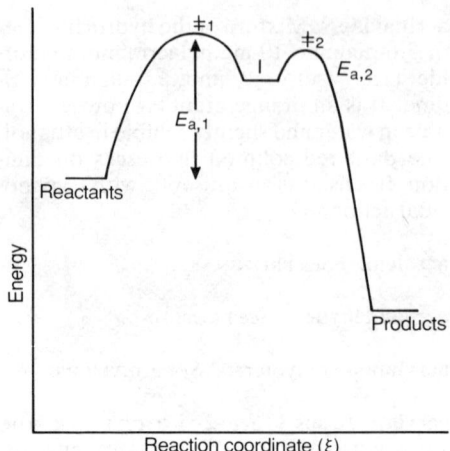

activated complex
Energy of a system during a chemical reaction: ·
$\ddagger_1$, activated complex for reactants → intermediate (I);
$\ddagger_2$, activated complex for intermediate → products.

of the formation and decay of the activated
complex leads to the EYRING EQUATION
(*see* Figure).

activated complex theory.   *See* ABSOLUTE
RATE THEORY.

activating group.   Electron-releasing group
in a benzene ring which causes the ring to be
more susceptible to further substitution.
Such groups are *ortho-*, *para-* directing (*see*
ORIENTATION IN THE BENZENE RING).

activation analysis.   Techniques in which a
neutron, ion or gamma-photon is captured
by an atom to produce a RADIONUCLIDE
which is then measured. The method is
specific, sensitive and may be applied to
very small amounts of sample. Neutron
activation analysis is most widely used. A
nuclide capturing a neutron emits a prompt
gamma-ray and forms a radioactive intermediate, the activity (A) of which is related
to the number of target atoms (N) by

$$A = Nf\sigma \left[1 - \exp\left(-0.639t/\tau\right)\right]$$

where f is the flux of neutrons, σ is the cross-
section of the nuclide for neutrons and τ is
the half-life of the intermediate. The most
sensitive method of determining the activity
is by measurement of the gamma-ray spectrum. The source of neutrons may be a
nuclear reactor, californium-252 or a small
accelerator known as a 14-MeV neutron
generator. Activation analysis by charged

particles (He^{3+} or H^+ in proton-induced X-ray emission, PIXE) is used for particular applications in which the prompt X-ray emission spectrum gives better discrimination between similar elements. Disadvantages of charged particle methods are the low flux of particles and the small penetration depth.

activation energy. Minimum energy that reactants must acquire to allow a reaction to occur (*see* ACTIVATED COMPLEX, Figure). A GIBBS FUNCTION of activation is defined from the EYRING EQUATION

$$k_2 = \kappa(kT/h) \exp(-\Delta G^{\ddagger}/RT)$$

from which the entropy and enthalpy of activation are given by

$$k_2 = \kappa(kT/h) \exp(\Delta S^{\ddagger}/R)$$
$$\exp(-\Delta H^{\ddagger}/RT)$$

For a bimolecular collision (*see* MOLECULARITY OF REACTION).

$$\Delta H^{\ddagger} = E_a - 2RT$$

See also ARRHENIUS EQUATION.

active carbon. *See* CARBON.

active mass. Old name for the ACTIVITY of a substance. *See also* MASS ACTION, LAW OF.

active site. Site on an ENZYME molecule to which the substrate is bound. The properties of the site, determined by the three-dimensional arrangement of the polypeptide chains and the amino acid sequence (*see* PROTEINS), are responsible for the nature of the interaction and hence substrate specificity.

activity. (1) In radiation, activity (A) is the number of atoms of a radioactive substance which disintegrate per unit time. *See also* RADIATION UNITS, SPECIFIC ACTIVITY. (2) In thermodynamics, activity (a_i) is a function introduced by G.N. Lewis to aid the treatment of real systems. Changes in CHEMICAL POTENTIAL can be correlated with experimentally measured quantities as follows

$$\mu_i - \mu_i^{\ominus} = RT \ln a_i = RT \ln m_i \gamma_i$$

where $\mu_i^{\ominus}$ is the chemical potential when $a_i = 1$ (i.e. pure state). Since there is no

method for determining individual ionic activities (a_+ and a_-), the mean ionic activity $a_{\pm}$ for the dissolved electrolyte is defined as

$$a_{\pm} = a_2^{1/\nu} = (a_+^{\nu+} a_-^{\nu-})^{1/\nu}$$

Activity values can be calculated from the results of any method for the measurement of ΔG: vapour pressure; solubility of sparingly soluble salts; emf of galvanic and concentration CELLS; depression of FREEZING POINT.

activity coefficient (γ_i). Ratio of the ACTIVITY to the concentration (or pressure) of the component i in the given state; in solution $\gamma_i = a_i/m_i$ and for gases $\gamma = f/p$, where f is the FUGACITY. γ_i, which approaches unity as the concentration (or pressure) approaches zero, is a measure of the departure of the system from ideal behaviour. For electrolytes, the mean ionic activity coefficient, $\gamma_{\pm} = (\gamma_+^{\nu+} \gamma_-^{\nu-})^{1/\nu} = a_{\pm}/m_{\pm}$, decreases with increasing concentration (*see* Figure) and the steepness of the slope increases with increasing valence type of the electrolyte. For a given valence type, γ_i is independent of the constituent ions at $m < 0.01$ mol kg^{-1}. In dilute solutions, activity coefficients can be calculated from the DEBYE–HÜCKEL ACTIVITY COEFFICIENT EQUATION.

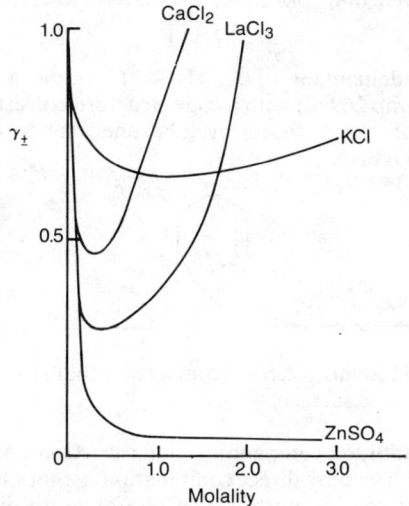

activity coefficient
Activity coefficients of electrolytes of different valence types.

acyclic compounds. Organic compounds with no rings (*see* ALIPHATIC COMPOUNDS).

acylation. Introduction of an ACYL GROUP into a molecule.

acyl group. The group RCO–, named from the CARBOXYLIC ACID (RCOOH) with the suffix –yl replacing –oic.

acyloins (1-hydroxyketones). Prepared by treating esters with sodium in an inert solvent followed by working up with acid: for example

$$2EtCOOEt \xrightarrow{2Na} \begin{array}{c} EtCO^-Na^+ \\ | \\ EtCO^-Na^+ \end{array} \xrightarrow{H^+}$$

$$\begin{array}{c} EtCOH \\ | \\ EtCOH \end{array} \longrightarrow \begin{array}{c} EtCO \\ | \\ EtCHOH \end{array}$$

Acyloins are converted to ALKANES by the CLEMMENSEN REDUCTION. 3-Hydroxy-2-butanone (acetoin) is produced by the action of yeasts on ethanol. When distilled 2,3-butanedione (*see* DIKETONES) is formed. Benzoin (PhCOCH(OH)Ph) is prepared from benzaldehyde by the action of sodium cyanide. It is oxidized to BENZIL by nitric acid and is reduced by sodium to hydrobenzoin   (PhCH(OH)CH(OH)Ph) and by zinc amalgam to STILBENE (*trans*-1,2-diphenylethene). The name benzoin is also given to a resin obtained from *Styrax benzoin. See also* CONDENSATION REACTION.

adamantane   ($C_{10}H_{16}$). HYDROCARBON (mp 269°C) with a cage structure consisting of three fused cyclohexane chairs (*see* Figure).

adamantane

ad atom. ATOM from a solid that sits on a plane surface.

addition compound (adduct). Compound formed by direct combination, generally in simple proportions, of two or more different compounds (e.g., aldehyde–bisulphite compound—RCH(OH)SO$_3$Na)—the product of an aldehyde and sodium bisulphite; the product of a diene and a dienophile, CLATHRATE COMPOUNDS.

addition polymerization.   *See* POLYMERIZATION.

addition reaction. Reaction in which a molecule adds to an unsaturated compound containing double or triple bonds (e.g., >C=C<, –C≡C–, –C≡N). Examples include the addition of bromine to an ALKENE, and of hydrogen cyanide to an ALDEHYDE. *See also* DIELS–ALDER REACTION, ELECTROPHILIC ADDITION, NUCLEOPHILIC ADDITION.

additive property. Property which, for a given system, may be represented by the sum of the corresponding properties of the components. MASS is exactly additive; MOLAR VOLUME and IONIC RADIUS are only approximately additive. *See also* COLLIGATIVE PROPERTY, CONSTITUTIVE PROPERTY.

adduct.   *See* ADDITION COMPOUND.

adenine.   *See* PURINES.

adenosine.   *See* NUCLEOSIDES.

adenosine diphosphate (ADP). Precursor of ADENOSINE TRIPHOSPHATE (ATP). It is formed on hydrolysis of ATP.

adenosine monophosphate (AMP). Formed on hydrolysis of ADENOSINE DIPHOSPHATE. It is an important structural component of NUCLEIC ACIDS, NAD$^+$, NADP$^+$ and FAD (*see* COENZYMES).

adenosine triphosphate (ATP). Primary energy source for numerous biological reactions (e.g., protein synthesis, ion transport, muscle contraction, electrical activity in nerve cells). The energy required is derived from its hydrolysis

$$ATP^{4-} + H_2O \rightarrow ADP^{3-} + P_i + H^+$$

where P_i is inorganic phosphate for which $\Delta G^{\ominus'}$ (pH 7) = -29.3 kJ mol^{-1}. The large decrease in the Gibbs function is available to power endergonic reactions. The hydrolysis of adenosine diphosphate (ADP) to adenosine monophosphate is accompanied by a similar large decrease in the Gibbs function, whereas for the hydrolysis of AMP to adenine and phosphate $\Delta G^{\ominus'} = -12.5$ kJ mol^{-1}. ATP is regenerated from AMP and ADP using the chemical energy

obtained from OXIDATIVE PHOSPHORYLA-
TION in the ELECTRON TRANSPORT CHAIN
ultimately from the oxidation of nutrients.

adenosine triphosphate

5'-adenylic acid.　See NUCLEOTIDES.

adhesion.　State in which two surfaces are
held together by interfacial forces (e.g.,
valence forces, interlocking forces). The
work of adhesion ($w_{A,B}$), the work required
to pull apart two immiscible liquids A and B
of unit cross-sectional area and so create
two liquid–air interfaces, is given by the
Dupré equation.

$$w_{AB} = \gamma_A + \gamma_B - \gamma_{AB}$$

See also COHESION.

adhesive.　Substance that is capable of
bonding other substances together by surface attachment. Examples include soluble
silicates, ceramics (silica/boric acid), cements, natural glues (starches, proteins,
rubber latex, resins, mucilages), asphalt,
thermoplastic resins, thermosetting resins
(epoxy resins), silicone polymers. Adhesives are used in construction, packaging
and furniture manufacture.

adiabatic demagnetization.　Process for
obtaining temperatures close to 0 K, in
which a paramagnetic salt (potassium
chrome alum) is first magnetized between
the poles of an electromagnet cooled in
liquid helium. The field is switched off and
the salt demagnetized adiabatically when
the temperature falls. The demagnetized
state is less ordered than the magnetized
state; the change requires energy which is
withdrawn from the internal energy of the
system.

adiabatic process (isentropic process).
Process in which no heat enters or leaves the
system (i.e. $q = 0$, $\Delta S = 0$). If work is done
by the system during such a process, $w < 0$,
$\Delta U < 0$ and the system cools (see FIRST LAW
OF THERMODYNAMICS). For reversible adiabatic expansion of an ideal gas

$$P_1 V_1^\gamma = P_2 V_2^\gamma = K$$

$$\gamma = C_p/C_v$$

$$T_2/T_1 = (P_2/P_1)^\gamma$$

See also ISOBARIC PROCESS, ISOCHORIC PRO-
CESS, ISOTHERMAL PROCESS, REVERSIBLE
PROCESS.

adipic acid.　See HEXANEDIOIC ACID.

ADP.　See ADENOSINE DIPHOSPHATE.

adrenaline.　See HORMONES.

adsorbent.　Material that will adsorb one
substance in preference to another from
solution (e.g., SILICA GEL, alumina see
ALUMINIUM INORGANIC COMPOUNDS, CEL-
LULOSE, KIESELGUHR).

adsorption.　Accumulation of a substance
at an interface. This is quantified by: n, the
number of molecules adsorbed per unit surface area or gram; V_{stp}, the equivalent
volume of the gas adsorbed at stp; θ, the
fractional coverage ($\theta = 1$ corresponds to
monolayer coverage); Γ, the surface excess,
which is the extra number of molecules per
unit area at the surface.

adsorption chromatography.　Form of
CHROMATOGRAPHY which uses an ADSOR-
BENT as a solid stationary phase and either a
liquid mobile phase (see THIN-LAYER
CHROMATOGRAPHY) or a gas mobile phase
(GAS–SOLID CHROMATOGRAPHY). Separation depends on the different extents to
which solutes are adsorbed by the solid;
least strongly adsorbed materials travel
faster.

adsorption indicator.　See INDICATORS.

adsorption isostere.　Relationship between the pressure of a gas above a surface
and the temperature at constant coverage.
See also HEAT OF ADSORPTION.

adsorption isotherm.　Equation that
relates the amount of a substance adsorbed

adsorption isotherm

Isotherm	Model	Equation[a]
Langmuir	Chemisorption with q = constant	$\theta = KP/(1 + KP)$
Temkin	Chemisorption with $q = q_0(1 + \alpha\theta)$	$\theta = a \ln bKP$
Freundlich	Chemisorption with $q = k \ln b\theta$	$\theta = aP^{1/n}$
Brunauer–Emmett–Teller	Multilayer physical adsorption	$\theta = 1/(1 - P/P_0) - 1/[1 + (C-1)P/P_0]$
Gibbs	Adsorption of a solute at an interface	$-d\gamma = \sum_i \Gamma_i \, d\mu_i$ For a single solute $\Gamma = (-c/RT)\,(d\gamma/dc)$

[a] q = heat of adsorption, P_0 = saturated vapour pressure, θ = fractional adsorption, Γ = surface excess, c = concentration, C = BET constant, K = equilibrium constant, α, a, b, c = constants, μ = chemical potential, γ = surface tension.

on a surface to the concentration of the substance at constant temperature. The form of the isotherm (*see* Table) is determined by the model of ADSORPTION chosen (e.g., if CHEMICAL ADSORPTION or physical adsorption).

aerosol. Colloidal dispersion of a solid or liquid in a gas. Examples include fog, liquid sprays, smokes and airborn dusts.

AES. (1) Atomic emission spectroscopy (*see* ATOMIC SPECTROSCOPY). (2) Auger electron spectroscopy (*see* AUGER EFFECT, ELECTRON SPECTROSCOPY).

affinity chromatography. Technique that depends on the highly biospecific interaction between an immobilized ligand and the macromolecule of interest. The ligand (e.g., enzyme, nucleic acid, hormone, antigen, dye) is bound by a hydrophilic spacer arm to a matrix (silica). Only molecules that bind with the immobilized ligand are retarded and held on the column to be eluted later under conditions in which the binding affinity is reduced. Immobilized boronic acid, which reversibly interacts with vicinal *cis*-diols, has an extremely broad specificity for the separation of nucleosides, nucleotides, glycoproteins, catecholamines, carbohydrates and tRNA. Anti-immunoglobulin ligands are used for antibody purification, triazine dyes for enzymes and lectins (e.g.,

concanavalin A) for specific sugars, carbohydrate residues or sequences in glycoproteins.

aflatoxins. Toxic metabolites of the fungus *Aspergillus flavus* found in grain crops, peanuts, etc. Aflatoxins are highly carcinogenic.

aflatoxins
Aflatoxin B$_1$.

Ag. *See* SILVER.

agar. *See* SEAWEED COLLOIDS.

agate. Crystalline, hard form of silica.

aggregate. (1) Mineral particles bonded together by cements or glasses; used in the construction industry. (2) Any macroscopic particle formed by the coming together of smaller particles.

aglucone. Non-sugar-like portion of a glucoside (*see* GLYCOSIDES).

aglycone. Non-sugar moiety of a GLYCO-
SIDE.

AH. *See* ALTERNANT HYDROCARBON.

air. Mixture of gases, the composition of
which varies with the altitude. The volume
composition of dry air at sea level is: N_2
78.08 percent, O_2 20.95 percent, Ar 0.93
percent, CO_2 0.03 percent, Ne 0.0018 percent, He 0.0005 percent, Kr 0.0001 percent,
Xe 8×10^{-6} percent, Rn 6×10^{-18} percent;
the CO_2 content can vary slightly; $\rho = 1.185$
$kg\,m^{-3}$ at 25°C and 1 atm. Air is noncom-
bustible, but will support combustion.
Liquid air (*see* LIQUEFACTION OF GASES) is
used as a cryogenic agent (liquid nitrogen is
safer in use).

Al. *See* ALUMINIUM.

Ala (alanine). *See* AMINO ACIDS.

alane. *See* ALUMINIUM INORGANIC COM-
POUNDS.

α-alanine (Ala). *See* AMINO ACIDS.

β-alanine. *See* 3-AMINOPROPANOIC ACID.

albumins. *See* PROTEINS.

alcohol. *See* ETHANOL.

alcoholates. *See* ALKOXIDES.

alcohol meter. Alcohol–air FUEL CELL
constructed so that the peak current is proportional to the amount of ethanol in the
vapour at the cathode. The electrodes are of
platinized graphite, and the electrolyte is
dilute acid.

alcohols. Organic compounds with one or
more hydroxyl groups (–OH) attached
directly to a carbon atom. Aromatic compounds with –OH attached directly to a
ring carbon are PHENOLS, which have distinctive properties. Compounds with one
–OH are MONOHYDRIC ALCOHOLS (primary, RCH_2OH; secondary, $RR'CHOH$;

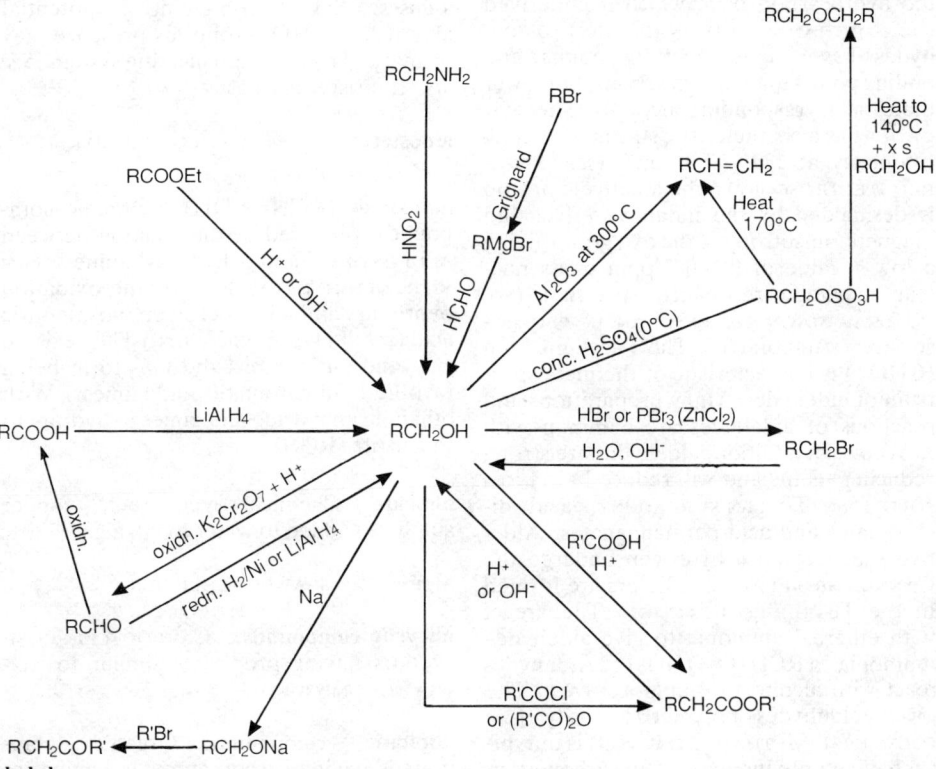

alcohols
Preparations and some reactions of primary alcohols.

tertiary, RR′R″COH), those with two –OH groups are dihydric alcohols (see DIOLS, GLYCOLS) and with three or more –OH groups POLYHYDRIC ALCOHOLS or polyols. The number of isomers increases with chain length: propanol, three; butanol, four; pentanol, eight; etc. Many alcohols and their esters are used as plasticizers and as solvents for paint and lacquers. Many of the commercial higher alcohols (C_6 upwards) are used for the manufacture of SURFACE-ACTIVE AGENTS. Fatty alcohols are those derived from natural fats and waxes (coconut oil, C_{12} and C_{14}; tallow, C_{16} and C_{18}) which are triglycerides. The esters are hydrogenated directly to the corresponding alcohol and glycerol. (For preparations and some reactions of primary alcohols—see Figure.)

aldehyde polymers.   See POLYMERS.

aldehydes ($C_nH_{2n}O$).   Organic compounds containing the group >CO, designated by the suffix –al added to the name of the hydrocarbon from which it is derived (i.e. METHANAL, ETHANAL, etc.). Aldehydes have higher melting points and boiling points than hydrocarbons, but lower than the corresponding alcohols. The OXO GROUP has a characteristic strong stretching frequency at 1720–1740 cm^{-1} (see INFRARED SPECTROSCOPY). The aldehydic proton is deshielded by the inductive effect and magnetic anisotropy of the oxo group giving a low τ value of 0.1–0.7 ppm in its nuclear magnetic resonance spectrum (see NUCLEAR MAGNETIC RESONANCE SPECTROSCOPY, SHIELDING). The acylium ion (CHO^+) is characteristic of the mass spectrum of aldehydes. Many preparations and reactions of aldehydes are common with KETONES. In addition, aldehydes are strong reducing agents and will reduce FEHLING'S SOLUTION, TOLLEN'S REAGENT, acid dichromates and acid permanganates. Aldehydes having no α-hydrogen undergo the CANNIZZARO REACTION. Esters are formed in the TISCHENKO REACTION. They react with ethereal ammonia to give aldehyde–ammonia ($RCH(OH)NH_2$). Aldehydes react with alcohols forming ACETALS. The lower aldehydes polymerize easily (see POLYMERS). METHANAL (HCHO) is untypical having only the oxo group carbon atom in the molecule. See also BENZALDEHYDE.

aldohexoses.   See HEXOSES.

aldoketens.   See KETENS.

aldol.   See 3-HYDROXYBUTANAL.

aldol condensation.   CONDENSATION between two ALDEHYDES or KETONES (or combination thereof) that have α-hydrogen atoms to give a β-hydroxyaldehyde or β-hydroxyketone. The absence of the α-hydrogen atom leads to the CANNIZZARO REACTION. ETHANAL in the presence of dilute sodium hydroxide, potassium carbonate or hydrochloric acid gives 3-HYDROXYBUTANAL (aldol)

$$2CH_3CHO \rightarrow CH_3CH(OH)CH_2CHO$$

With two different aldehydes all four possible products are found, their relative proportions being determined by the conditions used.

aldopentoses.   See PENTOSES.

aldoses.   Sugars containing a potential aldehyde (–CHO) group; its presence may be masked by inclusion in a ring system. See also HEXOSES, PENTOSES.

aldosterone.   See CORTICOSTEROIDS.

aldoximes (RCH=NOH).   Organic compounds, prepared by the reaction between ALDEHYDES and hydroxylamine (see NITROGEN HYDRIDES), or by the oxidation of primary amines by peroxomonosulphuric acid (see SULPHUR OXOACIDS). They exist in syn- and anti- forms, the syn- form being favoured in aromatic aldoximes. With ethanoic anhydride aldoximes dehydrate to the nitrile (RCN).

algicide.   Chemical agent (e.g., copper sulphate) added to water to kill algae.

algin.   See SEAWEED COLLOIDS.

alicyclic compounds.   CARBOCYCLIC COMPOUNDS having properties similar to ALIPHATIC COMPOUNDS.

aliphatic compounds.   Organic compounds having carbon atoms in chains (as opposed to rings).

alizarin ($C_{14}H_8O_4$). Orange–red MOR-DANT dye (*see* Figure); synthesized from anthraquinone-2-sulphonic acid, sodium hydroxide and potassium chlorate.

alizarin

Alk.   *See* ALKYL GROUP.

alkali.   (1) Hydroxide of one of the ALKALI METALS or any substance (e.g., calcium oxide, sodium carbonate) which when dissolved in water gives a solution of pH greater than 7. (2) Any aqueous solution having a pH of greater than 7.

alkali metals.   Elements LITHIUM, SODIUM, POTASSIUM, RUBIDIUM, CAESIUM and FRAN-CIUM forming group Ia of the PERIODIC TABLE. Their electronic structure is an inert gas configuration plus a single outer shell s-electron (e.g., Na is $[Ne]3s^1$). They are ELECTROPOSITIVE metals, easily forming the M^+ ion, the reactivity increasing down the group. Ionization energy, melting point and boiling point decrease down the group, whereas ionic and atomic radius and density increase. In aqueous solution, however, the mobilities of the ions increase down the group (i.e. implying a smaller radius) because of the greater solvation of the smaller cations. All standard electrode potentials of $M^+|M$ are highly negative.

Each element reacts with oxygen to give an oxide, and with water to give alkaline solutions and hydrogen. The salts (halides, sulphates, nitrates, etc.) are ionic, stable and soluble in water. The carbonates do not decompose on heating, and the nitrates, with the exception of lithium, give the nitrites not the oxides. Lithium, the first member of the group, shows similarities to magnesium (*see* DIAGONAL RELATIONSHIP) (e.g., it is the only alkali metal to react with nitrogen giving Li_3N).

alkaline cell.   Primary cell using sodium or potassium hydroxide as the electrolyte. Alkaline cells based on a manganese dioxide cathode and zinc anode have a lower internal resistance and consequently better performance than the LECLANCHÉ CELL. In particular at high current drain, a higher potential is maintained for longer. Also, the shelf life of this cell is longer and a higher energy density may be achieved. The potential given by the cell is about 1.9 V. A button cell may be constructed and, with some changes in the structure of the battery, alkaline manganese(IV) oxide–zinc cells may be recharged. Alternative anode materials are cadmium, indium, aluminium or sodium. Alternative cathode oxides are those of copper, iron, lead, mercury, nickel or silver. (For stoichiometry—*see* Figure.)

alkaline earth metals.   Elements BERYL-LIUM, MAGNESIUM, CALCIUM, STRONTIUM, BARIUM and RADIUM forming group IIa of the PERIODIC TABLE. Their electronic structure is a noble gas configuration plus two s-electrons in the outermost shell. They are ELECTROPOSITIVE metals readily forming M^{2+}, the reactivity increasing down the group. Although the second ionization potential is much greater than the first, the high lattice and hydration energies of M^{2+} compensate for this, and unipositive ions are not known. Beryllium has atypical properties, which resemble those of aluminium (*see* DIAGONAL RELATIONSHIP), but from magnesium to radium the atomic and ionic radii increase together with melting and boiling points and density. They are less reactive than the ALKALI METALS, but all react with oxygen and water (beryllium and magnesium form protective oxide films). The oxides become more ionic and thus more basic down the group. Beryllium hydroxide is amphoteric, whereas barium hydroxide is a reasonably strong base. The oxides are stable and are formed on heating salts of the oxoacids. Salts of singly charged anions are generally soluble in water and salts of doubly charged anions are insoluble.

cathode	$2MnO_2 + 2H_2O + 2e \rightleftharpoons Mn_2O_3.H_2O$		$E^{\ominus} = +0.118$
anode	$Zn + 2OH^- \rightleftharpoons Zn(OH)_2 + 2e$		$E^{\ominus} = -1.245$
overall	$Zn + 2MnO_2 + 2H_2O \rightleftharpoons Mn_2O_3.H_2O + Zn(OH)_2$	$E^{\ominus} = 1.433$	

alkaline cell
Stoichiometry of
an alkaline cell.

alkaloids

Compound	Structural formula	Occurrence/physical properties/pharmacology/uses
Ephedrine	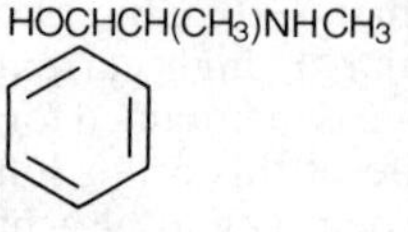 HOCHCH(CH₃)NHCH₃	l-Form obtained from *Ephedra vulgaris*; available as synthetic product. Very strong base; mp 34°C. Vasoconstrictor; activity resembles that of adrenaline (*see* HORMONES). Relieves catarrh, asthma, hay fever
Mescaline (3,4,5-trimethoxy-β-phenylethyl-amine)	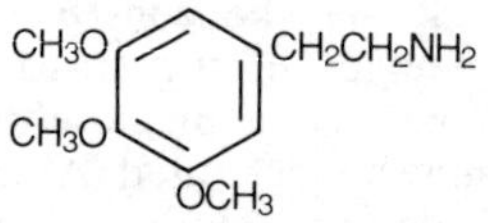 CH_3O ... $CH_2CH_2NH_2$ CH_3O OCH_3	Obtained from *Lophophora williamsii* (Mexican cactus); mp 35−6°C. Takes up CO_2 from atmosphere to give carbonate. CNS depressant; hallucinogenic
Ricinine (1-methyl-2-pyri-done-3-carbo-nitrile)	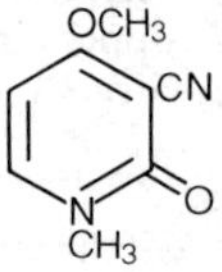 OCH_3 CN O CH₃	Constituent of seeds of *Ricinus communis L.* Optically inactive compound; does not form salts with acids; mp 201°C. Poisonous; causes convulsions, respiratory depression, hypotension
Nicotine (1-methyl-2-(3-pyridyl)-pyrrolidine)	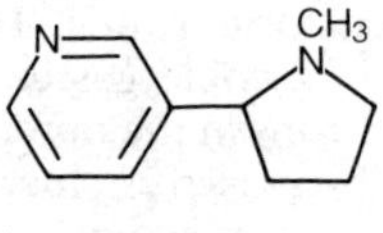 CH₃	Constituent of *Nicotiana tabacum*. Manufactured as byproduct of tobacco industry. When pure, oily colourless hygroscopic liquid (bp 247°C). Darkens on exposure to air. Forms partially miscible mixture with water; alkaline solution (pK_a 3.13, 8.02). Sulphate widely used as agricultural insecticide; source of nicotinic acid
Piperine (1-piperyl-piperidine)	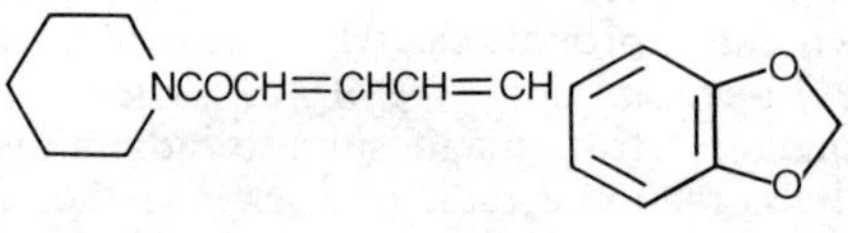 NCOCH=CHCH=CH	Obtained from fruit of various peppers (e.g., *Piper nigrum*, *Piper longum*); mp 130°C, pK_a 1.98. Acid hydrolysis gives piperidine and piperic acid. Not toxic to humans; responsible for sharp taste of pepper
Coniine (2-propyl-piperidine)	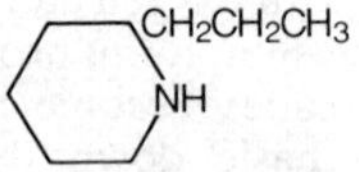 CH₂CH₂CH₃ NH	(−)-Form obtained from *Conium maculatum*. Alkaline, steam volatile liquid (bp 167°C), mousy odour. Darkens and polymerizes in air. Reduced at high temperature with HI to n-octane; oxidized to pyridine-1-carboxylic acid. Hydrochloride used as antispasmodic
Cocaine (benzoyl methyl ecognine)	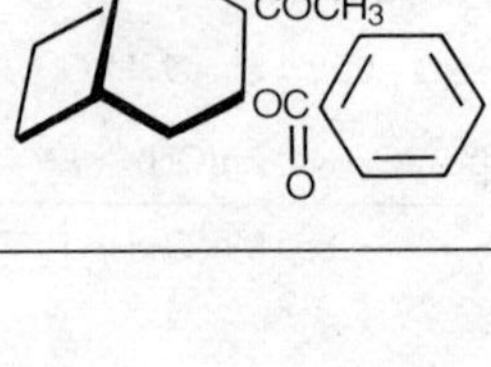 CH₃ N O COCH₃ OC O	Extracted from dried leaves of *Erythroxylon coca*, mp 98°C, pK_a 8.37. Decomposes on sterilization. Used as local anaesthetic; addictive

Compound	Structural formula	Occurrence/physical properties/ pharmacology/uses
Tropan alkaloids	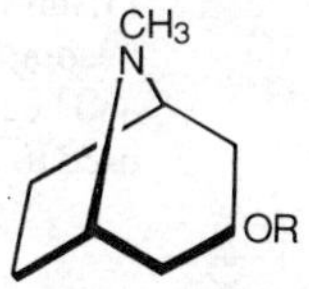	Constituents of such plants as *Atropa belladonna* (deadly nightshade), *Hyoscyamus niger* (henbane) and *Datura strammonium* (thorn apple). Many are esters of tropine
Tropine (3-tropanol)	R = H	Obtained by hydrolysis of atropine. A *meso*-compound (mp 63°K, pK_a 10.33), with secondary hydroxyl (oxidizable to ketone), tertiary amine with no double bonds
Atropine ((±-hyoscy-amine)	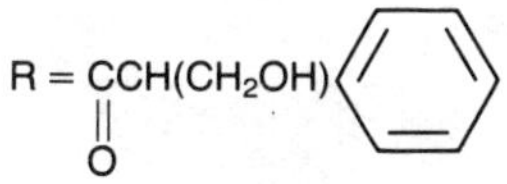 $R = \overset{\|\|}{\underset{O}{C}}CH(CH_2OH)$	Prisms (mp 114−16°C, pK_a 9.65). Used as pupil dilator, antidote for nerve gases
(1)-Hyoscy-amine		Obtained by resolution of atropine, easily racemized; mp 108.5°C
Hyoscine((−)-scopolamine)	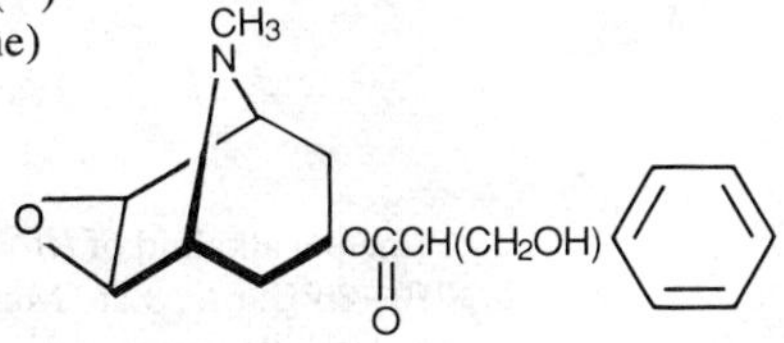	Obtained as viscous liquid from plants of Solanaceae family. Forms crystalline monohydrate (mp 59°C). Freely soluble in hot water and organic solvents. Affects CNS, vasoconstrictor. Used to prevent motion sickness
Ergot alkaloids	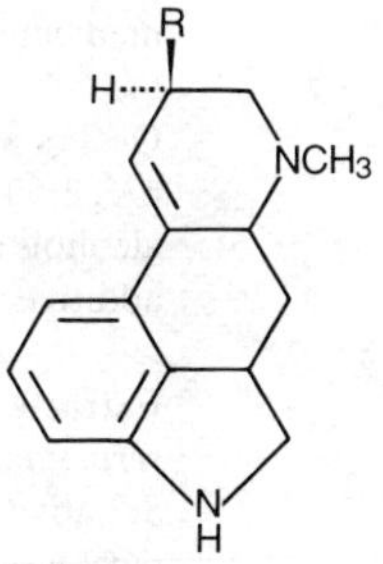	Constituents of *Claviceps purpurea*, a fungus parasitic on rye
Lysergic acid	R = COOH	Obtained by alkaline hydrolysis of ergot alkaloids (e.g., ergotamine). Scales (decomp. 238°C). Behaves as acid and base (pK_a 3.44, 7.68). Diethylamide (LSD, R=CONEt$_2$) is hallucinogenic
Ergometrine (ergometrin-ine)	$R = CONHCH(CH_2OH)CH_3$	Amide of lysergic acid with 2-amino-1-propanol. *l*-Form is water-soluble (decomp. 196°C, pK_a 7.3). Induces uterine contraction

continued overleaf

Compound	Structural formula	Occurrence/physical properties/ pharmacology/uses
Ergotamine	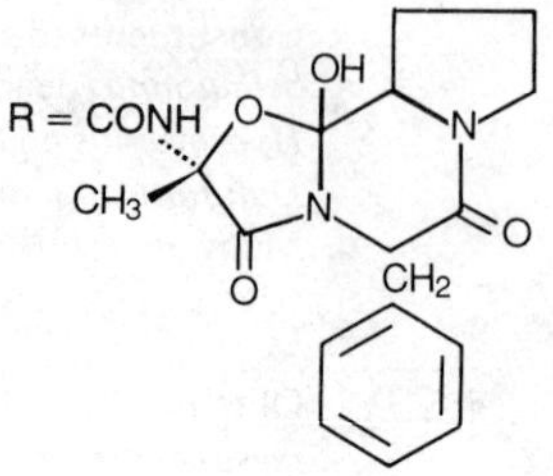 R = CONH	Prisms (decomp. 213°C). Darkens and decomposes on exposure to air, light and heat. Analgesic, vasoconstrictor; used to relieve pain of migraine
Nux vomica alkaloids	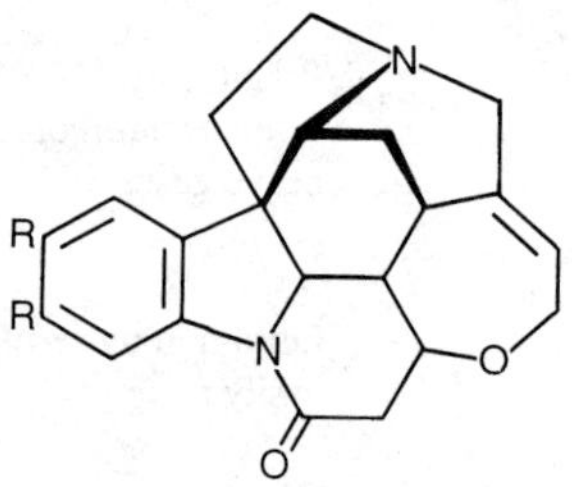	
Strychnine	R = H	Principal alkaloid of *Nux vomica*; mp 270−80°C; pK_a 8.26. Muscle relaxant; used to kill vermin and in homeopathic medicine
Brucine	R = OCH$_3$	Occurs with strychnine; mp 178°C, pK_a 2.50; 8.16. Used to denature alcohols and oil, and as lubricant additive
Reserpine	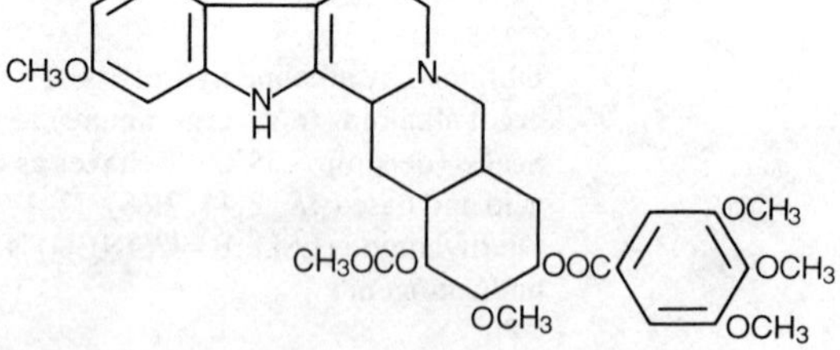	Extracted from root of *Rauwolfia serpentina*; mp 264°C. Darkens in air. Sedative, tranquillizer; used to treat nervous and mental disorders

Compound	Structural formula	Occurrence/physical properties/ pharmacology/uses
Yohimbine	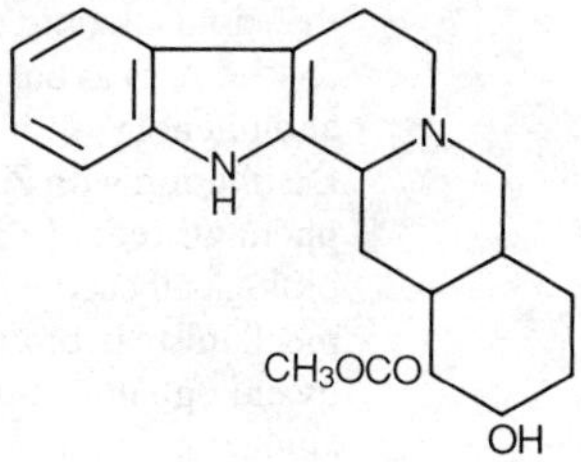	Obtained from leaves and bark of *Corynanthe johimbe*. Formerly used as aphrodisiac in veterinary medicine
Cinchona alkaloids	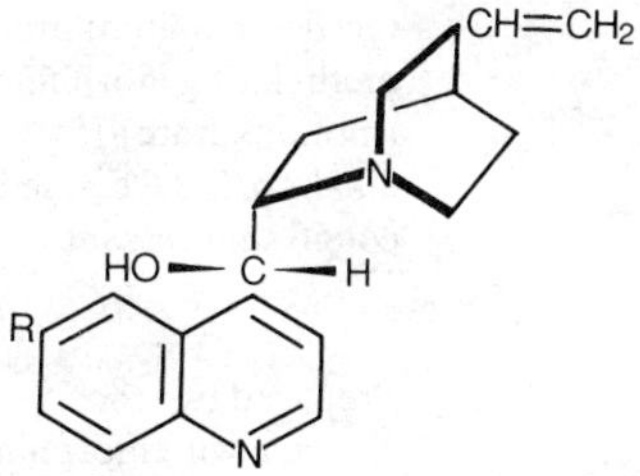	Obtained from bark of *Cinchona officinalis L.*
Cinchonine	R=H	Mp 255°C
Quinine	R=OCH$_3$	Principal alkaloid; mp (trihydrate) 57°C. Shows blue fluorescence especially in presence of H_2SO_4. Forms additive compounds with alcohols and benzene; oxidized (hot alkaline $KMnO_4$) to pyridine-2,3,4-tricarboxylic acid, CO_2 and NH_3; oxidized (CrO_3) to quinone and quininic acid. Base and salts are antimalarials; used as antipyretic for colds, influenza, cramp. Present in small amounts in tonic water
Quinidine	R=OCH$_3$	Isomeric with quinine; mp (anhyd.) 174−5°C. Chemical properties as quinine; salts more soluble than those of quinine
Opium alkaloids	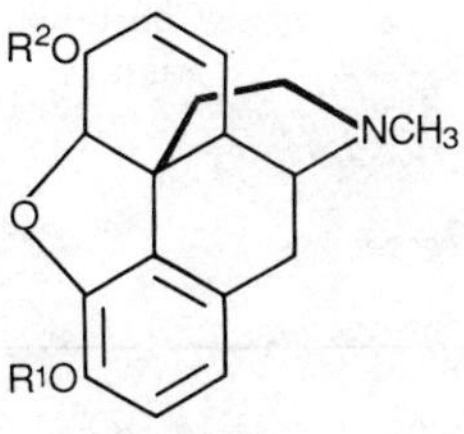	Extracted from exudates of unripe capsules of *Papaver somniferum*

continued overleaf

Compound	Structural formula	Occurrence/physical properties/pharmacology/uses
Morphine	$R^1=R^2=H$	Principal alkaloid of opium; decomp. 254°C. Acts as base (pK_a 8.07, 9.85) and phenol (soluble in alkali). Distillation with Zn dust gives phenanthrene. (+)-Stereoisomer biologically active; activates neuronal mechanism in brain normally activated by enkephalins. Sulphate valuable analgesic; highly addictive
Codeine (*O*-methyl-morphine)	$R^1=CH_3, R^2=H$	Occurs in opium; prepared by methylating morphine; mp (monohydrate) 155°C. Insoluble in alkali, pK_a 8.21. Used as analgesic and cough suppressant; highly addictive
Heroin (diacetyl-morphine, diamorphine)	$R' = R^2 = COCH_3$	Prepared by reacting ethanoyl chloride with morphine; mp 173°C. Used as analgesic; highly addictive
Papaverine	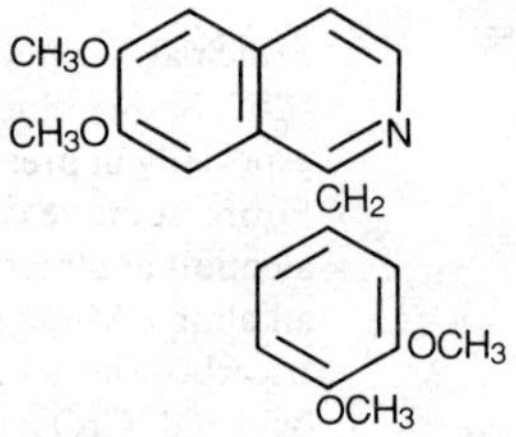	Optically inactive; mp 147°C, pK_a 6.4. Pharmacology resembles morphine (smooth muscle relaxant); non-addictive
Caffeine	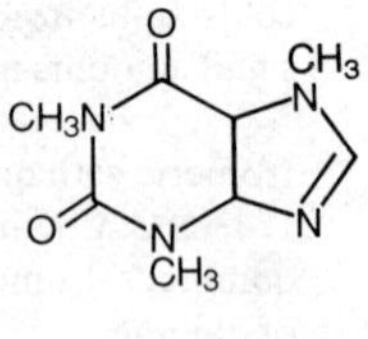	Constituent of tea, coffee. *See also* PURINES
Theobromine	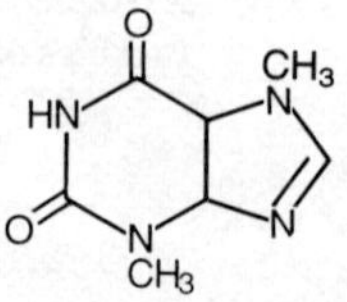	Constituent of cacao seeds. *See also* PURINES

alkaloids. Organic nitrogenous bases produced (with the exception of ergot alkaloids) by dicotyledonous plants. They occur as salts with organic hydroxy acids such as hydroxybutanedioic, 2-hydroxy-1,2,3-propanetricarboxylic, tannic and quinic (1,3,4,5-tetrahydroxycyclohexane-carboxylic) acids. They have potent pharmacological activity and form the basis of many drugs. The majority are very toxic both by inhalation and ingestion. They vary considerably in their chemical properties and constitution depending on the parent base: aryl-substituted amines, indole, pyridine, quinoline and isoquinoline. Most alkaloids are crystalline solids with a very bitter taste; they are sparingly soluble in water, but usually soluble in organic solvents such as ethanol, ether and trichloromethane. They are optically active, most being dextrorotatory; they have been used for the resolution of racemic acids into their enantiomorphs. They are basic, forming crystalline salts with acids; these are water-soluble. Drug preparations are usually based on the salts (e.g., hydrochloride, bromide, sulphate). Alkaloids give precipitates with such reagents as phosphomolybdic acid, potassium mercury(II) iodide and potassium triiodide. Their structures have been established as a result of chemical reactions: acetylation (for the number of hydroxyl groups), treatment with hydroiodic acid (for the CH_3O- and CH_3N- groups), determination of $>NH$ and $-NH_2$, hydrolysis, oxidation, exhaustive methylation and degradation to simple substances by heating with sodium hydroxide or zinc powder. Commercially the alkaloids are extracted from powdered plant material with alcohol, water or dilute acid, and then precipitated out on the addition of base. The crude extract, often containing a range of alkaloids, is purified by physical methods of separation including fractional crystallization, countercurrent distribution,

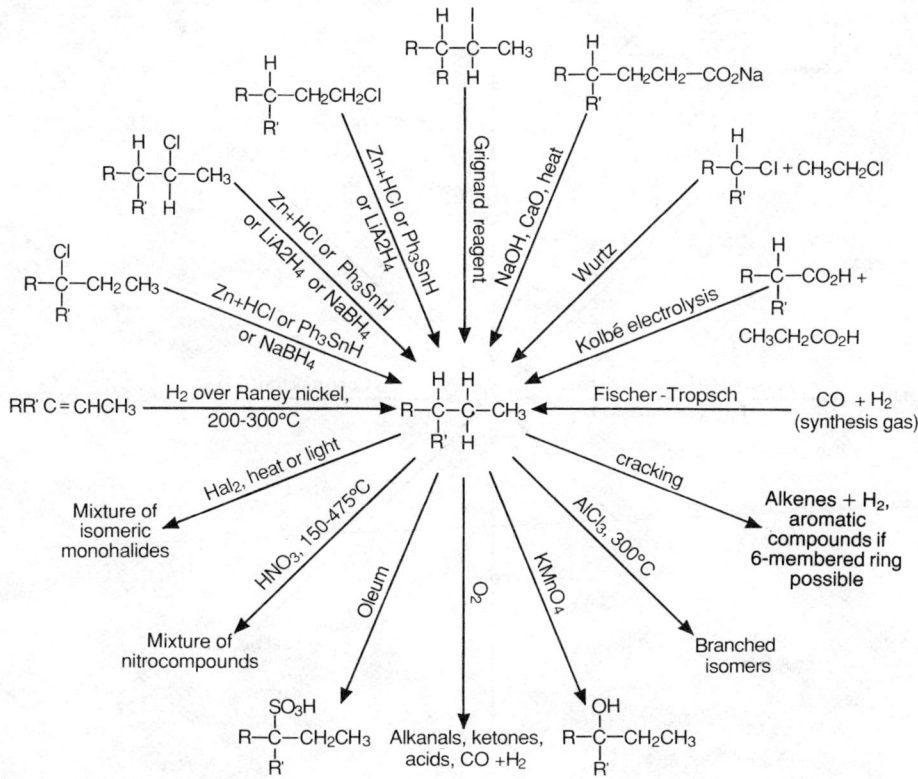

alkanes
Preparation and reaction of alkanes.

adsorption and partition chromatography. There are a very large number of plant alkaloids, the more important ones are tabulated.

alkanals.   *See* ALDEHYDES.

alkanes (paraffins, C_nH_{2n+2}).   Saturated aliphatic HYDROCARBONS, designated by the suffix –ane. ISOMERISM occurs for C_4 and above. C_1–C_4 (i.e. methane, ethane, propane and butanes) are gases, constituents of NATURAL GAS and PETROLEUM GAS, C_5–C_{15} are mostly liquids from PETROLEUM, and C_{16} and above are waxy solids. n-Compounds (unbranched compounds) always have the highest melting and boiling points. They are insoluble in water, but are soluble in organic solvents (e.g., alcohols, ethers, benzene). Uses are as fuels, chemical feedstocks, refrigerants and solvents. Chemical properties show a range of reactions (*see* Figure), the reactivity of C–H being tertiary>secondary>primary.

alkanolamines.   Amino alcohols, prepared by the reaction between an ALKENE OXIDE and aqueous ammonia at 50–60°C. They are used as emulsifiers, in the manufacture of detergents, as absorbents for acidic gases and as catalysts for POLYMERIZATION. *See also* ETHANOLAMINES.

alkanols (ROH).   ALCOHOLS, where R is an ALKYL GROUP.

alkene oxides (epoxides, $C_nH_{2n}O$).   Organic compounds derived from ALKENES with an oxygen atom bridged between the carbon atoms of the double bond. The formation of the oxide is known as epoxidation and may be effected via an OZONIDE or PRILESCHAIEV'S REACTION with a PERACID.

alkenes (olefins, C_nH_{2n}).   Unsaturated HYDROCARBONS containing a single double bond, designated by the suffix –ene. The carbon atoms of the double bond are sp^2-hybridized (*see* HYBRIDIZATION) and may

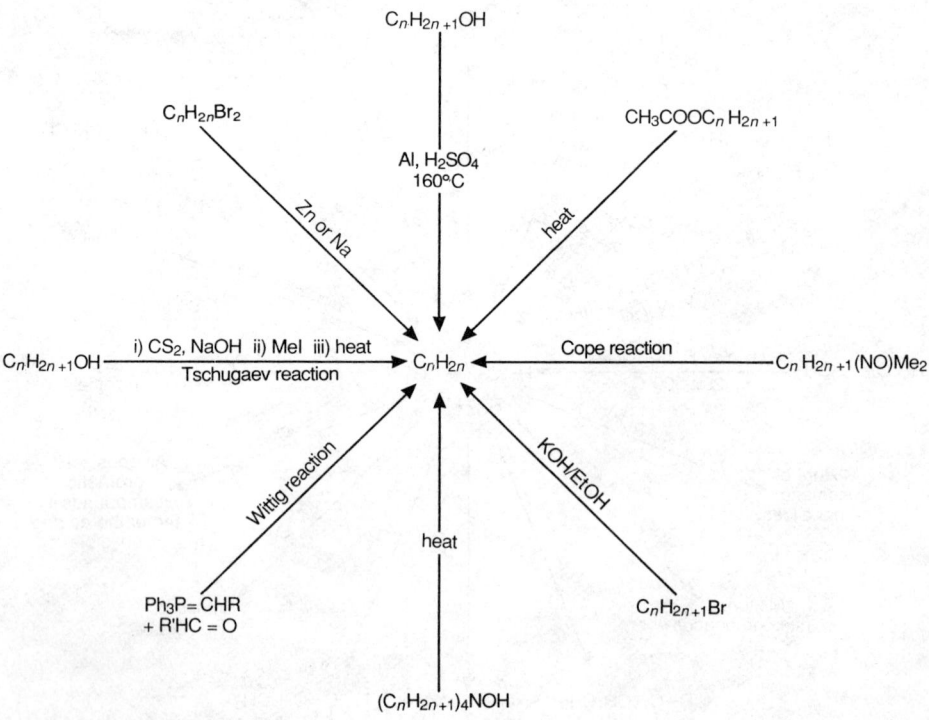

alkenes
Figure 1. Preparation of alkenes.

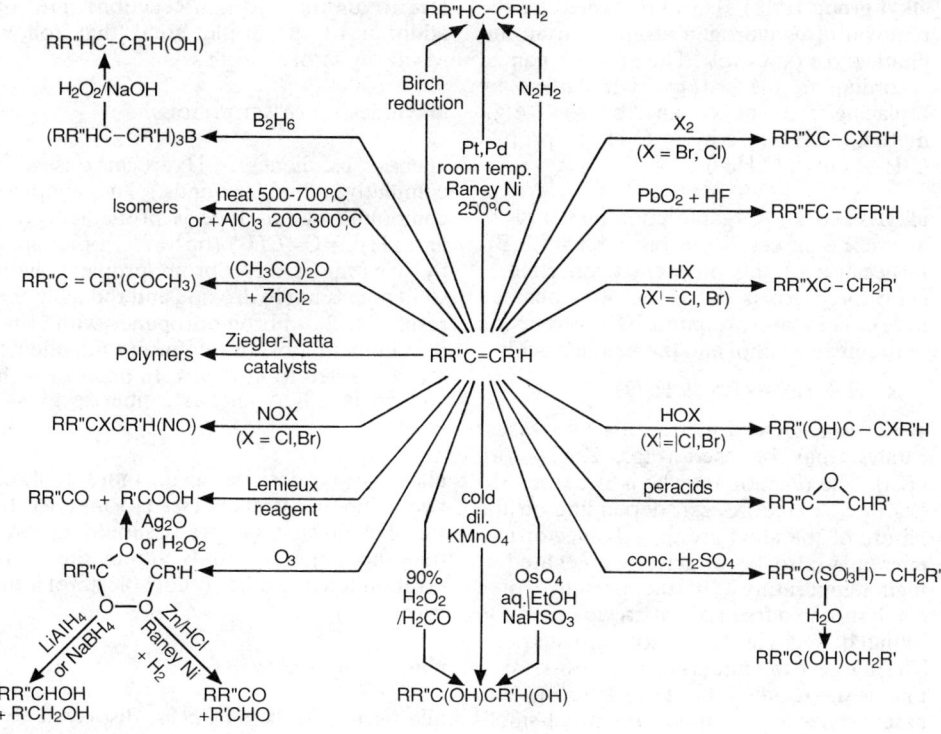

alkenes
Figure 2. Reactions of alkenes.

show *cis–trans* (geometrical) isomerism (*see* ISOMERISM). Their physical properties resemble ALKANES: C_2–C_4 are gases, C_5–C_7 liquids, and C_8 and beyond waxy solids. They are insoluble in water, but soluble in trichloromethane and benzene. Alkenes occur in PETROLEUM and are manufactured from petroleum by CRACKING (*see* Figure 1). ETHENE, PROPENE and butenes are important starting materials for industrial PETROCHEMICALS. Chemical properties are mostly of addition across the double bond (*see* MARKOWNIKOFF'S RULE) (*see* Figure 2).

alkenynes. Organic compounds containing both double and triple bonds (*see* ALKENES, ALKYNES).

alkoxides (alcoholates). Metal derivatives of an alcohol (e.g., CH_3OK, $[(CH_3)_3CO]_3$-Al), prepared by the solution of an alkali metal in the alcohol or indirectly (e.g., $TlCl_4$ + ROH + tertiary amine). Some are volatile, whereas others are polymeric.

Alkoxides are used as reagents in the formation of ETHERS and ALKENES, as catalysts and as paint additives.

alkoxyalkanes. *See* ETHERS.

alkoxyarenes. *See* AROMATIC ETHERS.

alkoxy group (OAlk). The group –OR, where R is an ALKYL GROUP. *See also* ETHERS.

alkyd resins. *See* POLYMERS, RESIN.

alkyl aryl ethers (phenolic ethers). *See* AROMATIC ETHERS.

alkylation. Introduction of an ALKYL GROUP into a molecule. Examples are the FRIEDEL–CRAFTS REACTION and in reactions of GRIGNARD REAGENTS. *See also* METHYLATION.

alkylbenzenes. *See* ARENES.

alkyl group (Alk). Group formed by the removal of a hydrogen atom from an aliphatic HYDROCARBON. The group is named according to the parent hydrocarbon by replacing the suffix –ane by –yl (e.g., methyl, CH_3–; ethyl, C_2H_5–; propyl, C_3H_7–, butyl, C_4H_9–).

alkyl halides. Organic compounds (RX, where R is an ALKYL GROUP and X = Cl, Br, I; because of the different properties of FLUOROCARBONS, X=F is not included here). They are prepared by the reaction between an alcohol and the halogen acid

$$ROH + HX \rightarrow RX + H_2O.$$

A Lewis acid (see LEWIS ACIDS AND BASES) catalyst may be used (e.g., $ZnCl_2$ with HCl). The reaction may be a S_N1 MECHANISM or S_N2 MECHANISM, depending on the nature of the alkyl group. Alkanes or alkenes may also be directly halogenated at high temperatures, in the presence of a catalyst or in a free radical CHAIN REACTION initiated by light (see PHOTOCHEMISTRY). The lightest members—CHLOROMETHANE, BROMOMETHANE and CHLOROETHANE—are gases; other alkyl halides are sweet-smelling liquids. The boiling points increase for a given alkyl group as Cl>Br>I, and for a given halogen tertiary<secondary<primary alkyl. Reactivity goes as Cl< Br< I; for example, iodomethane is decomposed by light to ethane and iodine. Alkyl halides react with nucleophiles. They are reduced to alkanes (see WURTZ REACTION) and give alkenes on boiling with ethanolic potassium hydroxide (see ELIMINATION REACTION). Alkyl halides are the precursors for amines, cyanides, carbylamines, esters, ethers, thioethers and thiols. They are used in the FRIEDEL–CRAFTS REACTION and to prepare GRIGNARD REAGENTS.

alkylidene group. The group RCH<, where R is an ALKYL GROUP. Thus 1,1-dibromoethane is also known as ethylidene dibromide. See also CARBENE.

alkylphenols. See HYDROXYTOLUENES, HYDROXYDIMETHYLBENZENES, PHENOLS.

alkynes (C_nH_{2n-2}). Unsaturated HYDROCARBONS containing a single triple bond, designated by the suffix –yne. Carbon atoms of the triple bond are sp-hybridized (see HYBRIDIZATION). Reactions are of addition to the triple bond that follow MARKOWNIKOFF'S RULES.

alkynides. See ACETYLIDES.

allenes (1,2-dienes). Hydrocarbons with cumulative double bonds. The simplest compound of this type is propadiene (allene, $H_2C=C=CH_2$) (bp –32°C), prepared by treating 1,2,3-tribromopropane with solid potassium hydroxide and reducing the resultant 2,3-dibromopropene with zinc dust in methanol. In sulphuric acid, allenes are converted to KETONES. In bases or with sodium in ether, allenes isomerize to ALKYNES.

allo-. Prefix designating the more stable of two geometric isomers (see ISOMERISM). In sterol chemistry, the prefix should indicate that the rings A and B are in the *trans* position to each other (allocholesterol is an exception).

allose. See HEXOSES.

allosteric effects. Effects displayed by enzymes with two or more receptor sites, in which occupation of one site can alter the specificity of the other site. The binding of the first molecule changes the conformation of the protein, thereby altering the activity of the enzyme.

allotropy. Existence of an element in more than one crystalline form; the various forms are termed allotropes. Examples include: sulphur and tin (ENANTIOTROPY); carbon and phosphorus (MONOTROPY).

alloy. Homogeneous molten mixture of two or more METALS and the solid-phase material that crystallizes from it. Alloys are usually harder and stronger than the pure metals and have been used to advantage in producing materials with particular properties (see BRASS, BRONZE, STEEL). Alloys may be solid solutions of one metal in another with a range of compositions and thus are distinguished from INTERMETALLIC COMPOUNDS. Alloys of mercury are known as AMALGAMS. See also HUME–ROTHERY'S RULE.

Allred–Rochow electronegativity scale. *See* ELECTRONEGATIVITY.

allyl alcohol. *See* 2-PROPEN-1-OL.

allyl group. *See* PROPENYL GROUP.

allylic rearrangement. Migration of a double bond in a PROPENYL GROUP (allyl group) during NUCLEOPHILIC SUBSTITUTION. For example bromination of 2-buten-1-ol under conditions favouring a CARBOCATION gives not only $CH_3CH=CHCH_2Br$, but also $CH_3CHBrCH=CH_2$. The positive charge in the propenyl cation is delocalized over the group leading to the possibility of attack at each end (e.g., $[CH_2=CH=CH_2]^+$) (*see* DELOCALIZATION).

allylic termination. *See* INHIBITION.

alpha-helix. *See* NUCLEIC ACIDS, PROTEINS.

alpha-particle (^4_2He). Helium nucleus emitted at a speed of $1–2 \times 10^7$ m s^{-1} by a RADIONUCLIDE during alpha-ray decay, resulting in a decrease in the NUCLEON NUMBER of the parent nucleus by two and of the nuclear mass by four. Generally there are two or more discrete and characteristic groups in the spectra of alpha-rays from a particular radionuclide; for uranium-238, 77 percent of the alpha-rays have an energy of 4.2 MeV and the remainder an energy of 4.15 MeV. Alpha-particles are used as bombarding agents in nuclear disintegration reactions. *See also* RADIOACTIVE DECAY SERIES, RADIOACTIVITY.

alternant hydrocarbon (AH). HYDROCARBON in which it is possible to divide the atoms of the resonating part of the molecule into two groups (sometimes called starred and unstarred) such that no atom of one group is adjacent to another atom of the same group, but is always adjacent to one or more atoms of the other group. Thus benzene, naphthalene and anthracene are alternate hydrocarbons (AHs), but fulvene is not (*see* Figure); no odd-numbered ring can be alternant. All the carbon atoms in an alternant hydrocarbon have a charge density of one. Due to asymmetric distribution of charge a non-alternant hydrocarbon has a dipole moment, whereas an alternant hydrocarbon does not.

altrose. *See* HEXOSES.

alumina (aluminium oxide). *See* ALUMINIUM INORGANIC COMPOUNDS.

aluminates. *See* ALUMINIUM INORGANIC COMPOUNDS.

aluminium (Al). Group IIIa element (*see* GROUP III ELEMENTS). It is the most abundant metal, being found naturally as aluminosilicates (*see* SILICATES) (clays, kaolin, micas, feldspars), as bauxite (mixture of hydrated oxides) and as cryolite (Na_3AlF_6). The metal is extracted by the HALL–HEROULT PROCESS. Aluminium is a silvery-white reactive metal that resists corrosion by the formation of a protective oxide film. If the film is broken by scratching or by amalgamation (*see* AMALGAMS), the metal may be attacked by water. It dissolves in dilute mineral acids and hot alkalis and is rendered passive by concentrated nitric acid. Aluminium has high thermal and electrical conductivity and high reflectivity. It becomes superconducting at 1.2 K. Aluminium reacts with most non-metals (e.g., carbon, nitrogen, phosphorus, halogens) and is attacked by mercury and the salts of the noble metals. The major use of aluminium is in the building and construction industry, and is alloyed in aircraft frames, in cables and food packaging. Aluminium powder is used in explosives, rocket fuels and as a reducing agent (*see* THERMITE REACTION). The gems ruby and sapphire are CORUNDUM coloured by chromium and cobalt, respectively.

benzene   naphthalene   anthracene   fulvene

alternatant hydrocarbon

$A_r = 26.98154$, $Z = 13$, ES [Ne] $3s^2 3p^1$, mp 660°C, bp 2470°C, $\rho = 2.70 \times 10^3$ kg m^{-3}. Isotopes: ^{27}Al, 100 percent.

aluminium inorganic compounds. Aluminium usually forms compounds in oxidation state +3 (Al(I) is known in some high-temperature gas-phase compounds) in octahedral or tetrahedral coordination; $E^{\ominus}_{Al^{3+},Al} = -1.66$ V. The structure of Al^{3+} in aqueous solution is extremely complex, as revealed by ^{27}Al nuclear magnetic resonance. $[Al(H_2O)_6]^{3+}$ is quite acid, and the hydroxide $Al(OH)_3$ is amphoteric, dissolving in both acids and alkalis. Simple and mixed salts of most strong oxoacids are known (see ALUMS). Some complexes are formed, especially with halogens (e.g., $[AlF_n]^{3-n}$, where $n=1$–6).

Aluminates, formally written as $[Al(OH)_4]^-$, are considered as hydroxy- or oxy-anionic complexes. They are prepared by the treatment of aqueous solutions of aluminium salts with excess base. Mixed metal aluminates (e.g., the spinel $MgAl_2O_4$) are formed by fusing the metal oxide with alumina (Al_2O_3). β-Alumina, used as a solid electrolyte, is the aluminate $Na_2O.11Al_2O_3$. Calcium aluminate is a constituent of Portland cement.

Aluminium oxide, alumina, has two crystal phases: α-alumina occurs naturally as CORUNDUM or emery and is an hcp structure (see CLOSE-PACKED STRUCTURES) of oxide ions with Al^{3+} in the octahedral holes. Pure α-alumina is prepared by firing hydrated alumina (AlO(OH)) at 1200°C. If heated at 600°C, γ-alumina with a defect spinel structure is formed. γ-Alumina, also called activated alumina, has catalytic properties due to its porosity and variable OH$^-$ content. For example, it catalyzes the dehydration of organic molecules. It is also used as a catalyst support for dispersed metals.

Aluminium halides may be made by direct reaction of the elements, or for aluminium trifluoride (AlF$_3$) by the reaction between hydrogen fluoride and Al_2O_3. The bonding in aluminium trichloride (AlCl$_3$) has considerable covalent character. Solid AlCl$_3$ contains octahedral aluminium, and in the gas phase Al_2Cl_6 with chlorine bridges is formed. AlCl$_3$ is a Lewis acid (see LEWIS ACIDS AND BASES) and is used as a catalyst for CRACKING petroleum. At high temperatures, the monohalides are formed from AlX$_3$ + Al.

Aluminium hydride (AlH$_3$) is a strong reducing agent that forms complexes containing $[AlH_4]^-$, which act as a source of H$^-$. The most important is lithium tetrahydridoaluminate (LiAlH$_4$) (LiH + AlCl$_3$ in ether) (e.g., in the preparation of ALKANES and ALCOHOLS). The hydride is obtained as alane (AlH$_3$)$_n$, a white polymer resulting from the addition of AlCl$_3$ to LiAlH$_4$.

aluminium organic derivatives. Aluminium has an extensive organometallic chemistry. Aluminium alkyls (AlR$_3$) are prepared from organomercury compounds and aluminium

$$2Al + 3R_2Hg \rightarrow 2AlR_3 + 3Hg$$

or from GRIGNARD REAGENTS and aluminium chloride

$$3RMgCl + AlCl_3 \rightarrow AlR_3 + 3MgCl_2$$

On a large scale, aluminium hydride directly reacts with alkenes, or Al + H$_2$ + AlR$_3 \rightarrow$ AlR$_2$H which in turn may be reacted with an alkene. The series of halides RAlCl$_2$ and R$_2$AlCl are important reagents made by reaction of aluminium with an alkyl halide. All organoaluminium compounds are reactive, they are easily oxidized and the lower alkyls inflame in air and explode in water. The lower alkyls also dimerize with bridging alkyl groups (cf AlCl$_3$, see ALUMINIUM INORGANIC COMPOUNDS). Like inorganic compounds of aluminium, the alkyls are Lewis acids (see LEWIS ACIDS AND BASES) which form adducts with electron donors such as amines, phosphines, ethers and thioethers. With lithium alkyls compounds LiAlR$_4$ are formed containing $[AlR_4]^-$. Organoaluminium compounds are components of ZIEGLER–NATTA CATALYSTS for the polymerization of alkenes, and catalyze the TISCHENKO REACTION for the preparation of esters.

aluminosilicates. See SILICATES.

aluminothermic reduction. See THERMITE REACTION.

alums. Mixed metal sulphates of general formula $M^I M^{III}(SO_4)_2.12H_2O$. Three structures are known all containing

$[M^I(H_2O)_6]^+$, $[M^{III}(H_2O)_6]^{3+}$ and $2[SO_4]^{2-}$. M^I may be any common univalent ion, except lithium. M^{III} is any trivalent ion including many transition element ions. Originally M^{III} was aluminium, but now these are referred to as aluminium alums, and the term alum covers all compounds of the general type. Chemically alums act as a mixture of the constituent sulphates. In compounds of similar structure, the sulphate group has been replaced by SeO_4^{2-}, BeF_4^{2-} and $ZnCl_4^{2-}$. Alums are used in leather tanning.

alvite. *See* HAFNIUM.

Am. Americum (*see* ACTINIDE ELEMENTS).

amalgam electrode. *See* ELECTRODE.

amalgams. Alloys of mercury and other metals. Most metals, except iron and platinum, dissolve in mercury to give liquids or solids, depending on the amount of mercury. Sodium amalgam ($NaHg_2$) has a definite composition. Amalgams are used in dental fillings, as electrodes of batteries and as convenient sources of reactive metals (e.g., sodium) in chemical reactions. ELECTROPOSITIVE metals, which normally could not be extracted from aqueous solution by electrolysis, are discharged at a mercury cathode and incorporated as an amalgam from which the metal may be recovered (e.g., SODIUM, RADIUM).

amatol. *See* EXPLOSIVES.

americum (Am). *See* ACTINIDE ELEMENTS.

amethyst. *See* CORUNDUM.

amides. Organic compounds containing the amido group (–$CONH_2$). As with AMINES, amides can be primary ($RCONH_2$), secondary (($RCO)_2NH$) or tertiary (($RCO)_3N$), and thus can be seen as acyl derivatives of ammonia. Primary amides are named by replacing the suffix –oic of the parent acid by –amide. Amides are prepared by: (1) heating the ammonium salt of a carboxylic acid or by heating the acid with urea; (2) AMMONOLYSIS; (3) partial hydrolysis of alkyl cyanides by, for example, polyphosphoric acid or concentrated hydrochloric acid; (4) the BECKMANN REARRANGEMENT. All amides, with exception of liquid METHANAMIDE, are white crystalline solids. The lower members are soluble in water with which they form HYDROGEN BONDS. Amides are hydrolyzed in acids and bases, and so their salts of strong inorganic acids (e.g., $CH_3CONH_2.HCl$) are unstable. Amides act as weak acids towards mercury(I) oxide forming ($RCONH_2)_2Hg$. They are reduced to amines by sodium in ethanol, and to amines containing one fewer carbon atom by the HOFMANN CONVERSION OF AMIDES. Dehydration to alkyl cyanides results from treatment with P_2O_5, PCl_5, $POCl_3$, $SOCl_2$ or by heating with ammonia. Amides evolve nitrogen when reacted with nitrous acid or alkyl nitrites. *N,N*-Dimethylamides (*see* *N,N*-DIMETHYLMETHANAMIDE) are good solvents for both polar and non-polar compounds. Inorganic amides (e.g., $NaNH_2$) contain the anion NH_2^- and are prepared by the action of ammonia on the metal or the ammonolysis of nitrides. Heavy metal amides, which can be explosive, may be prepared from the more stable alkali metal amides by a metathetical reaction (*see* DOUBLE DECOMPOSITION) with, for example, a thiocyanate, in liquid ammonia.

amidines ($RC(NH)NH_2$). Imido derivatives of amides, prepared from an imidic ester (RCN + $R'OH$ + HCl → $RC(NH)OR'$) and ammonia. They are strong bases and are readily hydrolyzed to amides.

amido group. The group –$CONH_2$. The term was formerly used for the AMINO GROUP.

amidol. *See* AMINOPHENOLS.

amidone. *See* METHADONE.

amination. Insertion of an AMINO GROUP into a molecule to form an AMINE.

amine oxides (R_3NO). Bases formed by the oxidation of tertiary AMINES. They exist as $R_3NH^+OH^-$ in aqueous solution. Amine oxides form addition compounds with hydrogen and alkyl halides (e.g., $[R_3NOR]^+Cl^-$, $[R_3NOH]^+Br^-$).

amines. Organic derivatives of ammonia in which one (primary), two (secondary) or three (tertiary) hydrogen atoms are replaced by alkyl or aryl groups. A fourth group may be added to give a quarternary ammonium salt (*see* QUATERNARY AMMONIUM COMPOUNDS). The lone pair on the nitrogen atom makes amines bases. Alkylamines are all stronger bases than ammonia ($pK_b > 4$), with $R_2NH > RNH_2 > R_3N$. Aromatic amines (*see* PYRIDINE) are very weak bases ($pK_b > 9$). The amino group is found widely in biologically important molecules (e.g., AMINO ACIDS, ALKALOIDS and some VITAMINS). The smell of rotting fish is due to amines produced by bacteria. NYLON is an amine derivative. Amines may be distinguished from one another by their reaction with nitrous acid (see below).

primary amines. For preparation and properties—*see* Figure.

secondary amines. Prepared from primary amines by reaction with an alkyl halide, also from sodium cyanamide and an alkyl halide. They undergo most of the reactions of primary amines. In addition they react with nitrous acid to give NITROSAMINES. The formation of these oils is a test for secondary amines. Enamines ($R_2C=CHNR_2$) are formed with aldehydes containing an α-hydrogen. Oxidation with Caro's acid (*see* SULPHUR OXOACIDS) gives only a ketone via a ketimine ($R_2C=NH$).

tertiary amines. Dissolve in cold nitrous acid to form the nitrite salt ($R_3NH^+NO_2^-$). Compounds containing the OXO GROUP are formed on warming if the alkyl group has an α-hydrogen. They are not affected by potassium permanganate, but are oxidized to the AMINE OXIDE ($R_3N^+O^-$) by Caro's acid, ozone or hydrogen peroxide. Tertiary amines may be degraded to secondary amines by reaction with BrCN and hydrolysis

$$R_3N + BrCN \rightarrow [R_3NCN]^+ Br^- \rightarrow$$

$$RBr + R_2NCN \overset{H^+}{\rightarrow} R_2NH$$

See also METHYLAMINES.

amino acid analysis. Amino acids in protein and peptide hydrolyzates can be analyzed, qualitatively and quantitatively, by ION EXCHANGE CHROMATOGRAPHY, partition chromatography (*see* CHROMATOGRAPHY) or ELECTROPHORESIS. Amino acid sequences can be obtained by specific enzymic treatment followed by specific reactions (e.g., C-terminal groups by reduction to –CH₂OH using lithium tetra-

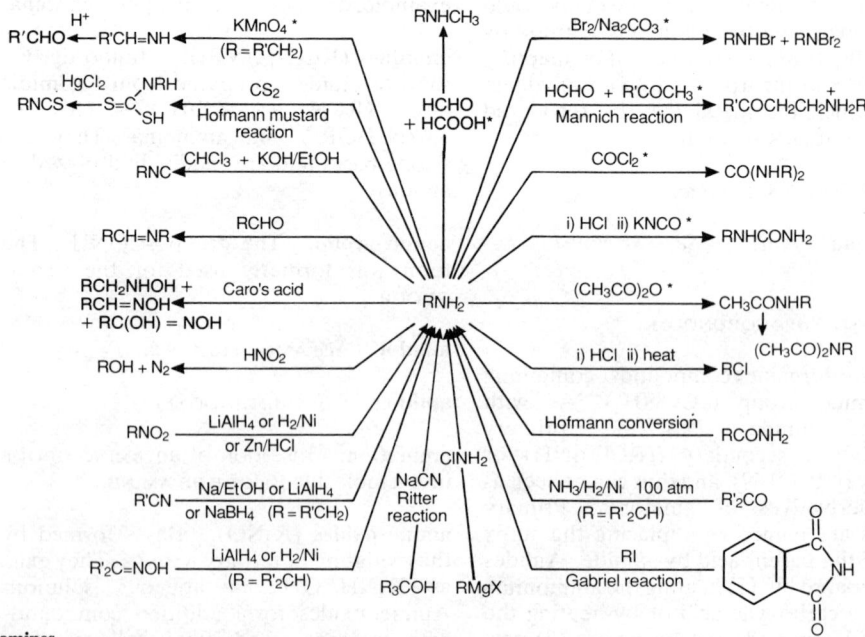

amines
Preparation and reactions of primary amines (* also given by secondary amines).

hydridoaluminate, N-terminal groups with 1-fluoro-2,4-dinitrobenzene). *See also* FORMOL TITRATION.

amino acids. Organic compounds (*see* Table) containing both carboxyl (–COOH) and amino (–NH$_2$) groups. Over 100 amino acids have been isolated and identified from natural sources, but only about 20 α-amino acids (–NH$_2$ and –COOH attached to the same carbon atom) form the building blocks of PROTEINS. These contain at least one asymmetric carbon atom (except glycine) with the L-configuration. Condensation of –NH$_2$ of one amino acid with –COOH of another gives an amide bond (–CONH–), hence leading to a protein. Amino acids are obtained by: (1) hydrolysis of proteins (acidic or enzymic) followed by chromatographic separation; (2) substitution (RCH(X)COOH + NH$_3$ + H$^+$); (3) STRECKER SYNTHESIS (RCHO + NH$_3$ (–H$_2$O) + HCN + H$^+$); (4) reductive amination (RCOCOOH + H$_2$,NH$_3$/Pd); (5) some can be prepared by fermentation. Amino acids are colourless, crystalline substances, melting with decomposition (>200°C), and are soluble in water, but practically insoluble in organic solvents. They are used as a food additive (e.g., monosodium glutamate (MSG, 621)) or dietary supplement. Essential amino acids are necessary nutritional factors for survival; they are not synthesized in the body and must be supplied from external sources. They are amphoteric, forming hydrochlorides and sodium salts; in solution the ZWITTERION is predominant.

Amino acids undergo the normal reactions associated with the –NH$_2$ and –COOH groups. Dehydration forms diketopiperazines; deaminases and decarboxylases remove the –NH$_2$ and –COOH groups, respectively. Treatment with methanal converts the amino group to –NHCH$_2$OH group leaving the –COOH group free for titration (FORMOL TITRATION). With ninhydrin, amino acids (except proline and hydroxyproline) form blue complexes. Fluoronitrobenzene (FDNB) blocks the –NH$_2$ group yielding yellow dinitrophenyl (DNP) derivatives. They are good ligands for a wide range of metals. Some amino acids, as opposed to PEPTIDES, exhibit potent biological activity (e.g., thyroxine — L-O-(4-hydroxy-3,5-diiodo-phenyl)-3,5-diiodotyrosine — a HORMONE that stimulates metabolism).

aminoacridines. When an amino group is substituted in the 3-, 6-, or 9- positions of an ACRIDINE molecule, a strong base is obtained as a result of delocalization of the positive charge on the cation. Only acridines that are highly ionized at physiological pH (e.g., PROFLAVINE, 9-aminoacridine) and that have a critical minimum area of flatness are active as ANTISEPTICS. The flat acridine ring is believed to be held by van der Waals forces to corresponding flat purine and pyrimidine rings in microbial DNA. Interaction of the positively charged amino groups with the phosphate anions results in stiffening of the helical structure of DNA.

4-aminoazobenzene (C$_6$H$_5$N=NC$_6$H$_4$NH$_2$). Orange–yellow needle-like solid (mp 126°C), prepared by diazotizing (*see* DIAZONIUM COMPOUNDS) aniline with an insufficiency of nitrous acid or by the rearrangement of diazoaminobenzene (C$_6$H$_5$N=NNHC$_6$H$_5$) when warmed with hydrochloric acid or aniline hydrochloride.

aminobenzene. *See* ANILINE.

4-aminobenzenesulphonamide. *See* SULPHONAMIDES.

2-aminobenzoic acid (anthranilic acid, H$_2$NC$_6$H$_4$COOH). Colourless crystalline solid (mp 145°C), prepared by the reduction of 2-nitrobenzoic acid. It is manufactured by the oxidation of phthalamide with alkaline sodium hypochlorite (*see* HOFMANN CONVERSION OF AMIDES). It behaves as an acid and an amine. On distillation, it decarboxylates to aniline and is used in the preparation of dyes and indigo (*see* INDIGOTIN). Decomposition of the ZWITTERION gives BENZYNE as intermediate.

4-aminobenzoic acid (*p*-aminobenzoic acid, vitamin B complex). *See* VITAMINS.

ε-aminocaprolactam. *See* CAPROLACTAM.

aminoglycosides. *See* ANTIBIOTICS.

amino group. The group –NH$_2$. *See also* AMIDO GROUP, AMINES.

amino acids

Properties of naturally occurring amino acids (RCH(NH$_2$)COOH).

Name (abbreviation)	R	M_r $[\alpha]_D^t$	$t/°C$	pK values (pI)	Occurrence[a]	General comments[b]
Monoaminomonocarboxylic acids						
Glycine (Gly)	H	75.05 —		2.34, 9.60 (5.97)	M: silk, gelatin	Only amino acid without asymmetric carbon
L-(+)-Alanine (Ala)	CH$_3$	89.10 +2.8°	25	2.35, 9.69 (6.02)	M: silk, gelatin, zein	N
L-(+)-Valine (Val)	(CH$_3$)$_2$CH	117.15 +6.42°	20	2.32, 9.62 (5.97)	S: range of fibrous proteins	E, N
L-(−)-Leucine (Leu)	(CH$_3$)$_2$CHCH$_2$	131.18 −10.8°	25	2.36, 9.60 (5.98)	M: haemoglobin, zein	E, N. One of the most common amino acids
L-(+)-Isoleucine (Ile)	CH$_3$CH$_2$CH(CH$_3$)	131.18 +11.29°	20	2.36, 9.68 (6.02)	S: casein, zein	E, N
L-(−)-Serine (Ser)	HOCH$_2$	105.10 −6.87°	20	2.21, 9.15 (5.68)	S: silk and range of proteins	H
L-(−)-Threonine (Thr)	CH$_3$CH(OH)	119.12 −28.3°	26	2.09, 9.10 (5.60)	S: casein	E, H
Monoaminodicarboxylic acids						
L-(+)-Aspartic acid (Asp)	HOOCCH$_2$	133.11 +4.70°	20	2.09, 3.86, 9.82 (2.98)	S: range of proteins	
L-(+)-Glutamic acid (Glu)	HOOC(CH$_2$)$_2$	147.13 +11.5°	18	2.19, 4.25, 9.67 (3.22)	M: prolamines, globulins, casein	
L-(−)-Asparagine (Asn)	H$_2$NOCCH$_2$	132.13 −5.42°	20	2.02, 8.80 (5.41)		D. Widely distributed in plants
L-(+)-Glutamine (Gln)	H$_2$NOC(CH$_2$)$_2$	146.15 +32.5°	20	2.17, 9.13 (5.70)	Range of plants, roots, seedlings	D

Name (abbreviation)	R	M_r $[\alpha]_D$	$t°C$	pK values (pI)	Occurrence[a]	General comments[b]
Diaminomonocarboxylic acids						
L-(+)-Lysine (Lys)	$H_2N(CH_2)_4$	146.19 +14.6°	20	2.18, 8.95, 10.53 (9.74)	M: albumins, globulins	E
L-(+)-Arginine (Arg)	$H_2NC(NH)NH(CH_2)_3$	174.20 +12.5°	20	2.17, 9.04, 12.48 (10.76)	M: protamines, histones	E. Hydrolysis gives ornithine
Diaminopimelic acid (Dap)	$HOOCCH(NH_2)(CH_2)_3$	190.05				Only found in bacterial proteins
L-(+)-Ornithine	$H_2N(CH_2)_3$	132.16 +11.5°	25	1.94, 8.65, 10.76 (9.7)		Isolated from proteins after alkaline hydrolysis
Sulphur-containing amino acids						
L-(+)-Cysteine (Cys)	$HSCH_2$	121.16 +9.8°	30	1.71, 8.33, 10.78 (5.02)	S: range of proteins	Breakdown product of cystine in body; soln. oxidized to cystine
L-(-)-Cystine (Cys–Cys)	SCH_2 $\mid$ SCH_2	240.30 −214.4° (in HCl)	21	1.0, 2.1, 8.02, 8.71 (5.06)	Abundant in skeletal tissue, hair	−S−S−bonds play important role in protein structure stabilization
L-(-)-Methionine (Met)	$CH_3S(CH_2)_2$	149.21 −8.2°	25	2.28, 9.21 (5.75)	Egg and vegetable proteins	E. Important role in biological methylations
Aromatic amino acids						
L-(-)-Phenylalanine (Phe)	$C_6H_5CH_2$	165.19 −35.14°	20	1.83, 9.13 (5.48)	Ovalbumin, zein, fibrin	E
L-(-)-Tyrosine (Tyr)	p-$HOC_6H_4CH_2$	181.19 −8.64° (in HCl)	20	2.20, 9.11, 10.07 (5.67)	Wide range of proteins	H. Least soluble of all amino acids

continued overleaf

Name (abbreviation)	R	M_r $[\alpha]_D^t$	$t/°C$	pK values (pI)	Occurrence[a]	General comments[b]
Heterocyclic amino acids						
L-(−)-Tryptophan (Trp)	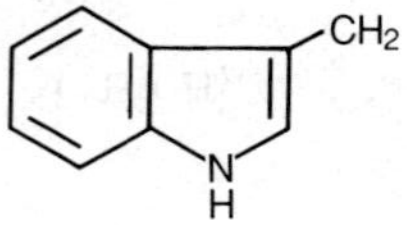	204.22 −31.5°	23	2.38, 9.39 (5.88)	Globulins, albumins, casein	E, H. Not synthesized by human body
L-(−)-Histidine (His)	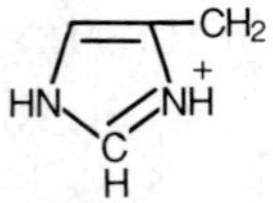	155.16 −39.01°	25	1.78, 5.97, 8.97 (7.59)	Histones, animal and vegetable globulins	E (rats)
Imino acids						
L-(−)-Proline (Pro) 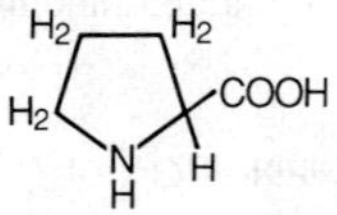	—	115.13 −85.0°	23	1.99, 10.6	Wide range of proteins	Not strictly an amino acid. Extracted from protein hydrolyzates
L-(−)-Hydroxyproline (Hyp) 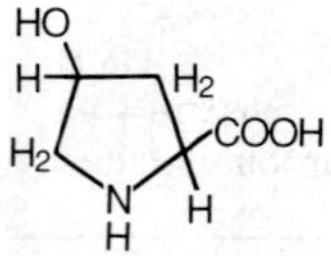	—	131.13 −75.2°	22.5	1.82, 9.65	Collagen, gelatin	

[a] M, major component; S, small amounts.
[b] E, essential amino acid; N, non-polar side chain acts as spacer in proteins; H, −OH group forms bonds with −NH and −CO group of main protein chain; D, −CONH$_2$ acts as a donor or acceptor of hydrogen bonds.

aminomethane.   See METHYLAMINES.

aminonaphthalenes (naphthylamines).   1-
and 2-compounds are prepared by reacting
the corresponding hydroxyl derivative (see
HYDROXYNAPHTHALENES) with ammonia in
the presence of ammonium sulphite
(Bucherer reaction). 1-Amino- and especially 2-aminonaphthalene are carcinogenic.
1-aminonaphthalene (α-naphthylamine).
Colourless crystals (mp 50°C), turning red
on exposure to air, unpleasant smell. Oxidation (CrO_3) gives α-naphthaquinone and
($KMnO_4$)benzene-1,2-dicarboxylic acid. It
is readily diazotized and coupled to aromatic hydroxyl or basic compounds, and couples
with diazonium salts in the 4-position.
Sulphonation at 130°C gives first the 4-sul-
phonic acid (used in the manufacture of the
dye Congo red), prolonged sulphonation
yields the 5- and 6-sulphonic acids.
2-aminonaphthalene (β-naphthylamine).
Colourless, odourless solid (mp 112°C).
Oxidized ($KMnO_4$) to benzene-1,2-dicarb-
oxylic acid, and couples with diazonium
salts only in the 1-position. Heating with
concentrated sulphuric acid yields 2-amino-
5-, -6-, -7- and -8-sulphonic acids according
to the temperature.

aminophenols.   Compounds containing
a phenolic hydroxyl and an amino group
attached to the benzene ring; crystalline
compounds, soluble in acids and alkalis,
readily oxidized in air. They undergo redox
reactions of the type exhibited by 1,4-quin-
ones. Aminophenols are widely used as
photographic developers and in the synthesis of dyestuffs.
2-aminophenol (2-$HOC_6H_4NH_2$). Solid
(mp 174°C), prepared by the reduction of
2-nitrophenol with sodium sulphide. With
ethanoic anhydride and sodium ethanoate it
gives 2-methylbenzoxazole. It is used as a
hair and fur dye.
4-aminophenol (rodinol, 4-$HOC_6H_4NH_2$).
Solid (mp 186°C), prepared by the reduction of 4-nitro- or 4-nitrosophenol.
4-hydroxyphenylglycine (glycin, 4-HO-
$C_6H_4NHCH_2COOH$). Solid (decomp.
246°C), prepared by the condensation of 4-
aminophenol with sodium chloroethanoate.
4-methylaminophenol (metol, HOC_6H_4-
$NHCH_3$). Solid (mp 260°C), prepared by

the decarboxylation of 4-hydroxyphenyl-
glycine or by heating a solution of methylamine and 1,4-dihydroxybenzene (hydroquinone) under pressure at 100°C. The
sulphate, mixed with hydroquinone, is the
widely used developer MQ.
2,4-diaminophenol (amidol, HOC_6H_3-
$(NH_2)_2$). Photographic developer; prepared by the electrolytic reduction of 1,3-
dinitrobenzene in acid solution.
   The methyl (or ethyl) ethers of the
amino-phenols, 2-methoxyaniline (*o*-anisi-
dine) and 4-methoxyaniline (*p*-anisidine) (or
ethoxyanilines, phenetidines) are prepared
by the reduction of the corresponding nitro-
anisole (ethyl phenyl ether, phenetole).
They are used in the preparation of azo
DYES.

3-aminopropanoic acid (β-alanine, H_2N-
CH_2CH_2COOH).   Naturally occurring
AMINO ACID that is not found in proteins;
white solid (mp 198°C (decomp.)), soluble
in water, $pK_1 = 3.6$, $pK_2 = 10.19$, prepared
by heating propenoic acid under pressure
with ammonia. It is used in the *in vivo* and *in
vitro* synthesis of pantothenic acid (see
VITAMINS).

β-aminopropylbenzene.   See   AMPHET-
AMINE.

6-aminopurine (adenine).   See PURINES.

amino sugars.   Compounds in which one
or more hydroxyl groups of a carbohydrate
are replaced by an amino group (see
Figure); very common in nature. The two
most important amino sugars are glucosamine (2-amino-2-deoxy-D-glucose) and
galactosamine. Glucosamine occurs in the
PEPTIDOGLYCANS of bacterial cell walls and
TEICHOIC ACIDS and in the shells of crabs
from which it can be extracted by boiling
with concentrated hydrochloric acid. The
N-acetyl derivative of muramic acid (3-*O*-
carboxyethyl-D-glucosamine) is one of the
components in the repeating unit of the
peptidoglycans of bacterial cell walls. Polymers of amino sugars are common in
nature. Many antibiotics (e.g., the aminoglycosides) form other classes of amino
sugars.
chitin. Polysaccharide, composed of units of
N-acetylglucosamine linked in a β-1,4
manner, with properties similar to those of

cellulose. It is found in the exoskeletons of insects and crustaceans.

chondroitin sulphate. Polysaccharide of repeating units of sulphated *N*-acetylgalactosamine and glucuronic acid which occurs in cartilage, tendons, bones and umbilical cord.

heparin. Polysaccharide similar to hyaluronic acid (*see below*) except that the glucosamine is a sulphamic acid derivative (the ethanoate group is replaced by –SO₃H) and some of the other hydroxyl groups are in the form of sulphates. It plays a vital role in the regulation of the clotting of blood. It is widely used clinically as an anticoagulant in the treatment of thrombosis.

hyaluronic acid. Polysaccharide of repeating units of *N*-acetylglucosamine and glucuronic acid. It is a component of animal connective tissue and of the synovial fluid; it is also present in the microcapsule of some bacteria.

2-amino-2-deoxy-D-glucose (glucosamine, β– form)

chitin

hyaluronic acid

neuraminic acid

amino sugars

neuraminic acid. Unusual amino sugar that is widely distributed in animal tissues and secretions, largely in the form of its *N*-acyl derivatives, the sialic acids.

aminotoluenes (toluidines, $H_2NC_6H_4CH_3$). 2-Aminotoluene (liquid, bp 200°C) and 4-aminotoluene (solid, mp 45°C) are prepared by the reduction (e.g., Fe + HCl) of the respective nitrotoluene. They are basic and form salts with mineral acids. They are used to prepare diazo DYES.

aminourea.   *See* SEMICARBAZIDE.

ammines.   Complexes of ammonia with metals, particularly transition metals. They are prepared by addition of aqueous ammonia to a solution of a soluble metal salt, or by the action of gaseous or liquid ammonia on an anhydrous salt.

ammonia.   *See* NITROGEN HYDRIDES.

ammonia soda process.   *See* SOLVAY PROCESS.

ammonium salts.   Many stable, crystalline salts of the tetrahedral NH_4^+ ion, exist which are prepared by the action of ammonia on the corresponding acid. Most are water-soluble and very similar in chemical properties to the ALKALI METAL salts. Salts of strong acids are fully dissociated and the solutions acidic. Many ammonium salts volatilize with dissociation at 300°C. Salts containing oxidizing anions decompose when heated (e.g., ammonium dichromate to Cr_2O_3 and N_2; ammonium nitrate to N_2O). Tetraalkylammonium ions (R_4N^+) (R_3N + RI) are of use when large univalent cations are required (e.g., in the determination of ionic size from transport number measurements).

ammonium carbonate (($NH_4)_2CO_3$). Used in baking powder and smelling salts (sal volatile).

ammonium chloride (NH_4Cl). Used in dry cells, as a MORDANT and as a flux.

ammonium nitrate (NH_4NO_3). Used mainly as a fertilizer; it is a component of many explosives.

ammonium sulphamate ($NH_4NH_2SO_3$). Non-selective HERBICIDE.

ammonium sulphate (($NH_4)_2SO_4$). Forms a series of ALUMS with the sulphates of iron and aluminium. Its main use is as a fertilizer.

ammonium thiocyanate (NH_4CNS). Colourless solid ($CS_2 + NH_3$ in EtOH); on heating forms the isomeric THIOUREA. It is used in photography.

ammonolysis. Splitting of compounds containing an OXO GROUP by ammonia to form AMIDES. For example

$$RCOOR' + NH_3 \rightarrow RCONH_2 + R'OH$$

$$RCOCl + 2NH_3 \rightarrow RCONH_2 + NH_4Cl$$

$$(RCO)_2O + 2NH_3 \rightarrow RCONH_2 +$$
$$RCOONH_4$$

amobarbital. *See* BARBITURATES.

amorphous. Describing a solid that is not crystalline, with no long-range order in its lattice and no cleavage planes. Many apparently amorphous powders have a microcrystalline structure as revealed by X-ray diffraction. Glasses are true amorphous solids.

AMP. *See* ADENOSINE MONOPHOSPHATE.

ampere (A; dimensions: $\varepsilon^{1/2} m^{1/2} l^{3/2} t^{-2}$). Basic SI unit; defined as that CURRENT which, if maintained in two parallel conductors of infinite length, of negligible cross-section and placed 1 metre apart in a vacuum, would produce a force between the two conductors equal to 2×10^{-7} newton per metre of length. The international ampere is the current which deposits 1.11800 milligrams of silver per second from a standard solution of silver nitrate.

amperometric titration. *See* TITRATIONS.

amphetamine (β-aminopropylbenzene, benzedrine, $C_6H_5CH_2CH(CH_3)NH_2$). Colourless liquid (bp 200°C (decomp.)); prepared by the reduction of phenylpropanone oxime. It is a vasoconstrictor and central nervous stimulant and is very addictive.

amphiprotic solvent. *See* PROTOPHILIC SOLVENT.

ampholines. Trade name for a series of polyamino–polycarboxylic acids with a range of p*I* values. They are used as carrier AMPHOLYTES to establish pH gradients in ISOELECTRIC FOCUSING.

ampholyte (amphoteric electrolyte). Substance that carries both positive and negative charges. An ampholyte (e.g., amino acid, protein, synthetic polyamino–polycarboxylic acid) can behave as both an acid and as a base.

amphotericin B. *See* ANTIBIOTICS.

amphoteric oxides. OXIDES that possess both an acidic and a basic function. An example is zinc oxide (ZnO) which dissolves in bases to give zincates (e.g., Na_2ZnO_2) and in acids to give zinc salts.

amu. *See* ATOMIC MASS UNIT.

amygdalin. *See* GLYCOSIDES.

amyl acetate (3-methylbutyl ethanoate). *See* ESTERS.

amylase. *See* ENZYMES.

amyl groups. Alkyl groups containing five carbon atoms.

n-amyl
pentyl ($CH_3CH_2CH_2CH_2CH_2-$)

sec-amyl
1-methylbutyl ($CH_3CH_2CH_2CH(CH_3)-$)

tert-amyl
1,1-dimethylpropyl (($CH_3)_2C(CH_2CH_3)-$)

iso-amyl
3-methylbutyl (($CH_3)_2CHCH_2CH_2-$)

amylopectin. *See* STARCH.

amylose. *See* STARCH.

amytal. *See* BARBITURATES.

An. *See* ACTINON.

anabolic agent. Compound that promotes storage of proteins and stimulates tissue metabolism. Such agents are often used in convalescence. Androgenic steroid HORMONES (e.g., testosterone) have marked anabolic properties.

anabolism. Sum of the processes involved in the biosynthesis of proteins, lipids, polysaccharides and nucleic acids from their constituents or simple precursors.

anaerobic process. Cellular process (e.g., fermentation) in which chemical energy is produced by a series of reactions in which molecular oxygen is not involved. The organic substrate is never completely oxidized. All fermentations are types of anaerobic respiration. In animals, plants and most bacteria, glucose is anaerobically broken down to lactic acid by glycolysis with the production of ADENOSINE TRIPHOSPHATE (ATP).

anaesthetic. Compound that induces loss of sensation in a specific part (local anaesthetic) or all of the body (general anaesthetic). Examples of general anaesthetics include cyclopropane (*see* CYCLOALKANES), HALOTHANE, ether, tribromoethanol, dinitrogen oxide (*see* NITROGEN OXIDES), BARBITURATES. Local anaesthetics include ALKALOIDS (e.g., cocaine), procaine, quinine, which all directly affect the nervous system, and chloroethane, which acts indirectly by freezing tissue.

analgesic. Compound that reduces pain perception and that augments subsequent anaesthetic agents. Examples include narcotics (e.g., morphine and pethidine), benzodiazepines (e.g., DIAZEPAM), ASPIRIN and 4-ACETAMIDOPHENOL.

analyte. Sample or solution being analyzed.

anatase. *See* TITANIUM.

anchimeric assistance. *See* NEIGHBOURING GROUP PARTICIPATION.

Andrew's isothermals. *See* REAL GAS.

androgens. Group of C_{19} HORMONES (e.g., testosterone and androsterone) with masculinizing properties biosynthesized from ethanoate and cholesterol in the testes of all vertebrates. Synthetic anabolic steroids are used to promote the growth of muscles and bones.

androsterone. *See* HORMONES.

anethole. *See* PHENOLS.

aneurine (thiamine, vitamin B_1). *See* VITAMINS.

angström (Å). Unit of length equal to 1×10^{-10} metre.

angular momentum. *See* MOMENTUM.

anharmonicity. Effect due to the weakening of a bond at large amplitudes which deviates from a pure HARMONIC OSCILLATOR. The frequency of vibration (ω_v) is dependent on the amplitude where

$$\omega_v = \omega_e \left[1 - x_e (v + 1/2)\right]$$

ω_e is the frequency for small oscillations, x_e the anharmonicity constant and v the vibrational quantum number. The vibrational energy (E) corrected for anharmonicity is

$$E = (v + 1/2)h\omega_e - (v + 1/2)^2 h\omega_e x_e$$

anhydride. Substance formed by the elimination of water from an acid or base. Organic anhydrides are important synthetic reagents (*see* ETHANOIC ANHYDRIDE, PHTHALIC ANHYDRIDE). In inorganic chemistry, the oxides of non-metals that form OXOACIDS in water are the anhydrides of the acid.

anhydrite. *See* SULPHATES.

anhydrous. Having the elements of water removed. Anhydrous salts (i.e. without WATER OF CRYSTALLIZATION) are used to detect the presence of water by a colour change (e.g., $CoCl_2$, blue to red; $CuSO_4$, white to blue) or to act as a drying agent (*see* DESICCANT).

anilides. *N*-Phenyl derivatives of ANILINE. *See also* ACETANILIDE.

aniline (aminobenzene, $C_6H_5NH_2$). Colourless oily liquid (bp 184°C), prepared by the reduction (Sn/HCl) of nitrobenzene. Aniline is steam distilled from alkaline solution. It is manufactured by vapour-phase hydrogenation of nitrobenzene over copper. Its uses are in the manufacture of dyes, pharmaceuticals, antioxidants and in the rubber industry. Aniline is a weaker base than aliphatic amines ($pK_b = 4.58$). It forms crystalline salts with mineral acids and forms the anion $PhNH^-$ with alkali metals. Aniline reacts with alkyl halides to give ultimately quaternary ammonium salts and is ethanoylated to acetanilide. It condenses with aldehydes to give SCHIFF'S BASES, undergoes the CARBYLAMINE REACTION, and with sodium nitrite in dilute mineral acid at 0°C gives the benzene DIAZONIUM COMPOUND. The amino group is strongly *ortho*- and *para*-directing (*see* ORIENTATION IN THE BENZENE RING) and brominates to 2,4,6-tribromoaniline. With glycerol and sulphuric acid, quinoline is formed (*see* SKRAUP REACTION), and phenylglycine is the product of reaction with trichloroethene.

anils (*N*-arylimides). *See* SCHIFF'S BASES.

animal charcoal. *See* CARBON.

anion. Negatively charged ion. When current flows anions are attracted to the ANODE of an electrochemical cell. *See also* CATION.

anion exchanger. *See* ION EXCHANGER.

anisidines. Methyl ethers of AMINO-PHENOLS.

anisole. *See* AROMATIC ETHERS.

anisotropic. Denoting a medium in which certain physical properties (e.g., thermal and electrical conductivity, refractive index) are different in different directions. In NUCLEAR MAGNETIC RESONANCE SPECTROSCOPY and ELECTRON SPIN RESONANCE SPECTROSCOPY, the tumbling motion of molecules in solution leads to ISOTROPIC

spectra. In the solid state, anisotropic effects may become apparent. Diamagnetic anisotropy occurs within molecules in which SHIELDING and deshielding of magnetic nuclei depend on the orientation of the nucleus with respect to the applied magnetic field. Crystals, other than of the cubic system, are anisotropic.

annealing. Controlled heating and cooling of a metal, glass or other solid to relieve stress.

annihilation. Destruction of a particle and its ANTIPARTICLE on collision; the energy released is carried away by photons or mesons (*see* ELEMENTARY PARTICLES), for example

$$\beta^+ + e^- \rightarrow 2 \text{ photons}$$

Each photon has an energy of 0.511 MeV.

annulenes (C_nH_n). Simple conjugated cyclic polyalkenes (*see* Figure), prepared to test the $(4n + 2)$ rule (*see* HÜCKEL'S RULE). The prefix [*n*]- is added to indicate the number of carbon atoms in the ring, benzene being [6]-annulene. Annulenes containing 12, 14, 16, 18, 20, 24 and 30 carbon atoms have been prepared; of these only [14]-, [18]- and [30]-annulenes have $(4n + 2)$ π-electrons, whereas the rest have $4n$. The latter molecules have olefinic properties; the [14]-annulene, although complying with the rule, does not show the magnetic properties required for aromatic character and further it is not planar. [18]-Annulene is planar and at low temperatures shows two broad peaks in the ^1H NMR spectrum (*see* NUCLEAR MAGNETIC RESONANCE SPECTROSCOPY) at δ 8.9 and δ 1.8 with an intensity

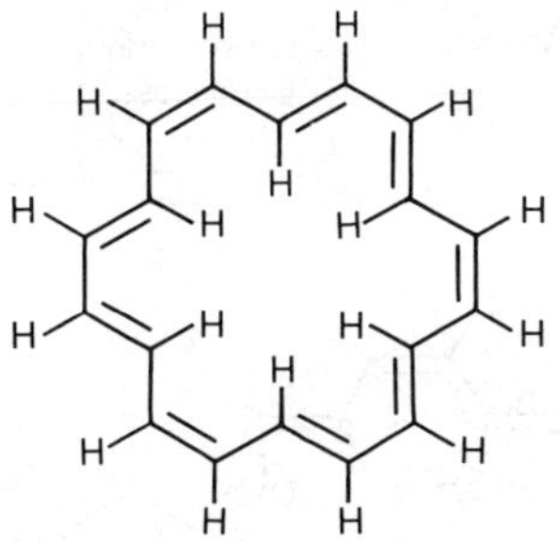

18 - annulene 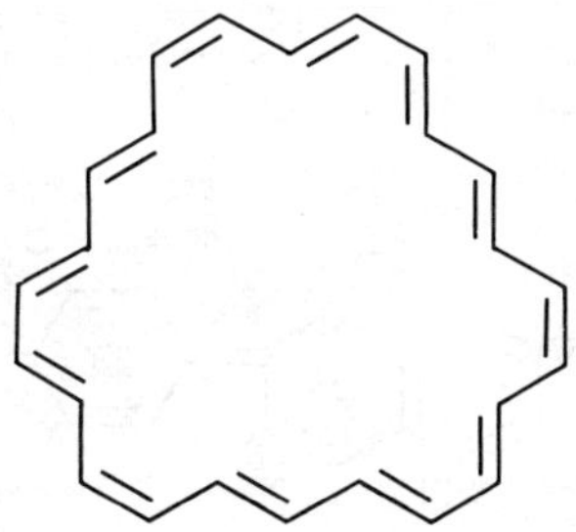   24 - annulene

annulenes

ratio of 2:1. The large peak is in the typical aromatic region and results from the protons outside the ring which are deshielded, the peak at the negative δ-value results from the strongly shielded protons inside the ring. At higher temperatures, the spectrum is a single resonance showing that the inside and outside hydrogen atoms are equilibrating rapidly; this can only occur if *cis–trans*-interconversion occurs about the double bonds. Annulenes are not stable compounds; the [4n + 2]-annulenes show typical aromatic properties (e.g., [18]-annulene has been converted to the nitro, ethanoyl, bromo and aldehyde derivatives by ELECTROPHILIC SUBSTITUTION reactions).

anode. Electrode at which OXIDATION occurs. *See also* CATHODE.

anodic protection. *See* CORROSION, PASSIVITY.

anodizing. Formation of an oxide or chloride film or coating on a metal acting as an ANODE by electrolysis in a suitable solution. Anodizing may increase resistance to CORROSION, abrasion or may alter the reflectance or absorbance properties of the metal.

anolyte. Electrolyte surrounding the ANODE in an electrochemical cell.

anomers. α- and β-isomers of a cyclic carbohydrate, differing only in the configuration at the carbonyl (hemiacetal, *see* ACETALS) carbon atom (e.g., glucose, *see* HEXOSES).

anorthite. *See* SILICATES.

antarafacial. *See* SUPRAFACIAL.

anthocyanins. Group of soluble FLAVONOID pigments responsible for the blue, red and purple colour of flowers, fruits and leaves (contributing to the autumn shades). Many change colour with acidity, generally red to blue as the pH increases. Their colour also depends on whether the anthocyanin is complexed with metal ions; in roses the red pigment is not complexed, whereas the blue of the cornflower is due to a metal complex (*see* Figure). They are 3,5-glycosides, which on hydrolysis yield one or more sugars and a coloured AGLYCONE, an anthocyanidine.

anthracene ($C_{14}H_{10}$). POLYCYCLIC HYDROCARBON showing aromatic character, obtained from the anthracene oil fraction of coal tar (0.5 percent). It is synthesized by a FRIEDEL–CRAFTS REACTION (2 mol benzyl chloride or benzene + dichloromethane). It is a colourless solid (mp 216°C), with a blue fluorescence; insoluble in water, but sparingly soluble in organic solvents. It is very reactive in the 9,10-positions: for example, chlorine adds to give anthracene dichloride, it reacts with maleic anhydride to form a Diels–Alder (*see* DIELS–ALDER REACTION) adduct (I) and with BENZYNE to give triptycene (II); in the presence of light it reacts with oxygen to give a photooxide (III); when irradiated in the absence of oxygen it dimerizes (IV) (*see* Figure). With aqueous nitric acid or chromic oxide it is oxidized to 9,10-anthraquinone (*see* QUINONES), but with nitric acid in glacial ethanoic acid 9-mono- and 9,10-dinitro-

red rose pigment
Gl=glucose

blue cornflower pigment
M=Fe^{3+} or Al^{3+}, Y=some anion

anthocyanins

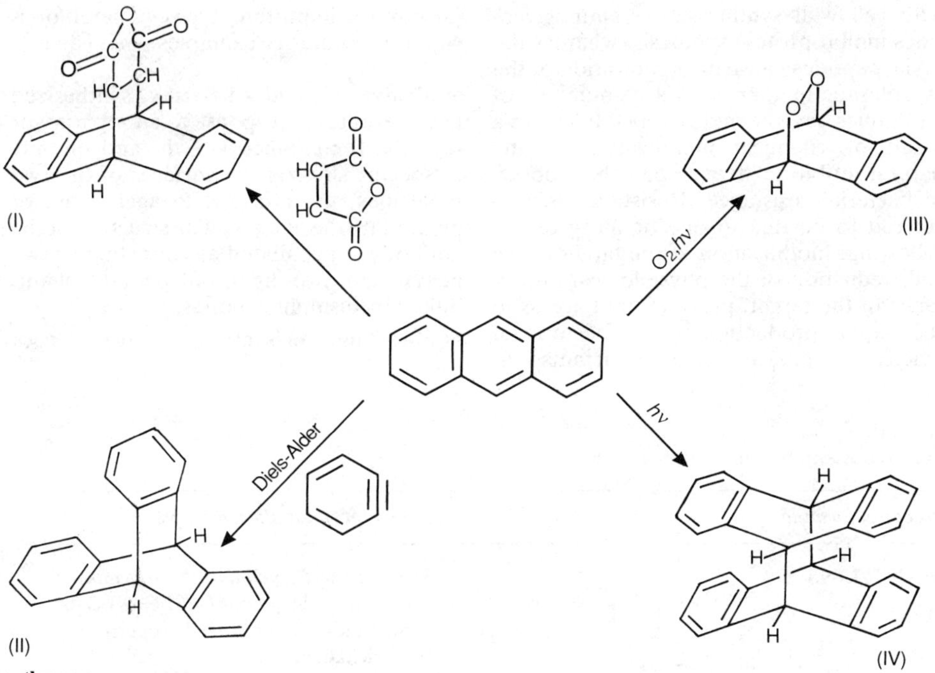

anthracene
Some reactions of anthracene.

anthracenes are formed. It is more basic than naphthalene or benzene and forms charge-transfer or π-complexes with Lewis bases. It is used in the manufacture of anthraquinone and many dyestuffs.

anthracite.   *See* COAL.

anthranilic acid.   *See*   2-AMINOBENZOIC ACID.

anthranol.   *See* HYDROXYANTHRACENES.

9,10-anthraquinone.   *See* QUINONES.

anthrol.   *See* HYDROXYANTHRACENES.

anthrone (10-keto-9,10-dihydroanthracene, $C_{14}H_{10}O$). Colourless needles (mp 155°C), prepared by heating 2-benzylbenzoic acid with concentrated sulphuric acid or the reduction of 9,10-anthraquinone (Sn + HCl in ethanoic acid). It is soluble in dilute alkali, which on acidification precipitates the enol form (10-hydroxyanthracene). It shows no phenolic properties and does not couple with diazonium salts. It is used in the

colorimetric determination of sugars and starches.

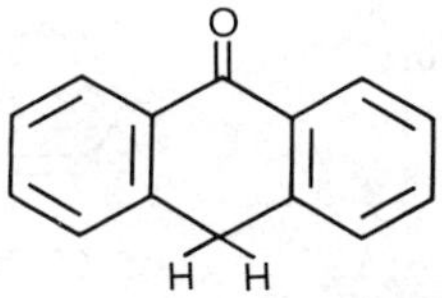

anthrone

anti-**addition.**   *See syn*-ADDITION.

antiaromatic. Referring to conjugate cyclic systems which are thermodynamically less stable than the corresponding acyclic analogues (e.g., cyclobutadiene and cyclooctatetraene, *see* CYCLOALKENES).

antibiotics. Organic compounds produced by one microorganism (fungi and some bacteria) that are lethal to another organism. The targets of antibiotic therapy are usually bacteria. The toxicity is selective because the antibiotic acts against specific targets in the microorganism; for example the penicillins and cephalosporins interfere with cell wall synthesis, the aminoglycosides inhibit protein synthesis, whereas the cyclic peptides cause disorganization of the cytoplasmic membrane. Large numbers of antibiotics have been developed (some as a result of chemical modification of the natural antibiotic) to overcome the problem of bacterial resistance. Resistance is considered to be due to one or more of the following: modification of the target in the cell, reduction of the physiological importance of the target, prevention of access to the target, production by the bacteria of inactivating enzymes (e.g., β-lactamases).

(For more important types of antibiotics with representative examples—*see* Table.)

antibody. Globular PROTEIN, synthesized in vertebrates in response to an ANTIGEN or HAPTEN. It combines with the antigen only at specific sites on the surfaces of the two molecules; the effect is to agglutinate or precipitate the antigen. The structure of the antibody is postulated as consisting of two heavy and two light polypeptide chains linked by disulphide bonds.

antibonding orbital. *See* MOLECULAR ORBITAL.

antibiotics
Some representative antibiotics.

Structural formula	Source/properties/activity/uses
PENICILLINS R COHN—[S (CH₃)₂ / N—COOH / O]	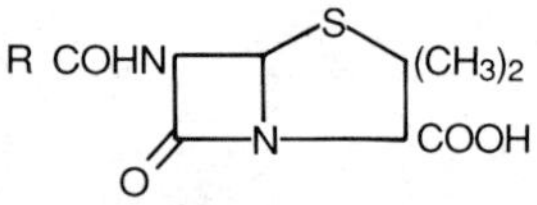 Produced by *Penicillium chrysogenum*, chemical modifications (see below) yield compounds with improved properties; bactericidal, act by inhibiting cell wall synthesis
Benzylpenicillin (penicillin G) R = (benzene)CH₂–	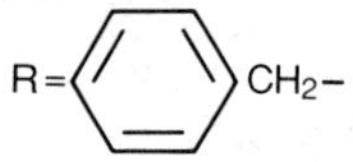 K salt is acid-labile, non-toxic and mainly effective against gram-positive organisms; resistance due to presence of penicillinase (β-lactamase) which opens the β-lactam ring giving penicilloic acid
Carbenicillin R = (benzene)CH– / COOH	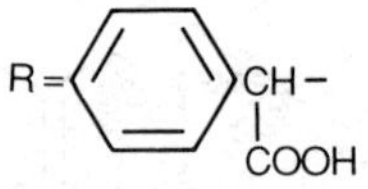 Active against gram-negative bacilli (e.g. *Pseudomonas aeruginosa*), useful in treatment of burns
Methicillin R = (benzene)OCH₃ / OCH₃	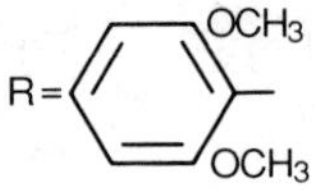 Active against penicillinase-producing organisms, slightly toxic; some organisms (e.g., some *Staphylococcus aureus* strains) are intrinsically resistant to methicillin
CEPHALOSPORINS	Formed by *Cephalosporium* sp.; broad-spectrum antibiotics, inhibitors of cell wall synthesis; active against pencillinase-producing organisms
Cephalosporin C H₂N \ HC(CH₂)₃COHN—[S / N—CH₂OCOCH₃ / O / COOH] / HOOC	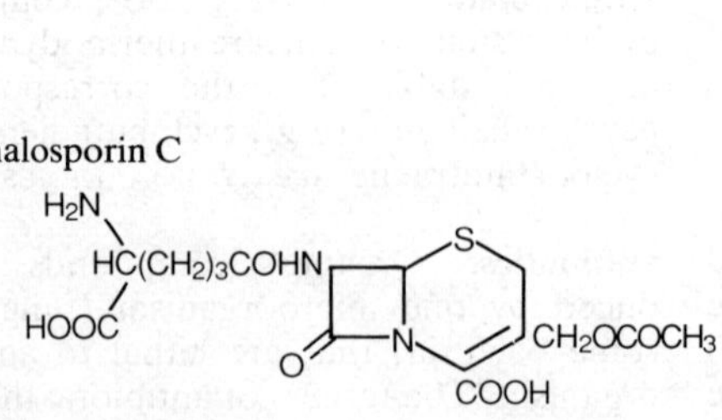 High degree of resistance to penicillinase, with low toxicity
AMINOGLYCOSIDES	Bactericidal, active against some gram-positive and many gram-negative bacteria; inhibitors of protein synthesis

Structural formula	*Source/properties/activity/uses*

Streptomycin

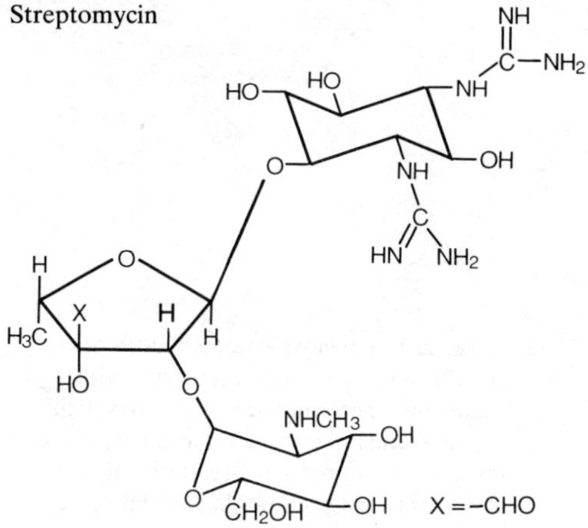

X = −CHO

Isolated from *Streptomyces griseus*; toxic, solutions (of sulphate); stable at pH 3−7 for long periods; active against *Mycobacterium tuberculosis*

Dihydrostreptomycin

X=CH$_2$OH

Similar antibacterial properties to streptomycin

Gentamicins

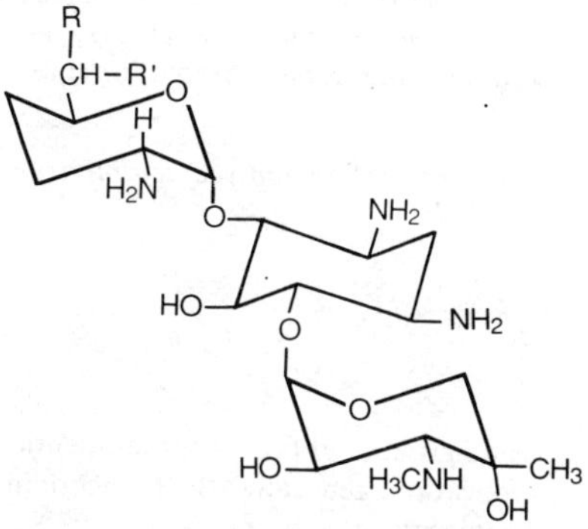

C$_{1a}$: R = H, R′ = NH$_2$
C$_1$: R = CH$_3$, R′ = NHCH$_3$
C$_2$: R = CH$_3$, R′ = NH$_2$

Produced by *Micromonospora purpurea*, consists of 3 closely related gentamicins (C$_1$, C$_{1a}$, C$_2$); broad-spectrum activity, useful against *Pseudomonas aeruginosa*, but no use against anaerobic bacteria; ototoxic, high doses can lead to renal failure

TETRACYCLINES

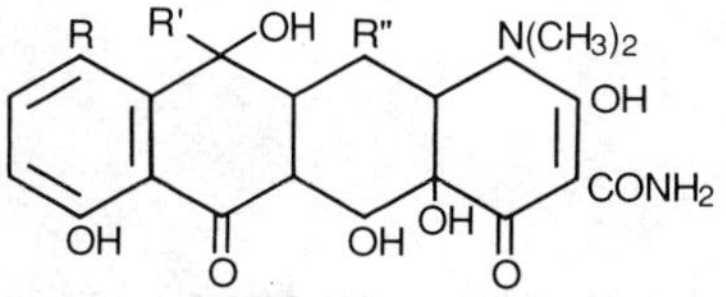

Isolated from *Streptomyces* spp; generally administered as hydrochloride or sulphate; broad-spectrum antibiotics against gram-positive and negative bacteria and against Rickettsiae; usefulness has declined because of increasing bacterial resistance; deposited in growing bone and teeth, causing staining; can cause liver damage

Tetracycline

R=H, R′=CH$_3$, R″=H

continued overleaf

Structural formula	*Source/properties/activity/uses*

Chlortetracycline

R=Cl, R'=CH$_3$, R''=H

Oxytetracycline

R=H, R'=CH$_3$, R''=OH

Demethylchlortetracycline

R=Cl, R'=R''=H

MACROLIDES

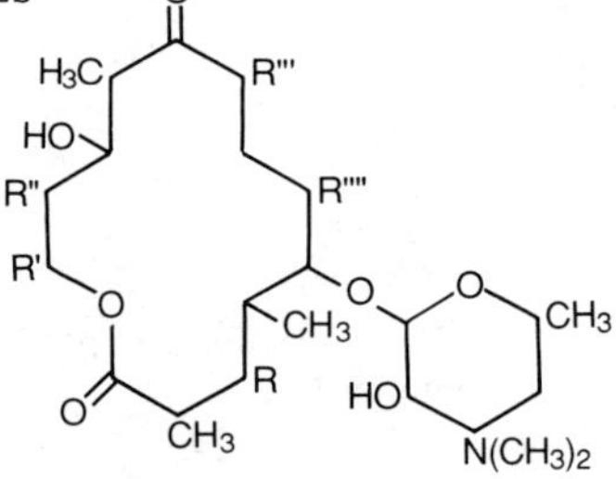

Large group of closely related antibiotics; consist of macrocyclic lactone ring with sugars attached; similar activity to penicillins, active against pneumococci and streptococci; low concentrations bacteriostatic, high concentrations bactericidal; inhibit protein synthesis

Erythromycin A

R=L-cladinose, R'=C$_2$H$_5$, R''=R''''=CH$_3$(OH), R'''=CH$_3$

Most important macrolide; produced by *Streptomyces erythreus*, weak base, not very soluble in water, neutral solution stable at 5°C

Oleandomycin

R=L-oleandrose, R'=R''=R''''=CH$_3$,

R''' = CH−CH$_2$ (epoxide)

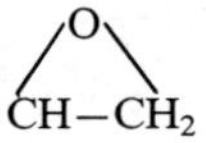

Produced by *Streptomyces antibioticus*

PEPTIDES

Large group, but few have therapeutic application; generally affect functioning of cytoplasmic membrane

Bacitracin

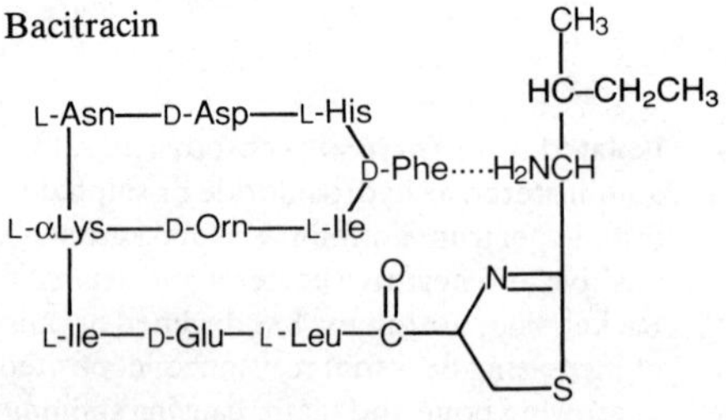

Produced by *Bacillus subtilis*; highly active against many gram-positive bacteria (useful against penicillin-resistant organisms) especially for skin and eye infections; nephrotoxic

Gramicidin S

L-Val–L-Orn–L-Leu–D-Phe–L-Pro

L-Pro–D-Phe–L-Leu–L-Orn–L-Val

Produced by *Bacillus brevis*; active against most species of aerobic and anaerobic gram-positive bacteria including mycobacteria; toxic against RBC, liver, kidney

Structural formula	*Source/properties/activity/uses*

Polymyxins

L-DABNH₂–R–L-Leu–L-DABNH₂
   └ L-DAB–L-Thr–L-DABNH₂┘
     |
   L-DABNH₂–L-Thr–L-DABNH₂–R′

Group of basic peptides, containing L-α,γ-diaminobutyric acid (DAB), L-threonine, 2 other amino acids and long-chain fatty residue; selectively active against gram-negative bacteria; neuro- and nephrotoxic; now replaced by gentamicin

B₁: R=D-Phe, R′=6-methyloctanoyl
B₂: R=D-Phe, R′=6-methylheptanoyl
E (colistin A): R′=D-Leu, R′=6-methyloctanoyl

POLYENE ANTIBIOTICS

Nystatin

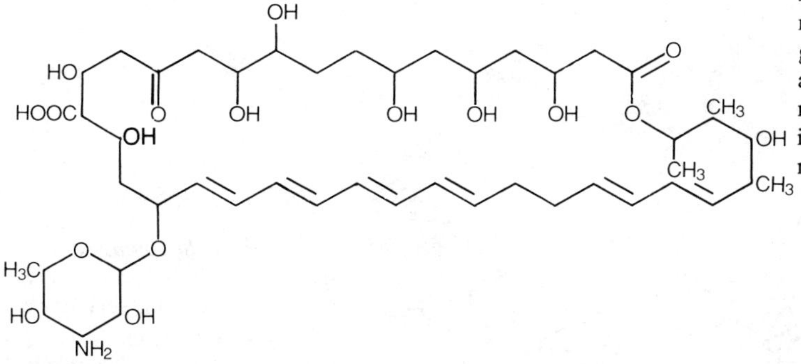

Polyene antibiotic; produced by *Streptomyces noursei*; not absorbed when given orally; active against yeast and fungi, mainly *Candida albicans* infections of skin and mucous membranes

Amphotericin B

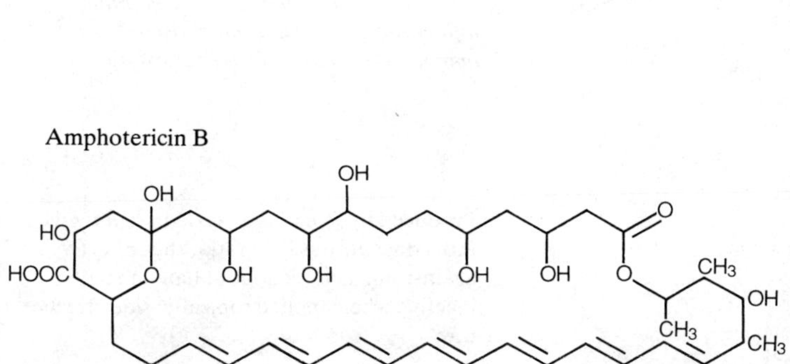

Polyene antibiotic; produced by *Streptomyces nodosus;* insoluble in water; forms low-solubility salts; inhibits growth of yeast-like fungi that cause systemic mycoses; ineffective against bacteria

continued overleaf

Structural formula	Source/properties/activity/uses
OTHER ANTIBIOTICS Chloramphenicol 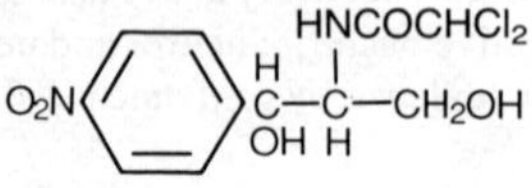	First broad-spectrum antibiotic discovered; produced by *Streptomyces venezuelae*; large-scale synthesis from 4-nitroacetophenone; *Salmonella typhii*, *Haemophillus influenzae* and *Bacillus pertussis* more susceptible to chloramphenicol than other antibiotics; strictly bacteriostatic; toxic, causing soreness in mouth (due to suppression of oral flora)
Fusidic acid and salts 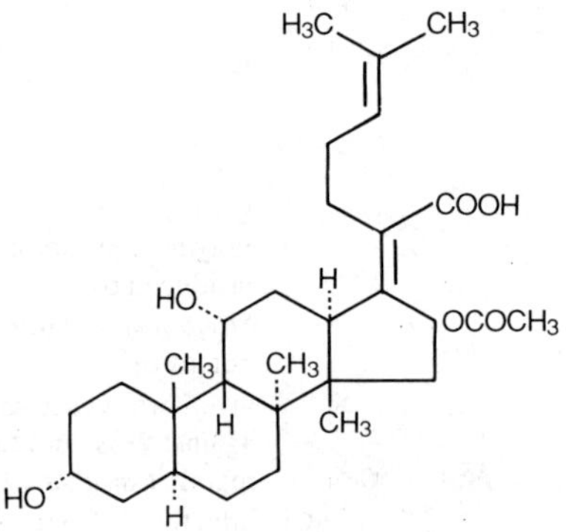	Narrow-spectrum steroid antibiotics, effective against penicillin-resistant streptococci; isolated from *Fusicium coccineum*; resistance unstable, reverts to sensitive state on subculture without antibiotic; no serious side effects
Novobiocin 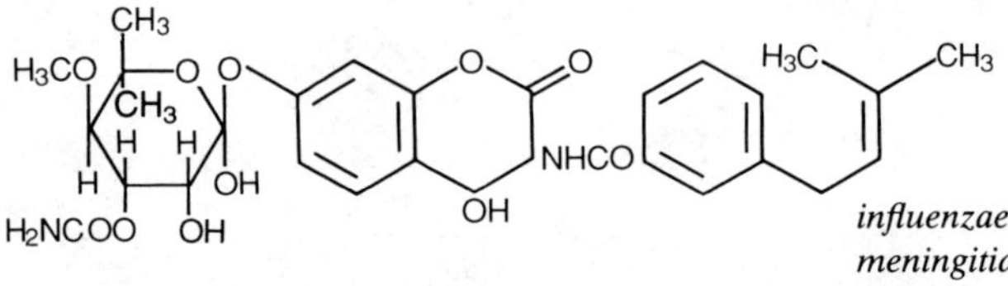	Isolated from *Streptomyces spheroides*, dibasic acid usually used as calcium salt; very active against *Streptococcus pneumoniae*, *Corynebacterium diphtheriae*, *Haemophilus influenzae*, *Neisseria gonorrhoeae*, *N. meningitidus*; primarily bacteriostatic
Griseofulvin 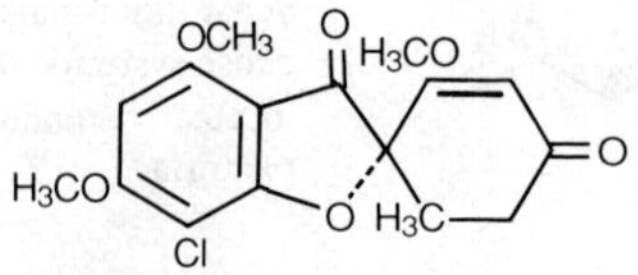	Produced by *Penicillium griseofulvum*; given p.o. concentrates in keratin, thus effective against fungal infections of hair and nails; inactive when applied topically; side effects rare

antichlor. Substance used for removing chlorine from a solution, clothing, etc. (e.g., sodium thiosulphate, sodium sulphite).

anti-**conformation.** *See* CONFORMERS.

antiferromagnetism. Magnetism in a material that has two interpenetrating sublattices magnetized in different directions. At low temperatures, the lattices are aligned antiparallel and so the material is diamagnetic (*see* DIAMAGNETISM) if the two lattices are equivalent and ferrimagnetic (*see* FERRIMAGNETISM) if they are not. At the Néel temperature (T_N), a cooperative effect occurs when the magnetic dipoles become aligned and the material exhibits PARAMAGNETISM. Chromium ($T_N = 475$ K,) α-manganese ($T_N = 100$ K) and nickel oxide ($T_N = 510$ K) are antiferromagnetic.

antifluorite structure. Structure adopted by compounds M_2X (e.g., K_2S, K_2O) in which the cations occupy the F^- positions and the anions the Ca^{2+} positions in the fluorite lattice (*see* CRYSTAL FORMS).

antifoaming agent. Substance that is dissolved in the liquid phase of a FOAM that destroys the foam or inhibits its formation: for example, polyamides and silicones are used in water boilers, and diethyl ether and octanol are sprayed on existing foams. Other uses are in electroplating and papermaking.

antifreeze. Additive that lowers the freezing point of a solution in order to prevent freezing at ambient temperatures; used specifically to refer to additives to water used to cool internal combustion engines. 1,2-DIHYDROXYETHANE is the most common antifreeze.

antigen. Substance that stimulates an animal to form a specific ANTIBODY. The antigenicity of a substance (e.g., bacterium, virus, pollen, protein) depends on the size and configuration of the antigen molecule. Proteins of relatively low molecular mass ($M_r < 10000$) are not very effective at producing an antibody (i.e. are not very antigenic). *See also* HAPTEN.

antihistamine. Synthetic compound, structurally related to HISTAMINE (usually a complex amine) whose presence in minute amounts counteracts the physiological and pharmacological effects of excess histamine in body fluids. Examples include cimetidine, diphenhydramine and mepyramine.

anti-**isomer.** *See* ISOMERISM.

antiknock additive. Substance added to PETROL (gasoline) to improve the OCTANE NUMBER. A typical mixture includes lead alkyls (*see* LEAD ORGANIC DERIVATIVES), to inhibit the precombustion chain reaction, and chloro- and bromoalkanes, to produce volatile lead compounds, which are carried out in the exhaust.

antimalarial agent. Natural or synthetic drug of the ALKALOID-type specific to combating malaria (e.g., quinine and chloroquine).

antimatter. *See* ANTIPARTICLE.

antimonates. *See* ANTIMONY OXIDES AND OXOACIDS.

antimony (Sb). METALLOID of group V (*see* GROUP V ELEMENTS). It occurs as stibnite (Sb_2S_3) (0.00005 percent of earth's crust). The ore is reduced with iron or roasted to antimony trioxide (Sb_2O_3) followed by reduction with carbon. Antimony has a bluish–white, metallic appearance. It burns in air to Sb_2O_3, is attacked by oxidizing acids (nitric acid to antimony pentoxide, and sulphuric acid to antimony sulphate), by halogens to SbX_3 and combines with metals to give antimonides (e.g., Zn_3Sb_2). Antimony is used in alloys (type metal, solder) and semiconductors. The chemistry is dominated by the +5 and +3 oxidation states, which are generally covalent; some cationic species (SbO^+ and Sb^{3+}) are known. Antimony compounds, which are very toxic, are used in flameproofing, paints, ceramics, enamels and glass. $A_r = 121.75$, $Z = 51$, ES [Kr]$4d^{10}5s^25p^3$, mp 630.74°C, bp 1750°C, $\rho = 6.68 \times 10^3$ kg m^{-3}. Isotopes: ^{121}Sb, 57.25 percent; ^{123}Sb, 42.75 percent.

antimony cationic species. Species present in a solution of antimony sulphate $(Sb_2(SO_4)_3)$ $(Sb_2O_3 + $ conc. $H_2SO_4)$ depend on the concentration of the acid: at low acid concentrations SbO^+ and $[Sb(OH)_2]^+$ are present; at higher concentrations the species are $[SbOSO_4]^-$ and $[Sb(SO_4)_2]^-$.

antimony halides. All the trihalides $(Sb_2O_3 + HF, Sb + Cl_2, Br_2, I_2)$ are known. They are hydrolyzed by water to the basic halide. They form many complexes (e.g., $[SbCl_4]^-$, $[SbF_4]^-$, $[SbCl_5]^{2-}$, $[SbCl_6]^{3-}$. The only pentahalides are the fluoride and chloride.
antimony trichloride $(SbCl_3)$. Forms complexes with neutral donors (e.g., $(PhNH_2)_2SbCl_3)$.
antimony pentafluoride (SbF_5). $(SbCl_5 + HF)$, composed of trigonal bipyramidal molecules in the solid and vapour states. It forms many complexes and complex ions (e.g., $[SbF_6]^-$, $[Sb_2F_{11}]^-$). It is a very powerful fluoride ion acceptor and is used to enhance the acidity of hydrofluoric acid and fluorosulphonic acid (HSO_3F). It is used as a fluorinating agent.
antimony pentachloride $(SbCl_5)$. Liquid $(SbCl_3 + Cl_2)$, is readily hydrolyzed and forms complexes (e.g., $[SbCl_6]^-$). It is used extensively as a chlorinating agent.
mixed (III) (V) complexes. Occur as salts (e.g., $M_2SbCl_6.SbCl_5)$.
oxohalides. Formed by the partial hydrolysis of the halides (e.g., $SbOCl$, $Sb_4O_5X_2)$.

antimony hydride (stibine, SbH_3). Unstable, poisonous and colourless gas $(Zn + Sb$ compound $+ HCl)$ that readily decomposes to antimony.

antimony organic derivatives. Range of alkyl and aryl derivatives (SbR_3) (Grignard reagent $+ SbCl_3)$ exist. They are colourless liquids, soluble in organic solvents, but not in water and are oxidized spontaneously in air. They form adducts with halogens, sulphur, and selenium (e.g., R_3SbS). With RCl tetraorganostibonium ions (R_4Sb^+) are formed.

antimony oxides and oxoacids
antimony trioxide (Sb_2O_3). White solid (burn Sb or Sb_2S_3 in $O_2)$, that contains Sb_4O_6 molecules (compare P_4O_6, *see* PHOS-

PHORUS OXIDES). At 540°C, Sb_2O_3 is converted to the macromolecular valentinite form containing infinite chains. At 900°C, it decomposes to antimony dioxide (see below). It is reduced to antimony with hydrogen or carbon at red heat; it is insoluble in water, but soluble in hydrochloric acid to give a solution of antimony trichloride $(SbCl_3)$. Boiling Sb_2O_3 with a solution of potassium hydrogen tartrate gives potassium antimonyl tartrate ('tartar emetic', $KSbO(C_4H_4O_6).H_2O)$, which is used as a parasiticide. Sb_2O_3 is soluble in bases to give solutions of antimonates(III) (e.g., $NaSbO_2)$. The free acid is unknown.
antimony dioxide (SbO_2). Obtained as a white powder when either Sb_2O_3 or antimony pentoxide (*see below*) is heated in air; it contains a 1:1 mixture of Sb(III) and Sb(V).
antimony pentoxide (Sb_2O_5). Yellow solid (Sb + conc. $HNO_3)$, which is soluble in bases to give antimonates(V) containing the $[Sb(OH)_6]^-$ ion (e.g., $NaSb(OH)_6)$, formerly called sodium pyroantimonate.
mixed oxides. Oxides based on SbO_6 octahedra (e.g., M^ISbO_3; $M^{II}Sb_2O_6$, $M^{III}SbO_4)$ are known; these are not antimonates.

antimony sulphides. Antimony trisulphide (Sb_2S_3) is formed either by direct combination or by precipitation with hydrogen sulphide from an acid solution of antimony(III) and is soluble in excess sulphide to give anionic complexes, mainly SbS_3^{3-}. The so-called pentasulphide is a NON-STOICHIOMETRIC COMPOUND and according to Mössbauer spectroscopy contains only antimony(III). Thioantimonates(V) (e.g., M_3SbS_4 $(Sb_2S_3 + S + alkali))$ are known.

antioxidant. Substance that retards oxidation, deterioration and rancidity. Rubber antioxidants include aromatic amines, substituted phenols and sulphur compounds. Some SEQUESTERING AGENTS inactivate metals that catalyze oxidation. Food antioxidants, effective at low concentrations, prevent rancidity and protect the nutritional value of the food. Permitted compounds include ascorbic acid and salts (E300–E304), tocopherols (E306–E309), propyl-, octyl- and dodecyl-3,4,5-trihydroxybenzoates (E310–E312), butylated anisole and hydroxytoluene (E320, E321), and lecithins (E322). Citric acid (E330) and phosphoric acid (E338) are used as synergists to enhance the effectiveness of the antioxidants.

antiparticle. Subatomic particle that has the same mass as another particle and equal, but opposite, value of some other property (e.g., charge). The antiparticle of the electron is the positron. The neutron and antineutron have magnetic moments opposite in sign to their spins. When a particle and its corresponding antiparticle collide ANNIHILATION occurs. Antimatter is postulated to be composed of antiparticles; so far no antimatter has been detected in the universe.

antipyretic. Drug used to reduce the temperature in the case of fever. Examples include ASPIRIN and salicylates.

antiseptic. Substance, applied to living tissue, to retard or stop the growth of bacteria. Examples include iodine, certain dyes (e.g., acriflavine), hydrogen peroxide, hypochlorites, hexachlorophene and quaternary ammonium compounds.

antistatic agent. Compound that imparts electrical conductivity to a surface to reduce the accumulation of charge. It may be a metal, conducting organic solid, water containing dissolved salts or a hygroscopic salt.

anti-Stokes. See RAMAN SPECTROSCOPY.

ao. See ATOMIC ORBITAL.

apatite. See PHOSPHORUS.

apoenzyme. Inactive form of an ENZYME. It must associate with a specific cofactor (see COENZYME) or ion before it can function.

aprotic solvent. Solvent (see SOLUTION) having no proton that can be donated to a solute (e.g., dimethyl sulphoxide, hexamethylphosphoramide). See also PROTOPHILIC SOLVENT.

Aquadag. Trade name for a dispersion of colloidal graphite in water. It is used as a lubricant for dies and tools.

aquamarine. See BERYLLIUM.

aqua regia. Mixture of concentrated hydrochloric and nitric acids (3:1, v/v) which dissolves noble metals (see PLATINIUM GROUP METALS). Its strong oxidizing power is due to the presence of free chlorine and nitrogen oxochloride.

aquation. Complexation of a metal ion by water molecules. This may involve replacement of an existing LIGAND.

aqueous. In which water is the solvent.

aquo ion. Hydrated (see HYDRATION) metal ion in aqueous solution. Water is the aquo ligand.

Ar. See ARGON, ARYL GROUP.

arabinose. See PENTOSES.

Arbusov reaction. See PHOSPHORUS ORGANIC DERIVATIVES.

arbutin. See GLYCOSIDES.

arc spectra. See ATOMIC SPECTROSCOPY, ELECTRONIC SPECTROSCOPY.

Arctons. See FREONS.

arenes (alkylbenzenes). Aromatic hydrocarbons, homologues of benzene. They are colourless liquids, with boiling points rising regularly for each additional methene group; the boiling point of a group of isomers are close. Arenes all burn with a smoky flame and are good solvents. A characteristic property of arenes is their rearrangement and/or disproportionation in the presence of a Lewis acid and halogen acid: ethylbenzene gives benzene and a mixture of diethylbenzenes. The order for the ease of disproportionation is $tert$-Bu> iso-Pr>Et$\gg$Me; the more stable the alkylcarbonium ion the easier the rearrangement. The benzene ring is very resistant to oxidation and when analogues of benzene are oxidized the side chain is attacked, the ultimate product being benzoic acid irrespective of the length of the side chain. The names for the substituted monocyclic aromatic hydrocarbons CUMENE, p-CYMENE, MESITYLENE, STYRENE, TOLUENE and XYLENE are retained; others are named systematically as derivatives of benzene.

arenols. *See* PHENOLS.

Arg (arginine). *See* AMINO ACIDS.

argentite. *See* SILVER.

arginase. *See* ENZYMES.

arginine (Arg). *See* AMINO ACIDS.

argol. *See* 2,3-DIHYDROXYBUTANEDIOIC ACID.

argon (Ar). Most abundant noble gas (GROUP O ELEMENT), obtained by the fractionation of liquid air (0.93 percent of air). It is slightly soluble in water forming CLATHRATE COMPOUNDS. Argon is used in welding operations requiring a non-oxidizing atmosphere and the absence of nitrogen, in gas-filled electric bulbs, thyratrons and fluorescent tubes.
$A_r = 39.948$, $Z = 18$, ES [Ne]$3s^2 3p^6$, mp $-189.2°C$, bp $-185.7°C$, $\rho = 1.78$ kg m^{-3}. Isotopes: ^{36}Ar, 0.34 percent; ^{38}Ar, 0.063 percent; ^{40}Ar, 99.6 percent.

argon detector. *See* DETECTION SYSTEMS.

Arkel–de Boer process. *See* CHROMIUM.

Arndt–Eistert synthesis. Reaction in which the hydrocarbon chain of a CARBOXYLIC ACID or its derivatives is extended by one METHENE GROUP. Diazomethane is used to form a keten which then gives the next higher homologue.

$$RCOOH + CH_2N_2 \xrightarrow[-H_2O]{Ag_2O} RCH=C=O$$
$$\xrightarrow{H_2O} RCH_2COOH$$

aromatic compounds. Unsaturated compounds containing a BENZENE ring or that have chemical properties similar to benzene that undergo ELECTROPHILIC SUBSTITUTION. *See also* HÜCKEL'S RULE.

aromatic ethers. Two recognized groups exist.
phenolic ethers (alkyl aryl ethers, alkoxyarenes, ArOR). Prepared by the reaction of an alkyl halide or sulphate with an alkaline solution of a phenol. Methoxybenzene (anisole, $C_6H_5OCH_3$) (bp 155°C) and ethoxy-

benzene (phenetole, PhOEt) (bp 172°C) are stable liquids unaffected by most acids and alkalis, but are decomposed by concentrated hydriodic (or hydrobromic) acid into phenol and alkyl iodide (bromide). Guaiacol (1-hydroxy-2-anisole), a catechol ether, occurs in beechwood tar from which it can be obtained by fractional distillation. It is synthesized from 1-amino-2-methoxybenzene via the diazonium salt. Treatment with hydriodic acid yields catechol (*see* DIHYDROXYBENZENES).
diaryl ethers (ArOAr). Conveniently prepared by the ULLMANN REACTION (e.g., refluxing a mixture of bromobenzene, phenol and KOH with Cu as catalyst gives diphenyl ether). Diphenyl ether is a solid (mp 28°C) with a geranium odour and is a valuable high-temperature heat transfer medium. It is not decomposed by hydriodic acid.

aromaticity. A monocyclic compound is aromatic if it has a reasonably planar cyclic structure, has $(4n + 2)$ π-electrons (*see* HÜCKEL'S RULE) and has unusual stability due to π-electron delocalization. An aromatic compound has the ability to sustain an induced ring current. Aromatic character is determined experimentally by physical properties which depend on the extent of delocalization of the π-electrons in the molecule, in a ^{1}H NMR spectrum (*see* NUCLEAR MAGNETIC RESONANCE SPECTROSCOPY) protons attached to the outside of an aromatic ring are highly deshielded and absorb far downfield from most other protons, usually beyond 7 ppm; thus for benzene $\delta = 7.3$, for pyridine $\delta = 7.1–8.5$ and for naphthalene $\delta = 7.3–7.8$ compared with a non-aromatic C–H of 5.3 ppm. Other methods include dipole moment measurements, X-ray analysis, and IR and UV spectroscopy.

aromatic substitution. *See* ELECTROPHILIC SUBSTITUTION, NUCLEOPHILIC SUBSTITUTION.

aromatization. HYDROFORMING of aromatic compounds from straight-chain aliphatic compounds.

Arrhenius dissociation theory. An electrolyte when dissolved in water dissociates into free ions. The degree of dissociation

(α) of weak electrolytes is given as Λ/Λ^0, where Λ is the MOLAR CONDUCTIVITY. *See also* CONDUCTANCE EQUATIONS, OSTWALD'S DILUTION LAW.

Arrhenius equation. For many reactions, the RATE CONSTANT increases exponentially according to the Arrhenius equation

$$k = A \exp(-E_a/RT)$$

A is the pre-exponential factor and E_a the ACTIVATION ENERGY. The exponential term in the above equation gives the fraction of collisions having energy of at least E_a. For simple gas-phase reactions, A may be identified with the frequency of collisions (*see* COLLISION THEORY).

arsenates. Arsenates(III) containing $[AsO_3]^-$ are formed from arsenic(III) oxide (As_2O_3) with a base; the free acid is unknown. A mixture of sodium and copper arsenates(III) is used as an insecticide for spraying fruit trees. Arsenic acid (H_3AsO_4), obtained as white crystals (As + conc. HNO_3), is tribasic and is a moderately strong oxidizing agent. The arsenates(v) resemble the orthophosphates and are generally isomorphous with them. Condensed arsenates, owing to their rapid hydrolysis, do not exist in solution. KH_2AsO_4 on dehydration at different temperatures gives rise to $K_2H_2As_2O_7$, $K_3H_2As_3O_{10}$ and $(KAsO_3)_n$.

arsenic (As). GROUP V ELEMENT. Principal sources of arsenic (0.0005 percent of earth's crust) are mispickel (FeAsS), orpiment (As_2S_3) and realgar (As_4S_4). The ores are roasted in air to arsenic(III) oxide (white arsenic, As_2O_3) and reduced to arsenic by carbon or hydrogen. Arsenic has a metallic appearance with a crystal structure similar to black phosphorus. An unstable, yellow allotrope (As_4) can be obtained by rapid condensation of the vapour. Arsenic combines on heating with oxygen (As_2O_3), sulphur (As_2S_3) and halogens (AsX_3), and reacts with concentrated sulphuric acid (As_2O_3) and dissolves in sodium hydroxide (Na_3AsO_3). It behaves as a metalloid; there is little evidence of cation chemistry, although the anions $[AsO_4]^{3-}$ and $[AsF_6]^-$ are well known. The element has few uses; it is added to alloys to give a hardening effect and is used as a

doping agent in semiconductors. Its compounds are used as insecticides. In large quantities arsenic compounds are poisonous.

$A_r = 74.9216$, $Z = 33$, ES [Ar] $3d^{10} 4s^2 4p^3$, sublimes 613°C, $\rho = 5.7 \times 10^{-3}$ kg m^{-3}. Isotope: ^{75}As, 100 percent.

arsenical pyrites. *See* ARSENIC.

arsenic halides. Arsenic forms two fluorides: the trifluoride (AsF_3) (heat As_2O_3 + CaF_2 + H_2SO_4), used as a mild fluorinating agent; the pentafluoride (AsF_5) (As + F_2). Both are readily hydrolyzed by water and act as acceptors (e.g., AsF_6^-). Arsenic trichloride ($AsCl_3$) (As + Cl_2), the only stable chloride, exists as a yellow oil. It forms many complexes in which it can act as an electron donor or acceptor. It has an appreciable electrical conductance, and is the starting material for many organic arsenicals. White crystalline arsenic tribromide ($AsBr_3$) and red crystalline arsenic triiodide (AsI_3) (As + halogen in carbon disulphide) have similar properties to $AsCl_3$. Hydrolysis of halides becomes more difficult in the order: F, Cl, Br, I. There are no oxohalides of arsenic.

arsenic hydride (arsine, AsH_3). Unstable, poisonous gas (metal arsenide + HCl; Zn + As compound + HCl) with a pyramidal structure. It is soluble in water and is a strong reducing agent giving arsenides with metal salt solutions. It decomposes to arsenic on heating (Marsh's test). AsH_3 is used in manufacture of n-type semiconductors.

arsenic organic derivatives. Trialkyl and triaryl arsines (R_3As) (AsX_3 + RMgX) are good donors towards *d*-group transition metals. They react with alkyl halides forming quaternary arsonium salts ($R_3R'As^+X^-$); in the presence of phenyllithium these form an arsenic YLIDE. The arsines ($RAsH_2$, R_2AsH and R_3As) are oxidized by air to alkylarsonic acid, dialkylarsonic acid and arsine oxide, respectively. Many compounds of the cacodyl radical (Me_2As-) are known; its chloride with zinc gives the spontaneously flammable cacodyl ($Me_2As-AsMe_2$). The most important aromatic compounds of arsenic are the arsonic acids ($ArAsO_3H_2$) (diazotized amine + Na_2HAsO_3); dehydration gives the anhydrides ($ArAsO_2$), analogous to the nitro compounds. Derivatives of these aromatic compounds (e.g., salvarsan, 3,3'-diamino-4,4'-dihydroxyarsenobenzene hydrochloride) are used in the treatment of protozoal diseases.

arsenic oxides.
arsenic(III) oxide (white arsenic, As_2O_3). Formed when arsenic burns in air; contains As_4O_6 molecules. It is soluble in water giving 'arsenious acid', and with bases forms arsenates(III). It is used as a standard reducing agent with iodine solutions in volumetric analysis and is employed in the glass industry.
arsenic(v) oxide (As_2O_5). Ill-defined compound (As_2O_3 + conc. HNO_3) that loses oxygen on heating. It is very soluble in water giving a solution of arsenic acid.

arsenic sulphides. As_4S_3, As_4S_4, As_2S_3 and As_2S_5 can be prepared by direct interaction; the last two can be precipitated from acidified solutions of As(III) and As(v) solutions with hydrogen sulphide. As_2S_3 and As_2S_5 are insoluble in water, but dissolve in alkali sulphide solutions to give thioanions. As_4S_3 and As_4S_4 have cage-like structures with As–As and As–S–As linkages. As_4S_4 is used in PYROTECHNICS.

arsenides. Binary compounds between a metal and arsenic.

arsine. *See* ARSENIC HYDRIDE.

arylation. Introduction of an ARYL GROUP (aromatic group) into a compound.

aryl group (Ar–). The group formed when a hydrogen atom is removed from a compound of the benzene series.

N-arylimines. *See* SCHIFF'S BASES.

arynes. Transient aromatic species in which two adjacent carbon atoms in a ring lack substituents, with two orbitals each lacking an electron. They undergo rapid addition reactions with nucleophiles; their existence is inferred from, for example, the identity of the products resulting from their trapping in DIELS–ALDER REACTIONS.

Benzene, naphthalene and pyridine have been shown to produce such species. *See also* BENZYNE.

As. *See* ARSENIC.

asbestos. *See* CALCIUM, MAGNESIUM, SILICATES.

ascorbic acid (vitamin C). *See* VITAMINS.

Asn (asparagine). *See* AMINO ACIDS.

Asp (aspartic acid). *See* AMINO ACIDS.

asparagine (Asn). *See* AMINO ACIDS.

aspartame. *See* SWEETENING AGENTS.

aspartic acid (Asp). *See* AMINO ACIDS.

asphaltenes. High-molecular-mass compounds extracted from BITUMEN by petroleum spirit.

aspirin (2-acetoxybenzoic acid, acetylsalicylic acid, $2-CH_3COOC_6H_4COOH$). White solid (mp 135–8°C), manufactured by the reaction between ethanoic anhydride and 2-hydroxybenzoic acid (salicylic acid). Widely used analgesic that reduces inflammation (antiinflammatory) and fever (antipyretic), due to the inhibition of PROSTAGLANDIN synthesis at the damaged tissue.

assembly. Thermodynamic system; collection of microscopic bodies or microstates not dependent on the number of phases or constituents. *See also* PARTITION FUNCTION, STATISTICAL MECHANICS.

associated liquid. Liquid in which the molecules are loosely aggregated: for example, by HYDROGEN BONDS. Such a liquid shows anomolously high melting and boiling points.

astatine (At). Radioactive GROUP VII ELEMENT, occurring by radioactive decay from uranium and thorium isotopes (*see* RADIOACTIVE DECAY SERIES) (total astatine in earth's crust is less than 25 grams). Astatine is prepared by alpha-bombardment of bismuth-209; 20 isotopes are known of which astatine-210 is the most stable (half-life: 8.3 hours). It is more metallic

than iodine with at least five oxidation states in aqueous solution. Little is known of its chemistry, although compounds such as CH_3At, HAt, AtI, $AtBr$ and $AtCl$ have been detected using mass spectroscopy. The existence of At_2 has not been established.
$A_r \sim 210$, $Z = 85$, ES [Xe] $4f^{14} 5d^{10} 6s^2 6p^5$, mp 305°C; bp 377°C.

ASTM. American Society for Testing and Materials; publishes the definitive index of powder X-ray diffraction data, and classification for ELASTOMERS.

asymmetry. *See* CHIRALITY.

At. *See* ASTATINE.

atactic polymer. *See* TACTICITY.

atmosphere (atm). Unit of pressure equal to 101 325 pascals (Nm^{-2}) or 760.0 mmHg. It is not coherent with SI units, and is usually used to express pressures well in excess of standard atmospheric pressure.

atom. Smallest possible unit of an element, consisting of protons and neutrons located in the nucleus and one or more electrons in surrounding shells or orbits. The numbers of electrons and protons are the same, and the atom is electrically neutral. Atoms of a given element are identical, but an element may have atoms of slightly different masses (*see* ISOTOPES). The ELECTRONIC CONFIGURATION describes the way in which the electrons are arranged around the nucleus.

atomic absorption spectroscopy (AAS). *See* ATOMIC SPECTROSCOPY.

atomic emission spectroscopy (AES). *See* ATOMIC SPECTROSCOPY.

atomic fluorescence spectroscopy. *See* ATOMIC SPECTROSCOPY.

atomic heat. Product of the atomic mass and heat capacity of an element. *See also* DULONG AND PETIT'S LAW.

atomic hydrogen. At high temperatures, molecular hydrogen is largely dissociated

into atoms (95 percent at 5000 K). Atomic hydrogen is also produced in an electric discharge in H_2, or in an atmosphere of H_2 and mercury vapour when irradiated by the 253.7 nm mercury line. Atomic hydrogen reduces the oxides and halides of many metals, and the nitrates(v) and nitrates-(III) of sodium and potassium to the metallic state. With elements HYDRIDES are formed. The heat evolved on recombination of atomic hydrogen (436.0 kJ mol^{-1}) provides the heat in atomic hydrogen welding. Hydrogen atoms are intermediates in radical chain reactions such as that between H_2 and O_2.

atomicity. Number of atoms in a molecule (e.g., oxygen, O_2, has an atomicity of two; ozone, O_3, an atomicity of three).

atomic mass unit (amu; 1.66033×10^{-27} kg). Unit of mass used to express relative atomic mass; 1/12 of the mass of an atom of carbon-12. It is sometimes called the dalton or unified mass unit.

atomic nucleus. *See* NUCLEUS.

atomic number (proton number, Z). Number of protons contained in the nucleus of an atom; this is equal to the number of orbiting electrons. Its value determines the position of the element in the PERIODIC TABLE.

atomic orbital (ao). WAVEFUNCTION of an atom. Atomic orbitals are described by the three QUANTUM NUMBERS n, l (values: 0, 1, 2, . . ., n–1) and m_l (values: $+l$, $+l$–1, . . ., 0, . . ., $–l$–1, $–l$). The spin quantum number ($s = +1/2, –1/2$) is required to describe fully the state of an electron in an orbital.

The principal quantum number is written followed by a letter, corresponding to the value of l (i.e. s, p, d, f, g, h, for $l = 0, 1, 2, 3, 4, 5, 6$), having a subscript denoting the orientation of the orbital and/or the value of m_l (e.g., x, y, z for the p-orbitals $m_l = +l$, $–l$, 0; z^2, $x^2–y^2$, xy, yz, xz for the five d-orbitals). In the absence of an orienting field, orbitals having common n and l quantum numbers are degenerate (*see* DEGENERACY). In molecules and complexes, this degeneracy may be split by neighbouring atoms or groups (*see* LIGAND FIELD

THEORY). (For distribution of electron density for some hydrogen atomic orbitals —*see* Figure.)

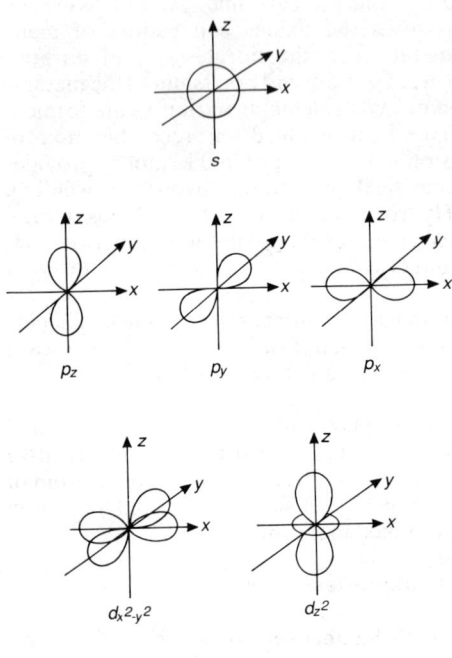

atomic orbital

atomic oxygen. Powerful oxidizing agent ($E^{\ominus} = 2.2$ V), produced by passing an electric discharge through gaseous oxygen at 1 mmHg pressure, by radiofrequency excitation at high pressures or by UV irradiation at short wavelengths (less than 190 nm).

atomic radius. Half the distance of closest approach of atoms in the structure of the element; this is not easily defined for elements with irregular structures. *See also* BOND LENGTH, IONIC RADIUS.

atomic spectroscopy. Analytical method utilizing the emission, absorption or FLUORESCENCE of light by atoms for determinating the elemental composition of a sample. The sample is introduced into a flame (e.g., oxy–ethyne), electric discharge (carbon arc) or furnace, where it is atomized. Atomic emission spectroscopy (AES) relies on the emission of light from a small fraction of excited atoms at wavelengths characteristic of the atom (e.g., BALMER SERIES). In atomic absorption spectroscopy (AAS), a monochromatic source of light (usually a hollow cathode discharge lamp of the element) at a specific absorption wavelength is passed through the atomized sample. ABSORBANCE is proportional to the concentration of the element in the sample and may be calibrated against known standards. The method is very sensitive, many elements may be determined to a few parts per million. Atomic fluorescence spectroscopy has a detector at right angles to the source which, in contrast to AAS, can be a broadband continuous source. Intensity of fluorescence is proportional to the intensity of the source and so electrodeless discharges or tunable lasers are used.

atomic units (au). Units defined such that the electronic mass, charge, $h/2\pi$ and $4\pi\varepsilon_0$ are unity. The energy of a hydrogen orbital is now $-1/2n^2$. The unit of energy is the hartree ($E_H = 27.21$ eV $= 2626$ kJ mol^{-1}). The unit of length is the bohr ($a_0 = 5.297 \times 10^{-11}$ m). In terms of fundamental constants, $a_0 = h^2\varepsilon_0/\pi me^2$. The atomic unit of time now becomes 2.419×10^{-17} s.

atomic weight. *See* RELATIVE ATOMIC MASS.

ATP. *See* ADENOSINE TRIPHOSPHATE.

ATPase. *See* ENZYMES.

ATR. *See* ATTENUATED TOTAL REFLECTANCE.

atropine. *See* ALKALOIDS.

attenuated total reflectance (ATR). Also known as frustrated internal reflectance (FIR) and internal reflectance spectroscopy (IRS). Enhanced absorption of ELECTRO-

MAGNETIC RADIATION by a sample is achieved by causing many reflections through a thin layer sandwiched between two prisms of different optical density. The method is used to study the INFRARED SPECTROSCOPY of adsorbed layers, paints, coatings, etc.

attractive forces. *See* INTERMOLECULAR FORCES.

Au. *See* GOLD.

aufbau principle. The electronic configuration of a multielectron atom or molecule is obtained by feeding electrons into atomic orbitals or molecular orbitals of increasing energy, taking into account the PAULI EXCLUSION PRINCIPLE and HUND'S RULES. The aufbau principle is the basis of the periodic classification of the elements (*see* PERIODIC TABLE).

Auger effect. Phenomenon in which, following the ejection of a core electron by an X-ray photon or energetic electron, a second electron is emitted by the energy released as an outer shell electron falls into the core. The effect competes with X-ray fluorescence and is more probable from atoms of low atomic number. Auger electrons are described by the shells between which the transitions occur (e.g., KL_1L_2) (*see* Figure). Auger electron spectroscopy is predominantly a surface technique, usually found in conjunction with electron spectroscopy for chemical analysis (ESCA) (*see* ELECTRON SPECTROSCOPY).

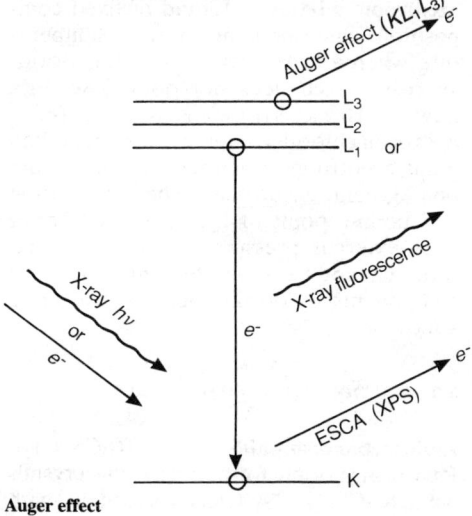

Auger effect

Auger electron spectroscopy. *See* AUGER EFFECT.

auramine. Yellow diphenylmethane basic DYE.

autocatalysis. Reaction in which the products promote the reaction itself. The general form of an autocatalytic reaction is $A + X \rightarrow P + 2X$.

autoinhibition. *See* INHIBITION.

autolysis. Self-destruction of a tissue, cell or part of a cell. It occurs just prior to and after the death of a cell and is mainly due to the release of hydrolytic enzymes from the lysosomes.

autooxidation. Oxidation by molecular oxygen that occurs in a one-electron radical process and which may be catalyzed by transition metal ions (*see* HOMOGENEOUS CATALYSIS). The metal ions go through an oxidation/reduction cycle. For example the decomposition of alkyl hydroperoxides is catalyzed by cobalt and manganese salts. The oxidation of *p*-xylene to 1,4-phthalic acid (*see* BENZENEDICARBOXYLIC ACIDS) in the manufacture of Terylene (*see* POLYMERS) uses a cobalt bromide catalyst.

autoprotolysis. Equilibrium between two molecules of water, one acting as an acid and one as a base

$$H_2O + H_2O \rightleftharpoons H_3O^+ + {}^-OH$$

The equilibrium constant for this reaction $K_w = a_{H_3O^+} \cdot a_{OH} = 1.008 \times 10^{-14}$ at 25°C.

autoradiography. Technique for the detection of compounds labelled with radionuclides, particularly suited to partition chromatography (*see* CHROMATOGRAPHY, PAPER CHROMATOGRAPHY, THIN-LAYER CHROMATOGRAPHY).

auxins. Plant hormones produced by growing shoot and root apices that have diverse effects, such as the stimulation of cell elongation, promotion of cell division. The two most common naturally occurring auxins are β-indoylacetic acid (IAA) and β-indoylacetonitrile (IAN). IAA is the active principle of horticultural 'rooting hormone' preparations. Synthetic auxins, widely used in weed control include:

2,4-dichlorophenoxyacetic acid (2,4-D). Selective growth regulator (mp 138°C), very toxic to broad-leaved species and used as weedkiller among grass crops and cereals.

2-methyl-4-chlorophenoxyacetic acid (MCPA). Solid (mp 120°C), prepared by the reaction of chloroethanoic acid with chlorinated *o*-cresol. Crude MCPA is a mixture of the 4- and 6-isomers. A selective weedkiller.

2,4,5-trichlorophenoxyacetic acid (2,4,5-T). Solid (mp 153°C), prepared from 2,4,5-trichlorophenol and sodium chloroethanoate. A selective herbicide for controlling many woody species and perennial weeds.

4-(4-chloro-2-methylphenoxy)butyric acid (MCPB). The compound itself is harmless to plants, but when absorbed and translocated it is a powerful herbicide. Other butanoic acid derivatives used commercially are 2,4-D$_B$ and 2,4,5-T$_B$, the analogues of 2,4-D and 2,4,5-T (*see above*).

1-naphthylacetic acid (NAA). Solid (mp 135°C) prepared by the chloromethylation of naphthalene, conversion to the nitrile and hydrolysis. Used to increase the ease of rooting of cuttings.

auxochrome.   Group in an organic molecule that modifies the absorption of light of a CHROMOPHORE. Typical auxochromes are methyl, chloro, amine and hydroxyl groups. Auxochromic effects are described in terms of an increase (hyperchromic) or decrease (hypochromic) in intensity of absorption, and of shifts to longer (bathochromic) or shorter (hypsochromic) wavelength. A hyperchromic effect is often associated with a bathochromic shift caused by an electron-donating auxochrome (e.g., methyl). Conversely, a hypochromic effect and hypsochromic shift may be caused by an electron-withdrawing auxochrome (e.g., chloro).

Avogadro constant (N_A).   Number of atoms or molecules in 1 MOLE of substance ($N_A = 6.022\,52 \times 10^{23}$ mol^{-1}).

Avogadro's hypothesis (Avogadro's law). Equal volumes of all gases (ideal) contain equal numbers of molecules at the same temperature and pressure.

axerophythol (vitamin A).   *See* VITAMINS.

axial form.   *See* CONFORMERS.

axial ratio.   (1) Ratio of the cell dimensions (a, b, c) in a crystal, b being taken as unity. In triclinic, monoclinic and orthorhombic systems, $a \neq b \neq c$, and the axial ratios are quoted as $a/b{:}1{:}c/b$. In hexagonal and tetragonal systems, $a = b \neq c$, and only the c/a ratio is needed. *See also* CRYSTAL LATTICE, CRYSTAL SYSTEMS. (2) For colloidal particles and macromolecules, the ratio L/D, where L is the length of the longest axis and D the length of the shortest axis. For a spherical particle $L/D = 1$, for pepsin 3, fibrinogen 18, myosin 100. Values of L/D are obtained from sedimentation, viscosity and streaming birefringence studies.

axis.   Line about which a figure, curve or body is symmetrical (axis of symmetry, *see* CRYSTAL SYMMETRY) or about which it rotates (axis of rotation). *See also* SYMMETRY ELEMENTS.

axis of rotation.   *See* SYMMETRY ELEMENTS.

axis of symmetry.   *See* CRYSTAL SYMMETRY.

azeotrope.   *See* AZEOTROPIC MIXTURE.

azeotropic mixture.   Liquid of fixed composition which boils at constant temperature when a COMPLETELY MISCIBLE LIQUID SYSTEM (which does not obey RAOULT'S LAW or HENRY'S LAW) is distilled (e.g., hydrogen chloride/water, maximum boiling point azeotrope; ethanol/water, minimum boiling point azeotrope). The composition and boiling point of the azeotrope varies with external pressure. Azeotropic mixtures cannot be separated into the pure components by distillation. *See also* REAL SOLUTION.

azides.   *See* NITROGEN HYDRIDES.

azidocarbondisulphide   (($SCSN_3$)$_2$). PSEUDOHALOGEN formed as white crystals when KSCSN$_3$ (CS$_2$ + KN$_3$) is oxidized with hydrogen peroxide. It decomposes at room temperature to $(SCN)_2$, sulphur and nitrogen.

azimuthal quantum number. *See* QUANTUM NUMBER.

azine. *See* PYRIDINE.

azine dye. *See* DYES.

azobenzene $(C_6H_5N=NC_6H_5)$. Orange–red solid, prepared by the reduction of nitrobenzene (Zn/NaOH/MeOH, LiAlH$_4$, Na/Hg or Fe/HCl). The *trans*-form (mp 68°C) is more stable; it may be converted to the *cis*-form (mp 71.4°C) by UV light. It is further reduced to hydrazobenzene $(C_6H_5NHNHC_6H_5)$ (Zn/NaOH, NaBH$_4$) or ANILINE (SnCl$_2$/HCl, TiCl$_2$/HCl). It is the most simple azo DYE containing the CHROMOPHORE $-N=N-$, but is not used as such because of its poor affinity for fibres.

azo compounds (ArN=NAr). Organic compounds containing the CHROMOPHORE $-N=N-$. They are prepared by coupling, usually at the *para*-position, a DIAZONIUM COMPOUND to a phenol or aniline derivative. The azo group is stable; the aromatic rings may be halogenated, nitrated and sulphonated. Sulphonic acid azo derivatives are important MORDANT DYES.

azo dye. *See* DYES.

azole. *See* PYRROLE.

azulene (bicyclo[5.3.0]deca-2,4,6,8,10-pentaene, $C_{10}H_8$). Intense blue-coloured solid (mp 98.5°C) benzenoid-type POLYCYCLIC HYDROCARBON (structure XXXI) containing a five-membered ring fused to a seven-membered ring with a dipole moment of 1 D. It is synthesized by dehydrogenation of cyclopentanocycloheptanol. It undergoes many typical aromatic substitution reactions, preferentially in the five-membered ring which is more electron-rich than the other ring. At 270°C it is transformed to its isomer NAPHTHALENE.

azurite. *See* COPPER.

B

B. *See* BORON.

β. *See* BOHR MAGNETON, COMPRESSI-
BILITY COEFFICIENT.

β-particle. Beta-particle (*see* BETA-
DECAY).

Ba. *See* BARIUM.

Babo's law. Addition of a non-volatile
solute to a volatile solvent (*see* SOLUTION)
lowers the VAPOUR PRESSURE of the solvent in proportion to the amount of solute
dissolved. *See also* BOILING POINT, FREEZ-
ING POINT, RAOULT'S LAW.

bacitracin. *See* ANTIBIOTICS.

back-bonding. Overlap of a filled orbital
of an ACCEPTOR with an unfilled orbital
(bonding or antibonding) of a donor. It
occurs in transition element complexes
especially in low oxidation states (e.g.,
METAL CARBONYLS). *See also* COORDINATE
BOND.

bacteria. Heterogeneous group of microorganisms, reproducing by binary fission,
characterized by a rigid cell wall that contains the polymer PEPTIDOGLYCAN. The
shape and chemical structure of the walls
are responsible for such features as the
GRAM STAIN and antigenic properties. The
complex chemical structure of the organism is classified broadly as either gram-
positive or gram-negative (*see* Figure). In
addition, some organisms possess flagella
or shorter appendages known as pili. Some
organisms (e.g., *Escherichia coli*) have
limited nutritional requirements, growing
on a simple salts/glucose medium, whereas
others are nutritionally very demanding,
requiring vitamins, amino acids, blood,
etc. for growth. Bacteria are widely distributed throughout the environment,
many are essential in the human body for
the breakdown of foodstuffs, etc., others,
the pathogens, are responsible for disease.
Bacteria in the soil are essential for recycling carbon and nitrogen (*see* CARBON
CYCLE, NITROGEN CYCLE). Bacteria are
used in sewage treatment and in industrial
fermentations for the production of many
organic chemicals, antibiotics, etc.

baddeleyite. *See* ZIRCONIUM.

Baeyer–Villiger oxidation. Reaction of
KETONES with PERACIDS to give ESTERS or
products of their hydrolysis (i.e. ALCOHOLS
and CARBOXYLIC ACIDS).

Baker–Nathan effect. In organic compounds, when a hydrogen atom is β to an
unsaturated carbon (i.e. H–C–C=) the
hydrogen–carbon σ-bond becomes less
localized by partial σ,π-CONJUGATION. *See
also* HYPERCONJUGATION.

Balmer series. When hydrogen molecules
dissociate in an electric discharge, the emission spectrum (*see* ATOMIC SPECTROSCOPY)
consists of series of lines. The frequency at
which the lines appear follows the relationship $R(1/n_1^2 - 1/n_2^2)$, where $n_2 = n_1+1$,
$n_1+2, n_1+3, \ldots$ for each series and $n_1 = 1$,
2, 3, 4, 5. R is the Rydberg constant (1.097
$\times 10^5$ cm^{-1}, 2.178 $\times 10^{-18}$ J). As n_2 increases the lines merge into a continuum at
a frequency of the ionization potential of
the particular state (R/n_1^2). The Balmer
series ($n_1 = 2$), which occurs in the visible
region (*see* ELECTROMAGNETIC SPECTRUM),
was discovered first, followed by the Lyman
($n_1 = 1$, ultraviolet), Paschen ($n_1 = 3$, infrared), Brackett ($n_1 = 4$) and Pfund ($n_1 = 5$)
series. Energy levels of atomic hydrogen
may be calculated by solving the SCHRÖ-
DINGER EQUATION, which identifies n as the
principal QUANTUM NUMBER.

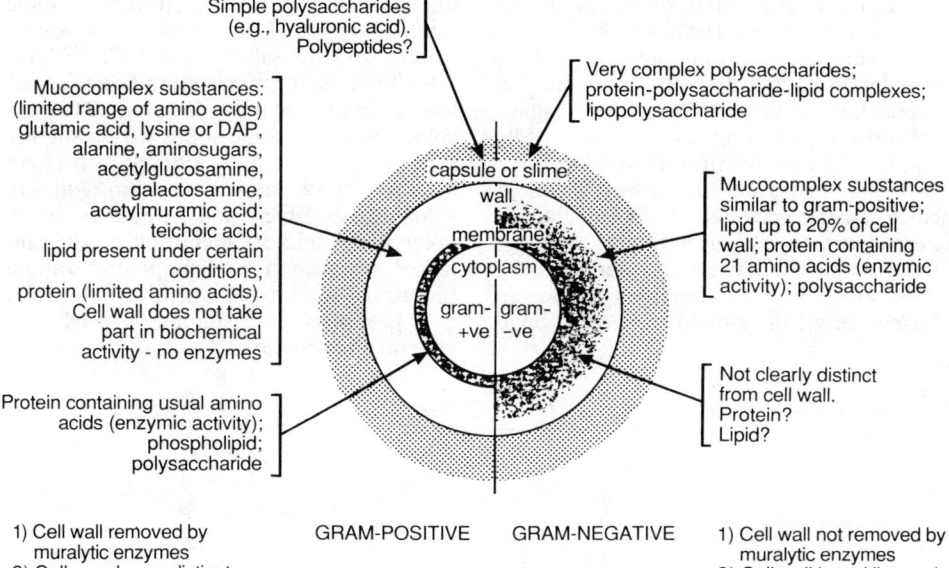

Simple polysaccharides
(e.g., hyaluronic acid).
Polypeptides?

Very complex polysaccharides;
protein-polysaccharide-lipid complexes;
lipopolysaccharide

Mucocomplex substances:
(limited range of amino acids)
glutamic acid, lysine or DAP,
alanine, aminosugars,
acetylglucosamine,
galactosamine,
acetylmuramic acid;
teichoic acid;
lipid present under certain
conditions;
protein (limited amino acids).
Cell wall does not take
part in biochemical
activity - no enzymes

Mucocomplex substances
similar to gram-positive;
lipid up to 20% of cell
wall; protein containing
21 amino acids (enzymic
activity); polysaccharide

capsule or slime
wall
membrane
cytoplasm

gram- gram-
+ve -ve

Protein containing usual amino
acids (enzymic activity);
phospholipid;
polysaccharide

Not clearly distinct
from cell wall.
Protein?
Lipid?

GRAM-POSITIVE   GRAM-NEGATIVE

1) Cell wall removed by
   muralytic enzymes
2) Cell membrane distinct
   structurally and
   biochemically

1) Cell wall not removed by
   muralytic enzymes
2) Cell wall is multilayered
   structure and may play
   role in metabolic activity

Examples:
Streptococcus pyogenes
Staphylococcus aureus

Examples:
Escherichia coli
Klebsiella aerogenes

bacteria
Summary of chemical structure of gram-positive and gram-negative bacteria.

$$E_n = me^4/8\varepsilon_0^2 h^2 (1/n^2)$$
$$= R/n^2$$

The theoretical value of the Rydberg constant is in good agreement with experiment. The series of lines seen in the emission spectrum of atomic hydrogen are thus transitions between energy levels of different principal quantum numbers.

balsam. Fragrant aromatic substance flowing spontaneously or by incision from certain plants; it does not necessarily remain liquid. It is commonly an OLEORESIN as in Canada balsam and is a source of volatile oils, resins and some TERPENES.

band broadening. Process common to all forms of CHROMATOGRAPHY, whereby the width of the zone slowly increases the longer the CHROMATOGRAM is run. The factors influencing band broadening are those corresponding to the terms in the VAN DEEMTER EQUATION.

bandhead. *See* INFRARED SPECTROSCOPY.

band spectrum. *See* ELECTRONIC SPECTROSCOPY.

band theory of solids. Periodic arrangement of positive nuclei in a crystalline solid (unit cell length a) leads to a SCHRÖDINGER EQUATION for an electron in a periodic potential $V(r)$, where $V(r) = V(an + r)$, $n = 1, 2, \ldots$

$$-h^2/8\pi^2 m (d^2\psi/dr^2) + V(r)\psi = E\psi$$

This is the Bloch equation. Two approximate solutions are obtained by assuming almost free electrons which are perturbed by the nuclei or, at the other extreme, essentially tightly bound electrons around individual atoms whose atomic orbitals only overlap as the atoms come together. The solution of the periodic potential model leads to energy bands separated by gaps. Whether a material is a METAL, INSULATOR or SEMICONDUCTOR depends on the population of the valence band, that is the highest occupied energy band (if it is full or half full) or if there is a small enough gap to allow population of the first unfilled band (the conduction band) (*see* METALS, Figure; SEMICONDUCTOR, Figure). The limits of the bands in three dimensions are known as Brillouin zones, and the surface which encloses all the occupied states is called the Fermi surface (*see also* FERMI ENERGY). Bands in metals may also be identified with the ATOMIC ORBITALS from which they are formed (e.g., the *s*-band, *p*-band, *d*-band, etc.).

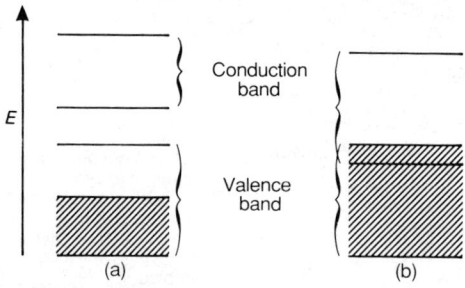

band theory of solids
Conduction in metals. (a) Half-filled valence band. (b) Overlapping valence and conduction bonds.

bar. CGS UNIT of pressure, equal to 10^6 dyne cm^{-2} or 10^6 Pa (N m^{-2}) (approximately 750 mmHg).

Barbier–Wieland degradation. Procedure for reducing the number of carbon atoms in a CARBOXYLIC ACID. An ESTER of the acid is reacted with phenyl magnesium bromide (*see* GRIGNARD REAGENTS) which when worked up in water gives an alkene. Chromium(VI) oxide catalyzes its decomposition to the product acid.

$$RCH_2COOR' + 2PhMgBr \longrightarrow$$

$$RCH_2C(OH)Ph_2 + R'OH \xrightarrow{H_2O}$$

$$RCH=CPh_2 \xrightarrow{CrO_3} RCOOH +$$

$$Ph_2CO$$

barbitone. *See* BARBITURATES.

barbiturates. Group of PYRIMIDINE derivatives, prepared by condensing urea or thiourea with disubstituted propanedioic esters. As cyclic diimides they form water-soluble sodium salts. Over 2000 barbiturates have been prepared and tested, but few are in general use (*see* Table). They are colourless, crystalline solids with a slightly bitter taste. They have a depressant effect on the central nervous system and are valuable sedatives and soporifics when taken orally and are anaesthetics when injected intravenously. Thiopental sodium (pentothal sodium) in which the oxygen at C–2 is replaced by sulphur is used as a general anaesthetic.

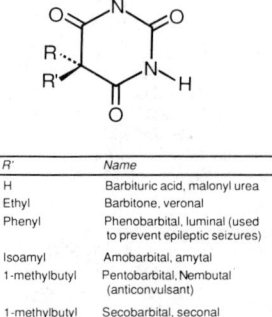

R	R'	Name
H	H	Barbituric acid, malonyl urea
Ethyl	Ethyl	Barbitone, veronal
Ethyl	Phenyl	Phenobarbital, luminal (used to prevent epileptic seizures)
Ethyl	Isoamyl	Amobarbital, amytal
Ethyl	1-methylbutyl	Pentobarbital, Nembutal (anticonvulsant)
Allyl	1-methylbutyl	Secobarbital, seconal (soporific)

barbiturates
Structures of some barbiturates.

Barff process. Protection of iron against CORROSION by heating the metal in steam. A layer of black iron(III) oxide is formed.

barite. *See* BARIUM, SULPHATES.

barium (Ba). Group IIa ALKALINE EARTH METAL (*see* GROUP II ELEMENTS). It occurs naturally as barite ($BaSO_4$) and is extracted by electrolysis of the molten chloride or by reduction of the oxide by aluminium. Barium is a silvery-white metal which reacts readily with water, oxygen, hydrogen, halogens and other non-metals. Its reactivity precludes the extensive use of the metal. It is alloyed with nickel in spark plug wires and with lead and calcium in Frary's metal. Soluble barium salts are poisonous.
$A_r = 137.33$, $Z = 56$, ES [Xe]$6s^2$, mp 714°C, bp 1640°C, $\rho = 3.51 \times 10^3$ kg m^{-3}, Isotopes: ^{130}Ba, 0.1 percent; ^{132}Ba, 0.10 percent; ^{134}Ba, 2.4 percent; ^{135}Ba, 6.6 per-

cent; ^{136}Ba, 7.8 percent; ^{137}Ba, 11.3 percent; ^{138}Ba, 71.7 percent.

barium compounds. Barium is an ELECTROPOSITIVE metal ($E^{\ominus}_{Ba^{2+},Ba}$ = −2.90 V) that forms ionic salts of Ba^{2+}. Two oxides (BaO and BaO_2) are known. BaO_2 is a bleaching agent giving hydrogen peroxide on reaction with water. BaO dissolves in water to give baryta ($Ba(OH)_2$), which is used in volumetric analysis, the manufacture of oils and greases, plastic stabilizers and pigments. Barium sulphate ($BaSO_4$) is the principal starting compound (via barium sulphide) for the synthesis of other barium salts. It is a component of the pigment LITHOPONE and is used in radiodiagnostics (barium meal). Barium titanate ($BaTiO_3$) is a perovskite (*see* PEROVSKITE STRUCTURE) and piezoelectric. Barium diphenylamine-4-sulphonate is a redox INDICATOR. Barium alkyls (RBaX and R_2Ba) act as GRIGNARD REAGENTS.

barn. Unit of measurement equal to 10^{-28} m^2, for the cross-sectional target area of the nucleus of an atom (e.g., for $^{55}_{25}$Mn, σ = 13.3 barn).

Barton reaction. PHOTOLYSIS of organic nitrites which contain a γ-hydrogen atom to give an oxime or nitroso dimer. The reaction is via a six-membered intermediate

$$ONO-RCHCH_2CH_2CH_2R' \rightarrow$$

$$HO-RCHCH_2CH_2CR'=NOH + dimer$$

baryon. *See* ELEMENTARY PARTICLES.

baryta. *See* BARIUM COMPOUNDS.

base dissociation constant (K_b). EQUILIBRIUM CONSTANT for the reaction $BH^+ \rightleftharpoons B$ + H^+, where BH^+ is a base. It may be expressed as pK_b = −lg K_b. *See also* ACID DISSOCIATION CONSTANT.

base exchange. Measure of the capacity of ZEOLITES and ION EXCHANGERS to exchange their cation for an equivalent of other cations without undergoing structural change. The term is not now in common use.

baseline. Constant signal from a detector before an experiment is commenced; a steady baseline is required for quantitative measurements. In chromatography, it is the signal when only the mobile phase is passing through the detector.

base metal. ELECTROPOSITIVE metal that dissolves in acid; the opposite of noble metal (*see* PLATINUM GROUP METALS).

base-pairing. Linking of complementary polynucleotide chains of NUCLEIC ACIDS by HYDROGEN BONDS between the opposite purine and pyrimidine bases (i.e. adenine–thymine and guanine–cytosine).

bases. *See* ACIDS AND BASES.

basic dyes. Water-soluble, cationic DYES; usually as the acid salts of amines or imines. Basic dyes are exceptionally bright, but have poor fastness except on acrylic fibres (*see* FIBRES). They may be used with cotton if a MORDANT has previously been applied.

basicity. Affinity for protons, that is the assessment of the position of equilibrium between an electron donor (Lewis base) and an acid (usually the proton). *See also* EQUILIBRIUM CONSTANT.

basic oxygen process (BOP). Method widely used for making STEEL. Pig iron and scrap iron are heated in a furnace across which oxygen is blown to remove impurities (carbon, silicon, phosphorus, manganese). The process has largely superceded the Bessemer and open hearth processes.

batch process. Chemical process in which a vessel is charged with reactants which are allowed to react to completion. The products are then removed and the process repeated.

bathochromic shift. *See* AUXOCHROME.

battery. CELL, or series of cells, from which the energy of an exoenergetic reaction is provided as electricity. In a primary battery, the reactants are held within the cell, and when consumed the life of the cell is over. A secondary battery, or ACCUMULATOR, may be recharged by passing a current through it from an external source. *See also* DANIELL CELL, ALKALINE CELL, LEAD–ACID BATTERY, LECLANCHÉ CELL.

bauxite. *See* ALUMINIUM.

Bayer–Hall–Heroult process. *See* HALL–HEROULT PROCESS.

bcc (body-centred cubic lattice). *See* CLOSE-PACKED STRUCTURES.

BCF (bromochlorodifluoromethane). *See* FREONS.

Be. *See* BERYLLIUM.

Beattie–Bridgeman equation. *See* EQUATION OF STATE.

Beckmann rearrangement. Intramolecular conversion of a KETOXIME to an AMIDE on treatment with phosphorus pentachloride, sulphuric acid, phosphoric acid, antimony(v) chloride or ethanoyl chloride

$$Ar_2C=NOH \rightarrow ArCONHAr.$$

ISOQUINOLINES are also formed by a Beckmann rearrangement from derivatives of 2-phenyl-1-1-*N*-amidoethane. In cyclic ketoximes, the nitrogen atom is incorporated into the ring. CAPROLACTAM is prepared in this way from cyclohexanone oxime.

becquerel (Bq). SI unit of ACTIVITY; the activity of a radionuclide decaying at a rate, on average, of one spontaneous nuclear transition per second (1 Bq = 27 pCi). *See also* RADIATION UNITS.

Beer–Lambert law. Expression of the linear relationship between the ABSORBANCE (A) of a sample at a given wavelength and the concentration (c) and path length (l) of the sample: $A = \varepsilon cl$. The constant of proportionality (ε) is the molar extinction (absorbance) coefficient.

Beilstein's test. Flame test for the presence of halogen in an organic compound. The material is heated on an oxidized copper gauze, a green flame indicating a halogen.

belladonna. *See* ALKALOIDS.

bending vibration. *See* NORMAL MODE.

Benedict's test. Test for reducing sugars. The addition of an alkaline solution of copper(II) sulphate and sodium 2-hydroxypropane-1,2,3-tricarboxylate (sodium citrate) to a sugar solution gives red–yellow colours or precipitates.

Benfield process. Process in which gases from the GASIFICATION of coal or oil are scrubbed with hot potassium carbonate solution to remove carbon dioxide.

benitoite. *See* SILICATES.

bent. *See* SHAPES OF MOLECULES.

bentonite. Clay-like material consisting largely of MONTMORILLONITE. Two varieties exist: one which swells considerably in water displaying THIXOTROPY, used as an emulsifying, suspending, anticaking and clarifying agent in the food industry (558) and the other a non-swelling form with strong absorptive properties which are enhanced by chemical activation.

benzaldehyde (oil of bitter almonds, C_6H_5CHO). Aromatic ALDEHYDE; colourless liquid (bp 180°C) with almond odour from which it receives its common name. It occurs in nature in amygladin (*see* GLYCOSIDES). It is manufactured by the oxidation of TOLUENE. It is used as a precursor of 3-PHENYLPROPENOIC ACID, dyes and perfumes. (For preparation and properties — *see* Figure.)

benzaldoxime ($C_6H_5CH=NOH$). OXIME that shows geometrical ISOMERISM. The *syn*-form (mp 34°C) is stable and is prepared from hydroxylamine (*see* NITROGEN HYDRIDES) and benzaldehyde. The *anti*-form (mp 127°C) is prepared by UV irradiation of a benzene solution of the *syn*-form.

benzal group. *See* BENZYLIDENE GROUP.

benzamide ($C_6H_5CONH_2$). White crystalline solid (mp 130°C), prepared by the action of concentrated aqueous ammonia on benzoyl chloride

$$PhCOCl + 2NH_3 \rightarrow PhCONH_2 + NH_4Cl$$

Benzamide undergoes general reactions of an AMIDE. It forms metallic salts (e.g., $PhCONHAg$, $(PhCONH)_2Hg$).

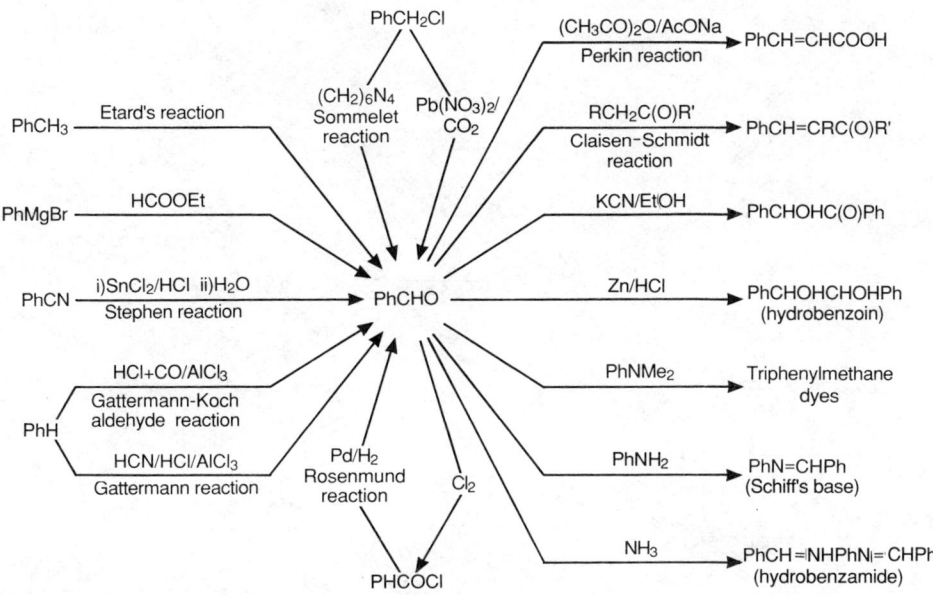

benzaldehyde
Preparation and reaction of benzaldehyde.

2-benzazine. *See* ISOQUINOLINE.

benzedrine. *See* AMPHETAMINE.

benzene (C_6H_6, PhH). Simplest stable structure with aromatic character (*see* HÜCKEL'S RULE). The molecule is a regular flat hexagon, with six hydrogen atoms lying in the plane of the ring (each $C\hat{C}H = 120°$); all C–C bonds are 13.97 pm long (compared with single C–C bond of 15.4 pm and double C=C bond of 13.3 pm) and have some double-bond character. Two equivalent Kekulé structures are used as the canonical forms of benzene (*see* Figure 1) and on this basis the resonance (stabilization) energy is 150.6 kJ mol^{-1}; it is thus more stable than the corresponding conjugated acyclic triene (hexa-1,3,5-triene).

(The double-headed arrow indicates that neither structure adequately represents the molecule, but that the structures together do.) Each carbon is in a state of trigonal HYBRIDIZATION; thus there are six σ C–H bonds, six σ C–C bonds and six $2p_x$-electrons which are all parallel and perpendicular to the plane of the ring (I). Each $2p_x$-electron overlaps its neighbours equally and all six can be treated as forming a molecular orbital embracing all six carbon atoms (II) and so the π-electrons are completely delocalized, thus stabilizing the molecule (*see* Figure 2).

Benzene is obtained as the lightest fraction from the distillation of coal tar hydrocarbons. It is manufactured by DE-HYDROGENATION and dealkylation of suitable PETROLEUM fractions (*see* PETRO-

benzene
Figure 1.   Kekulé structures of benzene.

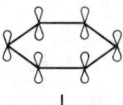 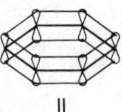 

I   II   III

benzene
Figure 2.   π-orbitals of benzene.

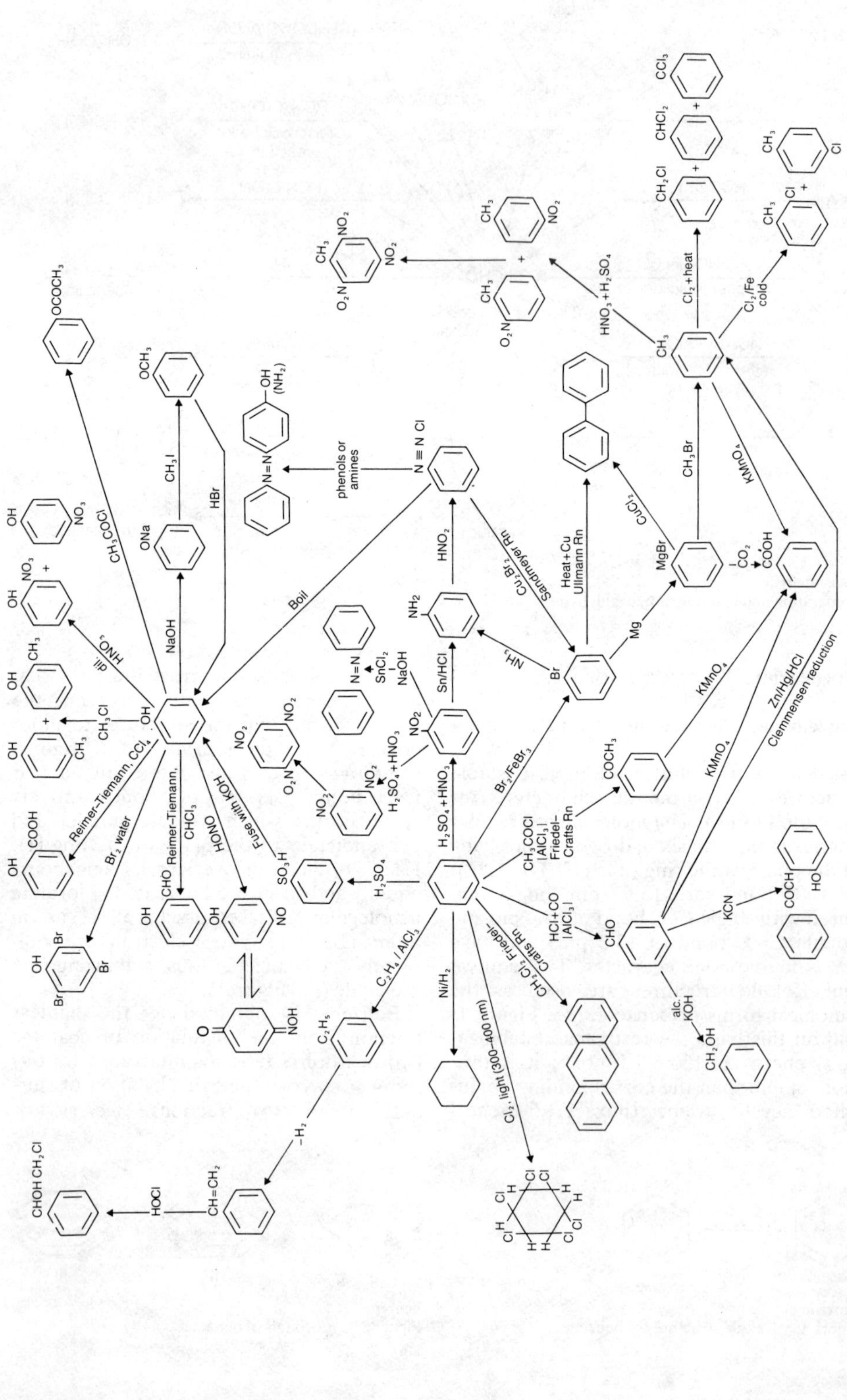
Reimer–Tiemann, CCl_4
Reimer–Tiemann, $CHCl_3$
Br_2 water
HONO
Fuse with KOH
CH_3COCl
NaOH
CH_3I
HBr
dil. HNO_3
CH_3Cl
Boil
phenols or amines
HNO_2
Sandmeyer Rn
Heat + Cu
Ullmann Rn
Sn/HCl
$SnCl_2$
NaOH
$H_2SO_4 + HNO_3$
H_2SO_4
$Br_2/FeBr_3$
NH_3
Mg
CO_2
CH_3Br
$KMnO_4$
$HNO_3 + H_2SO_4$
Cl_2 + heat
Cl_2/Fe cold
Friedel–Crafts Rn
HCl + CO
KCN
alc. KOH
Zn/Hg/HCl
Clemmensen reduction
C_2H_4 / $AlCl_3$
Ni/H_2
Cl_2, light (300–400 nm)
$-H_2$
HOCl

CHEMICALS). Benzene (mp 5.49°C, bp 80.2°C, ρ = 0.885 × 10^3 kg m^{-3}) is a thin, colourless, refractive, non-polar (ε_r = 2.284, 20°C) liquid with a characteristic smell (the vapour is very toxic when inhaled for long periods). It is highly flammable, burning with a yellow smoky flame. It absorbs strongly in the UV region, is an excellent solvent for fats and aromatic compounds, and is miscible with ethanol, ether and ethanoic acid. Benzene is a very stable compound, being slowly attacked by chromium(VI) oxide or acid potassium permanganate to carbon dioxide and water, and is reduced catalytically to cyclohexane (see CYCLOALKANES). In bright sunlight (radical reaction conditions) chlorine forms addition compounds up to $C_6H_6Cl_6$; in the absence of sunlight it undergoes ELECTROPHILIC SUBSTITUTION with halogens. The reaction is slow, but in the presence of iodine or iron(III) chloride substitution to C_6H_5Cl, $C_6H_4Cl_2$, etc. is rapid. Other electrophilic substitution reactions include mono-, di- and trinitration (see NITRONIUM ION), and sulphonation. It undergoes MERCURATION, and in the presence of aluminium trichloride (FRIEDEL–CRAFTS REACTION) it reacts with alkyl halides, acyl halides, carbon monoxide and oxygen to give alkylbenzenes, carbonyl derivatives, benzaldehyde and phenol, respectively. In liquid ammonia it is reduced by sodium (see BIRCH REDUCTION) to 1,4-dihydrobenzene. Nucleophilic substitution reactions of benzene are not possible, halogenobenzenes undergo nucleophilic displacement or elimination reactions. Disubstituted benzenes with groups located in the 1,2-positions are *ortho*-, in 1,3-positions are *meta*- and in 1,4-positions are *para*- (see Figure 3). Using π-bonding electrons, benzene forms many metallic complexes with transition metals (e.g., dibenzene chromium, $C_{12}H_{12}Cr$, see SANDWICH COMPOUND). Industrially benzene is the starting point for many chemicals (e.g., ETHYLBENZENE, CUMENE, cyclohexane, STYRENE, PHENOL, NYLON).

benzenediamines. *See* DIAMINOBENZENES.

benzenedicarboxylic acids   (C_6H_4-$(COOH)_2$). Three isomers are known, all prepared by the oxidation ($KMnO_4$) of the corresponding XYLENE.
benzene-1-2-dicarboxylic acid (phthalic acid). White crystalline solid (mp 231°C), which decomposes to the anhydride, manufactured from PHTHALIC ANHYDRIDE. It undergoes most reactions of DICARBOXYLIC ACIDS. It decarbonylates (KOH) to benzene and is reduced by sodium amalgam to di-, tetra- and hexahydrophthalic acids. A mercury salt derivative is produced by fusing with mercury(II) ethanoate (see Figure).
benzene-1,3-dicarboxylic acid (isophthalic acid). Powder (mp 346°C). It does not form an anhydride.
benzene-1,4-dicarboxylic acid (terephthalic acid). Colourless needles (sublimes 300°C), manufactured by the catalytic oxidation of *p*-xylene in the presence of a cobalt salt. It is used in the manufacture of the polyester Terylene (see POLYMERS, FIBRES).

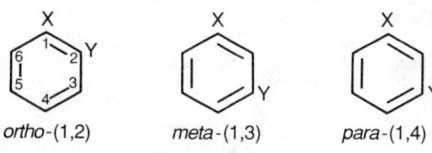

benzenedicarboxylic acids
Mercury salt derivative of benzene-1,2-dicarboxylic acid.

benzenediols. *See* DIHYDROXYBENZENES.

benzenehexacarboxylic acid (mellitic acid, $C_6(COOH)_6$). Colourless, crystalline solid (mp 287°C), formed by oxidation of charcoal using concentrated nitric acid. It occurs naturally as the aluminium salt, and is used in the preparation of DYES and POLYMERS. It is decomposed by heat to the anhydride (see CARBON OXIDES).

benzene hexachloride. *See* 1,2,3,4,5,6-HEXACHLOROCYCLOHEXANE.

benzenesulphonic acid ($C_6H_5SO_3H$). Colourless crystalline, deliquescent solid

X — 1,2 (*ortho*) / 1,3 (*meta*) / 1,4 (*para*) ring diagrams

ortho-(1,2)   meta-(1,3)   para-(1,4)

benzene
Figure 3. Nomenclature of disubstituted benzenes.

(mp (1.5H$_2$O) 44°C), its solutions in water are about as strong as sulphuric acid; prepared by the action of concentrated sulphuric acid on benzene at 80°C. A small amount of DIMETHYLSULPHONE is also produced. Benzenesulphonic acid is used as a catalyst in dehydration and esterification, and as a precursor of dyes and drugs. (For general properties—*see* SULPHONIC ACIDS.) *See also* TOSYL GROUP.

benzenetriols.   *See*   TRIHYDROXYBEN-ZENES.

benzenoid.   Describing an aromatic compound that conforms to HÜCKEL'S RULE.

benzenol.   *See* PHENOL.

benzfuran   (coumarone).   Heterocyclic ring system (*see* HETEROCYCLIC COMPOUNDS, structure XXX).

benzidine.   *See* 4,4'-DIAMINOBIPHENYL.

benzidine rearrangement.   Acid-catalyzed intramolecular conversion of hydrazo-arenes (I) to diaminobiaryls (II, III, IV) and aminodiarylamines (V, VI) (*see* Figure). The reaction is usually performed in aqueous or ethanolic solutions of hydrochloric acid or sulphuric acid.

benzil   (C$_6$H$_5$CO.COC$_6$H$_5$).   Yellow crystalline solid (mp 95°C), prepared by the oxidation of benzoin (*see* ACYLOINS) by nitric acid. It behaves as a typical α-DI-KETONE.

benzine.   Indefinite term roughly corresponding to a volatile gasoline. This name is not recommended for petroleum solvents since it may be confused with benzene.

benzocyclopentadiene.   *See* INDENE.

benzoic acid (C$_6$H$_5$COOH).   White crystalline carboxylic acid (mp 122°C), found naturally in resins (gums and balsam). Manufactured by: (1) air oxidation of toluene over cobalt and manganese ethanoates at 170°C; (2) hydrolysis of trichlorotoluene, from the chlorination of toluene, with calcium hydroxide in the presence of iron; (3) carbonylation of phthalic anhydride over cobalt phthalate. Benzoic acid readily forms ESTERS (benzoates) when refluxed with alcohols in acid solution. It is used in a variety of syntheses for POLYMERS, CAPROLACTAM, DYES and pharmaceuticals. The acid (E210), sodium benzoate (E211) and potassium benzoate (E212) are used as preservatives, and calcium benzoate in the preparation of BENZOPHENONE. The carboxyl group is *meta*-directing towards electrophilic substitution of the benzene ring.

benzoin.   *See* ACYLOINS.

benzol.   Mixture of predominantly aromatic hydrocarbons produced by the carbonization of coal; a fraction obtained from COAL TAR.

benzonitrile (phenyl cyanide, C$_6$H$_5$CN). Colourless oil (bp 191°C), prepared by the dehydrogenation of benzamide with phosphorus pentachloride, by the SANDMEYER REACTION with benzenediazonium chloride, or by refluxing benzaldehyde in methanoic acid containing hydroxylamine hydrochloride. It is used as a solvent and shows the usual reactions of NITRILES.

benzidine rearrangement

1,2-benzophenanthrene.   *See* CHRYSENE.

benzo[*def*]phenanthrene (pyrene).   *See* POLYCYCLIC HYDROCARBONS.

9,10-benzophenanthrene.   *See* TRIPHENYLENE.

benzophenone ($C_6H_5COC_6H_5$).   Solid KETONE (mp 49°C), with a geranium-like smell; prepared by the FRIEDEL–CRAFTS REACTION between benzoyl chloride or carbonyl chloride and benzene, or by heating calcium benzoate. It is reduced (Zn/EtOH, KOH) to benzhydrol (Ph_2CHOH) and is reduced (Zn/CH_3COOH) to benzopinacol ($Ph_2C(OH).C(OH)Ph_2$). Sodium dissolves in benzophenone to form the violet ketyl-containing radical anions

$$Ph_2CO + Na \rightarrow Ph_2C\dot{O}^- Na^+.$$

MICHLER'S KETONE is also a derivative of benzophenone.

benzo[*a*]pyrene (1,2-benzpyrene, $C_{20}H_{12}$). POLYCYCLIC HYDROCARBON (structure XV), occurring in coal tar, cigarette smoke and as a product of incomplete combustion. It has strong carcinogenic properties.

benzoquinone.   *See* QUINONES.

benzothiofuran.   *See* HETEROCYCLIC COMPOUNDS (structure XXXI).

benzoyl chloride (C_6H_5COCl).   Colourless liquid (bp 197°C) with pungent smell, prepared by distilling benzoic acid with phosphorus pentachloride or thionyl chloride and manufactured by chlorinating benzaldehyde. It is an important benzoylating agent used in the SCHOTTEN–BAUMANN REACTION.

benzoyl group.   The group C_6H_5CO- (PhCO–).

benzoyl peroxide (($C_6H_5CO)_2O_2$).   White, unstable solid (mp 104°C), prepared by adding benzoyl chloride and aqueous sodium hydroxide to cold hydrogen peroxide. It readily decarboxylates to give phenyl radicals and is used to initiate POLYMERIZATION.

benzoylsulphonic imide (saccharin).   *See* SWEETENING AGENTS.

1,2-benzpyrene.   *See* BENZO[*a*]PYRENE.

benzylalcohol (phenylmethanol, α-hydroxytoluene, $C_6H_5CH_2OH$). Simplest aromatic MONOHYDRIC ALCOHOL; colourless liquid (bp 204.7°C), prepared by the hydrolysis of benzyl chloride or by the catalytic reduction of benzaldehyde. It is used in perfumery in the form of its esters.

benzylamine.   *See* AMINOTOLUENES.

benzyl chloride ($C_6H_5CH_2Cl$).   Colourless, lachrymatory liquid (bp 179°C), prepared by the chlorination of toluene. It is hydrolyzed to benzyl alcohol by boiling water and reacts with ammonia in ethanol to benzylamine. The benzylcarbonium ion is stabilized by RESONANCE which leads to greater reactivity towards S_N1 and S_N2 reactions (*see* NUCLEOPHILIC SUBSTITUTION).

benzyl group.   The group $C_6H_5CH_2-$ (PhCH$_2$–).

benzylidine group (benzal group).   The group $C_6H_5CH<$ (PhCH<).

benzylpenicillin.   *See* ANTIBIOTICS.

benzyne (dehydrobenzene, C_6H_4).   Transient intermediate (life-time: 10^{-4}–10^{-5} s) generated in the DEHYDROHALOGENATION of a halobenzene with a strong base (e.g., amide ion or phenyllithium), and by the decomposition of benzenediazonium 2-carboxylate. It has not been isolated, but its existence is inferred from indirect evidence. Benzyne may be represented in two forms (*see* Figure). In the absence of other molecules, it reacts with itself to form the POLYCYCLIC HYDROCARBONS biphenylene (structure XXXIII) and triphenylene (structure XI). Both nucleophiles and electrophiles add to benzyne, thus the product of the reaction of lithium amide in liquid ammonia with chlorobenzene is aniline (i.e. it appears to be a substitution). It behaves as a dienophile in a DIELS–ALDER REACTION with, for example, FURAN or ANTHRACENE.

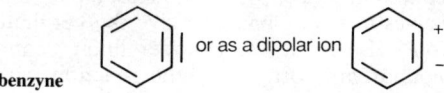

benzyne

Bergius process. Manufacture of liquid hydrocarbons by heating coal with heavy oil and hydrogen under pressure and with a catalyst.

berkelium (Bk). *See* ACTINIDE ELEMENTS.

Berlin green. *See* CYANOFERRATES.

Berthelot equation. *See* EQUATION OF STATE.

berthollide compound. Compound whose solid phase exhibits a range of compositions. *See also* NON-STOICHIOMETRIC COMPOUND.

beryl. *See* BERYLLIUM, SILICATES, SILICON.

beryllium (Be). ALKALINE EARTH METAL (*see* GROUP II ELEMENTS). It is found naturally as beryl ($3BeO.Al_2O_3.6SiO_2$) and in precious gems (e.g., emerald and aquamarine). The metal is extracted by electrolysis of fused BeX_2–NaX (X = F, Cl), or by reduction of BeF_2 by magnesium. Beryllium metal is alloyed with copper to give strong, non-magnetic alloys. It is also used as a moderator in nuclear reactors and as a window for X-rays. Beryllium resists attack by air, water, acids, etc. by forming an oxide layer. Its compounds are toxic.
A_r = 9.012 18, Z = 4, ES [He] $2s^2$, mp 1280°C, bp 2477°C, ρ = 1.85 × 10^3 $kg\,m^{-3}$. Isotope: ^9Be, 100 percent.

beryllium compounds. Beryllium is atypical of the ALKALINE EARTH METALS, showing a greater resemblance to aluminium (*see* DIAGONAL RELATIONSHIP) than to magnesium, calcium, strontium and barium. This stems from the small size and greater charge density of Be^{2+}, coupled with the high values of the ionization potentials (9.32 eV and 18.21 eV). Beryllium is therefore less ELECTROPOSITIVE than lithium ($E^{\ominus}_{Be^{2+}.Be}$ = −1.85 V). All compounds have some covalent character: for example, beryllium chloride ($BeCl_2$) exists in a tetrahedral polymeric structure. Beryllium compounds dissolve in water to give acidic solutions containing $[Be(OH)_3]^{3+}$. The metal dissolves in strong bases to give the beryllate ion (BeO_2^{2-}). Organoberyllium compounds are formed by the reaction with

GRIGNARD REAGENTS or with lithium alkyls or aryls. They are mostly liquids or low-melting solids which react violently with air and water. $Be(CH_3)_2$ is polymeric with tetrahedral coordination of beryllium and electron-deficient bonds (*see* ELECTRON-DEFICIENT COMPOUND) (*compare* ALUMINIUM). Highly coloured complexes are formed with ligands such as 2,2′-dipyridyl.

Bessemer process. Now obsolete method for the manufacture of steel. It has been replaced by the BASIC OXYGEN PROCESS.

beta-decay. Type of RADIOACTIVITY in which an unstable atomic nucleus changes into a nucleus of the same mass number (i.e. mass unchanged), but different proton number. Two types are recognized: negatron beta-decay in which a neutron is converted into a proton with the emission of a negatively charged electron (negatron) and an antineutrino ($\bar{v}$).

$$^1_0 n \longrightarrow {}^1_1 p + {}^{\,0}_{-1} e (\beta^-) + \bar{v}$$

as in the decay of carbon-14

$$^{14}_6 C \longrightarrow {}^{14}_7 N + {}^{\,0}_{-1} e + \bar{v}$$

and positron decay in which a proton is converted into a neutron with the emission of a positively charged electron (positron) and a neutrino (v):

$$^1_1 p \longrightarrow {}^1_0 n + {}^0_1 e (\beta^+) + v$$

The electrons or positrons emitted are called beta-particles; the energy distribution of the rays is continuous (*compare* ALPHA-PARTICLES) in the range 0–4 MeV, with a clear maximum value. They cause burns to the skin and are harmful if they enter the body. Thin sheets of metal provide protection. *See also* RADIOACTIVE DECAY SERIES, RADIOACTIVITY.

betaine (trimethylglycine, $(CH_3)_3\overset{+}{N}CH_2COO^-$). Very feeble base, which occurs in sugarbeet and can be obtained from beet molasses; prepared as a solid (mp 293°C) from trimethylamine and chloroethanoic acid.

betaines. Group of feebly basic compounds, resembling BETAINE, which occur in plants. They are intramolecular salts or ZWITTERIONS.

beta-particle. *See* BETA-DECAY.

beta-pleated sheet. *See* PROTEINS.

BET isotherm (Brunauer–Emmett–Teller isotherm). *See* ADSORPTION ISOTHERM.

bhang. *See* TETRAHYDROCANNABINOLS.

BHC (benzene hexachloride). *See* 1,2,3,-4,5,6-HEXACHLOROCYCLOHEXANE.

Bi. *See* BISMUTH.

biaxial crystal. Crystal, such as mica or selenite, with an optical axis in each of two directions. From a single incident ray two refracted rays are formed; the ordinary ray obeys the normal laws of refraction whereas the other ray, the extraordinary ray, has a refractive index that varies with its direction in the crystal. Both rays are plain polarized in directions at right angles to each other. The phenomenon is known as birefringence or DOUBLE REFRACTION.

bicarbonate. *See* CARBONIC ACIDS.

bicyclo compounds. CYCLOALKANES containing two rings with two or more atoms in common. They are named from the alkane containing the same total number of carbon atoms with the prefix bicyclo- and the number of carbon atoms in each of the three bridges in brackets. Numbering begins with one of the bridgeheads and goes on the longest path to the second bridgehead then the next longest path back and finally along the shortest path. An example is 2-methyl-7-ethylbicyclo[2.2.1]heptane. (For preparation and properties—*see* CYCLOALKANES.) *See also* BICYCLO[4.4.0] DECANE.

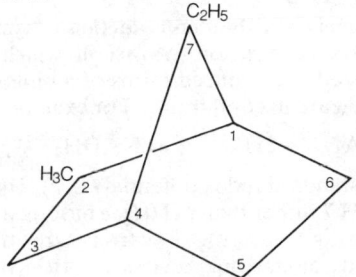

bicyclo compounds
Structure of 2-methyl-7-ethylbicyclo[2.2.1]heptane.

bicyclo[4.4.0]decane (decahydronaphthalene, decalin, $C_{10}H_{18}$). BICYCLO COMPOUND found in two geometrical structures (*see* Figure) with the fused cyclohexane rings in the chair conformation such that the bridged hydrogen atoms are *cis* (mp –45°C, bp 198°C) or *trans* (mp –32°C, bp 185°C). The *cis*-form is converted to the more stable *trans*-form over aluminium(III) chloride. Manufacture is by the hydrogenation of naphthalene at high pressures which yields a mixture of 90 percent *cis*-form, 10 percent *trans*-form. It is used as a solvent.

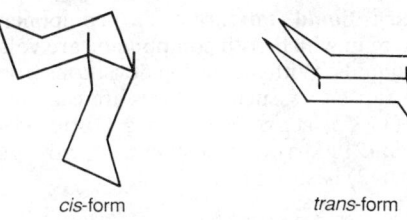

cis-form   *trans*-form

bicyclo[4.4.0]decane
Geometrical isomers.

bicyclo[5.3.0]deca-2,4,6,8,10-pentane. *See* AZULENE.

bidentate ligand. *See* CHELATE LIGAND.

bile acids. Products of the alkaline hydrolysis of bile salts formed in the liver and discharged into the small intestine as part of the bile. The acids (*see* Table) are combined by an amide linkage with glycine (glycocholate) or taurine (taurocholate). The bile salts assist in the emulsification of fat droplets, hence promoting the hydrolysis and

Compound	R	R'
Cholic acid	–OH	–OH
Deoxycholic acid	–H	–OH
Chenodeoxycholic acid	–OH	–H
Lithocholic acid	–H	–H

bile acids

absorption of fats from the intestinal tract. The four bile acids occurring in human and ox bile are cholic (mp 195°C), deoxycholic (mp 176°C), chenodeoxycholic (mp 143°C) and lithocholic acids (mp 186°C); the ratio of the first three (in humans) being 31:24:45.

bimolecular reaction. *See* MOLECULARITY OF REACTION.

binary compound. Chemical compound containing two elements only.

binary liquid mixture. Two-component system in which both components are volatile liquids. With increasing deviations from RAOULT'S LAW such mixtures are classified as: (1) COMPLETELY MISCIBLE LIQUID SYSTEMS; (2) PARTIALLY MISCIBLE LIQUIDS SYSTEMS; (3) IMMISCIBLE LIQUIDS.

binding energy. Energy equivalent to the MASS DEFECT when NUCLEONS bind together to form an atomic nucleus. When a nucleus is formed, some energy is released by the neutrons since they are in a lower energy state.

binodal curve. *See* THREE-COMPONENT SYSTEM.

biochemical oxygen demand (BOD). Quantity of dissolved oxygen ($mg\ dm^{-3}$) required during the aerobic breakdown of decomposable organic matter in water by microorganisms. A standardized means for estimating the degree of contamination of water supplies by industrial wastes and sewage, etc. The sample is diluted with water saturated with oxygen and the quantity of dissolved oxygen is measured immediately and after five days incubation at 20°C. A high BOD suggests the presence of a large number of organisms and hence high level of pollution.

biochemical standard state. Reactants and products are at unit concentration except that the hydrogen ion concentration is $10^{-7}\ mol\ dm^{-3}$ (i.e. pH 7.0). The change in the standard free energy ($\Delta G^{\ominus\prime}$) using this definition differs from $\Delta G^{\ominus}$. For the reaction

$$A + B \longrightarrow C + xH^+$$

$$\Delta G^{\ominus} = \Delta G^{\ominus\prime} + xRT \ln (1/10^{-7})$$
$$= \Delta G^{\ominus\prime} + 40.0\ kJ$$

(when $x = 1$ and $T = 298$ K)

The reaction is thus more spontaneous at pH 7.0 than it is at pH 0. For a reaction in which H^+ is a reactant the above equation becomes

$$\Delta G^{\ominus} = \Delta G^{\ominus\prime} - xRT \ln (1/10^{-7})$$

and the reaction is more spontaneous at pH 0 than it is at pH 7.0. The corresponding redox electrode potentials are related by

$$E^{\ominus} = E^{\ominus\prime} \pm 0.415/n$$

where the positive (negative) sign applies to reactions in which H^+ is a reactant (product). *See also* STANDARD STATE.

biocide. Substance that kills or inhibits the growth of bacteria, moulds, slime, fungi, etc. *See also* ANTISEPTIC, DISINFECTANT, FUNGICIDE.

biodegradable. Refers to a substance (e.g., sewage constituents) that can be decomposed by bacteria and other microorganisms. It is often specifically applied to detergents and pesticides; branched-chain alkylbenzene sulphonates are more resistant to decomposition than are the straight-chain compounds, the alcohol sulphate and non-ionic detergents are biodegradable. Organic phosphorus pesticides, although highly toxic, are more biodegradable than chlorinated biphenyls and DDT.

biofuel cell. FUEL CELL in which microorganisms provide the fuel for the reaction at the anode, either directly from reactions within the living cell or indirectly by the production of hydrogen, for example.

biological oxidation/reduction system. REDOX ELECTRODE SYSTEM in which the oxidized and reduced forms of a biological couple are in equilibrium. For example

$$NAD^+ + 2H^+ + 2e \rightleftharpoons NADH_2^+$$

The standard redox potential ($E^{\ominus\prime}$) is based on pH 7 rather than pH 0 (*see* BIOCHEMICAL STANDARD STATE). Electrons are transferred along an ELECTRON TRANSPORT CHAIN in the direction of increasing electrode potential.

bioluminescence. Production of light by living organisms, such as glow-worms, fireflies, bacteria, fungi and some deep-sea fish. The light is produced during the oxidation of LUCIFERIN, catalyzed by the enzyme luciferase. It may be continuous (in bacteria) or intermittent (in fireflies).

biomass. Total weight of living matter in a population, usually expressed as dry weight per unit area or volume.

biose. Carbohydrate with two carbon atoms; the only example is hydroxyethanal (HOCH$_2$CHO). It is not optically active.

biotin (vitamin B complex).   *See* VITAMINS.

biphenyl (diphenyl, phenylbenzene, C$_6$H$_5$–C$_6$H$_5$, Ph.Ph). Colourless crystalline solid (mp 71°C), insoluble in water, but soluble in ethanol and ether, prepared by the FITTIG REACTION, GOMBERG REACTION or ULLMANN REACTION. Industrially it is obtained by passing benzene mixed with superheated steam (>1000°C) into steel vessels coated with iron(II) oxide. It undergoes the usual nuclear substitution reactions, the phenyl group being *ortho-*, *para*-directing (*see* ORIENTATION IN THE BENZENE RING). On nitration, sulphonation and halogenation the first substituent enters mainly the 4-position (to lesser extent the 2-position), introduction of a second group usually takes place in the unsubstituted ring at 4'- and 2'-positions giving mainly 4,4'- (with some 2,4'- and 4,2'-) disubstituted biphenyls. In the solid phase the rings are coplanar, but in the gaseous phase one ring is rotated at 42° with respect to the other (i.e. the attainment of a position of minimum energy). If some or all the *ortho*-hydrogens are replaced by bulky groups, these interfere with one another so that the molecule cannot become planar (e.g., 2,2'-dicarboxy-6,6'-dinitrobiphenyl). This compound exists as two stable ENANTIOMERS which can be separated (*see* CHIRALITY). The 3,3'- and 4,4'-positions are too far apart for substituents to interfere with one another; the enantiomers are in equilibrium and cannot be separated. Biphenyl is used as a fungistat (E230) during the transport of citrus fruits and as a heat transfer medium.

biphenylene.   *See* POLYCYCLIC HYDROCARBONS (structure XXXIII).

bipy.   *See* DIPYRIDYL.

bipyridyl.   *See* DIPYRIDYL.

Birch reduction. Reduction of a terminal double bond by sodium or lithium in liquid ammonia in the presence of an alcohol (methanol, ethanol) (e.g., RCH=CH$_2$ → RCH$_2$CH$_3$). Aromatic compounds are reduced to the dihydro compound (e.g., C$_6$H$_6$ → C$_6$H$_8$). It is a useful reaction although relatively few substituents in the benzene ring are capable of withstanding the drastic conditions.

birefringence.   *See* DOUBLE REFRACTION.

Birkeland–Eyde process. Industrial method for the fixation of nitrogen by passing air through an electric arc (>3000°C) to produce nitrogen oxides. Since the reaction is endothermic, the equilibrium mixture must be rapidly cooled to reduce the dissociation of the oxides. The process is only economic if cheap electricity is available.

bis-. Prefix denoting two identical groups attached to a given atom.

bismuth (Bi).   GROUP V ELEMENT, which occurs (10^{-5} percent of the earth's crust) as the trisulphide (Bi$_2$S$_3$) associated with the sulphide ores of lead and copper. The ore is roasted to the trioxide (Bi$_2$O$_3$) and then reduced to the metal with hydrogen or carbon. Electrolytic purification is possible from a solution of bismuth(III) chloride (BiCl$_3$) in hydrochloric acid. The metal is brittle, with a reddish colour and a crystal structure similar to black phosphorus. Bismuth burns in air to the oxide and reacts with the halogens. It is only slowly attacked by acids, dissolving in concentrated nitric acid giving the nitrate and in sulphuric acid giving the sulphate. Bismuth exhibits oxidation states of +5 and +3, forms cluster ions and a range of complex cationic species (e.g., [Bi$_6$(OH)$_{12}$]$^{6+}$, [Bi$_6$O$_6$(OH)$_3$]$^{3+}$), the free Bi^{3+} ion does not exist. The metal is used in fusible alloys (e.g., WOOD'S METAL and type metal); the alloys expand on solidification and hence give a sharp impression.

The salts are used in pharmaceutical preparations, showing a marked contrast in toxicity to the arsenic compounds.
A_r = 208.9804, Z = 83, ES [Xe] $4f^{14} 5d^{10} 6s^2 6p^3$, mp 271.3°C, bp 1560°C; ρ = 9.8 × 10^3 kg m^{-3}. Isotope: ^{209}Bi, 100 percent.

bismuth compounds.
halides. White solid bismuth trifluoride (BiF$_3$) (Bi$_2$O$_3$ + HF), insoluble in water; with excess oxide the oxofluoride (BiOF) is formed. The remaining trihalides are made by direct combination of the elements. They are hydrolyzed by water and form complex ions (e.g., [BiCl$_4$]$^-$, [Bi$_2$Cl$_7$]$^-$). Bismuth pentafluoride (BiF$_5$) (Bi + F$_2$) is a very powerful fluorinating agent; it forms the complex ion [BiF$_6$]$^-$ (MF + Bi$_2$O$_3$ + BF$_3$).
oxides. The only well-established oxide is the trioxide (Bi$_2$O$_3$) (Bi + O$_2$ or bismuth salt + NaOH to give the hydroxide Bi(OH)$_3$ followed by dehydration), a yellow powder soluble in acids giving bismuth salts (e.g., nitrate, sulphate). It has no acidic properties, being insoluble in alkalis. Fusion with sodium hydroxide gives a non-stoichiometric buff-coloured powder called sodium bismuthate (NaBiO$_3$), which in nitric acid solution will oxidize Mn^{2+} to MnO$_4^-$ at room temperature. Bi$_2$O$_3$ is used in porcelain glazes and stained glass. The pentoxide (Bi$_2$O$_5$) does exist (Bi$_2$O$_3$ + powerful oxidizing agent), but has never been isolated in the pure state. It is extremely unstable, readily losing oxygen.
sulphides. The trisulphide (Bi$_2$S$_3$) is obtained as a dark brown precipitate when hydrogen sulphide is bubbled through an acidic solution of a bismuth salt. It is not acidic. The sulphide (BiS$_2$) (Bi + S at 1250°C) is known.
cationic bismuth compounds. Aqueous solutions contain well-defined hydrated cations, but there is no evidence for the simple [Bi(H$_2$O)$_n$]$^{3+}$ species. In neutral perchlorate solutions, the main species is [Bi$_6$O$_6$]$^{6+}$ and at higher pH values [Bi$_6$O$_6$(OH)$_3$]$^{3+}$. There is evidence for the association of bismuth ions with nitrate ions in solution (e.g., Bi(NO$_3$).(H$_2$O)$_n$). From acid solutions, crystalline salts such as Bi(NO$_3$)$_3$.5H$_2$O, Bi$_2$(SO$_4$)$_3$ and double salts of the kind M$_3$[Bi(NO$_3$)$_6$]$_2$.24H$_2$O can be obtained. Treatment of Bi$_2$O$_3$ with nitric acid gives bismuthyl salts such as BiO(NO$_3$) and Bi$_2$O$_2$(OH)(NO$_3$).

bisulphite addition compound. *See* ADDITION COMPOUND.

bittern. Solution of salts remaining after sodium chloride has been crystallized from seawater.

bitumen. Solid or viscous liquid residue from the distillation of PETROLEUM. When combined with small amounts of minerals, it is known as asphalt. *See also* ASPHALTENES.

biuret (H$_2$NCONHCONH$_2$). Solid (mp 193°C (decomp.)); formed by the loss of ammonia from two molecules of urea when heated.

biuret reaction. Substances containing at least two peptide (–CONH–) groups (e.g., BIURET, PROTEINS and peptides) on heating with sodium hydroxide and copper(II) sulphate give a violet or pink coloration. Amino acids do not give a reaction.

Bk. Berkelium (*see* ACTINIDE ELEMENTS).

black body radiation. Radiation emitted by a body capable of absorbing and emitting all frequencies with equal facility. The intensity of radiation plotted against wavelength goes through a maximum (λ_{max}), which moves to shorter wavelengths as the temperature increases. By the Wien displacement law $T\lambda_{max}$ = constant. The total energy per unit volume U follows Stefan's law $U = \sigma T^4$ where σ is Stefan's constant. A consideration of the nature of black body radiation led Planck to formulate his radiation law (*see* PLANCK CONSTANT).

bleaching agent. Chemical that removes colour from textiles, paper and/or other material. Oxidizing agents that are used as bleaches include bleaching powder (Ca(OH)$_2$/Cl$_2$), chlorine, chlorates(I), perborates, bleaching earths (*see* FULLER'S EARTH). In the bleaching of textiles, brighteners (e.g., diaminostilbenesulphonic acid) are added to absorb UV radiation and fluoresce in the visible. *See also* FOOD ADDITIVE.

Bloch equation. *See* BAND THEORY OF SOLIDS.

block copolymer. *See* COPOLYMERIZATION.

boat form. *See* CONFORMERS.

BOD. *See* BIOCHEMICAL OXYGEN DEMAND.

body-centred cubic lattice. *See* CLOSE-PACKED STRUCTURES.

bohr. *See* ATOMIC UNITS.

Bohr atom. Early model of the atom that equates the energy of an electron orbiting a positive nucleus at a particular distance with integral multiples of PLANCK CONSTANT. Although it predicts the correct form of the atomic spectrum of hydrogen (*see* BALMER SERIES), the use of an essentially classical model is wrong and has been superseded by the use of wave mechanics (*see* QUANTUM MECHANICS). *See also* SCHRÖDINGER EQUATION.

Bohr magneton (β; 9.2732×10^{-24} J T^{-1}).

Magnetic moment of a spinning sphere carrying unit electronic charge. From classical electromagnetic theory $\beta = eh/4\pi m = e\hbar/2m$.

Bohr radius. *See* BOHR ATOM.

boiling point. Temperature at which the VAPOUR PRESSURE of a liquid is equal to the external pressure. The normal boiling point is the temperature at which the vapour pressure is 1 atm (760 mmHg). A high (low) boiling point indicates large (small) inter-molecular forces of attraction. The boiling point of a solution is higher than that of the solvent (*see* Figure), the elevation of boiling point (COLLIGATIVE PROPERTY) for a solution of a non-volatile solute in a volatile solvent is related to the vapour pressure above the solution and its concentration

$$\Delta T_e = (RT^2/L_e) \ln p^\ominus/p$$
$$= RT_0^2 x_B/L_e$$
$$= (RT_0^2/L_e n_A) m_B$$
$$= k_e m_B$$

This is valid for dilute solutions of compounds of low molecular mass. The molal

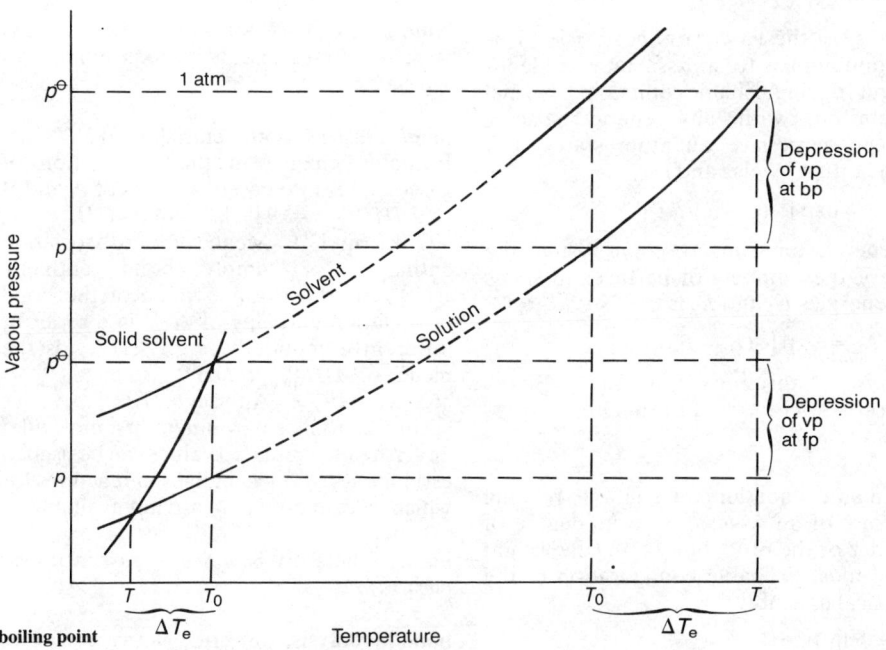

boiling point $\underbrace{}_{\Delta T_e}$ Temperature $\underbrace{}_{\Delta T_e}$

ebullioscopic constant (k_e) is a property of the solvent independent of the nature of the solute. For non-ideal solutions the ACTIVITY of the solvent is given by

$$\ln a_A = -(L_e/RT_0^2)\,\Delta T_e$$

The molecular mass of the solute (B) can be determined from measurements of ΔT_e for solutions of known molality; anomalous values are obtained for solutes which associate or dissociate. *See also* FREEZING POINT, RAOULT'S LAW.

boiling point diagram. *See* COMPLETELY MISCIBLE LIQUID SYSTEMS.

boiling point elevation. *See* BOILING POINT.

Boltzmann constant (k). Ratio of the universal GAS CONSTANT to the AVOGADRO CONSTANT. It may be regarded as the gas constant per molecule ($k = 1.380\,622 \times 10^{-23}$ J K^{-1}).

Boltzmann distribution. Equation derived from the BOLTZMANN EQUATION for the fraction of the total number of particles (molecules) in a particular energy level

$$n_i/N = \exp(-\varepsilon_i/kT)/Q$$

where Q is the PARTITION FUNCTION. This equation applies to an ASSEMBLY of identical but distinguishable units, in thermal equilibrium, with any energy spacing between successive quantum states. For two quantum levels i and j

$$n_i/n_j = \exp[-(\varepsilon_i - \varepsilon_j)/kT]$$

In general, for a macroscopic system the ratio of the numbers of particles in states with energies E_1 and E_2 is

$$N_1/N_2 = \exp[-(E_1 - E_2)/RT]$$

See also ARRHENIUS EQUATION, BOSE–EINSTEIN STATISTICS, FERMI–DIRAC STATISTICS.

Boltzmann equation. Equation relating ENTROPY of an ASSEMBLY to its degree of disorder or the probability (W) (the weight of the most probable configuration of the canonical assembly)

$$S = k \ln W$$

Its use involves an analysis of the distribution of all the microstates (systems) in the assembly among the available energy levels. *See also* PARTITION FUNCTION, SPECTROSCOPIC ENTROPY, STATISTICAL MECHANICS.

bomb calorimeter. Instrument for measuring ENTHALPY of combustion. The compound is ignited in a steel bomb in an atmosphere of oxygen under pressure, and the increase in temperature in the surrounding water is measured. Since the volume is constant, changes in the INTERNAL ENERGY are measured; these are converted to ΔH values given by $\Delta H = \Delta U + \Delta n_{gas}RT$. The method is only suitable for spontaneous reactions.

bond. Region of high electron density between atoms in a molecule. In simple COVALENT BONDS, the number of ELECTRON PAIRS that contribute to the bond determines whether the bond is single (one), double (two) or triple (three). In terms of MOLECULAR ORBITALS, a single bond is an electron pair in a σ-orbital to which may be added one or two pairs in π-orbitals to give double or triple bonds, respectively. *See also* BOND ORDER, COORDINATE BOND, HYDROGEN BOND, IONIC BOND.

bond angle. *See* SHAPES OF MOLECULES, VALENCE SHELL ELECTRON PAIR REPULSION THEORY.

bond dissociation enthalpy ($DH^{\ominus}_{A-B}$). Enthalpy change in the dissociation of gaseous AB into gaseous atoms of A and B (e.g. $DH^{\ominus}_{N=N} = 945$ kJ mol^{-1}; $DH^{\ominus}_{HO-H} = 492$ kJ mol^{-1}). Mean bond dissociation enthalpy, or simply bond enthalpy $EH^{\ominus}_{A-B}$, is the average value of the bond dissociation enthalpy of A–B in a series of different compounds (e.g., $EH^{\ominus}_{O-H}$ is the mean of $DH^{\ominus}_{HO-H}$, $DH^{\ominus}_{H-O}$, $DH^{\ominus}_{CH_3O-H} = 463$ kJ mol^{-1}). Although $EH^{\ominus}_{A-B}$ varies according to its environment in a molecule, nevertheless tabulated values can be used to estimate ENTHALPY changes in reactions for which calorimetric data are not available.

bond heterolysis. *See* HETEROLYTIC FISSION.

bond homolysis. *See* HOMOLYTIC FISSION.

bonding orbital. *See* MOLECULAR ORBITAL.

bond length. Internuclear distance between atoms in a molecule. Covalent bond lengths (*see* Table 1) may be assigned to pairs of atoms with similar bonds (*see* BOND ORDER, HYBRIDIZATION). From data of pairs of atoms, single-bond covalent radii (*see* Table 2) may be derived (e.g., the single-bond radius of F is half the bond length of F_2). Bond lengths are measured by X-RAY DIFFRACTION in solids and by

bond length
Table 1. Covalent bond lengths.

Bond	Length/ pm	Bond	Length/ pm
H–H	74	C–H	109
C–C	154	O–H	96
C=C	134	S–H	135
C≡C	120	C–O	143
C–C[a]	139	C=O	122
N–N	110	C–N	147
F–F	142	C≡N	116
Cl–Cl	199	C–Cl	177
Br–Br	228	C–Br	193

[a] Aromatic.

bond length
Table 2. Single-bond covalent radii.

Atom	Radius/ pm	Atom	Radius/ pm
H	37	C	77
N	74	O	74
F	72	Cl	99
Br	114	I	133
S	104	P	110
Si	117		

determination of properties such as IONIC MOBILITY, compressibility and entropy in the gas and liquid phases. *See also* ATOMIC RADIUS, IONIC RADIUS.

bond order. The π-bond order (p_{ab}) between atoms a and b is defined as $\sum_r n_r c_{ra} c_{rb}$. c_{ra} and c_{rb} are the Hückel molecular orbital coefficients (*see* HÜCKEL MOLECULAR ORBITAL THEORY) for the rth-orbital containing n_r electrons.

bond strength. *See* BOND DISSOCIATION ENTHALPY.

BOP. *See* BASIC OXYGEN PROCESS.

boranes (boron hydrides). *See* BORON INORGANIC COMPOUNDS.

borates. *See* BORON INORGANIC COMPOUNDS.

borax. *See* BORON.

borazine. *See* BORON INORGANIC COMPOUNDS.

borazole. *See* BORON INORGANIC COMPOUNDS.

borazon (boron nitride). *See* BORON INORGANIC COMPOUNDS.

borneol. *See* TERPENES.

Born–Haber cycle. Thermodynamic cycle (*see* Figure) based on HESS'S LAW for the calculation of the LATTICE ENERGY (U) of an ionic solid. For example, for sodium chloride, the lattice enthalpy is given by

$$\Delta H_{lattice} = \Delta_f H^\ominus - (L_s + 0.5D + I + E)$$

where L_s is the enthalpy of sublimation of

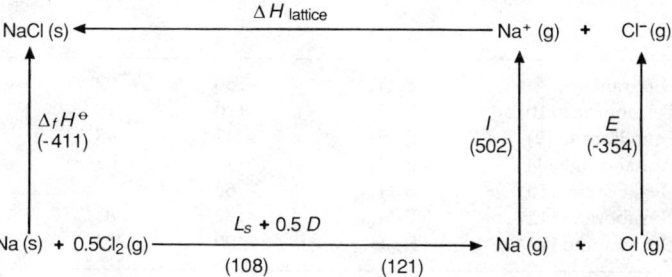

Born-Haber cycle

sodium to free atoms, D the dissociation energy of chlorine, I the ionization energy of sodium and E the electron affinity of chlorine. All the terms can be measured, except the lattice energy, which can thus be calculated. For sodium chloride, $\Delta H_{\text{lattice}} = -788$ kJ mol^{-1} corresponding to a lattice energy of -783 kJ mol^{-1}.

bornite. *See* COPPER.

Born–Landé equation. *See* LATTICE ENERGY.

Born–Oppenheimer approximation. Approximation that treats the nuclei of atoms in a molecule as immobile compared with the swiftly moving electrons. This is a great simplification when solving the SCHRÖDINGER EQUATION for the motion of electrons in the field of now static nuclei.

boron (B). Group IIIa element (*see* GROUP III ELEMENTS) widespread in nature, but occurring in only a limited number of places in minable quantity as borates (borax $Na_2B_4O_7.10H_2O$, rasorite and colemanite). The element is extracted by reduction of boron(III) oxide by magnesium, reduction of boron(III) chloride with hydrogen or by the electrolysis of fused potassium borofluoride. Crystalline boron is second only to diamond in hardness and is found in three modifications. The properties of elemental boron depends on the purity of the sample. For example a highly dispersed form with alkali impurities is pyrophoric, but amorphous boron oxidizes only slowly in air. Boron is generally inert, but it does

react with hot oxidizing acids and some non-metals at high temperatures. Boron is used as an oxygen scavenger in metallurgical processes and as an additive to STEEL and lightweight alloys. Boron-10 has a high cross-section for neutrons and thus is used as cladding and in control rods in nuclear reactors. Boron is an important component in borosilicate GLASSES and other ceramics. $A_r = 10.811$, $Z = 5$, ES [He] $2s^2 2p^1$, mp 2300°C, bp 3930°C, $\rho = 2.34 \times 10^3$ kg m^{-3}. Isotopes: ^{10}B, 18.7 percent; ^{11}B, 81.3 percent.

boron inorganic compounds. Boron mostly shows oxidation state +3 and is three- or four-coordinate in its compounds. All boron compounds are covalent, and boron–boron bonds are formed in many cluster compounds. Icosahedral B_{12} units exist in elemental boron. Boron compounds form Lewis acid (*see* LEWIS ACIDS AND BASES) adducts with amines. *boron oxides.* Boron(III) oxide is the major oxide formed by the ignition of boron in oxygen. It is acidic and dissolves in water to give boric acid ($B(OH)_3$). Boron(II) oxide (BO) is also known. Borates are anions containing boron and oxygen and are considered as salts of boric acid. Trigonal BO_3 and tetrahedral BO_4 units are found often linked into rings, cages and polymeric structures. In borax ($Na_2B_4O_7.xH_2O$) the borate ion has a complex structure (*see* Figure 1). The simpler perborate which shares an oxygen atom between two BO_3 units ($O_2B-O-BO_2$) is found in $Co_2B_2O_5$. Hydroxyl groups in borates may be replaced by alkoxy groups.

boron inorganic compounds
Some boranes.

Name	Formula	mp /°C	bp /°C	Stabiility to	
				Air	Hydrolysis
Diborane (6)	B_2H_6	−166	− 92.5	Inflames	Fast
Tetraborane (10)	B_4H_{10}	−120	18	Stable	Slow
Pentaborane (9)	B_5H_9	− 47	48	Inflames	By hot water
Pentaborane (11)	B_5H_{11}	−123	63	Inflames	Fast
Hexaborane (10)	B_6H_{10}	− 62	−	Stable	By hot water
Hexaborane (12)	B_6H_{12}	− 82	85	Stable	Fast
Decaborane (14)	$B_{10}H_{14}$	100	213	Stable	Slow

boron carbide (B_4C). Used extensively as an abrasive and in steel.

metal borides. Hard, electrically conducting. They are finding increasing use in cermets and high-temperature refractory materials.

boron–nitrogen compounds. Borazine (*see* Figure 2) also known as borazole, is isoelectronic with benzene but its polar nature renders it more reactive towards addition. A wide range of boron–nitrogen compounds and boron–carbon–hydrogen compounds exist (*see* CARBORANES).

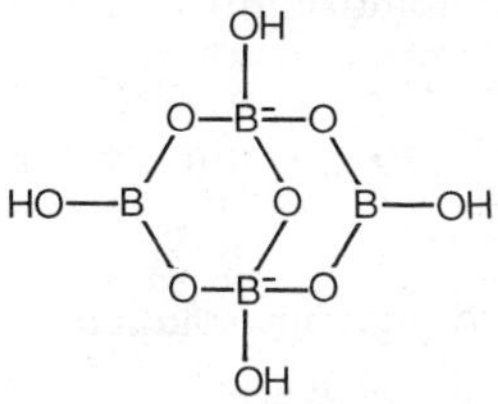

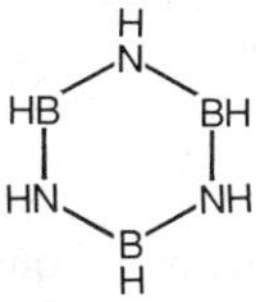

boron inorganic compounds
Figure 1. Structure of borate ion.   Figure 2. Structure of borazine.

boron hydrides (boranes). Electron-deficient compounds containing boron–hydrogen–boron bonds (*see* Table). Diborane (B_2H_6) is a reactive gas prepared by the reaction between lithium hydride and boron(III) fluoride. The higher boranes are more stable, their structures including complex cages of boron and hydrogen. They are prepared by thermal decomposition of diborane in a process similar to cracking of hydrocarbons. They are named according to the number of boron atoms with the number of hydrogen atoms in parentheses (e.g. hexaborane (12), B_6H_{12}). Boranes react with oxygen to give B_2O_3 and water, and with water to give $B(OH)_3$ and hydrogen. Diborane adds to unsaturated organic compounds in the process of HYDROBORATION, reacts with ammonia to give borazine and with metal hydrides to give tetrahydridoborates. Lithium and sodium tetrahydridoborates (MBH_4, where M = Li, Na) are widely used in organic synthesis as reducing agents (*see* ALKANES). Higher borane anions are known (e.g., $[B_3H_8]^-$, $[B_9H_9]^{2-}$, $[B_{10}H_{10}]^{2-}$).

boron halides. Colourless, volatile compounds that react readily with water and are synthesized by heating boron or borides with halogens. Boron(III) iodide (BI_3) is best prepared by the reaction between $NaBH_4$ and I_2. Halides are used as catalysts and as intermediates in the production of hydridoborates. Boron(III) chloride (BCl_3) is used in the refining of metals. Subhalides (e.g., B_2Cl_6, B_9Cl_9 and B_2F_4) are also known.

borosilicates. GLASSES formed by fusing silica and boron oxide and a metal oxide: for example, with sodium to give Pyrex.

boron organic derivatives. Organoboron compounds may be divided into those with direct C–B bonds and those linked through oxygen or nitrogen (borate esters and boroximes, respectively). Alkylboron compounds (R_3B) are reactive and spontaneously flammable in air. They are hydrolyzed to alkylboric acids: boronic ($RB(OH)_2$) and boronous ($R_2B(OH)$). Alkylboron compounds decompose in acids yielding alkanes and in alkaline peroxides yielding alcohols. They are used in synthesis in HYDROBORATION reactions. Boron-to-carbon bonds are also formed by reaction of boron(III) chloride (BCl_3) with a GRIGNARD REAGENT or in a FRIEDEL–CRAFTS REACTION with an aromatic hydrocarbon, $AlCl_3$ and BCl_3. Organic analogues of the borates ($[BR_4]^-$) are prepared from metal alkyls and R_3B. Organoborates readily react with electrophiles in synthetically useful reactions. Potassium tetraphenylborate is used in the gravimetric determination of potassium.

borosilicates. *See* BORON INORGANIC COMPOUNDS, GLASS, SILICATES.

Bose–Einstein statistics. Indistinguishable particles with zero or integral spin known as bosons (e.g., photons, phonons, ^{4}He) follow Bose–Einstein statistics. The probability (f) of a boson having energy E is

$$f = \{\exp[(E - \alpha)/kT] - 1\}^{-1}$$

where α is a normalizing parameter. *See also* BOLTZMANN DISTRIBUTION, FERMI–DIRAC STATISTICS.

boson. *See* BOSE–EINSTEIN STATISTICS.

Boyle's law. At constant temperature, the volume (V) of a given mass of gas is inversely proportional to its pressure (P) (i.e. *PV*

= constant). The law is only true for an IDEAL GAS. For a REAL GAS, the law is not obeyed because of the finite size of the molecules and the existence of INTERMOLECULAR FORCES. *See also* EQUATION OF STATE, GAS LAW, VAN DER WAALS EQUATION.

Boyle temperature (T_B). Temperature at which the value of PV for a REAL GAS is constant over an appreciable range of pressures (i.e. BOYLE'S LAW is valid). For a gas obeying VAN DER WAALS EQUATION, T_B = a/Rb. For hydrogen T_B = 110.04 K, compared with an experimental value of 106 K; for helium T_B = 22.64 K.

Bq.   *See* BECQUEREL.

Br.   *See* BROMINE.

Brackett series.   *See* BALMER SERIES.

bradykinin.   *See* PEPTIDES.

Brady's reagent.   Solution of 2,4-dinitrophenylhydrazine sulphate in methanol which reacts with aldehydes and ketones giving characteristic crystalline yellow to red derivatives.

Bragg's law.   When a beam of monochromatic X-rays of wavelength λ strikes a crystal surface in which the layers of atoms or ions are separated by a distance d, the maximum intensity of the diffracted ray occurs when

$$\sin \theta = n\lambda/2d$$

where θ is the angle which the incident beam makes with the crystal and n is an integer. *See also* X-RAY DIFFRACTION.

branched-chain reaction.   *See* CHAIN REACTION.

brass.   ALLOY of COPPER and ZINC. Six phases have been identified, the most important being the ductile α-phase (up to 30–35 percent zinc below 1000 K) and the stronger, more brittle β-phase (35–58 percent zinc, mixed with α- and γ-brass at extremes of composition). Brass is used for wiring, sheets, tubes and castings. The formation of identifiable INTERMETALLIC COMPOUNDS (e.g., $CuZn$, Cu_5Zn_8 and

$CuZn_3$) may be explained in terms of HUME–ROTHERY'S RULE.

Bravais lattice.   *See* CRYSTAL SYSTEMS.

breathalyzer.   *See* ALCOHOL METER.

breithamptite.   *See* NICKEL.

Bremsstralung.   Broad continuous spectrum obtained from an X-ray source with a maximum about two-thirds that of the primary energy source. This gives rise to an unwanted background in X-ray SPECTROMETRY and is usually removed by a monochromator.

Brillouin zone.   *See* BAND THEORY OF SOLIDS.

brine.   Concentrated aqueous solution of sodium chloride.

British anti-Lewisite.   *See* LEWISITE.

Brockmann scale.   Scale for the classification of the activity of alumina for use in ADSORPTION CHROMATOGRAPHY, in terms of increasing adsorption of different dyes.

bromates.   Salts of the oxoacids of bromine; they are all good oxidizing agents. Bromates(III) are unknown (*compare* CHLORATES.
bromates(I) (hypobromites, BrO^-). Prepared by the action of bromine on a base. The salts disproportionate to BrO_3^- at elevated temperatures. The weak acid, known only in solution, is prepared by the action of bromine on a suspension of HgO in water.
bromates(V) (bromates, pyramidal BrO_3^-). Prepared by the action of bromine on hot alkali solutions. They decompose on heating to the bromide and oxygen. They are reduced by bromides to give free bromine. The free acid ($Ba(BrO_3)_2$ + H_2SO_4), known only in solution, is a strong acid.
bromates(VII) (perbromates, tetrahedral BrO_4^-). Prepared by the oxidation of BrO_3^- electrolytically or with xenon difluoride. The oxidizing properties of perbromates increase with concentration. $KBrO_4$ is stable to 270°C, where it decomposes to $KBrO_3$ and O_2. The free acid can be crystallized as a dihydrate.

bromic acid. *See* BROMATES.

bromides. *See* HALIDES.

bromine (Br). GROUP VII ELEMENT present (0.01 per cent of the lithosphere) as bromides in seawater, natural brines and salt deposits, from which it is obtained by displacement with chlorine. Bromine is a dark red fuming liquid, giving off a red vapour which irritates the nose and throat. It is moderately soluble in water forming a hydrate and is miscible with non-polar solvents (e.g., carbon disulphide and tetrachloromethane). It is very reactive (less so than chlorine) with many other elements and is a strong oxidizing agent. It is a typical non-metallic element, the most stable oxidation state is –1, but it forms covalent compounds in oxidation states 1, 3, 5 and 7 mainly in combination with oxygen and other halogens. Bromine is used in the preparation of the petrol additive 1,2-DI-BROMOETHANE and in the manufacture of fumigants, flame-proofing materials, disinfectants and photographic chemicals.
$A_r = 79.904$, $Z = 35$, ES [Ar] $3d^{10} 4s^2 4p^5$, mp $-7.2°C$, bp $58.78°C$, $\rho = 3.12 \times 10^3$ kg m^{-3}, $E^{\ominus}_{Br^-,Br_2} = 1.065$ V. Isotopes: ^{79}Br, 50.54 percent; ^{81}Br, 49.46 percent.

bromine halides. *See* INTERHALOGEN COMPOUNDS.

bromine oxides. All have very low thermal stability. Dibromine monoxide (Br$_2$O) (Br$_2$ + HgO) is formally the anhydride of hypobromous acid. Bromine dioxide (BrO$_2$) and dibromine trioxide (Br$_2$O$_3$) are formed from mixtures of bromine and oxygen in an electric discharge at low temperatures. BrO$_2$ on hydrolysis gives bromide and bromate ions. Perbromyl fluoride (BrO$_3$F) (KBrO$_4$ + HF + SbF$_3$) is more reactive than is perchloryl fluoride and is hydrolyzed by bases to perbromate and fluoride ions.

bromine oxoacids. *See* BROMATES.

N-bromobutanimide (*N*-bromosuccinimide, NBS). Solid (mp 178°C), prepared by the action of bromine in sodium hydroxide on butanimide. NBS brominates olefins and unsaturated esters in the allyl and benzyl positions by a free radical

N-bromobutanimide

substitution reaction, thus RCH$_2$CH=CH$_2$ gives RCH$_2$CBr=CH$_2$. NBS also oxidizes secondary alcohols to ketones.

bromochlorodifluoromethane. *See* FREONS.

bromoethane (ethyl bromide, CH$_3$CH$_2$Br). ALKYL HALIDE (bp 38.5°C), manufactured by the FREE RADICAL addition of hydrogen bromide to ethene initiated by gamma-rays from cobalt-60. It is used in the preparation of BARBITURATES.

bromoform. *See* TRIHALOMETHANES.

bromomethane (methyl bromide, CH$_3$Br). Gaseous ALKYL HALIDE (bp 4.5°C), prepared from hydrogen bromide and methanol. It is used for METHYLATION and as a fumigant. It is extremely toxic (threshold limit value: 15 ppm).

N-bromosuccinimide. *See* N-BROMO-BUTANIMIDE.

bromothymol blue. *See* INDICATORS.

Brønsted–Lowry acids and bases. An acid is a substance that can donate a proton, and a base is one that can accept a proton. Thus

$$HA + B^- \rightleftharpoons HB + A^-$$
$$\text{acid}_1 \quad \text{base}_2 \quad \quad \text{acid}_2 \quad \text{base}_1$$

HA and A$^-$, and HB and B$^-$ are known as conjugate acid–base pairs. One of the pairs may arise from the solvent (e.g., H$_2$O + HCl → H$_3$O$^+$ + Cl$^-$). The equilibrium of water acting as both acid and base is AUTO-PROTOLYSIS. *See also* ACIDS AND BASES.

bronze. ALLOY of COPPER and TIN which was used extensively as a general-purpose metal. Now particular formulations have special uses (e.g., gun metal, phosphor bronze, aluminium bronze and beryllium bronze).

bronzes. Compounds of general formula A$_x$MO$_3$, where A is an ALKALI METAL,

ALKALINE EARTH METAL or hydrogen and M is niobium, tantalum or tungsten. For example, Na_xWO_3 is sodium tungsten bronze. They are metallic in appearance and are conducting. Blue hydrogen tungsten bronze is used in electrochromic devices.

brookite. *See* TITANIUM.

Brownian motion. Random motion imparted to particles of a COLLOID by collisions with molecules of the dispersing medium having translational kinetic energy ($3kT/2$). The phenomenon was first observed by the botanist R. Brown with pollen grains suspended in water. The average displacement $\langle x \rangle$ after time t is given by the Einstein equation $\langle x \rangle = (2Dt)^{1/2}$ where D is the diffusion coefficient (*see* FICK'S LAWS OF DIFFUSION).

brucine. *See* ALKALOIDS.

Brunauer–Emmett–Teller isotherm. *See* ADSORPTION ISOTHERM.

Brunswick green. Green PIGMENT comprising lead ethanoate, iron(II) sulphate, potassium hexacyanoferrate(II) and sodium chromate(VI) diluted with barium sulphate.

Bu. *See* BUTYL GROUP.

Bucherer reaction. *See* AMINONAPHTHALENES.

buckminsterfullerene (C_{60}). Truncated icosahedron of carbon atoms (*see* Figure) formed in material vaporized from a graphite surface by a high-intensity laser. The structure forms a 700-pm cage which may contain metal atoms (e.g., lanthanum).

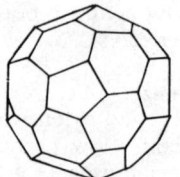

buckminsterfullerene

buffer solution. Solution that maintains nearly constant pH despite the addition of small amounts of acid or alkali. Most buffer solutions consist of a weak acid (HA) and its conjugate base (A–) (*see* ACIDS AND BASES). The pH of the solution may be calculated from the Henderson equation

$$pH = pK_a + \lg (c_A/c_{HA})$$

The buffer capacity (β) is the reciprocal of the slope of the pH titration curve at a given point and has a maximum of 2.303 $a/4$, where a is the total concentration of acid. *See also* TITRATIONS.

butadiene polymers. *See* POLYMERS.

butadienes (C_4H_4).
1,2-butadiene. An ALLENE.
1,3-butadiene. Gas (bp −4.4°C), conjugated DIENE prepared by passing cyclohexene over heated nichrome wire. 1,3-Butadiene polymerizes to a synthetic rubber (buna) over a sodium catalyst. Dimerization to 4-ethenylcyclohexene is a special case of the DIELS–ALDER REACTION.

2,5-Dihydrothiophene-1,1-dioxide (sulpholene) is produced by the reaction of 1,3-butadiene with sulphur dioxide.

butaldehydes (butanals).   *See* ALDEHYDES.

butanals. *See* ALDEHYDES.

butanedioic acid (succinic acid, HOOC-$(CH_2)_2$COOH). Saturated DICARBOXYLIC ACID (mp 185°C); found in amber, algae and other plants, and formed during the fermentation of sugars. It is manufactured by the reduction of hydroxybutanedioic acid (HIO_3 + red P) or of *cis*-butenedioic acid (H_2 + Ni), or from 1,2-dibromoethane by reaction with potassium cyanide and hydrolysis of the resultant $(CH_2CN)_2$. Heat (235°C) yields the cyclic anhydride (i.e. BUTANEDIOIC ANHYDRIDE), and the reaction with glycols gives polyesters (*see* POLYMERS). Butanedioyl chloride $(CH_2COCl)_2$ is the product of reaction with phosphorus pentachloride.

butanedioic anhydride (succinic anhydride). Prisms (mp 120°C); prepared by distilling BUTANEDIOIC ACID with ethanoic anhydride, ethanoyl chloride or phosphoryl chloride. It is converted back to the acid on boiling with water or mineral acids. Sodium in ethanol gives first γ-butyrolactone and then 1,4-dihydroxybutane. With alcohols (ROH), the half-ester of butanedioic acid ($ROOCCH_2CH_2COOH$) is formed. γ-Keto-acids ($RCOCH_2CH_2COOH$) are prepared from GRIGNARD REAGENTS.

butanedioic anhydride

butanediols. See DIHYDROXYBUTANES.

2,3-butanedione. See DIKETONES.

butanes. See ALKANES.

butanimide (succinimide). Crystals (mp 126°C), prepared by heating BUTANEDIOIC ANHYDRIDE in a stream of ammonia or by heating the diamide ((CH_2CONH_2)$_2$). Boiling in water gives BUTANEDIOIC ACID. Butanimide is used to prepare PYRROLE (distillation with zinc dust) and PYRROLIDINE (sodium in ethanol). Butanimide is weakly acidic; it reacts with potassium hydroxide to form the salt (CH_2CO)$_2$N⁻ K⁺. The important reagent *N*-BROMOBUTANIMIDE is prepared by treating butanimide with bromine at 0°C in the presence of sodium hydroxide.

butanimide

butanoic acids (butyric acids). See CARBOXYLIC ACIDS.

butanols (butyl alcohols, C_4H_9OH). There are four isomeric butanols. The alcohols and the butyl ethanoates are important solvents for resins and lacquers; other esters are used as flavouring essences and perfumes.
1-butanol (n-butyl alcohol, $CH_3(CH_2)_2$-CH_2OH). Manufactured by the base-catalyzed ALDOL CONDENSATION of ethanal to 2-butenal, followed by its reduction (Ni/H_2). It has the usual properties of a primary alcohol.
2-methyl-1-propanol (isobutyl alcohol, (CH_3)$_2$CHCH$_2$OH). Occurs in fusel oil. Shows typical properties of a primary alcohol: oxidized to 2-methylpropanoic acid and dehydrated to 2-methylpropene (isobutene). The presence of the branched chain next to the carbinol group makes rearrangement easy.
2-butanol (sec-butyl alcohol, methyl ethyl carbinol, $CH_3CH_2CH(CH_3)OH$). Manufactured by the acid-catalyzed hydration of butene (obtained from the cracking of petroleum). It possesses typical properties of a secondary alcohol, and is used in the preparation of BUTANONE.
2-methyl-2-propanol (tert-butyl alcohol, (CH_3)$_3$COH). Solid prepared by the hydration of 2-methylpropene in sulphuric acid followed by neutralization and steam distillation. Mild oxidizing agents are without effect, stronger ones split it into propanone, ethanoic acid and carbon dioxide. With bromine a dibromide is formed. Potassium *tert*-butoxide is a very strong base. See also MONOHYDRIC ALCOHOLS.

butanone (ethyl methyl ketone, MEK, $C_2H_5COCH_3$). KETONE; colourless liquid (bp 80°C) with a pleasant smell. It is manufactured by the dehydrogenation of 2-butanol and from butene and palladium chloride. MEK is used as a solvent particularly for vinyl and acrylic resins and other POLYMERS.

2-butenals (crotonaldehydes, $CH_3CH=$ CHCHO). Colourless pungent liquids that show geometrical isomerism (*see* ISOMERISM), prepared by the dehydration of 3-HYDROXYBUTANAL. 2-Butenals are α,β-unsaturated aldehydes having similar properties to PROPENAL, and are oxidized to 2-butenoic acids (*see* CARBOXYLIC ACIDS) or in the presence of vanadium pentoxide catalyst to BUTANEDIOIC ACID, which is an intermediate in the production of 1-butanol from ethanol.

butenedioic acid ($HOOCCH=CH$ COOH). Simplest unsaturated dicarboxylic acid (*see* Figure), found in the *cis*-form (maleic acid, I) and the *trans*-form (fumaric acid, II) (*see* ISOMERISM).

maleic acid (I)   fumaric acid (II)

cis-*butenedioic acid*. White crystalline solid (mp 130°C p$K_{a,1}$ = 1.92, p$K_{a,2}$ = 6.23), soluble in water; prepared by heating HYDROXYBUTANEDIOIC ACID at 250°C to give the anhydride III, which is then converted by hydrolysis in sodium hydroxide to the salt, which in turn is acidified by hydrochloric acid. Also formed by heating bromobutanedioic acid with aqueous potassium hydroxide. Industrially III is prepared by the oxidation of benzene over vanadium pentoxide (V_2O_5) at 420°C or 2-butene from cracked petroleum over V_2O_5 at 450°C. It is used to inhibit the decay of milks and fats and, with the anhydride, in the DIELS–ALDER REACTION. It is reduced catalytically or electrochemically to BUTANEDIOIC ACID, or is oxidized to *meso*-2,3-DIHYDROXYBUTANEDIOIC ACID by potassium chlorate/osmium tetroxide. Heating with sodium hydroxide at 100°C gives sodium (DL)-hydroxybutanedioic acid. III is used in the manufacture of polymers.

trans-*butenedioic acid*. White crystalline solid (mp 287°C), only slightly soluble in water; prepared from the *cis*-acid by prolonged heating at 150°C, by the condensation of PROPANEDIOIC ACID with OXO-ETHANOIC ACID in the presence of pyridine

$$HOOCCHO + CH_2(COOH)_2 \xrightarrow{\text{pyridine}}$$

$$HOOCCH=CHCOOH + H_2O + CO_2$$

or by the action of furfural with sodium chlorate. II occurs in nature in many plants. It is reduced to butanedioic acid and oxidized (KClO$_3$/OsO$_4$) to racemic 2,3-dihydroxybutanedioic acid. It is converted to III on heating or to hydroxybutanedioic acid by water under pressure.

butenedioic anhydride (III)

2-butene-1,4-diol (butenediol, $HOCH_2CH=$ $CHCH_2OH$). Colourless liquid obtained by the hydrogenation of 2-BUTYNE-1,4-DIOL over a catalyst. Both *cis*- and *trans*-forms are known. It is used in the manufacture of insecticides and pharmaceuticals.

butenes (butylenes, C_4H_8). Isomeric ALKENES formed by CRACKING PETROLEUM: 1-butene, $CH_2=CHCH_2CH_3$; 2-butene, $CH_3CH=CHCH_3$; 2-methylpropene (isobutene), $(CH_3)_2C=CH_2$. Butenes are precursors of POLYMERS and 2-butanol (*see* ALKENES, Figure).

2-butenoic acids (crotonic acids, $CH_3CH=$ $CHCOOH$). α,β-Unsaturated CARBOXYLIC ACIDS.

trans-*2-butenoic acid* (α-crotonic acid). Needles (mp 71.6°C), prepared by oxidizing *trans*-2-butenal with ammoniacal silver nitrate. It is used in resins, surface coating and pharmaceuticals.

cis-*2-butenoic acid* (isocrotonic acid). Liquid (bp 168–9°C), prepared from ethyl 3-oxobutanoate. It is converted to the *trans*-acid (*see above*) by heating at 180°C. *See also* KNOEVENAGEL REACTION.

2-buten-1-ol (crotyl alcohol, $CH_3CH=$ $CHCH_2OH$). Colourless liquid obtained by the reduction of 2-butenal. Phosphorus tribromide in the presence of pyridine gives 1-bromo-2-butene (crotyl bromide).

Butler–Volmer equation. Relationship between the cathodic current density (i) (*see* CURRENT) and OVERPOTENTIAL (η) for an electrochemical reaction in which the reaction at an electrode is rate-determining (*see* RATE-DETERMINING STEP). For a one-electron reaction

$$i = i_0 \{\exp(-\beta F\eta/RT) -$$
$$\exp[(1-\beta)F\eta/RT]\}$$

where i_0 is the exchange current density and β the symmetry factor ($0 < \beta < 1$), which is the fraction along the REACTION COORDINATE at the transition state (*see* ACTIVATED COMPLEX). For a multielectron reaction the analogous equation is

$$i = i_0[\exp(-\alpha_c F\eta/RT) -$$
$$\exp(\alpha_a F\eta/RT)]$$

α_c and α_a, the cathodic and anodic transfer

coefficients, respectively, may take any positive value which may be related to values of β for individual steps from a knowledge of the mechanism. At low values of η, linearization of the exponential terms of the equation for one-electron reactions gives

$$i = i_0 F\eta/RT$$

At high η the rate of the back reaction may be neglected, which leads to the Tafel equation

$$i = i_0 \exp(-\beta F\eta/RT)$$

A plot of $\ln i$ against η allows determination of β and i_0. The Tafel slope is of η against $\lg i$ and at 25°C is approximately $60/\beta$ mV per decade.

butyl alcohols. *See* BUTANOLS.

butylene glycols. *See* DIHYDROXYBUTANES.

butylenes. *See* BUTENES.

butyl groups (Bu groups). Alkyl groups containing four carbon atoms.
  n-butyl
  $CH_3CH_2CH_2CH_2-$
  sec-butyl
  (1-methylpropyl): $CH_3CH_2CH(CH_3)-$
  iso-butyl
  (2-methylpropyl): $(CH_3)_2CHCH_2-$
  tert-butyl
  (2-methyl-2-propyl): $(CH_3)_3C-$

The *tert*-butyl group has only alkyl groups attached to the tertiary carbon which gives stability to the carbonium ion R_3C^+. *tert*-Butyl halides undergo S_N1 reactions (*see* S_N1 MECHANISM) via this ion.

butyl rubber. *See* POLYMERS, RUBBER.

2-butynedioic acid (acetylenedicarboxylic acid, $HOOCC\equiv CCOOH$). Unsaturated DICARBOXYLIC ACID. It and its dimethyl ester undergo the DIELS–ALDER REACTION. For example

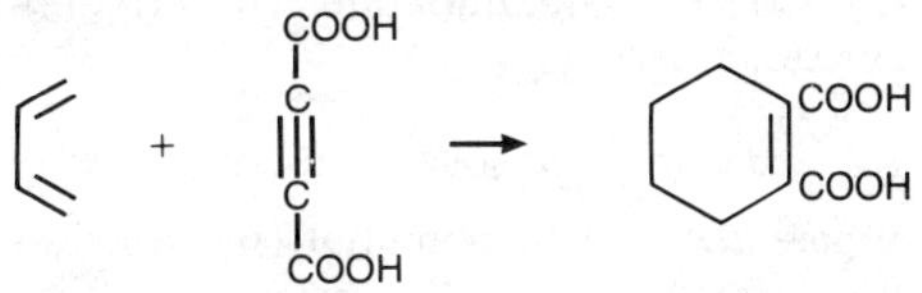

2-butyne-1,4-diol (butynediol, $HOCH_2C\equiv CCH_2OH$). White solid, prepared by the high-pressure reaction between ethyne and methanol. Its properties are typical of a di-primary alcohol (*see* ALCOHOLS, Figure); it is soluble in water and ethanol. It is used in electroplating as a brightner, as a corrosion inhibitor and as a polymerization accelerator.

butyric acid (butanoic acid). *See* CARBOXYLIC ACIDS.

C

C. *See* CARBON, COULOMB.

C. *See* CAPACITANCE, HEAT CAPACITY.

c. Concentration/mol dm^{-3} (*see* CONCENTRATION UNITS).

χ. *See* MAGNETIC SUSCEPTIBILITY.

χ-potential. Chi potential (*see* SURFACE POTENTIAL).

Ca. *See* CALCIUM.

cacodyl derivatives. Organoarsenic compounds (*see* ARSENIC ORGANIC DERIVATIVES) containing the Me$_2$As– grouping.

cadaverine. *See* POLYAMINES.

cadmium (Cd). Group IIb metal (*see* GROUP II ELEMENTS). It occurs naturally with zinc and as the sulphide ore greenockite. It is extracted by the reduction of cadmium oxide with carbon. Uses are in electroplating iron and steel for corrosion protection and in nickel–cadmium batteries; its compounds are used in pigments and phosphors. Cadmium also forms low-melting alloys employed as solders. Cadmium is a soft, bluish-white metal with properties very similar to those of zinc. Cadmium has a high cross-section for neutrons, which allows its use in nuclear reactors, but the extreme toxicity of the element's compounds precludes its widespread use. $A_r = 112.41$, $Z = 48$, ES [Kr] $4d^{10} 5s^2$, mp 321°C, bp 765°C, $\rho = 8.64 \times 10^3$ kg m^{-3}, Isotopes: ^{106}Cd, 1.2 percent; ^{108}Cd, 0.89 percent; ^{110}Cd, 12.4 percent; ^{111}Cd, 12.8 percent; ^{112}Cd, 24.1 percent; ^{113}Cd, 12.3 percent; ^{114}Cd, 28.8 percent; ^{116}Cd, 7.6 percent.

cadmium chloride lattice. *See* CRYSTAL FORMS.

cadmium compounds.
cadmium inorganic compounds. Cadmium is an ELECTROPOSITIVE metal ($E^{\ominus}_{Cd^{2+},Cd} = -0.402$ V) that forms Cd^{2+} in a range of salts with a coordination number generally of six. Some complexes are formed (e.g. [Cd(NH$_3$)$_4$]$^{2+}$, [Cd(CN)$_4$]$^{2-}$, [CdI$_4$]$^{2-}$. Cd$_2^{2+}$ (compare Hg$_2^{2+}$, *see* MERCURY INORGANIC COMPOUNDS) has been shown to exist in Cd/CdCl$_2$ melts. Cadmium oxide (CdO) is more reducible than zinc oxide and also more volatile, these properties being used to separate the elements. CdS and Cd(S, Se) are yellow and red pigments, respectively. CdS is also important in photovoltaic devices. Cadmium sulphate (CdSO$_4$) is used as an electrolyte in the Weston STANDARD CELL. All the halides are known. They are similar to the zinc halides, but are less soluble in water.
cadmium organic derivatives (R$_2$Cd and RCdX, where X=Cl, Br). Prepared from CdX$_2$ and RLi or Cd and RX. They are similar to, but more unstable than, GRIGNARD REAGENTS and are used specifically to prepare ketones from acyl chlorides (e.g. 2ROCl + R′$_2$Cd → 2RCOR′ + CdCl$_2$).

cadmium iodide lattice. *See* CRYSTAL FORMS.

caesium (Cs). Group Ia ALKALI METAL (*see* GROUP I ELEMENTS). It occurs with other alkali metals in the minerals carnallite (MgCl$_2$.KCl) and pollucite (CsAlSi$_2$O$_6$.-xH$_2$O). It is separated by ion exchange and extracted by electrolysis of the molten cyanide, or by reduction of molten CsAlO$_2$ with magnesium. Caesium is the most ELECTROPOSITIVE alkali metal ($E^{\ominus}_{Cs^+,Cs} = -3.02$ V), reacting violently with oxygen, water, halogens, etc. It forms mostly ionic compounds (Cs$^+$). The most important compound is caesium chloride (*see* CRYSTAL FORMS). The fluoride, carbonate and sulphate are also well known. Uses are in

80

photoelectric devices (caesium has the lowest ionization potential of 376 kJ mol^{-1}), as an oxygen GETTER and as a hydrogenation catalyst. ^{137}Cs is a gamma-ray source (half-life: 33 years). Caesium compounds are used in the ceramic industry and as windows in infrared spectroscopy. A_r = 132.9054, Z = 55, ES [Xe] $6s^1$, mp 28.7°C, bp 690°C, ρ = 1.90 × 10^3 kg m^{-3}. Isotope: ^{133}Cs, 100 percent.

caesium chloride lattice. *See* CRYSTAL FORMS.

caffeine. *See* ALKALOIDS, PURINES.

cage compound. *See* CLATHRATE COMPOUND.

Cal. *See* CALORIE

calamine. *See* ZINC.

calciferol (vitamin D$_2$). *See* VITAMINS.

calcination. (1) Process of heating minerals to high temperatures to form hard, insoluble aggregates (e.g., calcined bauxite is obtained by heating bauxite above 900°C). (2) Formation of a calcium carbonate precipitate from hard water. *See also* HARDNESS OF WATER.

calcite. *See* CALCIUM.

calcitonin. *See* HORMONES.

calcium (Ca). Group IIa ALKALINE EARTH METAL (*see* GROUP II ELEMENTS). It is the third most abundant metal, being found in several common minerals — marble, iceland spar, limestone, calcite (all CaCO$_3$), dolomite (CaCO$_3$.MgCO$_3$), fluorspar (CaF$_2$), gypsum (CaSO$_4$), asbestos (CaMg$_3$(SiO$_3$)$_4$) — and in seawater (0.15 percent CaCl$_2$). Calcium is extracted by electrolysis of the molten chloride obtained by the treatment of a carbonate ore with hydrochloric acid or from the SOLVAY PROCESS. The metal tarnishes in air, forming the oxide and nitride, and reacts violently with water. It also reacts directly with other non-metals (e.g., hydrogen and halogens). Calcium is used in alloys and in the preparation of some transition elements (e.g., vanadium, zirconium). It is a GETTER for oxygen and nitrogen in vacuum tubes. It is extremely important in living organisms, being present in bones, teeth, eggshells and soils.
A_r = 40.078, Z = 20, ES [A] $4s^2$, mp 850°C, bp 1487°C, ρ = 1.54 × 10^3 kg m^{-3}. Isotopes: ^{40}Ca, 96.9 percent; ^{42}Ca, 0.64 percent; ^{43}Ca, 0.14 percent; ^{44}Ca, 2.1 percent; ^{48}Ca, 0.18 percent (radioactive $t_{1/2}$ > 10^4 years).

calcium compounds. Calcium is an ELECTROPOSITIVE metal ($E^{\ominus}_{Ca^2,Ca}$ = –2.87 V) that forms ionic salts of Ca^{2+}.
lime (quicklime, CaO). Most important industrial base, produced by roasting the carbonate. It is used in the SOLVAY PROCESS, in the removal of sulphur dioxide, as a slag former in the manufacture of steel, as a dehydrating agent and as a refractory.
slaked lime (Ca(OH)$_2$). Relatively insoluble in water (0.2 g per 100 g water), but even so is used widely as a source of OH$^-$. Solutions of Ca(OH)$_2$ are used for detecting the presence of and removing carbon dioxide as the carbonate. Ca(OH)$_2$ is a constituent of mortar.
calcium hydride (CaH$_2$). Solid (Ca + H$_2$, 400°C), reducing agent (e.g., NaCl to Na, H$_2$O to H$_2$).
calcium carbide. Solid (CaO + C, 3000°C). Contains [C$_2$]$^{2-}$ (*see* CARBIDES), which hydrolyzes to ethyne and reacts with nitrogen to give calcium cyanamide (CaNCN), which is a source of fixed nitrogen for fertilizers.
calcium carbonate (CaCO$_3$). Widely found in nature (*see* CALCIUM). It is insoluble in water, but dissolves in solutions containing excess carbon dioxide to calcium hydrogen carbonate (Ca(HCO$_3$)$_2$).
calcium halides. All halides are known of which the chloride is the most important. It is used as a dehydrating agent, deicer for roads, antifreeze and refrigerant. Aqueous calcium chloride has a eutectic at –55°C.
calcium organic derivatives. Calcium does not have an extensive organometallic chemistry, but forms some Grignard-like reagents (*see* GRIGNARD REAGENTS) (e.g., in ether PhI + Ca → PhCaI).

Calgon. Phosphate detergent (*see* SURFACE-ACTIVE AGENT).

californium (Cf). *See* ACTINIDE ELEMENTS.

calomel (mercury(I) chloride). *See* MERCURY INORGANIC COMPOUNDS.

calomel electrode. *See* REFERENCE ELECTRODE.

Calor gas. Trade name for liquid petroleum gas (*see* PETROLEUM GAS) (butane). Calor propane is the trade name for liquified propane.

Calorie (kilogram calorie, Cal, kcal). Term still in limited use in estimating the energy value of foods. A value of 4.1 Cal for carbohydrates is 4100 cal g^{-1}.

calorie (cal). Amount of heat required to raise the temperature of 1 g of water 1 K at 288 K. It is not coherent with SI UNITS, and its use is discouraged (1 cal = 4.184 J).

calorific value. Amount of heat evolved during the complete combustion of unit weight of a solid or liquid or unit volume of a gas. It is usually used to express the energy values of fuels (MJ kg^{-1}) or energy content of foodstuffs (*see* CALORIE). *See also* CALORIMETRY.

calorimetry. Measurement of heat changes during physical changes or chemical reactions. The heat change is followed by measuring the change of temperature (ΔT), which is converted to energy supplied using the heat capacity (C) of the calorimeter assembly.

$$q = C\Delta T$$

Various types of calorimeter are available: (1) insulated vessel (vacuum flask) fitted with a temperature-detecting device using the method of mixtures, suitable for HEAT CAPACITY measurements and ENTHALPY changes during reactions in solution; (2) BOMB CALORIMETER, suitable for enthalpy of combustion of solids and CALORIFIC VALUE of fuels and foodstuffs; (3) flame calorimeter, suitable for enthalpy of combustion of gases and volatile liquids; (4) microcalorimeter (ampoule, batch, flow, titration, etc.), a very sensitive instrument for measuring enthalpy of reaction and enthalpy changes in biological systems, absorption on very small quantities of material.

camphene. *See* TERPENES.

camphor. *See* TERPENES.

candela (cd). SI UNIT of luminous intensity that is 1/0.6 mm^2 of a black body at the temperature of freezing platinum (2040 K).

Canderel. Trade name for a SWEETENING AGENT.

cane sugar. *See* SUCROSE.

Cannizzaro reaction. Reaction in alkaline solution of two ALDEHYDE molecules that do not have an α-carbon in which one is reduced to an alcohol and the other is oxidized to an acid (e.g., for BENZALDEHYDE 2PhCHO + NaOH → PhCH$_2$OH + PhCOONa).

canonical form. One of the possible structures of a molecule that together form a resonance hybrid (e.g., Kekulé structures of BENZENE).

capacitance (C; dimensions: $\varepsilon\, l$; units: F = A^{2}s^4 kg^{-1} m^{-2} = A s V^{-1} = C V^{-1}). Charge that must be added to a body to raise its potential by one unit. A capacitance of 1 farad requires 1 coulomb of electricity to raise its potential by 1 volt. The total capacitance (C) of a series of capacitances (C_1, C_2, . . .) is given by a reciprocal relationship ($1/C = 1/C_1 + 1/C_2 + \ldots$); capacitances in parallel are additive ($C = C_1 + C_2 + \ldots$). A conductor charged with Q coulomb to a potential V volt has a capacitance $C = Q/V$ farad, and differential capacitance $\tilde{C} = (\partial Q/\partial V)_{T,P,\mu}$.

capacity factor. *See* EXTENSIVE PROPERTY.

capillarity. Phenomenon of CAPILLARY RISE.

capillary column. Chromatography column for GAS–LIQUID CHROMATOGRAPHY made from glass, copper, stainless steel, Nylon, etc., of internal diameter 100–500 μm and lengths of up to 300 m, coated with a stationary liquid phase. Columns have a low resistance to gas flow and efficiencies of several thousand plates (*see* HEIGHT EQUIVALENT OF A THEORETICAL

PLATE, HETP). The eddy diffusion term of the VAN DEEMTER EQUATION no longer applies, and a modified equation is used to calculate the HETP.

capillary rise.   Height (h) to which a liquid rises in a tube of radius r

$$h = 2\gamma \cos \theta / \rho g r$$

caproic acid.   *See* n-HEXANOIC ACID.

caprolactam   (ε-aminocaprolactam,   II). Solid LACTAM (mp 69°C), prepared by the BECKMANN REARRANGEMENT of cyclohexanone oxime (I), which is obtained from cyclohexane and nitrosyl chloride. It is the precursor of Nylon-6,6 (*see* NYLON).

caprolactam

capryl alcohol.   *See* 2-OCTANOL.

caprylic acid.   *See* n-OCTANOIC ACID.

capsule. Mucilaginous   POLYSACCHARIDE material that completely surrounds some bacteria. The size, composition and amount of material depends on the species and the cultural conditions. In some species there is no capsule, and only cell-free slime is formed. Water is the principal component of capsules and slime. These extracellular polysaccharides may be composed of one or more sugar molecules; a range of HEXOSES, methylpentoses (*see* PENTOSES), AMINO SUGARS and URONIC ACIDS have been identified. In some organisms a protein with antigenic properties has been identified.

capture cross-section.   *See* CROSS-SECTION.

caramel. Brown substance, of unknown composition, obtained by the action of heat (or chemicals) on cane sugar (*see* SUCROSE) or other carbohydrates. It is used as a brown colouring and flavouring agent (E150) in foodstuffs.

carbamates ($NH_2COO^- M^+$). Salts of carbamic acid (not known in the free state).

Ammonium carbamate is formed by the reaction between ammonia and carbon dioxide, and hydrolyzes at 60°C to ammonium carbonate. Esters of carbamic acid are called URETHANES.

carbamides.   *See* UREA.

carbamido group.   The group –NH-$CONH_2$. *See also* UREA.

carbamyl group.   The group –$CONH_2$. *See also* AMIDES.

carbanion (:CR_3^-). Organic ion with an unshared pair of electrons carrying a negative charge on a carbon atom, produced by heterolytic cleavage of certain carbon–hydrogen, carbon–halide, carbon–carbon bonds, for example

$$CH_3COOEt + {}^-NH_2 \rightarrow$$

$${}^-CH_2COOEt + NH_3$$

The methyl anion (:CH_3^-), isoelectronic with ammonia, has the same fundamental type of geometry (pyramidal sp^3-HYBRIDIZATION). Carbanions that are stabilized by resonance are planar insofar as the delocalized electronic system is concerned. They exhibit a spectrum of stabilities; electronegative atoms or groups, depending on their electron-withdrawing ability, will lead to more or less stable carbanions; thus $-NO_2 > -SO_2R > -C{\equiv}N > {>}C{=}O > -COOR > Ph- > {>}C{=}C{<}$. Their existence, in many instances, can only be inferred from mechanistic studies (e.g., ALDOL CONDENSATION). The 1,4-addition of a stabilized carbanion to an α,β-unsaturated carbonyl compound, the MICHAEL CONDENSATION, is used in the creation of new carbon–carbon bonds

where Z = –CR, –CN, –NO_2.

carbazole ($C_{12}H_9N$). Nitrogen-containing HETEROCYCLIC   COMPOUND   (structure XXXV), comprising a pyrrole ring fused to two benzene rings. It occurs with ANTHRA-CENE in the solid that separates from anthracene oil and can be prepared from 2-biphenylamine. It is a weak base, which is insoluble in water and only sparingly soluble in most organic solvents. It is used in the preparation of dyestuffs.

carbene (methene, CH_2). Reactive bi-radical, the first member of the ALKENE series. It is formed by the photolysis of KETEN ($CH_2=C=O \rightarrow CH_2 + CO$). Carbene inserts into carbon–hydrogen bonds, also into oxygen–hydrogen and carbon–chlorine bonds (e.g., $CH_3CH(OH)CH_3$ gives $(CH_3)_3COH$, $CH_3CH_2CH(OH)CH_3$ and $(CH_3)_2CH(OH)CH_3$). Carbene also adds across double bonds to give cyclopropanes (*see* CYCLOALKANES).

carbenes (CR_2). Reactive intermediates in, for example, the ARNDT–EISTERT SYNTHESIS. Transition metal complexes are formed with carbenes.

carbenicillin.   *See* ANTIBIOTICS.

carbenium ion.   *See* CARBONIUM ION.

carbide (calcium carbide).   *See* CALCIUM COMPOUNDS, CARBIDES.

carbides.   Binary compounds of carbon with a more ELECTROPOSITIVE element. Three types of carbides are recognized. (1) Salt-like carbides formed with elements in groups I, II and III. With very electropositive elements, carbides are formed containing C^{4-} (methanides) and $[C_2]^{2-}$ (ACETYLIDES). They are prepared by reaction of carbon with the metal or metal oxide or, in the case of alkali metal carbides, by passing ethyne into a solution of the metal in liquid ammonia. These carbides are reactive, dissolving in water to give ethyne (e.g., $Na_2C_2 + 2H_2O \rightarrow 2NaOH + C_2H_2$). Calcium carbide, known commonly as 'carbide', was used as a source of ethyne for lamps. Aluminium carbide is hydrolyzed to methane. (2) Interstitial carbides are formed with carbon occupying octahedral holes in a transition metal lattice in which the metal radius is greater than 130 pm. Examples are TiC, ZrC, HfC, V_nC, Nb_nC, Ta_nC, Mo_nC, W_nC (where $n = 1$ or 2). This imparts great strength and hardness to the material, which also shows high melting points (3000–4800°C). For metals with atomic radii less than 130 pm distortion of the lattice accompanies incorporation of carbon and results in properties that are between those of ionic and true interstitial carbides. The carbides Cr_3C_2 and M_3C (where M = Mn, Fe, Co, Ni) hydrolyze in water and acids to hydrocarbons. (3) Covalent carbides of elements near carbon in the periodic table, notably SiC and B_4C. These are hard refractory materials (*see* BORON INORGANIC COMPOUNDS).

carbinol.   *See* METHANOL.

carbocation.   Positively charged ion in which the charge resides on a carbon atom or a group of carbon atoms. *See also* CARBONIUM ION.

carbocyclic compounds.   HOMOCYCLIC COMPOUNDS with closed rings of carbon atoms. *See also* ALICYCLIC COMPOUNDS.

carbodiimides ($RN=C=NR$). Prepared by heating disubstituted THIOUREAS with mercury(II) oxide

$$RNHCSNHR + HgO \rightarrow$$
$$RN=C=NR + HgS + H_2O$$

They are powerful dehydrating agents: for example, dicyclohexylcarbodiimide reacts with carboxylic acids to give anhydrides

$$2RCOOH + C_6H_{11}N=C=NC_6H_{11} \rightarrow$$
$$(RCO)_2O + (C_6H_{11}NH)_2CO$$

carbohydrates.   Substances with the general formula $C_x(H_2O)_y$, although not all carbohydrates conform to this general formula (e.g., 2-deoxyribose). Sugars, starches and cellulose, compounds with important structural energy functions in living materials, are all carbohydrates. They are produced in green plants by photosynthesis

$$6CO_2 + 6H_2O \underset{\substack{\text{animal}\\\text{metabolism}}}{\overset{\substack{\text{chlorophyll,}\\h\nu}}{\rightleftharpoons}} C_6H_{12}O_6 + 6O_2$$

Since many carbon atoms are asymmetric, carbohydrates can exist in many stereochemical and structural forms. Carbo-hydrates may be divided into several classes including MONOSACCHARIDES, DISACCHAR-IDES and POLYSACCHARIDES.

carbolic acid. *See* PHENOL.

carbomethoxy group. The group –CO.OCH$_3$.

carbon (C). GROUP IVB ELEMENT. It is found widely in nature (*see* Table) as the element (two crystalline allotropes: dia-mond and graphite) and combined (carbon dioxide, carbonates—limestone, dolomite, marble and chalk—organic compounds). Other forms of elemental carbon that are less well-defined are carbon black, charcoal and coke, which are produced by the partial

carbon
Forms of elemental carbon.

Name	Structure	Preparation	Uses
Graphite	Planes of covalently bonded hexagons with weaker van der Waals bonds between. C–C in rings = 141 pm, between planes = 335 pm. Hexagonal form with carbons in every other layer superposed. Rhombohedral form with every third layer superposed	Naturally occurring allotrope Purified from coke at high temperature	Lubricant, pencil leads, moderator in nuclear plant
Diamond	Tetrahedrally coordinated lattice	Naturally occurring allotrope Made from graphite at 3000 K and ~ 125 kbar	Gemstone, industrial cutting agent and abrasive
Charcoal	High-surface-area graphite with defect lattice	Slow combustion of wood, animal blood, sugar in absence of oxygen	Fuel, reducing agent, absorbent for gases
Coke	Graphite	Destructive distillation of coal (*see* CARBONIZATION)	Reducing agent, fuel
Carbon black (lampblack)	Microcrystalline graphite	Incomplete combustion of hydrocarbons	Filler, black pigment in rubbers, inks, polishers, decolorizer
Active carbon	Charcoal	Charcoal activated by treatment with steam to remove hydrocarbons	Absorbent, catalyst, catalyst support in sugar refining, dry cleaning, pollution control
Carbon fibre	Oriented graphite	Heating textile fibres in absence of air	Strengthener for polymers and other materials

combustion of organic and biological compounds. A major use of carbon is in the manufacture of STEEL and other metals. Carbon-12 is the standard of the atomic mass scale (*see* ATOMIC MASS UNIT), carbon-13 by virtue of its nuclear spin of 1/2 is observed in NUCLEAR MAGNETIC RESONANCE SPECTROSCOPY, and radioactive carbon-14 is used in CARBON-14 DATING. Graphite conducts electricity and because of its lamellar structure may intercalate atoms, molecules and ions (e.g., the higher alkali metals, halogens, $FeCl_3$, UCl_4, HSO^-_4—electrolysis of sulphuric acid with graphite anode). Diamond is the hardest element known. Its infinite lattice (*see* CRYSTAL FORMS) of tetrahedral carbon atoms gives it strength, makes it a thermal and electrical insulator, and renders it chemically inert. Carbon is insoluble in water, dilute acids and alkalis, and organic solvents. It burns in oxygen to carbon monoxide and carbon dioxide (*see* CARBON OXIDES) and dissolves in hot oxidizing acids. CARBIDES are formed with many metals.
A_r = 12.011, Z = 6, ES [He] $2s^2\ 2p^2$, mp 3730°C, bp 4830°C, $\rho_{graphite}$ = 2.25 × 10^3 kg m^{-3}, $\rho_{diamond}$ = 3.51 × 10^3 kg m^{-3}. Isotopes: ^{12}C, 98.9 percent; ^{13}C, 1.1 percent.

carbon-14 dating (radiocarbon dating). Method of estimating the age of archaeo-logical specimens of biological origin. Carbon-14 is produced by bombardment of atmospheric nitrogen by neutrons from cosmic rays (^{14}N + ^1n → ^{14}C + ^1H). The resulting atoms are incorporated into living systems. After death the carbon-14 decays (half-life: 5540 years), and so the age of the sample may be determined by combustion and analysis of $^{14}CO_2$. The method is accurate up to about 40000 years.

carbonates.　*See* CARBONIC ACIDS.

carbonation (carboxylation). Introduction of the CARBOXYL GROUP into a molecule, for example, by the reaction of carbon dioxide and a GRIGNARD REAGENT. The term is also used for the dissolution of carbon dioxide in a liquid under pressure.

carbon black.　*See* CARBON.

carbon cycle (1). Progress of carbon from air as carbon dioxide to plants by photosynthesis with the formation of sugars and starches, through the metabolism of animals to decomposition products which are ultimately returned to the atmosphere (*see* Figure 1). (2) Series of nuclear reactions, believed to be the source of energy in many stars, in which four hydrogen nuclei

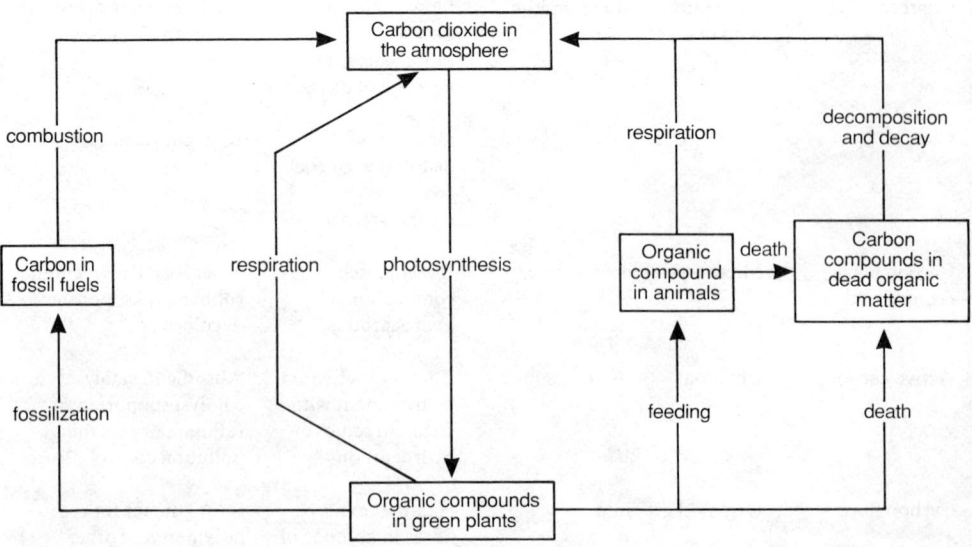

carbon cycle
Figure 1. Carbon cycle in nature.

combine to form a helium nucleus; carbon-12 acts as if it were a catalyst (*see* Figure 2).

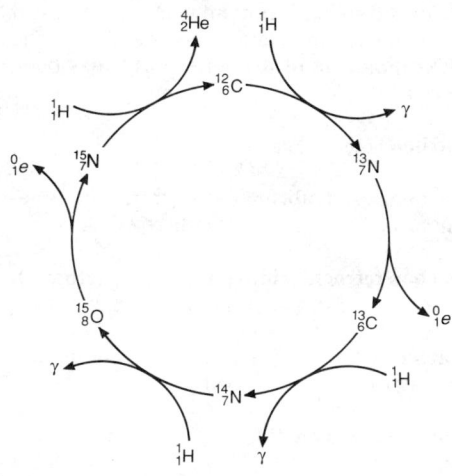

carbon cycle
Figure 2. Nuclear reactions.

carbon dioxide laser.   *See* LASER.

carbon dioxide lattice.   *See* CRYSTAL FORMS.

carbon disulphide (CS_2). Pale yellow, toxic, flammable liquid (bp 46°C), manufactured by direct reaction of the elements. It is used widely as a solvent and for the preparation of tetrachloromethane (*see* TETRAHALOMETHANES). CS_2 forms thiocarbonates ($[CS_3]^{2-}$) (*see* THIOCARBONIC ACID) with thioalkanes, thiocarbamates ($[R_2CNS_2]^-$) with amines and XANTHATES ($[RCOS_2]^-$) with alkoxides.

carbon fibre.   *See* CARBON.

carbonic acids.   Acids of CARBON OXIDES. METHANOIC ACID is the acid of carbon monoxide, PROPANEDIOIC ACID that of C_3O_2 and the most important carbonic acid H_2CO_3, the acid of carbon dioxide. When carbon dioxide dissolves in water, solvated molecules are formed before a slow conversion to H_2CO_3 ($pK_1 = 6.38$, $pK_2 = 10.32$). The acid salts (bicarbonates, HCO_3^-) are usually soluble in water; metal carbonates, with the exception of alkali metal carbon-

ates, are insoluble. Many carbonates are minerals (*see* CARBON) and of commercial importance as building materials and ingredients of CEMENT. Carbonates are converted to bicarbonates by the reaction with carbon dioxide and water which is reversible by heat. Metal carbonates, with the exception of alkali metal carbonates, are converted to the oxide with loss of carbon dioxide on heating. Sodium carbonate is manufactured by the SOLVAY PROCESS and is used in the glass, paper and textile industries.

carbonium ion (carbenium ion). Carbocation (positively charged) species containing a trivalent carbon atom (e.g., R_3C^+). Carbonium ions occur as reactive intermediates in many chemical reactions (e.g., solution of alkenes in strong acids; solvolysis reactions, *see* S_N1 MECHANISM). They show a wide range of stabilities in decreasing order of stability: TROPYLIUM ION ($C_7H_7^+$)>Ph_3C^+>$PhCH_2^+$, CH_2=CH-CH_2^+>$(CH_3)_3C^+$>$(CH_3)_2CH^+$>$CH_3CH_2^+$>CH_3^+, CH_2=CH^+, Ph^+. They may be stabilized by delocalization of the charge using solvation or by delocalization of the charge within the molecule using inductive and/or resonance effects. Bridged or non-classical carbonium ions are possible. They may be detected by NMR spectroscopy. Alkyl carbonium ions, prepared as a salt (alkyl fluoride + SbF_5 in liquid SO_2) are powerful alkylating agents. They have a marked affinity for NUCLEOPHILES.

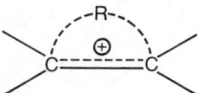

carbonium ion
Bridged or non-classical carbonium ion.

carbonization.   Destructive distillation of coal in the absence of air. High-temperature carbonization at 1100°C results in a pure, hard coke used in metallurgy. Smokeless fuel is the product of low-temperature carbonization at 900°C.

carbon oxides.   Five oxides are known: CO, CO_2, C_3O_2, C_5O_2 and $C_{12}O_9$. $C_{12}O_9$ is the anhydride of BENZENEHEXACARBOXYLIC ACID.

carbon monoxide (CO). Colourless, odourless, toxic gas (mp −205°C, bp −191°C); formed by the incomplete oxidation of carbon and its compounds. It is also produced industrially in the high temperature equilibria: $C + CO_2 \rightleftharpoons 2CO$ and $C + H_2O \rightleftharpoons CO + H_2$ (*see* WATER GAS) and $CH_4 + H_2O \rightleftharpoons CO + 3H_2$ (*see* SYNTHESIS GAS). CO is also synthesized by the dehydration of methanal by concentrated sulphuric acid. As a reducing agent CO reacts with oxygen, metal oxides and iodates. Although a weak Lewis acid (*see* LEWIS ACIDS AND BASES), it acts as a donor ligand with transition elements giving metal carbonyls. It is used in industrial syntheses; CO is often introduced into molecules via transition METAL CARBONYLS acting as catalysts.

carbon dioxide (CO_2). Colourless, dense gas (sublimes −78.5°C); present in the atmosphere (0.03 percent) as a result of biological activity and combustion. CO_2 is produced by the complete combustion of carbon-containing compounds and by the action of mineral acids on metal carbonates. Acidic (*see* CARBONIC ACIDS) and very inert, it is used as a refrigerant (solid CO_2 is 'dry ice'), as a carbonating agent for drinks, in fire extinguishers and as a carboxylating reagent with GRIGNARD REAGENTS. Industrially CO_2 is obtained as a byproduct of steam hydrocarbon REFORMING.

carbon suboxide (C_3O_2). Foul-smelling gas (bp 7°C), prepared by the vacuum dehydration of PROPANEDIOIC ACID by phosphorus-(v) oxide at 150°C. C_3O_2 polymerizes to coloured species containing lactone rings. It is a linear molecule O–C–C–C–O. The next molecule in the series C_5O_2 has been reported.

carbon steel. *See* STEEL.

carbon   tetrabromide (tetrabromomethane). *See* TETRAHALOMETHANES.

carbon tetrachloride (tetrachloromethane). *See* TETRAHALOMETHANES.

carbon tetrafluoride (tetrafluoromethane). *See* TETRAHALOMETHANES.

carbon   tetrahalides. *See* TETRAHALOMETHANES.

carbon   tetraiodide (tetraiodomethane). *See* TETRAHALOMETHANES.

carbon-to-metal bond. In organometallic compounds carbon may be directly linked to a metal by acting as a σ- or π-donor (*see* Figure).
σ-donor ligand. (1) Organic anionic donor such as in phenylsodium (*see also* ACETYLIDES, CARBIDES): a group in an ELECTRON-

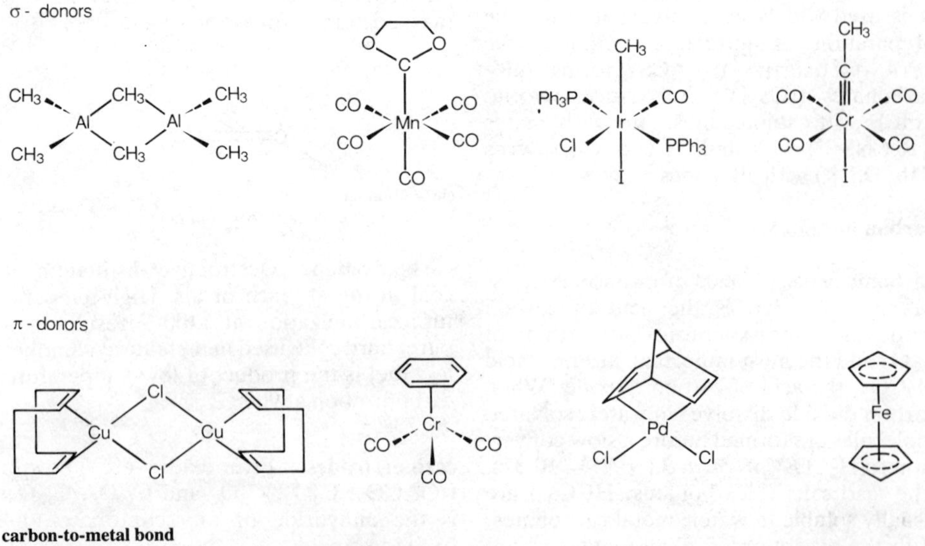

carbon-to-metal bond
Some complexes containing carbon-to-metal bonds.

DEFICIENT COMPOUND (e.g., $Al_2(CH_3)_6$, which has a bridging methyl group between two aluminium atoms(or which forms a direct σ-bond to a transition metal (e.g., CH_3 in $(h^5\text{-}C_5H_5)_2ZrCl(CH_3))$. (2) Neutral σ-donor and π-acceptor. The latter type is exclusively associated with transition elements in low oxidation states, for example, CO (*see* METAL CARBONYLS) and CARBENES.

π-donor complexes. Formed with a transition element in a low oxidation state, the ligand being an alkene, alkyne, π-allyl or cyclic-conjugated alkene (*see* SANDWICH COMPOUND). Metal–olefin bonds are thought to consist of direct overlap between the olefin π-orbital and an empty metal orbital and BACK-BONDING of d-electrons into a π^*-orbital on the ligand (*see* MOLECULAR ORBITAL).

carbonylation. Introduction of carbon monoxide into a molecule. Carbon monoxide carbonylates alkynes with a terminal hydrogen atom in the presence of nickel carbonyl (*see* METAL CARBONYLS) to give PROPENOIC ACIDS. With an iron pentacarbonyl catalyst cyclopentadienones and dihydroxybenzenes are formed. Industrially carbonylation occurs in HYDROFORMYLATION and in the OXO PROCESS.

carbonyl chloride (phosgene, $COCl_2$). Colourless, poisonous gas (bp 8°C), prepared from carbon monoxide and chlorine over active carbon. It is used to make toluenedicarbylamine, a precursor for polyurethane and polycarbonate POLYMERS and insecticides.

carbonyl derivatives. Compounds containing the OXO GROUP. *See also* ALDEHYDES, CARBONYL CHLORIDE, KETONES, METAL CARBONYLS.

carbonyl group. *See* OXO GROUP.

carboranes. Compounds consisting of carbon, boron and hydrogen. They are generally prepared by the reaction of ethynes with boron hydrides in the presence of a Lewis base. The structures are similar to those of boranes (*see* BORON INORGANIC COMPOUNDS). For example, the two icosohedral carboranes are $CB_{10}H_{12}^-$ and $C_2B_{10}H_{12}$ in which the BH group at the

vertex of the polyhedron is replaced by CH. Completely closed structures, such as octahedral $C_2B_4H_6$, are termed closo, and open structures such as pentagonal bipyramidal $C_2B_2H_6$ are nido. Carboranes are thermally stable and resistant to acids. Lithium derivatives may be made from lithium alkyls (e.g., $Li_2C_2B_{10}H_{10}$). Reactions of lithium carboranes with organic derivatives of silicon lead to stable silicon polymers.

carborundum (silicon carbide). *See* SILICON INORGANIC COMPOUNDS.

carboxy–. Prefix denoting the presence of the CARBOXYL GROUP.

carboxyhaemoglobin. *See* HAEMOGLOBIN.

carboxylation. *See* CARBONATION.

carboxyl group. The group –COOH (*see* CARBOXYLIC ACIDS).

carboxylic acids. Compounds containing the carboxyl group

$$-C\!\!\underset{\displaystyle OH}{\overset{\displaystyle O}{\diagup\diagdown}}$$

designated by the suffix –oic added to the name of the longest hydrocarbon chain.
saturated monocarboxylic acids (C_nH_{2n+1}-COOH). Fatty acids; pungent liquids (e.g., METHANOIC ACID, ETHANOIC ACID, PROPANOIC ACID), oils (C_4–C_9) and waxy solids (C_{10} and above). They are weak acids, the strength of which depends on the nature of the attached group, the more electron-withdrawing the group the stronger the acid (e.g., CH_3COOH, $pK_a = 4.8$; CCl_3COOH, $pK_a = 0.66$). The lower members have high melting and boiling points because of hydrogen bonding (*see* HYDROGEN BOND) between molecules. The melting points of n-acids show oscillations, members with an even number of carbon atoms in the alkyl chain have higher melting points than adjacent 'odd' acids. (For preparation and reaction—*see* Figure.)
unsaturated carboxylic acids. These acids have properties of ALKENES and saturated carboxylic acids if the double bond is far removed from the carboxyl group. An important class of compounds are α,β-

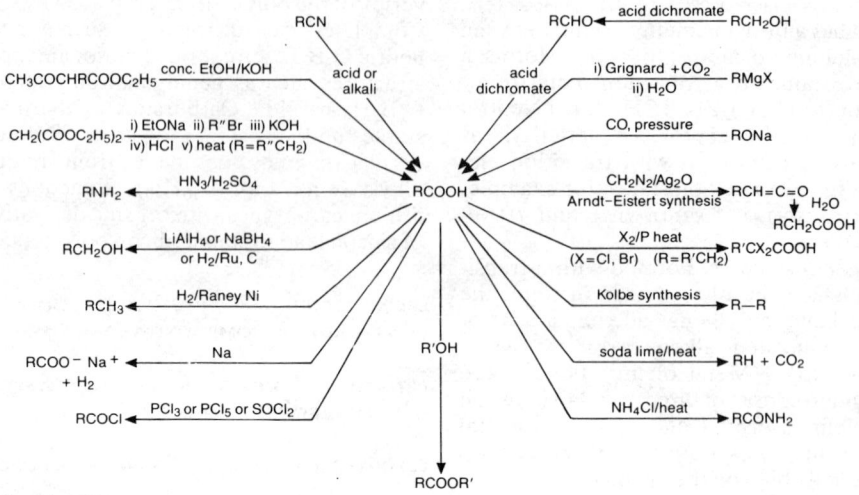

carboxylic acids
Preparation and reactions of monocarboxylic acids.

unsaturated acids, with the carboxyl group attached to an unsaturated carbon atom (e.g., 2-PROPENOIC ACID, acrylic acid). Hot alkali tends to move the double bond along the chain to give α,β-unsaturated acids. They may be fully reduced to the saturated alcohol by sodium tetrahydridoborate or to the unsaturated alcohol by lithium tetrahydridoaluminate. Long-chain unsaturated acids are found extensively in natural oils (e.g., linoleic acid, *cis,cis*-9,12-OCTA-DECADIENOIC ACID; linolenic acid, *cis,cis,-cis*-9,12,15-OCTADECATRIENOIC ACID). *See also* DICARBOXYLIC ACIDS, KNOEVENAGEL REACTION.

carboxymethylcellulose (CMC). CELLU-LOSE ethers in which some of the hydroxyl groups of cellulose are replaced by –OCH₂COOH groups by the reaction of alkaline cellulose with sodium chloroethanoate. The alkali metal salts are soluble in cold and hot water; other metallic salts are insoluble. Sodium salt is used as a thickener (E466) in food products, as a protective colloid and as an additive for synthetic detergents.

carbylamine reaction. Reaction of trichloromethane (*see* TRIHALOMETHANES) with ethanolic potassium hydroxide and a primary amine

$$CHCl_3 + 3KOH + RNH_2 \rightarrow$$
$$RNC + 3KCl + 3H_2O$$

This is a method for preparing CARBYL-AMINES and also a sensitive test for trichloromethane when R = phenyl.

carbylamines (isocyanides, isonitriles). Organic compounds containing the group –NC. These colourless, poisonous and most unpleasant liquids with lower boiling points than the isomeric cyanides are prepared by heating an alkyl iodide with AgCN in aqueous ethanol, by heating a primary amine and trichloromethane with ethanolic KOH (*see* CARBYLAMINE REACTION), or by dehydrating (POCl₃) *N*-substituted methanamides. They are reactive liquids, being hydrolyzed by acid to a primary amine and methanoic acid. They are reduced by nascent hydrogen to secondary amines and add to halogens (X₂) to give RNCX₂, sulphur to give isothiocyanates (RNCS) and form complexes with transition metals (e.g., Ni(CNPh)₄). On prolonged heating they rearrange to the NITRILE.

carbyne derivatives. Metal complexes containing the group ≡CR. *See also* ACE-TYLIDES.

carcinogen. Agent capable of causing the development of malignant cells in animals. Chemical carcinogens include benz[*a*]-

pyrene (occurring in tar, soot, tobacco smoke), aflatoxin B_1 (produced by certain moulds), MUSTARD GAS, many aromatic and polynuclear nitrogen compounds (AMINONAPHTHALENES, 4,4′-DIAMINOBIPHENYL, dibenzocarbazole, AZO COMPOUNDS, N–NITROSO COMPOUNDS), azacyclopropane (ethyleneimine), methoxychloromethane and β-propiolactone. UV radiation causes skin cancer, whereas other ionizing radiations (X-rays) may cause leukaemia. The agents are probably all mutagens.

cardiac glycosides. *See* GLYCOSIDES.

Carius method. Quantitative analysis of sulphur and halogens by complete oxidation of the compound with concentrated nitric acid and the gravimetric determination of the elements as AgX (X = halogen) and barium sulphate.

carnellite. *See* CAESIUM, CHLORINE, MAGNESIUM, POTASSIUM.

Carnot cycle. Cycle performed in a reversible heat engine, which working without friction, obtains the maximum amount of WORK from the passage of HEAT from a hotter to a colder body. Such engines consist of a heat source at a constant high temperature (T_2) (e.g., furnace, nuclear reactor) and a heat sink or cold reservoir maintained at a lower temperature (T_1) (e.g., atmosphere, cooling bath). The engine operates by carrying a working substance (e.g., ideal gas) through a reversible cycle as follows: (1) isothermal expansion at T_2; (2) adiabatic expansion to T_1; (3) isothermal compression at T_1; (4) adiabatic compression to T_2. At the end of each cycle the working substance is returned to its original state.

efficiency $\eta = -w_{max}/q_2$

$$= (q_2 - q_1)/q_2$$

$$= (T_2 - T_1)/T_2$$

Carnot theorem. Every perfect heat engine working reversibly between the same temperature limits has the same EFFICIENCY, whatever the nature of the working substance. The Carnot theorem provides a working equation (*see* CARNOT CYCLE), embodying the principles of the SECOND LAW OF THERMODYNAMICS.

β-carophyllene. *See* TERPENES.

Caro's acid (peroxomonosulphuric acid). *See* SULPHUR OXOACIDS.

carotenes. *See* CAROTENOIDS.

carotenoids. Yellow to red, fat-soluble pigments that occur in leaves, flowers and fruit, and in certain animal tissues. In plants, they are often masked by CHLOROPHYLL, but with ANTHOCYANINS they contribute to the colour of autumn foliage, when the chlorophyll is lost. They demonstrate the effect of long chains of conjugated carbon–carbon double bonds on the light-absorbing properties of organic compounds, showing electronic transitions of the π–π* type at wavelengths well into the visible region. Two types are recognized.
carotenes (tetraterpene hydrocarbons, $C_{40}H_{56}$). Chief colouring matter of carrots, butter and yolk of egg. The natural material consists of three isomers, α-, β- and γ-carotenes, which can be separated (and isolated from chlorophyll) by adsorption chromatography. β-Carotene (mp 181°C) is the most abundant (*see* Figure). α-Carotene is unsymmetrical, one ring being of α-IONONE type and the other β-ionone. Carotene is a precursor of vitamin A, a transformation that occurs in the liver. They are used as colouring agents in the food industry (E160 (a)).
xanthophylls ($C_{40}H_{56}O_2$). Yellow oxygenated pigments. The most common isomers are lutein and zeaxanthin, the dihydroxy derivatives of α- and β-carotenes, respec-

carotenoids
β-Carotene.

tively. They are used as yellow colouring materials in the food industry (E161).

carrageenan.   *See* SEAWEED COLLOIDS

carrier.   Water-insoluble material (polymer, alumina, glass, etc.) used to carry the active ligand in AFFINITY CHROMATOGRAPHY or the stationary phase in paper or thin-layer chromatography. *See also* SUPPORT MEDIUM.

carrier gas.   Mobile phase in GAS–LIQUID CHROMATOGRAPHY. The most common gases are hydrogen, helium and nitrogen, although argon, neon, air and carbon dioxide are used under special conditions. Gases supplied from cylinders at constant pressure and flow rate may require further purification and drying by molecular sieves. Flow rates are normally in the range 20–100 $cm^3 min^{-1}$.

CARS (coherent anti-Stokes Raman scattering).   *See* RAMAN SPECTROSCOPY.

carvene (limonene).   *See* TERPENES.

cascade liquefier.   *See* LIQUEFACTION OF GASES.

cascade process.   *See* LIQUEFACTION OF GASES.

case hardening.   Treatment of iron and steel articles by immersion in a bath of molten sodium or potassium cyanide containing sodium chloride or sodium carbonate. The cyanide decomposes at 750°C giving carbon, which combines with the metal. This and nitrogen, which forms nitrides, hardens the surface.

casein.   Phosphoprotein (*see* PROTEINS) in milk; main ingredient of cheese. It is an amorphous white solid, insoluble in water, soluble in alkalis, and is obtained from milk by treatment with acid or rennet. Casein is used for coating paper and textile sizing, as an adhesive, and in foods and feedstock.

cassiopeium (Cp, lutetium).   *See* LANTHANIDE ELEMENTS.

cassiterite.   *See* TIN.

cast iron.   ALLOY of iron with a high silicon content (up to 4 percent) and up to 4.5 percent carbon which is usually embedded as nodular graphite particles. Cast iron is used in many engineering applications, and modern heat treatments (e.g., austempering) give improved hardness and toughness. White cast iron containing 1 percent silicon is hard but not ductile. The carbon content is in the form of CARBIDES. Grey cast iron, the most used cast material, contains about 3 percent silicon and also contains graphite in the form of flakes. Nodular graphite irons are produced by the addition of small amounts of cerium or magnesium.

castor oil.   Natural oil from *Ricinus communis*; composed mostly of the glyceryl esters of ricinoleic acid (*cis*-12-hydroxy-9-octadecenoic acid). It is used for the manufacture of dehydrated castor oil and polyurethanes (*see* URETHANES).

catabolism.   *See* METABOLISM.

catalase.   *See* ENZYMES.

catalysis.   Kinetic phenomenon in which a small amount of a substance (catalyst) causes an increase in the rate of a chemical reaction without itself being consumed. The opposite process, which retards the rate of a reaction, is known as INHIBITION. By the SECOND LAW OF THERMODYNAMICS, if a catalyst increases the rate of a reaction it must, to preserve the EQUILIBRIUM CONSTANT, increase the rate of the back reaction. In a simple reaction, this may be envisaged as a lowering of the activation barrier over which the system must pass (*see* ACTIVATED COMPLEX, ACTIVATION ENERGY). Three categories of catalysis are recognized: (1) HOMOGENEOUS CATALYSIS, in which the catalyst is in the same phase as the reactants (*see also* ACID–BASE CATALYSIS); (2) HETEROGENEOUS CATALYSIS, usually in which a solid catalyzes a reaction in the gas or liquid phase; (3) enzyme catalysis (*see* ENZYME KINETICS, ENZYMES).   *See also* AUTOCATALYSIS, LANGMUIR–HINSHELWOOD MECHANISM, RIDEAL–ELEY MECHANISM, SUPPORTED CATALYST.

catalyst poison. Substance that destroys the catalytic activity of a surface. If, for example, a molecule adsorbs preferentially at an active site, it will poison the surface for other reactions. Sulphur compounds are effective poisons and are a particular problem in the REFORMING and CRACKING of PETROLEUM.

catalytic cracking. *See* CRACKING.

catalytic hydrogenation. *See* HYDROGENATION.

catalytic promoter. Substance that augments or preserves a catalyst without itself showing activity. One action that has been identified is to prevent the growth of particles with attendant loss of surface area (structural promotion). The promoters in the iron catalyst for the HABER PROCESS include oxides of potassium, calcium, aluminium, silicon and magnesium. Potassium oxide acts as an electronic promoter. Nickel steam-reforming catalysts are also promoted by oxides (e.g., silica).

catalytic reforming. *See* HYDROFORMING, REFORMING.

cataphoresis. *See* ELECTROPHORESIS.

catechol. *See* DIHYDROXYBENZENES.

catecholamines. Amine derivatives of catechol (*see* DIHYDROXYBENZENES) found in urine and plasma which act as HORMONES or neurotransmitters. The most important physiologically active compounds are adrenaline, noradrenaline and dopamine (an intermediate in the synthesis of adrenaline).

catenanes. Large CARBOCYCLIC COMPOUNDS with interlocking rings.

cathode. Electrode at which REDUCTION occurs. *See also* ANODE.

cathode ray. Stream of electrons emitted at the cathode in an evacuated tube containing an anode and a cathode. In cathode ray tubes, the electrons are produced by thermionic emission from a hot cathode and are accelerated down the tube to the anode.

cathodic protection. *See* CORROSION, PASSIVITY.

catholyte. Electrolyte surrounding the CATHODE in an electrochemical cell.

cation. Positively charged ion. When current flows cations are attracted to the CATHODE of an electrochemical cell. *See also* ANION.

cation exchanger. *See* ION EXCHANGER.

cationic surfactant. *See* SURFACE-ACTIVE AGENT.

caustic. Corrosive; refers to alkaline solutions of sodium hydroxide (caustic soda) and potassium hydroxide (caustic potash).

ccp (cubic close-packed structure). *See* CLOSE-PACKED STRUCTURES.

CD. *See* CIRCULAR DICHROISM.

Cd. *See* CADMIUM.

Ce. *See* CERIUM.

celestine. *See* STRONTIUM.

cellobiose. DISACCHARIDE (structure III) (mp 225°C); the repeating unit of CELLULOSE from which it can be obtained by partial hydrolysis with acid. It is dextrorotatory and is a reducing sugar. It forms an OXIME and an OSAZONE, and undergoes mutarotation. Cellobiose has a 1,4-β-glycosidic linkage (*compare* MALTOSE, which has a 1,4-α-glycosidic linkage). It is hydrolyzed by acid or β-glucosidase (*see* ENZYMES) to two molecules of glucose.

cellophane. Transparent sheet plastic made by extruding cellulose xanthate into acid. It is used as a wrapping material, but is highly flammable and now has been replaced by polypropene (*see* POLYMERS).

Cellosolve. Trade name for ETHYLENE GLYCOL MONOETHYL ETHER.

Cellosolve acetate. Trade name for the ethylene glycol monoethyl ether ethanoate (*see* ETHYLENE GLYCOL MONOETHYL ETHER).

cells. Series of conducting phases in contact, the electrodes are electronic conductors connected by electrolytes. At any phase boundary there is a difference in potential. The emf of the cell is the algebraic sum of these phase boundary potentials. Several types of cell are recognized.

primary cell. Device in which a spontaneous chemical reaction is used to produce electrical energy (e.g., DANIELL CELL, LÉCLANCHÉ CELL).

electrolytic cell. Reverse arrangement to a primary cell in which electrical energy from an external source brings about the desired chemical reaction (i.e. ELECTROLYSIS). *See also* ELECTRODEPOSITION OF METALS.

secondary cell (ACCUMULATOR). Energy-producing device that may be recharged by passing a current through it from an external source. For example, *see* LEAD–ACID BATTERY.

Both primary and secondary cells may be connected in series to form a BATTERY.

fuel cell. Produces energy from reactants which are continuously supplied, and the products are continuously removed.

reversible galvanic cell. Used for making thermodynamic measurements under reversible conditions.

concentration cell. Reversible cell in which there is no overall chemical reaction, the emf results from differences in concentration of one or more of the reactants.

celluloid. One of the earliest plastics. It consists of partially nitrated cellulose (*see* CELLULOSE NITRATE) plasticized with camphor (*see* TERPENES); a very flammable material.

cellulose. Polymer of glucose with in excess of 3000 glucose units, joined by β-1,4-glycosidic linkages in straight chains with no branching (*see* Figure). Cellulose is a white solid, insoluble in water, but soluble in ammoniacal copper hydroxide

($[Cu(NH_3)_4]^{2+}$). It is the main constituent of the cell walls of plants. It is the most widely distributed organic compound; cotton is almost pure cellulose, whereas wood is about 50 percent cellulose. Careful hydrolysis gives CELLOBIOSE, and complete hydrolysis yields GLUCOSE. It forms a triethanoate (*see* CELLULOSE ACETATE PLASTICS), a trinitrate (*see* CELLULOSE NITRATE) and ethers with alcohols. Man does not have the necessary enzymes to break down cellulose; microorganisms in the digestive tract of herbivores have the necessary enzyme systems to break the β-glycosidic linkages. Its largest use is in the rayon industry (*see* FIBRES). Microcrystalline cellulose (E460) is used as a non-nutritive bulking agent, binder, emulsion stabilizer and thickening agent in the food industry.

cellulose acetate plastics (cellulose ethanoate plastics). Formed by the treatment of cellulose (cotton, purified wood pulp) with ethanoic anhydride, ethanoic acid and sulphuric acid. Complete ethanoylation gives the triethanoate; used as a plastic material.

cellulose nitrate (nitrocellulose). Prepared by treating cellulose (cotton, wood pulp) with nitric and sulphuric acids. It is very flammable and is used extensively as an EXPLOSIVE, as a lacquer (*see* COLLODION) and formerly as a base for photographic films (*see* CELLULOID).

Celsius scale. *See* TEMPERATURE SCALES.

cement. (1) General term for an ADHESIVE. (2) Material based on calcium silicates and aluminates that forms a plastic paste with water and that hardens on standing. Common cement is formed by heating limestone and clay to 1700°C after which gypsum (*see* CALCIUM, SULPHATES, SUL-

cellulose

cement

Type	Composition	Properties
Portland	Calcium carbonate + aluminium silicates	Hard general-purpose cement resembling Portland stone
High alumina	Bauxite + limestone + iron and silicon oxides	Resistant to attack by seawater and acids. May lose strength over long periods
Rapid-hardening	Cement to which (e.g., $CaCl_2$) is added	Hardens within 2–3 days
Supersulphated	Portland cement + blast furnace slag	Resistant to sulphate

PHUR) is added. Different forms of cement exist (*see* Table).

cementite (Fe_3C). Orthorhombic phase in iron that contains small amounts of carbon.

centigrade scale. *See* TEMPERATURE SCALES.

central force field model. *See* FORCE CONSTANT.

centre of inversion. *See* SYMMETRY ELEMENTS.

centre of symmetry. *See* CRYSTAL SYMMETRY, SYMMETRY ELEMENTS.

cephalins. *See* PHOSPHOLIPIDS.

cephalosporins. *See* ANTIBIOTICS.

ceramics. Inorganic materials, usually containing oxides, borides or carbides, formed at high temperatures and used widely in engineering and industry as REFRACTORY MATERIALS, GLASSES, CEMENTS, ENAMELS, ABRASIVES, pottery, china, porcelains, etc.

cerargyrite. *See* SILVER.

cerebrosides. Group of glycosidic compounds (*see* GLYCOSIDES) occurring abundantly in the myelin sheaths of nerves in association with sphingomyelins (*see* PHOSPHOLIPIDS).

cerium (Ce). Most abundant LANTHANIDE ELEMENT. Extracted from the mineral monazite (($Ce,La,Nd,Pr)PO_4.ThO_2$). Cerium is added in small quantities to alloys to improve mechanical properties. Cerium-(IV) oxide is used in glass polish, ceramic coatings and in the Welsbach gas mantle (*see* INCANDESCENCE).
$A_r = 140.12$, $Z = 58$, ES [Xe] $4f^1 5d^1 6s^2$, mp 798°C, bp 3257°C, $\rho = 6.66 \times 10^3$ kg m^{-3}.

cerium compounds. Oxidation states +3 and +4 are found. Cerium(IV) is the most stable +4 state of any lanthanide element. Cerium(IV) is strongly oxidizing ($E^\ominus_{Ce^{4+},Ce^{3+}} = 1.61$ V) and is used in analytical chemistry. Cerium(III) salts are typical LANTHANIDE COMPOUNDS. CeF_4 is the only known cerium(IV) halide. $Ce(SO_4)_2$ is obtained by dissolving CeO_2 in sulphuric acid. Complex nitrates (e.g., $K_2Ce(NO_3)_6$) are also stable. Cerium(IV) salts are usually yellow and those of cerium(III) colourless.

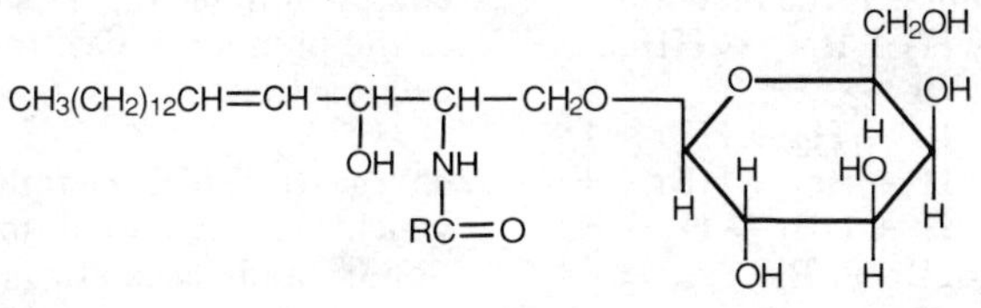

$$CH_3(CH_2)_{12}CH{=}CH{-}\underset{\underset{OH}{|}}{CH}{-}\underset{\underset{\underset{\underset{R\overset{\|}{C}{=}O}{|}}{NH}}{|}}{CH}{-}CH_2O{-}$$

cerebrosides

cerussite. *See* LEAD.

cesium. *See* CAESIUM.

cetrimide (hexadecanyltrimethylammonium bromide). *See* SURFACE-ACTIVE AGENT.

cetyl trimethyl ammonium bromide (hexadecanyltrimethylammonium bromide). *See* SURFACE-ACTIVE AGENT.

Cf. Californium (*see* ACTINIDE ELEMENTS).

CFM (chlorofluoromethanes). *See* FREONS.

CFT. *See* CRYSTAL FIELD THEORY.

cgs units. System of units based on the centimetre, gram and second. It was badly adjusted to use with thermal quantities (e.g., the inconsistently defined calorie) and with electrical units based on either unit permittivity (electrostatic units) or unit permeability (electromagnetic units) of free space. This system has now been superseded by SI UNITS (for cgs–SI conversion—see Appendix Table 2).

chabazite. *See* SILICATES.

chain carrier. *See* CHAIN REACTION.

chain reaction. Reaction in which intermediates (chain carriers) can initiate further reactions. In chemical chain reactions these intermediates are often FREE RADICALS or atoms, whereas in nuclear chain reactions they are neutrons. Chain reactions are initiated by the formation of radicals by light or heat from a normal molecule. The chain is propagated by a reaction between a molecule and radical, producing a product molecule and another radical which may continue the chain. Inhibition occurs when a radical attacks the product, and the chain terminates when two radicals combine, often in the reverse of initiation. An example is the reaction

$$H_2 + Br_2 \rightarrow 2HBr$$

initiation:	$Br_2 \rightarrow 2Br^{\cdot}$
propagation:	$Br^{\cdot} + H_2 \rightarrow HBr + H^{\cdot}$
propagation:	$H^{\cdot} + Br_2 \rightarrow HBr + Br^{\cdot}$
inhibition:	$H^{\cdot} + HBr \rightarrow H_2 + Br^{\cdot}$
termination:	$2Br^{\cdot} \rightarrow Br_2$

chair form. *See* CONFORMERS.

chalcocite. *See* COPPER.

chalcogens. GROUP VI ELEMENTS oxygen, sulphur, selenium, tellurium and polonium. They may complete the noble gas configuration by forming the chalconide (chalcogenide) ions (S^{2-}, Se^{2-}, Te^{2-}), that only exist in salts of the more electropositive elements.

chalcopyrite. *See* COPPER, IRON, SULPHUR.

chalk. Naturally occurring form of calcium carbonate (*see* CALCIUM INORGANIC COMPOUNDS) derived from the skeletal remains of diatoms and other marine creatures.

change of state. Change of matter in one physical state (solid, liquid or gas) into another. The change from the form stable at the lower temperature to that stable at the higher temperature is always accompanied by the absorption of energy. *See also* BOILING POINT, FREEZING POINT.

characteristic temperatures (θ). Temperatures at which the rotational or vibrational levels above the ground state first begin to be substantially populated; rotation $\theta_r = h^2/8\pi^2 Ik$, vibration $\theta_v = h\nu/k$. For oxygen $\theta_r = 2.079$ K and $\theta_v = 2273.64$ K. *See also* PARTITION FUNCTIONS.

charcoal. *See* CARBON.

charge density. Charge density at a point arising from a general distribution of charge as, for example, found in an ELECTROLYTE (*see* DEBYE–HÜCKEL ACTIVITY COEFFICIENT EQUATION) is related to the potential by Poisson's equation

$$\nabla^2 V = -\rho/\varepsilon_0$$

ε_0 is the permittivity of free space. *See also* ELECTRON DENSITY.

charge number (z). Positive number equal to the number of electrons transferred in a cell reaction.

charge–transfer complex. Complex in which there is a weak interaction between donor and acceptor (e.g., a complex between an aromatic derivative and a halogen). Such complexes are really indistinguishable from true complexes. *See also* CHARGE–TRANSFER SPECTRA.

charge–transfer spectra. Electronic spectra of complexes of TRANSITION ELEMENTS in which the transition causes the transfer of charge between an ORBITAL that is largely centred on a LIGAND and one on the metal. The bands are usually more intense than those arising from simple transitions between *d*-orbitals. The strong colours of chromates(VII) and manganates-(VII) are due to charge transfer.

Charles' law. At constant pressure, the volume (V) of a given mass of gas is directly proportional to its thermodynamic temperature (i.e. $V \propto T$). For an IDEAL GAS, $V = V_0 (1 + t/273.15)$ where V and V_0 are the volumes at t and $0°C$, respectively. *See also* GAS LAW.

chelate effect. Complexes of chelates (*see* CHELATE LIGAND) are more stable than those with an equivalent number of monodentate ligands. For example, the pK_s values for the reaction between $[Cu(H_2O)_4]^{2+}$ and $2NH_3$ and diaminoethene are 7.65 and 10.64, respectively. A large component of the effect arises from the changes in entropy.

chelate ligand. Multiple-coordinating LIGAND, which forms cyclic structures in COORDINATION COMPOUNDS. 1,2-Diaminoethane is an example of a bidentate chelating agent that coordinates through each nitrogen atom. Polydentate ligands are named according to the number the atoms capable of forming coordinate bonds (mono-, bi-, tri-, tetra-, pentadentate, etc.) (*see* Table). The ability of chelating agents to bind ions strongly has led to their use in analysis (e.g., EDTA, DIMETHYLGLYOXIME), water softening, metal scavenging and food preservation. Many biologically important compounds are chelating agents or have chelated metal ions (e.g., HAEMOGLOBIN, CHLOROPHYLL, VITAMINS B_6, B_{12} and C). Chelate complexes may show optical activity arising from the disposition of the ligands. For example, three bidentate ligands may coordinate octahedrally in two mirror image forms.

chemical adsorption. ADSORPTION at a surface may be classed according to the strength and nature of the interaction. Physical adsorption (physisorption) is characterized by long-range, weak van der Waals forces (*see* INTERMOLECULAR FORCES) with an enthalpy of adsorption (q_d) of less than -40 kJ mol^{-1} and with zero ACTIVATION ENERGY. Chemisorption occurs with the formation of covalent bonds with much greater q_d (-100 to -400 kJ mol^{-1}) but with zero or a small activation energy.

chemical combination, laws of. Three rules that define the way in which ELEMENTS may combine to form compounds (*see* CONSTANT COMPOSITION, LAW OF; EQUIVALENT PROPORTIONS, LAW OF; MULTIPLE PROPORTIONS, LAW OF).

chemical equivalent. Mass of a substance that will combine with or displace 1 g of hydrogen, 8 g of oxygen, 35.5 g of chlorine, etc. The equivalent weight represents the 'combining power' of the substance. For an atom it is the atomic mass divided by the valence; in redox reactions it is expressed as the mass equivalent to 1 mol of electrons.

chemical ionization mass spectrometry. *See* MASS SPECTROMETRY.

chemical potential (partial molar Gibbs function, μ_i, $\bar{G}_i$, units: J mol^{-1}). PARTIAL MOLAR QUANTITY defined by

$$\bar{G}_i = \mu_i = (\partial G/\partial n_i)_{T,P,n_j}$$

The chemical potential of a component in a system is the change in the total GIBBS FUNCTION at constant temperature and pressure when 1 mole of that component is added to an infinite amount of the system. For a change in composition

$$dG = \mu_1 dn_1 + \mu_2 dn_2 + \ldots + \mu_i dn_i$$

At equilibrium, $dG = 0$ and μ for a component is constant throughout the system (homogeneous or heterogeneous). For an ideal gas

$$\mu = \mu^\ominus + RT \ln P$$

For the ith component (1) in an ideal mixture of gases

$$\mu_i = \mu_i^\ominus + RT \ln p_i$$

and (2) in an ideal solution

chelate ligand
Some chelate ligands.

Name	Abbreviation	Formula/structure
2,2'-Dipyridyl	dipy bipy	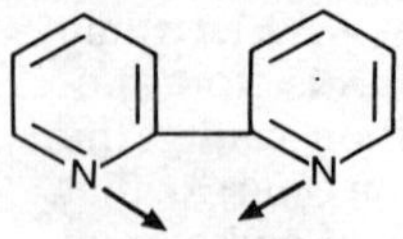
2,4-Pentanedionato	acac	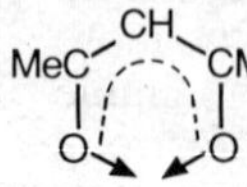
N,N-Diethylthiocarbamate		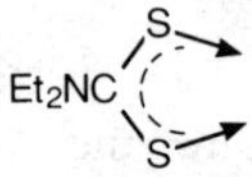
2-Phenylenebidimethyl-arsine	diars	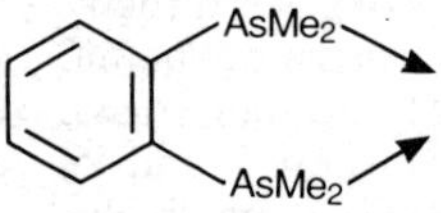
Diethenetriamine	dien	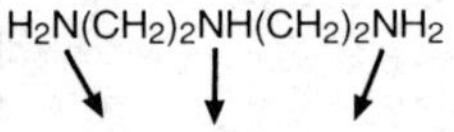 $H_2N(CH_2)_2NH(CH_2)_2NH_2$
Iminodiethanoate		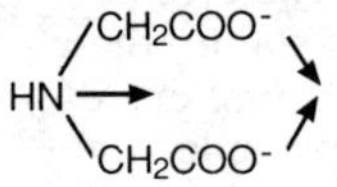
Triethenetetramine	trien	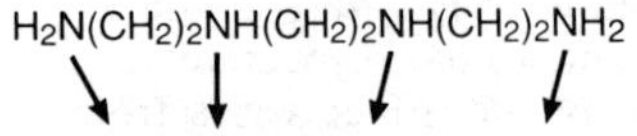 $H_2N(CH_2)_2NH(CH_2)_2NH(CH_2)_2NH_2$
Nitrilotriethanoate		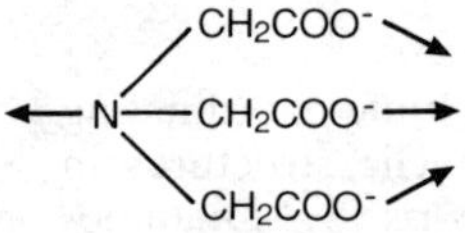
Ethenediaminetriethanoate		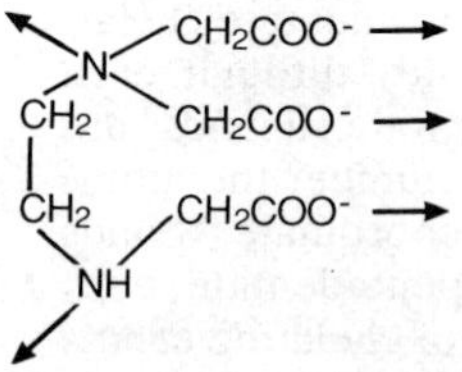
1,8-Bis(salicylideneamino)-3,6-dithiaoctane		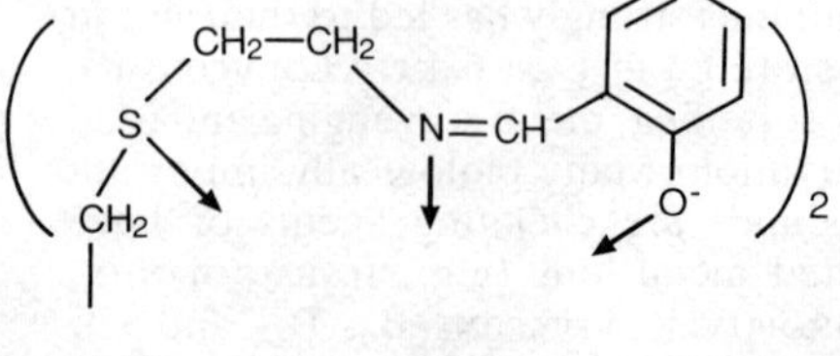
Ethylenediamine-tetraacetate	EDTA	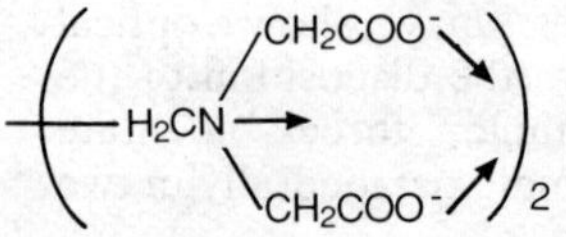

$$\mu_i = \mu_i^{\ominus} + RT \ln x_i$$

where $\mu_i^{\ominus}$ is the chemical potential when p_i or $x_i = 1$. When there are deviations from RAOULT'S LAW, p_i and x_i are replaced by FUGACITY (f_i) and ACTIVITY (a_i), respectively. In terms of molality

$$\mu_i = \mu_i^{\ominus} + RT \ln a_i$$
$$= \mu_i^{\ominus} + RT \ln \gamma_i m_i$$

where $\mu^{\ominus}$ is the chemical potential when $a_i = 1$. Although the chemical potential is closely identified with the Gibbs function, there are equivalent forms for all thermodynamic functions.

chemical shift. Small change in the energies of transitions arising from the chemical environment of an atom. This is found in NUCLEAR MAGNETIC RESONANCE SPECTROSCOPY, ELECTRON SPECTROSCOPY and MÖSSBAUER SPECTROSCOPY.

chemical ultrasonics (sonochemistry). Use of ultrasound (frequencies above 16 kHz) to induce chemical reactions. In solution three effects occur: (1) agitation of the solution as the sound waves pass through it; (2) microbubble formation and collapse (cavitation) which may create local pressures of several GPa and temperatures of 10^4–10^6 K; (3) microstreaming when a large amount of vibrational energy is channelled into small volumes. Free radicals may be produced, diffusion layers are reduced in size and chemical reactants are brought more quickly into contact. Reactions may be completed in shorter times with ultrasound than with conventional methods. For example, the hydrolysis of methyl benzoate goes to completion in 10 min, whereas a conventional reflux experiment requires 90 min. The REFORMATSKY REACTION is also accelerated by ultrasound. Ultrasound has also been used to promote heterogeneous reactions, particularly organometallic reactions.

chemical vapour deposition (CVD). Formation of thin films (usually metal) by deposition from the vapour involving a chemical reaction: for example, nickel may be deposited by the decomposition of nickel carbonyl.

chemiluminescence. Emission of light in a chemical reaction when one product is formed in an electronically excited state. An example is the oxidation of luminol by hydrogen peroxide. *See also* BIOLUMINESCENCE, FLUORESCENCE, LUMINESCENCE, PHOSPHORESCENCE.

chemisorption. *See* CHEMICAL ADSORPTION.

chemometrics. Use of computational and mathematical methods to extract information from analytical data. As well as statistical analysis, chemometrics includes pattern recognition, optimization techniques and robotics as applied to chemical problems.

chemotherapy. Treatment or prevention of a disease by the administration of a chemical: sulpha drugs, antibiotic, cisplatin, etc.

chenodeoxycholic acid. *See* BILE ACIDS.

Chile saltpetre. *See* SODIUM.

chi-potential. *See* SURFACE POTENTIAL.

chirality. Property of non-identity of an object with its mirror image, thus the right hand is the mirror image of the left hand. A molecule is chiral if it cannot be superimposed on its mirror image and exhibits OPTICAL ACTIVITY. The compound *Cabde* (e.g., 2-hydroxypropanoic acid), containing an asymmetric carbon atom, exhibits molecular chirality and the central carbon is said to be chiral. The two forms are optical isomers or enantiomers.

chirality
Figure 1.

In quartz there is chirality in the crystal structure. Molecular chirality is possible in compounds which have no asymmetric carbon atoms and yet possess non-superimposable mirror image structures. Restricted rotation about C=C=C bonds gives rise to the existence of optically active

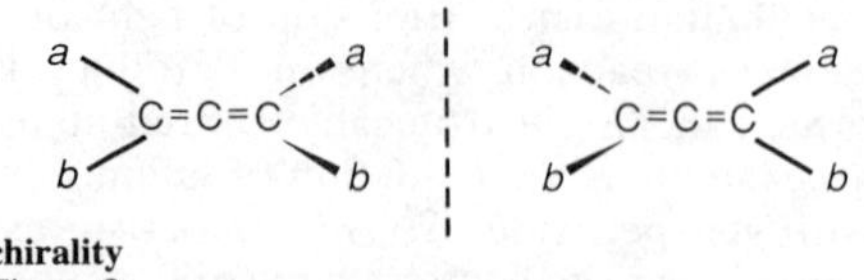

chirality
Figure 2.

forms as does restricted rotation about single bonds caused by bulky substituents in, for example, biphenyl compounds.

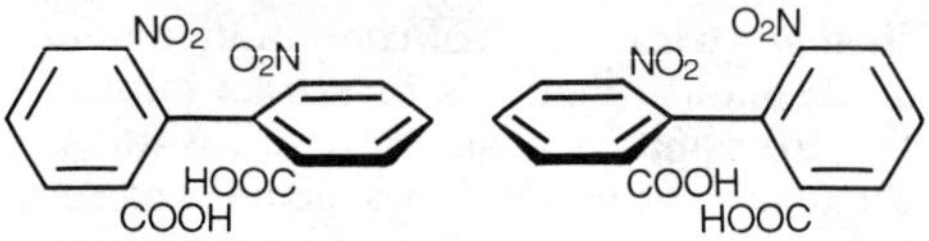

chirality
Figure 3.

chitin. *See* AMINO SUGARS.

chloral. *See* TRICHLOROETHANAL.

chloral hydrate. *See* TRICHLOROETHANAL.

chloramine ($ClNH_2$). Explosive liquid (mp $-66°C$), prepared from sodium hypochlorite and ammonia. Chloramine gives primary amines with GRIGNARD REAGENTS.

chloramines. Organic compounds with nitrogen–chlorine bonds, (e.g., CHLORAMINE-T).

chloramine-T (sodium N-chloro-4-toluenesulphonamide, Na^+ $^-ClN.O_2SC_6H_4CH_3$). Antiseptic and source of hypochlorous acid; prepared from 4-methylsulphonic acid (PCl_5, NH_3, $NaOCl$ + $NaOH$).

chloramphenicol. *See* ANTIBIOTICS.

chlorates. Salts of the oxoacids of chlorine. They are all good oxidizing agents.
chlorates(i) (hypochlorites, ClO^-). Prepared by passing chlorine through a solution of a base. Sodium hypochlorite, prepared commercially (in solution with sodium chloride) by electrolyzing brine and mixing the anode and cathode products, is used as a bleaching and sterilizing agent. The weak acid, $HOCl$, is prepared by passing chlorine into a suspension of mercury(ii) oxide in water.
chlorates(iii) (chlorites, bent ClO_2^-) (ClO_2 + peroxide). On heating, alkali chlorites give the chloride and chlorate(v). The free acid, $HClO_2$, is only known in solution.
chlorates(v) (chlorates, pyramidal ClO_3^-, Cl_2 + hot alkali solution). Alkali chlorates decompose on heating to the chloride and oxygen or perchlorate and chloride depending on the temperature. The free acid, $HClO_3$, ($Ba(ClO_3)_2$ + H_2SO_4), only known in solution, is a strong acid.
chlorates(vii) (perchlorates, tetrahedral ClO_4^-). Prepared by the electrolytic oxidation of chlorates. Perchlorates are the least polarized of anions and are much used for adjusting the ionic strength of solutions without the risk of forming complexes with the cations present. The free acid, obtained by treating a perchlorate with sulphuric acid, is a colourless oily liquid. It is used as an oxidizing agent; it is explosive on contact with organic materials.

chlordiazepoxide   (7-chloro-2-methylamino-5-phenyl-$3H$-1,4-benzodiazepine-4-oxide). Depressant of the central nervous system. The hydrochloride is Librium.

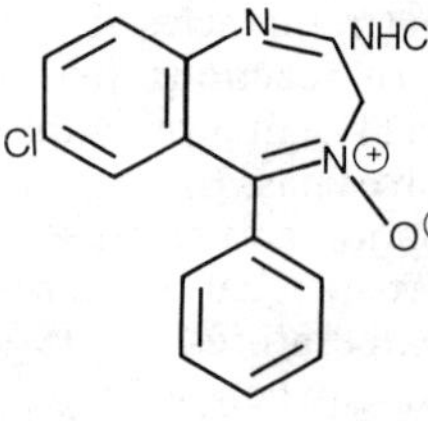

chlordiazepoxide

chloric acid. *See* CHLORATES.

chlorides. *See* HALIDES.

chlorine (Cl). GROUP VII ELEMENT that occurs (0.19 percent of the lithosphere) as sodium chloride in seawater and as carnallite and sylvite. It is manufactured by the electrolysis of brine with the simultaneous formation of sodium hydroxide. It is a greenish, poisonous diatomic gas which combines directly with most other elements; with hydrogen the reaction is slow in the dark but occurs rapidly by a chain reaction when irradiated. Chlorine is a strong oxidizing agent. It is soluble in water giving a hydrate ($Cl_2.7.3H_2O$) which has a clathrate structure (*see* CLATHRATE COMPOUND); a small amount of chloride and hypochlorite is formed. Chlorine is a typical non-metal exhibiting −1 oxidation state in ionic chlorides and covalent chloro-derivatives, and higher oxidation states mainly in combination with oxygen and other halogens. Industrially it is a very basic material used in the production of organo-chlorine compounds. Chlorine is used as a bleaching agent and bactericide in the food industry (925), and its derivatives are used as solvents, polymers (vinyl chloride), refrigerants and aerosol propellants. A_r = 35.453, Z = 17, ES [Ne] $3s^2\ 3p^5$, mp −100.98°C, bp −34.6°C, ρ = 3.214 kg m^{-3}, $E^{\ominus}_{Cl_2,Cl^-}$ = 1.3595 V. Isotopes: ^{35}Cl, 75.53 percent; ^{37}Cl, 24.47 percent.

chlorine halides.   *See* INTERHALOGEN COMPOUNDS.

chlorine oxides.
chlorine monoxide (Cl_2O). Anhydride of hypochlorous acid (HOCl); yellow–red explosive gas prepared by passing chlorine over mercury(II) oxide. It is unstable, dissociating into its elements.
chlorine dioxide (ClO_2). High explosive; usually prepared *in situ* ($NaClO_3 + H_2SO_4 + SO_2$). It is used as an oxidizing agent and as a bleaching agent for flour (926) and woodpulp. ClO_2 is soluble in water giving hypochlorous acid ($HClO_2$) and chloric acid ($HClO_3$).
dichlorine tetroxide ($ClO.ClO_3$) ($CsClO_4 + ClSO_3F$). Very unstable.
dichlorine hexoxide (Cl_2O_6) ($O_3 + ClO_2$). Very unstable.
dichlorine heptoxide (Cl_2O_7). Obtained by the dehydration of perchloric acid ($HClO_4$). Although explosive, Cl_2O_7 is more thermodynamically stable than are the other oxides of chlorine. It is a less powerful oxidizing agent.
perchloryl fluoride (ClO_3F) ($KClO_4 + HF + SbF_5$). Toxic gas that is thermally stable to 500°C and resistant to hydrolysis. At high temperatures, it is a powerful oxidizing agent with selective fluorinating properties (e.g., replacement of hydrogen by fluorine on methene groups).

chlorine oxoacids.   *See* CHLORATES.

chlorites.   *See* CHLORATES.

chloroacetic acids.   *See* CHLOROETHANOIC ACIDS.

2-chlorobenzaldicyanomethane.   *See* CS GAS.

chlorobenzene (C_6H_5Cl). Colourless liquid (bp 132°C), manufactured by the catalytic reaction (CuCl) of benzene vapour, air and hydrogen chloride

$$PhH + HCl + \tfrac{1}{2}O_2 \rightarrow PhCl + H_2O$$

It is used in the manufacture of aniline, phenol and DDT (*see* 1,1,1-TRICHLORO-2,2-BIS(4-CHLOROPHENYL)ETHANE).

2-chloro-1,3-butadiene   (chloroprene, $CH_2{=}CClCH{=}CH_2$). It is produced by passing 1-buten-3-yne (vinylacetylene) through concentrated hydrochloric acid in the presence of copper(I) chloride/ammonium chloride. Chloroprene (bp 60°C) polymerizes readily to the synthetic rubber NEOPRENE.

1-chloro-2-dichloroarsenoethene.   *See* LEWISITE.

1-chloro-2,3-epoxypropane   (epichloro-hydrin, $ClCH_2CH{-}CH_2$). Colourless liquid (mp −25.6°C); prepared by treating glycerol dichlorohydrins with solid sodium hydroxide at 25–30°C or by the chlorination of PROPENAL. Reactions are mainly those of the 1,2-epoxyethane ring. It is reduced by sodium amalgam to 2-PROPEN-1-OL and oxidized by nitric acid to 3-chloropropanoic acid. It forms glycerol diethers on treatment with alcohols and potassium hydroxide. It is used in the manufacture of glycerol, glycerol esters and epoxy resins.

chloroethane (ethyl chloride, C_2H_5Cl). Gaseous ALKYL HALIDE (bp 12.5°C), prepared by the chlorination of ethane at 400°C or by the addition of hydrogen chloride to ethene over aluminium chloride. It is used for the preparation of tetraethyllead (*see* ANTIKNOCK ADDITIVE, LEAD ORGANIC DERIVATIVES) and as a refrigerant.

chloroethanoic acids ($Cl_nCH_{3-n}COOH$, $n = 1$–3). α-Halogenocarboxylic acids.

chloroethanoic acid ($ClCH_2COOH$). Deliquescent solid (mp 161°C), prepared by treating ethanoic acid with chlorine in the presence of red phosphorus, or in sunlight. Industrially it is made by heating trichloroethene with 74 percent sulphuric acid at 140°C, or by direct chlorination of ethanoic acid. It is used in organic synthesis (for example, *see* REFORMATSKY REACTION), and in the manufacture of INDIGOTIN. It is converted to glycine (*see* AMINO ACIDS) by reaction with ammonia.

dichloroethanoic acid ($Cl_2CHCOOH$). Liquid (bp 194°C), prepared by adding calcium carbonate to a solution of trichloroethanal hydrate followed by sodium cyanide then heating the mixture. Hydrolysis gives OXOETHANOIC ACID.

trichloroethanoic acid (Cl_3CCOOH). Deliquescent solid (mp 58°C), which is one of the strongest organic acids ($pK_a = 0.66$), prepared by oxidation of trichloroethanal hydrate with concentrated nitric acid. The Cl_3C–$COOH$ bond is weak, trichloromethane being readily formed on hydrolysis. With concentrated alkali, methanoates are formed (*see* ESTERS, METHANOIC ACID.) The sodium salt is used as a herbicide.

chloroethene (vinyl chloride, $CH_2{=}CHCl$). Colourless, carcinogenic gas (bp −14°C), manufactured by: (1) hydrogen chloride and ethyne; (2) hydrolysis of 1,2-DICHLOROETHANE in sodium hydroxide; and (3) high-temperature chlorination of ethane. It forms the addition polymer polyvinyl chloride and many copolymers (*see* CO-POLYMERIZATION,   POLYMERIZATION, POLYMERS).

chlorofluoromethanes. *See* FREONS.

chloroform (trichloromethane). *See* TRI-HALOMETHANES.

chlorohydrins. Organic compounds containing the group ClC–COH, prepared by treating an alkene with chlorine water. They are oxidized by concentrated nitric acid to the α-halogenoacid. With alkalis 1,2-glycols are formed.

chloromethane (methyl chloride, CH_3Cl). ALKYL HALIDE (bp −24°C), manufactured by the direct reaction of methane and chlorine. It is used in the manufacture of aniline DYES, silicone rubbers (*see* POLYMERS, SILICON INORGANIC COMPOUNDS), tetramethyl-lead (*see* ANTIKNOCK ADDITIVE, LEAD ORGANIC DERIVATIVES), methylation of cellulose, as a refrigerant, local anaesthetic and fire extinguisher.

4-(4-chloro-2-methylphenoxy)butyric acid. *See* AUXINS.

chlorophylls. Green photosynthetic pigments of all plants, but not bacteria; magnesium complexes of the PORPHYRINS esterified with the long-chain alcohol phytol ($C_{20}H_{39}OH$). Chlorophyll *a* (*see* Figure), a black powder (mp 150–153°C) has a methyl group in position 3 (*); in chlorophyll *b*, a dark green powder (mp 120–130°C), an aldehyde group replaces the methyl group. Both are soluble in ethanol and ether. Chlorophyll acts as a catalyst in the PHOTOSYNTHESIS of carbohydrates from carbon dioxide and water with the release of oxygen. The porphyrin nucleus of both HAEM and chlorophyll is synthesized biologically from glycine and ethanoic acid; the steps involved in both the red blood cells and the chloroplasts of plant cells are identical. Chlorophyll, extracted from nettles, grass and lucerne with ethanol, is used as a food colouring agent (E140), as are the copper complexes (E141).

chlorophyll a

chloropicrin.   *See*   TRICHLORONITROMETHANE.

chloroplatinic acid. *See* PLATINUM COMPOUNDS.

chloroprene. *See* 2-CHLORO-1,3-BUTADIENE.

chlorosulphonic acid (chlorosulphuric acid). *See* SULPHUR HALOOXOACIDS.

chlorous acid. *See* CHLORATES.

2-chlorovinyldichloroarsine. *See* LEWISITE.

chlorpromazine (2-chloro-10-(3-dimethylaminopropyl)phenothiazine). Major antiemetic, antipsychotic tranquillizer (*see* Figure). The chlorine-free compound, promazine, is used as a tranquillizer and as an adjunct to anesthesia.

chlorpromazine

chlortetracycline. *See* ANTIBIOTICS.

cholane ring system. C_{24} skeleton of BILE ACIDS as depicted for the hydrocarbon 5β-cholane.

cholane ring system

cholesteric crystal. *See* LIQUID CRYSTALS.

cholesterol. *See* STEROLS.

cholic acid. *See* BILE ACIDS.

choline (vitamin B complex). *See* VITAMINS.

chondroitin sulphate. *See* AMINO SUGARS.

chromates(III), (IV), (V), (VI). *See* CHROMIUM COMPOUNDS.

chromatogram. Detector output recorded in the form of a continuous trace in GAS–LIQUID CHROMATOGRAPHY and HIGH-PERFORMANCE LIQUID CHROMATOGRAPHY. In PAPER CHROMATOGRAPHY and THIN-LAYER CHROMATOGRAPHY, it is the developed surface on which the separate zones

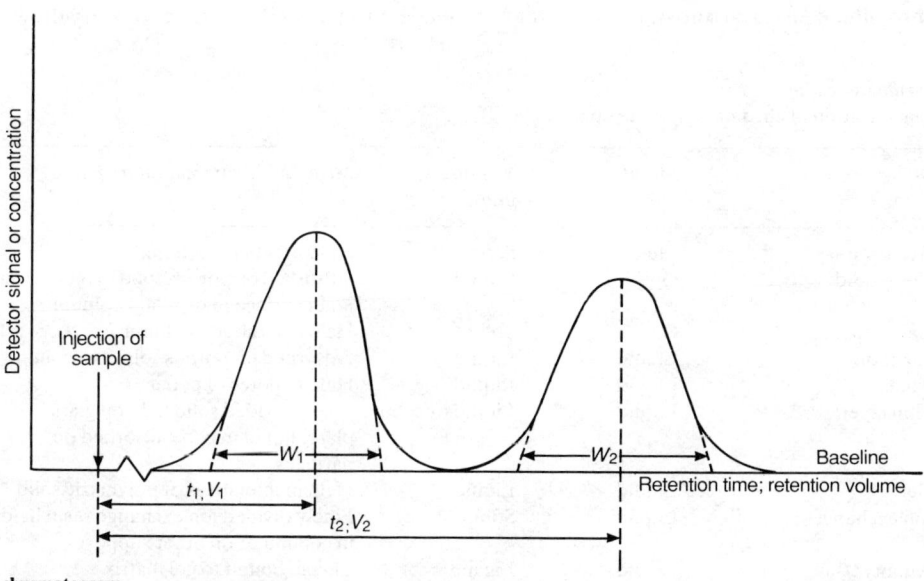

chromatogram
Typical chromatogram for the separation of a two-component mixture.

have been revealed; if the surface is scanned with a densitometer (or other suitable device) a similar trace can be obtained (*see* Figure). t_1, t_2, and V_1, V_2 are the retention times and volumes, respectively, for components 1 and 2. The resolution (R_s) is given by:

$$R_s = 2(t_2 - t_1)/(W_1 + W_2)$$

chromatography. Diverse group of separation processes based on differences in rates at which individual components of a mixture migrate through a STATIONARY PHASE under the influence of a MOBILE PHASE (*see* Table). All separations are based on differences in the extent to which solutes are partitioned between the mobile and stationary phase; described quantitatively by the temperature-dependent PARTITION COEFFICIENT (K)

$$K = c_s/c_m$$

where c_s and c_m are the total analytical concentrations of the solute in the stationary and mobile phases, respectively. The assumption that K is constant is generally satisfied provided the concentrations are low. Chromatographic methods are classified as elution analysis, FRONTAL ANALYSIS and displacement analysis.

chromite. *See* CHROMIUM, IRON, SPINELS.

chromium (Cr). Group VIb, first row TRANSITION ELEMENT. It occurs as the ore chromite ($FeCr_2O_4$), which also contains magnesium and aluminium, in seawater (0.05 μg dm^{-3}) and in moon rocks. Ores are reduced with carbon or silicon in an electric furnace to ferrochromium. Pure metal is obtained by thermal dissociation of chromium iodide *in vacuo* (the van Arkel–de Boer process) or by electroplating (*see* ELECTRODEPOSITION OF METALS) from chromic acid. The appearance, hardness and chemical resistance of an electroplated layer leads to an important use of chromium in finishing and protecting metal articles. It is a major component of stainless steel, nickel stainless STEEL, high-strength ferritic steels and THERMOCOUPLE (Chromel) and resistance wires (NICHROME). Compounds are used as pigments for colouring glass and ceramics and in the leather industry.
$A_r = 51.9961$, $Z = 24$, ES [Ar] $3d^5$ $4s^1$, mp $1857°C$, bp $2672°C$, $\rho = 7.19 \times 10^3$ kg m^{-3}. Isotopes: ^{50}Cr, 4.31 percent; ^{52}Cr, 83.76 percent; ^{53}Cr, 9.55 percent; ^{54}Cr, 2.38 percent.

chromium compounds. Oxidation states -1 to $+6$ are known, $+2$, $+3$ and $+6$ being of major importance. Chromium(II) is found as chromates (CrO_4^{2-}), yellow derivatives of the mis-named chromic acid (CrO_3). Barium chromate ($BaCrO_4$) and lead chromate ($PbCrO_4$, chrome yellow)

chromatography
Classification of chromatographic separations.

Name	Mobile phase	Stationary phase	Method of fixing stationary phase
Adsorption	Liquid	Solid	Held in tubular column
Gas–liquid	Gas	Liquid	Adsorbed on porous solid in tube or on inner surface of capillary column
Gas–solid	Gas	Solid	Held in a tubular column
Partition	Liquid	Liquid	Adsorbed on porous solid in a column
Paper	Liquid	Liquid	Held in pores of paper
Thin layer	Liquid	Liquid or solid	Finely divided solid held on glass plate, liquid may be adsorbed on particles
Gel	Liquid	Liquid	Held in interstices of polymeric solid
Ion exchange	Liquid	Solid	Finely divided ion exchange resin held in column or on other support
Affinity	Liquid	Ligand	Ligand bound to gel matrix

are used gravimetrically to determine chromium. The dichromate ion ($Cr_2O_7^{2-}$) (two CrO_3 units bridged with oxygen) is highly oxidizing ($E^{\ominus}_{Cr_2O_7^{2-},Cr^{3+}}$ = +1.33 V), as is chromyl chloride (CrO_2Cl_2) which is used in ETARD'S REACTION. Chromates(v) (CrO_4^{3-}) are dark green compounds which hydrolyze and disproportionate to chromium(III) and chromium(VI). CrF_5 is known (CrF_3+F_2, 350–500°C). Chromium(IV) is found in some mixed salts (e.g., $M_2^I CrO_4$, M^I = alkaline earth metal; $M_2^{II}CrF_6$, M^{II} = alkali metal) and as relatively unstable salts (e.g., CrF_4). Chromic salts (chromium-(III)) of Cr_2O_3 are known for most anions. Green aqueous solutions are formed containing octahedral complexes. Chromium-(II) (chromous) salts are reducing ($E^{\ominus}_{Cr^{2+},Cr}$ = –0.56 V. Chrome alum ($K_2Cr(SO_4)_2$.-$12H_2O$) is used as a gelatin hardener. Chromium in oxidation state –1 is found in the carbonyl $Cr(CO)_6^-$. Complexes of chromium(III) containing organic liquids are widely known (e.g., [$Cr(phen)_2$-$(OH)_2$]I_4.$4H_2O$, where phen = phenanthroline: $Na_4[Cr(ox)_2OH]_2$.$6H_2O$, where ox = oxonato group). SANDWICH COMPOUNDS of chromium include dibenzene chromium (h^6-C_6H_6)$_2$Cr and chromocene (h^5-C_5H_5)$_2$Cr.

chromocene. *See* CHROMIUM COMPOUNDS, METALLOCENES.

chromophore. Group in an organic molecule having empty π^*-orbitals (*see* MOLECULAR ORBITAL) into which electrons may be promoted. The energies of such transitions occur at wavelengths in the visible and UV regions of the ELECTROMAGNETIC SPECTRUM and thus give rise to colour in the molecule. Typical chromophores are C=O, C=C, N=N, NO_2. *See also* AUXOCHROME, ELECTRONIC SPECTROSCOPY.

chrysene (1,2-benzophenanthrene, $C_{18}H_{12}$). POLYCYCLIC HYDROCARBON (structure X) with aromatic character that occurs in COAL TAR. It fluoresces strongly under UV radiation.

chymotrypsin A. *See* ENZYMES.

CI. *See* CONFIGURATION INTERACTION.

CIDNP (chemically induced dynamic nuclear polarization). *See* NUCLEAR MAGNETIC RESONANCE SPECTROSCOPY.

CIMS (chemical ionization mass spectrometry). *See* MASS SPECTOMETRY.

cinchona alkaloids. *See* ALKALOIDS.

cinchonine. *See* ALKALOIDS.

cineole. *See* TERPENES.

cinnabar. *See* MERCURY.

cinnamic acids. *See* 3-PHENYLPROPENOIC ACIDS.

circular dichroism (CD). Elliptical polarization of light through a substance having different molar extinction coefficients (*see* BEER–LAMBERT LAW) for left ε_l and right ε_r circularly polarized light (*see* POLARIZED RADIATION). Molecular ellipticity (θ) is defined as $3300(\varepsilon_l - \varepsilon_r)$ for ε in dm^3 mol^{-1} cm^{-1} (*see* OPTICAL ROTATORY DISPERSION).

***cis*-isomer.** *See* ISOMERISM.

citrals. *See* TERPENES.

citrene (limonene). *See* TERPENES.

citric acid. *See* 2-HYDROXYPROPANE-1,2,3-TRICARBOXYLIC ACID.

Cl. *See* CHLORINE.

Claisen condensation. Formation of a β-ketoester, β-ketone or β-nitrile by the CONDENSATION REACTION of an ester with an ester, ketone or nitrile, respectively. It is an important synthetic reaction and is used, for example, to prepare ETHYL 3-OXO-BUTANOATE.

Claisen–Schmidt reaction. CONDENSATION REACTION (*see* BENZALDEHYDE, Figure) between an aromatic ALDEHYDE or KETONE and an aldehyde or ketone in the presence of dilute alkali to give an α,β-unsaturated compound. *See also* ALDOL CONDENSATION.

Clapeyron–Clausius equation. Relationship between the temperature dependence of VAPOUR PRESSURE and the ENTHALPY and volume changes in a phase transition.

$$\text{I} \quad\Longleftrightarrow\quad \text{II}$$

(form stable   (form stable
below T_{tr})   above T_{tr})

$$dp/dT = \Delta H/T\Delta V$$
$$= L_{tr}/T(V_{II} - V_{I})$$

For liquid–vapour equilibria, $V_g \gg V_l$ and assuming ideal behaviour equation, the above equation becomes

$$d \ln p/dT = L_e/RT^2$$

and similarly for the solid–vapour equilibrium. L_e and L_s can be determined from the slope of the plot of $\ln p$ against T^{-1}. For the solid–liquid equilibrium, the equation

$$dT/dp = T(V_l - V_s)/L_f$$

shows how the freezing point varies with the applied pressure; when V_l is greater (less) than V_s then increase of pressure causes a raising (lowering) of the freezing point. *See also* ONE-COMPONENT SYSTEM.

clarification. Removal from a liquid of small amounts of suspended matter. It is achieved by filtration, centrifugation or by the use of a clarifier (e.g., bentonite, *see* SILICATES).

Clark cell. *See* STANDARD CELL.

Clark oxygen electrode. *See* OXYGEN ELECTRODE.

Clark process. Removal of temporary HARDNESS OF WATER by the addition of calcium hydroxide (lime).

clathrate compound. Molecular compound formed by the inclusion or trapping of molecules of one type in the holes of a very open lattice of another type without the existence of any chemical bonds between the two components. Stable crystalline β-quinol clathrates containing O_2, N_2, CO, Ar, Xe and CH_3OH have been prepared; the entrapped gas is released on heating. Gas hydrates are formed by compressing water with such gases as Ar, Kr, Xe (e.g., $Xe.6H_2O$), Cl_2 (e.g., $Cl_2.7.3H_2O$) and Br_2 (e.g., $Br_2.8.5H_2O$). Complete filling of the cages is rarely achieved. A third type of clathrate compound—salt hydrates—is formed when tetraalkylammonium salts are crystallized from water with high water content (e.g., $C_6H_5COO\text{-}[N(n\text{-}C_4H_9)_4].39.5H_2O$).

Clausius–Mosotti equation. Equation to calculate the molecular POLARIZABILITY of a molecule

$$P_m = (M/\rho) \left[(\varepsilon_r - 1)/(\varepsilon_r + 2) \right]$$

clays. Natural aluminosilicates (*see* SILICATES) that occur as a wet, plastic material in soils. Their structure is of linked AlO_4 and SiO_4 tetrahedra with layers of magnesium and aluminium hydroxides. Particles of clay may be suspended in water without coagulation. Concentrated solutions form GELS. Clays are the primary constituents of many CERAMICS, and they are also used as fillers of rubbers, paints, polymers and paper. *See also* ZEOLITE.

cleavage planes. Directions in which a crystal splits along the planes of atoms or molecules in the crystal lattice.

Clemmensen reduction. Reduction of an OXO GROUP with amalgamated zinc and concentrated hydrochloric acid (i.e. $RCOR' \rightarrow RCH_2R'$).

closed shell molecule. Term used in MOLECULAR ORBITAL CALCULATIONS to describe a molecule or atom having a pair of electrons in each occupied orbital. An open shell molecule is one in which there is one or more unpaired electrons. *See also* MULTIPLICITY, RADICAL.

closed system. *See* SYSTEM.

close-packed structures. Most metals crystallize in one of three simple ways, two of which can be explained in terms of organizing spheres into the closest possible packing (*see* Figure). In a single plane, each sphere is surrounded by six close neighbours in a hexagonal arrangement (a). The spheres in the second layer fit into depressions in the first layer, each sphere now having 12 other touching spheres. A third layer can be laid on top of this in two different ways. (1) In hexagonal close-packed (hcp) structures (shown by Be, Cd, Co, He, Mg, Ti, Zn), the spheres in the third layer are directly over those in the first so the layer structure could be represented as ABABAB . . . (b). (2) In cubic close-packed (ccp) structures (shown in Al, Ar, Ca, Cu, Au, Pb, Ni, Ne, Pt, Ag, Xe), the spheres in the third layer occupy a different set of depressions to those in the first layer and do not reproduce the first layer; the layer structure is then ABCABCABC. . . . (a). The ccp arrangement gives rise to face-centred unit cells (fcc). Both hcp and ccp structures are close-packed, dense structures, with a coordination number of 12. In the third arrangement (shown by Ba, Cs, Cr, Fe, K and W), the atoms of the first layer are less closely packed than in other structures and the next layer lies in the depressions in the first layer. The third layer reproduces the first leading to a structure ABABAB . . . . This unit cell corresponds to a body-centred cubic (bcc) arrangement (c), with a coordination number of 8. In fcc and hcp cells the free volume is 26 percent, in a bcc cell it is 32 percent and in a simple cubic lattice 48 percent.

closo. *See* CARBORANES.

cloxacillin. *See* ANTIBIOTICS.

Cm. Curium (see ACTINIDE ELEMENTS).

CMC. *See* CARBOXYMETHYLCELLULOSE.

cmc (critical micelle concentration). *See* MICELLE.

CNDO (complete neglect of differential overlap). *See* MOLECULAR ORBITAL CALCULATIONS, SEMIEMPIRICAL MOLECULAR ORBITAL THEORY.

Co. *See* COBALT.

coagulation. Aggregation of a COLLOID at a primary minimum of the potential energy–distance curve. Coagulation occurs if the net energy of repulsion is less than the thermal energy of the particles. LYOPHOBIC COLLOIDS are coagulated by the addition of electrolytes which neutralize the sol, at the ISOPOTENTIAL POINT, by ultrasonic agitation or by the passage of an electric current. LYOPHILIC COLLOIDS are coagulated by high concentrations of added electrolyte (*see* SALTING OUT). *See also* FLOCCULATION.

coal. Natural carbonaceous material formed by the decay of vegetation. The carbon content varies between 50 and 95 percent and oxygen between 3 and 40 percent. The CALORIFIC VALUE is on average

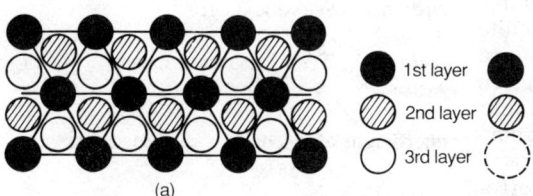
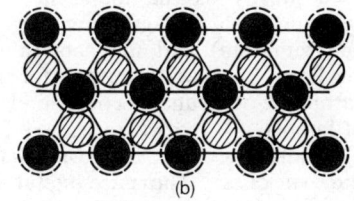
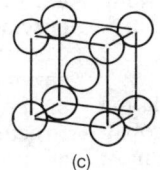

1st layer
2nd layer
3rd layer

(a)   (b)   (c)

close-packed structures
Close-packing of identical spheres. (a) Cubic close-packing (face-centred cubic packing); (b) hexagonal close-packing; (c) body-centred cube.

30 MJ kg^{-1} and increases with the carbon content along the series peat, lignites, bituminous coals, semibituminous coals, anthracite.

coal gas (town gas). Gas formed from coal by GASIFICATION that contains hydrogen (more than 50 percent), carbon monoxide and some hydrocarbons. It has a CALORIFIC VALUE of 18 MJ m^{-3}. Coal gas has been superceded as a fuel by NATURAL GAS.

coal tar. High-temperature carbonization of coal gives non-condensable gases, 'coal gas' (CO, H_2, CH_4, NH_3), coal tar and a non-volatile residue of coke and ash. Coal tar is an important source of many basic aromatic compounds, distillation yields four fractions: (1) light oil (bp < 200°C) containing benzene, toluene, xylenes, thiophene, pyrrole, phenol, cresols, pyridine, methylpyridines; (2) middle oil (bp 200–250°C) containing naphthalene, phenol, cresols, pyridine, methylpyridines; (3) heavy oil (bp 250–300°C) containing naphthalene, cresols, xylenols, quinoline; (4) anthracene oil (bp 300–350°C) containing phenanthrene, anthracene, carbazole.

cobalt (Co) First row TRANSITION ELEMENT. It is obtained from arsenides of silver, nickel, copper and lead, smaltite ($CoAs_2$) and cobaltite (CoAsS). Ore is roasted to the oxide, which is reduced by carbon. Cobalt is used mostly as the metal in alloys for machine tools and grinding (e.g., stellite and vitallium) and magnets (alnico and vicalloy), and as a binder for tungsten carbide. Cobalt and its compounds are also of importance in CATALYSIS (see FISCHER–TROPSCH PROCESS, HYDROFORMYLATION, OXO PROCESS). The pure metal is ferromagnetic (see FERROMAGNETISM), dissolves in mineral acids and oxidizes slowly in air. $A_r = 58.9332$, $Z = 27$, ES [Ar] $3d^7 4s^2$, mp 1495°C, bp 2870°C, $\rho = 8.92 \times 10^3$ kg m^{-3}. Isotope: ^{59}Co, 100 percent.

cobalt compounds. Common oxidation states in aqueous solution are +2 (cobaltous) and +3 (cobaltic). Cobalt(II) is most stable ($E^{\ominus}_{Co^{2+},Co} = -0.28$ V), showing a wide range of salts. Cobalt(III) is stabilized in the low-spin state by formation of complexes such as $[Co(NH_3)_6]^{3+}$ (see HIGH-SPIN COM-

PLEX). Cobalt forms three oxides: CoO, Co_2O_3 and the spinel Co_3O_4 (see SPINELS). Mixed oxides of this latter structure (e.g., $NiCo_2O_4$) show catalytic activity (see CATALYSIS). Cobalt oxides are used in the glass and ceramics industry to impart a blue colour. Cobalt chloride ($CoCl_2$) is blue in the anhydrous form but $CoCl_2.6H_2O$ is red, this colour change being used as a test for the presence of water. Cobalt carbonyl ($Co_2(CO)_8$) (Co + CO, 150°C, 300 atm) is an example of a metal–metal bond (see METAL CLUSTER COMPOUND). Further reduction to $Co_4(CO)_{12}$ occurs on heating. The ion $Co(CO)_4^-$ is produced in basic solution, the acid $HCo(CO)_4$ resulting from acidification. Cobalt carbonyls are important HYDROFORMYLATION catalysts. Cobalt is important biologically, being a necessary trace element and found in vitamin B_{12} (see VITAMINS). Examples of compounds exhibiting less common oxidation states are $[Co^{-I}(CO)_4]^-$, $Co_2^0(CO)_8$, $Co^IBr(PR_3)_3$, $[Co^{IV}F_6]^{2-}$. Organic derivatives include SANDWICH COMPOUNDS (e.g., dipentadienyl cobalt) and those with alkyl or alkenyl ligands (e.g., $CH_3Co(CO)_4$).

cobaltite. See COBALT.

cocaine. See ALKALOIDS.

COD (chemical oxygen demand). See OXYGEN CONSUMED.

codeine. See ALKALOIDS.

codon. See GENETIC CODE, NUCLEIC ACIDS.

coefficient of viscosity. See VISCOSITY.

coenzymes. Organic, non-protein molecules necessary for the activity of an ENZYME which is loosely associated with the enzyme (compare PROSTHETIC GROUP). Coenzymes participate in the substrate–enzyme interaction by donating or accepting protons or other groups. They may be structurally altered during the reaction, but are usually regenerated in subsequent reactions; thereby coupling two enzyme-catalyzed reactions (see Figure 1). The most important coenzymes are those involved in biological oxidation/reduction reactions

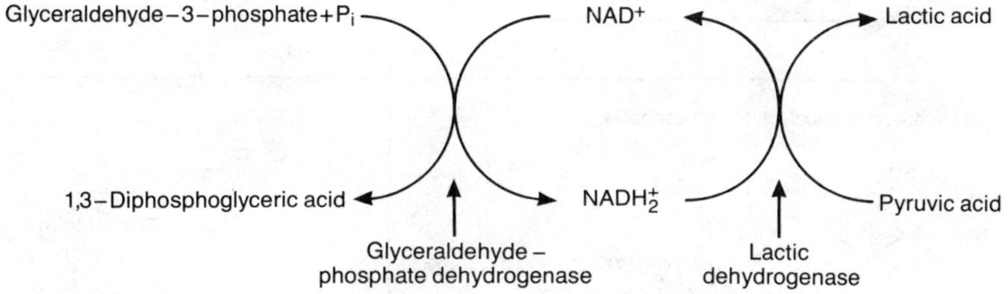

Glyceraldehyde–3–phosphate+P$_i$ — NAD$^+$ — Lactic acid

1,3–Diphosphoglyceric acid — NADH$_2^+$ — Pyruvic acid

Glyceraldehyde –
phosphate dehydrogenase

Lactic
dehydrogenase

coenzymes
Figure 1. Example of coupled enzyme-catalyzed reaction.

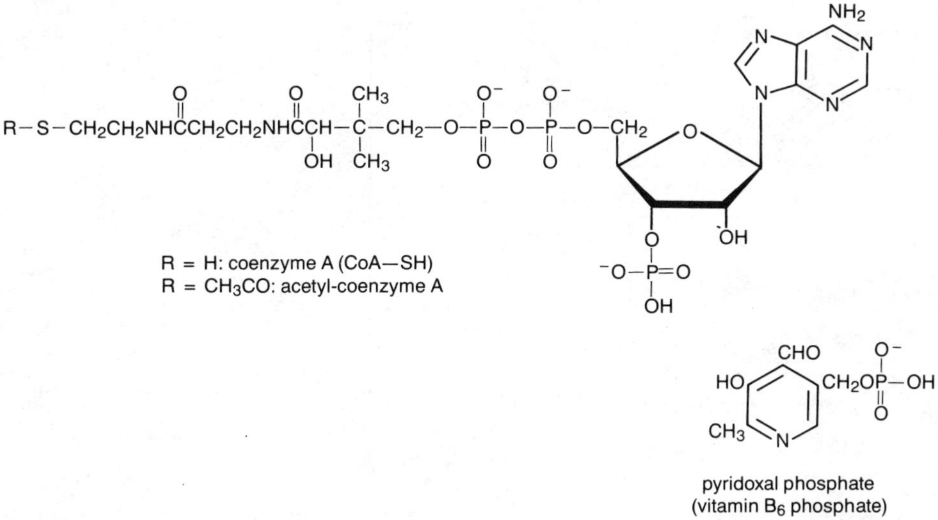

R = H: coenzyme A (CoA—SH)
R = CH$_3$CO: acetyl-coenzyme A

pyridoxal phosphate
(vitamin B$_6$ phosphate)

coenzymes
Figure 2. Structure of some coenzymes.

(i.e., hydrogen/electron transfer reactions) (*see* Table). The moiety actually involved in the reaction is indicated; the remaining part of the molecule is required for specificity. ADENOSINE TRIPHOSPHATE, biotin (*see* VITAMINS), coenzyme A and pyridoxal phosphate (*see* Figure 2) all act as coenzymes in various group transfer reactions (e.g., transaminations involving amino acids and α-keto acids).

coesite. *See* SILICA.

cofactor. *See* ENZYMES.

coherent anti-Stokes Raman scattering (CARS). *See* RAMAN SPECTROSCOPY.

coherent. Descriptive of ELECTRO-MAGNETIC RADIATION that has a single phase (e.g., as found in LASER radiation). Normal light, however, is not coherent.

coherent units. *See* SI UNITS.

cohesion. Work of cohesion is the work required to pull apart a column of a single liquid of unit cross-section and create two separate liquid–air interfaces ($w_{AA} = 2\gamma_A$). *See also* ADHESION.

coinage metals. Elements of group Ib (i.e. copper, silver and gold).

coke. *See* CARBON.

Coenzyme	Reaction

Nicotinamide adenine dinucleotide (phosphate)

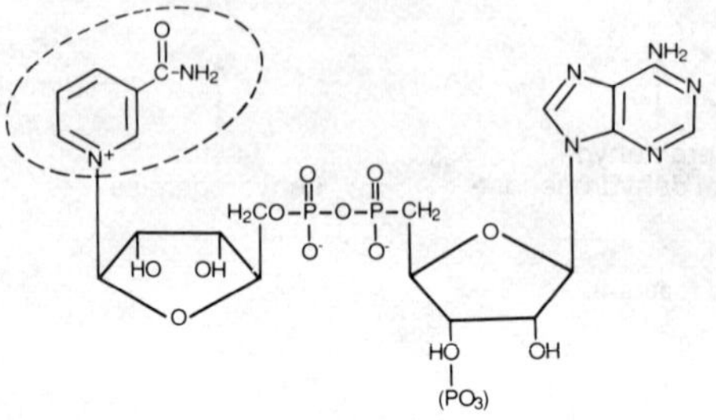

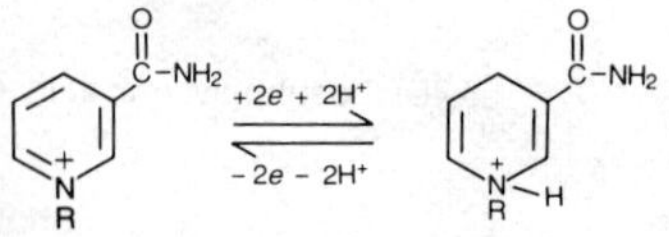

Flavin adenine nucleotide

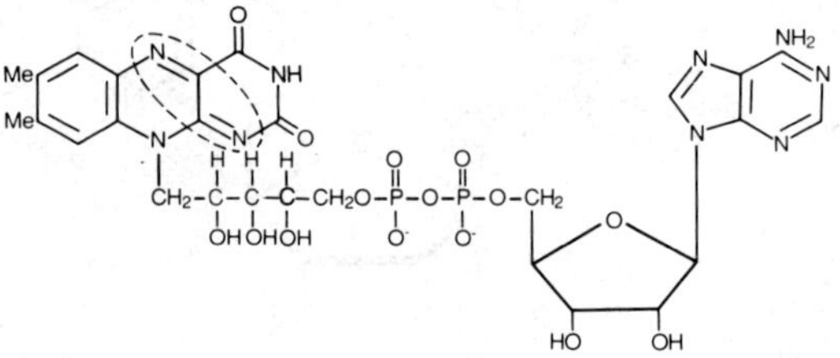

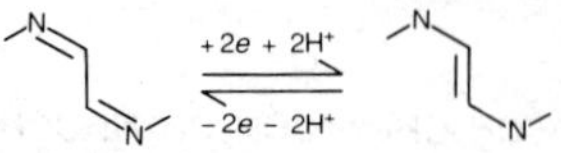

Coenzyme Q_{10}

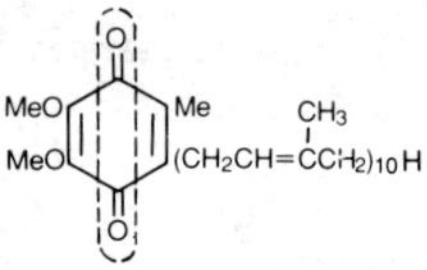

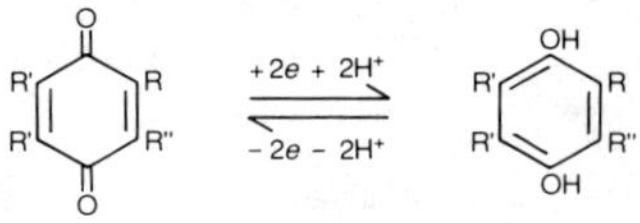

Cytochrome c

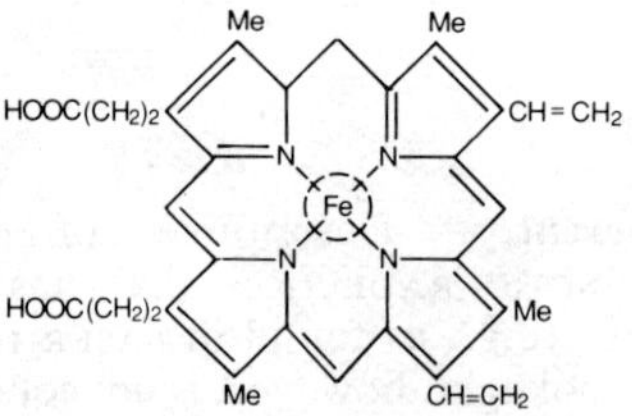

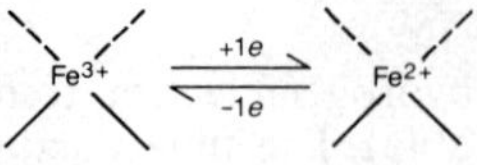

Lipoic Acid

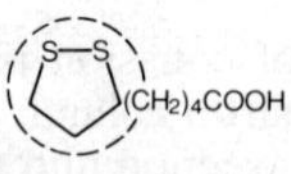

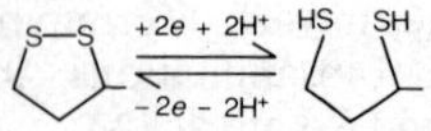

coenzymes
Coenzymes of electron transport and their redox reactions.

colemanite. *See* BORON.

colistin A. *See* ANTIBIOTICS.

collagen. *See* PROTEINS.

collidines. *See* METHYLPYRIDINES.

colligative property. Property that depends on the number of molecules or particles in the system rather than on their nature (e.g., the pressure of an ideal gas at constant temperature and pressure; elevation of BOILING POINT; depression of FREEZING POINT).

collision cross-section (σ; units: m^2). Area in which two molecules, radii (r_A, r_B) may collide

$$\sigma = \pi(r_A + r_B)^2$$

It is related to the MEAN FREE PATH. Typical values are: hydrogen 0.27 nm^2; benzene 0.88 nm^3. *See also* COLLISION THEORY.

collision frequency (Z; units: collisions s^{-1}). Number of collisions of a single molecule in 1 second.

$$Z = 2^{1/2}\pi d^2 \bar{c}(N/V)$$

where $\bar{c}$ is the average speed of the molecules. Numerical values for Z may be very high. *See also* COLLISION NUMBER.

collision number (Z_{11}, Z_{12}; units: collisions m^{-3} s^{-1}). If there are N_1 molecules in a volume V and each makes Z collisions per second (*see* COLLISION FREQUENCY), the total number of binary collisions between like molecules per unit volume per unit time is given by

$$Z_{11} = Z(N_1/V)/2$$
$$= \pi d^2 \bar{c}(N_1/V)^2/2^{1/2}$$
$$= \sigma \bar{c}(N_1/V)^2/2^{1/2}$$

The factor 1/2 is introduced to ensure that each collision between two molecules is only counted once. Substituting for the average MOLECULAR SPEED

$$Z_{11} = 2(N_1/V)^2 d^2 (\pi kT/m_1)^{1/2}$$

and for collisions between unlike molecules

$$Z_{12} = (N_1 N_2/V^2) r_{12}^2 (8\pi kT/\mu)^{1/2}$$

where $r_{12} = (d_1 + d_2)/2$ and μ is the REDUCED MASS. Numerical values for the collision number are very high: for example, nitrogen $d = 280$ pm at 298 K, $Z_{11} = 5 \times 10^{34}$ m^{-3} s^{-1}.

collision theory. Theory of the rates of bimolecular gas-phase reactions derived from first principles. The rate is the frequency of collisions multiplied by the fraction of collisions having an energy of at least the ACTIVATION ENERGY (E_a). For $A + B \rightarrow$ products, the RATE EQUATION is

$$\text{rate} = \sigma N_A(8kT/\pi\mu)^{1/2}[A][B] \times$$
$$\exp(-E_a/RT)$$

where μ is the REDUCED MASS and σ the COLLISION CROSS-SECTION. Experimentally determined rates of reactions are generally less than those calculated from the equation, a notable exception being the reaction $K + Br_2 \rightarrow KBr + Br$. The differences are thought to be due to steric factors which require a particular orientation of reactants to effect reaction. This is taken into account by replacing σ by σ^*, the reaction cross-section. The ratio σ^*/σ gives the steric factor P. *See also* ARRHENIUS EQUATION, RATE OF REACTION.

collodion. Solution of CELLULOSE NITRATE in ethanol or propanone. When applied to a surface and the solvent allowed to evaporate, a thin film remains.

colloid. Microheterogeneous system in which one component has dimensions in the region 1 nm–1 μm. Examples are AEROSOLS, inks, CEMENT, paints, DYES, FOAMS, PLASTICS, rubbers, soil. Three general classifications are recognized. (1) Colloidal dispersions. Thermodynamically unstable systems with high surface free energy which cannot be reconstituted after phase separation (*see* EMULSION, LYOPHOBIC COLLOID, SOL). (2) Solution of macromolecular materials (*see* POLYMERIZATION, POLYMERS). (3) Association colloids or COLLOIDAL ELECTROLYTES which form MICELLES (*see* LYOPHILIC COLLOID). Colloids are formed by the degradation of bulk matter (*see* COLLOID MILL) or aggregation of molecules and ions. Colloids cannot be filtered; they are separated by DIALYSIS (*see* ELECTRODIALYSIS, ULTRAFILTRATION). *See also* BROWNIAN MOTION, DONNAN MEMBRANE EQUILIBRIUM, SEDIMENTATION, TYNDALL EFFECT, ULTRACENTRIFUGE.

colloidal electrolyte. Solution of COLLOIDS that exhibits normal electrolyte behaviour (*see* CONDUCTANCE) until the critical micelle concentration is reached when deviation from the Debye–Hückel–Onsager relationship (*see* CONDUCTANCE EQUATIONS) is found. *See also* MICELLE.

colloid mill. Mechanical device that produces COLLOIDS by grinding, often in the presence of an inert dilutant which reduces the probability of aggregation. *See also* ELECTRODISPERSION.

colorimeter. *See* SPECTROPHOTOMETER.

Colour Index. Definitive reference work for DYES published by the Society of Dyers and Colourists (England) and the American Association of Textile Chemists and Colorists.

columbite–tantalite. *See* NIOBIUM.

columbium (Cb). *See* NIOBIUM.

column chromatography. Form of CHROMATOGRAPHY in which a column or tube is used to hold a solid phase or support a liquid stationary liquid phase. Gravity column chromatography is carried out in a glass column with diameters of several centimetres and length of up to 1 m. In GAS–LIQUID CHROMATOGRAPHY, stainless steel, aluminium, etc. tubes of internal diameter 2–6 mm and 1–4 m length and CAPILLARY COLUMNS are arranged in the form of a coil. Column temperature is an important variable that must be controlled.

column efficiency. *See* HEIGHT EQUIVALENT OF A THEORETICAL PLATE.

combination bands. Infrared absorption lines due to the sum, or difference, of two or more fundamental or harmonic vibrations. They are usually weak compared with the fundamental bands. *See also* INFRARED SPECTROSCOPY.

combustion. Rapid oxidation of a FUEL that occurs with, or without, a FLAME. Reactions that involve CHAIN REACTIONS of radicals, or in which the exothermicity causes runaway heating, may result in an

EXPLOSION. Gaseous mixtures of fuel and oxidant (usually, but not necessarily oxygen) may combust uniformly throughout or, if initiated in a small volume, pass as a wave through the volume (e.g., as in an internal combustion engine). Speeds of propagation of the wave range from about 0.3 m s^{-1} in hydrocarbon–air mixtures to 3 km s^{-1} in hydrogen–oxygen. Combustion temperatures are lowered in the presence of catalysts (e.g., platinum).

common ion effect. *See* SOLUBILITY PRODUCT.

completely miscible liquid systems. Binary liquid mixtures in which both components are volatile and completely miscible in the liquid phase. Two types of system exist. *zeotropic mixtures.* IDEAL SOLUTIONS in which RAOULT'S LAW is obeyed. The vapour pressure and boiling points of all mixtures lie between those of the pure components. Mixtures can be separated by distillation into the two pure components. Examples include benzene and toluene, 1,2-dibromoethane and 1,2-dibromopropane (*see* Figure 1).

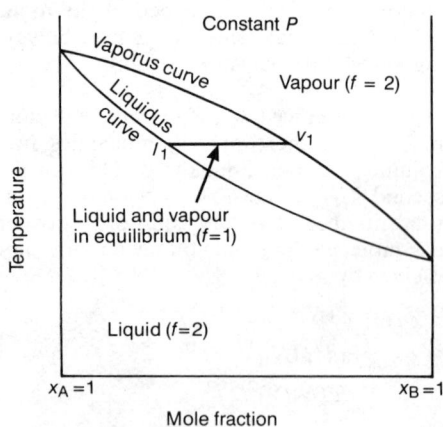

completely miscible liquid systems
Figure 1. Temperature–composition curves for zeotropic systems.

azeotropic mixtures. REAL SOLUTIONS in which the components are less similar, giving rise to mixtures which show positive or negative deviations from Raoult's law. Mixtures showing negative deviations (e.g., HCl/H_2O, HBr/H_2O) have a minimum in the vapour pressure–composition curve

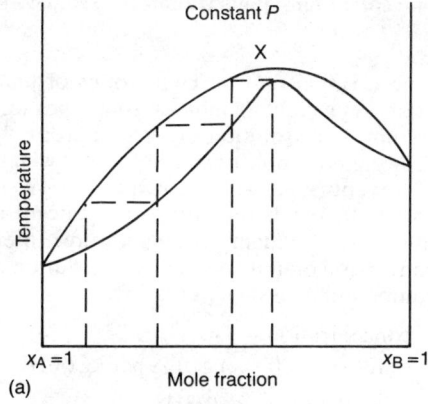

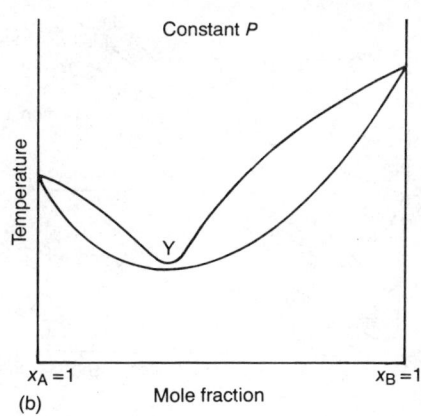

completely miscible liquid systems
Figure 2. Distillation diagrams for azeotropic mixtures.

(constant temperature) and a maximum in the temperature–composition curve (constant pressure). Fractional distillation gives a residue of azeotrope X (*see* Figure 2(a)); the distillate is the component present in excess. Mixtures showing positive deviations (e.g., ethanol/H_2O, dioxan/H_2O) have a maximum in the vapour pressure–composition curve and a minimum in the temperature–composition curve. Fractional distillation gives a residue of the component present in excess; the distillate is the azeotrope Y (*see* Figure 2(b)).

complex compound. Compound formed between chemically distinct species in which a COORDINATE BOND is formed. A complex may be charged, or a neutral compound or an intermediate in a chemical reaction (*see* ACTIVATED COMPLEX). The term is widely used to refer to a COORDINATION COMPOUND.

complexometric indicator. *See* INDICATORS.

complexometric titration. *See* TITRATIONS.

component. Number of components (c) of a system at equilibrium is the smallest number of independently variable constituents (molecular species), in terms of which the composition of each PHASE can be quantitatively defined (zero and negative quantities of components are permissible).

For the ice/water/water vapour system, $c = 1$ (i.e. the molecular species H_2O); for the equilibrium $CaCO_3 \rightleftharpoons CaO(s) + CO_2(g)$, $c = 2$. The number of components for a system is fixed; its value may change by varying the conditions (e.g., temperature).

composition–temperature diagrams. *See* COMPLETELY MISCIBLE LIQUID SYSTEMS, TWO-COMPONENT CONDENSED SYSTEM.

compressibility coefficient (β). Defined as

$$\beta = -(1/V)(\partial V/\partial P)_T$$

where $(\partial V/\partial P)_T$ is the rate of change of volume with pressure at constant temperature. *See also* HEAT CAPACITY.

compression factor (Z). Defined as

$$Z = PV_m/RT$$

For an IDEAL GAS Z is unity under all conditions; deviation from unity is a measure of imperfection. For real gases at very low pressures Z is close to unity. At high pressure $Z > 1$, which means they are harder to compress; this is the region where the repulsive forces are dominant. At lower pressures $Z < 1$, indicating that attractive long-range forces are dominating and favouring compression. Values of Z plotted as a function of the reduced pressure are, however, approximately the same for a wide variety of gases. The compression factor calculated from the CRITICAL CONSTANTS ($Z_c = P_cV_{m,c}/RT_c$) should be constant for all gases. Experimental values of

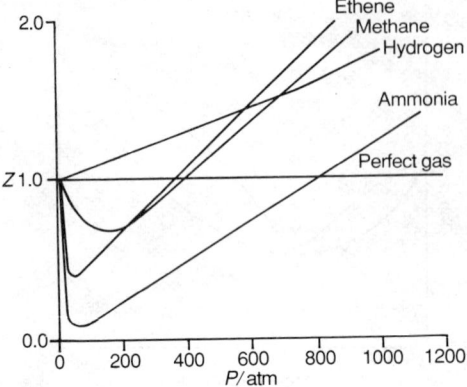

compression factor
Compression factor $Z = PV_m/RT$ for gases at 0°C.

Z_c are in the range 0.26–0.30; calculated values are listed elsewhere (*see* EQUATION OF STATE, Table).

concentration cells. CELLS in which there is no overall chemical reaction; the emf results from differences in concentration of one or more reactants. The general types of concentration cell recognised are as follows.
concentration cell with transport. Direct transfer of ions across a liquid junction occurs. The junction may be made with a SALT BRIDGE or the liquids may be allowed to be in direct contact. In the former, for a cell reaction involving univalent ions, the emf of the cell is given by

$$E_{cell} = RT/F \ln (a_1/a_2),$$

and in the latter by

$$E_{cell} = 2tRT/F \ln (a_1/a_2),$$

where a_1 and a_2 are the mean ionic activities at each side of the junction, and t is the TRANSPORT NUMBER of the ion to which the electrodes are not reversible.
concentration cell without transport. Transport of material occurs indirectly as a result of electrochemical reactions. The emf of such a cell is equal to $2RT/F \ln (a_1/a_2)$. Such cells are used to measure ACTIVITY values and test the validity of the DEBYE–HÜCKEL ACTIVITY COEFFICIENT EQUATION.
electrode concentration cell. Results from different partial pressures of gases at each electrode, or from different concentrations of metal in amalgam electrodes.

concentration polarization. *See* OVERPOTENTIAL.

concentration units. Two groups of units exist: (1) weight or mole of solute per unit volume of solution (e.g., $mol\,dm^{-3}$); (2) weight or mole of solute per unit weight of solvent (e.g., wt%, MOLAL, MOLE FRACTION). In the latter group the concentration is independent of temperature. Interconversion of units, for a w wt% solution of compound A in solvent B are:

concentration =
$1000w/(100 - w)$ g of A per kg of B

$c_A/mol\ dm^{-3} = w\rho/100M_{r,A}$

$c_A/mol\ m^{-3} = 1000w\rho/100M_{r,A}$

molality $(m_A) = 1000w/(100 - w)\,M_{r,A}]$

mole fraction $(x_A) =$
$(w/M_{r,A})/[w/M_{r,A} + (100 - w)/M_{r,B}]$

concerted reaction. Reaction that is completed in a single step through a single transition state. Thus the concerted enolization of propanone (*see* KETO–ENOL TAUTOMERISM) occurs by simultaneous attack of a base (B) and acid (A)

$$B \ldots H–CH_2–C=O \ldots HA \longrightarrow$$
$$\overset{|}{\underset{CH_3}{}}$$

$$BH^+ + CH_2=C–OH + A^-$$
$$\overset{|}{\underset{CH_3}{}}$$

S_N2 MECHANISMS are concerted, but S_N1 MECHANISMS, which involve an intermediate CARBONIUM IONS, are not. *See also* NUCLEOPHILIC SUBSTITUTION.

condensation. (1) Chemical reaction (e.g., CLAISEN CONDENSATION) in which two or more molecules combine to form a larger molecule with the elimination of a small molecule (e.g., water) (*see* CONDENSATION REACTION). (2) Change of state from the vapour to the liquid or solid form.

condensation polymerization. *See* POLYMERIZATION.

condensation reaction. Combination of two molecules, or two parts of the same molecule, usually with the elimination of a small molecule such as water, an alcohol or ammonia. ALDEHYDES and KETONES undergo a variety of condensation reactions (*see* ACYLOINS, ALDOL CONDENSATION, BENZALDEHYDE, CLAISEN CONDENSATION, CLAISEN–SCHMIDT REACTION, CYANO-ETHYLATION, 2,3-DIMETHYL-2,3-BUTANE-DIOL, MANNICH REACTION, MICHAEL CONDENSATION, PERKIN REACTION, POLY-MERIZATION, REFORMATSKY REACTION).

condensed ring system. Extension of the benzene structure in which two or more rings are fused together with two carbon atoms in common (e.g., ANTHRACENE, HETEROCYCLIC COMPOUNDS, NAPHTHA-LENE, POLYCYCLIC HYDROCARBONS).

condensed surface film. *See* SURFACE FILM.

condensed system. One composed of only liquid and/or solid phases. *See also* PHASE RULE, THREE-COMPONENT SYSTEM, TWO-COMPONENT SYSTEM.

conductance (G; dimensions: $\varepsilon l t^{-2}$; units: siemens, $S = \Omega^{-1} = kg^{-1} m^{-2} s^3 A^2$). Reciprocal of electric resistance. The resistance (R) of a conductor is proportional to its length (L) and inversely proportional to its cross-section (A), that is $R = \rho(L/A)$. Thus

$$G = 1/R = 1/\rho(L/A) = \kappa/(L/A)$$

where the CONDUCTIVITY (κ) is the reciprocal of the resistivity (ρ).

conductance equations. MOLAR CONDUC-TIVITY of strong electrolytes follows a square root relationship with concentration (i.e. $\Lambda = \Lambda^0 - Sc^{1/2}$). This was first noted by Kohlrausch and subsequently explained theoretically in terms of interionic attraction by Debye, Hückel and Onsager. The slope S is shown to be $A + B\Lambda^0$, where A and B are constants that depend on the nature of the solvent and temperature alone. The equation may be extended to higher concentrations by taking into account the finite size of ions when $\Lambda = \Lambda^0 - Sc^{1/2}/(1 + Cc^{1/2})$. The molar conductance of weak electrolytes is given by OSTWALD'S DILUTION LAW. In non-aqueous

solutions, an extension of the Onsager equation holds for completely dissociated electrolytes:

$$\Lambda = \Lambda^0 + Sc^{1/2} + Ec \ln c + J_1c - J_2c^{3/2}$$

where E, J_1 and J_2 are complex functions of the properties of the solutions.

conductance of aqueous solutions. *See* CON-DUCTANCE EQUATIONS, MOLAR CONDUCTIV-ITY, OSTWALD'S DILUTION LAW.

conductimetric titration. *See* TITRATIONS.

conduction band. *See* BAND THEORY OF SOLIDS.

conductivity (κ; dimensions: ε t^{-1}; units: $\Omega^{-1} m^{-1}$). Current flowing through unit area per unit potential gradient; formally called the specific conductance. It is the reciprocal of resistivity (ρ) (i.e. $\kappa = 1/\rho = (L/A)/R = JG$). In conductance cells, the term J is known as the cell constant and is determined using a solution of known conductivity. *See also* MOLAR CONDUCTIV-ITY.

configuration. Spatial arrangement of atoms or groups in a molecule. A change in configuration requires the breaking of a bond and reforming in a different way. In contrast, a change of conformation requires only rotation about a single bond. *See also* CONFORMERS, ISOMERISM.

configuration interaction (CI). Process of mixing electronic states performed in a MOLECULAR ORBITAL CALCULATION, usually after a SELF-CONSISTENT FIELD calculation. A commonly used set of states of a CLOSED SHELL MOLECULE is the 3×3 CI which is the GROUND STATE and first two excited singlet states. CI may be seen as a method of mixing in contributions from ionic forms as is done in VALENCE BOND ORBITAL.

conformers (conformational isomers). Molecules that differ from one another only by rotation about a single bond (rotational isomers). During the rotation of one methyl group in ethane relative to the other giving rise to the eclipsed and staggered conformations at 60° intervals (*see* NEWMAN PROJEC-TION FORMULA) the energy varies with the DIHEDRAL ANGLE in a sinusoidal manner; at

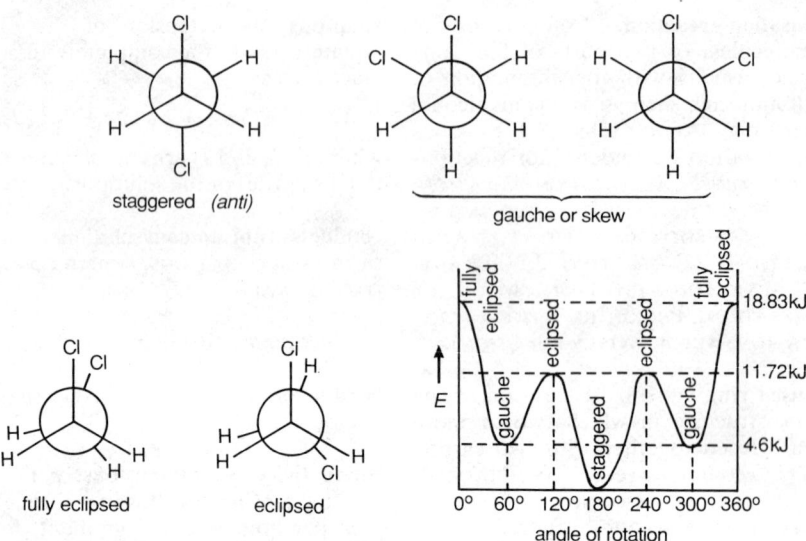

conformers
Figure 1. Rotational energy and conformations of 1,2-dichloroethane.

room temperature the molecules spend most of their time rocking in the potential energy trough (energy barrier = 11 kJ mol^{-1}). Neither conformation can be isolated.

The 60° and 180° conformations of, for example 1,2-dichloroethane are not identical; the 60° (and 300°) arrangement is known as the *gauche*- and the 180° the *anti*-conformation. The *anti*-conformation, with the lowest rotational energy, is the preferred one (i.e. the one in which the molecule largely remains) (*see* Figure 1).

Another example of conformational differences is shown by the CYCLOALKANES (C_nH_{2n}). Cyclohexane exists in two forms, the chair and the boat conformations; neither is completely strainless. The chair form is the stable conformer; it is rigid in the sense that to change the dihedral angle about any bond requires a simultaneous change in one or more bond angles in the molecule. Cyclohexane has only one important conformation in contrast to the normal alkanes. In the chair form there are two equivalent classes of hydrogen atoms: the six axial (*a*) carbon–hydrogen bonds are parallel to one another and to the symmetry axis of the molecule, the six equatorial (*e*) carbon–hydrogen bonds are roughly in the general plane of the ring. On alternate carbon atoms the axial hydrogen points upwards (downwards) and the equatorial hydrogen slightly downwards (upwards) (*see* Figure 2). A monosubstituted cyclohexane assumes the chair conformation in which the substituent occupies the equatorial position. *cis*-1,2-Disubstituted cyclohexanes are optically inactive by external compensation, whereas the *trans*-isomer exists as a pair of enantiomers.

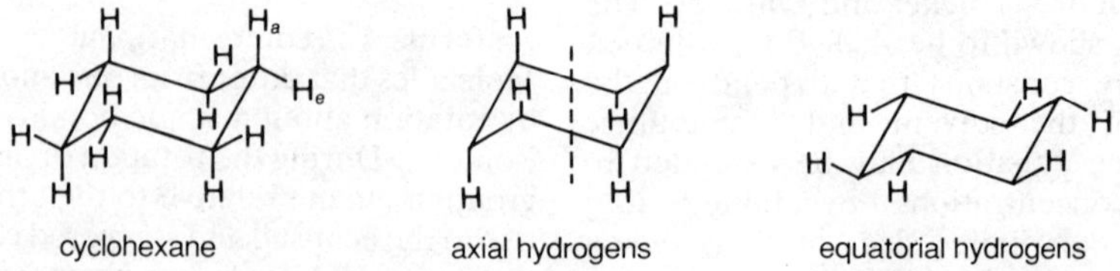

conformers
Figure 2. Conformation of cyclohexane.

congruent melting points. *See* TWO-COM-PONENT CONDENSED SYSTEM.

coniferin. *See* GLYCOSIDES.

coniferyl alcohol (4-hydroxy-3-methoxy-cinnamyl alcohol, $CH_3OC_6H_3(OH)CH=CHCH_2OH$). Constituent of the glucoside coniferin (*see* GLYCOSIDES). It is oxidized to VANILLIN.

coniine. *See* ALKALOIDS.

conjugate acids and bases. *See* ACIDS AND BASES.

conjugate solution. *See* PARTIALLY MISC-IBLE LIQUID SYSTEMS.

conjugate ternary solution. *See* THREE-COMPONENT SYSTEM.

conjugation. Alternating double (or triple) and single bonds (*see* Figure).

conrotation. In reactions of conjugated organic molecules, the rotation of π-orbitals in the same direction to form a σ-orbital. In disrotatory motion, the orbitals move in the opposite direction.

consecutive reactions. Reactions of a molecule that occur one after the other. The overall rate is determined by the slowest reaction (*see* RATE-DETERMINING STEP). An analytical solution for all but the most simple cases is impossible, and numerical solution of the INTEGRATED RATE EQUA-TION for two consecutive first-order reactions (*see* Table) is required.

conservation of energy. *See* FIRST LAW OF THERMODYNAMICS.

conservation of mass–energy. *See* FIRST LAW OF THERMODYNAMICS.

consolute temperature. *See* PARTIALLY MISCIBLE LIQUID SYSTEMS.

constantan. ALLOY of nickel (45 percent) and copper (55 percent) used in resistances and thermocouples.

constant boiling point solution. *See* AZEO-TROPIC MIXTURE.

constant composition, law of (constant proportions, law of; definite proportions, law of). A pure chemical compound always contains elements in the same proportions irrespective of the method of preparation. *See also* CHEMICAL COMBINATION, LAWS OF.

constant heat summation, law of. *See* HESS'S LAW.

constant proportions, law of. *See* CON-STANT COMPOSITION, LAW OF.

constititutive property. Property that depends mainly upon the structural arrangement of atoms or molecules (e.g., BOILING POINT, PARACHOR).

contact process. Manufacture of sulphuric(VI) acid by passing sulphur(IV) oxide (SO_2) and oxygen over vanadium–vanadium oxide catalysts at 450°C. Sulphur(VI) oxide (SO_3) so formed is dissolved in sulphuric acid to give oleum which is then diluted. *See also* SULPHUR OXIDES, SULPHUR OXOACIDS.

continuous spectrum. Broad continuous spectrum emitted by an incandescent solid. In some cases the spectrum approximates that of a black body, and such materials may be used as sources for INFRARED SPECTRO-SCOPY.

cooling curve. *See* THERMAL ANALYSIS.

coordinate bond (dative bond). Electron pair bond between two atoms (donor and ACCEPTOR) in which both electrons are contributed by one atom. Thus in the formation of the ammonium ion, the bond between ammonia and the proton (which

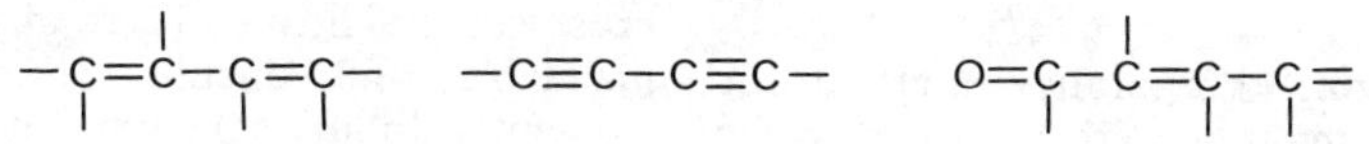

$$-\overset{|}{\underset{|}{C}}=\overset{|}{\underset{|}{C}}-\overset{|}{\underset{|}{C}}=\overset{|}{\underset{|}{C}}- \quad -C\equiv C-C\equiv C- \quad O=\overset{|}{\underset{|}{C}}-\overset{|}{\underset{|}{C}}=\overset{|}{\underset{|}{C}}-\overset{|}{\underset{|}{C}}=$$

conjugation

has no electrons) is formed from the lone pair of electrons on nitrogen. LEWIS ACIDS AND BASES are defined in terms of acceptors and donors of electron pairs. *See also* COVALENT BOND, IONIC BOND, VALENCE.

coordination compound. Inorganic COMPLEX COMPOUND containing a COORDINATE BOND. The bonding in coordination compounds may be described by LIGAND FIELD THEORY. They are named with the cation first followed by the anion. With complex ions, charged ligands are named before neutral ligands with the element or ion name having the ending -o (e.g., chloro, but water is aquo and NH₂ is ammine). The metal is named last, the element name is used for neutral or anionic complexes, the ending -ate is for cationic complexes. The OXIDATION STATE of the metal appears in parentheses after the name. The stability of complexes generally increases as the central ion increases in charge and electron affinity and decreases in size. This leads to an order of stability of transition metal divalent ions of manganese < iron < cobalt < nickel < copper. In terms of the ligand, stability increases with basicity, electron density (small ligand anions are favoured) and amount of chelation (*see* CHELATE LIGAND) (polydentate ligand complexes are more stable than monodentate ligand complexes); steric factors are also important.

Reactions of coordination compounds in which one ligand is replaced by another are classed as: A (association), which is a concerted reaction similar to S_N2 MECHANISMS in organic chemistry; I (interchange), in which a group is displaced directly by a group outside the coordination sphere; D (dissociation), in which a ligand leaves before the incoming group arrives (similar to S_N1 MECHANISM).

coordination isomerism. *See* ISOMERISM.

coordination number. Number of nearest neighbours of an atom or ion in a lattice, or the number of species bound to a COORDINATION COMPOUND.

Cope reaction. Preparative method for ALKENES from amine oxides by a cyclic elimination.

$$\begin{array}{c} \quad\quad H \\ \quad\quad / \\ RCH \quad O^- \\ \mid \quad\quad \mid \quad\quad \xrightarrow{150°C} RHC{=}CH_2 + \\ H_2C{-}N^+(CH_3)_2 \quad (CH_3)_2NOH \end{array}$$

copolymerization. Formation of a POLYMER from different monomers. ETHENE and PROPENE may be copolymerized to give a material which has properties different from each homopolymer. Four types of copolymer may be made. In random copolymers the different 'mers' occur in no particular order in the chain. In alternating copolymers (e.g., Nylon-6,6, *see* NYLON) the mers occur one after the other, whereas in block copolymers sections of the chain are of one polymer followed by a section of the other. Graft copolymers consist of a chain of one mer linked by chains of the second mer.

copolymers. *See* COPOLYMERIZATION.

copper (Cu). First row, group Ib TRANSITION ELEMENT. It occurs naturally in the metallic state, or as sulphide ores chalcocite (Cu_2S), covellite (CuS), chalcopyrite ($CuFeS_2$), bornite (Cu_5FeS_4) and enargite (Cu_3AsS_4). Semiprecious ores are green malachite ($CuCO_3.Cu(OH)_2$) and blue azurite ($2CuCO_3.Cu(OH)_2$). Ores are concentrated by flotation to 20–40 percent copper and may be roasted to lower the sulphur content. Heating in a reverbatory furnace produces a matte of copper suphide which is oxidized, yielding sulphur dioxide and crude copper. The impure copper is refined electrolytically. Copper is the most important non-ferrous metal. Its high electrical conductivity ($6.0 \times 10^7\,Sm^{-1}$) leads to its use in electrical wiring. Alloys (BRASS, copper–zinc; BRONZE, copper–tin) are valued for their corrosion resistance. Other uses include CATALYSIS, and as fungicides and pigments. Copper is a biologically important element, being present in some ENZYMES and in the oxygen carrier of some crustaceans. The metal is reddish, lustrous, malleable and ductile. It reacts with oxygen, sulphur, selenium and halogens, and dissolves in oxidizing acids such as nitric acid. On exposure to the atmosphere, copper or bronze is covered with a patina of verdigris, a green basic copper carbonate. $A_r = 63.546$, $Z = 29$, ES [Ar] $3d^{10} 4s^2$, mp 1083°C, bp 2567°C, $\rho = 8.92 \times 10^3$ kg m^{-3}. Isotopes: ^{63}Cu, 69 percent; ^{65}Cu, 31 percent.

copper compounds. Oxidation states +1, +2 and +3 are known. The aqueous chemistry is largely that of copper(II), the blue six-coordinate complexes of which show the JAHN–TELLER EFFECT. Copper(I) iodide (CuI) readily disproportionates (*see* DIS-PROPORTIONATION) to copper(II) and the metal in aqueous solution (K for 2Cu$^+ \rightarrow$ Cu^0 + Cu^{2+} is 1×10^7). Copper(I) (cuprous) salts are similar to those of Ag(I). The halides are pale yellow or colourless and insoluble; the oxide, sulphate and cyanide are known. Organic copper derivatives are of copper(I). Alkyls and aryls are known, as are complexes (e.g., (PhCu)$_2$, Cu$_4$I$_4$(PMe$_3$)$_4$). Copper(II) (cupric) forms stable salts of all the mineral acids, and many complexes are known, especially with N-containing ligands. The cuprammonium ion ([Cu(NH$_3$)$_4$]$^{2+}$) is formed by the addition of excess ammonia to a solution of a copper(II) salt. The solution dissolves cellulose and is used in the manufacture of rayon (*see* FIBRES). Basic copper ethanoate ([Cu(CH$_3$COO)$_2$.H$_2$O]$_2$) is the pigment verdigris (not to be confused with the corrosion patina, *see* COPPER).

copper phthalocyanin. *See* PHTHALOCYA-NINS.

copper pyrites (chalocopyrite). *See* COPPER, SULPHATES.

coprecipitation. Contamination of a precipitate by the inclusion or adsorption of material that is normally soluble in the MOTHER LIQUOR.

cordite. *See* EXPLOSIVES.

core electron. Electron occupying inner ORBITAL of an atom that is not normally involved in chemical bonding. *See also* ELECTRONIC CONFIGURATION, VALENCE.

coronene. *See* POLYCYCLIC HYDROCAR-BONS (structure XXV).

correlation diagram. *See* WALSH DIA-GRAM.

corresponding states. *See* REDUCED EQUATION OF STATE.

corrosion. Oxidation of a metal by the formation of a short-circuited electrochemical cell. The cathodic process, which also occurs on the metal surface, may be the reduction of oxygen (O$_2$ + 2H$_2$O + 4e $\rightarrow$ 4OH$^-$) or the evolution of hydrogen (2H$_2$O + 2e $\rightarrow$ H$_2$ + 2OH$^-$). The rate of corrosion is generally determined by the cathodic reaction (e.g., the availability of atmospheric oxygen). Corrosion may be prevented by coating the metal surface to exclude oxygen by painting, phosphating or electroplating, by applying a sufficiently negative potential to the metal in cathodic protection or by applying a positive potential to a metal that shows PASSIVITY (anodic protection). In aqueous media additives, such as arsenic compounds, may inhibit the cathodic reaction.

corticosteroids. Group of STEROIDS synthesized in the mammalian adrenal cortex from cholesterol; some exhibit hormone activity. They possess 21 carbon atoms with a double bond at C-4 and ketonic groups at C-3 and C-20 (*see* Table).
glucocorticosteroids. Principal action is on carbohydrate, fat and protein metabolism; promoting the deposition of glycogen in the liver, formation of glucose from tissue protein and mobilization of fat. Hydro-cortisone (17-hydroxycorticosterone, corti-sol), the main glucocorticosteroid hormone, is used extensively in the treatment of acute inflammation of various tissues. Cortisone (mp 215°C) is manufactured from the SAPO-NIN, diosgenin. It is used beneficially in the treatment of rheumatoid arthritis; the 9-fluoro derivative (dexamethasone) has a higher activity and fewer side effects than either cortisone or hydrocortisone.
mineralocorticoids. Group of hormones whose principal action is on water and electrolyte metabolism. Aldosterone causes the retention of sodium and is probably the principal hormonal factor

in maintaining electrolyte balance in mammals.

X	R	Compound
=O	-OH	Cortisone
-OH	-OH	Hydrocortisone
-H	-H	Deoxycorticosterone
-OH	-H	Corticosterone
=O	-H	11-Dehydrocorticosterone
-H	-OH	17-Hydroxydeoxycorticosterone
-OH	-H	Aldosterone[a]

a CH₃* replaced by -CHO.

corticosteroids

cortisol (hydrocortisone, 17-hydroxycorticosterone). *See* CORTICOSTEROIDS.

cortisone. *See* CORTICOSTEROIDS.

corundum. Rhombohedral form of aluminium oxide (*see* ALUMINIUM INORGANIC COMPOUNDS). Precious and semiprecious forms are oriental topaz (yellow), sapphire (blue, due to Co, Cr, Ti and Fe), ruby (red, due to Cr(III)), oriental amethyst (violet, due to Mn) and oriental emerald (green, due to Cr).

Cotton effect. *See* OPTICAL ROTATORY DISPERSION.

coulomb (C; dimensions: $\varepsilon^{1/2} m^{1/2} l^{3/2} t^{-1}$; units: C = A s). SI unit of charge defined as the quantity of electricity transported in 1 second by a current of 1 ampere.

Coulomb's law. Force between two charges ez_1 C and ez_2 C separated by r m in a medium of relative permittivity ε is

$$F(r) = z_1 z_2 e^2 / 4\pi\varepsilon_0 \varepsilon_r^2$$

where ε_0 is the permittivity of free space.

The potential r m away from a charge ez_1 is

$$V(r) = -z_1 e / 4\pi\varepsilon_0 \varepsilon_r$$

See also INTERMOLECULAR FORCES.

coulometer. Device for measuring the quantity of electricity passed. Formally a cell in which a reaction occurred quantitatively (e.g., the deposition of silver or copper, the reduction of iodine) was connected in series with the source of current. Now the amount of charge is determined by the electronic integration of the current.

coulometric titration. TITRATION in which the TITRANT is generated electrolytically, the end-point being detected either with an INDICATOR or other instrumental method. The amount of titrant generated is determined directly from the quantity of electricity passed (*see* FARADAY'S LAWS OF ELECTROLYSIS). For example, in the titration of a base, water is used for the generation of protons which immediately react with the base. Advantages are no standard solutions are required, unstable reagents (e.g., chlorine) can be generated and used immediately, microadditions of reagent can be made, and there is no dilution of the sample.

coumarin ($C_9H_6O_2$). Colourless solid (mp 68-70°C), prepared by heating 2-hydroxybenzaldehyde with ethanoic anhydride and sodium ethanoate. With GRIGNARD REAGENTS it gives benzopyrilium salts. Alkaline hydrolysis gives 2-coumaric acid (*trans*-2-hydroxycinnamic acid) which on irradiation produces the unstable coumarinic acid (*cis*-2-hydroxycinnamic acid), which spontaneously forms the δ-lactone, coumarin. Coumarin is widely distributed in plants as the glycoside. It is used as a natural perfume and flavouring material. WARFARIN is a coumarin derivative.

coumarin

coumarone. *See* BENZFURAN.

count. External indication given by a radioactive detector (e.g., a Geiger counter) of the amount of radioactivity to which the detector is exposed. Background counts are those from a source external to that being measured.

counter. In radiation, instrument for detecting and counting incident charged particles or photons. The particles cause ionization within the counter which then creates a current or voltage pulse. The pulses are counted electronically in ancillary equipment known as a scaler. Typical detectors include Geiger, proportional and scintillation counters.

countercurrent distribution. Parent of partition chromatography (*see* CHROMATOGRAPHY, PAPER CHROMATOGRAPHY) in which a mixture is shaken with equal volumes of two immiscible solvents and allowed to stand. The components distribute themselves between the two layers according to their PARTITION COEFFICIENTS. The two layers are separated, and each is further treated with more of the other pure solvent and repartitioning allowed to occur. If the process is repeated many times, material preferentially soluble in one solvent is concentrated in one set of fractions and that soluble in the other in another set of fractions. It is a tedious method when carried out manually; the Craig machine, developed to automate the procedure, has been superseded by partition chromatography.

counter ion. (1) Oppositely charged ion that balances the charge in an ELECTROLYTE. (2) Ion that compensates the charge in a LYOPHOBIC COLLOID.

coupling constant. *See* NUCLEAR MAGNETIC RESONANCE SPECTROSCOPY.

covalent bond. Bond formed from two atoms of similar ELECTRONEGATIVITY in which the electrons of the bond are shared equally between the atoms. *See also* BOND LENGTH, HYBRIDIZATION.

covalent crystal. *See* CRYSTAL.

covalent radius. *See* BOND LENGTH.

covellite. *See* COPPER.

covolume. *See* VAN DER WAALS EQUATION.

C_P. HEAT CAPACITY at constant pressure.

cp (cyclopentadienyl). *See* METALLOCENES.

Cr. *See* CHROMIUM.

cracking. Process that reduces the molecular mass of hydrocarbons from PETROLEUM. Thermal cracking is performed at increasing temperatures depending on the input oil and the desired products. Visbreaking at 450–500°C converts 20–25 percent to lighter fractions. Naphtha cracking (500–600°C) is more severe (45 percent conversion), and steam cracking (600–750°C) gives low-molecular-mass alkenes. A more important industrial process for producing high-octane petrol (*see* OCTANE NUMBER) is catalytic cracking in a FLUIDIZED BED REACTOR of molecular sieve ZEOLITES (e.g., zeolite Y). *See also* HYDROCRACKING, REFORMING.

cream of tartar. *See* 2,3-DIHYDROXYBUTANEDIOIC ACID.

creatine ((α-methylguanidino)ethanoic acid, I). Weakly basic nitrogenous compound present in the muscles of all vertebrates. Its phosphate, phosphocreatine, is an important energy reserve giving a large free energy decrease on hydrolysis. It is dehydrated to its internal anhydride creatinine (II), the end-product of creatine catabolism and a major nitrogenous constituent of urine.

$$H_2NC(NH)NMeCH_2COOH \xrightarrow{-H_2O}$$

(I)   (II)

creatine

creatinine. *See* CREATINE.

creep. Deformation of a material under constant load. In contrast, an ELASTIC solid only deforms when the load is being applied.

creosote. Fractions obtained in the distillation of coal tar; a dark liquid mixture of hydrocarbons, phenols, cresols and other aromatic compounds with a characteristic odour. It is used for preserving timber. Medical creosote, an almost colourless liquid mixture of phenols, mainly guaiacol (*see* AROMATIC ETHERS) and cresol (*see* HYDROXYTOLUENES), is obtained by the destructive distillation of wood. It is used as an antiseptic (less toxic than phenol) and as an expectorant.

cresol red. *See* INDICATORS.

cresols. *See* HYDROXYTOLUENES.

cresylic acids. Mixture of cresols, xylenols and other phenolic compounds obtained from the distillation of coal.

cristabolite. *See* SILICA.

critical constants. Values of pressure (P_c), volume (V_c) and temperature (T_c) when the gas and liquid phases of a fluid have the same DENSITY. The critical temperature is the temperature above which a gas cannot be liquefied by an increase in pressure. The critical volume is the volume of a fixed mass of fluid at T_c and P_c. The critical specific volume is its volume per unit mass in this state. Values of the critical constants can be calculated from the appropriate EQUATION OF STATE and determined experimentally from the P–V–T curves of a REAL GAS. When a liquid in a sealed tube is heated a temperature is reached at which the meniscus disappears, this is T_c; P_c can be measured if the tube is connected to a manometer. The critical volume can be obtained from measurements of the densities of the liquid and vapour in equilibrium (the orthobaric densities) at different temperatures; at T_c the densities are equal, and hence V_c can be calculated.

critical micelle concentration. *See* MICELLE.

critical solution temperature. *See* PARTIALLY MISCIBLE LIQUID SYSTEMS.

critical state. *See* CRITICAL CONSTANTS, REAL GAS.

cross-linked polymers. *See* POLYMERIZATION, POLYMERS.

cross-section (σ; units: BARN, m^2). Apparent or effective area presented by a target nucleus or particle to oncoming radiation. It is a measure of the probability that a nuclear reaction will occur. If N target atoms per m^3 are bombarded with I_0 particles the number transmitted is

$$I = I_0 \exp(-N\sigma x)$$

where x m is the thickness of the sample. *See also* COLLISION CROSS-SECTION.

crotonaldehydes. *See* 2-BUTENALS.

crotonic acids. *See* 2-BUTENOIC ACIDS.

crotyl alcohol. *See* 2-BUTEN-1-OL.

crown ethers. Macrocyclic polyethers. They are named as *n*-crown-*m*, where *n* is the total number of atoms and *m* is the number of oxygen atoms. The importance of crown ethers lies in their ability to solvate cations in non-aqueous solutions. For example, 12-crown-4 (*see* Figure) is specific for the lithium cation. The binding of a cation leaves the anion 'naked' which is then available for reaction. *See also* CRYPTANDS.

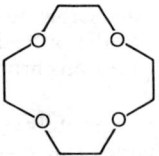

crown ethers
12-Crown-4.

crude oil. *See* PETROLEUM.

Crum–Brown's rule. Empirical rule predicting the course of aromatic substitution. A group X in a benzene ring will be *meta*-directing if the compound HX can be readily oxidized to HOX, otherwise a mixture of *ortho*- and *para*-isomers will be formed, thus HNO$_2$ → HONO$_2$ (–NO$_2$ is *meta*-directing), but HOH $\nrightarrow$ HO–OH (–OH is *ortho*,*para*-directing). This rule is not universally applicable. *See also* ORIENTATION IN THE BENZENE RING.

cryogenics. Study of very low temperatures and the techniques used for achieving them. A liquid gas provides a constant bath temperature from its triple point to its critical temperature. *See also* ADIABATIC DEMAGNETIZATION, REFRIGERATION.

cryohydrate. EUTECTIC MIXTURE of a TWO-COMPONENT CONDENSED SYSTEM in which one component is water (e.g., sodium chloride/water; cryohydric point, –22°C).

cryohydric point. *See* CRYOHYDRATE, EUTECTIC MIXTURE.

cryolite. *See* ALUMINIUM, FLUORINE.

cryometer. THERMOMETER designed to measure low temperatures. At temperatures down to 1 K, thermocouples can be used and to 0.01 K resistance thermometers. Below this magnetic or nuclear resonance detectors are required.

cryoscopic constant. *See* FREEZING POINT.

cryoscopy. Determination of RELATIVE MOLECULAR MASS by the FREEZING POINT depression method.

cryptands. Macropolycyclic polyaza–polyethers in which nitrogen atoms are the vertices of a three-dimensional structure containing $-O-CH_2CH_2-$ units (*see* Figure). They are named as (i,j,k)-cryptand, where i, j, k are the numbers of oxygen atoms in each branch between nitrogen atoms. It is possible to build molecules containing cavities of a variety of shapes and sizes which are specific receptors for given species. They have a greater ability than simple CROWN ETHERS to bind metal ions, particularly alkaline and alkaline earth metal ions and lanthanide ions. The bound metal complexes are referred to as cryptates.

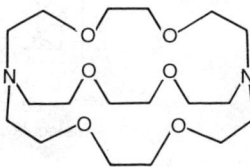

cryptands
(2,2,2)-cryptand.

cryptates. *See* CRYPTANDS.

crystal. Solid with regular polyhedral shape. All crystals of the same substance grow so that they have the same angles between the faces. They may not all have the same appearance, however, since different faces grow at different rates depending on the conditions.

crystal angles. *See* CRYSTAL STRUCTURE.

crystal field theory (CFT). Theory developed by H. Bethe in the 1930s for transition metal complexes in which the d-ATOMIC ORBITALS on the central metal ion are allowed to interact with point charges representing the LIGANDS. Although CFT gave a qualitative explanation of some properties, it has since been surpassed by the more rigorous treatment afforded by MOLECULAR ORBITAL theory known as LIGAND FIELD THEORY.

crystal forms. Some important common crystal forms are illustrated (*see* Figure). Compounds in parentheses have similar structures (isostructural).

crystal habit. Description of the external form of the crystal; this is due to the ordered arrangement of atoms, ions or molecules within the bulk of the solid.

crystal lattice (space lattice). Regular three-dimensional pattern of points in space such that the environment of any point is identical with any other. It is made up of the interpenetrating lattices of the constituent atoms (or ions) (*see* Figure), these may be the same (e.g., NaCl) or different (e.g., CaF_2, *see* CRYSTAL FORMS). The symmetry of the combined lattices determines the symmetry of the crystal. The points constitute the space lattice, for convenience they may be joined to provide three axes for the description of the points; OX, OY and OZ are the chosen axes, and with these the unit cell is a parallelopiped (edges OA, OB and OC). The unit cell is described by the three scalers a, b and c (primitive translations) and the angles α, β and γ. In the diagram, there is only one point to each unit cell (eight cells meet at one point and this is shared equally); there may be more than one point per cell in other lattices (e.g.,

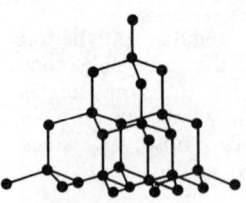

diamond
4-coordination

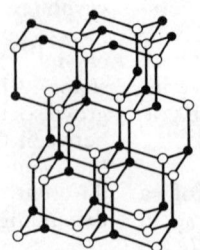

zinc blende (ZnS)
4:4-coordination
● Zn, ○ S
cubic form, ccp S with
Zn in tetrahedral holes.
(BN, SiC, AlP, CuCl)

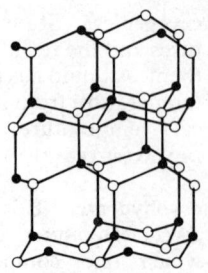

wurtzite (ZnS)
4:4-coordination
● Zn, ○ S
hexagonal structure, hcp S.
with Zn in tetrahedral holes
(ZnO, BeO, AlN, GaN)

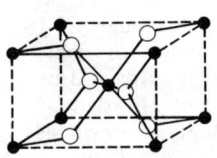

rutile (TiO_2)
6:3-coordination
● Ti, ○○ O
tetragonal structure
(PbO_2, MnO_2)

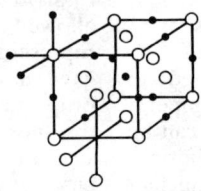

sodium chloride (NaCl)
6:6-coordination
● Na, ○ Cl
cubic structure, both Na
and Cl are ccp (alkali metal
halides except those of Cs)

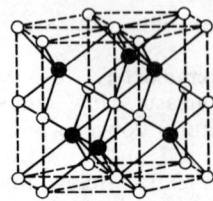

nickel arsenide (NiAs)
6:6-coordination
● As, ○ Ni
As is hcp with Ni in
octahedral holes
(CoTe, CrSb, FeS)

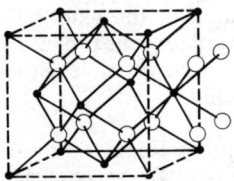

fluorite (CaF_2)
8:4-coordination
● Ca, ○ F
ccp of F^- with Ca in
tetrahedral holes
(many oxides, fluorides)

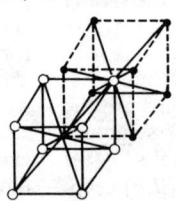

caesium chloride (CsCl)
8:8-coordination
● Cs, ○ Cl
large r_+/r_- ratio
(CsI, AgLi, HgTl)

graphite
layer structure
C-C distance in sheets =
142 pm, and between
sheets 340mm

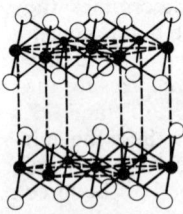

cadmium iodide (CdI_2)
layer structure
● Cd, ○ I
hcp of I^-

crystal forms

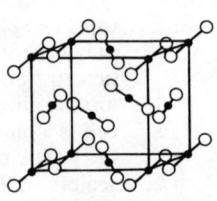

carbon dioxide (CO_2)
molecular structure
● C, ○○ O

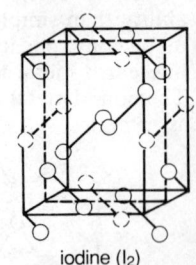

iodine (I_2)
molecular structure

face- or body-centred, *see* CLOSE-PACKED STRUCTURES). Special relationships between the axial lengths and ratios lead to the classification of CRYSTAL SYSTEMS.

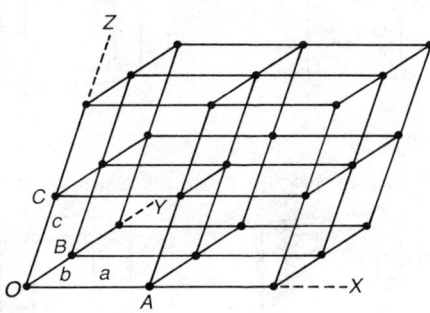

crystal lattice

crystallite (1) Small, imperfect crystal. (2) Single crystal in a microcrystalline substance.

crystallization. Process of forming crystals from a solution by increasing its concentration above the saturation point. Successive or fractional crystallization (the successive removal of the least soluble component) is a method of separation and purification.

crystallographic axis. *See* CRYSTAL.

crystallography. Study of crystal shape and form.

crystal plane. *See* CRYSTAL STRUCTURE.

crystal structure. Arrangement of atoms, ions or molecules in a crystal as revealed by X-RAY DIFFRACTION, in a three-dimensional CRYSTAL LATTICE. A characteristic parallelopiped, termed the unit cell, is described by the dimensions a, b and c, and the corresponding angles α, β and γ. The choice

crystal systems
Characteristics of the seven crystal systems.

System	Axial relationships	Essential symmetry	Number of point groups	Examples
Triclinic	$a \neq b \neq c$ $\alpha \neq \beta \neq \gamma \neq 90°, 120°$	No axes or planes	2	$CuSO_4.5H_2O$, $K_2Cr_2O_7$
Monoclinic	$a \neq b \neq c$ $\alpha = \gamma = 90°$, $\beta \neq 90°, 120°$	1 2-fold axis (C_2) or 1 plane	3	$CaSO_4.2H_2O$, $Na_2CO_3.10H_2O$, sucrose
Orthorhombic (rhombic)	$a \neq b \neq c$ $\alpha = \beta = \gamma = 90°$	3 2-fold axes (C_2) or 1 2-fold axis and 2 perpendicular planes intersecting in a 2-fold axis	3	$BaSO_4$, KNO_3, $PbCO_3$, α-S
Tetragonal	$a = b \neq c$ $\alpha = \beta = \gamma = 90°$	1 4-fold axis (C_4)	7	KH_2PO_4, SnO_2, TiO_2 (rutile)
Cubic (regular)	$a = b = c$ $\alpha = \beta = \gamma = 90°$	4 3-fold axes (C_3), 3 2-fold axes (C_2) or 3 4-fold axes (C_4)	5	NaCl, C (diamond), CaF_2, ZnS, $NaClO_3$, Pb, Hg, Ag, Au
Hexagonal	$a = b \neq c$ $\alpha = \beta = 90°, \gamma = 120°$	1 6-fold axis (C_6)	7	C (graphite), H_2O (ice), NiAs, Mg, Zn, Cd
Trigonal (rhombohedral)	$a = b = c$ $\alpha = \beta = \gamma \neq 90°, < 120°$	1 3-fold axis (C_3)	5	$CaCO_3$, $NaNO_3$, quartz, As, Sb, Bi

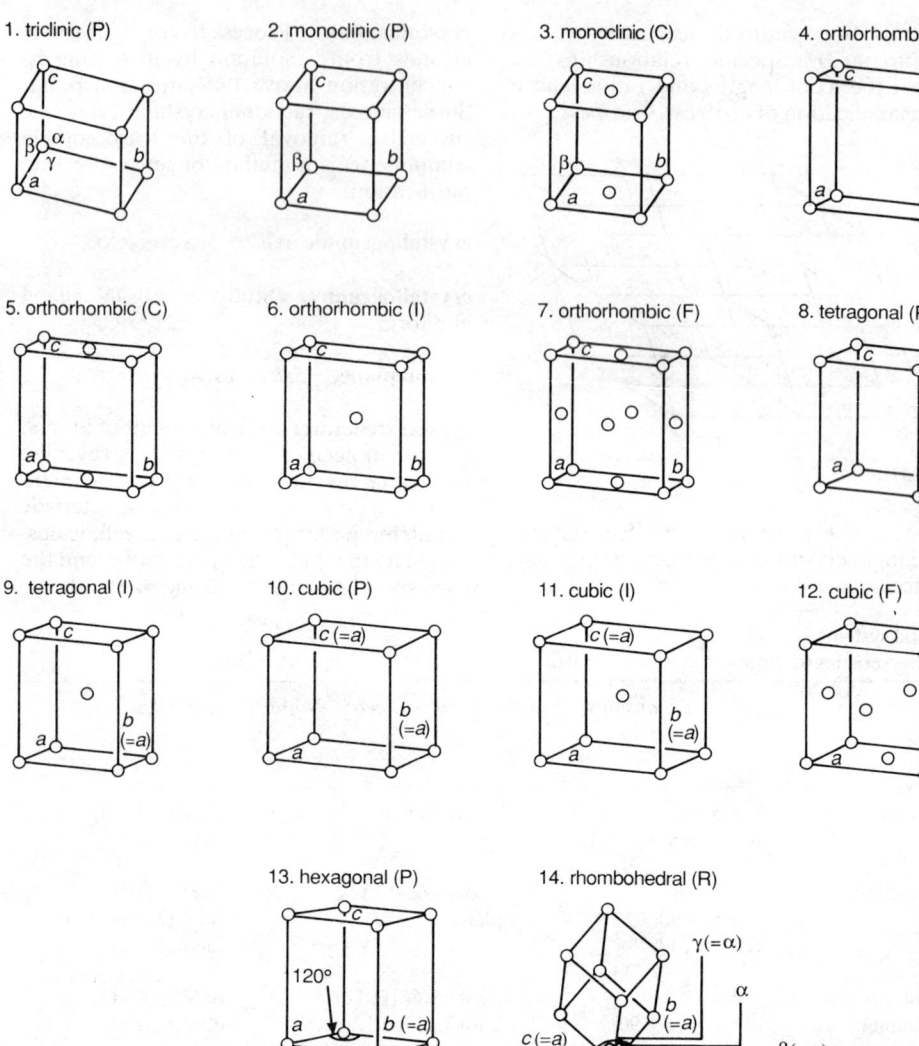

1. triclinic (P)   2. monoclinic (P)   3. monoclinic (C)   4. orthorhombic (P)

5. orthorhombic (C)   6. orthorhombic (I)   7. orthorhombic (F)   8. tetragonal (P)

9. tetragonal (I)   10. cubic (P)   11. cubic (I)   12. cubic (F)

13. hexagonal (P)   14. rhombohedral (R)

crystal systems
The 14 space or Bravais lattices. P, primitive unit; I (Innenzentrierte), body-centred; F, face-centred; C, 001 face-centred without symmetry demanding also centring of the prism faces. Interaxial angles are 90° unless otherwise indicated.

of axes, except in the triclinic system, is determined by the position of the SYM-METRY ELEMENTS. The shape and dimensions of the unit cell depend on the symmetry of the crystal. *See also* CRYSTAL SYSTEMS.

crystal symmetry. Term describing the regularities in the position and arrangement of the faces and edges of the crystal and also the position of the atoms within the crystal.

Such regularities are defined in terms of planes, axes and centres of symmetry (*see* SYMMETRY ELEMENTS). *See also* POINT GROUP.

crystal systems. Classification of crystalline substances on the basis of the shape and size of the unit cell. There are seven crystal systems (*see* Table) leading to 14 space or Bravais lattices (*see* Figure).

Cs. *See* CAESIUM.

CS gas (2-chlorobenzaldicyanomethane, $ClC_6H_4CH=C(CN)_2$). Riot gas prepared by the condensation of DICYANOMETHANE and 2-chlorobenzaldehyde.

CTAB (cetyl trimethyl ammonium bromide, hexadecanyltrimethylammonium bromide). *See* SURFACE-ACTIVE AGENT.

Cu. *See* COPPER.

cubic close-packed structure (ccp structure). *See* CLOSE-PACKED STRUCTURES.

cubic coordination. Coordination by eight ligands at the corners of a cube as in caesium chloride. *See also* CRYSTAL FORMS.

cubic system. *See* CRYSTAL SYSTEMS.

cumene ((1-methylethyl)benzene, C_6H_5-$CH(CH_3)_2$). Liquid (bp 152°C), manufactured by the reaction of benzene with PROPENE (H_2SO_4 as catalyst). Oxidation (air + basic catalyst) gives cumyl-α-hydroperoxide ($C_6H_5C(CH_3)_2OOH$) which on acidification gives phenol and propanone.

cumulative double bonds. Double bonds present in a chain in which at least three contiguous carbon atoms are joined by double bonds (e.g., $CH_2=C=C=C=CH_2$). Non-cumulative double bonds comprise every other arrangement of two or more double bonds in a single structure (e.g., $CH_3CH=CH–CH=CH–CH=CH_2$ or as in NAPHTHALENE). *See also* ALLENES.

cupellation. Refinement of precious metals (e.g., silver, gold) by melting in a shallow dish (cupel). Impurities are oxidized and adsorb on the walls of the cupel.

cuprammonium salts. *See* COPPER COMPOUNDS.

curare. South American indian name for a plant extract, which contains about 40 alkaloids; used as an arrow poison. A main component is tubocurarine which acts as a muscle relaxant; it is a bisbenzylisoquinoline alkaloid in which the nitrogen atoms are quaternary.

curie (Ci). Pre-SI unit of ACTIVITY chosen to approximate the activity of 1 g of radium-226. *See also* RADIATION UNITS.

Curie's law. Magnetic susceptibility per unit volume of a paramagnetic material is related to temperature by $\chi = C/T$. C is the Curie constant

$$C = \mu_0 N_v g^2 \, \beta^2 J(J + 1)/3k$$

where β is the BOHR MAGNETON, N_v is the number of spins per unit volume, g is the LANDE g-FACTOR and J is the quantum number associated with Russell–Saunders coupling (*see* TERM SYMBOLS).

Curie temperature. *See* FERROMAGNETISM.

curium (Cm). *See* ACTINIDE ELEMENTS.

current (I; dimensions: $\varepsilon^{1/2} \, m^{1/2} \, l^{3/2} \, t^{-2}$; units: $A = C \, s^{-1}$). Rate of transfer of electricity. The practical unit of current is the AMPERE. The ampere is one-tenth of the electromagnetic unit of current.

current density. *See* CURRENT.

current efficiency. Percentage of the current used in the electrolytic process under consideration.

$$\frac{\text{current}}{\text{efficiency}} = \frac{\substack{\text{actual yield of} \\ \text{element or compound}}}{\substack{\text{theoretical yield} \\ \text{(Faraday's laws)}}} \times 100$$

The deposition of foreign elements, evolution of extraneous gases, mechanical loss, vaporization of deposit at high temperatures, side reactions and leakage of current may all contribute to the reduction of the current efficiency.

Curtius reaction. Rearrangement of an acid azide (*see* NITROGEN HYDRIDES) to an N-alkyl URETHANE by refluxing with an alcohol

$$RCON_3 + R'OH \longrightarrow RNHCOOR' + N_2$$

C_V.   HEAT CAPACITY at constant volume.

CVD. *See* CHEMICAL VAPOUR DEPOSITION.

cyamelide. *See* CYANIC ACID.

cyanamide (NH_2CN). Colourless deliquescent solid (mp 42°C), prepared by the action of ammonia on cyanogen chloride, or water on calcium cyanamide. Cyanamide gives guanidine with ammonia, UREA with acids, and THIOUREA with hydrogen sulphide. On melting, the dimer dicyandiamide and trimer melamine are formed. Melamine is used to manufacture melamine–methanal plastics and resins (*see* POLYMERS). In the solid, cyanamide exists as a tautomer (*see* ISOMERISM) with a CARBODIIMIDE form

$$H_2N{-}C{\equiv}N \rightleftharpoons HN{=}C{=}NH$$

cyanic acid (HNCO).   Liquid (bp 23.5°C), prepared by the distillation of CYANURIC ACID. Above 0°C, it trimerizes to cyanuric acid and a higher solid polymer cyamelide (HNCO)$_n$. Aqueous solutions hydrolyze to carbon dioxide and ammonia. Cyanic acid reacts with alcohols to give URETHANES and forms cyanates (e.g., KCN + PbO $\rightarrow$ KNCO) and ISOCYANATES.

cyanide process.   Extraction of gold and silver from their ores by dilute sodium or potassium cyanide solution. The metal is complexed (*see* COMPLEX COMPOUND, CO-ORDINATION COMPOUND) and is recovered by reducing the complex with zinc dust.

cyanides (MCN). Salts of HYDROGEN CYANIDE. Organic, covalently bonded cyanides are also known as NITRILES. Potassium cyanide is manufactured by fusing potassium carbonate and carbon in an atmosphere of ammonia. Sodium cyanide is prepared by reacting hydrogen cyanide and aqueous sodium hydroxide. Other metal cyanides are prepared from sodium or potassium cyanides by DOUBLE DECOMPOSITION reactions. Aqueous solutions of cyanides hydrolyze slowly to the carbonate and ammonia. Cyanide is a common ligand in transition metal complexes (e.g., CYANO-FERRATES, [Au(CN)$_2$]$^-$). Uses of cyanides, which are extremely toxic, are in the CYANIDE PROCESS, CASE HARDENING, ELECTRO-DEPOSITION OF METALS and as a fumigant (Ca(CN)$_2$).

cyanine dyes.   *See* DYES.

cyanocobalamin (vitamin B$_{12}$). *See* VITAMINS.

cyanoethene.   *See* PROPENENITRILE.

cyanoethylation.   Introduction of the CNCH$_2$CH$_2$– group into a molecule with an active METHENE GROUP by a MICHAEL CONDENSATION. The best base catalyst is benzyltrimethylammonium hydroxide (e.g., CH$_3$COCH$_2$CH$_3$+CH$_2$=CHCN $\rightarrow$ CH$_3$COCH(CH$_2$CH$_2$CN)CH$_3$). A second group may then be introduced to give CH$_3$COC(CH$_2$CH$_2$CN)$_2$CH$_3$. In compounds with an active METHYL GROUP tri-cyanoethylation may be brought about.

cyanoferrates. Hexacyanoferrates, yellow [Fe(CN)$_6$]$^{4-}$ and red [Fe(CN)$_6$]$^{3-}$ are formed by the addition of alkali cyanides to Fe^{2+}. With Fe^{3+}, K$_4$Fe(CN)$_6$ gives a deep blue precipitate of Prussian blue (Fe$_4$[Fe(CN)$_6$]$_3$). Turnbull's blue, formed by Fe^{2+} and K$_3$Fe(CN)$_6$ is identical to Prussian blue. Soluble Prussian blue is KFe[Fe(CN)$_6$]. Berlin green is Fe[Fe(CN)$_6$]. Compounds are known with five CN ligands and a different sixth ligand (e.g., CO, NH$_3$, NO$_2{}^-$). The nitroprusside ion is [Fe(CN)$_5$NO]$^{2-}$.

cyanogen ((CN)$_2$).   Colourless, poisonous gas (bp −21°C), prepared by heating silver or mercury cyanides. It is hydrolyzed to hydrocyanic acid (*see* HYDROGEN CYANIDE) and CYANIC ACID, with small amounts of ethanediamide and ammonium methanoate. Between 400°C and 800°C, cyanogen polymerizes to paracyanogen. Cyanogen halides are prepared by the action of halogens on sodium cyanide.

cyanohydrins.   Organic compounds containing the group HOC–CCN, prepared by the reaction between compounds containing the OXO GROUP and hydrogen cyanide (as NaCN + dil. H$_2$SO$_4$). They are important reagents in synthesis, being hydrolyzed to α-hydroxyacids.

cyanuric acid (2,4,6-trihydroxy-1,3,5-triazine, C$_3$H$_3$N$_3$O$_3$). Solid, strong acid that is sparingly soluble in water; prepared by the dry distillation of UREA. Cyanuric acid decomposes on heating to CYANIC ACID vapour.

cyclamates.   *See* SWEETENING AGENTS.

cycle. Process in which a system in a given state passes through a sequence of changes back to its original state. For any cycle and any STATE FUNCTION, $\oint dX = \Delta X = 0$. In an isothermal cycle for a perfect gas, the WORK done by the gas is paid for by supplying the equivalent amount of energy to make the cycle operate (e.g., in the steam engine).

cyclic adenosine monophosphate (cyclic AMP, cAMP). Derivative of ADENOSINE TRIPHOSPHATE, widespread in animal cells acting as a messenger in many hormone-induced reactions. HORMONES (e.g., adrenaline) exert their effect by increasing or decreasing cellular cAMP levels. cAMP is involved in controlling gene expression and cell division and in immune responses.

cyclic compounds. *See* ACYCLIC COMPOUNDS, ALICYCLIC COMPOUNDS, CARBOCYCLIC COMPOUNDS, HETEROCYCLIC COMPOUNDS, HOMOCYCLIC COMPOUNDS.

cyclic peptides. *See* ANTIBIOTICS.

cyclic process. *See* CYCLE.

cyclic voltammetry. *See* LINEAR SWEEP VOLTAMMETRY.

cyclization. Formation of a cyclic compound from an open-chain compound.

cycloaddition. *See* WOODWARD–HOFFMANN RULE.

cycloalkanes (C_nH_{2n}). Saturated ALICYCLIC COMPOUNDS. Five- and six-membered rings are found in petroleum. Three-, four- and five-membered rings occur in TERPENES. Preparation methods include: (1) treatment of the α,ω-dihalogen with sodium or zinc (this is suitable for up to the 1,6-dihalogens); (2) conversion of cyclic ketones by the CLEMMENSEN REDUCTION; (3) condensation reactions between α,ω-dihalogenalkanes and sodium 3-oxobutanoate or sodium diethylpropanedioate. For example

(4) HOFMANN EXHAUSTIVE METHYLATION.

The rings of cycloalkanes are, apart from cyclopropane, not planar and are strained by different amounts. The thermodynamic stability is measured from the enthalpy of combustion per METHENE GROUP. The least strain occurs with six-membered cyclohexane (*see* RING STRAIN).

In many respects, the properties of cycloalkanes are similar to the ALKANES, but the lower members form addition products with ring fission. Thus cyclopentane halogenates maintaining the ring, whereas cyclopropane gives a mixture of the dihalogenopropanes. Cycloalkanes are largely unaffected by hydrogen over nickel, except cyclopropane (80°C to propane) and cyclobutane (200°C to n-butane). Cyclopropane (bp −34.5°C) is a powerful inhalatory anaesthetic. Cyclohexane (mp 6.5°C, bp 81.5°C), present in NATURAL GAS and PETROLEUM GAS, is also manufactured by the reduction of benzene by hydrogen over nickel. It is used to prepare CAPROLACTAM in the synthesis of Nylon (*see* FIBRES). The structure shows conformational ISOMERISM. *See also* CATENANES, CYCLOPHANES.

cycloalkenes (C_nH_{2n-2}). Unsaturated CARBOCYCLIC COMPOUNDS. Preparation is as for CYCLOALKANES and by the DIELS–ALDER REACTION. The smallest ring alkene is cyclopropene. Cyclobutadiene (I) is not stable, but may be prepared *in situ* from $(C_4H_4)Fe(CO)_3$. It undergoes Diels-Alder reactions with alkynes to give Dewar benzene. Cyclopentadiene (bp 41°C) is found in benzene from coal tar. The methene group is highly reactive, after losing H^+ the cyclopentadienyl anion (cp^-) is aromatic. Cyclopentadienylmagnesium bromide (*see* GRIGNARD REAGENTS) reacts with iron(III) chloride in THF to give the METALLOCENE h^5-dicyclopentadienyl iron (*see also* IRON COMPOUNDS, SANDWICH COMPOUND). On standing cyclopentadiene dimerizes by a Diels–Alder reaction to compound II and to further oligomers. Cyclopentadiene condenses with aldehydes and ketones in the

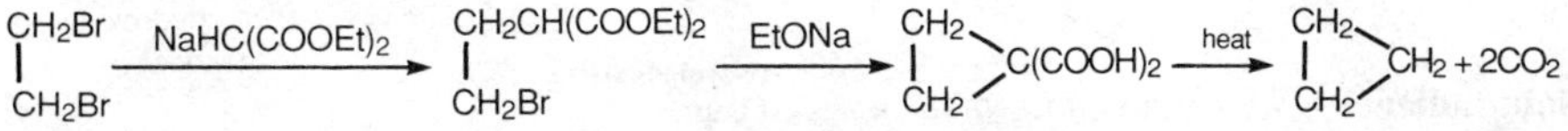

CH$_2$Br | CH$_2$Br →(NaHC(COOEt)$_2$) CH$_2$CH(COOEt)$_2$ | CH$_2$Br →(EtONa) CH$_2$ | CH$_2$ ⟩C(COOH)$_2$ →(heat) CH$_2$ | CH$_2$ ⟩CH$_2$ + 2CO$_2$

cyclobutadiene
(I)

cyclopentadiene
dimer (II)

fulvenes (III)

cycloheptatriene
(tropilidene, IV)

cycloheptatrienolone
(tropolone, V)

cycloheptatrienone
(tropone, VI)

1,3,5,7–
cyclooctatetraene
(VII)

cycloalkenes

presence of sodium ethoxide to form fulvenes (III). These are coloured compounds and behave as conjugated DIENES. Cyclohexatriene is BENZENE and has distinctive aromatic properties. Cycloheptatriene (tropilidene, IV) (bp 115.5°C) may be prepared by the photolysis of diazomethane in the presence of benzene. In turn, it may be used to prepared cycloheptatrienolone (tropolone, V). Cycloheptatrienone (tropone, VI) is prepared from methoxybenzene. V and VI show some aromatic character because of the possibility of forming the TROPYLIUM ION. 1,3,5,7-Cyclooctatetraene (VII) has a conjugated ring, but is not aromatic. It is prepared by the polymerization of four molecules of ethyne under pressure in the presence of $Ni(CN)_2$. Reactions frequently involve rearrangement to a six-membered ring which may be aromatic. Thus oxidation gives benzoic acid ($KMnO_4$ or O_2/V_2O_5) or benzene-1,4-dicarboxylic acid (CrO_3/CH_3-$COOH$). Bicyclo[4.2.0] ring compounds are formed with bromine or $Hg(CH_3$-$COO)_2$: for example

Br
Br

Large rings of cyclopolyenes and cyclopolyynes (e.g., cyclooctacosa-1,3,15,17-tetra-yne) have been prepared. *See also* ANNU-LENES.

cyclobutadiene. *See* CYCLOALKENES.

β-cyclodextrins. Macrocyclic carbohydrate molecules produced by bacterial degradation of starch, consisting of six to 12 glucose monomers arranged in the form of a torus (*see* Figure 1). The molecules have a rigid structure with a hollow interior cavity that is hydrophobic in nature. This contains one or more highly energetic polar water molecules which are expelled to include less polar molecules. These molecules form strong inclusion complexes with guest molecules of the correct size; aromatic groups have a high affinity for the cavity so that *ortho-*, *meta-* and *para-*isomers of aromatic compounds can be separated because of the different strengths of the complexes. The *para-*compound, which penetrates further into the cavity, is more strongly bound than the other isomers (*see* Figure 2). This forms the basis of INCLUSION COMPLEXING as a separation technique.

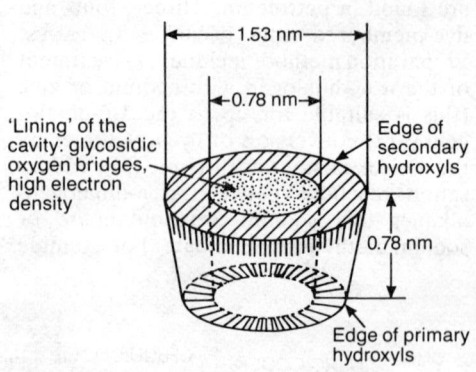

'Lining' of the cavity: glycosidic oxygen bridges, high electron density

1.53 nm

0.78 nm

Edge of secondary hydroxyls

0.78 nm

Edge of primary hydroxyls

β-cyclodextrins
Figure 1.

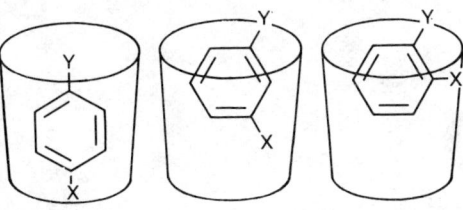

β-cyclodextrina
Figure 2. Simplified comparison of how *ortho-*, *meta-* and *para*-distributed benzenes sit in the cyclodextrin cavity. The X represents the more hydrophobic substituent, while Y represents the more hydrophilic one.

cycloheptatriene. *See* CYCLOALKENES.

cycloheptatrienyl cation. *See* TROPYLIUM ION.

cyclohexane. *See* CYCLOALKANES.

cyclohexanol ($C_6H_{12}O$). Colourless liquid (bp 161°C) with a camphor-like smell, prepared by the catalytic reduction (Ni/H_2) of phenol under pressure at elevated temperatures. It has typical properties of secondary MONOHYDRIC ALCOHOLS. Gentle oxidation (Cu, 250°C) or dehydrogenation yields CYCLOHEXANONE; more vigorous oxidation forms HEXANEDIOIC ACID. It is dehydrated with aluminum chloride to cyclohexene, and is used in the manufacture of NYLON, CELLULOID and detergents.

cyclohexanone ($C_6H_{10}O$). KETONE; colourless liquid (bp 157°C) with a smell of peppermints. It is prepared, with cyclohexanol, by the oxidation of cyclohexane by oxygen under pressure and with a cobalt catalyst. Cyclohexanone is oxidized to HEXANEDIOIC ACID. It is used in the manufacture of CAPROLACTAM and thus NYLON, lacquers, CELLULOID, artificial leathers and printing inks.

cyclooctatetraene. *See* CYCLOALKENES.

1,3-cyclopentadiene. *See* CYCLOALKENES.

cyclophanes. Ring systems containing benzene rings bridged between the *para-* or *meta*-positions (*see* ORIENTATION IN THE BENZENE RING) with METHENE GROUPS. Nomenclature is, for example, [*a,b,c,. . .*]-

paracyclophane, where *a,b,c,.* . . are the numbers of methenes between the rings. The compound having the trivial name cyclophane is [10,10]paracyclophane.

cyclopropane. *See* CYCLOALKANES.

cyclotron. Particle accelerator used to increase the kinetic energy of charged particles, by passing them in an outward spiral path through alternating electromagnetic fields. At the perimeter, the charged particles are directed to their target. Protons can achieve an energy of 10 MeV.

***p*-cymene** (4-methyl(1-methylethyl)benzene, $CH_3C_6H_4CH(CH_3)_2$). Colourless liquid, occurring in oils of thyme and eucalyptus and related to camphor from which it is prepared by heating with anhydrous zinc chloride. Oxidation gives a range of products including 4-methylbenzoic acid, 4-methylacetophenone, isopropyl benzoic acid, and benzene-1,4-dicarboxylic (terephthalic) acid. It is used as a paint thinner and in the manufacture of thymol (*see* PHENOLS).

Cys (cysteine). *See* AMINO ACIDS.

Cys–Cys (cystine). *See* AMINO ACIDS.

cysteine(Cys). *See* AMINO ACIDS.

cystine(Cys–Cys). *See* AMINO ACIDS.

cytidine. *See* NUCLEOSIDES.

3′-cytidylic acid. *See* NUCLEOTIDES.

cytochromes. Group of proteins, each with an iron-containing HAEM group, that form part of the ELECTRON TRANSPORT CHAIN. Electrons are transferred between Fe^{2+} and Fe^{3+} states. The terminal carrier cytochrome *a*–cytochrome a_3 is sometimes called cytochrome oxidase.

cytosine. *See* PYRIMIDINES.

cytotoxic agent. Substance that is injurious to living cells, usually with respect to tumour formation. Many such agents are bifunctional alkylating agents such as the NITROGEN MUSTARDS.

D

D. *See* DEBYE, DEUTERIUM.

D- (configurational symbol). *See* OPTICAL ACTIVITY.

D. Diffusion coefficient (*see* FICK'S LAW OF DIFFUSION), PARTITION COEFFICIENT.

d. Collision diameter (*see* COLLISION FREQUENCY, VISCOSITY).

Δ. Increase in thermodynamic function.

δ-scale. Delta-scale (*see* NUCLEAR MAGNETIC RESONANCE SPECTROSCOPY).

2,4-D (2,4-dichlorophenoxyacetic acid). *See* AUXINS.

Dacron. *See* FIBRES.

dalton. *See* ATOMIC MASS UNIT.

Dalton's law of partial pressures. Total pressure of a mixture of gases or vapours is equal to the sum of the partial pressures of the components. The principle is only strictly true for mixtures of IDEAL GASES. *See also* PARTIAL PRESSURE.

Daniell cell. Electrochemical CELL consisting of a zinc electrode dipping into a zinc sulphate solution and a copper electrode dipping into a copper sulphate solution. The two solutions are separated by a permeable membrane, such as a sintered glass disc or porous pot. The electrode reaction at the cathode is $Cu^{2+} + 2e \rightarrow Cu$ and at the anode is $Zn \rightarrow Zn^{2+} + 2e$. The theoretical standard emf of the cell is 1.100 V.

dansyl-. Prefix indicating a 1-dimethyl-aminonaphthalene-5-sulphonyl derivative of an amino acid. The derivatives exhibit intense yellow fluorescence under UV radiation, which permits the detection and determination of amino acids at very low levels. Such derivatives are used in end-group analysis of proteins.

dansyl chloride. *See* 1-DIMETHYLAMINO-NAPHTHALENE-5-SULPHONYL CHLORIDE.

Dap (2,6-diaminopimelic acid). *See* AMINO ACIDS.

dative bond. *See* COORDINATE BOND.

DDT (dichlorodiphenyltrichloroethane). *See* 1,1,1-TRICHLORO-2,2-BIS(4-CHLORO-PHENYL)ETHANE.

deactivating group. Electron-withdrawing group in a benzene ring that causes the ring to be less susceptible to further substitution; such groups are *meta*-directing (*see* ORIENTATION IN THE BENZENE RING).

de Broglie law. Wavelength of a particle is related to its momentum by Planck constant

$$\lambda = h/mv.$$

debye (D). Unit of a DIPOLE MOMENT; $1\,D = 3.336 \times 10^{-30}\,C\,m$.

Debye equation. Equation relating the dipole moment (p), the polarizability (α) and the relative permittivity of a substance

$$\alpha + p^2/3kT =$$
$$3\varepsilon_0(M/N_A\rho)[(\varepsilon_r - 1)/(\varepsilon_r + 2)]$$

Debye heat capacity equation. Equation used to evaluate the HEAT CAPACITY of a monatomic solid at temperatures approaching 0 K.

$$C_v = (12\pi^4/5)\,R(T/\theta_v)^3 = aT^3$$

where θ_v is the CHARACTERISTIC TEMPERATURE of vibration $(h\nu/k)$ and a is a constant for the particular solid. It is used to evaluate the ENTROPY change for a solid at temperatures below which heat capacity measurements are not possible.

Debye–Hückel activity coefficient equation. Theoretical equation for the calculation of the ACTIVITY COEFFICIENT of an ion in dilute solution.

$$\lg \gamma_i = -A\,|z_i^2|\,I^{1/2}/(1 + B\mathring{a}I^{1/2})$$

The second term in the denominator, in which $\mathring{a}$ is the effective diameter of the ion, allows for its finite size. As the IONIC STRENGTH I approaches zero this reduces to the limiting form

$$\lg \gamma_i = -A\,|z_i^2|\,I^{1/2}$$

The corresponding equations for the mean ionic activity coefficient of an electrolyte are

$$\lg \gamma_\pm = -A|z_+|\,|z_-|\,I^{1/2}/(1 + B\mathring{a}I^{1/2})$$

and

$$\lg \gamma_\pm = -A|z_+|\,|z_-|\,I^{1/2}$$

For aqueous solutions at 298 K, $A = 0.509$ mol$^{-1/2}$ kg$^{1/2}$, $B = 3.291 \times 10^9$ m^{-1} mol$^{-1/2}$ kg$^{1/2}$. The limiting equation gives values of $\gamma_\pm$ in excellent agreement with experimental values up to $I = 0.01$ mol dm^{-3}; large deviations are observed even at this ionic strength if $|z_+|\,|z_-| \geqslant 4$.

Debye–Hückel conductance equation. *See* CONDUCTANCE EQUATIONS.

Debye–Hückel–Onsager equation. *See* CONDUCTANCE EQUATIONS.

decahydronaphthalene. *See* BICYCLO-[4.4.0]DECANE.

decalin. *See* BICYCLO[4.4.0]DECANE.

decane. *See* ALKANES.

n-decanoic acid (capric acid). Solid (mp 31.4°C), rancid-smelling CARBOXYLIC ACID.

decantation. Removal of a liquid from a precipitate by pouring from the sediment.

decarbonylation. Elimination of carbon monoxide from a molecule. α-Ketoacids

and α-ketoesters undergo decarbonylation on warming with concentrated sulphuric acid (e.g., *see* 2-OXOPROPANOIC ACID).

decarboxylation. Elimination of carbon dioxide from a molecule, usually a carboxylic acid. SODA LIME is a common decarboxylating agent to yield ALKANES.

decay. Spontaneous transformation of one RADIONUCLIDE into a daughter nuclide, which may be radioactive, with the emission of alpha-, beta- or gamma-radiation. *See also* RADIOACTIVE DECAY SERIES, RADIO-ACTIVITY.

decay constant (λ; units: day^{-1}, min^{-1}, s^{-1}). First-order rate constant for the exponential decay of a RADIONUCLIDE

$$N = N_0 \exp(-\lambda t)$$

where N_0 and N are the number of atoms present at time zero and t, respectively. A more convenient quantity is the HALF-LIFE ($t_{1/2}$) related to λ by $\lambda = 0.693/t_{1/2}$. λ and $t_{1/2}$ are characteristic for a particular nuclide (e.g., for $^{32}_{15}$P, $\lambda = 0.0485$ day^{-1}; $t_{1/2} = 14.3$ day).

defect structure. Discontinuity in a CRYSTAL LATTICE. (1) A point or Schottky defect consists of a vacant site with the migrated atom at the surface. (2) A Frenkel defect consists of a vacant site in which the missing atom or ion has moved to an interstitial position. (3) A line defect (dislocation), a slip along the surface which occurs if there is more than one adjacent point defect. Defects are caused by strain, in some cases by irradiation or by introducing foreign ions into a structure. Crystals exhibiting defects show anomalous physical properties, for example electrical conductivity, which is made use of in the production of SEMI-CONDUCTORS.

definite proportions, law of. *See* CONSTANT COMPOSITION, LAW OF.

degeneracy (g). Number of independent wavefunctions associated with a given energy level; that is the number of distinct energy levels in an atom, molecule, etc. that have the same or nearly the same energy. In a rigid rotor with quantum number J, the degeneracy is $2J + 1$.

degree of dissociation. *See* CONDUCTANCE EQUATIONS.

degree of freedom
1) *phase rule.* Number of intensive state variables (f) that can be independently varied without changing the number of phases present in equilibrium; thus the state of a given amount of a gas is specified by any two of the variables P, V, T (i.e. $f = 2$).
2) *mechanical.* Number of independent 'square terms' (i.e. terms involving the square of a coordinate or velocity) required to express the total energy of a molecule. For a gaseous molecule of n atoms, the total number of coordinates (degrees of freedom) required to specify the position of all atoms is $3n$, of which 3 are for translation and 2 for rotation and $3n-5$ for vibration of a linear molecule, and 3 and $3n-6$, respectively, for a non-linear molecule. Thus for a non-linear molecule, assuming that all rotations and vibrations are active,

$$U_{max} = \tfrac{1}{2}m(v_x^2 + v_y^2 + v_z^2) +$$
$$\tfrac{1}{2}I(\omega_x^2 + \omega_y^2 + \omega_z^2) +$$
$$(3n-6)[\tfrac{1}{2}(mv^2 + kx^2)]$$

dehalogenation. Removal of a halogen atom from a compound.

dehydration. Removal of the elements of water from a compound. *See also* HYDRATION.

dehydroacetic acid. *See* 6-METHYLACETO-PYRANONE.

dehydrobenzene. *See* BENZYNE.

dehydrogenases. *See* ENZYMES.

dehydrogenation. Removal of hydrogen from a molecule resulting in an increase in the degree of unsaturation (e.g., alkanes to alkenes, cyclohexane derivatives to benzene derivatives). The process is accomplished at high temperatures and pressures in the presence of a catalyst (platinum, palladium, copper, selenium). In biochemical reactions the process, catalyzed by ENZYMES (i.e. dehydrogenases), is accompanied by electron transport (*see* ELECTRON TRANSPORT CHAIN). *See also* HYDROFORMING.

dehydrohalogenation. Removal of hydrogen and a halogen from a compound (e.g., formation of an ALKENE from an ALKYL HALIDE).

deionized water. Water that has been purified by DISTILLATION and by passing through an ION EXCHANGER column.

deliquescence. Absorption of water by a HYGROSCOPIC solid resulting in the formation of a saturated solution. It occurs when the partial water vapour pressure exceeds that of the aqueous saturated salt solution. The surface of the solid then becomes covered with a layer of a saturated solution. Salts forming hydrates with very low vapour pressure are deliquescent. *See also* EFFLORESCENCE.

delocalization. Sharing of electrons between several atoms in a molecule. This is found particularly in organic compounds containing multiple double bonds and gives its maximum effect in AROMATIC COMPOUNDS. In terms of MOLECULAR ORBITALS (mo), delocalization occurs when a filled mo extends over several atoms. The extra stability that is gained by this process is known as the delocalization energy and may be calculated simply by HÜCKEL MOLECULAR ORBITAL THEORY.

delta-scale. *See* NUCLEAR MAGNETIC RESONANCE SPECTROSCOPY.

demanding reaction. *See* SUPPORTED CATALYST.

demethylchlortetracycline. *See* ANTIBIOTICS.

demineralization of water. *See* ELECTRODIALYSIS.

demulsification. *See* EMULSION.

denaturation. *See* PROTEINS.

dendrite. Tree-like crystal formation arising from the different rates of crystal growth in different directions from the initial crystallization point. It is found in alloys and ice crystals.

denitrification. Return of fixed nitrogen (usually in the form of nitrates or ammonium salts) to the atmosphere. The process is brought about by such bacteria as *Pseudomonas denitrificans* which have the ability to use nitrates as their energy source. *See also* NITROGEN CYCLE.

density (ρ; dimensions: $m\ l^{-3}$; units: kg m^{-3}). Mass of a substance per unit volume. The density of a gas is measured by: (1) comparing the mass of an evacuated globe with that of the globe containing the gas at known temperature and pressure; (2) a buoyancy method using a microbalance. For an IDEAL GAS, $P = \rho RT/M_r$; for a REAL GAS, a correction for deviation from ideality must be applied. The limiting density (ρ_0) is the extrapolated value of the density at $P = 0$ (i.e. when the gas shows ideal behaviour). The molar mass can be obtained from the limiting density; $M = 22.414 \times 10^{-3}\rho_0$ kg. The density of a liquid is measured by: (1) comparing the mass of a pyknometer or density bottle filled with the liquid with that when filled with water at the same temperature; (2) measuring the length of stem of a hydrometer floating in the liquid using Archimedes' principle. Densities increase from gases through liquids to solids (e.g., hydrogen, 0.1; water, 1×10^3; mercury 13.6 $\times$ 10^3; osmium, 22.4 $\times$ 10^3 kg m^{-3}).

density gradient column. Tube filled with a mixture of solutions such that there is a gradual variation in density throughout its length. It is used for the determination of the density of small solid particles. It is also used to isolate protein or nucleic acid of specific molecular mass.

deoxycholic acid. *See* BILE ACIDS.

deoxyribonuclease. *See* ENZYMES.

deoxyribonucleic acid. *See* NUCLEIC ACIDS.

2-deoxyribose. *See* NUCLEIC ACIDS, PENTOSES.

depolarization. Minimization of POLARIZATION of an electrode by stirring, by an increase in temperature or by a chemical reaction.

depression of freezing point. *See* FREEZING POINT.

depth profiling. *See* SPUTTERING.

derivatization. Preparation of (1) a more volatile or thermally stable derivative of a compound which can be separated and identified by GAS–LIQUID CHROMATOGRAPHY (e.g., silylation, acylation, esterification) or (2) a compound that enhances selective detector response (e.g., in HIGH-PERFORMANCE LIQUID CHROMATOGRAPHY the use of such reagents as ω-bromoacetophenone (phenacyl bromide) and 2,4-dibromoacetophenone (4-bromophenacyl bromide) to form UV-active compounds).

derris. INSECTICIDE containing ROTENONE, obtained from the root of the shrub *Derris elliptica*; originally used as a fish poison.

Deryagin–Landau and Verwey–Overbeek theory. *See* DLVO THEORY.

desalination. *See* ELECTRODIALYSIS.

desiccant. Substance used to remove water. Desiccants may be used in non-aqueous solvents to dry organic reaction products, or in a desiccator to dry solids. The ability of common desiccants increases in the order $CaCl_2$, CaO, NaOH, MgO, $CaSO_4$, H_2SO_4, silica gel, $Mg(ClO_4)_2$, P_2O_5.

desorption. Removal of material from an INTERFACE; the opposite of ADSORPTION. Whereas adsorption occurs with zero or little activation energy (*see* ACTIVATED ADSORPTION), desorption requires energies which are about the HEAT OF ADSORPTION.

destructive distillation. Decomposition of complex organic liquids or solids by heat in the absence of air and the simultaneous distillation and condensation of volatile products (e.g., wood to charcoal, coal to coke, coal tar and other products).

detection systems. Requirements for a detector in GAS–LIQUID CHROMATOGRAPHY are: rapid and reproducible linear response at low concentrations; good stability over extended periods; a uniform response for a wide variety of compounds. No single detector meets all the requirements. Detectors (except mass spectrometers, *see below*) do not give a chemical identification; the substances are normally identified by the time they take to emerge from the column (*see* RETENTION TIME). A wide range of detectors has been described, often for specific purposes; the most commonly used ones are described.

thermal conductivity detector (katharometer). Records the change in thermal conductivity of the carrier gas around a heated platinum or tungsten wire when a solute is present in the gas. The change in conductivity is compared with a reference wire surrounded by pure carrier gas. Such detectors are simple, rugged, cheap, non-selective, accurate and do not destroy the sample. They are not as sensitive as other detectors.

flame ionization detector (FID). Components emerging in the effluent gas are ionized on entering the detector by burning in a stream of hydrogen. The gas is thus made conducting; the conductivity, measured by the current flow between the jet and a collector electrode, is a linear function of the ions formed and hence the concentration of that particular component. FID is a popular and highly sensitive detector which does not require a reference stream; it is destructive of the sample.

argon detector. Effluent from the column is passed over a beta-emitter (e.g., strontium-90 or tritium); this excites argon, the carrier gas, to a metastable state. The excited argon atoms cause ionization of the sample molecules by collision, and the resulting ion current is measured in essentially the same way as in the flame detector. It has a high sensitivity, but is easily overloaded.

photoionization detector (PID). Detector in which photons of suitable energy (e.g., UV radiation) are used to ionize solutes in the carrier gas. The resulting current is measured as in the argon and flame detectors.

mass spectrometry. Effluent gas is analyzed directly by passing part of the flow into a mass spectrometer; this gives both an identification and a check on the purity of the components.

detergent.   *See* SURFACE-ACTIVE AGENT.

detonation.   Exothermic chemical reaction accompanied by a very high percussion wave and marked increase in pressure. The explosion is important in engines where it can vary from pinking, an incipient stage, to the more severe condition of knocking, which gives reduced power output.

deuterated compound.   Organic compound in which some or all of the ^{1}H are replaced by ^{2}H (*see* DEUTERIUM). The compounds are named as d^n-compounds (e.g., d^6-benzene is C_6D_6). Deuterated compounds may be prepared from deuterated precursors (e.g., D_2O, D_2, DCl) and are used in the elucidation of mechanism (*see* ISOTOPE EFFECTS) and in the spectroscopic investigation of compounds (especially INFRARED SPECTROSCOPY and NUCLEAR MAGNETIC RESONANCE SPECTROSCOPY).

deuteriation.   Replacement of hydrogen in a molecule by DEUTERIUM.

deuteride (^{2}H$^-$).   Anion of DEUTERIUM.

deuterium (^{2}H, D).   Stable ISOTOPE of HYDROGEN consisting of one proton, one neutron and one electron. Deuterium has a relative atomic mass of 2.0144 and is found to the extent of 0.02 percent of naturally occurring hydrogen. D_2 is a mixture of 66 percent *ortho-* and 33 percent *para*-deuterium, the reverse of the proportions in PROTIUM because of the nuclear spin of 1 (*see ortho*-HYDROGEN). D_2 is obtained by the electrolysis of DEUTERIUM OXIDE (D_2O) or by distillation of liquid hydrogen. It is used as deuterium oxide in nuclear reactors. The chemical reactivity of deuterium in molecules is less than that of hydrogen; bonds between deuterium and elements are stronger largely because of the ZERO-POINT ENERGY. Therefore if deuterium is substituted for hydrogen, reactions in which the rate-determining step is the formation of a bond to hydrogen go more quickly and conversely more slowly if a bond is broken. This provides a useful tool for the elucidation of mechanisms (*see* ISOTOPE EFFECTS). Deuterated compounds (e.g., deuterotrichloromethane) are used in NUCLEAR

MAGNETIC RESONANCE SPECTROSCOPY where the spin 1 deuterons resonate at a much lower frequency than hydrogen.

deuterium isotope effect. *See* ISOTOPE EFFECTS.

deuterium oxide (heavy water, D_2O, ^2H_2O). Compared with ^1H_2O, D_2O has a higher melting and boiling point (3.81°C and 101.42°C, respectively) and is more viscous. HYDROGEN BOND formation is stronger and more extensive, and the IONIC PRODUCT is 10 times less. D_2O is separated from water in which it occurs to 0.0145 percent by distillation, electrolysis or chemical exchange. The only large-scale use of D_2O is as a MODERATE in heavy-water nuclear reactors.

deuteron ($^2H^+$, D^+). Cation of DEUTERIUM. It is generated in particle accelerators to bombard targets in the study of nuclear reactions. *See also* PROTON.

deuteronation. Transfer of a DEUTERON to a substrate.

Devarda's alloy. ALLOY of copper (50 percent), aluminium (45 percent) and zinc (5 percent) used in the analysis of nitrates.

devitrification. Crystallization of a GLASS.

dew point. Temperature at which a liquid starts to condense from the vapour, that is the pressure has attained the saturation VAPOUR PRESSURE at that temperature.

dexamethasone. *See* CORTICOSTEROIDS.

dextrins. POLYSACCHARIDES produced by the partial hydrolysis of starch by boiling with water under pressure at 250°C. They are white powders used for making adhesives and in confectionery.

dextrorotatory. *See* OPTICAL ACTIVITY.

dextrose. *See* GLUCOSE.

diacetone alcohol. *See* 4-HYDROXY-4-METHYL-2-PENTANONE.

diacetyl (2,3-butanedione). *See* DIKETONES.

diacetyldioxime. *See* DIMETHYLGLYOXIME.

diacetylmorphine (heroin). *See* ALKALOIDS.

diad axis. *See* CRYSTAL SYMMETRY.

diagonal relationship. Similarity in some properties between main group elements related diagonally in the PERIODIC TABLE. Thus lithium (group I) is similar to magnesium (group II) in forming a nitride and is different from the other members of group I. Beryllium and aluminium also show a diagonal relationship. The effect arises from a similarity in size of the elements in their compounds.

dialysis. Separation of particles of colloidal dimensions by a membrane. The membrane is chosen to allow small molecules and ions to pass while retaining the larger particles. Commonly used membranes are prepared from regenerated CELLULOSE (e.g., COLLODION, CELLOPHANE) and Visking, and these are usually in the form of a thimble or sausage skin. Dialysis is used to remove glucose added to stabilize the formation of some colloids. The kidneys remove urea and other impurities from the bloodstream by dialysis. ULTRAFILTRATION is the application of suction or pressure in dialysis; ELECTRODIALYSIS uses an electric potential to effect the separation.

diamagnetism. Tendency of magnetic dipoles in a material to oppose an applied magnetic field. All materials are diamagnetic to some degree, but this is much less than any paramagnetic effects (*see* PARAMAGNETISM).

3,6-diaminoacridine. *See* PROFLAVINE.

diaminobenzenes (benzenediamines, phenylenediamines, $C_6H_4(NO_2)_2$). Prepared by reduction (Zn/EtOH, NaOH or Fe/HCl) of the corresponding dinitrobenzenes or nitroanilines. They are dibasic. *1,2-diaminobenzene.* Yellow crystals (mp 102°C). Forms HETEROCYCLIC COMPOUNDS easily. With iron(III) chloride the dark red colour of 2,3-diaminophenazine is produced. Benztriazole (I) is formed with nitrous acid, benzimidazole (II) with organic acids and quinoxalines (III) with α-dioxo compounds. It is used as a dye precursor and photographic developer.

1,3-diaminobenzene. White crystals (mp 63°C). It is oxidized in air and forms a brown azo dye (Bismark brown) when treated with nitrous acid. It readily couples with DIAZO COMPOUNDS.

1,4-diaminobenzene. White crystals (mp 147°C). It is oxidized in air; stronger oxidation ($K_2Cr_2O_7$ or HNO_3) yields 1,4-benzoquinone (*see* QUINONES). It is used as a dye and photographic developer.

(I)   (II)

(III)

diaminobenzenes

4,4′-diaminobiphenyl (benzidine, H_2N-C_6H_4–$C_6H_4NH_2$). Colourless solid (mp 127°C), prepared by the BENZIDINE REARRANGEMENT of hydrazobenzene. It is used in the preparation of azo DYES.

1,4-diaminobutane (putrescine). *See* POLYAMINES.

1,2-diaminoethane (ethylenediamine, en, $H_2NCH_2CH_2NH_2$). Colourless liquid (bp 118°C), with a strong ammoniacal smell that fumes in air. It is manufactured by heating 1,2-dichloroethane with an excess of ammonia under pressure over copper(I) chloride. It absorbs carbon dioxide and water from the air to form the carbamate, and is used as a solvent for VAT DYES and for CASEIN and other resins, in adhesives and other coatings and in RUBBERS. PIPERAZINE is formed when 1,2-diaminoethane hydrochloride is heated. It forms complexes with transition metals bonding through each nitrogen atom (*see* CHELATE LIGAND).

1,6-diaminohexane (hexamethylene diamine). *See* POLYAMINES.

2,4-diaminophenol. *See* AMINOPHENOLS.

2,6-diaminopimelic acid (DAP). *See* AMINO ACIDS.

diamond. *See* CARBON.

diamond lattice. *See* CRYSTAL FORMS.

diamorphine (heroin). *See* ALKALOIDS.

diaryl ethers. *See* AROMATIC ETHERS.

diastereoisomers (diastereomers). STEREOISOMERS that are not related as object and mirror images of each other (i.e. they are not ENANTIOMERS). In the 1,2-dibromocyclopropanes, for example, the bromine atoms lie on the same side of the ring in the *cis*-isomer (I), whereas in the *trans*-isomers (II and III) they are on opposite sides of the plane of the ring. Compounds II and III are enantiomers. Compound I, which is superimposable on its mirror image, is diastereometrically related to both II and III (i.e. I and II and I and III are diastereoisomers). Two stereoisomers cannot be both enantiomers and diastereomers, a molecule can have only one pair of enantiomers, but it may (if valencies permit) have several diastereomers. The different diastereomers can be distinguished by X-ray analysis.

(I)

(II)   (III)

diastereoisomers

diazepam (Valium, 7-chloro-2,3-dihydro-1-methyl-5-phenyl-2*H*-1,4-benzodiazepin-2-one). White solid (mp 125°C), slightly soluble in water, soluble in ethanol and in dimethylmethanamide. One of several benzodiazepines widely used as tranquillizers, addictive. *See also* CHLORDIAZEPOXIDE.

diazepam

diazines. *See* HETEROCYCLIC COMPOUNDS, PYRAZINE, PYRIDAZINE, PYRIMIDINE.

diazo compounds. Organic compounds containing –N=N– groups which include diazoalkanes (*see* DIAZOMETHANE), DIAZONIUM COMPOUNDS and AZO COMPOUNDS.

diazomethane (CH_2N_2). Yellow, poisonous gas (bp –24°C), prepared from methylamine hydrochloride and nitroguanidine or by the distillation of *N*-nitroso-*N*-methyl-4-methylbenzenesulphonamide with ethanolic potassium hydroxide. It reacts as a NUCLEOPHILE with the loss of nitrogen and is a useful reagent for introducing the METHENE GROUP. Chloromethane is formed with hydrogen chloride, cyanomethane with hydrogen cyanide, epoxides and methyl ketones with aldehydes, PYRAZOLE

with ethyne and pyrazoline with ethene. Photolysis gives the highly reactive CARBENE. *See also* ARNDT–EISTERT SYNTHESIS.

diazonium compounds ($ArN=N^+ X^-$). Organic DIAZO COMPOUNDS, salts of the base ArN=NOH, named as aryldiazonium salts. They are prepared by the addition of cold sodium nitrite to an acidified solution of the aromatic amine. Diazonium salts are important synthetic reagents being the starting point for the manufacture of azo DYES, pharmaceuticals and other aromatic compounds. (For reactions—*see* Figure.)

dibenzene chromium. *See* SANDWICH COMPOUND.

dibenzocyclopentadiene. *See* FLUORENE.

5,6:14,15-dibenzoheptacene. *See* POLYCYCLIC HYDROCARBONS (structure VII).

dibenzyl. *See* DIPHENYLETHANES.

diborane. *See* BORON INORGANIC COMPOUNDS.

1,2-dibromoethane (CH_2BrCH_2Br). Dense liquid (bp 131°C), manufactured by passing ethene into cooled bromine. It is used in

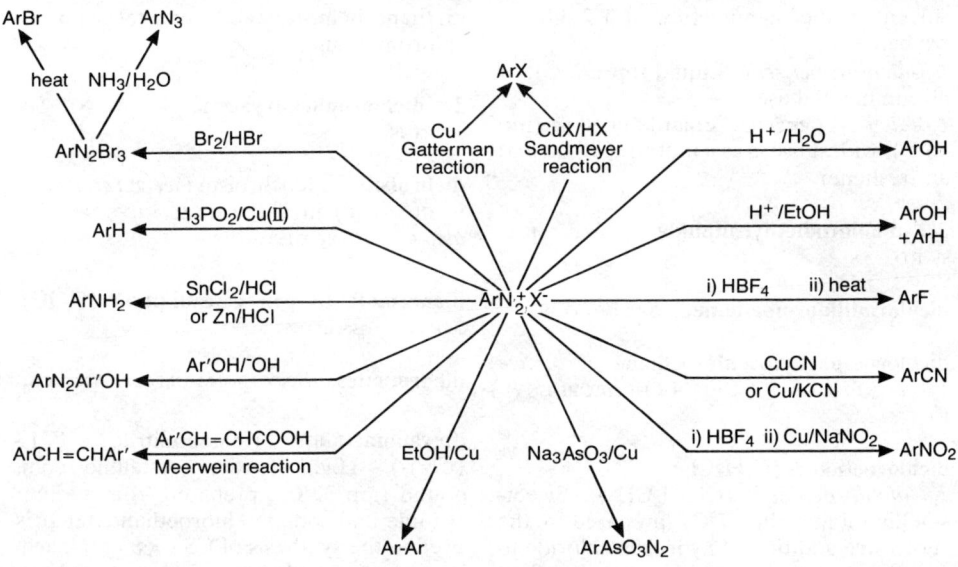

diazonium compounds
Reactions of diazonium compounds.

ANTIKNOCK ADDITIVES, as a fumigant and as a chemical reagent.

dicarboxylic acids. Compounds having two CARBOXYL GROUPS, designated by the suffix -dioic.
saturated dicarboxylic acids $(C_nH_{2n^-}(COOH)_2)$. Crystalline solids, of which the lower members are soluble in water. They form neutral and acid salts and occur widely in living systems. They are prepared by oxidation of the corresponding glycol, hydroxy acid or hydroxy aldehyde or hydrolysis of a dicyano compound or cyano acid. Diesters are prepared by heating the dicarboxylic acid with an alcohol, toluene and concentrated sulphuric acid. *See* BUTANEDIOIC ACID, ETHANEDIOIC ACID, HEPTANEDIOIC ACID, HEXANEDIOIC ACID, PENTANEDIOIC ACID, PROPANEDIOIC ACID.
unsaturated dicarboxylic acids $(C_nH_{2n-2}-(COOH)_2)$. Compounds showing geometrical ISOMERISM about the double bond. Values of $pK_{a,1}$ are less than $pK_{a,2}$, with a greater difference for the *cis*- than the *trans*-isomer. *See also* BUTENEDIOIC ACIDS.

dichlorobenzenes $(C_6H_4Cl_2)$. Prepared by the chlorination of benzene over iron (gives 1,2- and 1,4-isomers) and isomerization to the 1,3-derivative.
1,2-dichlorobenzene. Liquid (bp 179°C), used as a DYE intermediate, insecticide and solvent in the manufacture of 1,2-dihydroxybenzene.
1,3-dichlorobenzene. Liquid (bp 172°C), of no commercial use.
1,4-dichlorobenzene. Volatile crystals (mp 53°C), widest use is as a moth-repellant and air freshener.

2,2'-dichlorodiethylsulphide. *See* MUSTARD GAS.

dichlorodifluoromethane. *See* FREONS.

dichlorodiphenyltrichloroethane. *See* 1,1,1-TRICHLORO-2,2-BIS(4-CHLOROPHENYL)-ETHANE.

dichloroethanes $(C_2H_4Cl_2)$.
1,1-dichloroethane $(CHCl_2CH_3)$. Sweet-smelling liquid (bp 57°C), prepared by the successive addition of hydrogen chloride to ETHYNE. It is hydrolyzed by aqueous alkali to ethanal and is reduced by zinc dust in

methanol to ETHENE.
1,2-dichloroethane (CH_2ClCH_2Cl). Liquid (bp 84°C), manufactured by the reaction between ethene and chlorine in the presence of a catalyst. It is used to manufacture CHLOROETHENE, a precursor of polyvinyl chloride (*see* POLYMERS) and in ANTIKNOCK ADDITIVES. The chlorine atoms may be replaced in reactions by, for example, ethanoates, hydroxyl and amino groups.

dichloroethanoic acid. *See* CHLOROETHANOIC ACIDS.

dichloroethenes $(C_2H_2Cl_2)$.
1,1-dichloroethene $(CH_2=CCl_2)$. Colourless liquid (bp 37°C), prepared by heating 1,1,1-trichloroethane with lime at 80°C. It forms an addition polymer (*see* POLYMERIZATION).
1,2-dichloroethene (acetylene dichloride, CHCl=CHCl). Liquid (bp 55°C), prepared by the action of zinc dust on 1,1,2,2-tetrachloroethane. Two geometrical isomers are formed (80 percent *cis*, 20 percent *trans*). It is used as an industrial solvent.

dichloromethane (methylene chloride, CH_2Cl_2). Colourless liquid (bp 41°C), manufactured by the chlorination of methane. It is used widely as a solvent for paints, greases and cellulose acetate. Dichloromethane is more stable than tetra- or trichloromethane.

2,4-dichlorophenoxyacetic acid. *See* AUXINS.

dichroism. Pleochroism (*see* PLEOCHROMIC CRYSTAL) in a UNIAXIAL CRYSTAL. *See also* CIRCULAR DICHROISM.

dichromate oxygen consumption (DOC). *See* OXYGEN CONSUMED.

dichromates. *See* CHROMIUM COMPOUNDS.

dicyanomethane (malononitrile, $CH_2-(CN)_2$). Highly toxic, crystalline compound (mp 32°C), prepared from sodium cyanide and sodium chloroethanoate. It is used in the synthesis of CS GAS, as a leaching agent for gold and as a polar additive for lubricating oils.

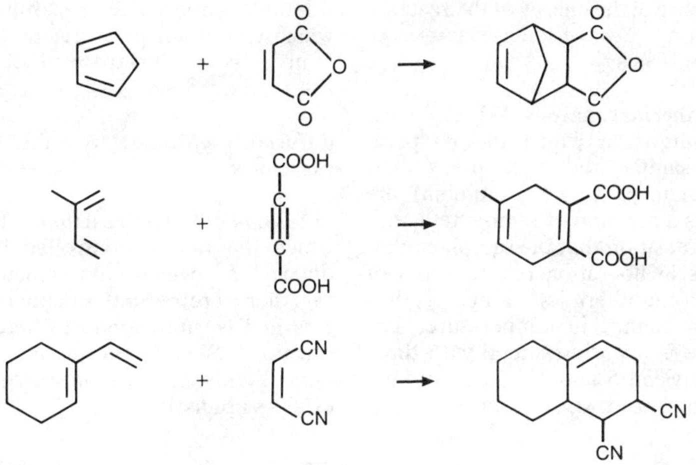

Diels–Alder reaction

di-π-cyclopentadienyl iron (ferrocene). *See* METALLOCENES.

Dieckmann synthesis. *See* RING CLOSURE.

dielectric constant. *See* PERMITTIVITY.

Diels–Alder reaction. 1,4-Cycloaddition of a dienophile (ALKENE or ALKYNE substituted with an electron-withdrawing group, *see* DEACTIVATING GROUP) across a conjugated diene (polyene, enyne or diyne). The reaction requires no catalyst and is stereospecific in that groups attached *cis* on the alkene maintain their stereochemical alignment in the product. Some examples are illustrated (*see* Figure).

dienes (C_nH_{2n-2}). Hydrocarbons with two double bonds. When the bonds are isolated each acts independently (*see* ALKENES). Molecules with cumulative double bonds are known as ALLENES. Double bonds arranged alternately with single bonds (1,3-dienes) are conjugated and have unique properties (*see* BUTA-DIENES).

Dieterici equation. *See* EQUATION OF STATE.

diethanolamine. *See* ETHANOLAMINES.

diethenyl ether (vinyl ether, ($CH_2=CH)_2O$). Unsaturated liquid ETHER (bp 39°C), prepared by heating ethene

chlorohydrin with sulphuric acid, then passing the product, 2,2′-dichlorodiethyl ether, over potassium hydroxide at 200–240°C. It is used as an anaesthetic, and it slowly polymerizes to a GEL on standing.

diethers. *See* ACETALS, KETALS.

1,4-diethylene dioxide. *See* 1,4-DIOXANE.

diethylene glycol. *See* 2,2′-DIHYDROXYDI-ETHYL ETHER.

diethyl ether (ether, ($C_2H_5)_2O$). Volatile liquid ETHER (bp 35°C), prepared by the continuous etherification process—by heating excess ethanol with concentrated sulphuric acid. It is soluble in water, miscible with ethanol and forms explosive mixtures in air due to the presence of the peroxide $CH_3CH(OOH)OC_2H_5$. It is a common solvent for organic synthesis (*see* GRIGNARD REAGENTS) and is also used as an anaesthetic.

diethyl malonate. *See* PROPANEDIOIC ACID.

differential scanning calorimetry (DSC). Analytical technique in which a sample and thermally inert reference material are subjected to a programmed temperature that is linearly ramped with time. The sample or reference is heated to maintain a zero temperature difference, and the energy is plotted as a function of the programmed temperature. This provides a direct calorimetric measure of the energy of the reaction or transition. *See also* DIFFERENTIAL THERMAL ANALYSIS.

differential thermal analysis (DTA). Analytical technique in which the temperatures of a sample and a thermally inert reference material (e.g., α-alumina) are measured as a function of sample temperature. Any transition that the sample undergoes results in liberation (exothermic) or absorption (endothermic) of heat with a concomitant change in temperature. The temperature is linearly ramped with time, typically between 5 and 12°C per minute. *See also* DIFFERENTIAL SCANNING CALORIMETRY.

differential titration. Direct method for the determining the first derivative plot of a potentiometric TITRATION, in which the emf is measured between two identical electrodes; one is shielded, immersed in, the ANALYTE. The addition of an aliquot of titrant produces an emf; this is reduced to zero on complete mixing of the solution. Further sequential additions of titrant give rise to an increasing emf which attains a maximum value at the END POINT.

diffraction. Spreading or bending of waves as they pass through an aperture or round the edge of a barrier. The diffracted waves subsequently interfere with each other producing regions of reinforcement and weakening. *See also* ELECTRON DIFFRACTION, X-RAY DIFFRACTION.

diffraction pattern. *See* X-RAY DIFFRACTION.

diffuse double layer. *See* ELECTRICAL DOUBLE LAYER.

diffusion. Process by which different substances mix as a result of random molecular motion. In any gaseous mixture or liquid solution, kept at uniform temperature, the composition eventually becomes uniform irrespective of the original distribution. The rate of diffusion of molecules in gels is practically the same as that in water, indicating the continuous nature of the aqueous phase. In solids, diffusion occurs very slowly at room temperature. The diffusion of gases into a stream of vapour is of con-

siderable importance in diffusion pumps which can attain pressures as low as 10^{-7} N m^{-2}. *See also* EFFUSION, FICK'S LAWS OF DIFFUSION.

diffusion coefficient. *See* FICK'S LAWS OF DIFFUSION.

diffusion-controlled reaction. Reaction in which the rate is controlled by the frequency at which the reactants come together. It represents an upper limit to the rate and is often found in heterogeneous reactions. *See also* DIFFUSION, ENCOUNTER PAIR, FICK'S LAWS OF DIFFUSION, LIMITING CURRENT DENSITY.

diffusion pump. *See* VACUUM.

digitalis. Dried leaves of the foxglove *Digitalis purpurea*. It is a valuable source of cardiac glycosides (*see* GLYCOSIDES), the main one of which is digitoxin ($C_{41}H_{64}O_{13}$). This is a valuable reagent in steroid chemistry since it forms molecular complexes with steroids having a 3-hydroxyl group of the β-configuration (as in cholesterol), but not those of the α-configuration. On hydrolysis, digitoxin yields digitoxigenen, a potent cardiac poison used (in preference to extracts of digitalis) in small doses to regulate heart action.

digitalis
Structure of digitoxigenen.

dihedral angle (ω). Angle between two parts of a molecule (e.g., hydrogen peroxide).

dihydric alcohols. *See* DIOLS, GLYCOLS.

2,3-dihydroindene. *See* INDAN.

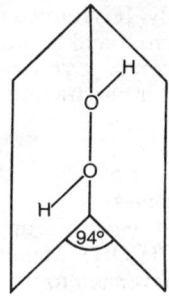

dihedral angle

2,3-dihydropyran (C_5H_8O). Six-membered oxygen HETEROCYCLIC COMPOUND, prepared by the dehydration and rearrangement of tetrahydrofurfuryl alcohol over alumina. It is readily protonated, the ion reacting with ROH to give an ether, which is easily cleaved with dilute acid. It is used as a protecting group for –OH, =NH, –SH and –COOH in organic syntheses.

4,5-dihydropyrazole. *See* 2-PYRAZOLINE.

dihydrostreptomycin. *See* ANTIBIOTICS.

dihydroxybenzenes (benzenediols, C_6H_4-$(OH)_2$).
1,2-dihydroxybenzene (catechol). Colourless solid (mp 105°C), soluble in water, ethanol and ether; prepared by alkaline fusion of 1,2-benzenedisulphonic acid or commercially by heating 2-chlorophenol or 1,2-dichlorobenzene under pressure with 20 percent sodium hydroxide solution at 190°C in the presence of copper sulphate. It is a powerful reducing agent, darkening on exposure to air and reducing silver nitrate and warm FEHLING'S SOLUTION. It is oxidized by silver oxide to 1,2-benzoquinone (*see* QUINONES) and condenses with phthalic anhydride in the presence of sulphuric acid to give ALIZARIN. Adrenaline (*see* HORMONES) is a derivative of catechol. Catechol is used as a photographic developer and antioxidant, and for the preparation of pharmaceuticals and dyes.
1,3-dihydroxybenzene (resorcinol). Colourless solid (mp 110°C), very soluble in water, ethanol and ether; prepared by alkaline fusion of 1,3- (and to some extent 1,2- and 1,4-) benzenedisulphonic acid. It is

not so powerful a reducing agent as the 1,2- and 1,4-compounds, but reduces silver nitrate and Fehling's solutions on warming. Fusion with phthalic anhydride gives fluorescein dyes. It is used extensively in the preparation of dyestuffs, plasticizers and resins.
1,4-dihydroxybenzene (quinol, hydroquinone). Colourless solid (mp 170°C), prepared by the reduction of 1,4-benzoquinone with sulphurous acid and commercially by the oxidation of aminobenzene ($MnO_2 + H_2SO_4$) followed by reduction (Fe) of the benzoquinone formed. Occurs in the glucoside arbutin (*see* GLYCOSIDES). It is a very powerful reducing agent used as a photographic developer. It is oxidized ($FeCl_3$) to 1,4-benzoquinone. Quinol forms CLATHRATE COMPOUNDS with a range of substances.

2,3-dihydroxybutanedioic acid (tartaric acid, CH(OH)COOH.CH(OH)COOH). Found in grapes and other fruit. With two identical asymmetric carbon atoms, two optically active forms (D(+)-, L(–)-) and two optically inactive forms (DL-, *meso*-) are possible (*see* ISOMERISM). It is prepared by the oxidation of BUTENEDIOIC ACIDS, the DL-form from the *cis*-isomer and the *meso*-form from the *trans*-isomer. D(+)-Tartaric acid is found in nature as reddish-brown crystals of potassium hydrogen tartarate (argol) which separate out during wine-making. Cream of tartar is a more pure white form from which the D-acid is obtained by treatment with calcium hydroxide, calcium chloride and then sulphuric acid. Sodium potassium D-tartarate (Rochelle salt) is used in the preparation of FEHLING'S SOLUTION. D-Tartaric acid is used in the food industry (E334) to prepare fizzy drinks. Racemic (DL-) tartaric acid crystallizes as the hemihydrate (mp 200°C). *meso*-Tartaric acid crystallizes as the monohydrate (mp 140°C).

dihydroxybutanes. There are five isomers, three of them chiral (*see* CHIRALITY). They are all colourless, rather viscous liquids. The important isomers are:
1,3-dihydroxybutane (1,3-butanediol, CH_3-CH(OH)CH_2CH_2OH). Manufactured by the reduction of 3-HYDROXYBUTANAL, or by the reaction of *cis*-2-butene with a peracid (*see* GLYCOLS). Properties are of a primary

and a secondary alcohol. It is used in the preparation of 1,3-BUTADIENE (by dehydration) and as an intermediate in plasticizers.

2,3-dihydroxybutane (2,3-butanediol, $CH_3(CHOH)_2CH_3$). Manufactured by the fermentation of potato mash or molasses using *Klebsiella* species, and synthetically from the dibromide. Reactions are of a disecondary alcohol.

1,4-dihydroxybutane ($HOCH_2(CH_2)_2CH_2OH$). Prepared by the reaction of ethyne and methanal, followed by reduction of the product. Reactions are of a typical primary alcohol. It is dehydrated to TETRAHYDROFURAN, and is used in the production of γ-butyrolactone and 2-pyrrolidone.

2,2′-dihydroxydiethyl ether (diethylene glycol, $(HOCH_2CH_2)_2O$). Colourless liquid ETHER (bp 244°C), prepared by the dehydration (phosphoric acid) of 1,2-ethanediol. It is used as a solvent for polymers (e.g., cellulose nitrate), dyes, lacquers, etc., and forms condensation POLYMERS.

4,4′-dihydroxy-α,β-diethylstilbene. *See* STILBOESTROL.

1,2-dihydroxyethane (ethylene glycol, $HOCH_2CH_2OH$). Colourless, odourless, viscous, hygroscopic, highly poisonous liquid (bp 197.6°C); manufactured by the hydration of 1,2-epoxyethane with sulphuric acid. Typical properties of GLYCOLS; it is oxidized to ETHANEDIOIC ACID and carbon dioxide, dehydrated to ETHANAL, reacts with halogen acids to give a halohydrin, and with PX_3 to the dihalide. It forms two series of esters. Glycol dinitrate (glycol + HNO_3/H_2SO_4) is incorporated into dynamite. It is widely used as 'antifreeze' in car radiators and in the manufacture of polyester fibres (e.g., Terylene).

di-(2-hydroxyethyl)amine. *See* ETHANOLAMINES.

3,4-dihydroxyphenylalanine (L-DOPA, $(HO)_2C_6H_3CH_2CH(NH_2)COOH$). L-Amino acid (mp 282°C (decomp.)); isolated from various plant sources, but not found in the human body. It is formed from tyrosine (*see* AMINO ACIDS) and is an intermediate in the synthesis of CATECHOLAMINES. L-DOPA is used in the treatment of Parkinson's disease.

dihydroxypropanes.
1,2-dihydroxypropane (propylene glycol, $CH_3CH(OH)CH_2OH$). Manufactured from propene by conversion to the chlorohydrin ($CH_3CH(OH)CH_2Cl$) followed by hydrolysis with sodium bicarbonate. Properties are typical of GLYCOLS, except that now a secondary alcohol group is involved. It is oxidized to ETHANOIC ACID and carbon dioxide, and rearranges in the presence of acid or on heating to PROPENAL. The molecule is asymmetric, but synthesis gives a RACEMIC MIXTURE and when fermented it becomes laevorotatory because the organism attacks the dextro-form preferentially. It is used as an 'antifreeze' (less poisonous than 1,2-DIHYDROXYETHANE), as a solvent and in the preparation of perfumes.

1,3-dihydroxypropane (trimethylene glycol, $HOCH_2CH_2CH_2OH$). Obtained by the fermentation of glycerol. Properties are of a diprimary alcohol, with the hydroxyl groups far enough apart not to influence each other.

diimides. *See* IMIDES.

diimine. *See* NITROGEN HYDRIDES.

diketones. Compounds containing two OXO GROUPS attached to carbon atoms. They are classed as α,β,γ, etc. according to the relative positions of the oxo groups (i.e. 1,2; 1,3; 1,4; etc.). They are an important class of compounds because of their ability to form CHELATE LIGANDS with TRANSITION ELEMENTS in complexes.

2,3-butanedione (diacetyl, dimethylglyoxal). Simplest α-diketone. A yellow oil (bp 88°C) prepared by the reaction of butanone with nitrous acid and hydrolysis of the resulting oxime.

$$CH_3COCH_2CH_3 + HNO_2 \rightarrow$$
$$CH_3COC(=NOH)CH_3 \rightarrow$$

$$CH_3COCOCH_3$$

Butanone may be oxidized directly by SeO_2, and 2,3-butanedione is also formed by the oxidation of 3-hydroxy-2-butanone. It shows all the typical reactions of KETONES, with first one and then the other oxo group reacting. It is reduced to the DIOL by lithium tetrahydridoaluminate. The dioxime (*see* DIMETHYLGLYOXIME) is used in the estimation of nickel.

2,4-pentanedione (acetylacetone). First β-diketone; colourless liquid (bp 139°C at 746 mmHg) that exhibits tautomerism (*see* ISOMERISM).

$$CH_3COCH_2COCH_3 \rightleftharpoons$$
$$CH_3C(OH)=CHCOCH_3$$

It is prepared by a CLAISEN CONDENSATION between ETHYL ETHANOATE and PROPANONE in DMSO and sodium hydride, or by the condensation of ETHANOIC ANHYDRIDE with propanone with a boron trifluoride catalyst. It is readily oxidized to ethanoic acid and propane by potassium hydroxide, and reacts with hydrazines (*see* NITROGEN HYDRIDES) to give PYRAZOLES. It is an important chelating ligand as the pentane-2,4-dionato group (acac = acetylacetonato).

diketones
Pentane-2,4-dionato ligand.

2,5-hexanedione (acetonylacetone). Simplest γ-diketone; liquid (bp 188°C), prepared from ETHYL 3-OXOBUTANOATE.

$$2(CH_3COCHCOOC_2H_5)^- Na^+ \xrightarrow{I_2}$$

$$\begin{array}{c} CH_3COCHCOOC_2H_5 \\ | \\ CH_3COCHCOOC_2H_5 \end{array} \xrightarrow[(ii)\,HCl]{(i)\,KOH}$$

$$CH_3COCHCOOH \xrightarrow{heat}$$
$$\quad | $$
$$CH_3COCHCOOH$$

$$\begin{array}{c} CH_3COCH_2 \\ | \quad\quad + 2CO_2 \\ CH_3COCH_2 \end{array}$$

With P_2O_5 and P_2S_5 it forms the five-membered ring compounds 2,5-dimethylfuran and 2,5-dimethylthiophen. *See also* BENZIL.

dilatancy. Increase in the apparent VISCOSITY with increasing shear rate. Dilatancy is shown particularly by pastes of densely packed deflocculated particles in which there is only sufficient liquid to fill the voids. Increase in shear rate causes dense packing to be broken down, permitting the particles to flow past one another.

dimensionally stable anode. Anode of titanium covered with platinum or ruthenium oxide used in a diaphragm cell for the production of chlorine by the electrolysis of brine.

dimensions. Expression of the relationship of physical quantities to the fundamental quantities mass, length and time by a formula involving the symbols m, l and t. Thus the dimensional formula for volume is expressed as l^3, velocity $l\,t^{-1}$ and force $m\,l\,t^{-2}$. Other fundamental quantities used in dimensional formulae are: θ, temperature; ε, the dielectric constant of a vacuum; and μ, the magnetic permeability of a vacuum (*see* Appendix Table 3).

dimer. Compound formed from two molecules of a monomer. *See also* OLIGOMER, POLYMERS.

dimesoperiodic acid. *See* IODATES.

dimethylamine. *See* METHYLAMINES.

1 - dimethylaminonaphthalene - 5 - sulphonyl chloride (dansyl chloride, $(CH_3)_2NC_{10}H_6$-SO_2Cl). Yellow–red crystalline solid (mp 70–73°C), used for fluorescent labelling of amines, amino acids and proteins. *See also* DANSYL-.

dimethylbenzenes. *See* XYLENES.

2,3-dimethyl-2,3-butanediol (pinacol, $(CH_3)_2C(OH)C(OH)(CH_3)_2$). GLYCOL (mp 38°C), prepared by reducing propanone with a magnesium amalgam in benzene. It crystallizes from water as the hexahydrate, and undergoes the PINACOL–PINACOLONE REARRANGEMENT and forms dimethylbutane when the vapour is passed over alumina at 400°C.

3,3-dimethyl-2-butanone (pinacolone, $CH_3COC(CH_3)_3$). KETONE; colourless liquid (bp 119°C), prepared by the distillation of 2,3-DIMETHYL-2,3-BUTANEDIOL, or its hydrate, with sulphuric acid. The reaction is an example of the PINACOL–PINACOLONE REARRANGEMENT. It is oxidized by bromine in sodium hydroxide to 2,2-dimethylpropanoic acid.

dimethyl ether (methyl ether, $(CH_3)_2O$). Gaseous ETHER (bp −24°C), manufactured by passing methanol vapour (350–400°C, 15 atm) over aluminium phosphate.

dimethylformamide. *See* N,N-DIMETHYL-METHANAMIDE.

dimethylglyoxal (2,3-butanedione). *See* DIKETONES.

dimethylglyoxime (diacetyldioxime, $CH_3C(=NOH)C(=NOH)CH_3$). OXIME of 2,3-butanedione (*see* DIKETONES); colourless needles (mp 234°C, sublimes 215°C), prepared by the condensation between hydroxylamine and 2,3-butanedione. It slowly polymerizes, condenses with 1,2-diaminobenzene and readily forms CHELATE LIGAND with metal ions. In particular, the dark red salts formed with nickel are used to estimate the metal.

N,N-dimethylmethanamide (dimethylformamide, DMF, $(CH_3)_2NCHO$). Tertiary AMIDE; liquid (bp 152°C) widely used as a NON-AQUEOUS SOLVENT for organic and inorganic compounds. It also acts as a catalyst in substitution, elimination and addition reactions, and as a methanoylating agent with phosphorus oxychloride.

dimethyl sulphate $((CH_3O)_2SO_2)$. Odourless, toxic liquid (bp 188°C), prepared from methanol and chlorosulphonic acid. It is used as a methylating agent (*see* METHYLATION).

dimethylsulphone (DMSO$_2$, $(CH_3)_2SO_2$). Solvent for polar and non-polar compounds, particularly useful at high temperatures.

dimethylsulphoxide (DMSO, $(CH_3)_2SO$). Polar ($\varepsilon_r = 46.7$) non-hydrogen-bonded (dipolar aprotic) solvent for polar and non-polar compounds. It is soluble in water and many organic solvents, forming stable coordination complexes with metals. DMSO is useful as a solvent for bimolecular nucleophilic reactions. It has a low toxicity and is used as a topical antiinflammatory agent.

1,3-dimethylxanthine (theophylline). *See* PURINES.

3,7-dimethylxanthine (theobromine). *See* ALKALOIDS, PURINES.

dimorphism.   Existence of a substance in two crystalline forms (e.g., zinc blende and wurtzite, *see* CRYSTAL FORMS). *See also* POLYMORPHISM.

dinitrobenzenes $(C_6H_4(NO_2)_2)$.
1,2-dinitrobenzene. Yellow crystals (mp 118°C), prepared from the nitration of acetanilide and its subsequent deethanoylation (H$^+$) and oxidation of the amino group ((1) NaNO$_2$, HBF$_4$ (2) NaNO$_2$, Cu). It differs from the 1,3-isomer by undergoing activated nucleophilic substitution with the displacement of one nitro group.
1,3-dinitrobenzene. White crystals (mp 90°C), formed by the nitration (fuming HNO$_3$ + conc. H$_2$SO$_4$) of nitrobenzene. Reduction with sodium sulphide gives first 3-nitroaniline, then 1,3-diaminobenzene. Hydrogen atoms *ortho* to a nitro group are activated towards nucleophilic attack: for example, sodium hydroxide and potassium ferricyanide give mostly 2,4-dinitrophenol.

1,4-dinitrobenzene. White crystals (mp 173°C); preparation and chemical reactions are similar to 1,2-dinitrobenzene.

2,4-dinitrofluorobenzene. *See* 1-FLUORO-2,4-DINITROBENZENE.

dinitrogen. *See* NITROGEN.

dinitrogen oxide (nitrous oxide). *See* NITROGEN OXIDES.

dinitrogen tetroxide. *See* NITROGEN OXIDES.

dinitrogen trioxide. *See* NITROGEN OXIDES.

diols (dihydric alcohols). Alcohols containing two hydroxyl groups (*see* GLYCOLS).

diopside. *See* SILICATES.

1,4-dioxane (1,4-diethylene dioxide). Six-membered oxygen HETEROCYCLIC COMPOUND (structure XXIII); colourless, very toxic liquid (bp 101°C), prepared by the acid-catalyzed condensation of 1,2-ethanediol followed by cyclization or by passing 2,2-epoxyethane over sodium hydrogen sulphate at 120°C. It gives oxonium salts with acids, halogens, mercury(II) salts and triiodomethane. Chlorination gives the di- and tetra-chloro derivatives. It is a valuable solvent in the pure state, being miscible with water to form a series of solvents of varying ε_r (2.24–80.4).

dioxin (2,3,7,8-tetrachlorodibenzo-*p*-dioxin). Byproduct of the production of 2,4,5-trichlorophenol from 1,2,4,5-tetrachlorobenzene, 1,2-dihydroxyethane and sodium hydroxide. Very toxic to humans.

dioxin

2,3-dioxobutane (2,3-butanedione). *See* DIKETONES.

dioxygen. *See* OXYGEN.

2,6-dioxypurine (xanthine). *See* PURINES.

2,6-dioxytetrahydropyrimidine (uracil). *See* PYRIMIDINES.

dipeptide. Compound formed by the condensation of two amino acids through a –CONH– link, with the elimination of water. For two amino acids there are two possible dipeptides; for example: glycylalanine ($H_2NCH_2CONHCH(CH_3)COOH$) and alanylglycine ($H_2NCH(CH_3)CONHCH_2COOH$).

diphenyl. *See* BIPHENYL.

diphenylacetylene. *See* DIPHENYLETHYNE.

diphenylamine (*N*-phenylbenzeneamine, $(C_6H_5)_2NH$). Secondary AMINE (mp 54°C), prepared by heating (260°C) phenol with aniline in the presence of zinc chloride, or by the ULLMANN REACTION; manufactured by heating (140°C, pressure) aniline and aniline hydrochloride. It is a weaker base than aniline ($pK_a = 0.9$) and is sufficiently acid to form salts with alkali metals. It is used to prepare azo DYES, and its derivatives are used in propellants and in the manufacture of RUBBER. It is a redox INDICATOR (colourless to violet at about 0.8 V versus DHE).

diphenylethanes
1,1-diphenylethane (($C_6H_5)_2CHCH_3$). Liquid prepared from benzene and ethyne in the presence of copper(I) chloride as catalyst.
1,2-diphenylethane (dibenzyl, $C_6H_5CH_2CH_2C_6H_5$). White solid (mp 52°C), prepared by the FRIEDEL–CRAFTS REACTION (PhH + CH_2ClCH_2Cl) or the condensation of benzyl bromide in the presence of sodium or copper. It is oxidized catalytically to benzaldehyde.

trans-**diphenylethene.** *See* STILBENE.

diphenyl ether. *See* AROMATIC ETHERS.

diphenylethyne (diphenylacetylene, $C_6H_5C{\equiv}CC_6H_5$). Colourless crystals (mp 61°C), obtained by treatment of stilbene dibromide with ethanolic potassium hydroxide. It shows the properties of a triple bond, and the substitution reactions of the benzene ring.

diphenylmethane ($C_6H_5CH_2C_6H_5$). Crystalline solid (mp 25.9°C), prepared by the FRIEDEL–CRAFTS REACTION ($PhCH_2Cl$ + PhH) or Grignard reaction (*see* GRIGNARD REAGENTS) ($PhMgBr$ + $PhCH_2Cl$). Reactions are similar to those of BIPHENYL, the first substituent entering the 4-position and the second the 4'-position. The METHENE GROUP is very reactive, bromination giving $C_6H_5CHBrC_6H_5$. Oxidation (CrO_3) yields BENZOPHENONE and when the vapour is passed through a red-hot tube FLUORENE is formed.

α,α-diphenyl-β-picrylhydrazyl radical. *See* 1,1-DIPHENYL-2-*p*-TRINITROPHENYLHYDRAZYL RADICAL.

1,1 - diphenyl - 2 - *p* - trinitrophenylhydrazyl radical (DPPH, α,α-diphenyl-β-picrylhydrazyl radical). Stable FREE RADICAL used as a standard in ELECTRON SPIN RESONANCE SPECTROSCOPY. It contains 1.63×10^{21} unpaired electrons per gram.

diphosphoric acid. *See* PHOSPHORUS OXO-ACIDS.

dipole. Combination of two electrically or magnetically charged particles of opposite sign separated by a short distance. Every covalent bond between unlike atoms constitutes a dipole. Electrons are drawn more closely to one of the atoms than to the other thereby conferring a negative charge on this atom. By convention the symbol ↦ indicates the direction of the dipole from the less to the more electronegative atom.

dipole–dipole interaction. *See* INTERMOLECULAR FORCES.

dipole–induced dipole interaction. *See* INTERMOLECULAR FORCES.

dipole moment (p; units: C m; $D = 10^{-18}$ esu cm = 3.336×10^{-30} C m). Product of one of the charges in a DIPOLE and the distance of separation of the charges ($p = qr$). For an isolated molecule $p = \sum_i q_i r_i$, where q_i are the charges on all nuclei and electrons, and r_i the position vectors referred to any origin. Dipole moments are vectoral in character. The dipole moment of a molecule as a whole can be regarded as the vector sum of the individual bond or group moments (*see* Figure).

dipole moment
θ = bond angle = 105°. Total dipole moment
$p = 2 \times 1.52 \cos 52.5 = 1.85$ D.

Perfectly symmetrical molecules such as hydrogen, tetrachloromethane, carbon dioxide and benzene have dipole moments of zero, whereas asymmetrical molecules, with an imbalance in the distribution of charge, are permanent dipoles. Dipole moments provide information about molecular structure and test calculated values of wavefunctions. Dipole moments can be determined by: (1) the vapour–temperature method, in which the relative permittivity and density of the vapour are measured at a series of temperatures; (2) the dilute solution method, in which relative permittivity, refractive index and specific volume of dilute solutions of a polar solute in a non-polar solvent are measured. They are important in determining the suitability of a solvent for a given solid and in the determination of INTERMOLECULAR FORCES.

dipole potential. *See* SURFACE POTENTIAL.

dipyridyls (2,2'-dipyridyl,I; 4-4'-dipyridyl, II). Both isomers are formed by the action of sodium on pyridine, with the subsequent oxidation of the disodium salt or by the oxidation of the corresponding phenanthroline followed by decarboxylation. 2,2'-Dipyridyl (dipy) (mp 72°C), formed by heating 2-bromopyridine with copper. It is an important chelating agent; when added to a solution of nickel(II) chloride the pink tris-2,2'-dipyridylnickel(II) chloride is formed. The derivatives, paraquat and diquat, are used as herbicides.

(I)   (II)

dipyridyls

direct dyes. DYES that colour fabric directly from aqueous solution, requiring no MORDANT or fixing agent. Direct dyes are usually sodium salts of sulphonic acids and are applied in baths containing sodium chloride or sodium sulphate.

disaccharides. GLYCOSIDES consisting of two sugar molecules; crystalline solids, soluble in water, which on hydrolysis yield two sugar molecules and which exhibit either reducing or non-reducing properties. Examples include SUCROSE, MALTOSE, CELLOBIOSE and LACTOSE (*see* Figure). Two other disaccharides that have been prepared are melibiose (from the trisaccharide RAFFINOSE) and gentiobiose (from the trisaccharide gentianose); these differ from the other disaccharides in that the two monosaccharide units are linked through the C-6 of the monosaccharide.

sucrose
α-D-glucopyranosyl-β-D- fructofuranoside
(I)

maltose (α-anomer)
4-O- α-D-glucopyranosyl-D-glucose
(II)

cellobiose (β-anomer)
4-O-β-D-glucopyranosyl-D-glucose
(III)

lactose (β-anomer)
4-O-β-D-galactopyranosyl-D-glucose
(IV)

gentiobiose (β-anomer)
6-O-β-glucopyranosyl-D-glucose
(V)

melibiose (β-anomer)
6-O-α-galactopyranosyl-D-glucose
(VI)

disaccharides
Naturally occurring compounds.

disilane. *See* SILICON INORGANIC COMPOUNDS.

disinfectant. Substance that kills or inhibits the growth of harmful microorganisms. Incomplete disinfectants only destroy the vegetative forms; complete disinfectants destroy the spores as well. Representative disinfectants are mercury compounds, oxidizing agents (e.g., hydrogen · peroxide), halogens and halogen compounds (Cl_2, I_2, NaOCl), PHENOLS and cresols (*see* HYDROXYTOLUENES), synthetic detergents, alcohols of low molecular mass (except methanol), gases (sulphur dioxide, methanal, 1,2-epoxyethane), heat and electromagnetic waves. Disinfectants are used to cleanse surgical equipment, sick rooms, drains, etc.

disintegration. Process in which an atomic nucleus breaks up into two or more fragments, spontaneously in a radioactive decay process or as a result of bombardment with high-energy particles.

dislocation. *See* DEFECT STRUCTURE.

disperse dyes. DYES applied to a material in a fine suspension. Such dyes are used particularly for dyeing cellulose triethanoate and other material that will not easily take soluble dyes.

disperse phase. Discontinuous phase of a two-phase system consisting of particles or droplets of one substance distributed through another continuous phase, the dispersion medium (e.g., the solid phase in a SOL, the gas phase in a foam). *See also* COLLOID.

dispersion. *See* COLLOID.

dispersion force. *See* INTERMOLECULAR FORCES.

displacement chromatography. Elution of components of a mixture on a column by a substance that is more strongly adsorbed than the components of the mixture. The components are eluted with sharp and distinct boundaries (e.g., ION EXCHANGE CHROMATOGRAPHY).

disproportionation. Reaction of the form $2A \rightarrow B + C$, often $2AX \rightarrow A + AX_2$. A may be a FREE RADICAL (e.g., $2C_2H_5^{\cdot} \rightarrow C_2H_4 + C_2H_6$ (X=H)) a transition element ion (e.g., $2Cu^+ \rightarrow Cu^{2+} + Cu$ (X=e)), or a compound (e.g., $2PF_4Cl \rightarrow PF_5 + PF_3Cl_2$).

disrotation. *See* CONROTATION.

dissociation. Splitting of a molecule into fragments (molecules, ions, atoms or free radicals). The EQUILIBRIUM CONSTANT for the reaction is known as the dissociation constant. The degree of dissociation (α) may be related to the dissociation constant: for example, for a simple acid HA $\rightleftharpoons$ $H^+ + A^-$

$$K = c\alpha^2/(1 - \alpha)$$

where c is the total concentration of HA.

dissociation constant. *See* EQUILIBRIUM CONSTANT.

dissociation energy. Thermodynamic dissociation energy (D_0) is the energy required to dissociate a diatomic molecule in its GROUND STATE. Spectroscopic dissociation energy (D_e) is the energy difference between the bottom of the potential well (*see* MORSE POTENTIAL) at the equilibrium internuclear separation and two atoms infinitely separated. D_0 and D_e differ by the ZERO-POINT ENERGY

$$D_0 = D_e - 1/2\, h\omega_e$$

dissolved oxygen (DO). Indicator of the condition of a water supply for biological, chemical and sanitary investigations. Adequate DO is necessary for the life of fish and aquatic animals.

distillation. Process by which volatile components may be purified and separated from other components in a mixture. It is usually carried out at constant pressure. The temperature of the mixture is raised until the total vapour pressure equals the external pressure; the vapour in equilibrium with the liquid is richer in the more volatile component. Fractional distillation is the separation of the two components of a binary liquid mixture by repeated distillation and condensation or by the use of a fractionating column. Efficiency of the column is determined by the number of theoretical plates to which the separation corresponds. For packed columns, efficiency is rated as the HEIGHT EQUIVALENT OF A THEORETICAL PLATE (HETP). A tube 30–40 cm high can be packed to the equivalent of about 100 theoretical plates, thus effecting practically perfect separation of all but the most difficult mixtures. *See also* COMPLETELY MISCIBLE LIQUID SYSTEMS, IMMISCIBLE LIQUIDS.

distillation diagrams. *See* COMPLETELY MISCIBLE LIQUID SYSTEMS.

distribution coefficient. *See* PARTITION COEFFICIENT.

disulphur dichloride. *See* SULPHUR HALIDES.

disulphuric acid. *See* SULPHUR OXOACIDS.

disulphurous acid (pyrosulphurous acid). *See* SULPHUR OXOACIDS.

diterpenes. *See* TERPENES.

dithiocarbonates. *See* XANTHATES.

dithionic acid. *See* SULPHUR OXOACIDS.

dithionous acid. *See* SULPHUR OXOACIDS.

DL- (configurational symbol). *See* OPTICAL ACTIVITY.

DLVO theory (Deryagin–Landau and Verwey–Overbeek theory). Theory that predicts the stability of LYOPHOBIC COLLOIDS by considering the energy of interaction (V_R) between particles. Overlap of double layers is repulsive, and this is balanced by attractive van der Waals and London forces (*see* INTERMOLECULAR FORCES). The attractive potential (V_A) at distance R may be approximated by $V_A = -aA/12R$, where a is the radius of the particles and A is the Hamaker constant, which has values between 3 and 20×10^{-20} J for different colloids. The repulsive potential varies exponentially with R. Variation in potential with distance leads to (1) maximum in V_R and a stable colloid, and (2) no maximum allowing coagulation to occur (*see* Figure). *See also* SCHULTZE–HARDY RULE.

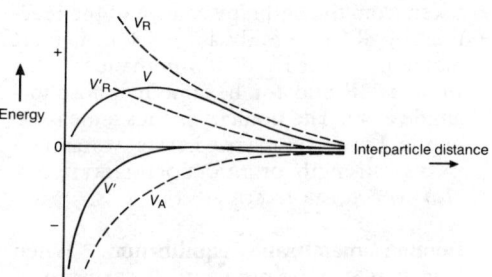

DLVO theory
Energy of intraction as a function of interparticle distance of a lyophobic colloid. $V = V_A + V_R$ for a stable colloid; $V' = V_A + V'_R$ for an unstable colloid.

DME. *See* DROPPING MERCURY ELECTRODE.

DMF (dimethylformamide). *See* N,N-DIMETHYLMETHANAMIDE.

DMSO. *See* DIMETHYLSULPHOXIDE.

DMSO₂. *See* DIMETHYLSULPHONE.

D_n. *See* DONICITY.

DNA (deoxyribonucleic acid). *See* NUCLEIC ACIDS.

DNF (2,4-dinitrofluorobenzene). *See* 1-FLUORO-2,4-DINITROBENZENE.

DNP-. Prefix denoting a 2,4-dinitrophenyl derivative, usually of an amino acid. Yellow DNP-amino acids (amino acid + 1-fluoro-2,4-dinitrobenzene) are used as reference standards in the determination of amino acid sequences in proteins.

DO. *See* DISSOLVED OXYGEN.

DOC (dichromate oxygen consumption). *See* OXYGEN CONSUMED.

dodecahedral coordination. *See* SHAPES OF MOLECULES.

dodecyl alcohol (lauryl alcohol). *See* 1-HYDROXYDODECANE.

dolomite. *See* CALCIUM, MAGNESIUM.

donicity (D_n). Measure of the ability of a solvent (S) to associate with an ionic solute, taken from the enthalpy change of the reaction $S + SbCl_5 \rightarrow S.SbCl_4^+ + Cl^-$ in an inert medium. D_n for 1,2-dichloroethane is 0, for water is 18 and for hexamethylphosphoramide is 40. The donicity is thus a measure of the Lewis base (*see* LEWIS ACIDS AND BASES) strength or NUCLEOPHILICITY. *See also* NON-AQUEOUS SOLVENTS.

Donnan membrane equilibrium. When ions that cannot permeate a SEMIPERMEABLE MEMBRANE are on one side only of the membrane the ions that can pass through the membrane do not have equal concentrations on either side. The results of osmotic pressure measurements on solutions of charged colloidal particles (*see* COLLOID) (e.g., proteins) need to be corrected for this effect or must be measured under conditions that minimize it (the addition of a large concentration of an indifferent electrolyte).

donor. *See* ACCEPTOR, COORDINATE BOND.

L-DOPA. *See* 3,4-DIHYDROXYPHENYL-ALANINE.

doping. Introduction of small amounts of substances to change the properties of a solid. In domain doping, the dopant does not form a SOLID SOLUTION, but remains as discrete aggregates. *See also* SEMICONDUCTOR.

d-orbital. *See* ATOMIC ORBITAL.

Dorn effect. *See* ELECTROKINESIS.

dose. Measure of the extent to which matter has been exposed to ionizing radiation.
absorbed dose. Energy per unit mass absorbed as a result of exposure (unit: GRAY, RAD).
maximum permissible dose. Recommended upper limit of absorbed dose that a person or organ should receive in a specified period of time (International Commission on Radiological Protection).
equivalent dose. Weighted absorbed dose equal to the product of the absorbed dose and a dimensionless factor (the quality factor, QF). The concept is introduced in an attempt to quantify the effects of different types of radiation. The SIEVERT is the SI unit of dose equivalent; for X-, gamma- and beta-rays $QF = 1$, for fast neutrons and protons $QF = 10$, for alpha-particles and fission fragments $QF = 20$. For radiation workers the annual whole body dose equivalent is 50 mSv and for the population at large 5 mSv.
exposure dose. Measure of X- or gamma-radiation to which a body is exposed. It is equal to the total charge collected on ions of one sign produced in unit mass of dry air (unit: ROENTGEN).

double decomposition (metathesis). Exchange reaction of the form $AB + CD \rightarrow AC + BD$. Usually one product is insoluble or volatile, and is therefore removed allowing the equilibrium to move in favour of the products.

double layer. *See* ELECTRICAL DOUBLE LAYER.

double refraction. Effects caused by having different refractive indices in different directions. This may be along crystallographic axes, or due to molecular alignment in electric or magnetic fields (e.g., $Fe(OH)_3$ sol) or in a rotating cylinder viscometer. This latter effect is known as streaming birefringence.

double salt. Salt formed by the crystallization from a solution of two or more components in a distinct lattice. *See also* ALUMS.

doublet. *See* MULTIPLET.

Down's cell. *See* SODIUM.

DPPH. *See* 1,1-DIPHENYL-2-p-TRINITROPHENYLHYDRAZYL RADICAL.

dropping mercury electrode (DME). Electrode in which a series of mercury drops are allowed to fall from a capillary tube into a pool of mercury that usually serves as the counter electrode. The hydrodynamics of the flow of electrolyte around the growing drop are known, and a fresh, smooth surface is presented to the electrolyte with each new drop. It is used extensively in polarographic analysis (*see* POLAROGRAPHY).

dyes
Colour Index classification of dyes.

Class	Chromophore
Nitroso (quinone oxime)	$-N=O$
Nitro	$-NO_2$
Azo	$-N=N-$
Azoic	$-N=N-$ formed in situ from amines, nitrosoamines, etc.
Stilbene	$ArCH=CHAr$ plus azo or azoxy
Diphenylmethane (ketone imine)	$Ar-\underset{\underset{NH}{\|}}{C}-Ar$
Triarylmethane	$(Ar_3C^+)\ X^-$ $Ar_2C=Ar=NH$ $Ar_2C=Ar=O$
Xanthene	
Acridine	
Quinoline	
Polymethine	$N-(C=C)_n-N^+$ (cyanine) $O-(C=C)_n-O^-$ (oxonol) $N-(C=C)_n-O$ (merocyanine)
Thiazole	
Indamine and indophenol	

continued overleaf

Class	Chromophore
Azine	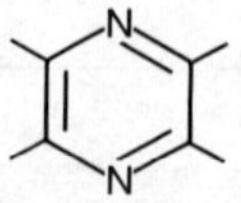
Oxazine	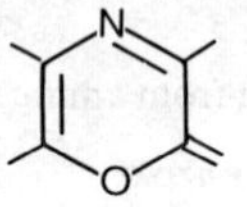
Thiazine	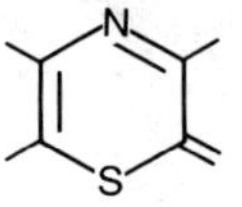
Sulphur	Unknown
Lactone	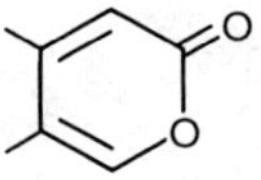
Amino ketone and hydroxy ketone	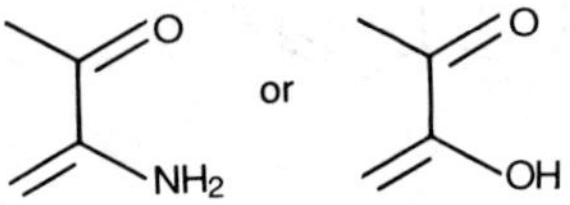
Anthraquinone	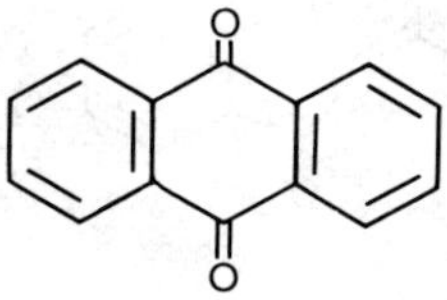
Indigoid	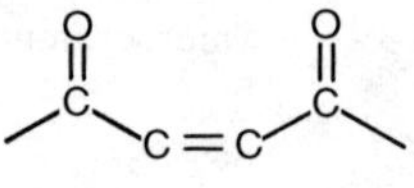
Phthalocyanine	*See* PHTHALOCYANINS

dry cell. *See* LECLANCHÉ CELL.

dry ice (solid carbon dioxide). *See* CARBON OXIDES.

DSC. *See* DIFFERENTIAL SCANNING CALORIMETRY.

DTA. *See* DIFFERENTIAL THERMAL ANALYSIS.

Duhem–Margules equation. Equation relating the vapour pressure to the composition of an ideal BINARY LIQUID MIXTURE. If the vapour behaves ideally, the

GIBBS–DUHEM EQUATION can be written as

$$(\partial \ln p_A/\partial \ln x_A)_{T,P} = (\partial \ln p_B/\partial \ln x_B)_{T,P}$$

which, if the solution is ideal, becomes

$$(\partial p_A/\partial x_A)_{T,P} = (p_A^{\ominus}/p_B^{\ominus})(\partial p_B/\partial x_B)_{T,P}$$

Thus plots of p_A and p_B against x_A and x_B, respectively, slope in opposite directions.

dulcitol. *See* HEXAHYDRIC ALCOHOLS.

Dulong and Petit's law. For a solid element, the product of the atomic mass and the specific HEAT CAPACITY (i.e. the molar heat capacity) is approximately $3R$ (25 J K^{-1} mol^{-1}). For certain 'soft' elements (those with a low FORCE CONSTANT, e.g., lead, copper) the molar heat capacity approaches a limiting value of $3R$ at room temperature. For elements such as carbon (diamond), silicon and boron with a high FORCE CONSTANT and/or low REDUCED MASS, the molar heat capacity does not approach the limiting value even at very high temperatures.

Dupré equation. *See* ADHESION.

Dy. Dysprosium (*see* LANTHANIDE ELEMENTS).

dyes. Coloured substances that impart colour to other materials. According to the theory of O. N. Witt, a dye molecule must possess a CHROMOPHORE. The colour may also be mediated by the presence of AUXOCHROMES. Natural dyes include saffron, henna, cochineal, woad, Tyrian purple and indigo. These are generally of the MORDANT-type and require fixing. The first synthetic dye was ANILINE. Twenty-two structural classes are recognized by the COLOUR INDEX (*see* Table). The method of application also gives rise to a different classification (*see* ACID DYES, BASIC DYES, DIRECT DYES, DISPERSE DYES, MORDANT, REACTIVE DYES, VAT DYES). The largest use of dyes is for colouring textiles. Other uses are in foodstuffs, wood, leather, polymers and PIGMENTS.

dynamic nuclear polarization. *See* NUCLEAR MAGNETIC RESONANCE SPECTROSCOPY.

dynamic viscosity. *See* VISCOSITY.

dynamites. *See* EXPLOSIVES.

dyne (dimensions: $m\,l\,t^{-2}$; units: g cm s^{-2}). CGS UNIT of force; defined as that force which when applied to a body of mass 1 gram gives it an acceleration of 1 centimetre per second squared (1 dyne $= 10^{-5}$ N).

dysprosium (Dy). *See* LANTHANIDE ELEMENTS.

E

E. *See* ELECTRODE POTENTIAL, ELECTRO-MOTIVE FORCE.

E- (stereochemical descriptor). *See* ISO-MERISM.

e. *See* ELECTRON.

ε. *See* PERMITTIVITY.

ε_0. *See* ZERO-POINT ENERGY.

η. *See* HAPTO-, OVERPOTENTIAL, VISCOS-ITY.

E1 mechanism (elimination unimolecular mechanism). Mechanism in which the RATE-DETERMINING STEP is ionization to a CARBONIUM ION (*compare* S_N1 MECHAN-ISM), followed by the elimination of a β-proton to form an alkene

$$(CH_3)_3Br \rightleftharpoons (CH_3)_3C^+ + Br^- \quad \text{(slow)}$$

$$HO^- + H\text{-}CH_2\text{-}C^+(CH_3)_2 \longrightarrow$$
$$H_2O + CH_2=C(CH_3)_2 \quad \text{(fast)}$$

The relative amounts of elimination or substitution processes (i.e. E1 or S_N1) from the same carbonium ion is mainly determined by the temperature, E1 being favoured at higher temperatures. E1 mechanism is more prevalent from *tert*-alkyl halides than *sec*-alkyl halides.

E2 mechanism (elimination bimolecular mechanism). Mechanism that does not involve an intermediate CARBONIUM ION; a typical reaction is

$$EtO^- + H\text{-}CH_2\text{-}CH(CH_3)\text{-}Br \longrightarrow$$
$$EtOH + CH_2=CHCH_3 + Br^-$$

In the presence of a strong base and not too polar solvent, S_N2 and E2 reactions occur in preference to S_N1 and E1 pro-cesses. The conditions for a substitution reaction with a *tert*-alkyl halide are a polar solvent, low temperature and weak base (S_N1 conditions) and for an elimination reaction a strong base, less polar solvent and higher temperature (E2 conditions). E2 reactions, which are stereospecific, give good yields of alkenes with secondary and primary bromides and sulphonates.

EAN. *See* EFFECTIVE ATOMIC NUMBER.

earth's atmosphere. *See* AIR.

ebullioscopic constant. *See* BOILING POINT.

EC. *See* ELECTRON CAPTURE.

eclipsed conformer. *See* CONFORMERS, NEWMAN PROJECTION FORMULA.

Edman degradation. Method for the determination of N-terminal amino acids in a PEPTIDE, without its hydrolysis, leading to its amino acid sequence. The pure

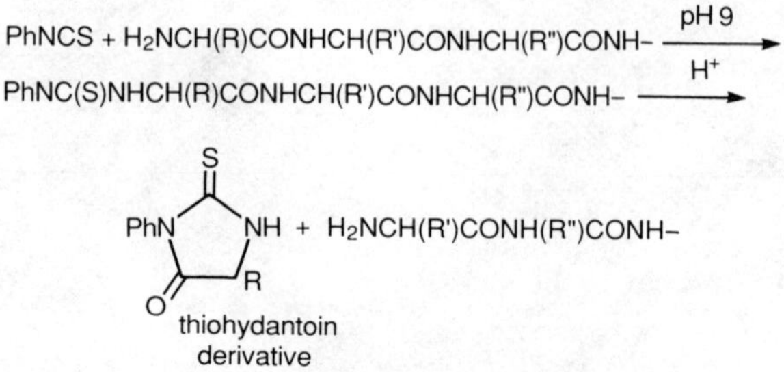

$$PhNCS + H_2NCH(R)CONHCH(R')CONHCH(R'')CONH- \xrightarrow{pH\,9}$$
$$PhNC(S)NHCH(R)CONHCH(R')CONHCH(R'')CONH- \xrightarrow{H^+}$$

thiohydantoin derivative

Edman degradation

156

peptide is treated with phenyl isothio-cyanate (pH 9) to give an *N*-phenylthio-carbamyl derivative. Treatment of the derivative with hydrogen chloride in nitro-methane yields a thiohydantoin and a peptide with one less amino acid residue (*see* Figure). By comparison with hydant-oins prepared from known amino acids the thiohydantoin, and hence the N-terminal amino acid, can be identified. Up to 30 successive Edman degradations can be performed automatically on polypeptides or small proteins.

EDTA (ethylenediaminetetraacetic acid, $(HOOCCH_2)_2NCH_2CH_2N(CH_2COOH)_2$). Important chelating or SEQUESTERING AGENT. EDTA is a hexadentate ligand (*see* CHELATE LIGAND, Table) and is used in titrations to determine transition metals.

EELS. *See* ELECTRON ENERGY LOSS SPECTROSCOPY.

E_F. *See* FERMI ENERGY.

EFA. *See* ESSENTIAL FATTY ACIDS.

effective atomic number (EAN). Number of electrons in the valence shell of a transi-tion metal complex. For a stable complex EAN = 18 (*see* EIGHTEEN-ELECTRON RULE), although some complexes of, for example, platinum(II) and iridium(I) are formed with EAN = 16.

effective nuclear charge. Charge due to the nucleus experienced by an electron in an ATOMIC ORBITAL. SHIELDING of the nucleus by inner shell electrons causes the effective nuclear charge to fall with increasing prin-cipal QUANTUM NUMBER and with the more directional orbitals.

efficiency (η). For a reversible Carnot heat engine, fraction of HEAT absorbed by the machine that is converted into WORK (*see* CARNOT CYCLE)

$$\eta = -w_{max}/q_2 = (T_2 - T_1)/T_2$$

where q_2 is the heat absorbed at the higher temperature (T_2). This is the maximum efficiency that could be expected if there were no inefficiencies in the operation of the machine.

efflorescence. Production of a powdery solid from crystals by the spontaneous loss of water of crystallization. It occurs when the partial water vapour pressure of the surroundings falls below the dissociation pressure of the hydrate (e.g., Na_2SO_4. $10H_2O$). *See also* DELIQUESCENCE.

effusion. Passage of a gas through an orifice (in a thin wall) that has a smaller diameter than the MEAN FREE PATH of the gas molecules. At constant pressure, the rate of effusion is proportional to the area of the orifice and to the root mean square speed of the molecules (and hence inversely proportional to the square root of the mole-cular mass). Thus

$$\text{rate of effusion/m s}^{-1} = (p/2\pi\rho)^{1/2}$$

This principle is the basis of Knudsen's method for determination of the molecular mass or, if that is known, the vapour pressure of a solid

$$p/\text{N m}^{-2} = w(2\pi RT/M_r)^{1/2}$$

where $w/\text{kg m}^{-2} \text{ s}^{-1}$ is the rate of effusion. *See also* GRAHAM'S LAW OF EFFUSION.

eigenfunction. *See* EIGENVALUE, WAVE-FUNCTION.

eigenvalue. Energy associated with an exact solution of the SCHRÖDINGER EQUA-TION.

eighteen-electron rule. Stable organo-metallic compounds of TRANSITION ELE-MENTS should have 18 valence electrons about the metal, that is they have the elec-tronic structure of the next highest noble gas. *See also* EFFECTIVE ATOMIC NUMBER.

eight-*N* rule. *See* HUME–ROTHERY'S RULE.

EIMS (electron impact mass spectrometry). *See* MASS SPECTROMETRY.

einstein. Amount of energy absorbed by one mole of a substance undergoing a photochemical reaction (*see* PHOTO-CHEMISTRY).

Einstein coefficients (A, B). Rate of emission of a photon from an electronically EXCITED STATE is $A + B.E'(\lambda)$, where A is the coefficient of spontaneous emission, B is the coefficient of stimulated emission and $E'(\lambda)$ is the energy density of radiation wavelength λ. Stimulated emission is more important at higher wavelengths as $A/B = 8\pi h(\lambda c)^{-3}$.

Einstein equation. *See* BROWNIAN MOTION.

einsteinium (Es). *See* ACTINIDE ELEMENTS.

Einstein photoelectric equation. Relationship between the energy E of electrons emitted by the action of electromagnetic radiation of frequency v. For a METAL $E = hv - e\phi$, where ϕ is the work function of the metal. For a gas $E = hv - I$, where I is the IONIZATION ENERGY of the gas.

Einstein–Stark law. Law of photochemistry which states that one quantum is absorbed by the molecule responsible for the primary photochemical process. This does not imply that only one molecule of product is formed when one photon is absorbed. If the subsequent reaction is a CHAIN REACTION, then the absorption of one photon might lead to several product molecules. *See also* QUANTUM EFFICIENCY.

Einstein's mass–energy equation. Relationship presenting the concept that energy (E) possesses mass (m)

$$E = mc^2$$

See also FIRST LAW OF THERMODYNAMICS.

Einstein's viscosity equation. Equation relating the VISCOSITY (η) of a suspension of particles to the viscosity of the solvent (η_0)

$$[(\eta/\eta_0) - 1]/\phi = \eta_{sp}/\phi = v + \kappa\phi$$

where ϕ is the volume fraction of the particles, and v and κ constants which depend on the shape of the particles. For spheres, $v = 2.5$, a value that increases with the AXIAL RATIO of the particles.

elastic. Descriptive of materials having the capacity to recover rapidly after a large deformation. *See also* ELASTOMER, POLYMERS, RUBBER.

elastin. *See* PROTEINS.

elastomer. Compound that is ELASTIC. Natural and synthetic RUBBERS comprise the largest class of elastomers, and recently polyurethane and thermoplastic elastomers based on block copolymers (*see* COPOLYMERIZATION) have been synthesized. Elastomers are classed by ASTM as M (saturated carbon chain), O (oxygen in chain), R (unsaturated chain rubbers), Q (silicon in chain), T (sulphur in chain) and U (oxygen and nitrogen in a carbon chain).

electrical double layer. Asymmetrical arrangement of ions at the interface between an electrolyte and another phase (e.g., ELECTRODE, MICELLE) (*see* Figure). The double layer acts like a capacitor which, in the most simple theory of Helmholtz and Perrin, has a constant capacitance due to a fixed layer of ions at the so-called OUTER HELMHOLTZ PLANE (OHP). A theory that is based on the approach of Debye and Hückel assumes a more diffuse layer spreading out exponentially into the electrolyte (Gouy–Chapman). The Stern model brings these extremes together in a

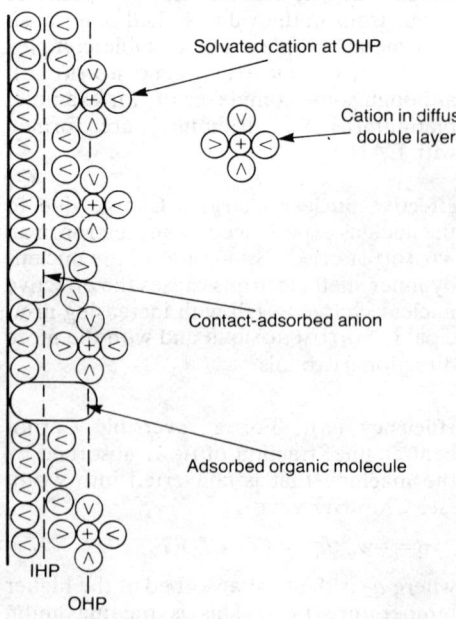

Solvated cation at OHP

Cation in diffuse double layer

Contact-adsorbed anion

Adsorbed organic molecule

IHP

OHP

electrical double layer
Possible structure of the electrode/electrolyte interface.

theory which can explain most of the experimental observations. Species may also adsorb on the surface of a solid phase (contact or specific adsorption), creating an inner Helmholtz layer which may have a different polarity to the OHP. *See also* ZETA-POTENTIAL.

electric dipole moment. *See* DIPOLE MOMENT.

electric polarization. *See* POLARIZATION.

electric susceptibility (χ_e). Dimensionless quantity, relating the electric POLARIZATION (P) and the electric field strength (E) producing it.

$$\chi_e = P/\varepsilon_0 E = \varepsilon_r - 1$$

electrocatalysis. Effect of the nature of an electrode surface on electrochemical reactions. For example, for a given OVERPOTENTIAL the hydrogen evolution current is some 10^{10} times greater on a platinum cathode than on mercury.

electrochemical cell. *See* CELLS.

electrochemical equivalent. Weight of a substance liberated or consumed by 1 coulomb of electricity. *See also* FARADAY'S LAWS OF ELECTROLYSIS.

electrochemical sensor. Device that can determine the presence and concentration of a particular substance by the measurement of an electrochemical property. Potentiometric devices measure equilibrium potentials (e.g., pH electrode, or other ION-SELECTIVE ELECTRODES). In galvanic and polarographic sensors, the substance is consumed electrochemically, the current passing is proportional to the amount of substance present (*see* OXYGEN PROBE).

electrochemical series. Arrangement of metals in decreasing order of their tendency to lose electrons and become ions. The tabulation of standard ELECTRODE POTENTIALS allows extension to non-metallic electrochemical reactions.

electrocyclic reaction. *See* WOODWARD–HOFFMANN RULE.

electrode. Electronic conductor in contact with an ELECTROLYTE that acts as a source of or sink for electrons. The electrode may be inert or take part in the electrochemical reaction. The term electrode is also used as a synonym for HALF-CELL.

electrodeposition of metals. Cathodic reduction of ions to the metal. Most metallic cations are discharged at potentials near to their equilibrium potential, but with nickel, iron and other transition element ions a significant OVERPOTENTIAL is required. When the electrolyte contains more than one ion, simultaneous discharge results in alloy deposition. In electrorefining, metals are produced with high purity. In some cases (e.g., aluminium), this is the only method of production (*see* HALL–HEROULT PROCESS). Metals may be recovered from dilute solutions by electrolysis in electrowinning. Electroplating is the coating of surfaces by a thin layer of metal to provide a decorative or protective finish (e.g., tin, nickel and chromium).

electrode potential (E; units: V). Potential of a HALF-CELL relative to that of a standard HYDROGEN ELECTRODE.
standard electrode potential ($E^{\ominus}$) of the half-cell ox + $ne \rightleftharpoons$ red is the potential that would be obtained with ox and red present at unit activity. All standard electrode (reduction) potentials refer to the reaction written as a reduction.

The electrode potential is related to the standard electrode potential and the activities of the ions by the NERNST EQUATION. The electrode potential is proportional to the free energy change for the reaction

$$\Delta G = -nFE$$

and thus is related to the equilibrium constant

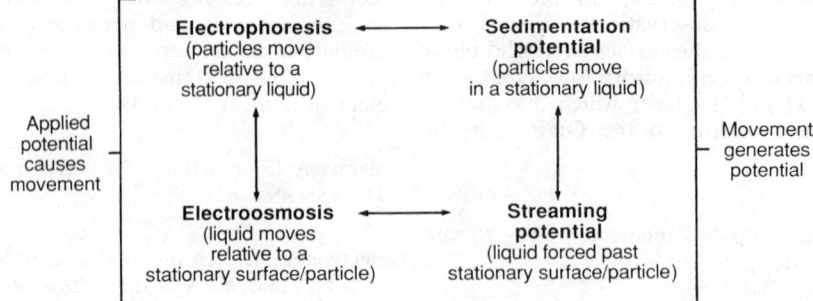

electrokinesis

$E^{\ominus} = (RT/nF) \ln K_{\text{therm}}$

and through the temperature-dependence of the potential to the ENTROPY change.

electrodialysis. Accelerated form of DIALYSIS carried out in a three-compartment apparatus in which the central compartment is separated by a SEMIPERMEABLE MEMBRANE. The colloid is contained in the central compartment, and a potential is applied between platinum gauze electrodes mounted in the surrounding solvent or dilute electrolyte. Electrodialysis is widely used in the desalination of brackish water.

electrodics. Study of the dynamics of electrochemical reactions. *See also* ELECTROCATALYSIS.

electrodispersion. Method of making a colloidal dispersion (*see* COLLOID) of a metal by striking an arc between metal electrodes in an inert dispersion medium.

electroflotation. Method of separating dilute suspensions into slurries and clear liquid. A fine stream of electrochemically generated bubbles is passed through the suspension carrying the particulate matter to the top of the cell where it is skimmed off. Electroflotation is employed in the treatment of steel rolling-mill wastes, paper-mill effluents, scouring liquors from yarn and for dealing with abattoir wastes.

electroforming. Process in which a thick layer of metal is electrodeposited (*see* ELECTRODEPOSITION OF METALS) on to a mould which may be of metal or other material that has been given a conducting coat.

electrokinesis. Group of phenomena (*see* Figure) that have a common origin in the asymmetrical distribution of ions at an interface, the ELECTRICAL DOUBLE LAYER. Movement of the phases at the interface depends on or determines the ZETA-POTENTIAL.

electrokinetic potential. *See* ZETA-POTENTIAL.

electrolysis. Use of electrical energy to bring about a chemical reaction. Important industrial processes are the electrolysis of brine to produce chlorine and chlorates and sodium hydroxide, and of water to give hydrogen. The voltage required to effect the reaction is the sum of the thermodynamic potential ($= -\Delta G/nF$), the OVERPOTENTIALS for each electrode reaction and a voltage to overcome the ohmic resistance of the electrolyte. ELECTRODEPOSITION OF METALS is an example of electrolysis.

electrolyte. Ionically conducting medium in an electrochemical cell. This is usually a solution of ions, but solid electrolytes (e.g., β-alumina) are used in some high-temperature batteries. An indifferent electrolyte carries almost all the current in a cell, but takes no part in the electrode processes. This ensures that the electroactive materials are transported to the electrodes by DIFFUSION. An electrolyte may also be classed as strong or weak according to its degree of dissociation (*see* CONDUCTANCE EQUATIONS).

electrolytic cell. *See* CELLS.

electromagnetic radiation. Sinusoidally fluctuating orthogonal electric and magnetic fields. It is characterized by its frequency (ν) or wavelength (λ), which are related to the velocity of propagation (c)

electromagnetic radiation
Units used in spectroscopy to define the energy of electromagnetic radiation.

	1 Hz	1 cm⁻¹	1 J mol⁻¹	1 eV
Hz	1.0	2.9979×10^{10}	2.5053×10^{9}	2.4182×10^{14}
cm⁻¹	3.3356×10^{-11}	1.0	8.3567×10^{-2}	8.0663×10^{3}
J mol⁻¹	3.9915×10^{-10}	11.9660	1.0	9.6485×10^{4}
eV	4.1353×10^{-15}	1.2397×10^{-4}	1.0364×10^{-5}	1.0

(the speed of light) by $c = \lambda v$. The energy (E) of the radiation is directly proportional to its frequency by PLANCK CONSTANT (h): $E = hv$. Electromagnetic radiation also has a particulate nature, the energy of a PHOTON being related to its momentum by DE BROGLIE LAW. A summary of the different non-SI units that are used in spectroscopy is listed (*see* Table). *See also* POLARIZED RADIATION.

electromagnetic spectrum. Range of energies of ELECTROMAGNETIC RADIATION. The extent of the electromagnetic spectrum of interest to chemical spectroscopists is illustrated (*see* Figure).

electrometric titration (potentiometric titration). *See* TITRATIONS.

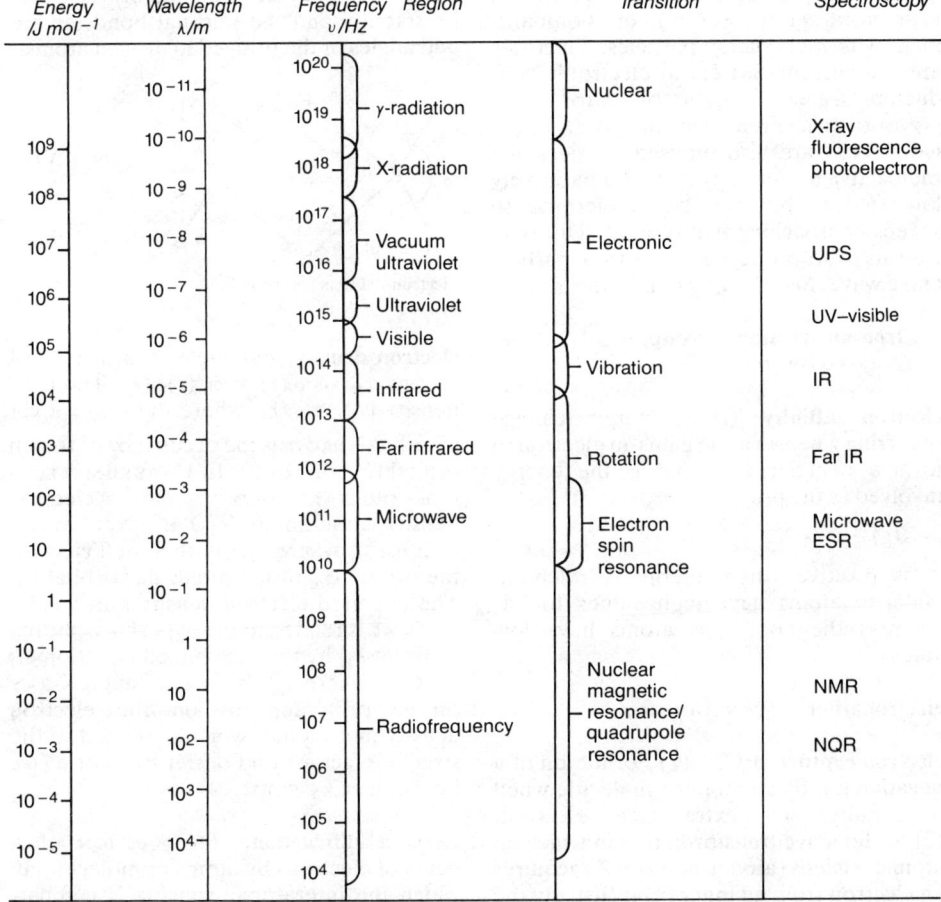

electromagnetic spectrum

electromotive force (emf, dimensions: $\varepsilon^{-1/2} m^{1/2} l^{1/2} t^{-1}$; units: $V = kg\ m^2\ s^{-3}\ A^{-1} = J\ A^{-1}\ s^{-1}$). Force that causes a current to flow between two points. The unit of electric potential is the VOLT. The sign of the electromotive force is defined so that a positive charge will tend to flow from a higher to a lower potential. Applied to a CELL, the emf is the electric potential between two pieces of metal with identical composition, the ends of the chain of conducting phases. Digital voltmeters have replaced potentiometers for the measurement of emf values.

electron ($_{-1}^{0}e$; mass 9.109 534 × 10^{31} kg; charge: 1.602 189 2 × 10^{19} C). Fundamental particle that can exist as a constituent of an atom (see ELECTRONIC CONFIGURATION) or in the free state. Electrons can be removed from some elements by heat, light (see ELECTRON SPECTROSCOPY) or bombardment with high-energy particles. Electrons are the current carriers in electronic conductors (see BAND THEORY OF SOLIDS, SEMICONDUCTOR). Free electrons (i.e. BETA-PARTICLES) are also emitted by decaying radioactive nuclei; they have comparatively low energy, but can be accelerated to speeds approaching that of light. The electron has the properties of both a particle and a wave. See also QUANTUM THEORY.

electron-abstracting group. See DE-ACTIVATING GROUP.

electron affinity (A). Energy change occurring when an atom gains an electron to form a negative ion, that is the energy involved in the process

$$X(g) + e \rightarrow X^-$$

A is positive when energy is released. Halogen atoms have high values for A, whereas the noble gas atoms have low values.

electronation. See REDUCTION.

electron capture (EC). (1) Formation of a negative ion by an atom or molecule when it acquires an extra free electron. (2) Radioactive transformation in which an atomic nucleus (atomic number Z) acquires an electron from an inner orbit (usually the K-shell) of the atom transforming it to a

nucleus of atomic number Z–1 (e.g., decay of $_{29}^{64}Cu$ to $_{28}^{64}Ni$). The disintegration energy is carried away by an emitted neutrino (ν)

$$_{1}^{1}p + _{-1}^{0}e \longrightarrow _{0}^{1}n + \nu$$

This type of capture is accompanied by emission of an X-ray, photon or Auger electron as the vacancy in the inner orbit is filled by an outer electron.

electron-deficient compound. Compound that has too few valence electrons to provide a pair for each pair of atoms close enough to be regarded as covalently bonded. This is found in boranes (see BORON INORGANIC COMPOUNDS), aluminium alkyls (see ALUMINIUM ORGANIC DERIVATIVES, CARBON-TO-METAL BOND) and bridged CARBONIUM IONS. A MOLECULAR ORBITAL description of diborane (for structure, see Figure) adequately describes its stability and the unusual bond lengths and angles of the bridged hydrogen atoms.

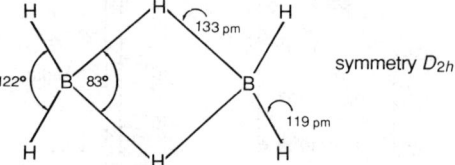

electron-deficient compound

electron density. One-electron density of an ORBITAL is $\rho_i(r) = \phi^*(r)\phi(r)$. The total density is $\sum_i n_i\rho_i(r)$, where the sum is over all orbitals and n_i is the occupancy of the ith orbital ($i = 0$, 1 or 2). In HÜCKEL MOLECULAR ORBITAL THEORY, the π-electron density at atom m (q_m) is given by $q_m = \sum n_r c_{rm}^2$ where c_{rm} is the coefficient of the orbital ϕ_m in the molecular orbital ψ_r. The unpaired electron density ρ in a FREE RADICAL is related to the hyperfine splitting constant a_H by the McConnell equation $a_H = Q\rho$, where Q is a constant. X-RAY DIFFRACTION depends on the electron density in a crystal, which is related to the structure factor by a Fourier transform (see FOURIER TRANSFORM ANALYSIS).

electron diffraction. DIFFRACTION of a beam of electrons by atoms or molecules in which the interatomic spacing is comparable to the wavelength of the beam. Electrons have the advantage over X-rays that their wavelength can be set by adjusting the accelerating voltage, V ($\lambda = h/(2eVm)^{1/2}$). Unlike X-rays, they have low penetrating power and so cannot easily to used to investigate crystal structure. Electron diffraction is used to study the structure (bond lengths and angles) of molecules in the gas phase and extensively in the study of solid surfaces (*see* LOW-ENERGY ELECTRON DIFFRACTION).

electronegativity. Measure of the ability of an atom to attract electrons to itself. The Mulliken electronegativity scale (χ^M) takes the mean of the ionization potential (*see* IONIZATION ENERGY) and the ELECTRON AFFINITY. From the Pauling electronegativity scale (χ^P), which has a more complex derivation, it is possible to calculate the DIPOLE MOMENT of a diatomic bond ($\Delta\chi^P$), the percent ionicity of the bond ($16\Delta\chi^P + 3.5(\Delta\chi^P)^2$) and the covalent/ionic resonance energy ($(\Delta\chi^P)^2$) eV, where $\Delta\chi^P = \chi_A^P - \chi_B^P$.

electron energy loss spectroscopy (EELS). Analytical technique associated with electron microscopy. A monochromatic beam of electrons causes ionization of inner-shell electrons of the sample, losing energy in the process. An EEL spectrum is a plot of the intensity of the electrons against energy loss. The peaks are weak and broad (most electrons are elastically scattered, which leads to an intense peak at zero loss), and the spectrum becomes increasingly complex with increasing atomic number. The technique is therefore most useful for light elements (e.g., carbon, oxygen and nitrogen) and thus complements techniques such as X-ray fluorescence (*see* X-RAY SPECTROMETRY). The fine structure of an EEL spectrum is studied in the technique extended energy loss fine structure (EXELFS) in the electron analog of EXTENDED X-RAY ABSORPTION FINE STRUCTURE (EXAFS).

electronic configuration. Arrangement of electrons in atoms and molecules. The number of electrons in an atom is the same as the atomic number, ranging from one for hydrogen to 103 for lawrencium. These are arranged in one to seven shells around the nucleus. The tendency for electrons to form complete outer shells accounts for the

valence of the element; these 'valence' electrons play an essential role in chemical bonding (*see* ATOMIC ORBITAL, PERIODIC TABLE, QUANTUM NUMBER). In molecules a LINEAR COMBINATION OF ATOMIC ORBITALS is used to construct MOLECULAR ORBITALS. *See also* ELECTRON PAIR, MOLECULAR ORBITAL CALCULATIONS, SHAPES OF MOLECULES.

electronic formula. *See* FORMULA.

electronic partition function. *See* PARTITION FUNCTION.

electronic spectroscopy. Absorption of light in the visible (400–750 nm) or UV part of the ELECTROMAGNETIC SPECTRUM causes transitions between electronic levels. Bands are seen in molecular spectra corresponding to the envelope of many vibrational transitions, each of which may have rotational fine structure. Atomic spectra (*see* ATOMIC SPECTROSCOPY, BALMER SERIES) consist simply of a series of sharp lines, also known as line spectra.

The SELECTION RULES for electronic transitions are, for ideal Russell–Saunders coupling $\Delta S = 0$, $\Delta L = 0$, ± 1 (*see* TERM SYMBOLS). (For experimental details, *see* SPECTROPHOTOMETER.) Electronic spectra give information about vibrational levels of molecules, dissociation energies (from the onset of the continuum), ionization potentials from atomic spectra, the structures of organic molecules (*see* AUXOCHROME, CHROMOPHORE) and the electronic structures of COORDINATION COMPOUNDS. Through the BEER–LAMBERT LAW, ABSORBANCE in this region is proportional to concentration that leads to the quantitative use of electronic spectra.

electron impact mass spectrometry. *See* MASS SPECTROMETRY.

electron nuclear double resonance spectroscopy. *See* ELECTRON SPIN RESONANCE SPECTROSCOPY.

electron pair. Two electrons with opposed spins (*see* PAULI EXCLUSION PRINCIPLE) in an ATOMIC ORBITAL or MOLECULAR ORBITAL. The electron pair may be in a bond or a lone pair (*see* LONE-PAIR ORBITAL).

electron paramagnetic resonance spectroscopy. *See* ELECTRON SPIN RESONANCE SPECTROSCOPY.

electron probe analysis. Analytical method in which an electron beam is scanned across a small area (diameter: 1 μm) of a sample and the characteristic X-rays detected. The method is capable of high spacial resolution and may detect as little as 10^{-14} g of an element.

electron spectroscopy. Techniques that measure the energies of electrons ejected from solids or gases after excitation by photons—known as photoelectron spectroscopy (generally PES), UV (UPS) or X-ray (XPS)—or energetic electrons (used in some forms of Auger electron spectroscopy AES, *see* AUGER EFFECT). The kinetic energy (E) of the primary electron is

$$E = h\nu - I - \Delta E_{vib} - \Delta E_{rot}$$

where $h\nu$ is the energy of the incident photon, I the ionization energy and ΔE_{vib} and ΔE_{rot} the changes in vibrational and rotational energy, respectively. The sample — a low-pressure gas or solid — is confined in an evacuated chamber with a source of exciting radiation. Emitted electrons are passed into an electrostatic analyzer from which a spectrum of the number of electrons against energy is produced. In UPS, the source is a helium discharge lamp ($\lambda = 58.4$ nm, $h\nu = 21.22$ eV) which excites outer or conduction band electrons. Ionization potentials (*see* IONIZATION ENERGY) of one or more transitions may be measured together with vibrational frequencies for GROUND STATE or EXCITED STATE molecules (*see* KOOPMAN'S THEOREM). Hyperfine structure due to rotational transitions is usually beyond the resolving power of electron spectrometers. Transitions to very high energy MOLECULAR ORBITALS give rise to Rydberg series of lines which tend to a limit of complete ionization. Information from UPS allows direct testing of theories of bonding, hybridization, etc. Irradiation by X-rays (e.g., MgK$_\alpha$, $h\nu = 1253.6$ eV; AlK$_\alpha$, $h\nu = 1486.6$ eV) causes the emission of core electrons. XPS is also known as electron spectroscopy for chemical analysis (ESCA), particularly when used to study solids. As core electrons do not take part in bonding, their energies are quite specific to the element, allowing ESCA to be used as a sensitive analytical technique. In addition, small CHEMICAL SHIFTS are seen due to differences in chemical environment, OXIDATION STATE, etc. The binding energy for an electron increases linearly with the charge on the atom. In ESCA of solids, the electrons can only escape from depths of about 5 nm, which makes PES a very surface-specific method. The filling of core levels by an outer electron is accompanied by the emission of X-rays (*see* X-RAY SPECTROSCOPY) or an Auger electron.

electron spectroscopy for chemical analysis (ESCA). *See* ELECTRON SPECTROSCOPY.

electron spin. Property of an electron that gives rise to its angular momentum. The spin angular momentum is quantized, the component along a given axis being $\pm (1/2)h/2\pi$. The spin QUANTUM NUMBER (M_s) takes values $\pm 1/2$. The spin angular momentum should not be confused with orbital angular momentum. Splitting of spin levels in a magnetic field is the basis of ELECTRON SPIN RESONANCE SPECTROSCOPY.

electron spin resonance spectroscopy (ESR spectroscopy, electron paramagnetic resonance (EPR) spectroscopy). Species that have an odd electron, which interacts weakly with the orbital angular momentum, particularly organic FREE RADICALS, have two energy levels in a magnetic field (B), E

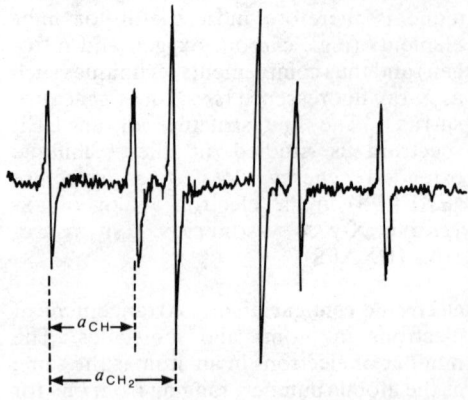

electron spin resonance spectroscopy
ESR spectrum of the radical $CH_3CH_2\dot{C}HOH$.
The signal is split into a doublet (by CH) of triplets (by –CH$_2$–). The methyl protons do not couple at all.

$= M_s g\beta B$. M_s is the spin quantum number (*see* ELECTRON SPIN), g is the LANDÉ g-FACTOR or spectroscopic splitting factor, having a value around two, and β is the BOHR MAGNETON. The transition between the levels is thus of energy $g\beta B$, which for a field of 0.33 T corresponds to a frequency of 9 GHz in the microwave region of the ELECTROMAGNETIC SPECTRUM. The microwave signal is modulated at 100 kHz, which leads to a first derivative spectrum (*see* Figure).

Measurement of g-values for paramagnetic ions of TRANSITION ELEMENTS shows deviations from the free electron value of 2.00232. For organic radicals coupling between NUCLEAR SPINS and electron spins leads to hyperfine splitting of the lines in the ESR spectrum. The energy of coupling between electrons and protons in a neutral radical is about 50 MHz (or 20 gauss). Further splitting occurs if the atom has a nuclear spin greater than one-half (e.g., nitrogen-14 has a nuclear spin of 1 which splits a line into three ($M_I = -1, 0, 1$)).

In electron nuclear double resonance (ENDOR), the sample is irradiated at one of the ESR frequencies as the spectrum is recorded as a function of a scanned frequency. The fixed frequency decouples the effect of that atom and leads to simplified spectra (compare with INDOR in NUCLEAR MAGNETIC RESONANCE SPECTROSCOPY). *See also* SPIN TRAPPING.

electron transport chain. Sequence of coupled reversible oxidation–reduction systems (in order of increasing redox potential) by which electrons are passed from the reduced COENZYME ($NADH_2^+$ or $FADH_2$) to molecular oxygen which is thereby reduced to water. As a result NAD^+ (FAD) is reformed and three (two) molecules of adenosine triphosphate produced from adenosine diphosphate and inorganic phosphate. It is the final stage of aerobic respiration in mitochondria involving several cytochromes (cyt)

$$NAD^+ \underset{\downarrow}{\rightarrow} FAD \rightarrow Co\ Q_{10} \rightarrow cyt\ b \underset{\downarrow}{\rightarrow}$$
$$\qquad ATP \qquad\qquad\qquad\qquad ATP$$

$$cyt\ c \rightarrow cyt\ a \underset{\downarrow}{\rightarrow} cyt\ a_3 \rightarrow O_2$$
$$\qquad\qquad ATP$$

The overall reaction is

$$NADH_2^+ + 3ADP + P_i + 0.5O_2 \longrightarrow$$
$$NAD^+ + 3ATP + H_2O$$

for which $\Delta G^{\ominus\prime} = -220\ kJ\ mol^{-1}$.

electron volt (eV). Energy acquired by one electron accelerated through one volt; equal to 1.602×10^{-19} J or 96.485 kJ mol^{-1}. It is used as an energy unit when discussing electrons in solids (*see* FERMI ENERGY), where typical energies are a few electron volts, and in high-energy particle physics.

electron-withdrawing group. *See* DEACTIVATING GROUP.

electroosmosis. Movement of a liquid phase through a capillary or plug of finely divided material when a potential difference is applied across the electrolyte (*see* ELECTROKINESIS). It is the reverse of ELECTROPHORESIS.

electrophile (electrophilic reagent). Electron-deficient reagent which is capable of accepting electrons. Examples include reducing agents, Lewis acids, positive ions (e.g., NITRONIUM ION) or molecules with a positive charge on a particular atom (e.g., sulphur(VI) oxide). In organic reactions they tend to attack negatively charged parts of a molecule. *See also* NUCLEOPHILE.

electrophilic addition. Characteristic reaction of ALKENES in which the first, RATE-DETERMINING STEP is attack by an ELECTROPHILE on the electron-rich part of the molecule, followed by a fast nucleophilic attack on the CARBONIUM ION:

$$>C=C< + H^+\ Z^- \longrightarrow -\overset{|}{C}-\overset{|}{\underset{H}{\overset{+}{C}}}< + Z^-\quad (slow)$$

$$-\overset{|}{C}-\overset{+}{\overset{|}{C}}< + Z^- \longrightarrow -\overset{|}{\underset{H}{C}}-\overset{|}{\underset{Z}{C}}-\quad (fast)$$

where $H^+\ Z^-$ is a strong proton acid (e.g., sulphuric, hydrochloric, hydrobromic or trifluoroethanoic acid).

electrophilic reagent. *See* ELECTROPHILE

electrophilic substitution. Reaction in which the first step is the attack by an ELEC-TROPHILE and the displacement of a group or atom. It is a feature of the reactions of benzene, in which an electrophile approaches the delocalized π-electrons in the ring (e.g., sulphonation or nitration)

$$PhH + NO_2^+ \longrightarrow PhNO_2 + H^+$$

electrophoresis. Migration of ions, colloidal particles, etc. through a stationary liquid under an applied electric field. Separation and identification of components in a mixture by this method depend upon the electrophoretic mobilities of the components at a fixed pH and ionic strength. The different experimental techniques available depend on the nature and state of dispersion of the material. These are the moving boundary or Tiselius method, particulate microelectrophoresis and zone electrophoresis in which the separation is carried out in a stabilizing or supporting medium such as filter paper, silica gel, etc. This may be combined with PAPER CHROMATOGRAPHY in two-dimensional electrochromatography. Polymers and paints may be deposited by an electrophoretic method by making the surface to be coated an electrode in a cell. *See also* ELECTROKINESIS, ISOELECTRIC FOCUSING.

electrophoretic mobility. *See* IONIC MOBILITY.

electroplating. *See* ELECTRODEPOSITION OF METALS.

electropositive. Descriptive of elements that have a tendency to lose electrons to give a positive ion.

electrorefining. *See* ELECTRODEPOSITION OF METALS.

electrorheology. Phenomenon of increased viscosity of a dispersion in an electric field. This is seen, for example, with poly(lithium 2-methylpropenoate) in hexane.

electrostatic precipitation. Removal of fine particles suspended in a gas by an electric field. Ions produced in the gas become attached to the particles and carry them to the electrodes.

electrostriction. Volume change caused by an electric field, especially an inhomogeneous electric field, for example, in the vicinity of an ion.

electrovalent bond. *See* IONIC BOND.

electrowinning. *See* ELECTRODEPOSITION OF METALS.

element. Substance that cannot be further divided by chemical means. It is defined by its ATOMIC NUMBER. ISOTOPES may have different numbers of neutrons. There are 92 naturally occurring elements plus the synthetic transuranic elements (*see* ACTINIDE ELEMENTS, TRANSACTINIDE ELEMENTS). *See also* PERIODIC TABLE.

element 108. *See* TRANSACTINIDE ELEMENTS.

elementary particles. Fundamental constituents of all matter in the universe. Some 200 short-lived elementary particles have been discovered and described. Two main classes are recognized: leptons (ELECTRONS, muons, NEUTRINOS, tau-particles) with weak interactions and no apparent internal structure; and hadrons (NUCLEONS, pions) with strong interactions and complex internal structure. Hadron structure is currently based on the concept of quarks, with charges that are $+2/3$ or $-1/3$ of the electronic charge. In this model, hadrons are divided into baryons (consisting of three quarks) which decay into protons, and mesons (consisting of a quark and an antiquark) which decay into leptons and photons. On this theory the only true elementary particles are leptons and quarks; the proton being a baryon consists of three quarks of different 'flavours' ($+2/3$, $+2/3$, $-1/3 = 1$) and the neutron ($2/3$, $-1/3$, $-1/3 = 0$). For each flavour there are equivalent antiquarks. The theory is well established by circumstantial evidence, but so far no quarks have been isolated.

elevation of boiling point. *See* BOILING POINT.

Eley–Rideal mechanism. *See* RIDEAL–ELEY MECHANISM.

α-elimination. Special type of elimination in which both groups are lost from the same carbon atom: for example, in the reaction between *tert*-butoxide ion and tribromomethane to give dibromocarbene (i.e. loss of hydrogen and bromine atoms from the same carbon), a very reactive intermediate.

$$(CH_3)_3C-O^- + H-CBr_3 \longrightarrow$$

$$(CH_3)_3CO-H + \bar{C}Br_3$$

$$Br-\underset{\underset{Br}{|}}{\bar{C}}-Br \longrightarrow [Br-\ddot{C}-Br] + Br^-$$

β-elimination. *See* E1 MECHANISM, E2 MECHANISM.

elimination bimolecular mechanism. *See* E2 MECHANISM.

elimination unimolecular mechanism. *See* E1 MECHANISM.

Elinvar. Commercial name for a nickel–chromium steel (36 percent nickel, 12 percent chromium + tungsten, manganese) with a small temperature coefficient of elasticity.

Ellingham diagram. Graphical method (*see* Figure) of representing the variation of the standard GIBBS FUNCTION with temperature for oxidation reactions

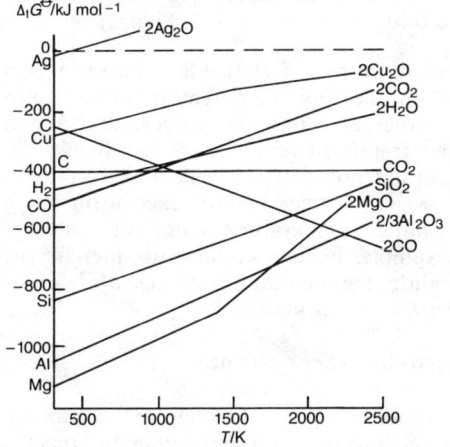

Ellingham diagram

involved in metal extraction. The linear $\Delta G^{\ominus}/T$ plots which, with the exception of carbon monoxide, have positive slopes meaning that the oxides (except carbon monoxide) become less stable at higher temperatures. A metal oxide will be reduced by a metal (including carbon and hydrogen) for which the $\Delta G^{\ominus}/T$ plot for oxide formation lies lower in the diagram (i.e. a lower metal will reduce an upper oxide). The reducing power of hydrogen is limited and does not increase with temperature; carbon is the most versatile reducing agent and will reduce most oxides provided the temperature is sufficiently high. Gold and precious metals occur as the metal, their oxides are reduced merely by heating at a temperature above that at which $\Delta G^{\ominus} = 0$.

ellipsometry. Study of surfaces by the change of POLARIZED RADIATION on reflection. Electrochemically generated oxide films have been studied by this method.

eluate. *See* CHROMATOGRAPHY.

eluant. Mobile phase used to carry out a chromatographic separation (*see* CHROMATOGRAPHY).

elution chromatography. Most common method for chromatographic separations, in which the mixture is applied at the head of the column or as a spot on paper or thin layer, and the individual components are separated by transportation along the STATIONARY PHASE by the continual addition and movement of the MOBILE PHASE. Rate of migration of a solute depends on the fraction of time it spends in the mobile phase (i.e. on its partition ratio). This type of development in a column allows the various components to be collected as separate fractions, in paper and thin layer chromatography separate spots are produced, whereas in gas–liquid and high-performance liquid chromatography a series of peaks on chart paper is recorded.

elution volume (V_s). Volume of solvent, measured between the placing of the solute at the head of the column and its elution at maximum concentration at the bottom; a term used exclusively in GEL FILTRATION CHROMATOGRAPHY, comparable to RETENTION VOLUME in other forms of chromatography.

elutriation. Process of separating a suspension of finely divided particles into fractions of various sizes by allowing it to settle against an upward stream of air or water. It is used in ore flotation.

emerald. *See* BERYLLIUM, CORUNDUM.

emery. *See* ALUMINIUM INORGANIC COMPOUNDS.

emf. *See* ELECTROMOTIVE FORCE.

emission spectra. *See* ATOMIC SPECTROSCOPY, ELECTRONIC SPECTROSCOPY.

empirical formula. *See* FORMULA.

emulsion. Colloidal dispersion (*see* COLLOID) of one liquid in another immiscible or partially miscible liquid. The particles are typically larger (0.1–10 μm) than in sols. Micro-emulsions (0.01–0.1 μm) are regarded by some as swollen MICELLES. Emulsions are termed 'oil-in-water' (O/W) or 'water-in-oil' (W/O), depending on whether the dispersed phase is oil or water, respectively. Emulsions are stabilized by emulsifying agents (emulsifiers) which are SURFACE-ACTIVE AGENTS, naturally occurring materials and sols. Emulsions are used widely in foodstuffs, especially dairy products, pharmaceuticals, cosmetics, agricultural sprays, paints and bituminous sprays. Emulsions are broken by mechanical agitation, filtering, thermal treatment and addition of substances that promote the formation of the opposite type of emulsion (i.e. O/W to destroy W/O). An example of demulsification is the formation of butter from milk.

emulsoid. *See* SOL.

en. *See* 1,2-DIAMINOETHANE.

enamel. GLASS fused to the surface of a metal, used as a protective or decorative coating.

enamines. *See* AMINES.

enantiomer. Molecule that possesses neither a plane nor a centre of symmetry; it cannot be superimposed on its mirror image. The molecule and its mirror image (enantiomers) rotate the plane of polarized light equally in opposite directions, one being the (+)-form and the other the (–)-form. The ISOMERS differ in their configuration at a chiral carbon (*see* CHIRALITY). Enantiomers cannot be distinguished by ordinary X-ray crystallography. The exact assignment of the spatial distribution of all the atoms in enantiomers can be made if certain types of atoms are present in a chiral molecule. The absolute configuration assigned to (+)-2,3-dihydroxybutanedioic acid (tartaric acid) (using the sodium rubidium salt) by anomalous X-ray scattering is

$$
\begin{array}{c}
\text{COOH} \\
\text{H} \!-\!\!\!\mid\!\!\!-\! \text{OH} \\
\text{HO} \!-\!\!\!\mid\!\!\!-\! \text{H} \\
\text{COOH}
\end{array}
$$

This is the configuration assigned using glyceraldehyde as the arbitrary reference compound. *See also* OPTICAL ACTIVITY.

enantiomorph. *See* ENANTIOMER.

enantiotopic. Two identical ligands *a* which are attached to a PROCHIRAL centre (e.g., *Caabd*) are enantiotopic if their separate replacement by some other group *e* gives rise to a pair of ENANTIOMERS (*Cabde*) and diastereotopic if they give rise to DIASTEREOISOMERS (*see* Figure).

enantiotropy. Reversible transformation of an allotropic or polymorphic form into another at a definite temperature. When the transition temperature lies below the melting point of the solid (*see* ONE-COMPONENT SYSTEM, Figure) each form has a definite temperature range of stability. Examples include sulphur, tin, mercury(II) iodide, ammonium nitrate. *See also* MONOTROPY, PHASE RULE.

enargite. *See* COPPER.

encounter pair (encounter complex). Reactants that diffuse together and react in a solvent cage.

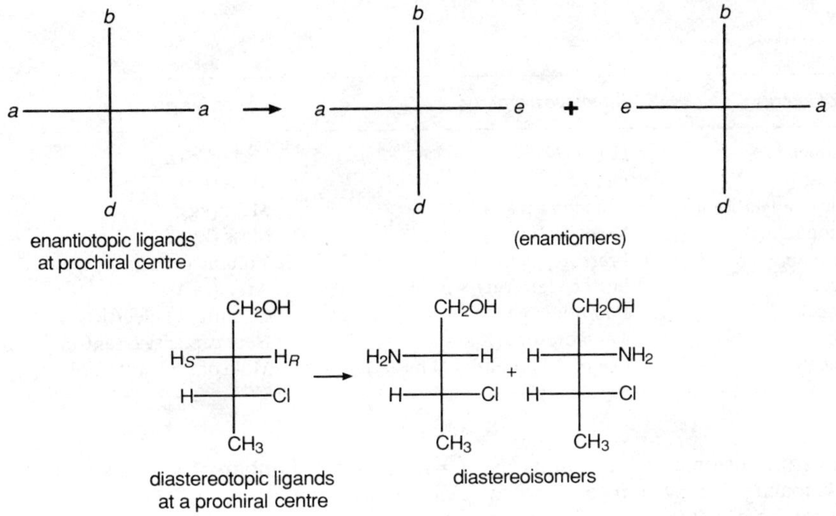

enantiotopic ligands
at prochiral centre

(enantiomers)

diastereotopic ligands
at a prochiral centre

diastereoisomers

enantiotopic

endergonic process. Process in which there is an increase in the GIBBS FUNCTION (i.e. $\Delta G > 0$).

endo-. Prefix used, with *exo-*, to describe the configuration in bridged or polycyclic ring systems. With reference to the bridge atoms, a substitutent is *exo-* (*cis-*) if it is on the same side as the bridge or *endo-* (*trans-*) if on the opposite side (*see* Figure).

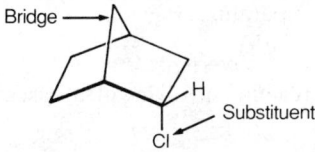

endo-2-Chlorobicyclo-
[2.2.1] heptane
(*endo*-norbornyl chloride)

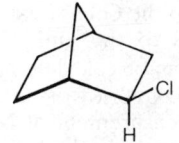

exo–2-Chlorobicyclo-
[2.2.1] heptane
(*exo*-norbornyl chloride)

endo-
Configuration of *endo-* and *exo*-norbornyl chlorides.

ENDOR (electron nuclear double resonance). *See* ELECTRON SPIN RESONANCE SPECTROSCOPY.

endorphins. Peptide neurotransmitters found in pituitary gland and hypothalamus. They have pain-relieving properties similar to the ENKEPHALINS.

endothermic process. Physical or chemical process that is accompanied by the absorption of heat (i.e. the ENTHALPY of the reactants is less than that of the products) (see Table). *See also* EXOTHERMIC PROCESS.

endothermic process

Conditions of process	Endothermic ($\Delta H > 0$)	Exothermic ($\Delta H < 0$)
Isothermal	Heat enters	Heat exits
Isobaric	$q > 0$	$q < 0$
Adiabatic	T falls	T rises

end-point. Stage in a titration when the indicator (or other form of indicator) undergoes maximum change (colour, potential, etc.) for a small amount of added titrant. It corresponds to a definite pH, ionic concentration or redox potential; it does not necessarily correspond to the EQUIVALENCE POINT.

energy
Different types of energy.

Type of energy	Intensive property	Extensive property
Mechanical	Force (N)	Distance (m)
Potential	gh (m^2 s^{-2})	Mass (kg)
Kinetic (translational)	Velocity (m s^{-1})^2	Mass (kg)
Rotational	Velocity (m s^{-1})^2	Mass (kg)
PV	Pressure (N m^{-2})	Volume (m^3)
Surface	Surface tension (N m^{-1})	Area (m^2)
Electrical	Potential (V)	Quantity of electricity (C)
Thermal	Temperature (K)	Heat capacity or entropy (J K^{-1})
Chemical	Chemical potential (J mol^{-1})	Mole of reactant (mol)

energy (dimensions: $m\,l^2\,t^{-2}$; units: N m = J). Capacity of a system to do WORK. All forms of energy (*see* Table) can be considered as the product of an EXTENSIVE PROPERTY and an INTENSIVE PROPERTY. *See also* FIRST LAW OF THERMODYNAMICS, INTERNAL ENERGY.

energy barrier. *See* ABSOLUTE RATE THEORY, ACTIVATED COMPLEX.

energy level. Given state of an atom or molecule, which is determined by the values of QUANTUM NUMBERS associated with the system. The lowest energy level is called the GROUND STATE. *See also* ELECTRONIC SPECTROSCOPY, PARTICLE IN A BOX, SCHRÖDINGER EQUATION.

enkephalins. Pentapeptides—Met-enkephalin (H-Tyr-Gly-Gly-Phe-Met-OH) and Leu-enkephalin (H-Tyr-Gly-Gly-Phe-Leu-OH)—that occur naturally in the brain. They have similar analgesic and other properties to the opiates morphine, codeine, etc. (*see* ALKALOIDS).

enols. *See* KETO–ENOL TAUTOMERISM.

enrichment. Process by which the amount of one or more isotopes in a given material is increased above that occurring in nature.

ensemble. Collection of a very large number of systems each constructed to be a replica on a thermodynamic level of the system of interest. In a canonical ensemble, the number of molecules and the volume of the individual systems are fixed, as is the total number of systems and the total energy.

enstatite. *See* SILICATES.

enthalpy (H; dimensions: $m\,l^2\,t^{-2}$; units J mol^{-1}). STATE FUNCTION defined as

$$H = U + PV$$

Absolute values of H cannot be obtained from classical thermodynamics, the enthalpy of an element in its standard state is arbitrarily taken as zero. For an endothermic process $\Delta H > 0$, and for an exothermic process $\Delta H < 0$. The enthalpy change in a reaction at constant pressure and temperature, is given by

$$\Delta H = \Sigma H_{products} - \Sigma H_{reactants}$$

For a reaction involving ideal gases

$$\Delta H = \Delta U + \Delta n_{gas}\,RT$$

$$\Delta n_{gas} = n_{gas,products} - n_{gas,reactants}$$

For a series of reactions, ΔH values are additive (*see* HESS'S LAW), ΔH varies with temperature (*see* KIRCHHOFF'S EQUATION) and is related to the GIBBS FUNCTION and in a galvanic cell to the emf of the CELL. Values of $H^\ominus$ can be calculated from values of the PARTITION FUNCTION using spectroscopic data. Measurement of ΔH values is by: (1) calorimetry (method of mixtures or BOMB CALORIMETER) and (2) determining the temperature coefficient of emf for reactions in cells (*see* THERMODYNAMICS OF CELLS).
standard enthalpy change ($\Delta H^\ominus$). Enthalpy change, at a specified temperature, during a

enthalpy

For a reaction

$$\Delta H^\ominus = \Sigma\, \Delta_f H^\ominus_{products} - \Sigma\, \Delta_f H^\ominus_{reactants} = \Sigma\, \Delta_c H^\ominus_{reactants} - \Sigma\, \Delta_c H^\ominus_{products}$$

Standard enthalpy of	Symbol	Enthalpy change during
Formation	$\Delta_f H^\ominus$	Formation of 1 mole of compound from its elements (tabulated values are available)
Combustion	$\Delta_c H^\ominus$	Complete combustion of 1 mole of substance in excess oxygen
Solution	$\Delta H^\ominus \cdot x H_2O$ $\Delta H^\ominus \cdot aq$	Solution of 1 mole of substance in x mole of water Solution of 1 mole of substance in excess water
Fusion Vaporization Sublimation Transition	L_f L_e L_s L_{tr}	Isothermal change of 1 mole of the substance from the form stable at the lower temperature to that stable at the higher temperature s→1, l→g, s→g, allotropic form I→II (normally measured at the corresponding transition temperature)

reaction or process when the reactants and products are in their STANDARD STATES.

enthalpy function (units: $J\,K^{-1}\,mol^{-1}$). For an IDEAL GAS the enthalpy function is given by

$$H_T^\ominus - H_0^\ominus = 2.5RT + RT^2\,(\partial \ln Q_i/\partial T)_P$$

Enthalpy of formation of a compound at a specified temperature can be calculated from tabulated values of the enthalpy function without the use of empirical HEAT CAPACITY equations. *See also* PARTITION FUNCTION.

entropy (S; dimensions: $m\ l^2\ t^{-2}\ \theta^{-1}$; units: $J\,K^{-1}\,mol^{-1}$). STATE FUNCTION and an EXTENSIVE PROPERTY of a system. For an isothermal reversible process

$$dS = dq_{rev}/T$$

and for a non-isothermal process

$$\Delta S = \Sigma\, dS = \int dq_{rev}/T$$

For an irreversible process: $q/T < \Delta S$, since $q < q_{rev}$. In an irreversible (spontaneous) process

$$\Delta S_{total} = \Delta S_{system} + \Delta S_{surroundings} > 0$$

If $\Delta S_{total} = 0$, there is no spontaneous process and if $\Delta S_{total} < 0$ the process is spontaneous in the reverse direction. An in-

crease in entropy is accompanied by an increase (decrease) in the disorder (order) of the system (e.g., the melting of a solid, vaporization of a liquid). Unlike U, H, G, etc., absolute values of S can be calculated from the variation of the heat capacity with temperature using the THIRD LAW OF THERMODYNAMICS ($S_{thermal}$). S is related to the GIBBS FUNCTION and to the temperature coefficient of $\Delta G^\ominus$, $E^\ominus$ and K

$$S = -(\partial G/\partial T)_P$$

$$\Delta S^\ominus = -[\partial(\Delta G^\ominus)/\partial T]_P = R[\partial(T\ln K)/\partial T]_P$$

$$= nF(\partial E^\ominus/\partial T)_P$$

Measurement of S or ΔS values is by: (1) isothermal change (fusion, vaporization, etc.)

$$\Delta S = \Delta H_{tr}/T$$

(2) for expansion of n mole of ideal gas from p_1, V_1, T_1 to p_2, V_2, T_2

$$\Delta S = nC_V \ln(T_2/T_1) - nR \ln(p_2/p_1)$$

(3) for non-isothermal process

$$\Delta S = S(T_2) - S(T_1) = \int_{T_1}^{T_2} dH/T$$

$$= \int_{T_1}^{T_2} C_P\, dT/T$$

that is the area under the C_P/T against T plot. If there are any transition points in the temperature range, then the isothermal entropy change must be added; (4) $S^{\ominus}$ values from spectroscopic data (*see* PARTITION FUNCTION). Contributions from translational, rotational, vibrational and electronic energies can be evaluated; (5) $\Delta S^{\ominus}$ values for a chemical reaction from tabulated values of $S^{\ominus}$; (6) from a knowledge of the temperature coefficients of E or K.

molar entropy values. Hard abrasive substances (e.g., diamond, silicon carbide) with infinite three-dimensional lattices have small entropy values, with increasing softness and increased thermal disorder (liquids and gases) the entropy increases.

standard entropy of an ion ($\bar{S}_i^{\ominus}$). This is the partial molar entropy of an ion in solution relative to the conventional standard entropy of the hydrogen ion taken as zero. $\bar{S}_i^{\ominus}$ increases with relative atomic mass and decreases with increasing charge. The values provide a measure of the ordering effect produced by the ions on the surrounding water molecules.

enzyme kinetics. Study of the rates of enzyme reactions, which show biphasic kinetics. They are first-order with respect to the substrate at low substrate concentrations and zero-order at higher concentrations; they are always first-order with respect to the enzyme. Applying the STEADY-STATE APPROXIMATION to the reaction between an enzyme (E) and a substrate (S)

$$E + S \underset{k_{-1}}{\overset{k_1}{\rightleftharpoons}} ES \overset{k_2}{\longrightarrow} E + P$$

the initial rate of reaction (v) is given by the Michaelis–Menten equation

$$v = V[S]/(K_m + [S])$$

where V is the maximum rate at high substrate concentrations. K_m, the Michaelis constant (the dissociation constant of the enzyme–substrate complex), is defined as $(k_{-1} + k_2)/k_1$. K_m and V are characteristic for a given enzyme and are independent of the enzyme concentration, but are dependent on the nature of the substrate. As the plot of v against [S] is hyperbolic, the Michaelis–Menten equation is rearranged into a linear form. The most common equation (Lineweaver–Burk) for this purpose is obtained by taking reciprocals of the Michaelis–Menten equation

$$1/v = (K_m/V)\,1/[S] + 1/V$$

From the double reciprocal plot of experimental values of v and [S], K_m and V may be obtained from the measured slope and intercept (*see* Figure). In the presence of a

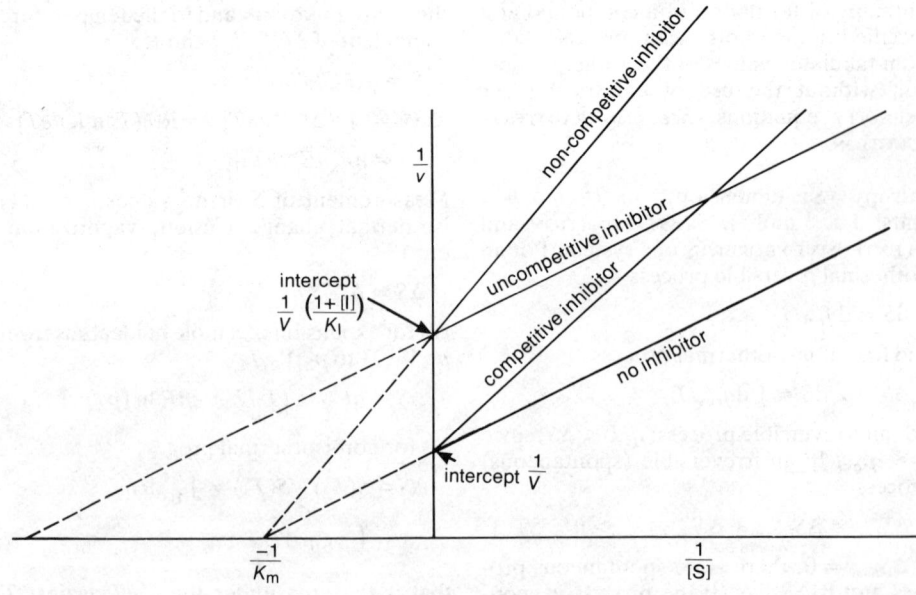

enzyme kinetics
Lineweaver–Burk plots for uninhibited and inhibited reactions.

fixed concentration of inhibitor [I], the above equation takes on different forms according to the type of inhibition (*see* ENZYMES):

competitive

$$1/v = (K_m/V)(1 + [I]/K_I) \, 1/[S] + 1/V$$

uncompetitive

$$1/v = (K_m/V) \, 1/[S] + (1 + [I]/K_I) \, 1/V$$

non-competitive

$$1/v = (K_m/V) \, (1 + [I]/K_I) \, 1/[S] +$$
$$(1 + [I]/K_I) \, 1/V$$

where K_I is the dissociation constant of the enzyme–inhibitor complex (*see* Figure). For multireactant enzyme systems the kinetics are more complex. The majority of such reactions fall into a few categories.
1) Ping-pong mechanism, where one product is formed prior to reaction with the second substrate

$$A + E \rightleftharpoons AE \rightleftharpoons P + E'$$
$$E' + B \rightleftharpoons E'B \rightleftharpoons E + Q$$

for example, dehydrogenation reactions catalyzed by flavoproteins.
2) Ordered mechanism of the type

$$A + E \rightleftharpoons AE + B \rightleftharpoons AEB \rightleftharpoons$$
$$QEP \rightleftharpoons P + QE \rightleftharpoons Q + E$$

A and Q are obligatory reactants and are usually coenzymes, for example, dehydrogenases requiring NAD^+.
3) Random mechanism

$$A + E \rightleftharpoons AE \searrow^B$$
$$\qquad AEB \rightleftharpoons QEP$$
$$B + E \rightleftharpoons BE \nearrow^A$$
$$Q \nearrow PE \rightleftharpoons P + E$$
$$\qquad \searrow^P$$
$$QE \rightleftharpoons Q + E$$

for example phosphotransferases.

enzymes. PROTEINS, present in all living organisms, responsible for catalyzing most cellular reactions with extraordinary specificity and power (*see* Table). These characteristics (at least in those enzymes not requiring a PROSTHETIC GROUP for activity) are determined solely by the specific spatial relationships between amino acid residues in the polypeptide chains. No enzyme-catalyzed reaction has been documented which would not occur (even-

tually) without catalysis, but the rate of an uncatalyzed reaction can be increased by many powers of 10. The enzyme is not used or destroyed in the reaction. The presence of a non-protein component is often required for enzymic activity; if combined with the protein it is known as a prosthetic group (e.g., the haemin group of catalase) and if loosely attached it is a COENZYME e.g., NAD^+). Certain metals may also be required as activators (cofactors) in a combined form (e.g., copper, magnesium).

Enzymes are highly specific, particularly those acting on carbohydrates and peptides, where a slight change in the stereochemical configuration is sufficient to render the enzyme incompatible and unable to effect catalysis.

Enzymes act by lowering the overall energy of activation of the reaction which involves a series of intermediates or a mechanism, different from that of the uncatalyzed reaction. They are active within a narrow pH range (less than two pH units); outside this range, the activity falls off rapidly due to changes in structure brought about by adverse conditions. Most enzymes have greatest activity at about 37°C; at higher temperatures the protein is denatured and all activity is irreversibly lost.

Enzymes are susceptible to a wide variety of inhibitors. (1) Noncompetitive inhibitors react with functional groups on the enzyme causing chemical transformation of the enzyme (e.g., heavy metal ions, alkylating agents). (2) Competitive inhibitors, which are molecules similar in shape and size to the substrate, block the active sites and prevent access by the substrate (e.g., sulphonamides block p-aminobenzoic acid sites in the synthesis of folic acid). This type of inhibition is reversible on the addition of more substrate. (3) Uncompetitive inhibitors bind reversibly to the enzyme–substrate complex to yield an inactive enzyme–substrate–inhibitor complex (*see* ENZYME KINETICS). This type of inhibition is uncommon with one-substrate systems.

Enzymes can be obtained from plants, animals and microorganisms by solvent extraction and are purified by precipitation with salts or methanol or by ultrafiltration, chromatography and/or electrophoresis. Many enzymes are available in the crystalline form.

Classification of enzymes is according to

enzymes

Properties of some common enzymes.

Enzyme	Source	Approx. M_r [a]	Optimum pH [a]	Reaction catalyzed or bonds broken
Amylase	Animal, plant	15 500	6.5–7.0	Breaks down α-1,4-glucosidic bonds in starches and glycogen to maltose
ATPase	Beef heart, heart myosin, mitochondria	225 000–750 000	7.4–8.0	Splits ATP into ADP and inorganic phosphate
Catalase	Liver, bacteria	225 000–250 000	7.0	Contains haematin, breaks down hydrogen peroxide to oxygen; prevents accumulation of toxic peroxides
Chymotrypsin A	Intestine	20 000–25 000	7.5–8.0	Hydrolysis of peptide linkages adjacent to aromatic L-amino acids
Deoxyribonuclease	Ox pancreas	42 000–60 000	4.5–5.0	Hydrolysis of native or denatured DNA to range of oligo- and mononucleotides
α-Glucosidase	Yeast	68 000	4.0–7.0	Splits α-1,4-glucosidic bonds (e.g., maltase)
β-Glucosidase	Rat kidney	50 000	4.0–5.0	Splits β-1,4-glucosidic bonds (e.g., emulsin)
Invertase (β-fructo-furanosidase)	Bakers' yeast	270 000	4.0–5.5	Hydrolysis of sucrose and fructose (invert sugar)
Lysozyme	Egg white, nasal mucosa	14 600	5.0–6.0	Splits β-1,4-glycosidic linkages between N-acetylglucosamine and N-acetylmuramic acid. Used to digest bacterial cell walls
Papain	Papaya latex	23 000	6.0–6.8	Hydrolysis of peptide linkages at bonds involving basic amino acids leucine or glycine
Penicillinase (β-lactamase)	*Bacillus leicheniformis*	23 000–35 000	7.0	Hydrolysis of β-lactam ring of penicillin yielding inactive penicilloic acid
Pepsin	Hog stomach	36 000	2.0	Hydrolysis of peptide linkages at bonds adjacent to aromatic or dicarboxylic L-amino residues
Peroxidase	Plants, horseradish	40 000	6.0–7.0	Contains haematin, destroys peroxides by transferring oxygen to oxidizable substrates
Phosphatases (acid) (alkaline)	Beef spleen Calf intestinal mucosa	23 000 80 000–100 000	5.0–6.0 8.0–10.0	Range of enzymes which hydrolyze the linkages between phosphoric acid residues and organic compounds (e.g., glucose-1-phosphate)
Rennin	Stomach of ruminants	34 000	2.0–4.0	Milk-clotting enzyme (as rennet used in cheese-making and for making junkets). Hydrolyzes peptides
Ribonuclease	Bovine pancreas	13 700	7.0–7.5	Hydrolysis of RNA to oligo- or mononucleotides
Trypsin	Bovine pancreas	23 800	7.0–9.0	Hydrolysis of peptide linkages at bonds adjacent to L-arginine or L-lysine residues
Urease	Many plants (not vertebrates)	42 000–60 000	6.5–7.0	Hydrolysis of urea to ammonia and carbon dioxide

[a] The range of M_r values and optimum pH values are for the enzyme extracted from different sources.

the reactions which they catalyze; they were formerly named by the addition of the suffix -ase to the substrate or process of the reaction. The confusing nomenclature has been clarified in a recommendation of IUB and IUPAC in which each enzyme has a specific code number. The old nomenclature is still retained and trivial names are usually given in discussion of a particular enzyme. Six main types of enzyme are recognized.

oxidoreductases. Enzymes catalyzing redox reactions in which the reduced form of the substrate (MH_2) is oxidized by an electron transfer process, often involving the presence of a coenzyme (e.g., NAD^+)

$$MH_2 + NAD^+ \xrightarrow{\text{dehydrogenase}} M + NADH_2^+$$

(e.g., dehydrogenases, oxidases, peroxidases, hydrogenases, and hydrolyases).

transferases. Enzymes catalyzing the transfer of a group from one compound to another (e.g., transaminases, transpeptidases, kinases).

hydrolases. Enzymes catalyzing the hydrolytic cleavage of C–O, C–N and C–C bonds (e.g., esterases, glucosidases, proteinases, lipases) and also some other bonds (e.g., sulphatases, phosphatases).

lyases. Enzymes that cleave C–C, C–O, C–N and some other bonds by elimination leaving double bonds or conversely which add groups across a double bond (e.g., decarboxylases, lyases, anhydrases, synthases).

isomerases. Enzymes catalyzing structural or geometrical changes within a molecule (e.g., epimerases, rotases, isomerases, mutases, racemases).

ligases. Enzymes catalyzing the reaction of two molecules with the elimination of a pyrophosphate bond (e.g., synthetases).

Enzymes are used extensively in baking, wine-making, brewing and in the manufacture of pharmaceuticals, biodetergents, leather and paper, and in sewage disposal.

eosin. *See* INDICATORS.

ephedrine. *See* ALKALOIDS.

epichlorohydrin. *See* 1-CHLORO-2,3-EP-OXYPROPANE.

epimerization. Changing of the configuration at a single asymmetric centre in a

molecule containing more than one asymmetric centre (i.e. the interconversion of DIASTEREOISOMERS). ALDOSES that produce the same OSAZONE must have identical configurations on all the asymmetric carbon atoms except the α-carbon (only the aldehyde group and the α-carbon are involved in osazone formation). Such sugars are known as epimers.

epinephrine (adrenaline). *See* HORMONES.

epitaxy. Growth of one solid phase on another. The structure of the lower phase can force a particular orientation on the growing (epitaxial) layer. *See also* CHEMICAL VAPOUR DEPOSITION.

epoxides. *See* ALKENE OXIDES.

epoxy group. The group $>\!C\!-\!C\!<$ with an O bridging the two carbons.

epoxy resins. *See* POLYMERS.

EPR (electron paramagnetic resonance spectroscopy). *See* ELECTRON SPIN RESONANCE SPECTROSCOPY.

Epsom salts (magnesium sulphate). *See* MAGNESIUM COMPOUNDS.

equation of state. Equation that relates the pressure, volume and temperature for a given amount of substance. Criteria of success of such an equation are an accurate representation of the observed pressure, volume, temperature relationship over a moderate range of conditions, a simplicity of form and the use of only a few adjustable parameters. The equations (*see* Table) conform to the first criterion with increasing success down the table, but the number of parameters increases from zero to an infinite number.

equatorial form. *See* CONFORMERS.

equilibrium. A system is in equilibrium when it has no further tendency to change its properties. The fundamental criterion for equilibrium is: in a system of constant energy and volume, the total ENTROPY (S) is a maximum (i.e. $\Delta S_{H,V} = 0$). The move-

equation of state

Name	Equation	Critical constants				Reduced form
		P_c	$V_{m,c}$	T_c	Z_c	
Ideal gas equation	$P = RT/V_m$					
van der Waals equation	$P = RT/(V_m - b) - a/V_m^2$	$a/27b^2$	$3b$	$8a/27Rb$	$3/8$ 0.375	$P_r = 8T_r/(3V_r - 1) - 3/V_r^2$
Berthelot equation	$P = RT/(V_m - b) - a/TV_m^2$	$(2aR/3b^3)^{1/2}/12$	$3b$	$2(2a/3Rb)^{1/2}/3$	$3/8$ 0.375	$P_r = 8T_r/(3V_r - 1) - 3/T_rV_r^2$
Dieterici equation[a]	$P = [RT/(V_m - b)]\exp(-a/RTV_m)$	$a/4e^2b^2$	$2b$	$a/4Rb$	$2/e^2$ 0.2706	$P_r = [e^2T_r/(2V_r - 1)]\exp(-2/T_rV_r)$
Beattie–Bridgeman equation	$P = (1 - \gamma)RT(V_m + \beta)/V_m^2 - \alpha/V_m^2$ where $\alpha = a_0(1 + b/V_m)$ $\beta = b_0(1 - b/V_m)$ $\gamma = c_0/V_mT^3$					
Virial equation (Kammerlingh Onnes)	$P = (RT/V_m)[1 + B(T)/V_m + C(T)/V_m^2 + \ldots]$					

[a] e is the exponential 2.71, not a parameter.

ment of a system towards equilibrium is compounded of two parts: the achievement of (1) minimum energy and (2) maximum entropy. Frequently it is impossible to achieve these simultaneously and a compromise occurs. The most useful criterion under normal working conditions at constant temperature and pressure is that the GIBBS FUNCTION (G) is a minimum and $\Delta G_{T,P} = 0$. At the macroscopic level, when equilibrium is attained, all activity apparently ceases, but at the microscopic level it is the balancing of opposing forces (in chemical reactions the equality of the rates of the forward and back reaction, i.e. a position of dynamic and not static equilibrium). The position of equilibrium is unchanged by the presence of a catalyst, but can be changed by altering the temperature and pressure (only for reactions in which there is a change in the number of molecules) and in solution by a change of IONIC STRENGTH.

equilibrium constant (K). For the equilibrium, at constant temperature and pressure

$$aA + bB \rightleftharpoons lL + mM$$

the thermodynamic equilibrium constant is defined by

$$K_{\text{therm}} = (a_L^l a_M^m / a_A^a a_B^b)_e$$
$$= (c_L^l c_M^m / c_A^a c_B^b)_e (\gamma_L^l \gamma_M^m / \gamma_A^a \gamma_B^b)$$

K is related to the standard GIBBS FUNCTION

$$\Delta G^\ominus = -RT \ln K$$

or

$$K = \exp(-\Delta G^\ominus / RT)$$

In practice (assuming ideal behaviour), ACTIVITY is often replaced by concentration (K_c, $K_{c/c\ominus}$); mole fraction (K_m); partial pressure (K_p, $K_{p/p\ominus}$). K_{therm}, K_c, and K_m have units of (concentration)$^{\Delta\nu}$, K_p has units of (pressure)$^{\Delta\nu}$, whereas K_x, $K_{c/c\ominus}$, $K_{p/p\ominus}$, etc. are dimensionless ($\Delta\nu$ is the difference between the number of product and reactant molecules). For reactions in solution

$$K_{\text{therm}} = K_c \prod_i \gamma_i^i$$

For gas phase reactions

$$K_p = K_c(RT)^{\Delta\nu} = K_N(RT/N_A)^{\Delta\nu}$$
$$= K_x P^{\Delta\nu}$$

K_{therm} is a true constant for a reaction at a given temperature; because of departure from ideal behaviour K_p, K_c and K_m vary with pressure or ionic strength, approaching K_{therm} as $P \rightarrow 0$ or $I \rightarrow 0$. The value of K varies with the temperature (see VAN'T HOFF ISOCHORE). The magnitude of K determines the extent to which a reaction can proceed, if K is large, ($\Delta G^\ominus \ll 0$) formation of products is favoured; a small value of K ($\Delta G^\ominus > 0$) indicates that the reaction does not proceed to any appreciable extent. Methods of determination include: direct chemical analysis (e.g., SOLUBILITY PRODUCT), measurement of equilibrium pressure (gas equilibria), conductance (e.g., dissociation constant), electrometric titrations (e.g., dissociation constant and STABILITY CONSTANT), emf measurements (e.g., ionic product), absorption spectrometry (e.g., acid and indicator constants). Values of K may be calculated from tabulated values of $\Delta_f G^\ominus$ and standard ELECTRODE POTENTIALS and from the PARTITION FUNCTION. See also ACID DISSOCIATION CONSTANT.

equipartition of energy. When quantum effects can be ignored, the average energy of every quadratic term in the energy expression has the same value, $kT/2$ per molecule or $RT/2$ per mole. The total energy of a molecule is divided among the different DEGREES OF FREEDOM. A theoretical value for the HEAT CAPACITY (C_V) of a gas can be calculated

$$C_V = R(3 + \nu_r + 2\nu_v)/2$$

where ν_r and ν_v are the number of active rotations and vibrations, respectively.

equivalence point. Stage in a titration when the reactants and products are present in equivalent amounts according to the stoichiometric equation. It may not correspond exactly with the END-POINT.

equivalent conductivity (Λ_E; units: Ω^{-1} cm^2 equiv^{-1}). CONDUCTIVITY per unit CHEMICAL EQUIVALENT; an obsolete term superseded by MOLAR CONDUCTIVITY.

equivalent proportions, law of (reciprocal proportions, law of). If two elements A and B each form a compound with a third element C, then a compound of A and B will contain A and B in their relative proportions in which they react with C. This does not hold generally for compounds with variable oxidation states (e.g., TRANSITION ELEMENTS) or for NON-STOICHIOMETRIC COMPOUNDS. *See also* CHEMICAL COMBINATION, LAWS OF.

equivalent weight.   *See* CHEMICAL EQUIVALENT.

Er.   Erbium (*see* LANTHANIDE ELEMENTS).

erbium (Er).   *See* LANTHANIDE ELEMENTS.

erg (dimensions: $m\ l^2\ t^{-2}$).   CGS UNIT of energy; defined as the work done when a force of 1 dyne is moved through a distance of 1 cm in the direction of the force (1 erg = 10^{-7} J).

ergometrine (ergometrinine).   *See* ALKALOIDS.

ergosterol.   *See* STEROLS.

ergot alkaloids.   *See* ALKALOIDS.

ergotamine.   *See* ALKALOIDS.

Erichrome Black T.   *See* INDICATORS.

erionite.   *See* SILICATES.

erythritols ($C_4H_{10}O_4$).   All four tetrahydric alcohols are known but only *meso-*erythritol (erythritol, *see* Figure) occurs naturally (in lichens and some algae). It is a white solid (mp 121.5°C), very soluble in water, and is about twice as sweet as sucrose. On heating with methanoic acid it gives 2,5-dihydrofuran. It is used as a coronary vasodilator.

*meso-*erythritol

erythro-.   Prefixes *erythro-* and *threo-* are used to describe the configuration of DIASTEREOMERS. If, in the NEWMAN PROJECTION FORMULA, at least two sets of similar groups are in the eclipsed configuration the configuration is *erythro-*, as shown.

erythro-
(a) *meso-*2,3-dichlorobutane, (b) L-*erythro-*2,3-dichloropentane, (c) D-*threo-*2,3-dichloropentane.

erythromycin.   *See* ANTIBIOTICS.

erythrose.   *See* TETROSES.

erythrulose.   *See* TETROSES.

Es.   Einsteinium (*see* ACTINIDE ELEMENTS).

ESCA (electron spectroscopy for chemical analysis).   *See* ELECTRON SPECTROSCOPY.

ESR.   *See* ELECTRON SPIN RESONANCE SPECTROSCOPY.

essential amino acids.   AMINO ACIDS that cannot be synthesized by an organism in sufficient quantities; they must be supplied in the diet. In man, they are arginine, histidine, lysine, threonine, methionine, isoleucine, leucine, valine, phenylalanine and tryptophan.

essential elements.   Elements required by living organisms to ensure normal growth, development and maintenance. Apart from the elements found in organic compounds (carbon, hydrogen, oxygen and nitrogen), plants, animals and bacteria require a range of elements in the inorganic form in varying amounts depending on the type of organism. The major elements present in tissues (> 0.005 percent) are sodium, potassium and chlorine (chief electrolytic components of cells and body fluids), calcium, magnesium and phosphorus (present in bone, calcium is essential for nerve and muscle activity, phosphorus is a constituent of energy carriers, ATP and the nucleic acids) and sulphur (for amino acid synthesis). The trace elements which occur at much lower concentrations may serve as cofactors (*see* ENZYMES) or as constituents of complex molecules (iron in haem and cobalt in vitamin B_{12}). The important ones are iron, manganese, zinc, copper, iodine, cobalt, selenium, molybdenum, chromium and silicon.

essential fatty acids (EFA). Unsaturated linoleic (*see cis,cis*-9,12-OCTADECADIENOIC ACID), linolenic (*see cis,cis,cis*-9,12,15-OCTADECATRIENOIC ACID) and arachidonic acids, abundant in vegetable oils, which must be in the diet for healthy growth; the body cannot synthesize them fast enough to meet demand. As constituents of PHOSPHO-LIPIDS and GLYCERIDES, they are vital for membrane formation and fat metabolism. A deficiency leads to lack of growth, skin lesions and necrosis.

essential oils. Volatile oils derived from leaves, stems or twigs of plants usually carrying the odour or flavour of the plant. Chemically they are often TERPENES, but many other types occur. The oils, except those containing esters, are non-saponifiable (*see* SAPONIFICATION). Some are nearly pure single substances (e.g., oil of wintergreen (methyl salicylate)); others are mixtures (e.g., oil of turpentine (pinene, dipentene) and oil of bitter almonds (benzaldehyde and hydrocyanic acid)). They have a pungent taste and odour; are usually colourless when fresh, but become darker and thicker on exposure to air; are insoluble in water, but are soluble in organic solvents. Methods of extraction include: steam distillation; pressing fruit rinds; solvent extraction; maceration of flowers and leaves followed by solvent extraction. Typical essential oils are those obtained from cloves, roses, lavender, citronella, eucalyptus, peppermint, camphor, sandalwood, cedar and turpentine. They are widely used

essential oils

Oil of	Natural source	Main component(s)	Uses
Anise	*Illicium verum*	Anethole	Carminative, ingredient of cough lozenges, flavouring
Bay	*Laurus nobilis*	Myrcene	Astringent, perfume
Bergamot	*Citrus bergamia*	Linalool, limonene	Perfume, flavouring
Camphor	*Cinnamonum camphora*	Cineole, safrole	Medicinally in linaments
Caraway	*Carum carvi*	Carvone	Relief of flatulence
Cinnamon	*Cinnamonum zeylanicum*	Cinnamic aldehyde	Carminative, antiseptic, flavouring perfume
Cloves	*Eugenia caryophillus*	Eugenol	Antiseptic, local anaesthetic in dentistry
Coriander	*Coriandrum sativum*	Coriandrol	Stimulative
Dill	*Anethum graveolens*	Carvone	Relief of flatulence in infants
Eucalyptus	*Eucalyptus* sp.	Cineole, α-pinene	Medicinal, perfume
Jasmine	*Jasminum officinale*	Jasmone, benzyl ethanoate	Perfume
Peppermint	*Mentha piperita*	Menthol, menthyl esters	Aromatic carminative, relief of flatulence
Thyme	*Thymus vulgaris*	Thymol	Perfume, antiseptic
Turpentine	Variety of pines	α-Pinene	Paint thinner

as food flavourings, solvents, and in per-fumery and medicines.

esterification. Formation of an ESTER by the acid-catalyzed reaction between a carboxylic acid and an alcohol.

esters. Compounds formed by the replacement of the hydroxyl hydrogen atom in an oxoacid by an alkyl group. The most important esters are formed by CARB-OXYLIC ACIDS. They are named as the alkyl salts (suffix: -ate) of the acid (e.g., $CH_3COOC_2H_5$ is ethyl ethanoate). Carboxylic esters and water are formed in equilibrium with an acid and an alcohol. High yields of esters are obtained by using excess alcohol or by removing the product water in a constant boiling mixture with benzene. The reaction is slow at room temperature, but may be increased by rais-ing the temperature and by the addition of a mineral acid which acts as a catalyst. Other methods of preparation include the reaction of alcohols with acid chlorides, anhydrides and amides (in the presence of boron tri-fluoride). Silver salts of acids react with alkyl halides to give esters. Methyl esters may be prepared from the acid and diazo-methane. Ethyl ethanoate is made com-mercially by the TISCHENKO REACTION. New esters are formed by the exchange of an ester with an ester, carboxylic acid or alcohol in the process of transesterification. Esters are pleasant-smelling liquids that occur naturally in fruits and are used to flavour foods (e.g., 3-methylbutyl ethano-ate (amyl acetate) in pear drops). Their major use is as solvents, particularly methyl and ethyl ethanoates. Esters of cellulose (*see* CELLULOSE ACETATE PLASTICS) are used as photographic film. ASPIRIN is a widely used analgesic. Esters are hydro-lyzed to acids and alcohols by acids or alkalis, converted to alcohols by lithium tetrahydridoaluminate, lithium tetrahydrido-borate and hydrogen gas over nickel, and to AMIDES with ammonia (*see* AMMONOLYSIS). Sodium in ether followed by treatment with acid gives ACYLOINS. Pyrolysis of esters, particularly ethanoates gives ALKENES. Esters of inorganic oxoacids, alkyl sul-phates, nitrates, nitrites, etc. may be pre-pared by reaction of the acid or gaseous oxide with an alcohol.

estrogens. *See* OESTROGENS.

estrone (oestrone). *See* HORMONES.

Et. *See* ETHYL GROUP.

Etard's reaction. Oxidation of toluene by chromyl chloride to give benzaldehyde

$$PhCH_3 + 2CrO_2Cl_2 \rightarrow PhCH(OCrCl_2OH)_2$$

$$\xrightarrow{H_2O} PhCHO$$

When more than one methyl group is present only one is oxidized. Aromatic compounds with longer alkyl chains give ketones.

ethanal (acetaldehyde, CH_3CHO). ALDE-HYDE; colourless, pungent liquid (bp 20.8°C), miscible with water, ethanol and ethers. It is manufactured by: (1) the air oxidation or dehydrogenation of ethanol over a silver catalyst; (2) direct oxidation of ethene under pressure in a solution of copper(II) and palladium(II) chlorides at 50°C. It is used in the preparation of ethanoic acid, ethanol, paraldehyde and POLYMERS. (For general preparation and properties for aldehydes—*see* KETONES, Figures.) Peculiar to ethanal is the polymer-ization to the cyclic trimer paraldehyde, which is formed when ethanal is treated with concentrated sulphuric acid at room temperature, and metaldehyde, a solid tetramer precipitated at 0°C by concen-trated sulphuric acid.

ethanamide (acetamide, CH_3CONH_2). AMIDE, colourless deliquescent needles (mp 82°C), prepared by the dry distillation of ammonium ethanoate. Ethanamide is a weak base. (For general properties—*see* AMIDES.)

ethane (C_2H_6). ALKANE (mp −183.3°C, bp −88.6°C); minor constituent of NATU-RAL GAS. It is used as a fuel, precursor of ETHENE and refrigerant.

ethanedioic acid (oxalic acid, $(COOH)_2$). Crystalline, poisonous solid (mp 101°C (dihydrate), 189.5°C (anhydrous), sublimes 157°C, $pK_{a,1} = 1.271$, $pK_{a,2} = 4.266$), found in rhubarb, spinach, sorrel and members of the *Oxalis* genus. Industrial preparation is by: (1) heating sodium methanoate and treating the sodium ethanedioate so formed with sulphuric acid; (2) oxidation of sucrose by nitric acid in the presence of vanadium pentoxide. Ethanedioic acid is decomposed to methanoic acid and carbon dioxide and then carbon monoxide and water (heat to 200°C), carbon monoxide, carbon dioxide and water (conc. H_2SO_4, 90°C), and carbon dioxide and water (manganate(VII)). The ESTER ethyl ethanedioate is produced by refluxing the anhydrous acid with ethanol and gives the diamide when reacted with ammonia. When heated with 1,2-DIHYDROXYETHANE the cyclic compound (I) is formed. Reduction of ethanedioic acid is to HYDROXYETHANOIC ACID ($Zn + H_2SO_4$) or OXOETHANOIC ACID ($Mg + H_2SO_4$). Ethanedioic acid is used as a cleaning agent and as an intermediate in synthesis.

(I)

ethanenitrile (acetonitrile, methyl cyanide, CH_3CN). Liquid NITRILE (bp 82°C), prepared from ETHYNE and ammonia or by the dehydration of ETHANAMIDE. Ethanenitrile is an important NON-AQUEOUS SOLVENT.

ethanethioic acid (thioacetic acid, CH_3COSH). Fuming liquid (bp 87°C), made by the action of hydrogen sulphide on ethanoic anhydride. It exists exclusively in the thiol form. It is more reactive than ethanoic acid, undergoes free radical addition to unsaturated compounds and ethanoylates aromatic amines.

ethanoates (acetates, CH_3COOR). ESTERS of ETHANOIC ACID. Methyl and ethyl ethanoates are solvents.

ethanoic acid (acetic acid, CH_3COOH). CARBOXYLIC ACID; colourless pungent liquid (mp 16.6°C, bp 119°C, $pK_a = 4.75$). It is manufactured by: (1) catalytic (Mn(II), pressure, 200°C) oxidation of butane; (2) catalytic (Pd–Cu(I)) oxidation of ethene; or (3) reaction between carbon monoxide and methanol over a cobalt or rhodium catalyst. It is also produced by the fermentation of ethanol to give vinegar. Two-thirds of ethanoic acid produced is used in the manufacture of cellulose triethanoate (see CELLULOSE ACETATE PLASTICS) and ETHENYL ETHANOATE. Other uses are as an industrial solvent and in the manufacture of BENZENEDICARBOXYLIC ACIDS and a variety of ESTERS (see ETHANOATES).

ethanoic anhydride (acetic anhydride, $(CH_3CO)_2O$). ANHYDRIDE of ETHANOIC ACID; colourless pungent liquid (bp 139.5°C). It is prepared by distilling sodium ethanoate and ethanoyl chloride

$$CH_3COCl + CH_3COO^- \longrightarrow$$

$$CH_3CO.O.COCH_3$$

Manufacture is by: (1) distillation of ETHYNE in glacial ethanoic acid with a mercury(I) salt catalyst; or (2) reaction of KETEN and glacial ethanoic acid. It is slightly soluble in water in which it hydrolyzes and is soluble in organic solvents. It is rapidly hydrolyzed to ethanoic acid by alkalis. It is used for ETHANOYLATION in pyridine, when it reacts less vigorously than ethanoyl chloride, and in the FRIEDEL–CRAFTS REACTION. It is reduced to ethanol by lithium tetrahydridoaluminate, sodium tetrahydridoborate, diborane and hydrogen over nickel. The anhydride reacts with nitrogen pentoxide to give ethanoyl nitrate (CH_3COONO_2). It is used for the production of cellulose triethanoate (see CELLULOSE ACETATE PLASTICS) and aspirin.

ethanol (ethyl alcohol, C_2H_5OH). Colourless liquid (bp 78.3°C) with a pleasant odour. It is miscible with water with the evolution of heat and contraction of volume forming a minimum boiling AZEOTROPIC MIXTURE (95.6 percent ethanol); 100 percent ethanol is obtained by azeotropic distillation in the presence of benzene. Ethanol is obtained from the fermentation of molasses, grain or starch and is synthesized by the hydration of ethene in 96 percent sulphuric acid followed by the hydrolysis of the mono- and diethylsulphates with steam. It has all the properties of primary MONOHYDRIC ALCOHOLS: oxidation gives ETHANAL and ETHANOIC ACID; ethoxides are formed with sodium, calcium and aluminium; it reacts with acids to give esters; it is dehydrated with sulphuric acid to diethyl ether or ethene. Bleaching powder ($CaOCl_2$) converts it to trichloromethane, whereas chlorine gives trichloroethanal. It is the active principle of intoxicating drinks; pharmacologically it acts as a central depressant, low doses have a stimulating effect. Its uses are as a solvent for many organic compounds, and in the manufacture of ethanal and ethyl ethanoate.

ethanolamines　(2-hydroxyethylamines, $(HOCH_2CH_2)_nNH_{3-n}$,　$n = 1, 2, 3$). Aminoalcohols prepared by the reaction between ethene oxide and ammonia. The mixture is separated by fractional distillation. They are widely used as emulsifying agents, detergents and in the preparation of cosmetics, toiletries and herbicides. Monoethanolamine is used to remove acid gases from NATURAL GAS.

ethanoylation (acetylation). Introduction of the ethanoyl group (CH_3CO-) into a compound containing an active hydrogen atom. ETHANOYL CHLORIDE and ETHANOIC ANHYDRIDE are used extensively as ethanoylating agents.

ethanoyl chloride (acetyl chloride, CH_3COCl). Colourless, fuming, pungent liquid (bp 55°C), prepared from ethanoic acid or its sodium salt and $POCl_3$, PCl_5 or $SOCl_2$. It readily hydrolyzes to ethanoic acid and hydrochloric acid. It is used to ethanoylate compounds containing active hydrogen atoms (*see* ETHANOYLATION).

ethanoyl group (acetyl group). The ACYL GROUP CH_3CO-. *See also* ETHANOYLATION.

ethene (ethylene, $CH_2{=}CH_2$). ALKENE (mp –169°C, bp –104°C); colourless gas occurring in NATURAL GAS and PETROLEUM, and manufactured by CRACKING petroleum. It is the most important synthetic organic chemical. Most ethene is used in the production of POLYMERS.

ethene oxide. *See* ALKENE OXIDES.

ethenol (vinyl alcohol, $CH_2{=}CHOH$). End form of ETHANAL. It cannot be isolated in pure state.

ethenone. *See* KETENS.

ethenylation. Addition of compounds containing an active hydrogen atom to ETHYNE forming the ETHENYL GROUP (vinyl group)

$$HC{\equiv}CH + HX \longrightarrow H_2C{=}CHX$$

ethenylbenzene. *See* STYRENE.

ethenyl ethanoate (vinyl acetate, $CH_2{=}CHOOCCH_3$). Colourless, lachrymatory liquid (bp 73°C), manufactured by passing a mixture of ethyne and ethanoic acid vapour over zinc ethanoate on charcoal at 170°C, or ethene, sodium ethanoate and palladium chloride in aqueous copper(II) chloride:

$$C_2H_4 + PdCl_2 + 2CH_3COONa \longrightarrow$$
$$CH_2{=}CHOOCCH_3 + 2NaCl + Pd +$$
$$CH_3COOH$$

It is oxidized in air and polymerizes and copolymerizes (with ethenol (vinyl alcohol)), catalyzed by ZIEGLER–NATTA CATALYSTS. POLYMERS are used as general plastics and adhesives. It is an intermediate in the manufacture of polyethenol and poly(ethenyl ethanoates).

ethenyl group (vinyl group). The group $CH_2{=}CH-$. *See also* ETHENYLATION.

ether.　Common name for DIETHYL ETHER.

ethers. Organic compounds (ROR'), where R and R' are alkyl or aryl (*see* AROMATIC ETHERS); named as the dialkyl or diaryl ether if $R = R'$, or as an alkoxyalkane (or alkoxyarene). Lower-molecular-mass ethers are liquids with pleasant smells. They are insoluble in water, but soluble in organic solvents. (For general methods of preparation and properties of alkyl ethers— see Figure.) DIETHYL ETHER is commonly known as 'ether'. Cyclic ethers are CROWN ETHERS. Polyethers (*see* POLYMERS) are found as epoxy resins, polyacetals, poly(oxyphenylenes), phenoxy resins and poly(aldehydes). DIETHENYL ETHER is an example of an unsaturated ether.

ethoxybenzene. *See* AROMATIC ETHERS.

2-ethoxyethanol. *See* ETHYLENE GLYCOL MONOETHYL ETHER.

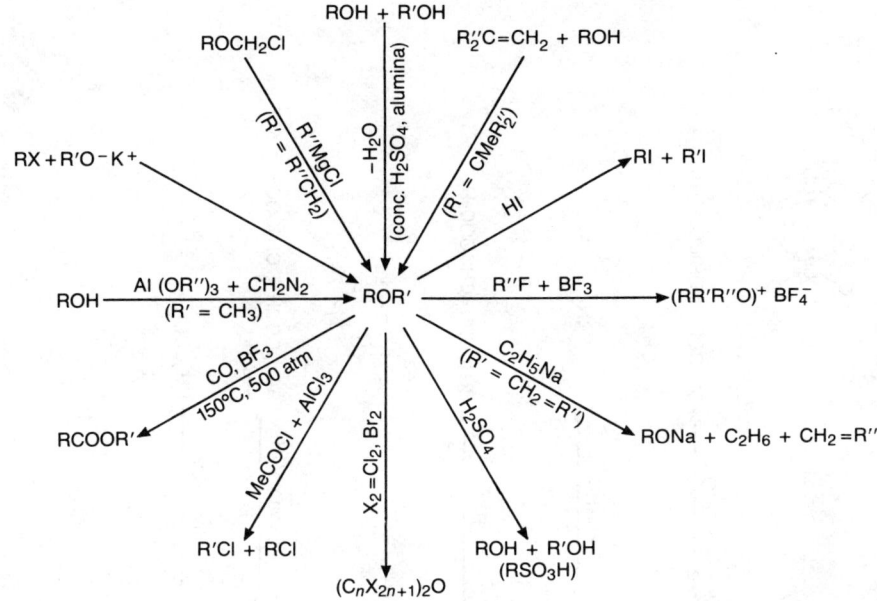

ethers
Reactions and preparation of alkyl ethers.

ethoxy group. The group C_2H_5O- (EtO–). *See* ETHERS.

ethyl acetate. *See* ETHYL ETHANOATE.

ethyl acetoacetate. *See* ETHYL 3-OXO-BUTANOATE.

ethyl alcohol. *See* ETHANOL.

ethylamines. *See* AMINES.

ethylbenzene ($C_6H_5C_2H_5$). Isomer of the XYLENES. Colourless liquid (bp 136°C), prepared industrially by the alkylation of benzene with ethene in the presence of a catalyst ($AlCl_3/HCl$ or H_3PO_4). It is easily oxidized to benzoic acid and undergoes catalytic DEHYDROGENATION in the presence of steam to STYRENE and catalytic hydrogenation to ethylcyclohexane.

ethyl bromide. *See* BROMOETHANE.

ethyl carbamate. *See* URETHANES.

ethyl chloride. *See* CHLOROETHANE.

ethylene. *See* ETHENE.

ethylenediamine. *See* 1,2-DIAMINOETHANE.

ethylenediaminetetraacetic acid. *See* EDTA.

ethylene dibromide. *See* 1,2-DIBROMO-ETHANE.

ethylene dichloride (1,2-dichloroethane). *See* DICHLOROETHANE.

ethylene glycol. *See* 1,2-DIHYDROXY-ETHANE.

ethylene glycol monoethyl ether (2-ethoxy-ethanol, Cellosolve, $C_2H_5OCH_2CH_2OH$). Colourless liquid (bp 135°C) with pleasant smell, manufactured by heating ethene (ethylene) oxide and ethanol in the presence of a catalyst or by treating 1,2-dihydroxy-ethane (ethylene glycol) with diethyl sulphate and sodium hydroxide. Reaction with ethanoic acid gives the ethanoate (Cellosolve acetate). Cellosolve and Cellosolve acetate are used extensively as solvents in cellulose nitrate lacquers.

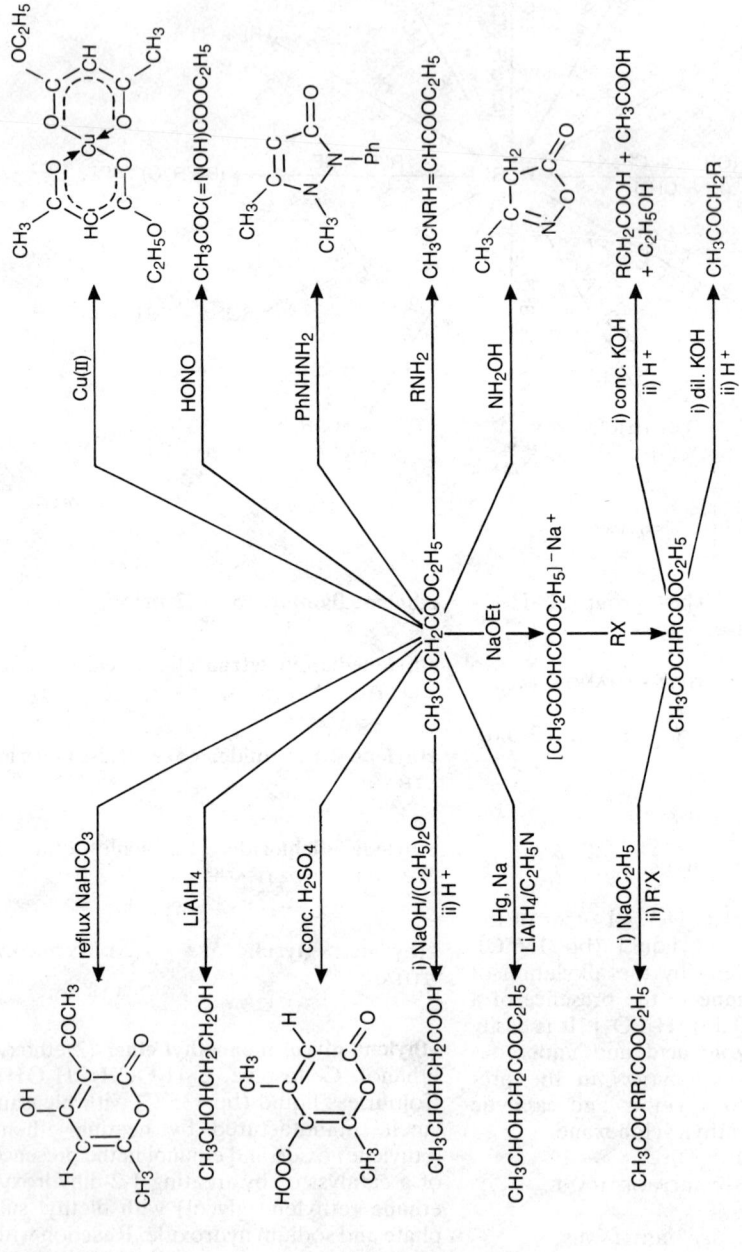

ethyl 3-oxobutanoate
Some reactions of ethyl 3-oxobutanoate.

ethyl ethanoate (ethyl acetate, acetic ester, CH$_3$COOC$_2$H$_5$).   ESTER; colourless liquid (bp 77°C), manufactured from ETHANOL and ETHANOIC ACID or by the TISCHENKO REACTION. It is used extensively as a solvent and artificial essence, and for the synthesis of ETHYL 3-OXOBUTANOATE.

ethyl group.   The ALKYL GROUP C$_2$H$_5$– (Et–).

ethylidene group.   *See* ALKYLIDENE GROUP.

ethyl iodide (iodoethane).   *See* ALKYL HALIDES.

ethyl 3-oxobutanoate (ethyl acetoacetate, acetoacetic ester, CH$_3$CH(OH)CH$_2$COO-C$_2$H$_5$).   β-Ketonic ESTER; colourless, pleasant-smelling liquid (bp 181°C), prepared by the CLAISEN CONDENSATION of two molecules of ethyl ethanoate in the presence of sodium or sodium ethoxide. It exhibits KETO–ENOL TAUTOMERISM, in which 93 percent is in the keto-form at room temperature. It is an important synthetic reagent (*see* Figure), especially of ketones containing the ETHANOYL GROUP and of monocarboxylic acids.

ethyne (acetylene, HC≡CH).   Most simple ALKYNE; colourless gas (bp –84°C) having an ethereal smell. Ethyne is sparingly soluble in water and is soluble in propanone, and is explosive in air. (For preparation and general reactions—*see* Figure.) Ethyne is a precursor for a variety of industrial chemicals including propanenitrile, ethenyl chloride, ethenyl ethanoate, ethanal and ethanoic acid, and is used in oxy-ethyne welding.

ethyne complexes.   *See* CARBON-TO-METAL BOND.

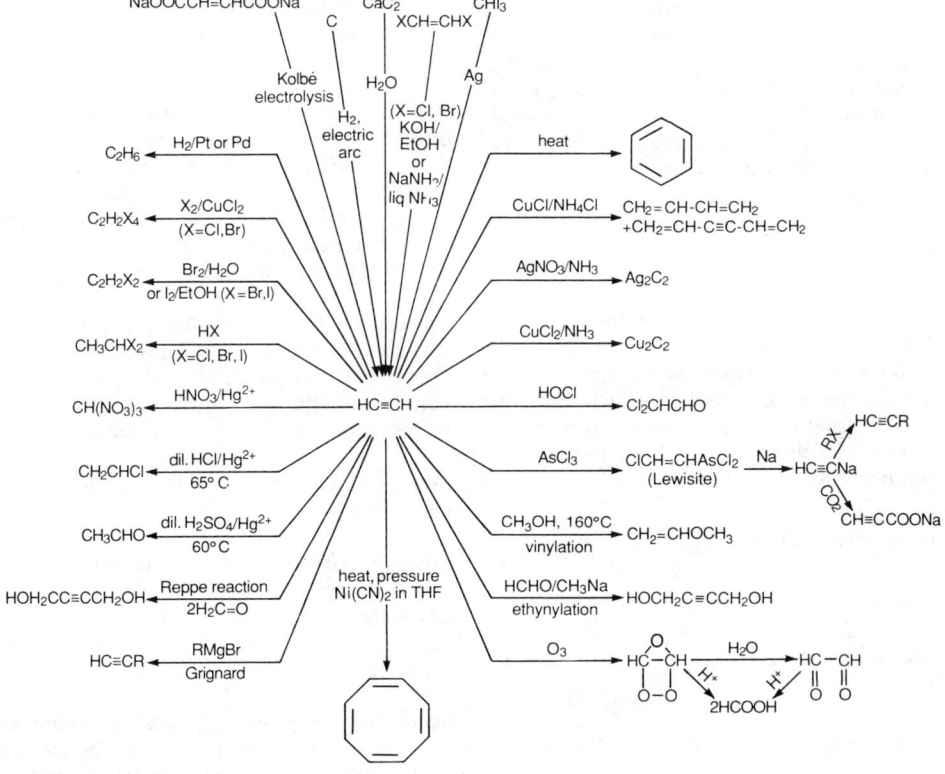

ethyne
Reactions and preparations of ethyne.

ethynylation. Reaction of ETHYNE that preserves the ethynic bond.

ethynyl group. The group HC≡C–.

Eu. Europium (see LANTHANIDE ELEMENTS).

eugenol. See PHENOLS.

europium (Eu). See LANTHANIDE ELEMENTS.

eutectic arrest. Duration of the constant temperature period at the eutectic temperature during the cooling of a TWO-COMPONENT CONDENSED SYSTEM. The arrest is maximum for a melt having the eutectic composition.

eutectic mixture. Solid mixture of constant proportion of components A and B, in a TWO-COMPONENT CONDENSED SYSTEM, which separates out from the liquid of the same composition at a constant (eutectic) temperature. It has a fixed melting point, but is not a pure compound and cannot usually be represented by a simple empirical formula. The two visibly distinct crystalline forms can be separated mechanically. The eutectic point is an invariant point ($f' = 0$).

eV. See ELECTRON VOLT.

evaporation. Change of state of a liquid into a vapour at a temperature below its BOILING POINT. Evaporation occurs at the surface of a liquid when molecules with higher kinetic energies escape into the gas phase, resulting in a lowering of the temperature of the liquid.

even–even nucleus. See NUCLEAR SPIN.

even–odd nucleus. See NUCLEAR SPIN.

EXAFS. See EXTENDED X-RAY ABSORPTION FINE STRUCTURE.

excess functions. See REAL SOLUTION.

exchange current. See BUTLER–VOLMER EQUATION.

exchange current density (i_0, j_0). Anodic and cathodic current density at an electrode when no net current flows, that is at equilibrium (i.e. zero OVERPOTENTIAL). For a multi-step reaction ox + $ne \rightarrow$ red

$$i_0 = nFk^0[\text{ox}]^{1-\alpha}[\text{red}]^\alpha$$

k^0 is a rate constant defined at the standard electrode potential.

exchange energy. Short-range repulsive energy between atoms in a molecule which becomes important when there is sufficient overlap between the two sets of atomic orbitals. See also LENNARD–JONES (12,6) POTENTIAL.

excimer. Dimeric complex in which one molecule is in the ground electronic state and the other in an excited electronic state. This has no vibrational structure as a return of the excited molecule to the ground state causes dissociation. Excimer formation is shown at high concentrations by molecules having long fluorescent lifetimes (e.g., pyrene).

excited state. Stationary state of a particle or system of particles of higher energy than the GROUND STATE: for example, electronically excited state (electron promoted to higher energy ORBITAL), vibrationally excited state (population of higher vibrational states within a given electronic state).

exclusion limit. In GEL FILTRATION CHROMATOGRAPHY, the molecular mass above which molecules are incapable of penetrating the gel pores and are eluted with the interstitial volume of the column.

exclusion principle. See PAULI EXCLUSION PRINCIPLE.

exergonic process. Process in which there is a decrease in the GIBBS FUNCTION (i.e. $\Delta G < 0$).

exo-. See ENDO-.

exothermic process. Physical or chemical process that is accompanied by the evolution of HEAT (i.e. the ENTHALPY of the reactants is greater than that of the products). See also ENDOTHERMIC PROCESS.

explosives   Some conventional explosives.

Name/structure	Preparation/description/properties/uses
Amatol	Mixture of NH_4NO_3 and TNT
Cordite	Mixture of 60% cellulose nitrate, 40% nitroglycerin; smokeless powder
Cyclonite (RDX, 1,3,5-hexahydro-1,3,4-triazine) O_2NN⌐⌐NNO_2 N NO_2	Prepared by nitration of hexamethylenetetramine; very powerful explosive with greater brisance (shattering effect) than TNT; used in mixture with TNT in large bombs and in plastic bombs
DINA (*N*-nitrodiethanolamine nitrate)	Powerful explosive; its low mp (52°C) permits it to mix better with TNT and to plasticize cellulose nitrate
Dynamites	Mixtures of about 55% NH_4NO_3, 15% $NaNO_3$, 15% mixture of glycol nitrate and nitroglycerin
Gelatin dynamite (gelignite)	Mixture of wood pulp, $NaNO_3$ and nitroglycerin, gelatinized with 2−6% cellulose nitrate; plastic and can be loaded into bore holes, water resistant; useful for work in wet places
Gun cotton (cellulose nitrate)	12.5−13.5% N; smokeless powder; propellant
Gun powder	Mixtures of about 75% KNO_3, 15% charcoal, 10% sulphur; as propellant has been replaced by smokeless powders, but still used in primers, fuses and pyrotechnics
Lead azide $Pb(N_3)_2$	Prepared from lead ethanoate and NaN_3; very sensitive to shock; used as detonator
Mercury fulminate $Hg(OCN)_2$	Prepared by reaction between Hg, EtOH and HNO_3; highly explosive, detonated by heat or shock; used as detonator
Nitroglycerin (glycerol trinitrate) CH_2ONO_2 | CH_2ONO_2 | CH_2ONO_2	Made by nitration of glycerol with HNO_3 and H_2SO_4
1-Nitroguanidine $HN=CNH_2NHNO_2$	Prepared by action of conc. H_2SO_4 on guanidinium nitrate; component of some explosives, about as powerful as TNT, explodes without producing a flash
Nitrostarch	Contains 11−13% N; used to replace TNT as demolition explosive
Pentaerythritol tetranitrate $C(CHONO_2)_4$	Mixed with TNT used as a high-explosive charge for bombs, torpedos and mines, and for demolition purposes
Tetracene (1-(5-tetrazolyl)-4-guanyl tetrazene) N—N N⌐ ⌐$CN=NNHNHC(=NH)NH_2$ N H	Formed from HNO_2 and aminoguanidine; primary explosive that can replace $Pb(N_3)_2$ or mercury fulminate

continued overleaf

Name/structure	Preparation/description/properties/uses
Tetryl (2,4,6-trinitrophenyl-methylnitramide) 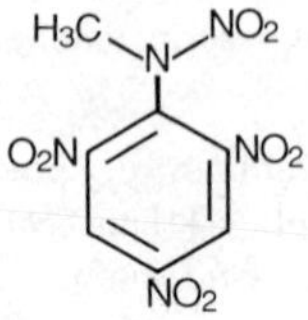	Manufactured by nitration of dimethylaniline in conc. H_2SO_4 (one of Me groups is oxidized during the reaction); booster charge for high-explosive shells initiated by detonator
1,3,5-Trinitrobenzene 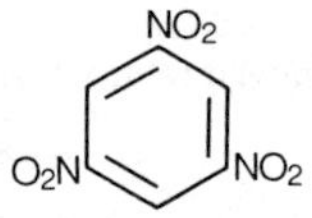	More powerful explosive than TNT, but not made in satisfactory yields because introduction of 3rd nitro group in absence of activating group is difficult
Trinitrotoluene (TNT) 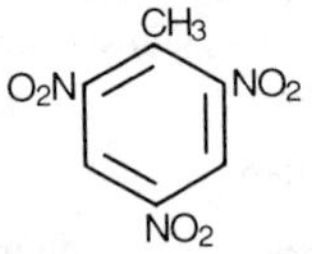	Prepared by nitration of toluene; melts at 81°C and does not explode until 250°C, can be poured into shells in liquid form; insensitive to shock, requires detonator; used for filling bombs, shells and hand grenades, either alone or mixed with other explosives

expectation value. Average value of a quantity x written as $\langle x \rangle$ and defined as

$$\int \psi^* x \psi \, d\tau \Big/ \int \psi^* \psi \, d\tau$$

Of importance to quantum mechanics in that the expectation value of the Hamiltonian operator is the eigenvalue of the Hamiltonian. *See also* SCHRÖDINGER EQUATION.

explosion. An explosion may be thermal, caused by the exponential dependence of reaction rate on temperature (*see* ARRHENIUS EQUATION) or be caused by a CHAIN REACTION which has steps that increase the number of radicals in the system. In the case of the reaction between hydrogen and oxygen, the radical reactions branch by

$$H^\cdot + O_2 \longrightarrow {}^\cdot OH + O^\cdot$$

and

$$O^\cdot + H_2 \longrightarrow {}^\cdot OH + H^\cdot$$

Whether or not a mixture of hydrogen and oxygen will explode depends upon both temperature and pressure, the particular values of which determine the explosion limits.

explosion limits. *See* EXPLOSION.

explosives. Substances, or mixtures, (*see* Table) which when submitted to shock, friction, sparks, etc. undergo rapid exothermic decomposition or reaction with the production of large volumes of gas. Three main classes are recognized.
propellants. Compounds which burn steadily and can be detonated only under extreme conditions.
initiators. Compounds that are very sensitive to mechanical shock; used as detonators to initiate the explosion of larger amounts of less sensitive material.
high explosives. Compounds that burn without undue violence but can be detonated by a sudden large mechanical or explosive shock.

extended X-ray absorption fine structure (EXAFS). Near the absorption edge (i.e. the threshold energy required to emit a photoelectron) the X-ray absorption spectrum shows fine structure (König structure) which may be related to the environment of the atom. Analysis of this structure at greater than 100 eV above the edge (hence the term 'extended') allows calculation of interatomic distances (to $\pm$ 1 pm), coordination numbers and the relative displacements of the surrounding atoms. *See also* X-RAY SPECTROMETRY.

extender. Additive to natural or synthetic RUBBER, often an inert filler. *See also* VULCANIZATION.

extensive property (capacity factor). Property, the value of which is proportional to the quantity of material in the system (e.g., mass and volume); such properties are additive. *See also* INTENSIVE PROPERTY.

extinction coefficient. *See* BEER–LAMBERT LAW.

extraction. (1) Process of obtaining a metal from its ore. (2) Separation of soluble material from a solid mixture by a solvent. (3) Separation of one or more compounds from a liquid mixture using a solvent with which the liquid is immiscible. *See also* LEACHING, LIQUID–LIQUID EXTRACTION, PARTITION COEFFICIENT.

Eyring equation. Second-order RATE CONSTANT (k_2) for a bimolecular reaction

$$k_2 = \kappa(kT/h)\,\bar{K}^{\ddagger}$$

where κ, the transmission coefficient, is the fraction of vibrations of the ACTIVATED COMPLEX along the REACTION COORDINATE which leads to products (κ is usually about unity). $\bar{K}^{\ddagger}$ is an equilibrium constant for the formation of the activated complex with the contribution to the PARTITION FUNCTION made by the vibrational mode which leads to decomposition subtracted out.

E,Z- (stereochemical descriptor). *See* ISO-MERISM.

F

F. *See* FARAD, FLUORINE.

F. *See* FARADAY CONSTANT.

f. *See* ACTIVITY COEFFICIENT, FUGACITY, OSCILLATOR STRENGTH.

FAC. *See* FAST AFFINITY CHROMATOGRAPHY.

face-centred cubic lattice (fcc). *See* CLOSE-PACKED STRUCTURES.

facile reaction. *See* SUPPORTED CATALYST.

FAD (flavin adenine dinucleotide). *See* COENZYMES.

Fahrenheit scale. *See* TEMPERATURE SCALES.

Fajan's rules. Ionic bonds (rather than covalent) are favoured between ions if: (1) the ions have small charges; and (2) the cation is large and the anion small.

farad (F; $F = C\ V^{-1} = A\ s\ V^{-1} = A^2\ s^4\ kg^{-1}\ m^{-2}$). SI-derived unit of electrical CAPACITANCE. A capacitance of 1 farad requires 1 coulomb of electricity to raise its potential by 1 volt.

Faraday constant (F). Charge carried by 1 mole of electrons (i.e. F = electronic charge $\times N_A = 96\ 484.56\ C\ mol^{-1}$).

Faraday's laws of electrolysis. (1) In any electrochemical process, the amount of chemical reaction is proportional to the quantity of electricity passed through the electrolyte. (2) Masses of different substances deposited or dissolved by the same quantity of electricity are proportional to their chemical equivalents. The quantity of electricity that will discharge 1 gram

equivalent is known as the FARADAY CONSTANT.

α-farnesene. *See* TERPENES.

farnesol. *See* TERPENES.

fast affinity chromatography (FAC). Extension of AFFINITY CHROMATOGRAPHY, based on a macroporous silica to which is bonded a hydrophilic spacer terminating in an active epoxide function. This increases the potential ligands over those used with soft gels. Problems associated with leaching of ligands are minimized. Elution times are 3–40 min for flow rates of 1 $cm^3\ min^{-1}$.

fast ion conductor. *See* SOLID ELECTROLYTE.

fats. Main form in which lipids, a potential energy source, are stored in higher animals and some plants. Fat is composed chiefly of triglycerides, in which the GLYCEROL is substituted with one or more different fatty acids, mainly oleic (*see cis*-9-OCTADECENOIC ACID), palmitic (*see* HEXADECANOIC ACID) and stearic acids (*see* n-OCTADECANOIC ACID). They are hydrolyzed to glycerol and the acid with acids, alkalis and by the action of lipases. Alkaline hydrolysis (*see* SAPONIFICATION) produces soaps. At temperatures above 250°C, they decompose to acrolein (*see* PROPENAL). Fats can be extracted from tissues with ether or other solvents.

fatty acids. Alkyl CARBOXYLIC ACIDS, so-called because of their occurrence in natural fats (*see* GLYCERIDES). They may be saturated or have one (*see cis*-9-OCTADECENOIC ACID), two (*cis,cis*-9,12-OCTADECADIENOIC ACID), three (*cis,cis,cis*-9,12,15-OCTADECATRIENOIC ACID) or more double bonds.

190

fcc (face-centred cubic lattice). *See* CLOSE-PACKED STRUCTURES.

FDMS (field desorption mass spectrometry). *See* MASS SPECTROMETRY.

FDNB. *See* 1-FLUORO-2,4-DINITROBEN-ZENE.

Fe. *See* IRON.

Fehling's solution. Solution of copper(II) sulphate, sodium potassium 2,3-dihydroxy-butanedioate and sodium hydroxide which in the presence of reducing sugars gives a red deposit of copper(I) oxide. It can be used qualitatively or quantitatively.

feldspar. *See* ALUMINIUM.

feldspars. Family of anhydrous alumino-silicates, the most common constitutents of igneous rocks, which include such minerals as orthoclase $(K[(AlO_2)(SiO_2)_3])$ with aluminium or silicon in tetrahedral co-ordination. They are used in ceramics and enamelling. *See also* SILICATES.

FEM. *See* FIELD EMISSION MICROSCOPY.

fenchone. *See* TERPENES.

Fenton's reagent. Aqueous solution of an iron(II) salt and hydrogen peroxide used for oxidizing POLYHYDRIC ALCOHOLS.

ferberite. *See* TUNGSTEN.

fermentation. Chemical processes induced by living organisms (bacteria, yeasts, moulds, fungi) or enzymes. Fermentation is used in the preparation of breads and other foodstuffs, in the manufacture of alcoholic beverages (wine, beer, etc.), lactic acid, ethanoic acid, gluconic acid and many amino acids, synthetic biopolymers, ANTI-BIOTICS and single-cell proteins from petroleum. The activated sludge process for sewage digestion is a form of fermentation.

Fermi–Dirac statistics. Indistinguishable particles with spin one-half known as fermions (e.g., electrons) are subject to the PAULI EXCLUSION PRINCIPLE and follow Fermi–Dirac statistics. The probability (f) of a particle having energy E is given by

$$f = \{ \exp [(E - E_F)/kT] + 1 \}^{-1}$$

E_F is the FERMI ENERGY. *See also* BOLTZMANN DISTRIBUTION, BOSE–EINSTEIN STATISTICS.

Fermi energy (E_F). Energy of the top of the filled band of a METAL at 0 K

$$E_F = h^2/8\pi^2 m_e(3\pi^2 N_c)^{2/3}$$

where N_c is the number of conduction band electrons (mass: m_e) per unit volume. It is therefore the negative of the ionization energy of the metal. *See also* BAND THEORY OF SOLIDS, FERMI–DIRAC STATISTICS.

fermion. *See* FERMI–DIRAC STATISTICS.

Fermi resonance. Accidental DEGENERACY between vibrational states in polyatomic molecules. It is commonly found between fundamental and overtone bands.

fermium (Fm). *See* ACTINIDE ELEMENTS.

ferrates (II), (III), (IV), (VI). *See* IRON COMPOUNDS.

ferredoxins. Group of red–brown non-haem proteins, containing iron in association with sulphur, found in some bacteria and plants. They are strong reducing agents and are components of ELECTRON TRANSPORT CHAINS (e.g., in photosynthesis and nitrogen fixation).

ferricyanides. *See* CYANOFERRATES.

ferrimagnetism. Magnetism associated with FERRITES.

ferrites. Materials that have two interpenetrating magnetic sublattices which are not equivalent (*see* ANTIFERROMAGNET-ISM). When an antiparallel alignment of the magnetic dipoles occurs, there is residual magnetism. Magnetite (Fe_3O_4) with a SPINEL structure is the most common ferrite. Ferrites have high electrical resistivity coupled with high magnetic permeability and show HYSTERESIS. They are used as cores of inductive elements, in memory devices and in microwave equipment.

ferroalloy. Alloy of iron and other elements made by smelting mixtures of ores. It is used for manufacturing special STEELS (e.g., ferrochromium, ferrovanadium, ferromanganese). *See also* SIEMENS–MARTIN PROCESS.

ferrocene. *See* METALLOCENES.

ferrocyanides. *See* CYANOFERRATES.

ferroelectricity. Dielectric HYSTERESIS shown by compounds with non-equivalent dipolar sublattices (e.g., the perovskite barium titanate).

ferromagnetism. Materials that exhibit PARAMAGNETISM at high temperatures and spontaneously acquire a permanent magnetic dipole below a characteristic temperature (the Curie temperature, T_c) are ferromagnetic. Iron, cobalt and nickel are ferromagnetic elements. Magnetization of a ferromagnet requires individual magnetic domains in the bulk material to be aligned. Ferromagnets typically show HYSTERESIS in the curve of magnetic susceptibility against applied magnetic field.

ferrosoferric oxide. *See* IRON COMPOUNDS.

fertilizer. Substance or mixture that contains one or more of the primary plant nutrients: nitrogen (free ammonia, ammonium nitrate or urea), phosphorus (superphosphate) and potassium (potassium chloride, sylvite ore) and sometimes secondary nutrients (calcium, magnesium, sulphur) and/or trace elements (iron, copper, boron, manganese, zinc, molybdenum).

fibres. Thread-like materials that may be spun and which are used in textiles, insulation, as fillers, etc. Natural fibres include cotton, wool and asbestos. The most important synthetic fibres are polyesters (e.g., Terylene, Dacron), NYLONS, cellulose acetate (rayon, viscose), acrylic resins. *See also* POLYMERS.

Fick's laws of diffusion. (1) Flux J (amount/unit area/unit time) of substance is proportional to its concentration gradient:

$$J = -D(\partial c/\partial x)_t$$

(2) Rate of change of concentration with time is given by

$$(\partial c/\partial t)_x = D(\partial^2 c/\partial x^2)_t$$

The constant D (units: $m^2 \ s^{-1}$) is called the diffusion coefficient and for many substances in water has values of the order of $10^{-10} \ m^2 \ s^{-1}$

fictile. Describing compounds that rearrange very easily. *See also* FLUXIONAL COMPOUNDS.

FID (flame ionization detector). *See* DETECTION SYSTEMS.

field desorption mass spectrometry. *See* MASS SPECTROMETRY.

field emission microscopy (FEM). Imaging of areas of constant work function on a metal. A strong negative electric field applied to a metal cathode in a vacuum causes the emission of electrons by quantum mechanical tunnelling from the Fermi level of the metal (*see* TUNNEL EFFECTS). In order to produce fields of sufficient magnitude, the metal is usually in the form of a sharp point or thin wire. The emission current (I) is given by the Fowler–Nordheim equation which in its most simple form is $I = A/F^2 \exp(-B\phi^{3/2}/F)$, where A and B are constants and ϕ is the work function of the emitting site. A phosphored screen acting as an anode gives an image of the emitter with a magnification of the order of the ratio of the tip to screen distance to the radius of the tip.

field ionization mass spectrometry. *See* MASS SPECTROMETRY, TIME-OF-FLIGHT SPECTROMETER.

field ion microscopy (FIM). Technique related to FIELD EMISSION MICROSCOPY in which the metal tip is subjected to a strong positive field in an atmosphere of an imaging gas, usually helium. The gas atoms are ionized, the cation produced travels to the cathode screen and creates an image of the surface at which it was ionized. Cooling the emitter minimizes lateral movement of the ions, and atomic resolution is easily achieved. The method is confined to refractory metals, which can withstand the high stresses associated with the electric field. In atom probe FIM, the image of an atom of interest is allowed to fall on a small probe hole in the screen. A pulse of voltage on the emitter or the action of a laser (laser atom probe FIM) causes field desorption of atoms at the surface which pass through the probe hole and are analyzed by a TIME-OF-FLIGHT SPECTROMETER.

filler. Inert material incorporated into POLYMERS (plastics and rubbers) to improve properties such as wear and friction resistance, hardness, impact resistance and to reduce the cost. Examples of fillers are FIBRES (e.g., wood, glass fibre, asbestos) and minerals (micas, slate).

film. *See* SURFACE FILM.

film balance. *See* SURFACE FILM.

FIM. *See* FIELD ION MICROSCOPY.

FIMS (field ionization mass spectrometry). *See* MASS SPECTROMETRY.

FIR (frustrated internal reflectance). *See* ATTENUATED TOTAL REFLECTANCE.

first law of thermodynamics (conservation of energy). Total amount of energy, including that equivalent to mass (mc^2) is constant in an isolated SYSTEM. It is a law of nature consistent with all known experimental results. For a closed system

$$\Delta U = U_2 - U_1 = q + w$$

where q is the HEAT absorbed by the system and w the WORK done on the system. q and w can have different values depending on the path taken, but their algebraic sum is invariable.

$$\Delta U_{system} + \Delta U_{surroundings} = 0$$

If $w = 0$, $\Delta U = q$; for ADIABATIC PROCESSES $q = 0$ and $\Delta U = w$. The first law is concerned only with changes in the energy of material systems and not with the absolute energy of any system.

first-order reaction. *See* ORDER OF REACTION.

first-order transition. *See* PHASE TRANSITION.

Fischer indole synthesis. Formation of substituted indoles by heating aryl hydrazones of an aldehyde or ketone with a catalyst (e.g., fused zinc chloride, Lewis acids). INDOLE itself cannot be prepared from ethanal by this method.

Fischer projection formula. Method of representing a three-dimensional structure (e.g., the projection formulae I and II) in two dimensions in which the chiral atom(s) lie in the plane of the paper and, by convention, bonds drawn horizontally are above the plane of the paper and those drawn vertically are below it. In a Fischer projection, III and IV refer to the same molecule, whereas V and VI are ENANTIOMERS.

perspective formulae

projection formula
(I)

Fisher projection
(II)

(III)

(IV)

(V)

(VI)

To tell whether or not two projections are identical or are a pair of enantiomers, interchange any two groups, a pair at a time. Each such interchange corresponds to converting the molecule to its enantiomer; after two interchanges the original molecule is again obtained. Thus

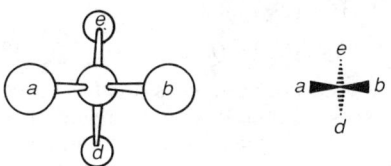

(VII)   interchange e and d enantiomer of (VII)   interchange a and b (VII)

The Fischer projection does not adequately display the spatial relationships between groups on adjacent atoms. The NEWMAN PROJECTION FORMULA (sawhorse projection) portrays this more clearly.

Fischer–Tropsch process. Synthesis of hydrocarbons and alcohols by the catalytic HYDROGENATION of carbon monoxide. The typical classes of reactions are

$$(2n + 1)H_2 + nCO \rightarrow C_nH_{2n+2} + nH_2O \quad (1)$$

$$(n + 1)H_2 + 2nCO \rightarrow C_nH_{2n+2} + nCO_2 \quad (1a)$$

$$2nH_2 + nCO \rightarrow C_nH_{2n} + nH_2O \quad (2)$$

$$nH_2 + 2nCO \rightarrow C_nH_{2n} + nCO_2 \quad (2a)$$

$$2nH_2 + nCO \rightarrow C_nH_{2n+1}OH + (n-1)H_2O \quad (3)$$

$$(n + 1)H_2 + (2n-1)CO \rightarrow$$
$$C_nH_{2n+1}OH + (n-1)CO_2 \quad (3a)$$

Methanation of SYNTHESIS GAS (reaction 1) occurs over nickel. In general, reactions 1, 2 and 3 are catalyzed by cobalt and 1a, 2a, 3a by iron. Iron catalysts produce more alkenes. Some branching occurs and small amounts of aromatics are formed. The mechanism of catalysis is thought to be via METAL CARBONYLS. *See also* OXO PROCESS.

fission. *See* NUCLEAR FISSION.

Fittig reaction. Condensation of two molecules of an aryl halide in the presence of sodium in ethereal solution to give the diaryl

$$ArX + 2Na + Ar'X \longrightarrow Ar–Ar' + 2NaX$$

where Ar and Ar' may be the same or different; some Ar–Ar and Ar'–Ar' will also be formed. *See also* WURTZ–FITTIG REACTION, WURTZ REACTION.

fixation. Removal of unwanted silver halides from photographic films after development. Thiosulphates (*see* SULPHUR OXOACIDS), which form water-soluble complex salts with silver halides, are most widely used as fixatives. Fixing baths often contain hardeners (chrome or potassium alum) to toughen the gelatin film.

fixed bed reactor. Catalytic reactor in which the catalyst is held in a fixed matrix, usually on a SUPPORT MEDIUM. The reacting gas or liquid flows through the bed. In some processes, fixed bed reactors have been superseded by FLUIDIZED BED REACTORS.

fixed points. *See* TEMPERATURE.

Flade potential. *See* PASSIVITY.

flame. Hot, luminous region in which gases undergo COMBUSTION. The light produced arises from FLUORESCENCE of molecules formed in the excited state and from INCANDESCENCE of solid particles (e.g., carbon). The flame front is the region between the luminous zone and the dark zone of unburnt gas.

flame emission spectroscopy. *See* ATOMIC SPECTROSCOPY.

flame ionization detector. *See* DETECTION SYSTEMS.

flame retardant. Compound that is used to treat materials to give them resistance to COMBUSTION. Examples are borates, phosphates, bromine, chlorine, antimony(III) oxide and aluminium(III) oxide.

flash photolysis. Relaxation method for measuring the rates of fast reactions of radicals, excited species, etc. An intense flash of light absorbed by the reaction mixture creates the species, the reactions of which are followed with time by spectrophotometry, measurement of electrical conductance or any technique that can operate on the time-scale of the experiment. The use of LASERS allows the study of picosecond reactions.

flash point. Lowest temperature at which the application of a small flame causes the vapour above a flammable, volatile liquid to ignite when the liquid is heated in a standard cup under defined conditions. At the flash point the application of a naked flame gives a momentary flash, rather than sustained combustion.

flavanone. *See* FLAVONE.

flavin adenine dinucleotide (FAD). *See* COENZYMES.

flavone (2-phenylchromone). Solid (mp 97°C), insoluble in water, soluble in organic solvents. Flavone (*see* Figure) occurs naturally as the dust on the leaves and flowers of primulas. It is synthesized from 2-hydroxyacetophenone and benzaldehyde. The reduction product is flavanone.

flavone

flavonoid. ACETOGENIN containing the $C_6-C_3-C_6$ skeleton (*see* Figure), classified according to the structure of the C_3 portion. This usually forms a ring containing oxygen as in FLAVONE. Flavonoids and their glycosides are widely distributed in nature as water-soluble pigments: ANTHOCYANINS (red, blue and purple), aurones and chalcones (yellow), flavones and flavonols (pale yellow and ivory).

flavonoid

flavoproteins. Group of conjugated proteins in which a derivative of riboflavin (e.g., FAD, *see* VITAMINS) is bound as the prosthetic group. They function as dehydrogenases in the ELECTRON TRANSPORT CHAIN.

flint. *See* SILICA.

flocculation. Weak, reversible aggregation of a COLLOID at large interparticle separations. This may occur at a secondary minimum in the potential energy–distance curve, or by bridging of long-chain additives, such as GELATIN. The term is often used synonymously with aggregation and COAGULATION.

flotation of ores. Separation of ores from a suspension by creating an aerated FOAM.

The rising bubbles of gas carry the lighter ores to the top of the tank where they are skimmed off. *See also* ELECTROFLOTATION.

flow process. Chemical process in which the reactants are continuously fed into a chamber, for example, containing a catalyst, and the products continuously removed. *See also* BATCH PROCESS, FUEL CELL, STOPPED-FLOW METHOD.

fluid. Any substance that can flow (e.g., liquid, gas).

fluidity (ϕ; units: $m^2 \, N^{-1} \, s^{-1}$). Reciprocal of VISCOSITY.

fluidized bed reactor. Catalytic reactor in which the catalyst bed is maintained in continuous agitation by an upstream of gas or liquid. A balance is achieved when the buoyancy of the fluid is equal to the gravitational force on the particles. The turbulence produced allows good thermal and physical contact in the system. It is used in the petroleum industry (e.g., in HYDROCRACKING). *See also* FIXED BED REACTOR.

Fluon. *See* POLYTETRAFLUOROETHENE.

fluorapatite. *See* FLUORINE, PHOSPHORUS.

fluorene (dibenzocyclopentadiene, $C_{18}H_{10}$). POLYCYCLIC HYDROCARBON (structure XXXVI), present in coal tar, prepared by passing diphenylmethane through a red-hot iron tube or by the reduction of fluorenone (diphenylene ketone). It is a colourless solid (mp 116°C), with a blue fluorescence. Fluorene is less acidic than INDENE, but forms a sodium derivative, liberates hydrocarbons from Grignard reagents and condenses with aldehydes and ketones. When halogenated or nitrated the first substituent enters the 2-position and the second the 4-position. It is oxidized by chromium(III) oxide to fluorenone and by potassium permanganate to 1,2-benzenedicarboxylic acid.

fluorescein. *See* INDICATORS.

fluorescence. Relaxation of a molecule in an electronically EXCITED STATE to the GROUND STATE of the same MULTIPLICITY by

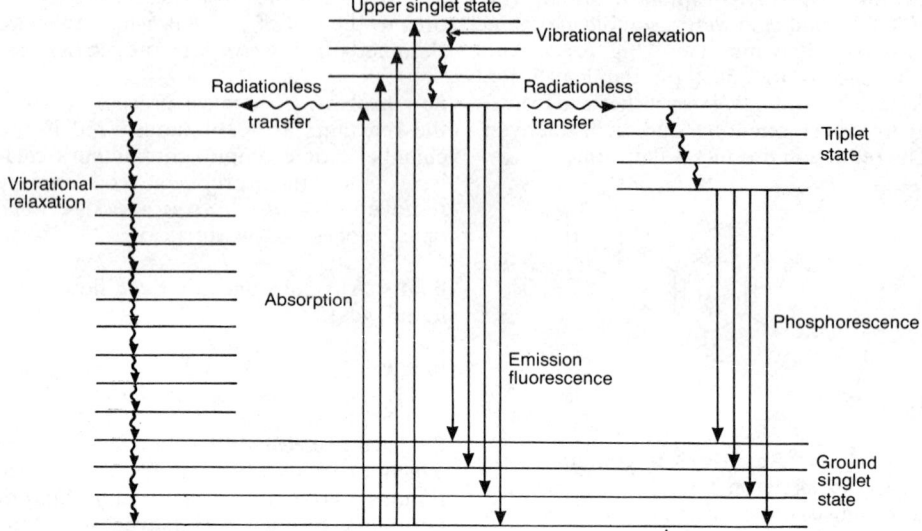

fluorescence
Possible transitions between electronic and vibrational states of a molecule.

the emission of a photon. The relationship of fluorescence to other modes of relaxation (radiationless transitions, INTERSYSTEM CROSSING to a triplet state, and PHOSPHORESCENCE) is shown in the Figure. Fluorescent life-times are 1–100 ns. Fluorescence is measured in a SPECTROPHOTOMETER similar to that used for absorption ELECTRONIC SPECTROSCOPY, but with the detector at right angles to the source. The intensity of fluorescence is small so the technique requires a high-intensity light source, such as a xenon discharge lamp. The quantum yield of fluorescence (ϕ) is the ratio of the number of photons fluoresced to the number incident on the sample

$$\phi = k_f/(k_f + k_i + k_q[Q])$$

where k_f and k_i are the rate constants for fluorescence and internal radiationless transition, respectively, and $k_q[Q]$ is the rate constant for external quenching by a molecule Q (e.g., molecular oxygen, iodide, transition metal ions). k_q may be determined by a STERN–VOLMER PLOT. *See also* EXCIMER.

fluorescence indicator.   *See* INDICATORS.

fluorides.   *See* HALIDES.

fluorine (F).   GROUP VII ELEMENT which occurs (0.08 percent of the lithosphere) as fluorspar, cryolite and fluorapatite. The greenish gas, obtained by the electrolysis of a melt of KHF_2 and HF using carbon anodes, is stored in metal (copper, steel or monel) cylinders or in the liquid state. Fluorine is the most chemically reactive of all elements, combining directly with all the elements other than oxygen and the lighter noble gases, tending to bring out the highest oxidation state of the element. It attacks many compounds forming the fluoride; organic compounds often inflame in the gas. It forms a single series of compounds in –1 oxidation state, high oxidation state compounds (e.g., SF_6) are covalent, low oxidation state compounds are ionic (e.g., NaF). Fluorine, hydrogen fluoride and fluorides are toxic in large amounts, but fluorides are essential to life and beneficial in the prevention of dental caries. Fluorine compounds are used in the preparation of metals (e.g., uranium-235), inert plastics, refrigerants, propellants and in water treatment.

$A_r = 18.998403$, $Z = 9$, ES [He] $2s^2\,2p^5$, mp –219.62°C, bp –188.14°C, $\rho = 1.7$ kg m^{-3}, $E^{\ominus}_{F^-,F_2} = 2.82$ V. Isotope: ^{19}F, 100 percent.

fluorine halides. *See* INTERHALOGEN COMPOUNDS.

fluorite lattice. *See* CRYSTAL FORMS.

fluorocarbons. HYDROCARBONS in which some or all of the hydrogen atoms are replaced by fluorine. Fluorocarbons have different properties from the other halogenated hydrocarbons, largely due to the strength of the carbon–fluorine bond (448 kJ mol^{-1}); for example, the alkyl fluorides do not undergo the WURTZ REACTION. Fluorocarbons are prepared by: (1) direct fluorination of hydrocarbons producing mixtures, but these may be controlled by the choice of catalyst; for example, silver fluoride gives predominantly perfluoro compounds; (2) addition of hydrogen fluoride to alkenes and alkynes; (3) exchange of the fluorine of inorganic fluorides (e.g., AsF_3, AgF, Hg_2F_2) with organic halides. Industrially hydrocarbons are electrochemically fluorinated at nickel anodes in anhydrous hydrogen fluoride with an inorganic fluoride electrolyte. The lower alkyl fluorides are stable, but fluoropentane and above spontaneously decompose into an alkene and hydrogen fluoride. 1,2-Difluoroalkanes are also unstable. However because of the increasing strength of the carbon–fluorine bond with the numbers of fluorine atoms, perfluorocompounds are chemically inert. POLYTETRAFLUOROETHENE (PTFE, *see* POLYMERS) is widely used where resistance to chemical attack is required. Inert fluorocarbon oils, greases and dielectrics are also of importance. Other uses are as refrigerants, fire extinguishers and propellants for aerosols. *See also* FREONS.

1-fluoro-2,4-dinitrobenzene (FDNB, 2,4-dinitrofluorobenzene, DNF, Sanger's reagent, $C_6H_3F(NO_2)_2$). Yellow liquid (mp 26°C; bp 137°C, 2 mmHg); reagent used to identify the terminal amino group of a PEPTIDE or PROTEIN. The fluorine atom is activated (*see* DINITROBENZENES) and the compound condenses with the amino group to form *N*-2,4-dinitrophenyl derivatives.

Fluorolube. *See* HEXACHLORO-1,3-BUTADIENE, MULL.

fluorophosphates
monofluorophosphoric acid (H_2PO_3F). Colourless, viscous and highly toxic liquid (P_2O_5 + conc. aq. HF). The salts resemble the sulphates; the esters are exceedingly toxic, inhibiting cholinesterase. Derivatives are used as insecticides and proposed as nerve gases.
hexafluorophosphoric acid (HPF$_6$). Colourless, fuming, toxic liquid stable in neutral and alkaline solution (PF_5 + HF). The salts (BrF_3 + MF + P_2O_5) are similar to the perchlorates (*see* CHLORATES). The free acids are used as metal cleaners.

fluorspar. *See* CALCIUM, FLUORINE.

fluothane. *See* HALOTHANE.

fluxional compounds. Stereochemically non-rigid compounds which can rearrange from one structure to an entirely equivalent structure with extraordinary ease; not to be confused with tautomers (*see* ISOMERISM) which involve rearrangements between non-equivalent structures. The rate of rearrangement between the forms (energies of activation: 25–100 kJ mol^{-1}) can be controlled by a suitable choice of temperature, so that it is slow enough to allow detection of individual molecules by NMR spectroscopy. At −30°C, the molecule $Ti(C_5H_5)_4$ exhibits two sharp signals showing the presence of two types of protons (10 of one kind in rings a,a' and 10 of a second kind in rings b,b') (*see* Figure). An increase in the temperature results in a process in which rings of type a and b exchange roles more rapidly until at 60°C they are no longer distinguishable (i.e. a single resonance peak is exhibited). A common example of fluxional behaviour is the interconversion of pyramidal molecules (e.g., PF_5).

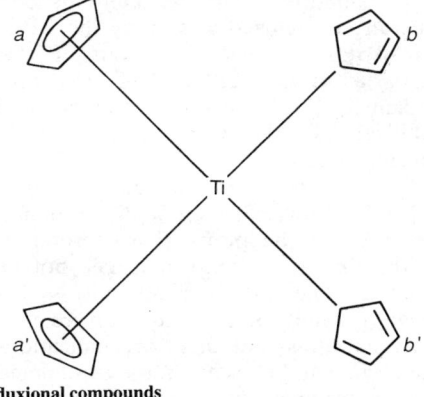

fluxional compounds

FMO theory. *See* FRONTIER MOLECULAR ORBITAL THEORY.

foam.   Coarse, colloidal dispersion (*see* COLLOID) of a gas in a liquid or solid. A typical example is the bubbles caused by agitating a solution of a SURFACE-ACTIVE AGENT. Solid foams (e.g., expanded polystyrene, foam rubber) are used as fillers and materials in furniture. Liquid foams are used in fire-fighting, gas adsorption and the separation of proteins. *See also* ANTIFOAMING AGENT.

folic acid (vitamin B_6). *See* VITAMINS.

follicle-stimulating hormone (FSH). *See* HORMONES.

food additive.   Substance added to food to perform a specific function (e.g., preservation or improvement of flavour and appearance). Typical additives (with examples) include ANTIOXIDANTS, anticaking agents ($Ca_3(PO_4)_2$, E341(c)), antifoaming agents (dimethylpolysiloxane, 900), bleaching agents ($KBrO_3$, 924; Cl_2 925), colouring agents (natural, synthetic and coal tar dyes, tartrazine, E102; cochineal, E120; CHLOROPHYLL, E140; carotene, E160(a)), emulsifying agents (CARBOXYMETHYLCELLULOSE, E466; glycerol mono- and distearates, E471), gelling agents (PECTIN, E440(a)), HUMECTANTS (glycerol, E422), thickening agents (calcium alginate, E404; agar, E406), firming agents (calcium lactate, E327), flavouring modifiers and enhancers (sodium glutamate, E621), nutrients (vitamins), preservatives and antimicrobials (benzoic acid, E210; sulphur dioxide, E220).

The additives with E numbers are generally recognized as safe by the EEC. E100–E180 are permitted colours, E200–E290 preservatives, E300–E321 antioxidants, E322–E494 emulsifiers and stabilizers, E420–E421 sweeteners, E442 solvents, E905–E907 mineral hydrocarbons, E1400–E1442 modified starches, E170–E927 miscellaneous additives. Numbers without the prefix E are proposals (at the time of writing) and need not be adopted. The list is continually being updated. Many of the additives are perfectly harmless, but there are indications that certain additives are non-essential and perhaps even harmful to sensitive people.

forbidden transition. *See* SELECTION RULES.

f-orbital. *See* ATOMIC ORBITAL.

force constant (k; units: N m^{-1}). Measure of the restoring force for vibration of a bond. For a simple HARMONIC OSCILLATOR, the force constant is the constant of proportionality in Hooke's law

$$\mu \, d^2x/dt^2 = -kx$$

where μ is the REDUCED MASS and x the displacement. For complex molecules, the force constants of each bond are related to the potential energy of the molecule. According to the central force field model the energy U is given in terms of the force constants and displacements δx

$$U = \tfrac{1}{2} \sum_{i}^{bonds} \sum_{j>i}^{bonds} k_{ij} \, (\delta x_{ij})^2$$

A better model that takes account of the molecular structure is the valence bond force field model which includes cross terms such as $1/2[k_\alpha x e_{ij} x e_{jk}(\delta\alpha)^2]$, where $\delta\alpha$ is the displacement of the angle between atoms ijk and xe is the equilibrium bond distance. Typical values for force constants range from 50 N m^{-1} for bond bending in benzene to 2000 N m^{-1} for the carbon–oxygen stretch in carbon monoxide.

force field calculation (molecular mechanics).   Method of calculating the shapes and energies of molecules that treats a molecule as an array of atoms whose positions are governed by a set of classical mechanical potential functions.

formaldehyde. *See* METHANAL.

formal electrode potential. *See* REDOX ELECTRODE SYSTEM.

formalin. *See* METHANAL.

formamide. *See* METHANAMIDE.

formation constant (K_f). EQUILIBRIUM CONSTANT for the reaction of a metal ion with an electron pair donor or LIGAND (L) to form a coordination compound

$$M + nL \rightleftharpoons ML_n$$

(charges on M and L determine the charge on the complex).

$$K_f = a_{ML_n}/a_M a_L^n$$

The overall stability constant (β) is the product of all the step-wise constants. K_f values can be obtained from emf measurements of concentration cells without transport and from potentiometric TITRATIONS.

formic acid. *See* METHANOIC ACID.

formol titration. Method for estimating carboxyl groups in amino acids. Methanal reacts with amino groups to form mono- and dimethylol compounds (–NH(CH$_2$-OH) and –N(CH$_2$OH)$_2$, respectively), thereby lowering the pK_a of the amino group. Thus, in the presence of methanal, the carboxyl groups of an amino acid can be estimated by titration with an alkali.

formula. In chemistry, written representation, using symbols, of a chemical entity. There are several kinds of formulae.
empirical formula. Expression, in its simplest form, of the relative numbers and kinds of atoms in the molecule (i.e. composition, not structure), thus for hydrogen H, and glucose CH$_2$O.
molecular formula. Derived from the empirical formula and molecular mass, gives the actual number of different kinds of atoms in the molecule, thus for hydrogen H$_2$ and for glucose C$_6$H$_{12}$O$_6$.
structural formula. Indication of the location of the atoms relative to one another, thus for glucose:

electronic formula. Structural formula in which the bonds are replaced by dots indicating the sharing of electrons between the atoms, thus for methane

See also ISOMERISM, STEREOISOMERS.

formula weight. Mass in grams of a substance numerically equal to the formula (molar mass).

formylation. *See* METHANOYLATION.

formyl group. *See* METHANOYL GROUP.

fossil fuel. Organic FUEL derived from fossilized plant material as solid COAL, liquid PETROLEUM or gaseous NATURAL GAS.

Fourier transform analysis (FT analysis). Powerful method for improving the signal-to-noise ratio in types of instrumental analysis. A Fourier transform pair relates the amplitude of a signal in the time domain $f(t)$ to that in the frequency domain $F(\omega)$.

$$F(\omega) = \int_{-\infty}^{\infty} f(t) \exp(-2\pi i \omega t) \, dt$$

$$f(t) = \int_{-\infty}^{\infty} F(\omega) \exp(2\pi i \omega t) \, 2\pi \, dt$$

In spectroscopic methods, the experiment is conducted such that data are rapidly collected in the time domain and converted by an FT to the frequency domain of the usual spectrum. In INFRARED SPECTROSCOPY, the use of an interferometer (*see* SPECTROPHOTOMETER) yields the FT of the spectrum. In NUCLEAR MAGNETIC RESONANCE SPECTROSCOPY, a pulse of radio-frequency energy applied for 1–1000 μs causes a disturbance of the magnetic moments of all the nuclei. An FT of the resulting free induction decay is the conventional frequency spectrum. Any set of data may be Fourier transformed in order to filter out noise or a background signal. The most efficient algorithm is known as a fast Fourier transform. However, the hardware methods described above are faster.

Fowler–Nordheim equation. *See* FIELD EMISSION MICROSCOPY.

Fr. *See* FRANCIUM.

fractional crystallization. *See* CRYSTALLIZATION.

fractional distillation. *See* DISTILLATION.

fragment ions. *See* MASS SPECTROMETRY.

francium (Fr). Radioactive group Ia ALKALI METAL. It occurs in the decay series of neptunium and actinium (*see* RADIO-

ACTIVE DECAY SERIES), and therefore is found in uranium and thorium ores. The most stable isotope is francium-233 (β^- half-life: 22 minutes). Its chemistry is known from tracer studies. Francium forms salts of Fr^+.
$A_r = 223$, $Z = 87$, ES [Rn] $7s^1$, mp 27°C, bp 680°C, $\rho = 2.4 \times 10^3$ kg m^{-3}.

Frank–Condon principle. Internuclear distance is not altered during the course of an electronic transition. On a Morse curve (*see* MORSE POTENTIAL) such a transition is drawn as a vertical line.

Frasch process. Process for obtaining sulphur from underground deposits by passing superheated steam down a shaft to liquefy the sulphur, which is then forced up a concentric tube by compressed air.

free electron molecular orbital. *See* PARTICLE IN A BOX.

free energy. *See* GIBBS FUNCTION.

free energy diagram. *See* ELLINGHAM DIAGRAM.

free energy function. Values of the function $(G^\ominus - H_0^\ominus)/T$ calculated from the equation

$$(G^\ominus - H_0^\ominus)/T = 30.471 -$$
$$19.144[(3/2)\lg M_r + (5/2)\lg T] -$$
$$19.147\lg Q$$

can be used to obtain $\Delta G^\ominus$ values for reactions, avoiding the use of empirical heat capacity equations

$$\Delta_f G^\ominus/T\,(T,K)/T =$$
$$[\Delta_f H^\ominus/T + (G_T^\ominus - H_0^\ominus)/T]_{compound} -$$
$$\Sigma\,[(G_T^\ominus - H_0^\ominus)/T]_{element}$$

free energy of activation. *See* ABSOLUTE RATE THEORY.

free radical. Molecule or ion with unpaired electrons. These may be stable (e.g., O_2 a diradical, NO, $Ph_3C^\cdot$) or extremely short-lived and reactive (e.g., $^\cdot CH_3$, $^\cdot OH$). Alkyl free radicals are formed by photolysis or pyrolysis of metal alkyls (e.g., tetra-

methyllead gives methyl radicals, *see* LEAD ORGANIC DERIVATIVES, ANTIKNOCK ADDITIVE). Free radical substitution of aromatic molecules usually leads to *ortho-* or *para-* attack irrespective of the nature of the substituents (*see* ELECTROPHILIC SUBSTITUTION) because of the possibilities of stabilizing the intermediate radical. The TRIPHENYLMETHYL radical is exceptionally stable because the unpaired electron is delocalized (*see* DELOCALIZATION) over the phenyl rings (*see* TRIPHENYLCARBINOL, TRIPHENYLMETHANE). Free radicals are common intermediates in radical CHAIN REACTIONS. The reactions of organic free radicals show the attempts of the radical to pair the odd electron. This may be accomplished by the abstraction of an atom (often a hydrogen atom) from a molecule forming another free radical (e.g. $^\cdot CH_3 + H–CH_2CH_3 \rightarrow CH_4 + CH_3{}^\cdot CH_2$), disproportionation (e.g., $2\,^\cdot C_2H_5 \rightarrow C_2H_6 + C_2H_4$) or by coupling with another radical (e.g., $2\,^\cdot C_6H_5 \rightarrow C_6H_5–C_6H_5$). Free radicals may be studied by ELECTRON SPIN RESONANCE SPECTROSCOPY if they are sufficiently long-lived, or they may be trapped to give stable free radicals (*see* SPIN TRAPPING) or frozen in a glass (*see* MATRIX ISOLATION).

freeze-drying. Method of dehydrating biological or chemical samples without the use of excessive heat. The material is frozen and water sublimed under vacuum. Mixed oxide catalysts of high surface area may be prepared from freeze-dried precursor salts.

freezing mixture. Mixture used for small-scale refrigeration. For example, mixture of salt and ice gives a temperature of –22°C; solid carbon dioxide ('dry ice') and propanone –78°C. *See also* CRYOHYDRATE.

freezing point. Temperature at which the solid and liquid forms of a substance have the same VAPOUR PRESSURE. The freezing point of a solution of a non-volatile solute in a volatile solvent is lower than that of the pure solvent (*see* BOILING POINT, Figure). The depression of the freezing point of an ideal solution is related to the vapour pressure above the solution and to its concentration by

$$\Delta T_c = (RT_0^2/L_f) \ln p^\ominus/p$$
$$= RT_0^2 x_B/L_f$$
$$= (RT_0^2/L_f n_A) m_B$$
$$= k_c m_B$$

This equation is valid for dilute solutions of low molecular mass (*see* BOILING POINT), when pure crystals of solvent separate from the dilute solution. The molar cryoscopic constant (k_c) is a property of the solvent and is independent of the nature of the solute. If the solution is non-ideal the ACTIVITY of the solvent is given by

$$\ln a_A = -(L_f/RT_0^2) \Delta T_c$$

Small temperature changes (ΔT_c) are measured with a Beckmann thermometer, thermocouple or thermistor. Anomalous values of ΔT_c, and hence $M_{r,B}$, are obtained for solutes which undergo association or dissociation. *See also* COLLIGATIVE PROPERTY, RAOULT'S LAW.

Frenkel defect. *See* DEFECT STRUCTURE.

Freons (Arctons). Polyhalogenated derivatives of methane and ethane containing fluorine and, in most examples, chlorine and bromine (*see* Table). Freons are prepared by fluorination of chlorinated hydrocarbons in the presence of antimony halide catalysts (e.g., $SbCl_4F$). They combine chemical and thermal stability, low boiling points, low viscosity, low surface tension and low toxicity. They are used as refrigerants, propellants for aerosols, fire extinguishers, solvents and as intermediates in POLYMERIZATION. *See also* FLUOROCARBONS.

Freons
Some common Freons.

Freon	Structure
11	Cl_3CF
12	Cl_2CF_2
13	$ClCF_3$
13B1	$BrCF_3$
22	$HClCF_2$
113	$ClCF_2CClF_2$
113B2	$BrCF_2CBrF_2$
114	$ClCF_2CCl_2F$
142	CH_3CClF_2

frequency (v; units: $Hz = s^{-1}$). Number of times a wave reaches its maximum or minimum amplitude per second. For an electromagnetic wave (*see* ELECTROMAGNETIC RADIATION) the frequency is related to the speed of light and the WAVELENGTH by $v = c/\lambda$, and to the energy by $E = hv$, where h is PLANCK CONSTANT.

Freundlich adsorption isotherm. *See* ADSORPTION ISOTHERM.

Friedel–Crafts reaction. Alkylation (a) or acylation (b) of aromatic (and certain aliphatic) compounds by substitution of one or more hydrogen atoms in the presence of a Lewis acid ($AlCl_3 > BF_3 > SbCl_5 > FeCl_3 > SnCl_4 > ZnCl_2$) (*see* Figure). Alkylating agents include alkyl halides, alkenes, alcohols; acylating agents (aliphatic or aromatic) include acid halides, acids, esters, anhydrides. Common solvents, if the substrate is solid, are nitrobenzene and carbon disulphide; the orientation may be affected by the nature of the solvent.

frontal analysis. Infrequently used chromatographic separation technique in which a solution of the mixture is continuously added to the column; only the least adsorbed component is obtained in a pure state as it moves fastest. The second fastest component is then eluted mixed with the first, and eventually the eluate is of identical composition to that added at the column head.

frontier molecular orbital theory (FMO). Description of bonding and reactions between chemical entities by a consideration of the relative energy and symmetry of the highest occupied molecular orbitals and LOWEST UNOCCUPIED MOLECULAR ORBITALS —the so-called frontier molecular orbitals.

fructose (laevulose, $C_6H_{12}O_6$). Most common ketohexose (*see* HEXOSES). D-(–)-Fructose, a white crystalline power (mp 102°C), is laevorotatory ($[\alpha]_D^{20} = -92°$) and occurs in fruits and honey and in sucrose. Commercially, D-(–)-fructose is prepared by the hydrolysis of the polysaccharide inulin which occurs in dahlia tubers and Jerusalem artichokes. In natural products it is always in the FURANOSE form,

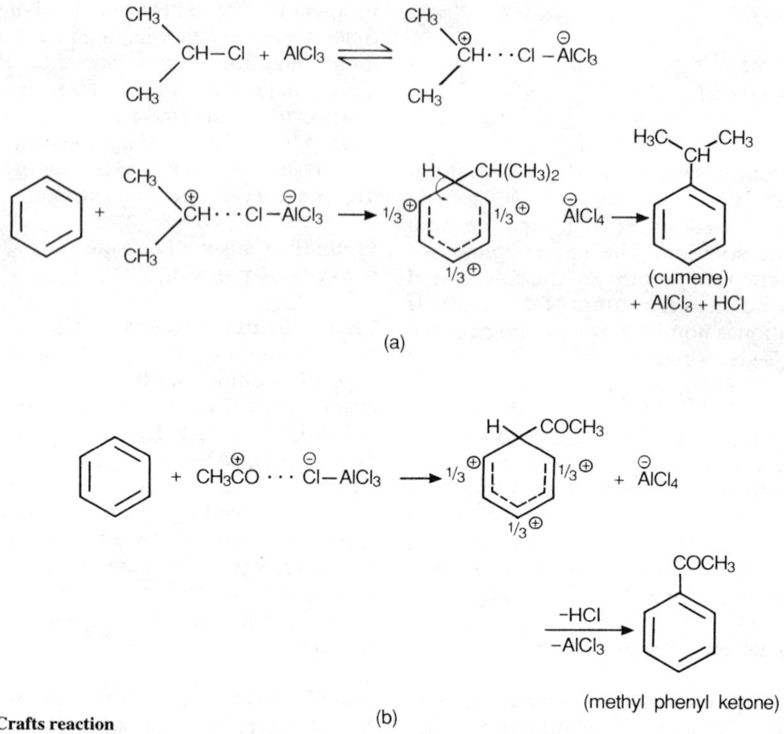

(a)

(methyl phenyl ketone)

(b)

Friedel–Crafts reaction
(a) Alkylation, (b) acylation.

but it crystallizes in the PYRANOSE form. It is very soluble in water, is twice as sweet as glucose, which it resembles in many of its properties, giving the same osazone with phenylhydrazine. Differences are due to the fact that fructose is a ketone. It is not oxidized by bromine water.

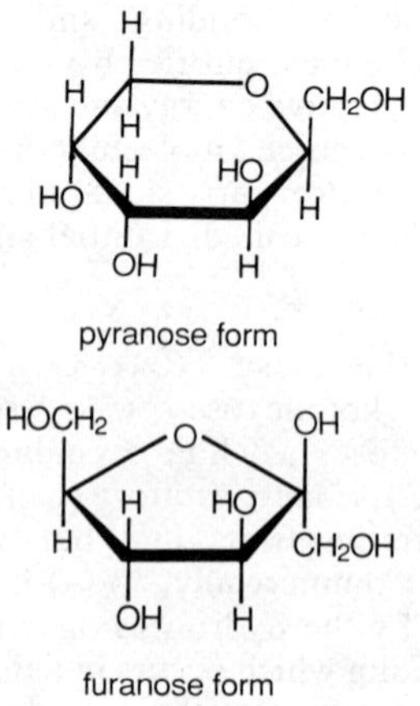

pyranose form

furanose form

fructose

frustrated internal reflectance (FIR). *See* ATTENUATED TOTAL REFLECTANCE.

FSH (follicle-stimulating hormone). *See* HORMONES.

FT analysis. *See* FOURIER TRANSFORM ANALYSIS.

FTIR (Fourier transform infrared). *See* FOURIER TRANSFORM ANALYSIS, INFRARED SPECTROSCOPY.

fucose (6-deoxygalactose, $C_6H_{12}O_5$). Methyl PENTOSE. L-Fucose is present in gum tragacanth, blood polysaccharides and seaweeds.

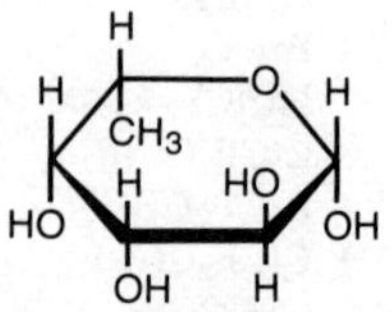

fucose

fuel. Organic fossil fuel. COAL, PETROL-EUM and NATURAL GAS are the most widely used fuels, being employed in combustion engines, power plants and heating systems. All rely on the COMBUSTION of the fuel in air. Special fuels are used to provide high specific impulse values (force × time/mass of fuel); for example, liquid hydrogen (*see* PROPELLANT). Other liquid fuels are methanol, ethanol and ammonia (with oxygen as oxidant) and aniline, 2-furaldehyde (oxidant: fuming nitric acid). Uranium and plutonium, used to generate heat in nuclear power stations, are termed 'nuclear fuels'. *See also* NUCLEAR FISSION.

fuel cell. Electrochemical CELL in which the reactants of an exoenergetic reaction are continuously supplied and the products continuously removed so that electricity may be supplied indefinitely. The fuel is usually hydrogen, but hydrazine, hydrocarbons, alcohols, etc. are possible. The cathode reaction is invariably the reduction of oxygen which may be from air ($O_2 + 4H^+ + 4e \rightarrow 2H_2O$). The electrolyte is concentrated acid or alkali, an ion-selective membrane, or for high-temperature fuel cells molten alkali carbonates or solid-doped zirconium oxide. Electrode materials are chosen for their high electrocatalytic properties and may be made of precious metal, silver or nickel. The Bacon fuel cell, a medium-temperature (100–200°C) alkali cell with nickel electrodes, gave 1 A at about 0.75 V in the Gemini spacecraft. Water produced from the hydrogen/oxygen reaction was used for drinking.

fugacity (f_i; units: N m^{-2}, atm). Variable for a real gas which replaces the (partial) pressure in each equation where the latter appears. For a real gas,

$$\mu \neq \mu^\ominus + RT \ln P/P^\ominus$$

(*see* CHEMICAL POTENTIAL), and to allow for deviations from ideal behaviour (without the use of such equation as VAN DER WAALS EQUATION) this is written as

$$\mu = \mu^\ominus + RT \ln f/P^\ominus$$

where f is an effective pressure, but its value ensures that the chemical potential is given by the above equation whatever the pressure. Since all gases tend to ideal behaviour as $P \rightarrow 0$, the limiting value of f is P. The fugacity coefficient ($\gamma = f/P$) has a limiting value of unity as $P = 0$. Values of f are calculated from the variation of the COMPRESSION FACTOR (Z) with pressure

$$\ln f/P = \int_0^P (Z-1) \, dP/P$$

The value of f depends on the pressure and the temperature. *See also* ACTIVITY.

Fuller's earth. Naturally occurring porous colloidal aluminium silicate (clay) with high absorptive properties. It is widely used in the decolorizing of oils, and as an insecticide carrier, rubber filler and oil well drilling mud.

fulminic acid (C=NOH). Unstable, volatile isomer of CYANIC ACID. It polymerizes to metafulminic acid. Fulminates of mercury(I) and silver(I) (metal + excess HNO_3 + EtOH) are explosive and are used as detonators.

fumaric acid (*trans*-butenedioic acid). *See* BUTENEDIOIC ACIDS.

function. *See* STATE FUNCTION.

functional group. Atom or group of atoms, acting as a unit, which replaces a hydrogen atom in an hydrocarbon molecule and thereby imparts characteristic properties on the molecule (e.g., $-NH_2$, $>CO$, $-Cl$, $-COOH$).

fundamental particles. *See* ELEMENTARY PARTICLES.

fundamental thermodynamic equations. Four fundamental equations relating a thermodynamic or CHEMICAL POTENTIAL to its appropriate independent variables

$$dU = T \, dS - P \, dV + \sum_i \mu_i \, dn_i$$

$$dA = -S \, dT - P \, dV + \sum_i \mu_i \, dn_i$$

$$dH = T \, dS + V \, dP + \sum_i \mu_i \, dn_i$$

$$dG = -S \, dT + V \, dP + \sum_i \mu_i \, dn_i$$

Using these equations, U, H, A and G can be expressed in terms of a chosen chemical potential and its derivatives with respect to the corresponding independent variables S, T, V and P.

fungicide. Compound that controls fungal growth. Two types exist: (1) those designed to protect against the growth of the fungus, such as copper compounds (for fruit and vegetables), organic mercurials (for seed treatment, and as a paint additive), metal organic compounds (wood preservatives); (2) those designed to eradicate fungus already present, such as lime sulphur, dithiocarbonates, organic mercurials, methanal, dinitro compounds. Fungicides are highly toxic.

2-furaldehyde (furfural, C_4H_3OCHO). Colourless liquid (bp 162°C), which darkens on exposure to air and light, prepared by hydrolysis of vegetable products (pentose-containing material) with steam at 180°C. It undergoes the CANNIZZARO REACTION with alkalis to give furoic acid and furfuryl alcohol and reacts with ammonia to yield furamide. It forms resins with phenol and aniline, and is used as a decolorizing solvent and as a chemical intermediate.

furan (C_4H_4O). Five-membered oxygen-containing HETEROCYCLIC COMPOUND (structure II) obtained as a colourless liquid (bp 32°C) by the catalytic decarbonylation of 2-furaldehyde with steam. Substituted furans are synthesized by the dehydration of 1,4-diketones (i.e. dienols) with sulphuric acid or phosphorus pentoxide. Furan has properties intermediate between those of benzene and a simple cyclic diene which is also an enol ether. On reduction (Ni/H_2) TETRAHYDROFURAN is formed. Furan readily undergoes aromatic substitution reactions and is very sensitive to strong acids, which cause exothermic polymerization. With chlorine at low temperatures 2-mono-, 2,5-di- and 2,3,5-trichlorofurans are formed.

furanose. Form of a sugar that contains the five-membered ring furan (see HETERO-CYCLIC COMPOUNDS, structure II) (e.g., ribose, 2-deoxyribose; see PENTOSES), exist-ing in the α- and β-forms. Furanose sugars are less stable than the PYRANOSE forms.

furazans (1,2,5-oxodiazoles). Substituted five-membered HETEROCYCLIC COMPOUND (structure XIII) formed by the action of sodium hydroxide on the dioximes of 1,2-diketones.

furfural. See 2-FURALDEHYDE.

furfuryl group. Univalent group (C_4H_3-OCH_2-) from 2-FURALDEHYDE. See also HETEROCYCLIC COMPOUNDS.

furyl group. The group C_4H_3O- (see HETEROCYCLIC COMPOUNDS).

fused ring. Molecular structure in which two or more aromatic rings have two carbon atoms in common. See also POLYCYCLIC HYDROCARBONS.

fused salts. Molten solutions of ionic compounds. Pure salts melt at high temperatures (in excess of 600°C). Eutectics (see EUTECTIC MIXTURE) of binary and ternary systems melt at much lower temperatures (e.g., LiCl–KCl, 355°C; AlCl$_3$–NaCl, 107°C). Being composed of ions, fused salt melts have high electrical conductivity, and thus admit an extensive electrochemistry. For example, the extraction of many electropositive metals (potassium, calcium, aluminium) is possible in fused salts. The ions in a melt may aggregate by ion association (see ION PAIR) or by HYDROGEN BOND formation (e.g., in NaHSO$_3$) and lead to lower conductivities than expected. Metals dissolved in salts of the metal have anomalously high conductivities arising from charge transfer between the metal atoms and ions. Molten salts are used in batteries, fuel cells and metal extraction, and as catalysts.

fusidic acid. See ANTIBIOTICS.

fusion. See NUCLEAR FUSION.

G

G. *See* CONDUCTANCE, GIBBS FUNCTION.

g. *See* GRAM.

g. *See* DEGENERACY.

γ. *See* ACTIVITY COEFFICIENT, NUCLEAR GYROMAGNETIC RATIO, RATIO OF HEAT CAPACITIES, SURFACE TENSION.

γ-rays. *See* GAMMA-RAYS.

Ga. *See* GALLIUM.

Gabriel reaction. Conversion of phthalamide via its potassium salt to an *N*-alkylphthalamide which may be hydrolyzed by boiling potassium hydroxide to the alkylamine. The primary amine is produced in good yield free from secondary and tertiary amines. *See also* AMINES.

gadolinite. *See* YTTRIUM.

gadolinium (Gd). *See* LANTHANIDE ELEMENTS.

galactaric acid. *See* 2,3,4,5-TETRAHYDROXYHEXANEDIOIC ACID.

galactosamine. *See* AMINO SUGARS.

galactose. Aldohexose (*see* HEXOSES). D-(+)-Galactose occurs in several polysaccharides (galactans), is combined with glucose in the disaccharide LACTOSE and with glucose and fructose in the trisaccharide RAFFINOSE. It is prepared as a white solid (monohydrate, mp 118°C) by the hydrolysis of lactose and separated from glucose by fractional crystallization. Chemically it is similar to its isomer glucose. Oxidation (HNO_3) gives the optically inactive dicarboxylic acid mucic acid.

galacturonic acid. *See* URONIC ACIDS.

galena. *See* LEAD, SULPHUR.

gallane. *See* GALLIUM COMPOUNDS.

gallia. *See* GALLIUM COMPOUNDS.

gallic acid. *See* 3,4,5-TRIHYDROXYBENZOIC ACID.

gallium (Ga). Group IIIb element (*see* GROUP III ELEMENTS) found with zinc and aluminium in low concentrations. The richest ore—germanite—contains only 0.6 percent gallium on average. Commercially gallium is extracted from the residues of zinc smelters by electrolysis of strongly alkaline solutions of gallium salts. Gallium is a soft, white metal that dissolves in acids, alkalis and reacts with air to form a protective oxide film. Because of an unusually low melting point gallium has the longest liquid range of any element. Gallium is an intrinsic SEMICONDUCTOR with a band gap of 1.4 eV. It is used in the electronics industry particularly in the binary alloys GaAs and GaP. A_r=69.723, Z=31, ES [A] $3d^{10} 4s^2 4p^1$, mp 29.8°C, bp 2400°C, ρ=5.91 × 10^3 kg m^{-3}. Isotopes: ^{69}Ga, 60.4 percent, ^{71}Ga, 39.6 percent.

gallium compounds. Chemistry of gallium is similar to that of ALUMINIUM. Its main oxidation state is +3 ($E^{\ominus}_{Ga^{3+},Ga} = -0.53$ V). Some apparently divalent chalogenides are known (e.g., GaS, GaSe, GaTe), but these contain (Ga^IGa^{III}) and not Ga^{2+}. Gallium is amphoteric, but more acid than aluminium. Gallia (Ga_2O_3) is formed in an α- and a γ-form by heating the hydroxide, nitrate, etc. Many simple and mixed salts of oxoacids are formed including a series of alums. The trihalides of gallium are prepared by direct combination of the elements. Gallium trifluoride (GaF_3) is ionic and readily forms complexes (e.g., $[GaF_6]^{3-}$). The other halides are more covalent; gallium trichloride ($GaCl_3$) forms dimers. A simple hydride has not been isolated, but adducts of gallane (GaH_3) (e.g., $(CH_3)_3$-$NGaH_3$) are known as is the anion $[GaH_4]^-$ in lithium tetrahydridogallate ($LiGaH_4$), which is a milder reducing agent than lithium tetrahydridoaluminate. Organogallium compounds are less stable than the corresponding aluminium ones. GaR_3 and complexes of $[GaR_2]^+$ are best known.

gallotannic acid. *See* TANNIC ACID.

galvanic cell. *See* CELLS.

Galvani potential. *See* INTERFACIAL POTENTIAL.

galvanizing. Method of corrosion protection in which steel is coated with zinc by dipping in molten metal or by electrodeposition. Zinc, being more electronegative, is preferentially corroded.

gamma-rays. ELECTROMAGNETIC RADIATION of extremely short wavelength (10^{-14}–10^{-10} m) and intensely high energy (10 keV–100 MeV) emitted by excited atomic nuclei during passage to a lower excitation state. They usually accompany alpha- and beta-emissions, as in the decay of radium, and always accompany fission. A common source is cobalt-60

$$^{60}_{27}Co \xrightarrow{\beta} {}^{60}_{28}Ni \xrightarrow{\gamma} {}^{60}_{28}Ni$$

in which the de-excitation of nickel-60 is accompanied by the emission of gamma-photons with an energy of 1.17 MeV. Isotopes emit gamma-rays of characteristic energy; these are detected (and counted) with semiconductor Ge(Li) detectors. They are extremely penetrating and best absorbed by dense materials such as lead. Exposure to gamma-rays may be lethal; complete protection is essential. They are used to initiate chemical reactions and polymerizations, and in food preservation.

gangue. Stony material found with metal ORES.

gas. One of the three states of matter, characterized by a very low density and viscosity, large changes in volume accom-panying change in temperature or pressure, the ability to diffuse readily into other gases and to occupy the whole of its container. A substance at temperatures above the critical temperature is known as a gas; below that temperature it is known as a vapour. *See also* IDEAL GAS.

gas absorption. Dissolution of a gas in a liquid. It is used to separate gas mixtures (e.g., a water-soluble gas from a mixture with air). *See also* ABSORPTION COEFFICIENT OF A GAS.

gas adsorption. *See* ADSORPTION.

gas analysis. Analysis of a mixture of gases by one or more of the following methods. (1) Volumetric (Orsat) method, selective absorption of constituents in succession (measuring the volume at constant temperature and pressure after each absorption) (e.g., CO_2 in KOH solution, O_2 in alkaline pyrogallol, ethyne in ammoniacal Cu_2Cl_2). (2) Burning gas in excess oxygen and measuring the change in volume of gases on absorption as in (1). (3) Titration method, absorption of a constituent in a solvent followed by suitable titration; useful for sulphur dioxide, ammonia and hydrogen chloride. (4) Gravimetric method, selective absorption of a gas onto a weighed solid. (5) Physical methods, such as thermal conductivity, UV or IR spectroscopy, vapour phase chromatography, magnetic susceptibility.

gas chromatography mass spectrometry. *See* GAS–LIQUID CHROMATOGRAPHY, MASS SPECTROMETRY.

gas constant (R; dimensions: $m^2\,l^2\,t^{-2}\,\theta^{-1}$; units: $J\,K^{-1}\,mol^{-1}$). R is related to pressure (P), volume (V) and temperature (T) of an ideal gas

$$R = PV/T$$
$$= 101\,325 \times 22.4 \times 10^{-3}/273.15$$
$$= 8.314\,J\,K^{-1}\,mol^{-1}$$

gas electrodes. *See* HALF-CELL.

gas hydrate. *See* CLATHRATE COMPOUND.

gasification. Conversion of solid (coal, coke, carbon) or liquid (petroleum) FUELS into fuel gases. Gasification of petroleum may be by hydrogen (hydrogasification) or steam (steam reforming) during which the higher hydrocarbons may undergo CRACKING. Gasification of coke may be via the reaction with oxygen to give PRODUCER GAS or steam (WATER GAS or SYNTHESIS GAS). In the high-pressure Lurgi process methane is formed

$$C + 2H_2 \rightarrow CH_4$$

or

$$CO + 3H_2 \rightarrow CH_4 + H_2O$$

gas law. The law relating the temperature (T), pressure (P) and volume (V) of an IDEAL GAS. BOYLE'S LAW and CHARLES' LAW can be combined in the universal gas equation

$$PV = nRT$$

where n is the number of moles of the gas and R the GAS CONSTANT. The law is only obeyed by real gases to a very limited extent and only at high temperatures and low pressures. *See also* EQUATION OF STATE.

gas–liquid chromatography (GLC). Chromatographic procedure in which the MOBILE PHASE is an inert gas and the STATIONARY PHASE is a liquid held on a solid surface. The sample is injected as vapour at the head of the column; the components partition themselves between the gas and liquid phases. The rate at which the components are eluted depends on their partition coefficients; a favourable PARTITION COEFFICIENT results in a low rate, components with low solubility in the liquid phase migrate rapidly. Qualitative identification of a component is based on the time (*see* RETENTION TIME) for it to appear in the DETECTION SYSTEM, quantitative data are obtained from the evaluation of peak areas on the chromatogram (*see* Figure). The apparatus, although simple in concept, requires sophisticated equipment for the control of gas flow, injection block, temperature control (constant or programmed) of the column, sample collection, and the detection system and recorder.

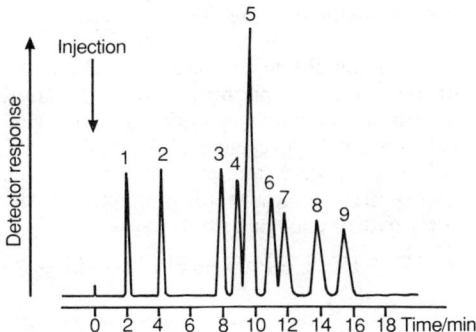

gas–liquid chromatography
Typical gas chromatogram for the separation of: (1) benzene; (2) toluene; (3) ethylbenzene; (4) *p*-xylene; (5) *m*-xylene; (6) *o*-xylene; (7) isopropylbenzene; (8) styrene; (9) *n*-propylbenzene.

gas–liquid equilibrium. *See* ONE-COMPONENT SYSTEM.

gas oil. Fraction obtained by the distillation of PETROLEUM. It is used as a fuel oil for heating systems and as a feedstock for CRACKING.

gasoline. Liquid fraction from the distillation of PETROLEUM used as a motor and aviation fuel. It is mostly obtained as NAPHTHA from the primary distillation of crude oil, but is also produced from CRACKING and REFORMING higher fractions. The fuel is characterized by its OCTANE NUMBER and may be combined with different additives (*see* ANTIKNOCK ADDITIVE).

gas-sensing membrane probe. Sensor consisting of an ION-SELECTIVE ELECTRODE combined with a REFERENCE ELECTRODE to form a complete cell. The emf of the cell is a function of the gas to be determined. Gases such as sulphur dioxide or ammonia are determined by the pH of the internal electrolyte solution which responds to changes in partial pressure of the gas.

gas–solid chromatography. Separation of mixtures by the use of a mobile gas phase and a solid adsorbent. With the use of special polymeric materials for column packing, the technique is almost as useful as GAS–LIQUID CHROMATOGRAPHY (GLC). The apparatus is similar to that for GLC.

gas–solid equilibrium. *See* ONE-COMPONENT SYSTEM.

gas thermometer. *See* THERMOMETER.

Gattermann aldehyde synthesis. Reaction of benzene (or phenol) with a gaseous mixture of hydrogen cyanide and hydrogen chloride in the presence of aluminium trichloride to give an arylimine which on hydrolysis gives low yields of benzaldehyde (or 4-hydroxybenzaldehyde).

$$HCl + HCN + PhH \rightarrow PhCH{=}NH \rightarrow$$
$$PhCHO$$

Gattermann–Koch aldehyde synthesis. METHANOYLATION of an aromatic hydrocarbon to the corresponding aldehyde by treatment of a solution of the hydrocarbon with a gaseous mixture of carbon monoxide and hydrogen chloride under pressure in the presence of aluminium trichloride and copper(I) chloride.

$$PhH + CO + HCl \rightarrow PhCHO + HCl$$

Gattermann reaction. Replacement of a diazonium group (*see* DIAZONIUM COMPOUNDS) in an aromatic compound by a halogen under the catalytic influence of copper powder and the corresponding hydrogen halide. It is a variant of the SANDMEYER REACTION.

$$PhN_2Cl \xrightarrow{Cu,HCl} PhCl + N_2$$

gauche **conformation.** *See* CONFORMERS.

Gay-Lussac's law. When gases combine they do so in volumes that are in a simple ratio to one another and to that of the product if gaseous (e.g., three volumes of hydrogen combine with one volume of nitrogen to give two volumes of ammonia). The law is strictly only valid for ideal gases. CHARLES' LAW is sometimes referred to as Gay-Lussac's law.

GCMS (gas chromatography mass spectrometry). *See* GAS–LIQUID CHROMATOGRAPHY, MASS SPECTROMETRY.

Gd. Gadolinium (*see* LANTHANIDE ELEMENTS).

Ge. *See* GERMANIUM.

gel. COLLOID containing a high concentration of the dispersed phase in which the particles aggregate to give a continuous network. The dispersion medium is immobilized and the gel may be relatively solid (e.g., SILICA GEL) or jelly-like (e.g., GELATIN). Reversible gels are converted to sols on heating.

gelatin. White or yellow crystalline solid produced by boiling a source of collagen (*see* PROTEINS), usually bones and skin of animals. Gelatin consists of the peptides of collagen and is rich in glycine, proline, hydroxyproline, lysine and hydroxylysine (*see* AMINO ACIDS). When dissolved in hot water, gelatin produces a viscous solution which sets to a jelly on cooling. Gelatin is used in cooking, photography and glues.

gel electrophoresis. *See* ELECTROPHORESIS.

gel filtration chromatography. Separation technique dependent on the shape and size of the solute molecules. The mixture, applied at the top of a column of gel is washed through the gel with water, buffer solution or organic solvent. Solutes with molecules larger than the largest pores of the swollen gel beads (above the exclusion limit) cannot penetrate the particles and pass through the bed in the liquid phase outside the gel and emerge from the bottom first. Smaller molecules can penetrate the particles to varying extents resulting in a partition between the liquid inside the gel and that outside. The gels are based on cross-linked dextrans, polyacrylamides, high-molecular-mass polysaccharides (e.g., agarose), etc.; the amount of cross-linking determines the swelling properties of the gel and the exclusion limits. The technique can be used not only for the separation of macromolecules, but also for an approximate determination of relative molecular mass (M_r); the elution volume is approximately a linear function of $\lg M_r$.

gelignite (gelatin dynamite). *See* EXPLOSIVES.

gel permeation chromatography. *See* GEL FILTRATION CHROMATOGRAPHY.

gem-. *See* GEMINAL.

geminal (*gem-*). Indicates that two substituents are attached to the same atom; thus *gem*-dibromoethane is Br_2CHCH_3.

general acid–base catalysis. *See* ACID–BASE CATALYSIS.

genetic code. Information for the synthesis of proteins coded in the structure of the DNA. The addition of a particular amino acid to a polypeptide chain is determined (coded) by a specific sequence(s) of three NUCLEOTIDES (codons) in the DNA molecule (*see* NUCLEIC ACIDS). As 64 (i.e. $4 \times 4 \times 4$) codons are possible using the two purine and two pyrimidine bases and only 20 amino acids are used, each amino acid may be specified by more than one triplet thus methionine is coded by AUG, phenylalanine by UUU or UUC, isoleucine by AUU, AUC or AUA, and proline by CCU, CCC, CCA or CCG (where A = adenine, C = cytosine, G = guanine and U = uracil). The initiation of protein synthesis is signalled by the codon AUG; in the middle of a message this codon also codes for the introduction of a methionyl residue in the normal way. Termination is signalled by UAA, UAG or UGA. The genetic code is universal and is the basis of the hereditary information of nucleic aids.

gentamicins. *See* ANTIBIOTICS.

gentiobiose. *See* DISACCHARIDES.

geometrical isomerism. *See* ISOMERISM.

geranial. *See* TERPENES.

geraniol. *See* TERPENES.

germanes. *See* GERMANIUM COMPOUNDS.

germanite. *See* GALLIUM.

germanium (Ge). GROUP IV ELEMENT, METALLOID. It is recovered from sulphide ores of silver, tin, copper and iron in a similar way to SILICON, an element it resembles in structure and general properties, although germanium is more reactive. Its main uses are as a SEMICONDUCTOR, PHOSPHOR and in special glasses. $A_r = 72.59$, $Z = 32$, ES [A] $3d^{10} 4s^2 4p^2$, mp 937°C, bp 2830°C, $\rho = 5.35 \times 10^3$ kg m^{-3}. Isotopes: ^{70}Ge, 20.6 percent; ^{72}Ge, 27.4 percent; ^{73}Ge, 7.8 percent; ^{74}Ge, 36.5 percent; ^{76}Ge, 7.7 percent.

germanium compounds. Germanium forms compounds in the +2 and +4 oxidation states. Germanium(II) compounds are strongly reducing and convert readily to germanium(IV). Germanium dichloride ($GeCl_2$) is prepared by passing the tetrachloride vapour over heated germanium. The difluoride is also known and gives $[GeF_3]^-$ with F^-. Germanium sulphide (GeS) is prepared from the disulphide and germanium. Hydrolysis of $GeCl_2$ gives germanium monoxide. Germanium dioxide (GeO_2) occurs in a high-temperature crystabolite form (*see* SILICON) in a rutile lattice (*see* CRYSTAL FORMS). Salts of GeO_2 (germanates) are similar to silicates. Metagermanates (e.g., Na_2GeO_3) and orthogermanates (e.g., Mg_2GeO_4) have silicon analogues. Strontium germanate ($SrGeO_3$) contains cyclic $[Ge_3O_9]^{6-}$. In aqueous solution, germanium(IV) exists as $[GeO(OH)_3]^-$. $[GeO_2(OH)_2]^{2-}$ and $[[Ge(OH)_4]_8(OH)_3]^{3-}$. All the tetrahalides are known. $GeCl_4$ is used to prepare semiconductor-grade germanium ($GeCl_4 + 2H_2 \rightarrow Ge + 4HCl$). $GeCl_4$ and GeF_4 are soluble in many organic solvents. Germanium hydrides (germanes) are more stable than silanes. Monogermane (GeH_4) is prepared from sodium tetrahydridoborate and GeO_2 in acid. Higher members (up to Ge_9H_{20}) are obtained from a high-pressure electric discharge in GeH_4. The thermal decomposition of GeH_4 to germanium and hydrogen at 280°C is used in the semiconductor industry to lay down a thin film of germanium. Some alkyl and aryl germanium compounds are known.

getter. Substance that removes gases from a VACUUM by adsorption or chemical reaction. Magnesium is used in thermionic valves and titanium in sputter ion pumps (*see* SPUTTERING).

gibberellins. Plant growth hormones; metabolites of the fungus *Gibberella fujikuroi* which cause marked elongation of the stems of most plants. Gibberellins are also known to be present in higher plants. They are tetracyclic lactone carboxylic acids, and although they contain only 19 carbon atoms they are related to the diterpenes (*see* Figure). Examples include gibberellic acid (mp 233°C). Gibberellins are widely used in agriculture.

gibberellins
Structure of gibberellic acid.

Gibbs adsorption isotherm. *See* ADSORPTION ISOTHERM.

Gibbs–Duhem equation. Equation relating the ACTIVITY values of the components of a binary liquid mixture

$$n_A \, d\bar{J}_A + n_B \, d\bar{J}_B = 0$$

where $\bar{J}_A$ and $\bar{J}_B$ are PARTIAL MOLAR QUANTITIES of A and B. For the GIBBS FUNCTION

$$(d \ln a_A / d \ln x_A)_{T,P} = (d \ln a_B / d \ln x_B)_{T,P}$$

Gibbs function (free energy) (G; dimensions: $m \, l^2 \, t^{-2}$; units: J mol^{-1}). STATE FUNCTION defined by

$$G = H - TS$$

For a pure substance, G depends on the nature of the substance, the state in which it exists, the temperature and, for gases, the pressure. For a mixture, G also depends on its composition

$$G = \mu_A n_A + \mu_B n_B$$

The decrease in G between specified states at constant temperature gives a measure of the net WORK (w') available from the process other than work of expansion. The equation

$$\Delta G = \Delta H - T\Delta S$$

shows the balance between the energy and entropy terms in determining the direction of the process; at low temperatures the value of ΔG depends mainly on ΔH, but at

higher temperatures the $T\Delta S$ term becomes dominant. If $\Delta G < 0$ the process is spontaneous; if $\Delta G > 0$ then the process as written will not proceed, but the reverse process will be spontaneous; if $\Delta G = 0$ and G is a minimum then the system is in a state of EQUILIBRIUM. G is related to other thermodynamic quantities

$$dG = V \, dP - S \, dT$$

leading to the identities

$$(\partial G / \partial T)_P = -S$$

and

$$(\partial G / \partial P)_T = V$$

and to the pressure or activity of a component

$$G = G^{\ominus} + RT \ln p/p^{\ominus}$$

and

$$G = G^{\ominus} + RT \ln a$$

(*compare* CHEMICAL POTENTIAL). The variation of ΔG of a reaction with temperature

$$\Delta G = \Delta H_0^{\ominus} - \Delta a T \ln T - \Delta b T^2 / 2 - \Delta c T^3 / 6 + \ldots + JT$$

is almost linear because variations in the heat capacity of the reactants and products are self-compensating in their effects on ΔH and ΔS. The mixing of two components is a spontaneous process, $\Delta G_{mixing} < 0$ (*see* IDEAL SOLUTIONS). Absolute values of G cannot be obtained from classical thermodynamics. Values of ΔG can be obtained: (1) from values of ΔH and ΔS; (2) from K or $\Delta G^{\ominus}$ and the initial and final ACTIVITY (concentration, pressure) values (*see* VAN'T HOFF ISOTHERM); (3) for ionic reactions from a knowledge of the emf of the cell; (4) from $\Delta H_0^{\ominus}$ and heat capacity data; (5) $\Delta G^{\ominus}$ from tabulated values of $\Delta_f G^{\ominus}$ for reactants and products; (6) for reactions between gases from the FREE ENERGY FUNCTION. *standard molar Gibbs function of formation* ($\Delta_f G^{\ominus}$). Value of ΔG for the formation of 1 mole of the compound from its elements with all the components in their standard states at a specified temperature (for elements, $\Delta_f G^{\ominus} = 0$). Compounds that have a large positive value for $\Delta_f G^{\ominus}$ (e.g., ethyne, benzene) are thermodynamically unstable with respect to their elements. *standard Gibbs function for reactions.* Can be calculated from $\Delta G^{\ominus}$ values for reactants and products

$$\Delta G^{\ominus} = \Delta_f G^{\ominus}_{products} - \Delta_f G^{\ominus}_{reactants}$$

$\Delta G^{\ominus}$ values are additive, whereas ΔG values are not (see VAN'T HOFF ISOTHERM).

Gibbs–Helmholtz equation. For reactions at constant pressure

$$\Delta H - \Delta G = - T [\partial(\Delta G)/\partial T]_P$$

Alternative forms are

$$[\partial(\Delta G/T)/\partial T]_P = - \Delta H/T^2$$

$$[\partial(\Delta G/T)/\partial(1/T)]_P = - \Delta H$$

The slope of the plot $\Delta G/T$ against T^{-1} at a given T is the ENTHALPY change. For reactions at constant volume

$$\Delta U - \Delta A = - T [\partial(\Delta A)/\partial T]_V$$

See also THERMODYNAMICS OF CELLS.

glass. Amorphous, rigid material formed from a melt that is cooled below its melting point without crystallizing. Inorganic glasses are formed from the oxides of the elements boron, silicon (see SILICA), germanium and phosphorus, known as 'glass formers'. Other oxides, 'conditional glass formers' (tellurium, selenium, molybdenum, tungsten, bismuth, aluminium, gallium and vanadium) form glasses in combination with a second oxide. Many organic glasses are known (e.g., candy floss is a sugar glass). Glasses, unlike crystals, are ISOTROPIC, do not show sharp melting points, but they are elastic (particularly glass fibres). Many glasses are transparent, they are resistant to chemical attack, and they are thermal and electrical insulators. Such electrical conductivity shown by glasses is by virtue of the movement of ions. Colour is imparted to glasses by the addition of transition element ions. The glassy state is also known as the vitreous state.
borosilicate glass. Based on boron and silicon oxides with a metal oxide (e.g., sodium oxide gives Pyrex glass used for most laboratory and kitchen ware).
crown glass. Contains barium or potassium oxides, and flint or lead glass contains lead oxide.
chalcogenide glass. Based on the sulphides or selenides of glass-forming elements.

glass electrode. HALF-CELL Ag,AgCl|HCl $(0.1 \text{ mol dm}^{-3})$|glass. In a cell with a saturated calomel electrode, the emf varies with the concentration of hydrogen ions

$$E_G = E'_G + (RT/F) \ln a_{H_3O^+}$$
$$= E'_G - 0.059 \text{ pH}$$

The emf of the glass electrode depends on the equilibrium between protons in solution and the univalent cations in the glass. It reaches equilibrium very quickly, suffers from no interfering species and may be miniaturized. Glass electrodes may only be used in solutions of pH less than 11.

GLC. See GAS–LIQUID CHROMATOGRAPHY.

glide plane. SYMMETRY OPERATION; the TRANSLATION of an object followed by a reflection across a plane containing the translation axis. See also SPACE GROUP.

Gln (glutamine). See AMINO ACIDS.

globulins. See PROTEINS.

Glu (glutamic acid). See AMINO ACIDS.

glucagon. See HORMONES.

glucocorticosteroids. See CORTICOSTEROIDS.

glucopyranose. See GLUCOSE.

glucosamine. See AMINO SUGARS.

glucose Aldohexose (see HEXOSES). D-(+)-Glucose (dextrose, α-D-(+)-glucopyranose) is the most common hexose, being found in ripe grapes, honey, sweet fruits and blood. It is a constituent of starch, cellulose, glycogen and sucrose from which it may be obtained by hydrolysis, and is a white solid (mp 146°C) with a sweet taste (not as sweet as cane sugar). It is a strong reducing agent, reducing FEHLING'S SOLUTION and ammoniacal silver nitrate; when heated with concentrated sodium hydroxide it turns brown and resinifies. It forms a cyanhydrin, an oxime, an OSAZONE with phenylhydrazine and glucosates with various metallic hydroxides (e.g., $C_6H_{12}O_6 \cdot CaO$). Mild oxidizing agents (bromine

α - form   aldehyde form   β - form
(I)   (II)

glucose
Anomers of glucose.

water) give D-gluconic acid, whereas nitric acid yields the dicarboxylic acid D-glucaric acid. With methanol and hydrochloric acid, it forms a mixture of the α- and β-D-glucopyranosides (*see* GLYCOSIDES). It is reduced (Ni/H_2 or electrolytically) to sorbitol (*see* HEXAHYDRIC ALCOHOLS) and fermented by yeast to ethanol. The specific rotation ($[\alpha]_D^{20}$) is +113.4°; in solution it undergoes mutarotation to a mixture of the α- and β-forms with $[\alpha]_D^{20} = 52.5°$. In the α-form (I) the C-1 hydroxyl is axial, whereas in the β-form (II) it is equatorial (the α- and β-forms are not optical isomers) (*see* Figure); these hemiacetals give rise to α- and β-methylglucosides. The particular diastereoisomers are known as anomers and the carbon atom is called the anomeric carbon. Glucose is used in the food industry.

glucosidases. *See* ENZYMES.

glucosides. *See* GLYCOSIDES.

glucuronic acid. *See* URONIC ACIDS.

glutamic acid (Glu). *See* AMINO ACIDS.

glutamine (Gln). *See* AMINO ACIDS.

glutaraldehyde ($OHC(CH_2)_3CHO$). Prepared from propenal and ethenylethyl ether. It is used as a cross-linking agent for proteins and resins.

glutaric acid. *See* PENTANEDIOIC ACID.

glutathione (glutamylcysteinylglycine). *See* PEPTIDES.

glutelins. *See* PROTEINS.

Gly (glycine). *See* AMINO ACIDS.

glycerides. Mono-, di- and triesters of GLYCEROL. Triglycerides are a major component of fats and oils found in plants and animals from which glycerol is formed on hydrolysis. One hydroxyl group may be esterified with a phosphate group giving a phosphoglyceride (*see* PHOSPHOLIPIDS) or to a sugar forming a glycolipid.

glycerin. *See* GLYCEROL.

glycerol (glycerin, 1,2,3-trihydroxypropane, $HOCH_2CH(OH)CH_2OH$). Simplest and most important trihydric alcohol; colourless, sweet-tasting, viscous liquid, miscible with water and ethanol, but insoluble in ether. It occurs (as GLYCERIDES) in combination with fatty acids in animal and vegetable fats and oils. It is obtained, as a byproduct, in the manufacture of soap by hydrolysis of the esters. It can be synthesized from propene by a variety of standard reactions.

$$CH_3CH=CH_2 \xrightarrow{Cl_2} ClCH_2CH=CH_2 \xrightarrow{HOCl}$$

$$ClCH_2CH(OH)CH_2Cl \xrightarrow{base}$$

$$\overset{O}{\underset{CH_2-CHCH_2Cl}{\triangle}} \xrightarrow{NaOH}$$

$$HOCH_2CH(OH)CH_2OH$$

Replacement of hydrogen by alkali metals occurs on heating with the metal or hydroxide. It reacts with hydrochloric acid to form GLYCEROL CHLOROHYDRINS and with nitric acid to give NITROGLYCERIN. Propenal is formed on dehydration with concentrated sulphuric acid. Oxidation yields 2,3-dihydroxypropanal, dihydroxypropanone, 2,3-dihydroxypropanoic acid and ethanedioic acid. It is used as a humectant and sweetner in confectionery (E422) and in the manufacture of explosives, medicines, perfumes, cosmetics, cellulose films and as a moistening agent for tobacco.

glycerol chlorohydrins
1,2-dihydroxy-3-chloropropane ($ClCH_2$-$CH(OH)CH_2OH$). Colourless, viscous liquid obtained by passing dry hydrogen chloride into glycerol containing 2 percent ethanoic acid at 105°C; some 1,3-dihydroxy-2-chloropropane (*see below*) is also formed. The dinitrate is used in the manufacture of low-freezing dynamite.
1,3-dichloro-2-hydroxypropane ($ClCH_2$-$CH(OH)CH_2Cl$). Obtained by passing excess dry hydrogen chloride into glycerol. Heating with potassium hydroxide yields α-epichlorohydrin (*see* 1-CHLORO-2,3-EPOXYPROPANE). It is used as a solvent for cellulose nitrate and resins.
2,3-dichloropropanol ($ClCH_2CHClCH_2$-OH). Obtained by the chlorination of 2-propenol. It is oxidized by nitric acid to 1,2-dichloropropanoic acid and reacts with sodium hydroxide to give α-epichlorohydrin.

glycerol trioleate. *See* OLEIN.

glycerol tripalmitate. *See* PALMITIN.

glycerol tristearate. *See* STEARIN.

glyceryl trinitrate. *See* NITROGLYCERIN.

glycin (4-hydroxyphenylglycine). *See* AMINOPHENOLS.

glycine (Gly). *See* AMINO ACIDS.

glycocholic acid. *See* BILE ACIDS.

glycogen. POLYSACCHARIDE, similar in structure to amylopectin (*see* STARCH) but more branched and with a higher relative molecular mass. It functions as the major carbohydrate reserve in animals, being stored in the liver and muscles. It is a white powder, soluble in water, giving a purplish–red colour with iodine. Complete hydrolysis gives D-(+)-glucose. In the digestive system it is broken down by α-glucosidase to glucose, and in cells phosphorylases, in the presence of inorganic phosphate, convert it to glucose-1-phosphate which is a starting material for many metabolic pathways.

glycol. 1,2-Ethanediol. *See also* GLYCOLS.

glycolipids. *See* GLYCERIDES.

glycollic acid. *See* HYDROXYETHANOIC ACID.

glycols ($C_nH_{2n}(OH)_2$). Aliphatic ALCOHOLS containing two hydroxyl groups. They are colourless liquids. 1,2-Glycols are obtained in the *cis*-form by the oxidation of an alkene with potassium permanganate, lead tetraethanoate or osmium tetroxide, by heating a chlorohydrin (e.g., $ClCH_2CH_2$-OH) with weak alkali or by the hydrolysis of alkane dihalides with sodium hydroxide. The epoxidation of alkenes (I) with peracids is stereospecific, the epoxide (II) formed undergoing hydrolytic attack *trans* to the ring forming a *trans*-1,2-diol (vicinal glycol, III).

The reaction of 1,2-epoxyethane in the presence of hydrogen ions and a trace of water leads to the formation of polyethylene glycols ($HOCH_2CH_2(OCH_2CH_2)_n$-OCH_2CH_2OH). In the presence of acid, 1,2-diols undergo dehydration and rearrangement (the PINACOL–PINACOLONE REARRANGEMENT). The bond between the carbon atoms in a 1,2-diol is cleaved by oxidizing agents; chromic acid and permanganate oxidize to the highest oxidation states, periodic acid and lead tetraethanoate oxidize only as far as the first aldehydes.

glycoproteins. *See* PROTEINS.

glycosides. Compounds formed by the acid-catalyzed reaction of the hemiacetal of a monosaccharide with an alcohol; on account of the equilibrium between the α- and β-forms (*see* GLUCOSE) of the monosaccharide a mixture of glycosides is formed (*see* Figure). These are stable compounds, with a bitter taste, which do not mutarotate and show no reducing properties. When glucose is the sugar the compound is known as a glucoside. The natural glycosides (starch, celluloses, nucleic acids, etc.) all contain monosaccharides which are incorporated via glycoside linkages; they nearly all belong to the β-series. They are hydrolyzed by the appropriate β-glycosidase. The non-β-D-glucopyranose   methyl-α-D-glucopyranoside   methyl-β-D-glucopyranoside

glycosides

sugar moiety of the glycoside is known as the aglycone; in many naturally occurring glycosides the C-1 substituent is a phenolic or other grouping.

amygdalin ($C_6H_{11}O_5.O.C_6H_{10}O_4.O.CH-(CN)C_6H_5$). Glycoside in which the sugar is gentiobiose (6-β-glucosylglucose) and the AGLYCONE is benzeneacetonitrile. It occurs in the kernels and leaves of many plants, such as apricot, pear and bitter almonds. Hydrolysis yields gentiobiose (*see* DI-SACCHARIDES) (and finally glucose), benzaldehyde and hydrogen cyanide.

arbutin (hydroquinone-β-D-glucoside, $C_6H_{11}O_5.O.C_6H_4OH$). It occurs in many plants; the black colouring of leaves of certain pear trees in autumn is due to its enzymic hydrolysis to 1,4-dihydroxybenzene which in air is oxidized to a black dye. It has pharmacological value, possessing diuretic properties.

coniferin ($C_6H_{11}O_5.O.C_6H_3(OCH_3)CH=CHCH_2OH$). D-Glucoside of CONIFERYL ALCOHOL, present in fir trees. VANILLIN is produced on oxidation.

salicin ($C_6H_{11}O_5.O.C_6H_4CH_2OH$). β-D-Glucoside of 2-hydroxybenzyl alcohol; colourless crystals with a bitter taste, soluble in water. It occurs in the leaves and bark of species of willow and poplar. β-Glucosidase hydrolyzes it to glucose and 2-hydroxybenzyl alcohol.

glycylglycine ($H_2NCH_2CONHCH_2-COOH$). Simplest of all PEPTIDES, prepared by the action of a base on 2,5-piperazinedione or the rection of glycyl chloride on glycine. It is soluble in water, slightly soluble in ethanol, but insoluble in ether; $pK_1 = 3.12$, $pK_2 = 8.17$.

glyoxylic acid. *See* OXOETHANOIC ACID.

gold (Au). Group Ib, third row TRANSITION ELEMENT. It occurs naturally as nuggets or embedded in quartz (*see* SILICA). It is extracted as $[Au(CN)_2]^-$ by dissolution in aerated cyanide solutions. Three-quarters of the gold produced is used in jewellery. Other uses are in electrical contacts and semiconducting devices, and as a monetary standard. Gold is a soft yellow metal with high thermal and electrical conductivity. Purity is expressed on the carat scale (parts of gold per 24 parts of total metal). Gold is inert chemically to oxygen, simple acids and akalis, but does dissolve in AQUA REGIA and selenic acid (*see* SELENIUM OXOACIDS). Gold COLLOIDS are formed by reduction of solutions of gold chloride, the colour of the colloid. being determined by the particle size. Reduction by $SnCl_2$ gives a colloid known as purple of Cassius. $A_r = 196.9665$, $Z = 79$, ES [Xe] $5d^{10}$ $6s^1$, mp 1064.4°C, bp 2807°C, $\rho = 19.3 \times 10^3$kg m^{-3}. Isotope: ^{197}Au, 100 percent.

gold compounds. Oxidation states +1 (aurous) and +3 (auric) are found in gold compounds. AuF_5 is the sole example of gold(v); gold(ii) exists with some sulphur ligands. No single Au^{3+} salts are formed, all are complexed, usually in square planar coordination (*see* SHAPES OF MOLECULES). AuF_3. has some ionic character, but with a three-dimensional polymeric structure. Other gold(iii) halides have a planar arrangement of atoms with halide bridges. Gold(i) halides (heat on gold(iii) halides) are linear. Organic gold compounds are formed readily. Alkyl gold mercaptans (AuRS) are used in thermal gold plating. Compounds with gold clusters have been synthesized (e.g., $Au_{11}[P(p\text{-}C_6H_4F)_3]_7I_3$). This compound has a central gold atom surrounded by 10 others.

gold number. Measure of the effectiveness of a PROTECTIVE COLLOID by its ability to prevent the COAGULATION of a gold SOL. *See also* COLLOID.

Goldschmidt process. Process for the manufacture of sodium methanoate (NaOH, CO, 200°C, pressure) and sodium ethanedioate (heat on sodium methanoate).

Goldschmidt radii. *See* IONIC RADIUS.

Goldschmidt reaction. Reduction of metal oxides by aluminium powder. *See also* THERMITE REACTION.

Gomberg reaction. Method for the preparation of diaryl compounds by adding an aromatic compound to an alkaline solution of a diazonium salt.

$$p\text{-}BrC_6H_4N_2^+ \ Cl^- + PhH \longrightarrow$$
$$p\text{-}BrC_6H_4Ph + N_2 + HCl$$

g-orbital. *See* ATOMIC ORBITAL.

Gouy balance. Chemical balance that can weigh a sample in and out of a magnetic field. It is used to determine the MAGNETIC SUSCEPTIBILITY of materials.

Gouy layer. *See* ELECTRICAL DOUBLE LAYER.

gradient elution. Elution technique in which the polarity of the initial solvent is progressively changed by programmed addition of increasing quantities of a second solvent. It is widely used in HIGH-PERFORMANCE LIQUID CHROMATOGRAPHY.

graft polymer. *See* COPOLYMERIZATION.

Graham's law of effusion. Rate at which a gas effuses through an orifice is inversely proportional to the square root of its density. The principle is used for the separation of isotopes (e.g., hydrogen and deuterium gas). *See also* KINETIC THEORY OF GASES.

gram (g). Fundamental unit of MASS in cgs units; one-thousanth of a KILOGRAM. Formerly it was used in such units as gram-atom, gram-molecule and gram-equivalent, which have now been replaced by the MOLE.

gramicidin S. *See* ANTIBIOTICS.

gram-molecular volume (V_m). Volume occupied by a gram-molecule of a substance in the gaseous phase. According to AVOGADRO'S HYPOTHESIS, all gases have the same value of V_m under the same conditions of pressure and temperature. At stp, $V_m = 2.241\ 383 \times 10^{-2}\ m^3\ mol^{-1}$.

Gram stain. Differential staining technique to distinguish the two main classes of BACTERIA. A heat-fixed smear of bacteria is stained with crystal voilet and then a dilute solution of iodine. The smear is then washed with either ethanol or propanone. Gram-negative bacteria are decolorized, whereas gram-positive cells retain a deep purple colour. The decolorized gram-negative cells are usually counterstained with safranin.

grand partition function. *See* PARTITION FUNCTION.

granite. Acid igneous rock, with varieties differing in colour and grain size, consisting of granular aggregates of quartz, FELDSPAR (usually potash feldspar) and mica containing 65–70 percent silica. It is used in building and paving. *See also* SILICA, SILICATES.

graphite. *See* CARBON.

graphite lattice. *See* CRYSTAL FORMS.

gravimetric analysis. Technique of QUANTITATIVE ANALYSIS in which a desired constituent is converted, usually by precipitation or combustion, to a pure compound or element of definite composition which is then dried and weighed.

gray (Gy). Derived SI unit of absorbed DOSE of ionizing radiation; the absorbed dose when the energy per unit mass imparted to matter by the radiation is $1\ J\ kg^{-1}$. *See also* RADIATION UNITS.

greenockite. *See* CADMIUM.

Grignard reagents. Aryl or alkyl magnesium halides (RMgX) formed by the reaction between corresponding halide (RX) and magnesium in ether. An equilibrium is established

$$R_2Mg + MgX_2 \rightleftharpoons 2RMgX$$

all species being solvated. Higher oligomers are formed in more concentrated solutions. Grignard reagents undergo two types of reaction: (1) addition across multiple bonds and (2) double decompositions with compounds containing active hydrogen or halogen atoms. Both types of reaction occur cleanly, in good yield and thus are of great importance in synthetic chemistry.

addition reactions. Grignard reagents add across C=O, C=S, C≡N, S=O, etc., with the alkyl group adding to the less electronegative atom of the multiple bond. This provides a useful synthetic method of introducing new carbon–carbon bonds. For example

$$RMgX + R'CHO \longrightarrow RR'CHOMgX$$

$$\xrightarrow{H_2O} RR'CHOH + Mg(OH)X$$

$$RMgX + CO_2 \longrightarrow RCOOMgX$$

$$\xrightarrow{H_2O} RCOOH + Mg(OH)X$$

Acid chlorides and anhydrides give ketones, which react further to tertiary alcohols. Two carbon atoms may be introduced by reaction with ethene oxide. *See also* ALCOHOLS, ALDEHYDES, AMINES, CARBOXYLIC ACIDS, KETONES.

double decomposition. Metal halides, alkyl halides, and water react with Grignard reagents to give ALKANES.

$$2RMgX + 2AgX \longrightarrow R_2 + 2Ag + 2MgX_2$$

$$RMgX + R'X \longrightarrow RR' + MgX_2$$

$$RMgX + H_2O \longrightarrow RH + Mg(OH)X$$

Such reactions are also used to prepare other Grignard reagents (e.g., by reaction with cyclopentadiene or ethyne). Organometallic compounds of boron, phosphorus, silicon, tin, beryllium, mercury, zinc and other elements are prepared in a similar way. For example

$$4RMgX + SiCl_4 \longrightarrow R_4Si + 4MgXCl$$

griseofulvin. *See* ANTIBIOTICS.

Grotthus–Draper law. Law of photochemistry which states that only the radiation actually absorbed by the reacting system can initiate reaction. Not all radia-

tion absorbed brings about reaction, some may be re-emitted as fluorescence or as heat.

Grotthus theory. Movement of protons through water by a proton jump mechanism. HYDROGEN BOND formation allows the net movement of a positive charge by rearrangement of bonds. This provides an explanation of the high mobilities of protons and hydroxyl ions in aqueous solution.

ground state. Lowest energy state of an atom or molecule. *See also* EXCITED STATE, SCHRÖDINGER EQUATION.

group. *See* PERIODIC TABLE.

group 0 elements. Monoatomic gaseous elements: HELIUM, NEON, ARGON, KRYPTON, XENON and RADON, each being the first element of a period (*see* PERIODIC TABLE). The closed-shell electronic structures are completely stable as shown by high ionization potentials and lack of chemical activity. They are all low boiling point gases, their physical properties varying systematically with atomic number. The gases are spherically symmetrical, with very weak interatomic interactions and consequent low enthalpies of vaporization. The behaviour of the lighter gases approaches that of an ideal gas. Argon, krypton and xenon form CLATHRATE COMPOUNDS with water and β-quinol under pressure. The ability of the gases to combine with other elements is limited; only krypton, xenon and radon form compounds and even then only bonds to fluorine and oxygen are stable. The threshold of chemical activity is reached at krypton, the activity of xenon is markedly greater and presumably that of radon is greater still if the compounds were not so short-lived.

group I elements. ALKALI METALS (group Ia) plus the coinage metals COPPER, SILVER and GOLD. The latter are usually classed with the transition elements by virtue of oxidation states which have semi-filled *d*-orbitals, +2 (copper, silver) and +3 (gold). *See also* PERIODIC TABLE.

group II elements. ALKALINE EARTH METALS (group IIa) plus ZINC, CADMIUM and MERCURY (group IIb). They are characterized by having two s-electrons outside a full shell, and thus tend to form M^{2+} in ionic compounds. The elements at the extremes of the group (i.e. BERYLLIUM and MERCURY) show atypical behaviour. Group IIb elements form covalent complexes, although not to the same extent as the TRANSITION ELEMENTS, which they follow and are sometimes classed with. *See also* PERIODIC TABLE.

group III elements. Group IIIa TRANSITION ELEMENTS comprise SCANDIUM, YTTRIUM, LANTHANUM and ACTINIUM and the main group elements (group IIIb) are BORON, ALUMINIUM, GALLIUM, INDIUM and THALLIUM; all have the outer shell electronic configuration $s^2 p^1$. Ionization potentials for boron are high enough to preclude an extensive ionic chemistry. Boron compounds are covalent and resemble those of SILICON (*see* DIAGONAL RELATIONSHIP), and the element is classed as a METALLOID. Going down the group the atomic radii are larger, and the elements show ionic and metallic character. The trend is also shown in the oxides, which become more basic. MX_3 compounds behave like Lewis acids (*see* LEWIS ACIDS AND BASES) with the semi-filled p-orbital being able to accept electrons. Thus complexes of the form $[MX_4]^-$ or M_2X_6 are formed. The univalent state becomes more important down the group; no univalent boron compounds are known, whereas aqueous thallium(I) is more stable than thallium(III). The existence of thallium(I) is an example of the INERT PAIR EFFECT. *See also* PERIODIC TABLE.

group IV elements. Main group elements CARBON, SILICON, GERMANIUM, TIN and LEAD (group IVb), together with the TRANSITION ELEMENTS TITANIUM, ZIRCONIUM and HAFNIUM. The elements are characterized by an outer shell electronic configuration $s^2 p^2$ and thus generally have compounds in the +4 oxidation state. In group IVb, there is a trend from electronegative non-metals (carbon) through semiconducting METALLOIDS (silicon, germanium) to ELECTROPOSITIVE METALS (tin, lead). Compounds of carbon are largely covalent, whereas those of lead are ionic. There is an increasing tendency to form stable divalent compounds descending the group (*see* INERT PAIR EFFECT). All elements react with oxygen and chlorine. The hydrides all exist, from the relatively stable methane (CH_4) to the very unstable PbH_4. *See also* PERIODIC TABLE.

group V elements. Usually refers to the main group elements: NITROGEN, PHOSPHORUS, ARSENIC, ANTIMONY and BISMUTH. These all have an outer electronic configuration $s^2 p^3$. Beyond the stoichiometries of the simpler compounds (e.g., NH_3, PH_3, NCl_3, $BiCl_3$), there is little resemblance between nitrogen and the remaining elements. Nitrogen and phosphorus are non-metals; the heavier elements are METALLOID or metallic in character. Phosphorus, like nitrogen, is essentially covalent in all its chemistry, whereas arsenic, antimony and bismuth show tendencies to cationic behaviour. Although the electronic structure of the next noble gas could be achieved by electron gain, considerable energies are required. To form P^{3-} from phosphorus requires about 1500 kJ mol^{-1}, thus ionic compounds such as Na_3P are few. Loss of valence electrons is similarly difficult to achieve because of the high ionization potentials. The 5+ ions do not exist, but for antimony(III) and bismuth(III) cationic behaviour does occur. Bismuth trifluoride (BiF_3) is predominantly ionic, and salts such as antimony sulphate ($Sb_2(SO_4)_3$) and salts of the oxo ions SbO^+ and BiO^+ occur. Some of the more important trends are shown by the oxides, which change from acidic or neutral for nitrogen, acidic for phosphorus to basic for bismuth and by the halides which show increasing ionic behaviour. Phosphorus trichloride (PCl_3) is instantly hydrolyzed by water to phosphorous acid ($HPO(OH)_2$), whereas the other trihalides are hydrolyzed to arsenic trioxide, antimony oxochloride (SbOCl) and bismuth oxochloride (BiOCl). There is an increase in the stability of the lower oxidation state with increasing atomic number, thus bismuth pentoxide is the most difficult to prepare and the least stable pentoxide. (For general types of compound and the stereochemical possibilities—*see* Table.) *See also* PERIODIC TABLE.

group V elements

Compounds of group V elements and their stereochemistries.

Number of bonds to other atoms	Geometry[a]	Typical examples
3	Pyramidal	NH_3, PH_3, $AsCl_3$, $SbPh_3$
4	Tetrahedral	NEt_4^+, PH_4^+, $PO(OH)_3$, Cl_3PO
	ψ-Trigonal bipyramidal	KSb_2F_7, SbOCl, $(PhNH_2)SbCl_3$, SbOCl
5	Trigonal bipyramidal	PF_5, $SbCl_5$, PPh_5, $AsPh_5$
	Square pyramidal	$SbPh_5$
	ψ-Octahedral	K_2SbF_5, Sb_2S_3, $(PhNH_2)_2SbCl_3$
6	Octahedral	PCl_6^-, AsF_6^-, $Sb(OH)_6^-$

[a] ψ indicates a central atom with lone pair(s).

group VI elements. OXYGEN, SULPHUR, SELENIUM, TELLURIUM and POLONIUM. The atoms are two electrons short of the configuration of the next noble gas, and the elements show essentially non-metallic covalent chemistry except for polonium and, to a very slight extent, tellurium. They may complete the noble gas configuration by forming: (1) the chalconide ions (see CHALCOGENS); (2) two electron-pair bonds (e.g., $(CH_3)_2S$, H_2S, SCl_2; (3) ionic species with one bond and one negative charge (e.g., RS^-); (4) three bonds and one positive charge (e.g., R_3S^+). In addition to such divalent species, the elements form compounds in formal oxidation states 4 and 6 with four, five or six bonds. Some examples of compounds of group VI elements are listed (see Table). Gradual changes of properties are evident with increasing size and decreasing electronegativity. These include: decreasing thermal stability of H_2X compounds: increasing metallic character of the elements; increasing tendency to form anionic complexes (e.g., $[TeBr_6]^{2-}$); decreasing stability of compounds of high oxidation states; emergence of cationic properties for polonium. See also PERIODIC TABLE.

group VII elements. The halogens: FLUORINE, CHLORINE, BROMINE, IODINE and ASTATINE. They are terminal members of respective periods and form a group in which the elements show a strong family relationship. The properties of the elements

group VI elements

Compounds of group VI elements and their stereochemistries.

Valence	Number of bonds	Geometry	Typical examples
2	2	Angular (V-shaped)	H_2S, S_8, H_2Te
	3	Pyramidal	Me_3S^+
	4	Square planar	$Te[SC(NH_2)_2]_2Cl_2$
4	2	Angular (V-shaped)	SO_2
	3	Pyramidal	SO_3^{2-}, SOF_2
	4	Trigonal bipyramidal	SF_4, RSF_3
		Tetrahedral	Me_3SO^+
	5	Square pyramidal	$[SF_5]^-$
	6	Octahedral	$[SeBr_6]^{2-}$
6	3	Trigonal planar	$SO_3(g)$
	4	Tetrahedral	SO_2Cl_2, $SO_3(s)$
	5	Trigonal bipyramidal	SOF_4
	6	Octahedral	SF_6, SeF_6, $Te(OH)_6$

and their compounds change progressively with increasing atomic number. The atoms are a single *p*-electron short of a noble gas configuration and thus readily form the anion X^- or a single covalent bond; consequently they are essentially non-metallic elements. They have high electron affinities, electronegativities and ionization energies which decrease down the group. None of the halogens occur in the elemental state because of their high reactivity; all exist as diatomic molecules. The halogens react with metals forming ionic halides and with non-metals to give covalent halides. Fluorine only exhibits a valency of one, the others have high oxidation states using the vacant *d*-electron levels. There is evidence of increasing metallic character down the group, only iodine forms positive ions. *See also* PERIODIC TABLE.

group theory. Theory of the properties of collections of mathematical elements. In chemistry, this is applied to the symmetry operations of molecules. Each structure may be assigned to a POINT GROUP which embodies the symmetry of a molecule. The character table (the tabulation of the traces of the matrices describing each symmetry operation in each irreducible representation) of a point group allows the exploitation of the symmetry in calculating molecular properties. Using group theory it is possible to generate MOLECULAR ORBITALS which have the symmetry of one of the irreducible representations of the group. These are orthogonal to orbitals of different symmetry, and their use can reduce the amount of calculation required to determine orbital energies, etc. This is, however, at the cost of losing the concept of a simple chemical bond as symmetry-adapted orbitals are usually delocalized over the molecule. Electronic transitions on the absorption of light must also preserve the symmetry in a molecule and so SELECTION RULES may be generated concerning the symmetries of ground and excited state orbitals. The symmetries of the NORMAL MODES of vibration are given by group theory and so allow determination of the IR or Raman activity.

guaiacol. *See* AROMATIC ETHERS.

guanine. *See* PURINES.

guanosine. *See* NUCLEOSIDES.

5′-guanylic acid. *See* NUCLEOTIDES.

gulose. *See* HEXOSES.

gums. Exudates from the bark of various plants or extracts from the stored food in the seeds of members of the pea family. They are produced in large amounts by certain algae, a process used commercially. Gums are soluble in water giving viscous colloidal solutions, but insoluble in organic solvents. They consist of complex hetero-polysaccharides containing a range of SUGAR and URONIC ACID residues. Ceratonia gum (E410), guar gum (E412), gum tragacanth (E413), gum arabic (E414), xanthan gum (E415) and karaya gum (416) are extensively used in the food industry as stabilizers, thickeners and emulsifiers, and in the pharmaceutical industry as hydrocolloids.

gun cotton. *See* EXPLOSIVES.

gun metal. *See* BRONZE.

gun powder. *See* EXPLOSIVES.

Gy. *See* GRAY.

gypsum. *See* CALCIUM, SULPHATES, SULPHUR.

gyromagnetic ratio. *See* NUCLEAR GYROMAGNETIC RATIO.

H

H. *See* HENRY, HYDROGEN.

H. *See* ENTHALPY.

H_0. *See* HAMMETT ACIDITY FUNCTION.

h. *See* PLANCK CONSTANT.

h-. *See* HAPTO-

Ha. Hahnium (*see* TRANSACTINIDE ELEMENTS).

Haber process. Industrial process for the synthesis of ammonia from atmospheric nitrogen and hydrogen in the presence of a catalyst (α-iron + oxides) at pressures of 10^2–10^3 atm at 450–550°C. Although an exothermic reaction, the elevated temperature is required to obtain a satisfactory rate of conversion.

habit. *See* CRYSTAL STRUCTURE.

hadron. *See* ELEMENTARY PARTICLES.

haem. Iron(II)–porphyrin complex with iron in the +2 oxidation state (*see* Figure). Haemin (($RN)_4Fe^+$), the oxidized form of haem, is the PROSTHETIC GROUP of a number of important enzymes (e.g., catalase).

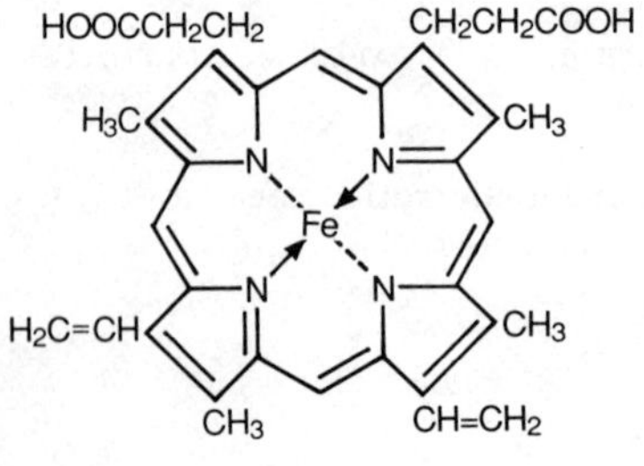

haem

haematite. *See* IRON.

haemin. *See* HAEM.

haemocyanin. Blue respiratory pigment of molluscs, arachnids and crustaceans, containing two copper atoms directly linked to a polypeptide chain.

haemoglobin. Complex molecule consisting of globin (a protein fraction) and HAEM. Haemoglobin in vertebrate blood transports oxygen to the tissues; carbon monoxide acts as a poison since it forms a more stable complex (carboxyhaemoglobin) than does oxygen.

hafnium (Hf). Group IVb, third row TRANSITION ELEMENT. It is found with ZIRCONIUM in baddeleyite and zircon and alvite $(Be,Hf,Th,Y,Zr)O_2.SiO_2.xH_2O$. Hafnium is separated from zirconium by ION EXCHANGE CHROMATOGRAPHY. $A_r = 178.49$, $Z = 72$, ES $[Xe]5d^2 6s^2$, mp 2227°C, bp 4602°C, $\rho = 13.31 \times 10^3$ kg m^{-3}. Isotopes: ^{174}Hf, 0.16 percent; ^{176}Hf, 5.21 percent; ^{177}Hf, 18.56 percent; ^{178}Hf, 27.1 percent; ^{179}Hf, 13.75 percent; ^{180}Hf, 35.22 percent.

hafnium compounds. Usual oxidation state is +4, but +1 and +3 are also found. The chemistry is almost identical to that of zirconium. $Hf(NO_3)_4.N_2O_5$ has been sublimed from solutions containing both elements. The carbide (Ta_4HfC_5) has the highest known melting point (4215°C).

hahnium (Ha). *See* TRANSACTINIDE ELEMENTS.

half-cell. Electrode in contact with electrolyte, at which an electrochemical equilibrium is established. Two half-cells in contact through the electrolyte (e.g., by a SALT BRIDGE) form a CELL. In representing half-cells a full vertical line indicates the junction of two phases (e.g., $Zn^{2+}|Zn$) and a broken vertical line a liquid junction.

Variation of the electrode potential with the activities of species in the half-cell is given by the NERNST EQUATION. The following types of half-cell are recognized. (1) Metal electrode. Solid element in contact with a solution of its ions (e.g., $Cu^{2+}|Cu$). (2) REDOX ELECTRODE SYSTEM. Inert metal in a solution containing its ions in different oxidation states (e.g., $Fe^{3+},Fe^{2+}|Pt$). (3) Inert electrode with neutral solutes in different oxidation states (e.g., quinone, hydroquinone|Pt). (4) Gas electrode. Inert electrode with gas in equilibrium with its ions in solution (e.g., HYDROGEN ELECTRODE $H^+|H_2,Pt$). (5) Metal, insoluble metal salt electrode. Metal coated with an insoluble salt in contact with a solution containing a common anion (e.g., Ag, $AgCl|Cl^-$). These are commonly used as REFERENCE ELECTRODES. (6) Cationic-responsive electrode. Electrodes whose potential depends on the concentration of a cation (e.g., glass electrode (H^+), $Ca^{2+}|CaC_2O_4(s),PbC_2O_4(s),Pb\,(Ca^{2+})$).

half-life ($\tau_{1/2}$, $t_{1/2}$). Time taken for the concentration of a reactant to fall to half its initial value. For a first-order reaction (see ORDER OF REACTION) $\tau_{1/2} = \ln 2/k_1$, and for the general reaction A $\rightarrow$ P of order n (where $n > 1$) and initial concentration $[A_0]$ (i.e. $d[A]/dt = k_n[A]^n$) $\tau^{1/2} = 1/k_n[A_0]^{n-1}$.

half-wave potential. *See* POLAROGRAPHY.

halides. Formally compounds of the –1 oxidation state of bromine, chlorine, fluorine and iodine. Two broad classes exist: (1) those produced by metals of low charge (e.g., alkali and alkaline earth metals) with three-dimensional, ionic lattices, high melting point and boiling point, and good conductors in the fused state; (2) those formed by non-metals and B-subgroup metals which are volatile and non-conducting with molecular lattices. Dry methods of preparation (obligatory when the product is easily hydrolyzed) include direct halogenation, and for chlorides the reaction between sulphur monochloride, chlorine and oxides and for fluorides treating chlorides with anhydrous HF or heating oxides with calcium fluoride, or calcium fluoride and sulphuric acid. Wet methods, feasible when the halide is not hydrolyzed, include the reaction of the metal, oxide, hydroxide or carbonate with the appropriate halogen acid and precipitation from solution.

Type AB halides have the sodium chloride lattice (6:6) or caesium chloride lattice (8:8) type structure (*see* CRYSTAL FORMS); of type AB_2, the fluorides have a fluorite structure or rutile structure, whereas most other dihalides form layer lattices; of type AB_3, the trifluorides of many lanthanides have slightly distorted rhenium trioxide structures whereas most of the trichlorides have a lead dichloride-type structure. The shape of molecules in the vapour state depends on the electronic configuration of the less electronegative element; phosphorus pentachloride is trigonal bipyramidal, arsenic triiodide is pyramidal, tetrachloromethane tetrahedral and aluminium trichloride exists as the dimer. Most alkali metal and barium halides are soluble in water with some hydration of ions; heavy metal halides (e.g., silver, lead) are insoluble. Covalent halides (except those of carbon) hydrolyze readily. All halide ions have the ability to function as ligands and form with metal ions, or covalent halides, such complexes as $[SiF_6]^{2-}$, $[HgI_4]^{2-}$ and $[Co(NH_3)_4Cl_2]^+$.

Hall–Heroult process (Bayer–Hall–Heroult process). Production of aluminium by ELECTROLYSIS of alumina in molten electrolyte (Al_2O_3, Na_3AlF_6, MgF_2, 950°C) between carbon electrodes. The overall reaction is $2Al_2O_3 + 3C \rightarrow 4Al + 3CO_2$ or $2Al_2O_3 + 6C \rightarrow 4Al + 6CO$.

haloforms. *See* TRIHALOMETHANES.

halogenation. Incorporation, by addition or substitution, of a HALOGEN into a compound. ALKENES add halogens and hydrogen halides across the C=C double bond; some aromatic compounds react with halogens by substitution in the presence of a Lewis acid catalyst.

halogen halides. *See* INTERHALOGEN COMPOUNDS.

halogens. *See* GROUP VII ELEMENTS, BROMINE, CHLORINE, FLUORINE, IODINE.

halothane (Fluothane, $CHBrClCF_3$). Heavy, volatile liquid (bp 50°C). It is used as a general anaesthetic.

Hamaker constant. *See* DLVO THEORY.

Hamiltonian operator. *See* SCHRÖDINGER EQUATION.

Hammett acidity function (H_0). Measure of the ability of an acid (*see* ACIDS AND BASES) to accept a proton from a weak base indicator ($B + H^+ \rightleftharpoons BH^+$).

$$H_0 = pK_{BH^+} + \lg [BH^+]/[B]$$

A large negative H_0 indicates a strong acid (*see* SUPERACID).

Hammett equation. Correlation of the structure and reactivity of side-chain derivatives of aromatic compounds. The dissociation constants (K) of a series of 4-(*para*-)substituted benzoic acids and the rate constant (k) for the alkaline hydrolysis of the 4-substituted ethyl benzoates are related by the equation

$$\lg k = \lg K + C$$

Similar results are obtained for 3- (meta)-substituted, but not for 2- (*ortho*)-substituted derivatives. When the ring substituent is hydrogen the equation becomes

$$\lg k_0 = \lg K_0 + C$$

where K_0 and k_0 are the dissociation constant of benzoic acid and the rate constant for the hydrolysis of ethyl benzoate, respectively. Subtracting the two equations gives the Hammett equation

$$\lg k/k_0 = \rho \lg K/K_0 = \rho\sigma$$

where $\sigma = \lg K/K_0$. The strength of a substituted benzoic acid relative to that of benzoic acid and hence the value of σ, the substituent constant, depends on the nature and position of the substituent in the ring. The value of σ corresponds to the polar character of the substituent; thus the more electron-attracting (donating) the more positive (negative) its value (relative to hydrogen at $\sigma = 0$). The ρ term, known as the reaction constant (i.e. the slope of the plot of $\lg k/k_0$ or $\lg K/K_0$ against σ) measures the sensitivity of a reaction to the electrical effects of substituents in the 3- or 4-positions. A large value of ρ (positive or negative) means a high sensitivity to substituent influence. The value of ρ is influenced by changes of temperature, solvent, etc.; these are usually not large unless there has been a change in the actual reaction mechanism.

Hammick and Illingworth's rules. Empirical set of rules governing the course of aromatic substitution. A group AB in a benzene ring (A attached to the nucleus) is: (1) 3- (*meta*-)directing if B is in a higher group of the periodic table than A or if in the same group B is of lower atomic mass (e.g., $-NO_2$, $-SO_3H$, $-CN$); (2) is 2,4-(*ortho*,*para*-)directing if B is in a lower group than A or if AB is a single atom (e.g., $-OH$, $-NH_2$, Cl). The rules are not universally applicable. *See also* ORIENTATION IN THE BENZENE RING.

(a)

(b)

Hantzsch synthesis
(a) Substituted pyridines, (b) substituted pyrroles.

Hantzsch synthesis. (1) Substituted pyridines are synthesized by condensation of a β-dicarboxyl compound (2 mol) with an aldehyde (1 mol) and ammonia (1 mol), and the subsequent oxidation of the dihydro product with nitric acid (*see* Figure (a)). The ester can be hydrolyzed and decarboxylated to 2,4,6-trimethylpyridine. (2) Substituted pyrroles are synthesized by the condensation of a β-diketo ester, chloropropanone and ammonia (or primary amine) (*see* Figure (b)).

hapten. Small chemical molecule that reacts specifically with an ANTIBODY, but which does not itself stimulate antibody production. When complexed with a large protein molecule (e.g., globulin) antibodies are produced that are specific to both the hapten and to the carrier protein.

hapto (*hⁿ* or *ηⁿ*). Designating the number of atoms in a ligand bonded to an acceptor; also referred to as hapticity. For example, dibenzene chromium is hexahapto and ferrocene is pentahapto (*see* SANDWICH COMPOUND).

hard and soft acids and bases (HSAB). Classification of LEWIS ACIDS AND BASES proposed by Pearson in 1963, which is based on the ease with which electrons may be distorted by the presence of other species. Attributes that may be possessed by hard acids are low polarizability, small size and high positive oxidation state (e.g., H^+, Li^+, Na^+, Be^{2+}, Al^{3+}, Co^{3+}, Si^{4+}, BF_3, AlH_3, SO_3, CO_2). Soft acids may have high polarizability, low or zero positive charge and be of large size (e.g., Cu^+, Ag^+, Hg^+, BH_3, I_2, Br_2). In a similar way, soft bases may have high polarizability, low electronegativity and may be able to form π-COMPLEXES. Hard bases have the opposite properties. Bases arranged in increasing order of hardness are: Ph_3P, $S_2O_3^{2-}$, SO_3^{2-}, SCN^-, I^-, PhSH, NO_2^-, NH_3, F^-, –OMe, –OH, H_2O. According to HSAB theory, hard acid–hard base pairs and soft acid–soft base pairs are more stable than any hard–soft combinations. Thus silver ions in aqueous ammonia forms $Ag(NH_3)^+$ (i.e. soft acid–soft base) and not AgOH (i.e. soft acid–hard base). The theory successfully explains many chemical systems: for example, metal complexes, charge transfer complexes and any reactions forming bonds between species. *See also* ACIDS AND BASES.

hardening. (1) Catalytic (Ni) HYDROGENATION of unsaturated fats and oils to give saturated products with higher melting points; used in the food industry to manufacture margarines. (2) Alloying of a metal to increase its hardness. *See also* CASE HARDENING.

hardness. *See* HARDNESS OF WATER, MOH'S SCALE.

hardness of water. Inability of water to form a foam with SOAPS due to the presence of salts of alkaline earth metals. Temporary hardness due to bicarbonates is removed by boiling. Permanent hardness is usually expressed as calcium carbonate equivalents in parts per million. Water is softened by hydrated lime and soda ash in a hot or cold process, and, in addition, with sodium phosphate in the hot lime soda and phosphate process. Ion exchange methods use ZEOLITES to exchange alkaline earth metals for sodium ions (Permutit process) or protons in an hydrogen ion exchange process. The spent ion exchanger is regenerated by washing with sodium chloride or sulphuric acid, respectively.

harmonic oscillator. Two masses that vibrate about an equilibrium point with a restoring force proportional to the displacement (Hooke's law). It is a model for vibrations within a molecule. Solution of the SCHRÖDINGER EQUATION leads to quantized energy levels $E_{vib} = (v + 1/2)h\omega$ (where $v = 0, 1, 2, \ldots$) is the vibrational quantum number and ω is a characteristic frequency $= (k/\mu)^{1/2}/2\pi$, where μ is the REDUCED MASS and k the force constant.

hartree. *See* ATOMIC UNITS.

Hartree–Fock method. *See* SELF-CONSISTENT FIELD.

hashish. *See* TETRAHYDROCANNABINOLS.

hcp (hexagonal close-packed structure). *See* CLOSE-PACKED STRUCTURES.

He. *See* HELIUM.

heat (q; dimensions: $m\ l^2\ t^{-2}$; units: J). Form of energy transferred between a SYSTEM and its surroundings by thermal conduction, convection or electromagnetic radiation. Heat is an EXTENSIVE PROPERTY and always flows from a body of higher temperature to one of lower temperature. Heat transfer is an irreversible process. *See also* FIRST LAW OF THERMODYNAMICS.

heat capacity (C; dimensions: $m\ l^2\ t^{-2}\ \theta^{-1}$; units: $J K^{-1}\ mol^{-1}$).   Amount of heat required to raise the temperature of a system by 1 K either at constant volume (C_V) or at constant pressure (C_P). Specific heat ($c = C/m$) is the heat capacity per unit mass

$$C_V = (\partial U/\partial T)_V$$

and

$$C_P = (\partial H/\partial T)_P$$

For any system

$$C_P - C_V = \alpha V[P + (\partial U/\partial V)]_T = \alpha^2 TV/\beta$$

where α is the coefficient of expansion and β the coefficient of compressibility. The difference increases as α increases (i.e. solid, liquid, gas) for gases $\alpha = 1/T$ and $\beta = 1/P$, hence

$$C_P - C_V = R$$

The variation of the heat capacity of gases with temperature can be expressed by an empirical equation of the type

$$C_P = a + bT + cT^2 + \ldots$$

where $a, b, c, \ldots$ are experimentally determined constants for each gas. This equation is used in the calculation of enthalpy changes at different temperatures (*see* KIRCHHOFF'S EQUATION) and ENTROPY changes. Heat capacity cannot be predicted by classical thermodynamics but can be calculated from spectroscopic data using the PARTITION FUNCTION. Heat capacities for gases are measured: (1) at constant pressure, by Callendar and Barnes continuous flow method; (2) at constant volume by the Joly differential steam calorimeter. Heat capacities for solids and liquids are determined by the method of mixtures.

heat content.   *See* ENTHALPY.

heat engine.   *See* CARNOT CYCLE.

heatite.   *See* SILICA.

heat of.   *See* ENTHALPY.

heat of adsorption.   Integral heat of adsorption is the heat given out per mole on adsorption. This is the negative of the enthalpy of adsorption (i.e. $-\Delta_{ads}\bar{H} = Q/n$). The differential heat of adsorption is $dQ/dn(q_d)$, thus $-\Delta_{ads}\dot{H} = dQ/dn$. Therefore

$$-\Delta_{ads}\dot{H} = -\Delta_{ads}\bar{H} + n d(\Delta_{ads}\bar{H})/dn$$

The isosteric heat is the heat adsorbed at constant coverage and is analogous to the heat of vaporization defined by the CLAPEYRON–CLAUSIUS EQUATION

$$(\partial \ln P/P_0/\partial T) = q_{st}/RT^2$$

The relationship between q_d and q_{st} is

$$q_{st} = q_d + RT.$$

heat of atomization.   Heat required (i.e. absorbed) to dissociate 1 mole of an element into its gaseous atoms.

heat of combustion, formation.   Heat evolved at constant temperature and pressure when 1 mole of the substance is completely combusted in oxygen or formed from its constituent atoms. *See also* HEAT OF REACTION.

heat of reaction.   Heat evolved at constant temperature and pressure during a chemical reaction, usually expressed for molar amounts of reactants or products in specified physical states. The negative value of the heat of reaction is the enthalpy change (ΔH) of the reaction. For an endothermic process the heat of reaction has a negative value and ΔH a positive value. For the reaction

$$C(s) + 0.5\ O_2(g) \rightarrow CO_2(g)$$

heat of reaction = heat of combustion
of carbon

= heat of formation
of CO_2

= 393.5 kJ mol^{-1}

$$\Delta H = \Delta_c H(C) = \Delta_f H(CO_2)$$

$$= -393.5\ kJ\ mol^{-1}$$

heavy hydrogen. *See* DEUTERIUM.

heavy water. *See* DEUTERIUM OXIDE.

height equivalent of a theoretical plate (HETP). Measure of the efficiency of a chromatographic column defined as $H = L/N$, where L is the length of the column packing and N the number of theoretical plates obtained from any peak on a CHROMATOGRAM by $N = 16(V/W)^2$, where V is the RETENTION VOLUME (time) and W the peak width at base. *See also* VAN DEEMTER EQUATION.

Heisenberg's uncertainty principle. If the momentum of a particle is known to lie within the range δp, then the uncertainty in the position δq is governed by the relationship $\delta q \, \delta p \geqslant h/4\pi$.

Heitler–London orbital. *See* VALENCE BOND ORBITAL.

helium (He). Noble gas (GROUP O ELEMENT) that occurs in the atmosphere (5.2 × 10^{-4} percent of air) and natural gases (up to 7 percent) in association with minerals containing alpha-emitters (e.g., pitchblende, monazite); it is more common on larger planets and stars. Helium is obtained by the fractionation of liquid air. Naturaly occurring helium is essentially all helium-4; helium-3 (10^{-7} percent of air) is obtained by beta-decay of tritium. Helium is the only known substance which has no TRIPLE POINT, and the only one that can be solidified only under pressure. Liquid He_{II} with a lower entropy than that of He_I (transition temperature: 2.2 K) exhibits the phenomenon of superconductivity. He_{II} has an immeasureably low viscosity and readily forms films that can flow, apparently without friction, over the edge of a vessel. Helium is completely inert with no normal chemistry. It is used as an inert gas for arc welding, for filling balloons and, with 20 percent oxygen, as an atmosphere for deep-sea divers. Liquid helium is used as a coolant in nuclear reactors and superconducting magnets.
A_r = 4.002602, Z = 2, ES $1s^2$, mp –272.2°C, bp –268.934°C; ρ = 0.1785 kg m^{-3}. Isotope: ^{4}He, 100 percent.

helix. *See* NUCLEIC ACIDS.

Helmholtz function (maximum work function) (A; dimensions: $m \ l^2 \ t^{-2}$; units: J mol^{-1}). STATE FUNCTION defined by
$$A = U - TS$$
It is related to the GIBBS FUNCTION by
$$G = A + PV$$
For an isothermal reversible process, $\Delta A = w$ (i.e. total WORK done on the system). At constant temperature and volume if $\Delta A < 0$, the process is spontaneous, if $\Delta A = 0$, the system is in equilibrium (A is a minimum), and if $\Delta A > 0$, the process as written is not spontaneous. For a system doing no work other than PdV
$$dA = -PdV - SdT$$
leading to the identities $(\partial A/\partial V)_T = -P$ and $(\partial A/\partial T)_V = -S$. A varies with temperature (*compare* GIBBS–HELMHOLTZ EQUATION). Since most chemical processes are carried out at constant pressure, G is a more useful function than A.

Helmholtz layer. *See* ELECTRICAL DOUBLE LAYER.

hemiacetals. *See* ACETALS.

hemicelluloses. Ill-defined group of alkali-soluble, easily hydrolyzed POLYSACCHARIDES (e.g., XYLANS, galactans, mannans) which are components of the cell wall matrix. The chain-like molecules are shorter than those of CELLULOSE and may have short branches. In some plants, they are used as food reserves.

hemimellitine (1,2,3-trimethylbenzene, $C_6H_3(CH_3)_3$). Colourless liquid (bp 175°C), found in small amounts in coal tar; an isomer of MESITYLENE.

hemimorphite. *See* SILICATES.

Henderson equation. *See* BUFFER SOLUTION.

Henderson–Hasselbach equation. *See* BUFFER SOLUTION.

henry (H; units: kg m^2 s^{-2} A^{-2} = Vs A^{-1}). SI unit of inductance.

Henry's law. At a constant temperature, mass of gas dissolved in a given volume of solvent is proportional to the partial pressure of the gas in equilibrium with the solution

$$p_B = k_B x_B$$

The law is only valid when there is no chemical reaction between the gas and the solvent. For IDEAL SOLUTIONS, the law becomes identical with RAOULT'S LAW; $k_A = p_A^\ominus$ and $k_B = p_B^\ominus$. *See also* REAL SOLUTION.

heparin. *See* AMINO SUGARS.

heptacene. *See* POLYCYCLIC HYDROCARBONS (structure VI).

heptanedioic acid (pimelic acid, HOOC-$(CH_2)_5$COOH). Saturated DICARBOXYLIC ACID (mp 104°C), prepared by the hydrolysis of 1,5-dicyanopentane (from KCN + 1,5-dibromopentane) or by the reduction of 2-hydroxybenzoic acid with sodium in 3-pentanol followed by the addition of water and HCl. Distillation at 300°C with ethanoic anhydride gives CYCLOHEXANONE. It is used in the manufacture of POLYMERS.

heptoses ($C_7H_{14}O_7$). CARBOHYDRATES with seven carbon atoms, obtained synthetically from an hexose; D-glucose gives rise to two epimeric glucoheptoses.

herbicide. Compound used to destroy unwanted vegetation, especially weeds, grasses and woody plants. Originally such inorganic compounds as sodium chlorate, ammonium sulphamate, arsenic and boron compounds were used; more recently specific organic compounds have been introduced.
selective herbicides. Includes AUXINS (2,4-D, 2,4,5-T), which overstimulate growth hormones, and phenols, carbamates, urea derivatives and SIMIZINE, which eliminate weeds without damage to the crop.
non-selective herbicides. Soil sterilants such as sodium chlorate and ammonium sulphamate, which kill woody plants and trees.
contact herbicides. An example is PARAQUAT.

heroin (diacetylmorphine). *See* ALKALOIDS.

hertz (Hz). SI unit of FREQUENCY; equal to 1 cycle per second.

Hess's law. If a reaction can be carried out in stages the algebraic sum of the ENTHALPY changes of the separate stages is equal to the enthalpy change of the complete reaction as carried out in one stage, and is independent of the route taken. It is a corollary of the FIRST LAW OF THERMODYNAMICS and is used to calculate enthalpy changes of reactions from tabulated values of $\Delta_f H^\ominus$. *See also* BORN–HABER CYCLE.

heterocyclic compounds (heterocycles). Cyclic compounds in which one or more of the ring carbons is replaced by another atom (*see* Figure). Common 'hetero' atoms are nitrogen, oxygen and sulphur, but ring compounds with arsenic, phosphorus, and selenium are known. The double bonds in the five-membered and some six-membered unsaturated rings are conjugated as in BENZENE; these compounds show aromatic properties to varying degrees.

heterogeneous. Referring to a mixture of phases (e.g., liquid–solid, liquid–solid–vapour). Heterogeneous reactions occur between reactants in different phases.

heterogeneous catalysis. CATALYSIS, in which the catalyst is in a different phase (usually a solid) from the reactants and products (usually gases or liquids). The reaction may be considered to occur in a series of steps: (1) DIFFUSION of reactants to the surface; (2) ADSORPTION; (3) surface reaction; (4) desorption of products; (5) diffusion of products away from the surface. Any of the steps may be limiting (*see* RATE-DETERMINING STEP). The nature of adsorption is critical to the path the reaction will take. For example, 2-propanol dehydrates to propene over an alumina catalyst, but dehydrogenates to propanone over copper. A theory of heterogeneous catalysis must take account both of the electronic and geometric structure of the surface (*see* LANGMUIR–HINSHELWOOD MECHANISM, RIDEAL–ELEY MECHANISM). Heterogeneous catalysis dominates industrial chemistry

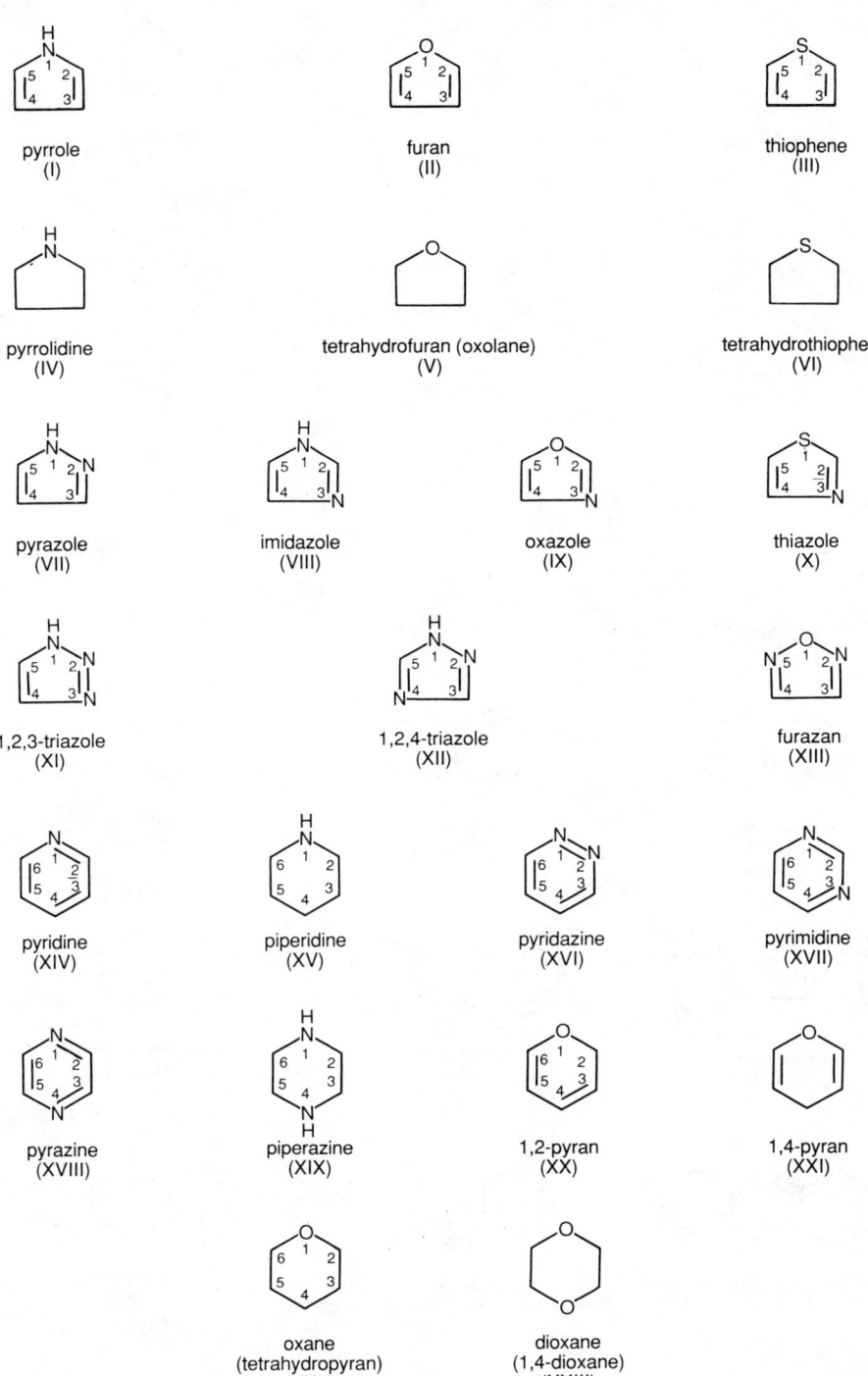

pyrrole
(I)

furan
(II)

thiophene
(III)

pyrrolidine
(IV)

tetrahydrofuran (oxolane)
(V)

tetrahydrothiophene
(VI)

pyrazole
(VII)

imidazole
(VIII)

oxazole
(IX)

thiazole
(X)

1,2,3-triazole
(XI)

1,2,4-triazole
(XII)

furazan
(XIII)

pyridine
(XIV)

piperidine
(XV)

pyridazine
(XVI)

pyrimidine
(XVII)

pyrazine
(XVIII)

piperazine
(XIX)

1,2-pyran
(XX)

1,4-pyran
(XXI)

oxane
(tetrahydropyran)
(XXII)

dioxane
(1,4-dioxane)
(XXIII)

heterocyclic compounds
Heterocyclic ring systems and numbering systems.

continued overleaf

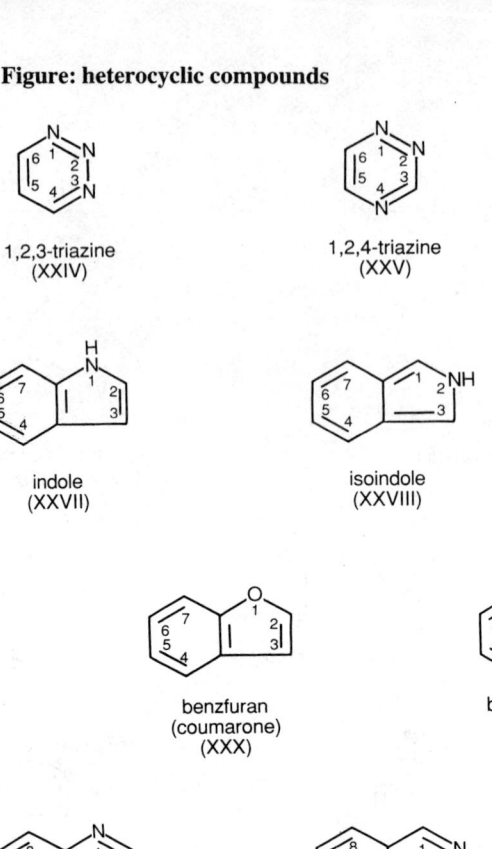

1,2,3-triazine
(XXIV)

1,2,4-triazine
(XXV)

1,3,5-triazine
(XXVI)

indole
(XXVII)

isoindole
(XXVIII)

purine
(XXIX)

benzfuran
(coumarone)
(XXX)

benzothiofuran
(XXXI)

quinoline
(XXXII)

isoquinoline
(XXXIII)

pteridine
(XXXIV)

carbazole
(XXXV)

acridine
(XXXVI)

1,10-phenanthroline
(XXXVII)

phenazine
(XXXVIII)

phenothiazine
(XXXIX)

phenoxazine
(XL)

heterocyclic compounds
Heterocyclic ring systems and numbering systems.

in which reactions may be made more selective or to proceed at lower temperatures. TRANSITION ELEMENTS, particularly the PLATINUM GROUP METALS, are used extensively for HYDROGENATION, POLYMERIZATION and OXIDATION. Expensive catalysts may be used in the form of SUPPORTED CATALYSTS. *See also* CRACKING, FISCHER–TROPSCH PROCESS, HABER PROCESS, HYDROFORMYLATION, OXO PROCESS, PETROCHEMICALS, REFORMING.

heterolytic fission. Reaction in which the electrons forming the broken bond are shared unequally among the fragments formed (e.g., $XR \rightarrow X^+ + R^-$). *Compare* HOMOLYTIC FISSION.

heteronuclear. Descriptive of a molecule consisting of different atoms or groups (e.g., hydrogen chloride, nitric oxide, carbon monoxide). *Compare* HOMONUCLEAR.

heteropoly acids and anions. Polymeric oxides or hydroxides of molybdenum or tungsten containing at least one heteroatom (e.g., phosphorus, silicon, iron). They are very soluble in water and oxygen-containing solvents, stable in acids, but decompose in strong bases. The structures are based on MoO_6 and WO_6 octahedra. Exmples are sodium 12-molybdophosphate ($Na_3PMo_{12}O_{40}$), dimeric potassium 6-tungstocobaltate ($K_8Co_2W_{12}O_{42}$).

HETP. *See* HEIGHT EQUIVALENT OF A THEORETICAL PLATE.

hexacene. *See* POLYCYCLIC HYDROCARBONS (structure V).

hexachlorobenzene (C_6Cl_6). Colourless crystalline solid (mp 227°C), prepared by the exhaustive chlorination of benzene with chlorine in the presence of iron(III) chloride, or chlorine on hexachlorocyclohexane in hexachloroethane. It is used in the preparation of hexafluorobenzene.

hexachloro-1,3-butadiene (Fluorolube, $CCl_2=CClCCl=CCl_2$). Colourless, viscous liquid (bp 210°C). It is used to disperse solids as MULLS for INFRARED SPECTROSCOPY.

1,2,3,4,5,6-hexachlorocyclohexane (benzenehexachloride, $C_6H_6Cl_6$). Product of the chlorination of benzene in the presence of sunlight. Of the eight possible STEREOISOMERS, seven are known (α, β, γ, δ, ε, η and θ), all in the chair form. The γ-isomer (lindane: *a, a, a, e, e, e*) (mp 112.5°C) is a powerful insecticide, more stable and active than DDT (*see* 1,1,1-TRI-CHLORO-2,2-BIS(4-CHLOROPHENYL)ETHANE).

hexachloroethane (perchloroethane, C_2Cl_6). Solid (mp 187°C), manufactured by the chlorination of 1,1,2,2-tetrachloroethane over aluminium trichloride, or by the chlorinolysis of propane. It is used as a moth-repellant.

hexacyanoferrates. *See* CYANOFERRATES.

hexad axis. *See* CRYSTAL SYMMETRY.

hexadecanoic acid (palmitic acid, CH_3-$(CH_2)_{14}COOH$). MonoCARBOXYLIC ACID (mp 63°C), occurring naturally as glycerides in animal and vegetable fats from which it may be extracted by steam distillation, it is a major component of cow's milk. It is used to make soap as the sodium or potassium salt (*see* n-OCTADECANOIC ACID) and candles (*see* PALMATIN, STEARIN).

hexadecanyltrimethyl ammonium bromide. *See* SURFACE-ACTIVE AGENT.

hexagonal close-packed structure (hcp). *See* CLOSE-PACKED STRUCTURES.

hexagonal system. *See* CRYSTAL SYSTEMS.

hexahelicene (phenanthro[3,4c]phenanthrene, $C_{28}H_{16}$). Organic compound containing six fused benzene rings (*see* Figure). It is not a planar molecule (steric overcrowding prevents the terminal rings A

hexahelicene

and B becoming coplanar) but has a helical, chiral structure with one turn. It exists as two enantiomeric forms which have been resolved using a π-acid complex. The specific rotation is 3700°.

hexahydric alcohols (hexitols, $C_6H_{14}O_6$). Alcohols with six hydroxyl groups; obtained by the reduction of the OXO GROUP of an ALDOSE or KETOSE with sodium tetrahydridoborate or by catalytic hydrogenation. There are 10 hexitols with the structure $HOCH_2(CHOH)_4CH_2OH$, eight exist as four pairs of ENANTIOMERS. They exhibit the properties of primary and secondary alcohols.

dulcitol. Reduction product of GALACTOSE, occurs widely in plants (mp 198°C).

D-mannitol (manna sugar). Reduction product of MANNOSE; occurs in seaweed, grass, fruits, fungi and in the exudate of *Fraxinus ornus* (mp 168°C). It is used as a texturizing agent, as a sweetener in sugar-free products and as an anticaking agent (E421).

D-sorbitol. Reduction product of GLUCOSE; occurs in red seaweed and fruits such as apples, pears, cherries and peaches (mp 110°C). It is metabolized by the body. It is used as a sweetening agent (E420) and substitute for glycerol in the confectionery industry and in the manufacture of ascorbic acid (vitamin C), pharmaceuticals and adhesives. *See also* SWEETENING AGENTS.

hexahydropyridine. *See* PIPERIDINE.

hexamethenetriazine. *See* HEXAMINE.

hexamethylene diamine. *See* POLYAMINES.

hexamine (hexamethenetriazine, 1,3,5,7-tetraazaadamantane). White crystals (sublimes 263°C), prepared from METHANAL and ammonia. It is used as a starter fuel for camping stoves, in vulcanizing rubber and to manufacture resins.

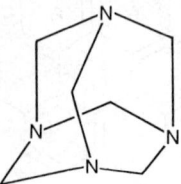

hexamine

hexanedioic acid (adipic acid, HOOC-$(CH_2)_4$COOH). Saturated DICARBOXYLIC ACID (mp 150°C), prepared industrially by the catalytic (e.g., cobalt ethanoate) oxidation of cyclohexane. Heating with ethanoic anhydride gives the cyclic ketone cyclopentanone. Major industrial uses are in the production of polyester resins, NYLON (*see* FIBRES) and other plastics.

2,5-hexanedione (acetonyl acetone). *See* DIKETONES.

hexanes. *See* ALKANES.

n-hexanoic acid (capric acid). Oily liquid (bp 205°C) CARBOXYLIC ACID, with goat-like smell.

hexitols. *See* HEXAHYDRIC ALCOHOLS.

hexosan. *See* POLYSACCHARIDES.

hexoses ($C_6H_{12}O_6$). Most important group of the simple sugars; almost all POLYSACCHARIDES are built up of hexose units. GLUCOSE, GALACTOSE and MANNOSE (commonly occurring in plants either in the free state or as components of a polysaccharide or as GLYCOSIDES) are the most important aldohexoses. With four structurally different asymmetric carbon atoms, the aldohexoses can exist in 16 optically active forms corresponding to the D- and L-isomers (*see* Figure 1). There is strong evidence that the sugars exist as six-membered rings in the solid state. FRUCTOSE is the only ketohexose of any importance; other ketohexoses are sorbose, tagatose and psicose (*see* Figure 2). With ethanoic anhydride, both the aldohexoses and the ketohexoses form a pentaethanoate (indicating five hydroxyl groups); with hydroxylamine both form oximes (indicating the presence of a carbonyl group); on reduction with red phosphorus and hydrogen iodide both form 2-iodohexane and n-hexane (indicating six carbons in a chain). Oxidation (bromine water) of the aldohexoses yields pentahydroxy (aldonic) acids ($C_6H_{12}O_7$), indicating the presence of an aldehyde group, whereas oxidation (HNO_3) of the ketohexoses yields a mixture of acids containing fewer carbon atoms than the original sugar (indicating the presence of a keto group). High-pressure catalytic reduction (Ni/H_2) converts the aldohexoses to the corresponding alcohol.

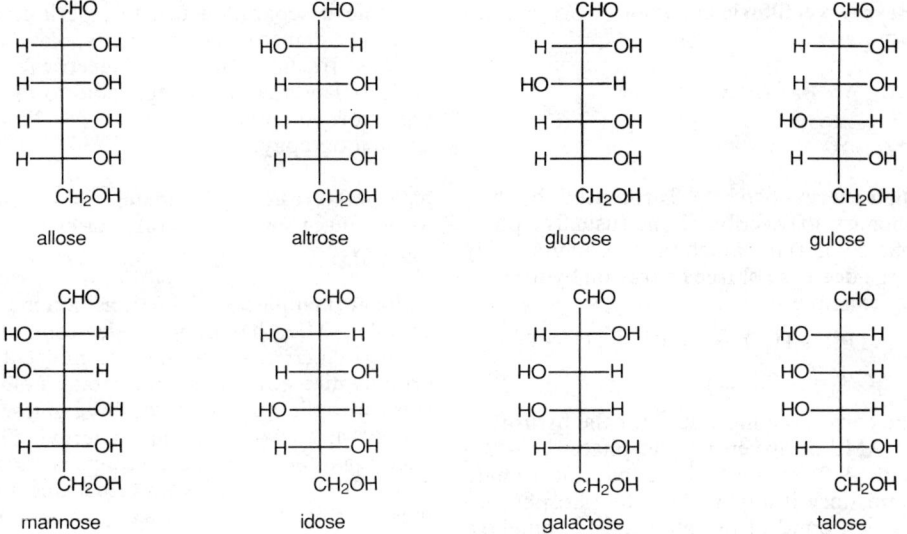

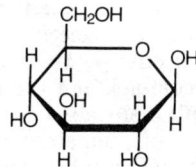

Haworth formulation
of β - D - glucose

D - galactose

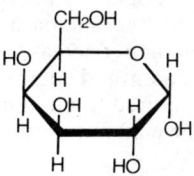

conformation formulation
of β - D - glucose

D - mannose

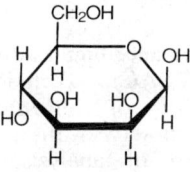

hexoses
Figure 1. Fischer and Haworth projection formulae for
D-aldohexoses.

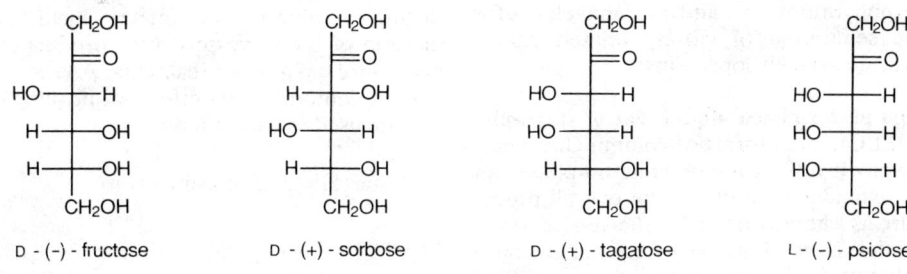

hexoses
Figure 2. Structure of 2-ketohexoses.

Heyrovsky–Ilkovic equation.   *See* POLAR-OGRAPHY.

Hf.   *See* HAFNIUM.

Hg.   *See* MERCURY.

high-energy bond.   Term used by bio-chemists to describe a bond (usually a phos-phate bond) for which there is an unusually large decrease of free energy on hydrolysis: for example

$$ATP^{4-} + H_2O \rightarrow ADP^{3-} + H^+ + P_i^{2-}$$
$$\Delta G^{\ominus\prime} = -29.3 \text{ kJ mol}^{-1}$$

the corresponding values for the hydrolysis of AMP or glycerol-1-phosphate are -13.8 and -9.2 kJ mol^{-1}. It is an unfortunate term since it implies that the phosphorus–oxygen bond in the molecule is somehow different from the normal covalent bond. The large amount of energy released is attributed to the increase in resonance stabilization and decrease in electostatic repulsion. The stored energy is released for use when required (e.g., in muscle contrac-tion or in the anabolic synthesis of essential metabolites).

highest   occupied   molecular   orbital (HOMO).   *See* LOWEST   UNOCCUPIED MOLECULAR ORBITAL.

high-performance liquid affinity chromato-graphy (HPLAC).   Technique combining the selectivity of conventional AFFINITY CHROMATOGRAPHY with the high resolu-tion, speed of analysis and sensitive detection of HIGH-PERFORMANCE LIQUID CHROMATOGRAPHY. Typical HPLAC columns, based on 10 μm spherical silica for high-dimensional stability are boronate for the separation of nucleotides, nucleosides, carbohydrates, etc. and concanavalin A for the separation of closely related carbo-hydrates and glycoproteins.

high-performance liquid chromatography (HPLC).   All forms of column chromato-graphy involving a mobile liquid phase can be extended to HPLC. The overall proce-dure is characterized by the use of small particle sizes (down to 5 μm), narrow bore columns (1–7 mm internal diameter) and high column inlet pressures (up to 600 atm)

to achieve separation in short periods of time. The detector system for observing the various fractions may be electrical or optical. The method is applicable to more types of compounds than is gas–liquid chromatography.

high-pressure liquid chromatography.   *See* HIGH-PERFORMANCE LIQUID CHROMATO-GRAPHY.

high-spin complex.   In the transition metal complex ML_6 having n d-electrons, by LIGAND FIELD THEORY the five molecular orbitals that will contain the metal d-elec-trons are split into a group of three of lower energy and two of higher energy. The energy gap between these groups is called Δ_0 (or $10D_q$). When feeding electrons into these orbitals (*see* AUFBAU PRINCIPLE) there comes a point at d^4 when a choice must be made between placing an electron in an already singly occupied, but lower-energy, orbital or to place it in an empty orbital of higher energy. The former leads to a low-spin complex and the latter to a high-spin complex, so-called after the rela-tive numbers of unpaired electrons. Whether a high- or low-spin configuration is adopted depends on the relative magni-tudes of Δ_0 and the energy required to pair two electrons. *See also* LIGAND FIELD STABILIZATION ENERGY.

Hill–de Boer equation.   *See* SURFACE PRESSURE.

His (histidine).   *See* AMINO ACIDS.

histamine   (4-(2-aminoethyl)imidazole, $C_5H_9N_3$).   Base (mp 83°C), formed by the bacterial degradation of histidine (*see* AMINO ACIDS). It is present in ergot and is a pharmacologically active compound, pro-ducing a number of effects including dilation of blood vessels. Overproduction plays a role in allergic reactions. ANTIHIST-AMINES counteract the effects of histamine and are used to treat allergies.

histidine (His).   *See* AMINO ACIDS.

histones.   *See* PROTEINS.

Hittorf method.   *See* TRANSPORT NUMBER.

Ho. Holmium (*see* LANTHANIDE ELEMENTS).

Hofmann conversion of amides. Preparation of a primary AMINE from an AMIDE by the action of alkaline bromine solution. The amine has one fewer carbon atoms than the amide. The stoichiometry of the reaction is

$$RCONH_2 + Br_2 + 4OH^- \rightarrow$$
$$RNH_2 + 2Br^- + CO_3^{2-} + 2H_2O$$

The mechanism is by a 1,2-shift via a bridged cation.

Hofmann elimination (Hofmann exhaustive methylation). Elimination reaction of –onium hydroxides which may be used to prepare ALKENES and to determine the arrangement of carbon atoms in ring compounds containing nitrogen.

$$[H\text{-}CR_2\text{-}CR_2\text{-}NR_3]^+ OH^- \xrightarrow{\text{heat}}$$
$$CR_2{=}CR_2 + R_3N + H_2O$$

Hofmann's rule states is that if the ethyl group is present the predominant alkene is ETHENE, or more generally the elimination of hydrogen occurs preferentially from the β-carbon joined to the largest number of hydrogen atoms. A compound without a β-carbon atom cannot undergo this reaction (e.g., tetramethylammonium hydroxide gives methanol with some dimethyl ether).

Hofmann's rule. *See* HOFMANN ELIMINATION.

Hofmeister series. *See* LYOTROPIC SERIES.

hole. Electronic vacancy in a solid. SEMICONDUCTORS having p-type dopants conduct by the movement of holes in the valence band.

holmium (Ho). *See* LANTHANIDE ELEMENTS.

HOMO (highest occupied molecular orbital). *See* LOWEST UNOCCUPIED MOLECULAR ORBITAL.

homo-. Prefix used in organic chemistry indicating: (1) a difference of –CH₂– in an otherwise similar structure: (e.g., serine, $HOCH_2CH(NH_2)COOH$; homoserine,

$HOCH_2CH_2CH(NH_2)COOH$); (2) a polymer made up of a single monomer.

homocyclic compounds. Compounds with closed rings of the same atom. Rings of carbon atoms, which form the majority of homocyclic compounds, are known as CARBOCYCLIC COMPOUNDS. *See also* ALICYCLIC COMPOUNDS.

homogeneous. System with a uniform composition or structure. A solution is a homogeneous mixture of solvent and solute. Homogeneous reactions occur between reactants in the same phase (e.g., between gases or liquids).

homogeneous catalysis. CATALYSIS in which the catalyst is in the same phase (usually liquid) as the reactants and products. A common type is ACID–BASE CATALYSIS. Of importance to industry is the use of the ions of TRANSITION ELEMENTS to effect reactions such as HYDROFORMYLATION, OXO PROCESS, oxidation (*see* AUTOXIDATION), the synthesis of adiponitrile, the WACKER PROCESS and POLYMERIZATION (*see* ZIEGLER–NATTA CATALYST). Homogeneous catalysts may also lead to stereospecific reactions. For example, optically active α-amino acids may be prepared from unsaturated precursors using rhodium complexes containing optically active ligands. *See also* WILKINSON'S CATALYST.

homologous series. Series of organic compounds that differ by a METHENE GROUP. Each member is known as an homologue.

homologue. *See* HOMOLOGOUS SERIES.

homolytic fission. Reaction in which the electrons forming the broken bond are shared equally among the fragments formed (e.g., $XR \rightarrow X^{\cdot} + R^{\cdot}$). *See also* HETEROLYTIC FISSION.

homonuclear. Descriptive of a molecule consisting of identical atoms (e.g., hydrogen, chlorine). *See also* HETERONUCLEAR.

homopolar bond. *See* COVALENT BOND.

homopolymer. *See* POLYMERS.

***h*-orbital.** *See* ATOMIC ORBITAL.

hormones   Some representative hormones.

Hormone	Structural formula	Occurrence, properties, uses
Adrenaline (epinephrine)	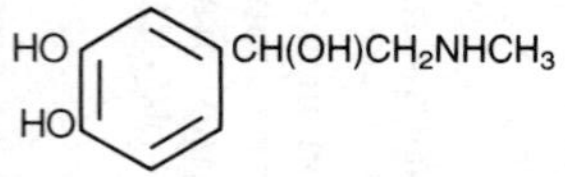 HO-C$_6$H$_3$(OH)-CH(OH)CH$_2$NHCH$_3$	Main hormone excreted by adrenal medulla in mammals. Secretion stimulated by nervous system under conditions of stress, pain, fear and fall in blood sugar. Most important effects are on dilation and constriction of blood vessels and carbohydrate metabolism, resulting in stimulated blood flow and high glucose levels. Synthesized from catechol, mp 212°C. Included in local anaesthetics to constrict blood flow locally. Powerful bronchodilator used in treatment of bronchial asthma. Naturally occurring (−)-form is about 20 times more active than (+)-form
Noradrenaline (norepinephrine)	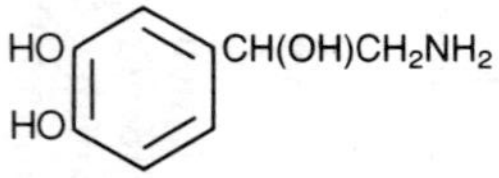 CH(OH)CH$_2$NH$_2$ (HO, HO)	Released in adrenal medulla with adrenaline. Precursor for adrenaline, mp 103°C. Acts to constrict small arteries thereby increasing blood pressure and causing contraction of smooth muscle. Present in pulp and peel of bananas
Oestrone (estrone)	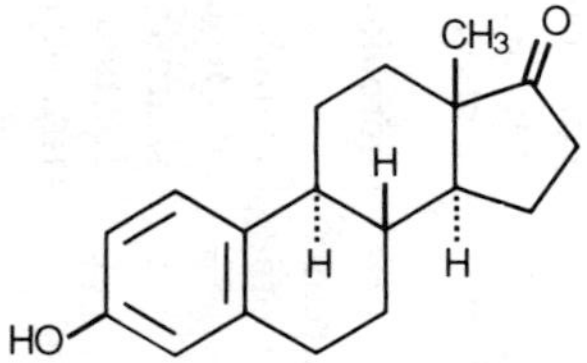	Human OESTROGEN isolated from urine of pregnant women (mp 260°C). Corresponding C-17 alcohol, oestradiol-17β (OH trans to C-13 methyl), is even more active
Progesterone (progestin)	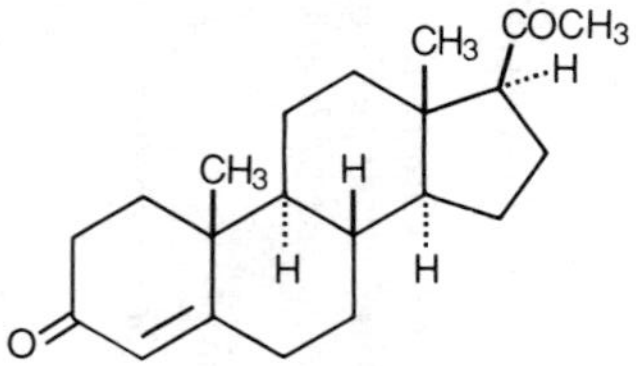	Secreted by corpus luteum, governing growth and development of uterus during pregnancy (mp 122°C). Exerts antiovulatory effect when administered during days 5−25 of menstrual cycle; this is basis for use of synthetic progestins as oral contraceptives (see also STILBOESTROL). Intermediate in biosynthesis of all steroid hormones. Can be manufactured from cholesterol
Testosterone	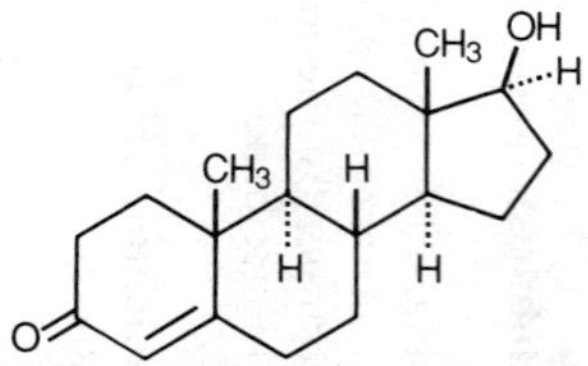	Human ANDROGEN formed in testes, controlling development and maintenance of male sex organs and secondary sex characteristics (mp 154°C). Used, together with synthetic analogues, in treatment of disorders due to impaired secretion of natural hormone

Hormone	Structural formula	Occurrence, properties, uses
Androsterone	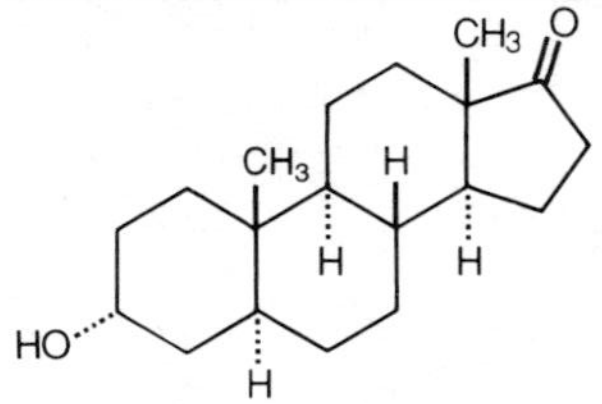	Typical urinary 17-oxosteroid (mp 183°C), formed by the metabolic reduction of testosterone. Less potent than testosterone
17α-Ethynyl-oestradiol	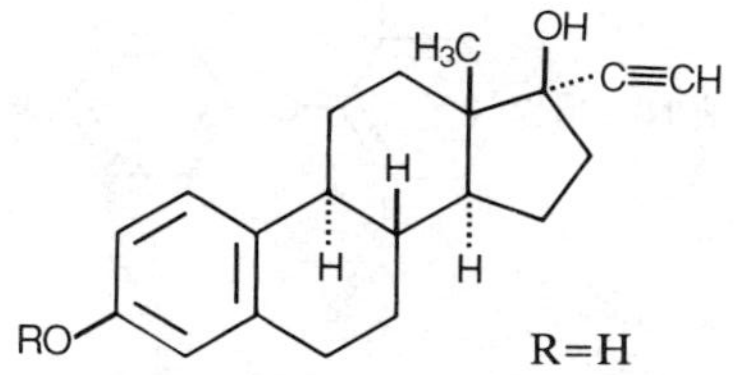 R=H	Synthetic oestrogen with potent activity (inhibition of ovulation), widely used in oral contraceptives. Manufactured from natural oestrogen, oestrone by reaction with potassium acetylide (HC≡CK) in liquid ammonia
3-Methoxyethynyl-oestradiol (mestranol)	R=CH₃	Similar to 17α-ethynyloestradiol. Manufactured from oestrone methyl ether
Insulin	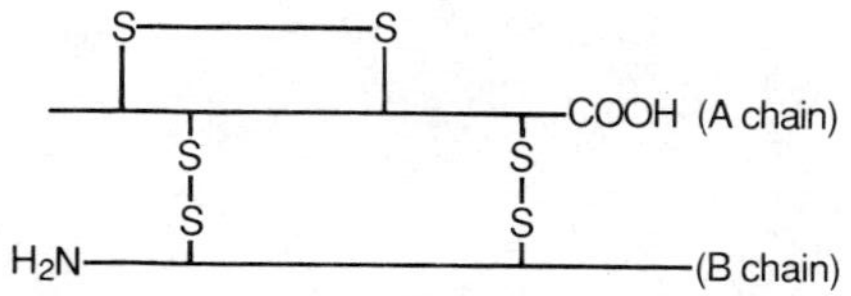 A and B chains are polypeptide chains (21 and 30 amino acid residues, respectively) linked by disulphide bridges between cystine residues	Protein ($M_r = 5800$), secreted in pancreas in response to high blood glucose. Stimulates glucose utilization and prevents glycogen breakdown in liver. Absence leads to diabetes mellitus
Glucagon		Straight-chain polypeptide (29 amino acid residues), synthesized and secreted in pancreas in response to low blood glucose. Increases glucose levels by stimulating metabolic breakdown of glycogen. Thus antagonizes effects of insulin

continued overleaf

Hormone	Structural formula	Occurrence, properties, uses
Luteinizing hormone (LH; interstitial cell-stimulating hormone, ICSH)		Glycoprotein, secreted by anterior pituitary gland. In males, stimulates sex hormone (androgen) production in testes; in females, stimulates ovulation, progesterone synthesis and corpus luteum formation
Follicle-stimulating hormone (FSH)		Glycoprotein, secreted by pituitary gland. In females, stimulates growth of specialized structures in ovary (graafian follicles); in males, stimulates formation of sperm in testes. FSH and LH are usually released together and act synergistically. Major component in treatment of infertility
Prolactin (lactogenic hormone, luteotrophic hormone, LTH)		Protein, consisting of single chain of 198 amino acid residues with three disulphide linkages, produced by anterior pituitary gland. Stimulates and maintains lactation, and secretion of progesterone in mammal; secretion is increased by suckling. In birds stimulates secretion of crop milk by the crop glands
Thyrotropin (thyroid-stimulating hormone, TSH)		Glycoprotein, consisting of two large peptide units (one of which is similar to LH and FSH), secreted by anterior pituitary gland. Stimulates growth of thyroid gland, its uptake of iodine and secretion of thyroxine and triiodothyronine
L-Thyroxine (3,5,3′,5′-tetraiodo-thyronine)	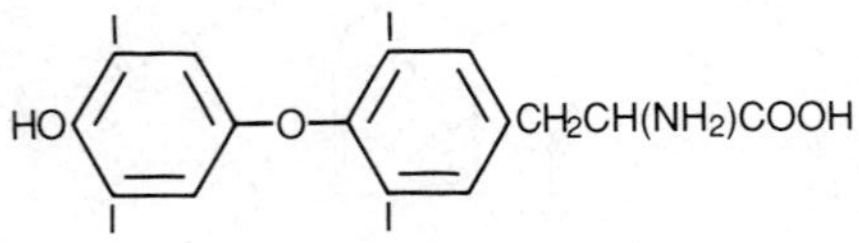 HO—⟨benzene ring⟩—O—⟨benzene ring⟩—CH₂CH(NH₂)COOH	Formed in thyroid glands from tyrosine and iodine (derived from blood) which occurs in protein-bound form thyroglobulin. Thyroxine and 3,5,3′-triiodothyronine are secreted into blood when gland is stimulated. Essential regulators of basic metabolic rate, in particular they increase oxygen consumption and energy production. Deficiency leads to goitre and cretinism; excessive secretion leads to Graves's disease
Parathyroid hormone (PTH)		Single-chain polypeptide of about 75 amino acid residues, secreted by parathyroid gland. Controls blood calcium levels. Low calcium levels stimulate hormone secretion which increases transfer of calcium from bones to blood
Calcitonin (thyrocalcitonin)		Straight-chain polypeptide containing 32 amino acid residues, produced by mammalian thyroid gland. Lowers blood calcium (and phosphate) levels. Operates in opposition to PTH
Oxytocin		Straight-chain peptide containing 9 amino acids secreted by posterior pituitary gland. Contracts uterus and stimulates lactation

hormones. Chemical substances excreted by an endocrine gland and transported by the blood to target tissues where they exert a specific physiological action. They do not fall into any particular class of compounds; they may be steroids (*see* CORTICO-STEROIDS), peptides or simple organic amines (*see* Table). AUXINS are plants' equivalent of hormones.

horn silver. *See* SILVER.

HPLAC. *See* HIGH-PERFORMANCE LIQUID AFFINITY CHROMATOGRAPHY.

HPLC. *See* HIGH-PERFORMANCE LIQUID CHROMATOGRAPHY.

HSAB. *See* HARD AND SOFT ACIDS AND BASES.

Hückel molecular orbital theory. Theory of conjugated molecules which deals only with bonding between neighbouring π-orbitals. The overlap integral S_{ab} is taken to be unity for a = b and zero for a ≠ b. The coulomb integral H_{aa} is called α, and the resonance integral (β) is only considered for a bonded to b. The general secular equation is

$$c_m(\alpha - E) + \sum_{n,m} c_n\beta = 0$$

The sum is taken over atoms n bonded to each atom m. Values for the energies of orbitals are in terms of α and β (e.g., butadiene has four orbitals of energy $\alpha \pm 1.62\beta$ and $\alpha \pm 0.62\beta$). Molecular orbital calculations based on this simple Hückel approach are styled as CNDO (complete neglect of differential overlap). *See also* MOLECULAR ORBITAL CALCULATIONS, SEMI-EMPIRICAL MOLECULAR ORBITAL THEORY.

Hückel's rule. For a monocyclic system C_nH_n, containing n π-electrons with each carbon atom providing one π-electron, to exhibit aromatic character (i.e. unusual stability) the molecule must have $(4n + 2)$ π-electrons, where n is an integer 0, 1, 2, . . . A further requirement for aromaticity is (reasonable) planarity of the ring; if the molecule is non-planar the overlap of the p-orbitals is diminished or absent. If $n = 1$, the molecule (i.e. BENZENE), has a closed shell of six π-electrons and shows aromatic character; the remaining hydrocarbons in the three to eight-ring systems are non-benzenoid. However, the cyclobutenyl dication, the cyclopentadienyl anion and the TROPYLIUM ION which conform to the $(4n + 2)$ rule exhibit aromatic character. The rule has been successfully applied to many bicyclic and polycyclic systems (e.g., naphthalene, anthracene and phenanthrene) and to various bicyclic non-benzenoid aromatic compounds, (e.g. AZULENE, where $n = 2$). *See also* AROMATICITY.

humectant. Substance (e.g., GLYCEROL) that absorbs water vapour from the atmosphere and prevents food from drying out and becoming hard.

Hume–Rothery's rule (eight-N rule). A metal in group N has a structure in which each atom has eight–N nearest neighbours. Thus carbon (group IV element) is four coordinate in diamond, and tellurium and selenium (group VI elements) crystallize in strings with two nearest neighbours. *See also* ALLOY, BRASS, INTERMETALLIC COMPOUND.

humic acids. Group of aromatic acids of high molecular mass and unknown structure present in soil. They can be extracted from peat and brown coal.

humidity
absolute humidity. Mass of water vapour per unit mass of dry air.
relative humidity. Ratio of the partial vapour pressure of the water vapour in the air to the partial pressure in the air when saturated at the same temperature (usually expressed as a percentage).

humus. Dark brown amorphous material in soil consisting of decomposed plant and animal remains. It provides the nutrients for plant growth and increases the water-absorbing properties of the soil.

Hund's rule. Electrons with parallel spins have a lower energy than a corresponding pair with opposed spins. Hund's rule in conjunction with the PAULI EXCLUSION PRINCIPLE and the AUFBAU PRINCIPLE is used to build up the electronic structure of atoms and molecules. *See also* MOLECULAR ORBITAL, PERIODIC TABLE.

hyaluronic acid.   *See* AMINO SUGARS.

hybridization. Superposition of atomic wavefunctions to create a hybrid orbital: for example, the sum of a 2s-orbital and one 2p-orbital leads to a linear sp-hybrid, a 2s-orbital and two 2p-orbitals to an sp^2-hybrid with maximum electron density towards the corners of an equilateral triangle and, thirdly, a tetrahedral sp^3-hybrid is formed from a 2s-orbital and all three 2p-orbitals. Each of the above combinations is found in compounds of carbon in alkynes, alkenes and alkanes, respectively. Hybridization involving d-orbitals has been used to describe the shapes of transition metal complexes; dsp^3-hybrids are trigonal bipyramidal and d^2sp^3 octahedral (*see* SHAPES OF MOLECULES).

hydrases.   *See* ENZYMES.

hydrate. Solid in which water is incorporated as WATER OF CRYSTALLIZATION or in a clathrate (*see* CLATHRATE COMPOUND). Zeolites and other silicates and clays crystallize with varying amounts of water.

hydration. Association of molecules of water with a species. In solutions (*see* SOLVATION), water usually associates with cations, the dipole of the water molecule being directed towards the ion. ANHYDROUS solids may form HYDRATES by taking up water in a process known as hydration.

hydrazides.   *See* NITROGEN HYDRIDES.

hydrazine.   *See* NITROGEN HYDRIDES.

hydrazoic acid.   *See* NITROGEN HYDRIDES.

hydrazones. Monoimines of the hydrazine series (*see* NITROGEN HYDRIDES), of general formula >C=NNR$_2$, formed by the condensation of hydrazine or substituted hydrazine with aldehydes or ketones, for example

$$(CH_3)_2CO + H_2NNH_2 \rightarrow (CH_3)_2C=NNH_2$$
propanone hydrazone

$$PhCOCH_3 + H_2NNHPh \rightarrow$$
$$Ph(CH_3)C=NNHPh$$
acetophenone phenylhydrazone

Acid hydrolysis breaks down hydrazones to the corresponding hydrazine and carbonyl compound. Unsubstituted hydrazones react with a further molecule of carbonyl compound to give an azine

$$(CH_3)_2C=NNH_2 + OC(CH_3)_2 \rightarrow$$
$$(CH_3)_2C=NN=C(CH_3)_2$$

Phenylhydrazones react with strong acids to give indoles (*see* FISCHER INDOLE SYNTHESIS). 2,4-Dinitrophenylhydrazine is used for the preparation of crystalline derivatives for the identification of keto compounds.

hydrides. Compounds of hydrogen and another element containing the hydride ion (H$^-$) or a covalently bonded hydrogen atom, named generally hydro (hydrido).
ionic hydrides (H$^-$). ENTHALPY of formation of H$^-$ is +151 kJ mol^{-1}, thus only the most ELECTROPOSITIVE elements form ionic hydrides. Hydrides of the ALKALI METALS have a sodium chloride lattice (*see* CRYSTAL FORMS) in which H$^-$ has a radius between that of F$^-$ and Cl$^-$ (135–150 pm). The saline hydrides (alkali metal and ALKALINE EARTH METAL hydrides) are white crystalline solids which are very powerful reducing agents $E^{\ominus}_{H_2,H^-} = -2.25$ V).
complex hydrides. Hydrides containing the ion [MH$_4$]$^-$, where M = B, Al, Ga, are widely used in synthesis as reducing agents, in particular sodium tetrahydridoborate (*see* BORON INORGANIC COMPOUNDS) and lithium tetrahydridoaluminate (*see* ALUMINIUM INORGANIC COMPOUNDS).
hydrido complexes of transition elements. Complexes in which hydrogen acts as a ligand are known for each transition element except scandium, yttrium and lanthanum. [ReH$_9$]$^{2-}$ and [TcH$_9$]$^{2-}$ are the only binary complexes. Hydrogen may act as a simple ligand or as a bridge between metal atoms as in [(CO)$_4$WHHW(CO)$_4$]$^{2-}$. It is thought that NITROGEN FIXATION occurs via Fe–H or Mo–H.
covalent hydrides. Hydrides with more electronegative elements in which hydrogen has an oxidation state of +1 (e.g., H$_2$O, CH$_4$) are rarely classed as such.

hydrobenzoin.   *See* ACYLOINS.

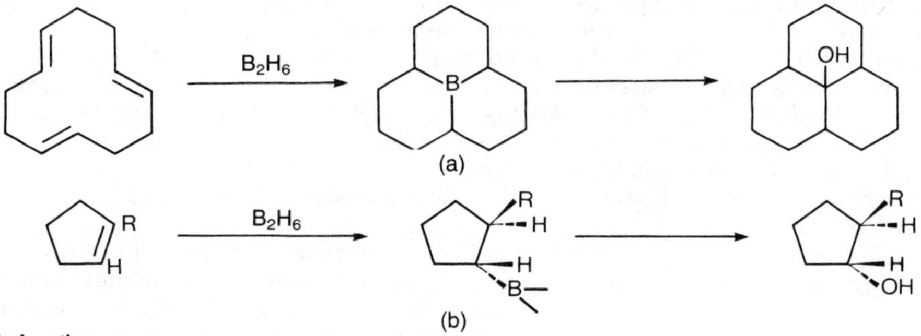

(a)

(b)

hydroboration

hydroboration. Formation of organoboron compounds (*see* BORON ORGANIC DERIVATIVES) by the addition of diborane (BORON INORGANIC COMPOUNDS) in tetrahydrofuran to unsaturated organic compounds (*see* Figure (a)). The subsequent decomposition, usually *in situ*, yields the organic product (e.g., alkanes by acid hydrolysis, alcohols by oxidation, amines from chloramine, 1,4-addition to α,β-unsaturated aldehydes). The importance of hydroboration is in the stereo-specificity of the reactions showing: (1) anti-Markownikoff addition (*see* MARKOWNIKOFF'S RULE): for example

$$RCH=CH_2 \rightarrow RCH_2CH_2B< \rightarrow$$
$$RCH_2CH_2OH$$

(2) *cis*-addition (*see* Figure (b))

(3) a high degree of steric control with no rearrangement. The reaction of organoboranes with carbon monoxide to give R_3CBO is a route to tertiary alcohols.

hydrobromic acid. *See* HYDROGEN HALIDES.

hydrocarbons. Compounds of carbon and hydrogen only.

hydrochloric acid. *See* HYDROGEN HALIDES.

hydrochlorides. Salts of weak organic bases and hydrogen chloride (e.g., aniline hydrochloride, $C_6H_5NH_2.HCl$).

hydrocortisone. *See* CORTICOSTEROIDS.

hydrocracking. Catalytic high-pressure process for the production of high-octane petrol or aviation fuel. Oil and hydrogen (310–60°C, 100–150 atm) are passed over a bifunctional catalyst consisting of an acidic support (e.g., zeolites) which promotes CRACKING and a hydrogenating metal (e.g., nickel, molybdenum, tungsten, palladium). *See also* REFORMING, HYDROGENATION.

hydrocyanic acid. *See* HYDROGEN CYANIDE.

hydrodealkylation. *See* HYDROGENOLYSIS.

hydrofluoric acid. *See* HYDROGEN HALIDES.

hydroforming (catalytic reforming). Method based on DEHYDROGENATION, cyclization and isomerization of aliphatic materials (containing at least six carbon atoms) to give aromatic compounds with the same number of carbon atoms. The process is carried out at high temperatures and pressures in the presence of a catalyst (oxides of chromium, molybdenum, and vanadium on alumina support, platinum); (e.g., n-hexane to benzene, n-heptane to toluene, n-octane to xylenes and ethylbenzene). The process is used to improve the OCTANE NUMBER of gasoline by increasing the proportion of aromatics.

hydroformylation. Process for manufacture of ALDEHYDES (C_2–C_4) in which hydrogen, carbon monoxide (SYNTHESIS GAS) and ALKENES are reacted in the liquid phase over cobalt (oxide or carbonyl) or rhodium–phosphine complexes. The METHANOYL GROUP and a hydrogen atom add across the double bond of the alkene, probably via a cobalt carbonyl–alkene complex ($Co(CO)_7RCH=CH_2$). Hydroformylation is the first step in the OXO PROCESS for the production of alcohols. *See also* FISCHER–TROPSCH PROCESS, HYDROGENATION.

hydrogasification. *See* GASIFICATION.

hydrogel. GEL in which the dispersion medium is water.

hydrogen (H). Lightest element. The atom of the most abundant isotope, PROTIUM, consists of one proton and one electron (*see also* DEUTERIUM, TRITIUM). The element occurs free as dihydrogen (H_2) in small quantities in NATURAL GAS, but is mostly found combined in water, hydrocarbons and other organic and biological compounds. H_2 is the lightest of all gases; it is colourless, odourless and practically insoluble in water. H_2 is manufactured by the ELECTROLYSIS of water, by steam reforming (*see* GASIFICATION) hydrocarbons and from SYNTHESIS GAS. It is prepared in the laboratory by the action of dilute acids on ELECTROPOSITIVE metals such as zinc and iron. Hydrogen reacts explosively with oxygen under certain conditions of temperature and pressure (*see* CHAIN REACTION, EXPLOSION) and with halogens (*see* HYDROGEN HALIDES). Metal oxides are reduced at high temperatures. The relatively poor reactivity of hydrogen at low temperatures is a reflection of the strength of the hydrogen–hydrogen bond (436.0 kJ mol^{-1}). ATOMIC HYDROGEN is much more reactive. HYDRIDES are formed with most elements. Bonding in the hydrogen molecule may be described by MOLECULAR ORBITAL or VALENCE BOND ORBITAL methods. H_2 is used in the manufacture of ammonia by the HABER PROCESS and also in the HYDROGENATION of vegetable oils and coal, and in the reduction of oxides. The 'hydrogen economy' is a concept in which hydrogen plays a central role in energy production and use. In this hydrogen would be generated by electrolysis from spare nuclear electricity, be stored or transported and then burnt or used to produce elec-

tricity in a FUEL CELL. *See also* ORTHOHYDROGEN.

$A_r = 1.00794$, $Z = 1$, ES $1s^1$, mp $-259.2°C$, bp $-252.9°C$, ρ ($0°C$, 1 atm) $= 0.08987$ kg m^{-3}. Isotopes: ^1H, 99.98 percent; ^2H (D) 0.02 percent.

hydrogenases. *See* ENZYMES.

hydrogenation. Addition of hydrogen to a compound, usually an unsaturated organic compound, at an elevated temperature, pressure and over a catalyst. ALKYNES may be hydrogenated to ALKENES and then ALKANES over a supported metal catalyst such as nickel, platinum or palladium. Aromatic compounds are also hydrogenated at high pressure (200 atm) over nickel. Compounds containing the OXO GROUP hydrogenate to ALCOHOLS then alkanes. Unsaturated vegetable and animal fats are hardened by hydrogenation under mild conditions which do not affect the ester groups. Acids and esters are hydrogenated to alcohols over a reduced copper–ammonium chromate catalyst. Carbon monoxide is hydrogenated in the FISCHER–TROPSCH REACTION (*see also* HYDROFORMYLATION, OXO PROCESS) and petroleum, coal and tar are hydrogenated to produce useful fuels and chemicals (*see also* HYDROCRACKING). Catalysts for these processes are transition elements, their oxides or sulphides, or homogeneous catalysts derived from transition metals especially platinum, cobalt, rhodium and copper. Hydrogenation accompanied by cleavage of a molecule is known as HYDROGENOLYSIS. *See also* BIRCH REDUCTION.

hydrogen bond. Special type of interaction between molecules that occurs whenever a polar bond containing the hydrogen atom (e.g., $-OH$, $-NH$) interacts with an electronegative atom such as oxygen, nitrogen, fluorine or chlorine. The nucleus of the hydrogen atom, when its electron forms a covalent bond with an electronegative atom, becomes partly unshielded. Consequently this proton can interact directly with another electronegative atom on a different molecule. Depending on the strength of the bond (usually about 10 percent of that of a covalent bond) such binding can exist in the gaseous phase, as well as in the liquid and solid phases. Normal hydrogen bond distance is too long for covalent character. Such bonds cause the self-association of water and alcohols and polymeric chains in liquid hydrogen fluoride and account for the unusual properties of water and the relatively high boiling point of water, hydrogen fluoride and ammonia (compared with hydrogen sulphide, hydrogen chloride and phosphine). Hydrogen bonding occurs between the bases in chains of DNA and between >CO and >NH groups in proteins, and is responsible for the secondary structure of proteins. Intramolecular hydrogen bonds (e.g., in *cis*-butenedioic acid (maleic acid) in contrast to the *trans*-butenedioic acid (fumaric acid)) occur between −COOH and −COO⁻ groups on the same side of the double bond.

hydrogen-bridged structure. *See* ELECTRON-DEFICIENT COMPOUND.

hydrogen bromide. *See* HYDROGEN HALIDES.

hydrogen chloride. *See* HYDROGEN HALIDES.

hydrogen cyanide (HCN). Covalent gas (bp 26°C), toxic with an almond-like smell. Manufactured by the oxidation of methane and ammonia (catalyst, 800°C)

$$2CH_4 + 3O_2 + 2NH_3 \rightarrow 2HCN + 6H_2O$$

It is also formed from cyanides and sulphuric acid and the dehydration of methanamide. Liquid hydrogen cyanide has a high relative permittivity ($\varepsilon_r = 107$ at 25°C) arising from hydrogen bonding. It easily polymerizes. Aqueous solutions (hydrocyanic acid, prussic acid) are weakly acidic ($K_a = 2.1 \times 10^{-9}$). Derivatives of hydrogen cyanide include CYANIDES, NITRILES, isocyanides and isonitriles (*see* CARBYLAMINES). It is used to manufacture adiponitrile and PROPENENITRILE (HCN + C_2H_2, CuCl + NH_4Cl catalyst).

hydrogen electrode. HALF-CELL H⁺|H₂,M, where M is a metal usually platinum. The electrode reaction is $H^+ + e \rightleftharpoons \frac{1}{2}H_2$, the ELECTRODE POTENTIAL of which is given by the NERNST EQUATION

$$E_{H^+,H_2} = E^{\ominus}_{H^+,H_2} + RT/F \ln [a_{H^+}/(p_{H_2}/\text{atm})^{1/2}]$$

The standard potential when $a_{H^+} = 1$ and $p_{H_2}/\text{atm} = 1$ is by convention the arbitrary zero of electrode potential. The simplest form of the hydrogen electrode is a piece of platinized platinum dipping into a solution containing H⁺, over which pure hydrogen is bubbled. In this form it is an important REFERENCE ELECTRODE.

hydrogen fluoride. *See* HYDROGEN HALIDES.

hydrogen halides (HX, where X = F, Cl, Br, I). Gases at room temperature (bp: HF, 19.5°C; HCl, −84.9°C; HBr, −66.8°C; HI, −35.4°C), prepared by treating a halide with sulphuric acid (e.g., $CaF_2 + H_2SO_4 \rightarrow CaSO_4 + 2HF$), by direct combination of the elements (very explosive in the case of HF) or by the hydrolysis of phosphorus trichloride, tribromide or triiodide. They are molecular species, the atoms being joined by a COVALENT BOND. HF is polymeric, held together by strong HYDROGEN BONDS. Their solutions in water are known as hydrohalic acids. All hydrogen halides are used to prepare and manufacture organic halides (*see* FLUOROCARBONS, FREONS, chloro-, bromo- and iodocompounds).

hydrogen fluoride. Used as a strongly dehydrating nonaqueous solvent. Hydrofluoric acid (36 percent HF in water) corrodes glass and so is handled in Teflon (*see* POLYTETRAFLUOROETHENE) or platinum apparatus.

hydrogen chloride. Used as a condensing agent. It reacts with many metals to give chlorides usually in a low OXIDATION STATE. Hydrochloric acid (concentrated acid is 43 percent HCl) is corrosive and is used in industry as a strong acid.

hydroiodic acid. Strongest acid in water.

hydrogen iodide. *See* HYDROGEN HALIDES.

hydrogen ion concentration. *See* pH.

hydrogenolysis. Cleavage of a molecule by HYDROGEN (*compare* AMMONOLYSIS, HYDROLYSIS). For example, toluene undergoes hydrodealkylation to methane and benzene, and methyl dodecanoate gives dodecanol and methanol.

hydrogen peroxide

hydrogen peroxide (H_2O_2). Colourless liquid (bp 150°C), prepared electrochemically from sulphuric acid or in a cyclic process involving derivatives of anthraquinone (*see* Figure).

The pure liquid is stable, but aqueous solutions decompose easily when catalyzed

$$H_2O_2 \rightarrow H_2O + \tfrac{1}{2}O_2$$

Hydrogen peroxide is a strong oxidizing agent ($E^{\ominus}_{H_2O_2,H^+} = 1.77$ V) and a bleach but may be reduced by potassium permanganate. Uses are in synthesis, as a bleach in the textile and paper industries, in pollution control and as a propellant.

hydrogen sulphide (H_2S). Extremely poisonous gas (mp −85.5°C, bp −60.3°C) with a revolting odour. It occurs in natural gas from North Sea and in volcanic emissions. H_2S is prepared by the action of acids on metal sulphides (e.g., FeS). Aqueous solutions are weakly acidic ($pK_1 = 7$, $pK_2 = 12$) containing the HS^- ion and traces of S^{2-}. In acid solution, it is a mild reducing agent. With heavy metal ions in solution insoluble sulphides are formed. It burns in air forming sulphur(IV) oxide (sulphur dioxide, *see* SULPHUR OXIDES).

hydrogen uranyl phosphate (HUP, HUO_2-$PO_4.4H_2O$). Solid proton conductor having $\kappa = 4$ S m^{-1} at 25°C. *See also* SOLID ELECTROLYTE.

hydroiodic acid. *See* HYDROGEN HALIDES.

hydrolysis. Reaction of a substance with water. In aqueous solution of electrolytes, hydrolysis of cations produces weak bases and hydrolysis of anions gives weak acids.

$$CH_3COO^- + H_2O \rightarrow CH_3COOH + OH^-$$

Thus the ethanoate ion is hydrolyzed giving alkaline solutions. In organic chemistry, hydrolysis refers to the displacement of a group by the HYDROXYL GROUP. For example,

$$RCH_2Cl + H_2O \rightarrow RCH_2OH + HCl$$

Hydrolysis is a particular example of SOLVOLYSIS.

hydrometalation. Addition of a metal HYDRIDE across an unsaturated bond with a hydrogen atom adding on one side and the remaining radical on the other. HYDROBORATION is an important synthetic example in which borane adds. Other metals whose hydrocompounds take part in hydrometalation are aluminium, zirconium and silicon, the last if a coordinately unsaturated transition metal cayalyst is used (e.g., Vaskà's compound, *see* IRIDIUM).

hydronation. Transfer of a hydrogen cation (without regard to the nuclear mass of the hydrogen entity) to a substrate. *See also* DEUTERONATION, PROTONATION, TRITONATION.

hydronium ion. *See* HYDROXONIUM ION.

hydrophilic colloid. *See* LYOPHILIC COLLOID.

hydrophobic colloid. *See* LYOPHOBIC COLLOID.

hydroquinone (1,4-dihydroxybenzene). *See* DIHYDROXYBENZENES.

hydroxides. Inorganic compounds containing a HYDROXYL ION. The hydroxides of the alkali metals are soluble in water, giving strongly alkaline solutions. Other metallic hydroxides are insoluble or are only sparingly soluble bases. Hydroxides of nonmetals (e.g., $B(OH)_3$) are acidic and better written as such (i.e. H_3BO_3). *See also* ACIDS AND BASES, ALKALI.

hydroxonium ion (hydronium ion, $(H_2O)_nH^+$). Hydrated PROTON. In water, the life time of an individual hydroxonium ion is only about 100 fs since protons migrate rapidly from one water molecule to the next. In an electric field, the apparently large value of the mobility of a proton arises from this effect (*see* GROTTHUS THEORY). In crystals of hydrated mineral acids (e.g., $H_3O^+ Cl^-$) the ion is a flat triangular pyramid which resembles NH_4^+. $H_5O_2^+$ is found in some crystals and in solution. H_3O^+ hydrogen-bonded to three other water molecules is $H_9O_4^+$.

hydroxy-. Prefix indicating the presence of a HYDROXYL GROUP.

1-hydroxy-2-anisole (guaiacol). *See* AROMATIC ETHERS.

hydroxyanthracenes ($C_{14}H_{10}O$) (anthrols). *1- and 2-hydroxyanthracenes* (anthrols). Prepared from the corresponding naphthalenesulphonic acid by alkaline fusion. These exhibit the characteristic properties of phenols.
9-hydroxyanthracene (anthranol). Unstable yellow solid (mp 120°C), prepared by acidifying an alkaline solution of its isomer ANTHRONE or the reduction of 9,10-anthraquinone (Al powder + conc. H_2SO_4). It shows the general properties of a hydroxy compound; it couples with diazonium salts and is rapidly oxidized to dihydrodianthrone. The tautomerism between anthrone and 9-hydroxyanthracene is known as *trans*-annular tautomerism; the equilibrium mixture consists mainly of the keto form.

hydroxybenzenes. *See* PHENOLS.

2-hydroxybenzoic acid (salicylic acid, HO-C_6H_4COOH). Aromatic acid (pK_a = 2.95) that occurs naturally as its methyl ester (*see* METHYL 2-HYDROXYBENZOATE) in many essential oils. It melts at 159°C and decomposes to phenol and carbon dioxide at 200°C, and is manufactured by the Kolbé–Schmitt reaction—sodium phenoxide and carbon dioxide are heated under pressure at 120°C. It is used as an antiseptic, for the preparation of DYES, ASPIRIN, methyl 2-hydroxybenzoate, and phenyl 2-hydroxybenzoate (salol).

3-hydroxybutanal (aldol, $CH_3CH(OH)$-CH_2CHO). Colourless oil (bp 83°C, 20

mmHg), prepared by the ALDOL CONDENSATION of ETHANAL. On heating aldols eliminate water to form unsaturated compounds (e.g., aldol gives 2-butenal, CH_3-CH=CHCHO).

hydroxybutanedioic acid (malic acid, HOOCCH(OH)CH_2COOH). Acid found in apples and other fruits. It is manufactured by heating *cis*-BUTENEDIOIC ACID (maleic acid) with sulphuric acid under pressure, and synthesized in the laboratory by heating bromobutanedioic acid with a suspension of silver oxide. With one asymmetric carbon, D- and L-forms are possible (*see* ISOMERISM); naturally occurring malic acid is in the L-(−)-form (mp 100°C), whereas synthetic malic acid is DL (mp 131°C). Malic acid behaves both as an acid and an alcohol. Water is eliminated from the molecule acting as a β-hydroxy-acid to form an unsaturated acid (*trans*-butenedioic acid, fumaric acid) or butenedioic anhydride.

3-hydroxy-2-butanone. *See* ACYLOINS.

17-hydroxycorticosterone (cortisol, hydrocortisone). *See* CORTICOSTEROIDS.

2-hydroxy-2,4,6-cycloheptatrien-1-one (tropolone). Solid, prepared by oxidation of 1,3,5-cycloheptatriene with potassium permanganate. It shows aromatic character, but does not exhibit any ketonic properties. The tropolonate ion forms five-membered chelate rings with many metals (e.g., $Fe(O_2C_7H_5)_3$). A number of tropolones occur naturally (e.g., cholchicine and thujaplicins present in some essential oils).

2-hydroxy-2,4,6-cycloheptatrien-1-one

hydroxydimethylbenzenes (xylenols, $C_6H_3(CH_3)_2OH$). Components of coal tar. The six possible isomers are low-melting solids with the general properties of PHENOLS.

1-hydroxydodecane (dodecyl alcohol, lauryl alcohol, $CH_3(CH_2)_{10}CH_2OH$). Solid obtained synthetically, with other homologues, by the polymerization of ethene using triethylaluminium (ZIEGLER–NATTA CATALYST) at 50–100 atm at below 130°C, followed by oxidation to the alkoxide and subsequent hydrolysis

$$AlEt_3 + nC_2H_4 \rightarrow Al[(C_2H_4)_nEt]_3 \xrightarrow{O_2}$$
$$Al[O(C_2H_4)_nEt]_3 \xrightarrow{H_2SO_4}$$
$$HO(C_2H_4)_nEt + Al_2(SO_4)_3$$

The sulphate, usually as the sodium salt, is used as a detergent.

hydroxyethanal. *See* BIOSE.

hydroxyethanoic acid (glycollic acid, CH_2-$OHCOOH$). Solid (mp 80°C), α-hydroxy-acid that is soluble in water. It occurs naturally in sugar cane. The acid is prepared by the hydrolysis of potassium chloroethanoate, or by the electrolytic reduction of ethanedioic acid. It undergoes reactions of both ALCOHOLS and CARBOXYLIC ACIDS, and forms the LACTIDE 1,4-dioxan-2,5-dione. Uses are in textile-processing and cleaning.

2-hydroxyethylamine. *See* ETHANOL-AMINES.

1-hydroxyketones. *See* ACYLOINS.

hydroxylamine. *See* NITROGEN HYDRIDES.

hydroxylases. *See* ENZYMES.

hydroxylation. Introduction of the HYDROXYL GROUP into a molecule.

hydroxyl group. The group –OH. It is attached to a carbon atom in ALCOHOLS and in CARBOXYLIC ACIDS.

hydroxyl ion (OH^-). Ion formed as the result of the removal of a proton from water. It is the anion found in HYDROXIDES and the basis of an early definition of a BASE. *See also* BRØNSTED–LOWRY ACIDS AND BASES.

4-hydroxy-3-methoxybenzaldehyde. *See* VANILLIN.

4-hydroxy-3-methoxycinnamyl   alcohol. *See* CONIFERYL ALCOHOL.

2-hydroxy-1-(1-methylethyl)-4-methylbenzene (thymol).   *See* PHENOLS.

4-hydroxy-4-methyl-2-pentanone (diacetone alcohol, $(CH_3)_2C(OH)CH_2COCH_3$). Liquid (bp 164°C), prepared by the condensation of PROPANONE in the presence of barium hydroxide. A molecule of water may be eliminated to form 4-hydroxy-4-methylpent-3-en-2-one when heated with acid or iodine. It is oxidized by bromine in sodium hydroxide to 3-hydroxy-3-methylbutanoic acid in a haloform reaction (*see* TRIHALOMETHANES).

hydroxynaphthalenes   (naphthols, $C_{10}H_8O$). 1- and 2-hydroxynaphthalenes present in coal tar, are prepared by alkaline fusion of the corresponding naphthalenesulphonic acid. They are solids (mp 96 and 122°C, respectively) with a faint phenolic odour, sparingly soluble in water but soluble in alkali solutions giving naphthoxides. They are oxidized ultimately to 1,2-benzenedicarboxylic acid and reduced (Na + 3-methyl-1-butanol) to tetrahydro-1- (or tetrahydro-2-) hydroxynaphthalene. They react with nitrous acid giving nitroso compounds (monoximes of quinones), are nitrated to the 4- and 1-nitro compounds, respectively, and couple with diazonium salts to give azo dyes. Direct sulphonation leads to a range of mono-, di- and trisulphonic acids which form the bases of intermediates for solubilized dyes. 2-Hydroxynaphthalene is used in the manufacture of antioxidants for the synthetic rubber industry; some 2-hydroxynaphthalene ethers (nerolins) are used in perfumery.

4-hydroxyphenylglycine. *See* AMINO-PHENOLS.

hydroxyproline (Hyp).   *See* AMINO ACIDS.

2-hydroxypropane-1,2,3-tricarboxylic acid (citric acid, II). Acid (mp 153°C (anhydrous)), found anhydrous in citrus fruits (lime, lemon, orange, etc.) and produced by the fermentation of sugars by various moulds and fungi. It is synthesized from glycerol (1,2,3-trihydroxypropane, I) (*see* Figure). Citric acid acts as a tribasic acid,

$$\begin{array}{c} CH_2OH \\ | \\ CHOH \\ | \\ CH_2OH \end{array} \xrightarrow{HCl} \begin{array}{c} CH_2Cl \\ | \\ CHOH \\ | \\ CH_2Cl \end{array} \xrightarrow{[O]} \begin{array}{c} CH_2Cl \\ | \\ CO \\ | \\ CH_2Cl \end{array} \xrightarrow{HCN} \begin{array}{c} CH_2Cl \\ | \\ C(OH)CN \\ | \\ CH_2Cl \end{array} \xrightarrow{KCN} \begin{array}{c} CH_2CN \\ | \\ C(OH)CN \\ | \\ CH_2CN \end{array} \xrightarrow{H^+} \begin{array}{c} CH_2COOH \\ | \\ C(OH)COOH \\ | \\ CH_2COOH \end{array}$$

(I)   (II)

2-hydroxypropane-1,2,3-tricarboxylic acid
Synthesis from glycerol.

forming salts with metals, and as an alcohol. When heated to 170°C a molecule of water is eliminated to give propene-1,2,3-tricarboxylic acid (aconitic acid). On pyrolysis a number of products are formed including 3-oxobutanoic acid. Fuming sulphuric acid yields 3-OXOPENTANEDIOIC ACID. Citric acid is used extensively in the food industry (E330).

2-hydroxypropanoic acid (lactic acid, $CH_3CH(OH)COOH$). Acid formed in the fermentation of sugars (D-form from lactose, *see* DISACCHARIDES), and in the breakdown of carbohydrates in muscles (L-form). It is prepared by the hydrolysis of cyanoethanal. It is a typical α-hydroxyacid. The acid gives 2-OXOPROPANOIC ACID with FENTON'S REAGENT. Three optical ISOMERS are known: L-(+) (mp 26°C); D-(−) (mp 26°C); and DL (mp 18°C). It is used as an additive in the food industry (E270).

hydroxypropanone (acetol, $CH_2(OH)-COCH_3$). Simplest hydroxyketone; liquid (bp 145°C), prepared by heating bromopropanone with potassium hydroxide in methanol and then adding ethyl methanoate. As a α-hydroxyketone it reduces TOLLEN'S REAGENT, FEHLING'S SOLUTION giving DL-2-HYDROXYPROPANOIC ACID and methanoic and ethanoic acids, and forms an OSAZONE with phenylhydrazine.

hydroxyquinol. *See* TRIHYDROXYBENZENES.

8-hydroxyquinoline (oxine, C_9H_7ON). Brown crystalline compound (mp 76°C), obtained by the alkali fusion of quinoline-8-sulphonic acid formed from QUINOLINE (H_2SO_4, 220°C). It is insoluble in water, soluble in ethanol, forms insoluble chelates with many metals and is widely used in gravimetric analysis. The copper derivative is used as a fungicide.

hydroxysuccinic acid (malic acid). *See* HYDROXYBUTANEDIOIC ACID.

hydroxytoluenes (cresols, methylbenzenols. $CH_3C_6H_4OH$). Components of the middle and heavy fractions of COAL TAR; colourless steam volatile liquids, (mp, bp: *o*- 31, 191°C; *m*- 11.5, 202.2°C; *p*- 34.8, 201.9°C). Pure isomers are prepared from the corresponding sulphonic acid by alkaline fusion or from the aminotoluene via the diazo compound. All are reduced to toluene with zinc dust. They are used as the starting materials for pesticides, herbicides, disinfectants and plasticizers (TRICRESYL PHOSPHATE).

hygroscopic. Referring to a substance that absorbs water from the atmosphere (e.g., NaOH, $MgCl_2$, P_2O_5). Hygroscopic substances are often used as desiccants. *See also* DELIQUESCENCE.

hyoscine. *See* ALKALOIDS.

hyoscyamine (atropine). *See* ALKALOIDS.

Hyp (hydroxyproline). *See* AMINO ACIDS.

hyperchromic effect. *See* AUXOCHROME.

hyperconjugation. 'No-bond resonance', which occurs when a $>CH_2$ or $-CH_3$ group is adjacent to a multiple bond. Propene is stabilized by resonance of this kind (*see* Figure). The hydrogen atoms are not free; the effect is to increase the ionic character of the carbon–hydrogen bond, the electrons of which become partially delocalized through conjugation. From the molecular orbital point of view, π-orbitals overlap with π-orbitals to produce conjugation. π-Orbitals, however, can overlap to a certain extent with adjacent σ-orbitals to form extended orbitals; this is hyperconjugation. Various physical data have been explained

$$\underset{\underset{H}{|}}{\overset{\overset{H}{|}}{H-C-CH=CH_2}} \longleftrightarrow \underset{\underset{H}{|}}{\overset{\overset{H^+}{\cdot\cdot^-}}{H-C-CH=CH_2}} \longleftrightarrow \underset{\underset{H}{|}}{\overset{\overset{H^+}{|}}{H-C=CH-\bar{C}H_2}} \longleftrightarrow \cdots$$

hyperconjugation
Stabilization of propene.

by hyperconjugation (e.g., bond lengths, the CH_3–C bond length in CH_3–C≡CH is shortened and has partial double bond character, and the greater stability of propene than is 'expected').

hyperfine structure. *See* ELECTRON SPIN RESONANCE SPECTROSCOPY.

hyperpolarizability (β). POLARIZABILITY (α) of a molecule, in general, depends on its orientation. In a very strong field, as found in laser beams, the induced moment is proportional to the square of the field strength (βE^2); the proportionality constant is the hyperpolarizability.

hypertonic. Referring to a solution which has a higher OSMOTIC PRESSURE than another solution.

hypo. Common name for sodium thiosulphate (*see* SULPHUR OXOACIDS).

hypobromites. *See* BROMATES.

hypochlorites. *See* CHLORATES.

hypochromic shift. *See* AUXOCHROME.

hypoiodites. *See* IODATES.

hyponitric acid. *See* NITROGEN OXOACIDS.

hyponitrous acid. *See* NITROGEN OXO-ACIDS.

hypophosphoric acid. *See* PHOSPHORUS OXOACIDS.

hypophosphorous acid. *See* PHOSPHORUS OXOACIDS.

hyposulphurous acid (dithionous acid). *See* SULPHUR OXOACIDS.

hypotonic. Referring to a solution which has a lower OSMOTIC PRESSURE than another solution.

hypoxanthine. *See* PURINES.

hypsochromic shift. *See* AUXOCHROME.

hysteresis. Phenomenon mostly of solids and surfaces in which a measured property has different values on increasing an independent variable from those when decreasing it. Ferromagnets (*see* FERRO-MAGNETISM) and FERRITES show magnetic hysteresis between the susceptibility and the external field. Hysteresis is also shown by the adsorption of water by textiles and the adsorption of gases by porous solids.

Hz. *See* HERTZ.

I

I. *See* IODINE.

I. *See* CURRENT, IONIC STRENGTH, ION-
IZATION ENERGY, MOMENT OF INERTIA.

i. Current density (*see* CURRENT); VAN'T
HOFF FACTOR.

***i*₀.** *See* EXCHANGE CURRENT DENSITY.

IAA (β-indoylacetic acid). *See* AUXINS.

IAN (β-indoylacetonitrile). *See* AUXINS.

ice. Solid WATER. The hydrogen-bonded
(*see* HYDROGEN BOND) structure formed at
0°C and 1 atm is similar to that of diamond
(*see* CRYSTAL FORMS), with tetrahedrally
coordinated oxygen atoms. At higher
pressures, more dense structures are
formed with fewer hydrogen bonds.

iceland spar. *See* CALCIUM.

ice-point. Temperature at which ice and
water are in equilibrium at standard atmospheric pressure; a fixed point on the
Celsius scale. *See also* TEMPERATURE
SCALES.

icosahedral coordination. *See* SHAPES OF
MOLECULES.

ICP. *See* INDUCTIVELY COUPLED PLASMA.

ICR. *See* ION CYCLOTRON RESONANCE.

ICSH (interstitial cell-stimulating hormone). *See* HORMONES.

IDA. *See* ISOTOPE DILUTION ANALYSIS.

ideal gas (perfect gas). Hypothetical gas
which obeys the GAS LAW exactly. An ideal
gas would consist of molecules of negligible
volume and have negligible forces between

them; all collisions between molecules and
between molecules and the walls of the
container would be perfectly elastic. In
practice, no gas shows ideal behaviour,
although helium, hydrogen and nitrogen
approximate to ideality at high temperatures and low pressures.

ideal solution. Characterized by: (1) a
vapour pressure–concentration curve
which conforms to RAOULT'S LAW; (2) zero
heat of mixing (i.e. $\Delta H_{\text{mixing}} = 0$); and (3) a
solution volume that is the sum of the
component volumes

$$\Delta G_{\text{mixing}} = x_A RT \ln x_A + x_B RT \ln x_B$$

$$\Delta S_{\text{mixing}} = -x_A R \ln x_A - x_B R \ln x_B$$

For a given mole fraction $\Delta G_{\text{mixing}} < 0$ and
$\Delta S_{\text{mixing}} > 0$ independent of the nature of the
components. The driving force for mixing is
the entropy factor rather than the enthalpy
factor. Although no solution is completely
ideal, some (e.g., benzene and toluene)
with similar molecules conform to ideality
reasonably well.

idose. *See* HEXOSES.

***i*₍L₎.** *See* LIMITING CURRENT DENSITY.

Ile (isoleucine). *See* AMINO ACIDS.

Ilkovic equation. *See* POLAROGRAPHY.

ilmenite. *See* TITANIUM.

imidazole (1,3-diazole, $C_3H_4N_2$). Five-
membered nitrogen HETEROCYCLIC COM-
POUND (structure VIII); solid (mp 90°C)
prepared from ethanedial (CHOCHO),
methanal and ammonia. It is related to
pyrrole, but is more stable towards oxidation, reduction and acids, and is a weak base
($pK_a = 6.9$). The most important derivative
is histidine (*see* AMINO ACIDS).

imides. Organic compounds containing the imido group –CONHCO–, formed by the reaction between dibasic acids or their anhydrides and ammonia. Imides are weakly acidic and are used in synthesis. *See also* BUTANIMIDE.

imines. Organic compounds containing the imino group =NH or –NH– in a cyclic compound. Imino complexes with transition metal ions may be formed by the reaction of an amine with an aldehyde complex. *See also* SCHIFF'S BASES.

imino group. The group =NH or –NH– in a ring.

immiscible liquids. Pairs of liquids for which the mutual solubilities are so small as to be insignificant. Two layers exist at all temperatures, and each component exerts its own vapour pressure independently of the presence of the other. At a given temperature, the total vapour pressure is constant for mixtures of all compositions. Assuming ideal behaviour

$$\frac{\text{mass of A in vapour}}{\text{mass of B in vapour}} = \frac{w_A}{w_B} = \frac{p_A M_{r,A}}{p_B M_{r,B}}$$

If a mixture is heated until the total pressure is atmospheric, distillation occurs at a temperature below the boiling point of either component; the relative masses in the vapour phase are given by the above equation. When water is one component, the process is known as steam distillation; this enables compounds of high molecular mass to be distilled over in relatively large amounts despite the fact that they have a low vapour pressure at the temperature of distillation (e.g., aniline/water).

imperfect gas. *See* IDEAL GAS.

IMS. *See* INDUSTRIAL METHYLATED SPIRITS.

incandescence. Emission of light from a hot body. Oxides of transition metals readily become incandescent. A mixture of thoria (98 percent) and ceria (2 percent) are the principle constituents of the Welsbach gas mantle.

inclusion complexing. Modified HIGH-PERFORMANCE LIQUID CHROMATOGRAPHY method which, although essentially chemical in nature, depends on all aspects of the structure, functionality, shape and chirality of the solute. An inclusion complex is formed when the hydrophobic part of a solute molecule is attached (dipole–dipole interaction, hydrogen bonding, etc., *see* INTERMOLECULAR FORCES) to the inner surface of a β-CYCLODEXTRIN molecule. The strength of bonding is governed primarily by the ability of the solute molecule to fit into the cavity of the cyclodextrin molecule. The stationary phase consists of cyclodextrins bonded to 5-μm silica beads. Typical mobile phases include methanol/water and ethanenitrile/water; the organic molecule in the solvent competes with the solute molecule in the cavity, an increase in the concentration of the organic solvent leads to a lower retention of the solute. The method can be used to separate enantiomers, diastereoisomers, structural and positional isomers, and polycyclic hydrocarbons.

incongruent melting point. *See* MELTING POINT, TWO-COMPONENT CONDENSED SYSTEM.

indan (2,3-dihydroindene, C_9H_{10}). Liquid (bp 176.5°C), product of the reduction of INDENE with sodium and ethanol.

indene (benzocyclopentadiene, C_9H_8). POLYCYCLIC HYDROCARBON (structure XXVIII), obtained from the light oil fraction of COAL TAR; colourless liquid (bp 181°C) which readily polymerizes. It is reduced (Na + EtOH) to INDAN (2,3-dihydroindene), oxidized (CrO_3) to homophthalic acid and combines with halogens to form 2,3-dihalogenoindene. With aldehydes and ketones in the presence of alkali highly coloured benzofulvenes (*see* CYCLO-ALKENES) are formed.

independent migration of ions. *See* MOLAR IONIC CONDUCTIVITY.

indican. Glucoside of INDOXYL obtained from the plant *Isatis tinctora*; the natural source of INDIGOTIN.

indicator electrode. Electrode, the potential of which is determined by the concentration of electroactive species in, for example, a TITRATION. *See also* REFERENCE ELECTRODE.

indicators. Substances that indicate, by a change in their colour, the presence or absence or concentration of some other substance or ion. More specifically indicators are used to mark a precise stage in a chemical reaction (e.g., the EQUIVALENCE POINT in a titration). Indicators are available for a range of different titrations (*see* Table). Universal or multiple range (acid–base) indicators (as solution or papers), obtained by mixing a number of indicators, give colour changes over a wide range of pH values. They are only useful for the determination of an approximate pH. Some titrations are self-indicating (e.g., the disappearance of the colour in titrations involving permanganate or iodine).

indifferent electrolyte. Added constituent to an electrolyte solution that takes no part in the electrode processes under study. Its functions are to reduce the resistance of the solution and, by carrying almost all the current, to ensure that the electroactive constituents (i.e. those taking part in the electrode reactions) reach the electrode surfaces by diffusion and not by electrolytic transport.

indigo. *See* INDIGOTIN.

indigotin (indigo). Important and long-known dyestuff (woad), a dark blue powder with a copper lustre; originally obtained from INDICAN by acid hydrolysis to give INDOXYL followed by atmospheric oxidation. It is now synthesized from 2-amino-benzoic acid and chloroethanoic acid via indoxyl. It is insoluble in water, but when the paste is agitated with sodium dithionate

the insoluble indigotin is reduced to the soluble, colourless *leuco*-base (*see* Figure). The material to be dyed is soaked in this solution and exposed to the air when the blue dye is regenerated in the cloth. Two geometric isomers are possible; the *trans*-form is the more stable.

indium (In). Group IIIb metal (*see* GROUP III ELEMENTS) found in small quantities associated with zinc, from which it is recovered by electrolysis of acid solutions. Indium is a soft metal which oxidizes slowly in air, dissolves in acids and reacts with halogens, sulphur, selenium and phosphorus. It is used in low-melting alloys, in electronic devices and in glass-sealing alloys.
$A_r = 114.82$, $Z = 49$, ES $[Kr]4d^{10}5s^25p^1$, mp 157°C, bp 2080°C, $\rho = 7.3 \times 10^3\,kg\,m^{-3}$. Isotopes: ^{113}In, 4.2 percent; ^{115}In, 95.8 percent.

indium compounds. Indium is an ELECTROPOSITIVE metal ($E^{\ominus}_{In^{3+},In} = -0.342$ V) which forms compounds of In^{3+}, which are similar to those of Al^{3+}, and of In^+, which disproportionate in aqueous solution.
indium inorganic compounds. Yellow indium oxide (In_2O) is formed by burning the metal in air or heating indium hydroxide ($In(OH)_3$). It readily oxidizes to In_2O_3. As with aluminium and gallium the trihalides, with the exception of the trifluoride, are known and are dimeric in the gas phase. Complexes are formed (e.g., $InCl_2$ is $In^I[In^{III}Cl_4]$). In_2Cl_3, In_4Cl_5 and In_4Cl_7 are also known. The trihalides hydrolyze in water to species such as $[In(OH)]^{2+}$, $[In_2(OH)_4]^{4+}$ and $[InCl_n]^{3-n}$, where $n = 1$ to 7. They are Lewis acids forming 1:2 adducts (*see* LEWIS ACIDS AND BASES).
indium organic derivatives. Indium alkyls and aryls, also Lewis acids, are prepared by reaction of indium with organomercury

Indigo
(water-insoluble)

Indigo white (*leuco*-base)
(water-soluble)

indigotin

indicators
Types of indicators.

Type	Description	Application/comments	Example
Acid–base	Weak organic acid or base, ionized and unionized forms have different colours $HIn \rightleftharpoons H^+ + In^-$	pH at which colour changes depends on pK_{In}. Useful range of indicator $pK_{In} \pm 1$, the equivalence point of titrations should be in this range	Methyl red, pK_{In} 5.1; cresol red, pK_{In} 7.9; phenolphthalein, pK_{In} 9.6
Redox (oxidation–reduction)	Substance that undergoes a definite colour change during its reversible oxidation or reduction $In_{ox} + ne \rightleftharpoons In_{red}$	Potential at which colour change depends on $E_{In}^{\ominus}$, useful range $E_{In}^{\ominus} \pm 0.0591/n$, the equivalence point of the titration should be in this range	1,10-Phenanthroline iron(II) complex, $E_{In}^{\ominus} = 1.11$ V (acid solution); methylene blue, $E_{In}^{\ominus} = 0.52$ V
Metallochromic (complexometric)	Multidentate chelating agent that exhibits different colours in the metallized and non-metallized forms $M^{2+} + HIn^{(m-1)} \rightleftharpoons MIn^{(m+1)} + H^+$ During most of titration indicator is in metallized form, only free M^{2+} being titrated with chelate (e.g., EDTA). At equivalence point chelate removes metal ion from dye complex and colour changes	Stability of metal chelate complex must be greater than that of the metal–indicator complex. pH of titrand must be carefully controlled	Erichrome Black T for Pb, Mg, Zn, pH controlled. Calgon for Ca
Precipitation (1) Adsorption	Indicator functioning by adsorption onto the surface of a precipitate. During the addition of $AgNO_3$ to NaCl, the AgCl is negatively charged, when there is excess Ag^+ the precipitate becomes positively charged, the indicator adsorbed and precipitate changes colour	Precipitate must be in colloidal form, the indicator ion must be of opposite sign to the ion of the precipitating agent, and must only be adsorbed immediately after complete precipitation	Acid dyes such as eosin, fluorescein and their derivatives
(2) Coloured precipitate	Substance that forms a coloured precipitate at the end-point (e.g., the use of Na_2CrO_4 in silver nitrate titrations)	Solubility product of the coloured precipitate (Ag_2CrO_4) must be less than that of the silver halide. Titration is in neutral solution	Specific example
(3) Coloured solution	Substance that forms a soluble coloured compound at the end-point (e.g., Volhard titration)	Titration of $AgNO_3$ (acid) with thiocyanate with iron(III) sulphate as indicator. Bright red coloration due to iron thiocyanate complex	Specific example

compounds. Solid trimethylindium has a tetrameric structure with unsymmetrical methyl–indium bonds. Indium–transition element bonds exist in compounds such as $In[Co(CO)_4]_3$.

INDO (intermediate neglect of differential overlap). *See* MOLECULAR ORBITAL CALCULATIONS, SEMIEMPIRICAL MOLECULAR ORBITAL THEORY.

indole (2,3-benzpyrrole, 1-benzo[*b*]-pyrrole, C_8H_7N). Nitrogen HETEROCYCLIC COMPOUND (structure XXVII), in which benzene is fused to pyrrole. It occurs in coal tar, various plants and in faeces (from bacterial degradation of tryptophan). The colourless crystals (mp 52°C), with a faecal odour at high doses, are prepared by the dehydration of *N*-methanoyl-2-amino-benzene on heating with a *tert*-butoxide. The FISCHER INDOLE SYNTHESIS is the most general method of forming the indole ring system. Like pyrrole, indole is neutral. It forms a potassium salt with the metal and reacts readily with electrophilic reagents. The 3-position is sufficiently nucleophilic to undergo the MANNICH REACTION. In very dilute solutions, indole has a pleasant odour and is used in perfumery.

indole-3-ethanoic acid (heteroauxin, $C_{10}H_9NO_2$). Solid (mp 169°C), obtained by the oxidative deamination of tryptophan (*see* AMINO ACIDS). Decarboxylation yields skatole (3-methylindole). It occurs in plants where it functions as a growth hormone (i.e. an AUXIN).

INDOR (internuclear double resonance). *See* NUCLEAR MAGNETIC RESONANCE SPECTROSCOPY.

indoxyl. Keto form (I) of 3-hydroxyindole (*see* Figure); yellow crystalline solid (mp 85°C) occurring in woad as the glucoside indican and in urine as a result of bacterial

decomposition of tryptophan (*see* AMINO ACIDS). There are several methods of synthesis, usually starting from aniline or 2-aminobenzoic acid and chloroethanoic acid. It is readily oxidized in alkaline solution to INDIGOTIN.

β-indoylacetic acid. *See* AUXINS.

β-indoylacetonitrile. *See* AUXINS.

induced dipole. *See* INTERMOLECULAR FORCES.

induced dipole–induced dipole interaction. *See* INTERMOLECULAR FORCES.

inductive effect. Effect that results when atoms of different electronegativities are bonded together; thus, since chlorine has a greater electronegativity than carbon, the electron pair in a carbon–chlorine covalent bond will be displaced towards the chlorine, resulting in the formation of a permanent dipole. This type of electron displacement along a carbon chain decreases rapidly as the distance from the source increases. Electron-withdrawing groups (*see* DEACTIVATING GROUP) (I–; e.g., –NO_2, –CN, –CHO, –COOH, –Cl, –Br) substituted in a benzene ring reduce the electron density on the ring and decrease its susceptibility to further ELECTROPHILIC SUBSTITUTION. Electron-releasing groups (I+; e.g., –OH, –NH_2, –OCH_3, –CH_3) have the opposite effect. This effect is a useful concept in explaining certain aspects of organic reactions and the relative strengths of acids and bases.

inductively coupled plasma (ICP). PLASMA produced by passing a radiofrequency discharge through an inert gas at atmospheric pressure. A sample solution is nebulized and passed in a nitrogen or argon atmos-

(I)

indoxyl

phere through a tube which is heated by radiofrequency radiation. The plasma has a temperature of 6000 K and may be used instead of a flame in the analysis of the sample by ATOMIC SPECTROSCOPY.

industrial methylated spirits (IMS). Solvent; 95 percent ethanol plus additives to render it non-potable.

inelastic neutron scattering. Slow NEUTRONS with kinetic energy that is comparable with thermal energy levels in solids are inelastically scattered by phonons. The energies of the scattered neutrons give information about interatomic forces and vibrational modes.

inert gases. *See* GROUP O ELEMENTS.

inert pair effect. Tendency of heavier elements of the main groups II, III, IV and V (*see* PERIODIC TABLE) to form compounds in oxidation states two less than the common oxidation state for that group. Thus mercury is difficult to ionize at all, and thallium(I), lead(II) and bismuth(III) are found. This effect is not due solely to the difficulty of removing *s*-electrons, but rather is due to the decreasing strengths of bonds formed descending a group.

infinite dilution. Term applied to a hypothetical limiting electrolyte having zero concentration of ions (i.e. an infinite dilution of an inert solvent). The MOLAR CONDUCTIVITY at infinite dilution (Λ^0, also Λ^∞) is the maximum conductivity attained by an electrolyte when the ions are free from intermolecular forces, and it is fully dissociated. *See also* CONDUCTANCE EQUATIONS, OSTWALD'S DILUTION LAW.

infrared radiation. *See* ELECTROMAGNETIC SPECTRUM.

infrared spectroscopy (IR spectrophotometry). Measurement of absorption of radiation in the range 1–300 μm (mostly 2.5–25 μm) caused by the excitation of molecular rotations or vibrations. The SELECTION RULE for IR activity requires a change of dipole on absorption. (For instrumental design—*see* SPECTROPHOTOMETER.) For simple molecules, rotational fine structure may be observed (*see* VIBRATION–

ROTATION SPECTRA). The IR spectra of polyatomic molecules is discussed in terms of NORMAL MODES and is usually too complex to determine from a given structure. However certain groups (e.g., carbonyl) have vibrational modes at characteristic frequencies whatever the molecule they are in. Thus the IR (vibrational) spectrum of an organic molecule may give an indication of the groups it contains, but also the spectrum as a whole provides a unique record of that molecule. The near IR region (1–3 μm) excites overtones of vibrations and is used in quantitative analysis of different functional groups. Restricted rotation and ring puckering in closed molecules are studied in the far IR region (beyond 25 μm). IR reflectance methods are used to investigate molecules adsorbed on surfaces (*see* ATTENUATED TOTAL REFLECTANCE).

inhibition. Retardation of a reaction that may be catalytic (*see* CATALYSIS). Important inhibition processes are the use of phenols and quinones in POLYMERIZATION and in CORROSION. Autoinhibition (allylic termination) is exhibited in the polymerization of monomers that contain reactive allylic hydrogen atoms. Oxygen, iodine and nitrogen monoxide are also FREE RADICAL inhibitors. *See also* QUENCHING.

initiation. *See* CHAIN REACTION, FREE RADICAL.

inner potential. *See* INTERFACIAL POTENTIAL.

inosine. *See* NUCLEOSIDES.

5′-inosinic acid. *See* NUCLEOTIDES.

inositol (vitamin B complex). *See* VITAMINS.

insecticide. Pesticide used to control insect life. General types are: (1) inorganic, including arsenic, lead and copper compounds and mixtures; (2) natural organic compounds such as ROTENONE, pyrethrins, nicotine (*see* ALKALOIDS); (3) synthetic organic compounds: (a) chlorinated hydrocarbons such as DDT (*see* 1,1,1-TRICHLORO-2,2-BIS(4-CHLOROPHENYL-ETHANE)), dieldrin, 4-dichlorobenzene, (b) organic esters of phosphorus such as para-thion $((C_2H_5O)_2P(S)OC_6H_4NO_2)$. Most are toxic to humans. The organic phosphorus types are biodegradable, whereas the chlorinated hydrocarbons resist biodegradation.

instability constant. *See* STABILITY CONSTANT.

insulator. Substance that does not conduct electricity. In practice this includes materials with conductivities of less than 10^{-4} S^{-1} m^{-1}. In terms of the BAND THEORY OF SOLIDS, insulators arise from a full valence band and empty conduction band separated by an energy gap which is large compared to the value of kT. Insulators include pure ionic solids, fully covalently bonded crystals (e.g., diamond) or molecular crystals in which there is no overlap between adjacent molecules (e.g., solid benzene). *See also* METAL, SEMICONDUCTOR.

insulin. *See* HORMONES.

integrated rate equations. Simple forms of the RATE EQUATION may be integrated to give an expression for the concentration of a species with time (*see* Table).

intensive property (intensity factor). Property, the value of which is independent of the amount of material in the system (e.g., density, pressure, temperature); such properties are not additive. In a SPONTANEOUS PROCESS, the intensive property always decreases. *See also* EXTENSIVE PROPERTY.

interatomic distance. *See* BOND LENGTH.

integrated rate equations

Reaction	Rate equation[a]	Integrated rate equation[a]
$A \xrightarrow{k_0} P$	$-d[A]/dt = k_0$	$[A] = [A_0] - k_0 t$
$A \xrightarrow{k_1} P$	$-d[A]/dt = k_1[A]$	$[A] = [A_0] \exp(-k_1 t)$
$A \xrightarrow{k_2} P$	$-d[A]/dt = k_2[A]^2$	$[A] = [A_0]/(1 + k_2[A_0]t)$
$A + B \xrightarrow{k_2} P$	$-d[A]/dt = k_2[A][B]$	$k_2 t = [1/([B_0] - [A_0])] \times$ $\ln([A](B_0)/[B][A_0])$
$A \xrightarrow{k_1} B \xrightarrow{k_1'} C$	$-d[A]/dt = k_1[A]$	$[A] = [A_0] \exp(-k_1 t)$
	$d[B]/dt = k_1[A] - k_1'[B]$	$[B] = [A_0][k_1/(k_1 + k_1')] \times$ $[\exp(-k_1 t) - \exp(-k_1' t)]$
	$d[C]/dt = k_1'[B]$	$[C] = [A_0] - [A] - [B]$
$A \underset{k_{-1}}{\overset{k_1}{\rightleftharpoons}} B$	$-d[A]/dt = k_1[A] - k_{-1}[B]$	$[A] = \{[A_0]/(k_1 + k_{-1})\} \times$ $\{k_{-1} + k_1 \exp[-(k_1 + k_{-1})t]\}$
	$d[B]/dt = -d[A]/dt$	$[B] = [A_0] - [A]$
$A \xrightarrow{k_1} B$ $\searrow^{k_1'}$ C	$-d[A]/dt = (k_1 + k_1')[A]$	$[A] = [A_0] \exp[-(k_1 + k_1')t]$
	$d[B]/dt = k_1[A]$	$[B] = [A_0][1 - \exp(-k_1 t)]$
	$d[C]/dt = k_1'[A]$	$[C] = [A_0][1 - \exp(-k_1' t)]$

[a] $k_n = n$th order rate constant, $[A_0]$ = initial concentration of A.

intercalation compound. Compound in which extra atoms, molecules or ions are introduced into an existing structure. Graphite (*see* CARBON) and titanium disulphide have lamellar lattices which may accommodate, for example, fluorine, potassium, sulphuric acid and iron(III) chloride between the layers. Sodium β-alumina is a sodium ion conductor (*see* SOLID ELECTROLYTE) in which sodium ions are intercalated between aluminosilicate layers. ZEOLITES have three-dimensional structures capable of intercalating a range of foreign atoms. Intercalation compounds may be prepared by solid-state reaction or by electrolysis from a melt.

interface. Region between two homogeneous phases. When one of the phases is a gas or vapour, this region is referred to as a surface. For some purposes an interface may be considered as an infinitely thin lamina, but the transition of properties from one phase to the other may extend over several nanometres when the region is more correctly called an interphase.

interfacial potential (Galvani potential, ϕ). Work required to move unit charge from infinity to the interior of the phase. The potential difference across an electrode–electrolyte interface is the difference between the Galvani potentials (i.e. $\Delta\phi = \phi_{solid} - \phi_{solution}$). The interfacial potential is the sum of the surface potential (work done crossing the surface dipole layer) and the outer, or Volta, potential (work done to bring a charge to just outside the range of image forces).

interfacial tension. *See* SURFACE TENSION.

interferon. Low-molecular-mass protein, synthesized by animal cells in response to viral infection. It specifically inhibits replication of the virus, and plays a significant role in recovery. Human interferon is produced by gene cloning in suitable bacterial hosts.

interhalogen compounds. Compounds formed by direct combination of the halogens; the product depends on the conditions (e.g., chlorine with fluorine in equal volumes gives ClF, but with excess fluorine ClF_3 is formed). For every class the boiling point increases as the difference between the electronegativities of the two halogens increases. Compounds with three or four halogens are not known, although POLYHALIDE IONS exist.

Compounds of type AX resemble the halogens in physical properties; the more polar the bond the greater the thermal stability of the molecule. AX-type compounds are more reactive than the halogens, they convert metals to mixed halides, are hydrolyzed by water, and form addition compounds with alkenes and some alkali metal halides (e.g., $NaIBr_2$).

Of the AX_3 compounds, ClF_3 is most reactive, but BrF_3 is more useful as a fluorinating agent (converting metals, oxides, chlorides and bromides to fluorides) and as a non-aqueous solvent which undergoes self-dissociation. Some metal fluorides dissolve in BrF_3 to give such compounds as $KBrF_4$ and $SbBrF_6$. ICl_3 is much less reactive; in the molten state it is a good conductor. Thermal stabilities are in the order $BrF_3 > ClF_3 > ICl_3$.

Of the AX_5 compounds, BrF_5 is most reactive, resembling ClF_3 in its violent reaction with water. Liquid IF_5 is a good electrical conductor which reacts with KI at its boiling point to give KIF_6.

IF_7, the only heptahalide, made by heating IF_5 with F_2 at 250°C, is comparable with ClF_3 and BrF_5 in its violent fluorinating action.

Several oxofluorides are known. ClO_2F ($BrF_3 + KClO_3$) is a very reactive compound which forms solid additives with BF_3 and SbF_5 regarded as chloronium salts $[ClO_2]^+[BF_4]^-$, $[ClO_2]^+[SbF_6]^-$. Perchloryl fluoride (ClO_3F, $KClO_4 + HFSO_4$) is a thermally inert and stable gas.

intermetallic compound. Material composed of two or more types of METAL atom that exists in distinct homogeneous phases with properties different from the individual metals. A wide range of crystal structures are found in intermetallic compounds. Many compounds are formed with one metal occupying interstitial positions in the close packed lattice (*see* CLOSE-PACKED STRUCTURES) of another. For example, the sodium chloride lattice (*see* CRYSTAL FORMS) is obtained by filling the octahedral holes in a face-centred cubic metal (e.g., CuSe, SrTe, UC). Other structures are determined by the relative sizes of the metal atoms or the electron-to-atom ratio (*see* HUME–ROTHERY'S RULE). Icosohedral co-ordination of large atoms about a smaller atom is also found (e.g., $MoAl_{12}$). Disordered structures are found in which lattice sites may be occupied by either metal. *See also* ALLOY.

intermolecular. Between molecules. *See also* INTRAMOLECULAR.

intermolecular forces. Weak forces occurring between molecules. Several types of attractive forces, which are due to electrostatic interaction, under the general name of van der Waals forces, are recognized.
dipole–dipole interaction. When two polar molecules are close to each other their dipoles interact giving rise to an attractive force.
dipole–induced dipole interaction (induction effect). The presence of a polar molecule in the vicinity of another molecule (polar or non-polar) has the effect of polarizing the second molecule. The induced dipole interacts with that of the first molecule giving rise to an attractive force.
induced dipole–induced dipole interaction (London or dispersion forces). Although non-polar molecules have no permanent dipoles, their electron clouds are fluctuating, giving rise to an instantaneous dipole moment which is continuously changing in magnitude and direction. If one such molecule suddenly acquires an electronic arrangement which gives it an instantan-

eous dipole, this can induce an instantaneous dipole in a second molecule leading to an attractive force between them. Dispersion forces play an important role in protein structure; tertiary structure is determined to a large extent by contact of non-polar groups.
The potentials (energy) vary as r^{-6} (*see* Table); only dipole–dipole interaction is temperature-dependent. In addition, other types of force operate in special cases.
ion–ion interaction This force is predominant in ionic crystals, molten metals, salts and ionic solutions; it is not important in non-polar liquids. Coulombic forces between ions are effective over longer distances than any other electrical force, they vary as r^{-2}.
ion–dipole interaction. For an ion and a dipolar molecule lying along the same axis the attractive force is proportional to r^{-4}.
ion–induced dipole interaction. If an ion is in the vicinity of a non-polar molecule (spherically symmetrical charge density), electrostatic interaction causes a redistribution of the charge density giving rise to an induced dipole which interacts with the ion.
hydrogen bonding. See HYDROGEN BOND.
See also LENNARD–JONES (12,6) POTENTIAL.

internal conversion. Process in which an electronically excited atom relaxes to its ground state, the energy released being used to eject another electron from an inner shell. The ion produced is in an excited state and may emit an X-ray photon or Auger electron (*see* AUGER EFFECT). *See also* ELECTRON SPECTROSCOPY.

intermolecular forces

Type	Energy of interaction[a]	Example	Magnitude/kJ mol^{-1}
Covalent bond	*	H–H	200–800
Ion–ion	$-Q_1Q_2/\varepsilon r$	Na^+Cl^-	40–400
Ion–dipole	$-Q\mu/\varepsilon r^2$	$Na^+(H_2O)_n$	4–40
Dipole–dipole	$-2(\mu_1\mu_2/4\pi\varepsilon_0)^2(1/3r^6kT)$	SO_2–SO_2	0.4–4
Dipole–induced dipole	$-2\alpha_2\mu_1^2/4\pi\varepsilon_0r^6$	HCl–C_6H_6	0.4–4
Dispersion	$-[3I_1I_2/2(I_1+I_2)]\alpha_1\alpha_2/r^6$	He–He	4–40
Ion–induced dipole	$-\alpha_2Q_1^2/\varepsilon r^4$	Na^+ $^-C_6H_6$	0.4–4
Hydrogen-bond	*	HF–HF	4–40

[a] Q = charge, μ = dipole moment, I = first ionization energy, α = polarizability, r = separation of the two species (1,2).

* No simple expression for this type of interaction.

internal energy (U; dimensions: $m\ l^2\ t^{-2}$; units: $J\,mol^{-1}$, $U = f(T,V)$). STATE FUNCTION. The value of U is determined by: (1) the number of molecules; (2) the kinetic, rotational and vibrational motion of the molecules; (3) the structure of the molecules; (4) the nature of the individual atoms; (5) the number and arrangement of the electrons; and (6) the nature of the nuclei. Monatomic gases possess translational or KINETIC ENERGY only ($3RT/2$), diatomic and polyatomic molecules may also possess rotational and vibrational energy, depending on the temperature ($RT/2$ per DEGREE OF FREEDOM). Absolute values of U cannot be obtained from classical thermodynamics alone.

When a quantity of heat q_V is transferred to a constant volume system: $q_V = \Delta U$. For an isothermal constant pressure system

$$q_P = \Delta U + P\Delta V = \Delta H$$

ΔU varies with temperature (*see* KIRCHHOFF'S EQUATION), and is related to the HELMHOLTZ FUNCTION, the HEAT CAPACITY and to the pressure and temperature.

$$(\partial U/\partial V)_S = -P$$

and

$$(\partial U/\partial S)_V = T$$

Values of U (total and for each degree of freedom) can be calculated from the PARTITION FUNCTION using spectroscopic data.

internal pressure. *See* VAN DER WAALS EQUATION.

internal reflectance spectroscopy. *See* ATTENUATED TOTAL REFLECTANCE.

internuclear double resonance (INDOR). *See* NUCLEAR MAGNETIC RESONANCE SPECTROSCOPY.

interphase. *See* INTERFACE.

interstitial compound. *See* INTERCALATION COMPOUND, SOLID SOLUTION.

interstitial volume. *See* VOID VOLUME.

intersystem crossing. Conversion of a molecule in a low vibrational level of an excited singlet state to a higher vibrational level of a triplet state (*see* ELECTRONIC SPECTROSCOPY, MULTIPLICITY). The process takes 10–100 ns. The decay of the long-lived triplet state gives rise to PHOSPHORESCENCE. *See also* FLUORESCENCE.

intramolecular. Within a molecule. An intramolecular hydrogen bond in 2-HYDROXYBENZOIC ACID stabilizes the anion and thus increases the strength of the acid. An example of an intramolecular reaction is the 1,2-shift of a methyl group and then a hydride ion in the reaction of 2,2-dimethyl-1-propanol with hydrogen bromide to give one product 1-bromo-2-methylpropane. *See also* INTERMOLECULAR.

intrinsic viscosity ($[\eta]$). *See* VISCOSITY.

Invar. ALLOY of nickel and iron.

invariant system. System in which there are sufficient algebraic equations to permit the complete definition of all variables. According to the PHASE RULE when $f = 0$, the system in equilibrium is completely defined. The invariant point for a ONE-COMPONENT SYSTEM occurs when three phases coexist in equilibrium (TRIPLE POINT), for a TWO-COMPONENT CONDENSED SYSTEM (at constant pressure) at the eutectic point and for a THREE-COMPONENT SYSTEM (at constant pressure) at the ternary eutectic point.

inversion. (1) Change of optical configuration (*see* STEREOCHEMISTRY, WALDEN INVERSION); (2) SYMMETRY OPERATION.

inversion temperature. *See* JOULE–THOMSON EFFECT.

invertase. *See* ENZYMES.

invert sugar. *See* SUCROSE.

in vitro. Describes biological processes that are made to occur outside the living organism (literally means 'in glass'). *See also IN VIVO*.

in vivo. Describes biological processes as they occur within the living organism.

iodates. Salts of the iodine oxoacids, they are all strong oxidizing agents. Iodates(III) are unknown (*compare* CHLORATES).

iodates(ı) (hypoiodites, IO⁻). Prepared by similar methods to the chlorates(ı) and with similar properties. The acid, HIO is very unstable.

iodates(v) (iodates, pyramidal IO_3^-). Alkali metal iodates when heated give periodate and iodide. Iodates are reduced by acid solutions of iodides to give free iodine. Iodic acid (HIO_3, I_2 + fuming HNO_3) is the only known halic acid to exist in the free state. Dehydration gives di-iodine pentoxide.

iodates(vıı) (periodates). They present a complicated picture on account of the stability of paraperiodic acid (H_5IO_6). Periodates are of four formula types: KIO_4 from metaperiodic acid (HIO_4); $Na_4I_2O_9$ from dimesoperiodic acid ($H_4I_2O_9$); $Pb_3(IO_5)_2$ from (hypothetical) meso-periodic acid (H_3IO_5); Ag_5IO_6 from para-periodic acid (H_5IO_6).

iodic acid.   *See* IODATES.

iodides.   *See* HALIDES.

iodimetry.   Use of iodine in titrations, generally titrated with sodium thiosulphate (I_2 reduced to I^-) using starch as indicator. It is used indirectly in estimation of oxidizing agents (e.g., H_2O_2, CIO^-, IO_3^-, Cu^{2+}, Cl_2) which liberate iodine from acidified potassium iodide solutions.

iodine (I).   GROUP VII ELEMENT, present (10^{-4} percent of the lithosphere) as iodide ions in brine and as iodates(v) in Chilean nitrate deposits. The element is obtained from iodides by displacement with chlorine and from iodates by reduction with sodium bisulphite. Iodine is a black solid with a metallic lustre, subliming to a violet vapour. It is insoluble in water, soluble in carbon disulphide and tetrachloromethane giving purple solutions and in unsaturated hydrocarbons, liquid sulphur dioxide, alcohols and ketones, where charge transfer occurs, giving brown solutions. It is less reactive than the other halogens and is the most electropositive (metallic) element in the group. It reacts with other halogens (*see* INTERHALOGEN COMPOUNDS) and with bases gives IODATES. Iodine exhibits oxidation states of −1, +1, +3, +5 and +7, showing some cationic character in the formation of oxosalts (e.g., IO_2SO_4). It is

required as a trace element in man, where it is concentrated in the thyroid gland. It is used as a mild antiseptic (tincture of iodine), in the manufacture of iodine compounds, photographic materials and quartz–halogen lamps.

$A_r = 126.9045$, $Z = 53$, ES [Kr] $4d^{10}5s^25p^5$, mp 113.5°C, bp 184.35°C, $\rho = 4.94 \times 10^3$ $kg\,m^{-3}$. Isotopes: ^{127}I only stable isotope, 14 radioisotopes.

iodine halides.   *See* INTERHALOGEN COMPOUNDS.

iodine lattice.   *See* CRYSTAL FORMS.

iodine number (iodine value).   Measure of the unsaturation in a fat or vegetable oil; the percentage, by weight, of iodine absorbed by an unsaturated substance under defined conditions.

iodine oxides.   The only true oxide of iodine, di-iodine pentoxide (I_2O_5, heat HIO_3) is a white polymeric powder decomposing to the elements above 300°C and hydrolyzed by water to iodic acid. It is a strong oxidizing agent.

iodine oxoacids.   *See* IODATES.

iodine value.   *See* IODINE NUMBER.

iodoethane.   *See* ALKYL HALIDES.

iodoform.   *See* TRIHALOMETHANES.

iodomethane (methyl iodide, CH_3I). Sweet-smelling ALKYL HALIDE (bp 42.5°C), prepared by heating methanol, red phosphorus and iodine. It is used as a methylating agent (*see* METHYLATION).

ion.   Atom or group of atoms that carry a positive (CATION) or negative (ANION) charge. *See also* IONIZATION.

ion association.   *See* ION PAIR.

ion chromatography.   Technique for the simultaneous determination of anions and cations, using in sequence a cation exchanger (*see* ION EXCHANGER) column, a conductivity detector, an anion exchanger column, an anion suppressor column and a detector (e.g., second conductivity detector, constant potential amperometric detector, optical detector). With a suitable eluant a range of anions and cations can be separated and detected. In modern instruments, alternating layers of high-capacity ion exchange screens and ultra-thin ion exchange membranes are used; this reduces the VOID VOLUME. The elutant and regenerant flow are countercurrent to maximize performance. The technique is of use in water analysis; the analysis time is considerably less than for the separate determinations and detection limits approach 1 ppm for most species.

ion cyclotron resonance (ICR). Ions (mass m, charge q) generated in a cubic cell of metal plates are confined to circular paths by an applied magnetic field (B) and repelling potentials on the plates at right angles to the field. The frequency (ω) of their orbit is qB/m. For $B = 1.2$ T, the mass range 10–1000 of M^+ corresponds to $\omega = 1.8$ MHz–18 kHz. Irradiation by an external electric field frequency ω causes absorption of energy in a manner similar to that in NUCLEAR MAGNETIC RESONANCE SPECTROSCOPY and ELECTRON SPIN RESONANCE SPECTROSCOPY. The technique may thus be used to mass analyze ions (see MASS SPECTROMETRY). Modern ICR spectrometers apply a pulse of radiofrequency radiation and Fourier transform the decaying induced currents on receiving plates in the cell. The long life-times of ions in the cell allows the investigation of ion–molecule reactions.

ion–dipole interaction. See INTERMOLECULAR FORCES.

ion exchange chromatography. Separations carried out on ion exchange resins (see ION EXCHANGER). Different ions in the sample are displaced from the appropriate ion exchanger one by one by changing the pH, ion type or concentration of the eluting liquid, or the temperature. Materials separated include simple inorganic and organic ions, amino acids and polyelectrolytes, such as proteins, nucleic acids.

ion exchanger. Synthetic, insoluble, cross-linked polymer carrying acidic or basic ionogenic groups that has high exchange capacity. Strong anion exchangers are quaternary amines) ($RN(CH_3)_3^+$ OH^-);

weak types contain secondary and tertiary amines (normally used in the Cl^- or OH^- form); strong cation resins contain sulphonic acid groups, whereas weak ones contain carboxyl groups (normally used in the H^+ form). The early naturally occurring ZEOLITES have been replaced by synthetic exchangers. Ion exchangers are used extensively in water treatment, extraction and analysis; a 'mixed bed' of anion and cation exchangers is used to produce 'ion-free' water (see DEIONIZED WATER).

ion exclusion. Process in which an ion exchange resin (see ION EXCHANGER) absorbs non-ionized solutes (e.g., sugar), but not ionized solutes present in solution. It is probably due to the Donnan membrane potential (see DONNAN MEMBRANE EQUILIBRIUM), whereby ionic solutes exist at higher concentration in the solution than in the resin, whereas non-ionic solutes are distributed uniformly between the solvent in the column and that within the resin matrix. As a result the ionic solute moves faster down an ion exchange column than does the nonpolar solute, leading to separation.

ionic atmosphere. See CONDUCTANCE EQUATIONS, ELECTRICAL DOUBLE LAYER.

ionic bond. Bond formed between ions in which an electron has been transferred completely from one atom to another (e.g., Na^+Cl^-).

ionic conductivity. See MOLAR IONIC CONDUCTIVITY.

ionic crystal. See CRYSTAL STRUCTURE.

ionic migration. See TRANSPORT NUMBER.

ionic mobility (u; units: m^2 V^{-1} s^{-1}). Velocity of an ion under a unit potential gradient. It is related to the MOLAR IONIC CONDUCTIVITY ($\Lambda = zuF$), the TRANSPORT NUMBER ($t_+ = u_+/(u_+ + u_-)$) and the diffusion coefficient ($D = uRT/zF$).

ionic product of water (K_w). EQUILIBRIUM CONSTANT of the AUTOPROTOLYSIS of water
$$H_2O \rightleftharpoons H^+ + {}^-OH$$

$$K_w = a_{H+}a_{-OH}$$

At 25°C $K_w = 1.0 \times 10^{-14}$ or $pK_w = 14$.

ionic radius
Ionic radii (/pm) calculated by Pauling (P) and Goldschmidt (G).

Ion	G	P	Ion	G	P	Ion	G	P
H^-	154	208	Li^+	68	60	B^{3+}	20	20
F^-	133	136	Na^+	98	95	Al^{3+}	45	50
Cl^-	181	181	K^+	133	133	La^{3+}	104	115
O^{2-}	145	140	Rb^+	148	148	C^{4+}	15	15
S^{2-}	190	184	Cs^+	167	169	Co^{2+}	70	72
			Ag^+	113	126	Fe^{2+}	76	75
			Be^{2+}	30	31	Ni^{2+}	68	69
			Mg^{2+}	65	65	Fe^{3+}	53	—
			Ca^{2+}	94	99	Ce^{4+}	87	101
			Ba^{2+}	129	135	Pb^{4+}	81	84

ionic radius. Effective size of an ion in a crystal. Ionic radii have been determined by Pauling who, using sodium chloride as standard, apportioned interionic distances (from X-RAY DIFFRACTION) of isoelectronic ions in the ratio of their EFFECTIVE NUCLEAR CHARGES (*see* Table). Goldschmidt also prepared a set of values from a wider set of data using a more empirical method.

ionic strength (I; units: $mol\,dm^{-3}$, $mol\,kg^{-1}$, etc.). Function which provides a method for the expression of the total ionic concentration for all types of electrolyte. For fully dissociated electrolytes

$$I = 1/2 \sum_i c_i z_i^2$$

or

$$I = 1/2 \sum_i m_i z_i^2$$

the summation being extended over all the different ionic species in solution.

ionic strength adjustor (isa). Concentrated solution of an indifferent electrolyte, added to a sample (or calibration) solution to give a solution of constant ionic strength. The ACTIVITY COEFFICIENT is thereby kept constant and the NERNST EQUATION can be used with concentrations instead of activities. It is used in potentiometric analytical methods.

ion–induced dipole interaction. *See* INTERMOLECULAR FORCES.

ion–ion interaction. *See* INTERMOLECULAR FORCES.

ionization. Gain or loss of an electron to create an ION. *See also* ELECTRON AFFINITY, IONIZATION ENERGY.

ionization energy (ionization potential, I). Energy required to remove an electron

ionization energy
First and second ionization energies of the first 18 elements in electronvolts.

H							He
13.599							24.588
—							54.418
Li	Be	B	C	N	O	F	Ne
5.392	9.323	8.298	11.260	14.53	13.618	17.423	21.565
75.641	18.211	25.38	24.38	29.602	35.118	34.98	40.964
Na	Mg	Al	Si	P	S	Cl	Ar
5.139	7.646	5.989	8.152	10.487	10.360	12.967	15.760
47.29	15.035	18.828	16.346	19.72	23.40	23.80	27.62

from a gaseous atom or ion. It is measured by photoelectron spectroscopy (*see* ELECTRON SPECTROSCOPY). The first and second ionization potentials for the first two rows of the periodic table are given in the Table. *See also* KOOPMANN'S THEOREM.

ionization isomerism. *See* ISOMERISM.

ionization potential. *See* IONIZATION ENERGY.

ionizing radiation. Radiation of sufficiently high energy to cause the production of ions in the medium through which it passes: for example, high-energy particles (electrons, protons, alpha-particles) or short-wave radiation (UV, X-, gamma-rays). Extensive damage to the molecular structure of the medium occurs either as a result of direct transfer of energy or of secondary electrons released. In biological systems, the effect is particularly dangerous due to the formation of free radicals which can have powerful oxidizing or reducing properties.

ionogenic surface. *See* SURFACE CHARGE.

α-, β-ionones (4-(2,6,6-trimethyl-2-cyclohexenyl)-3-buten-2-ones, $C_{13}H_{20}O$). α-Ionone (bp 120°C), β-ionone (bp 135°C); powerful odorants smelling strongly of violets (α- sweeter odour than β-) (*see* Figure). Condensation of citral (*see* TERPENES) with propanone gives pseudoionone (bp 143°C, 12 mmHg) which on treatment with alkali gives a mixture of the isomers.

(CH₃)₂

CH=CHCOCH₃

CH₃

(a)

(CH₃)₂

CH=CHCOCH₃

CH₃

(b)

α-, β-ionones
(a) α-Ionone, (b) β-ionone.

ion pair. Oppositely charged ions that associate in an electrolyte forming an overall neutral entity. Ion-pairing is promoted by high concentration of ions in a solvent of low relative permittivity. Three types of ion pairs are distinguished depending on whether each ion remains fully solvated, if solvent molecules are shared between ions or if the ions are in contact and most of the solvation shell is lost. The formation of ion pairs results in lower values of the conductivity than predicted by the Onsager equation (*see* CONDUCTANCE EQUATIONS).

ion pump. *See* SPUTTERING.

ion-selective electrode. MEMBRANE ELECTRODE that responds selectively to one (or several) ionic species in the presence of other ions. The membrane may be a single crystal or supported crystalline material, glass (*see* GLASS ELECTRODE), polymer or organic solution. Other types which use the membrane (e.g., NAFION) to establish an equilibrium concentration in the internal electrolyte are GAS-SENSING MEMBRANE PROBES and biosensors.

ion-selective membrane. Separator for electrochemical cells that transports only anions or cations (or certain anions or cations). Membranes may be inorganic solid electrolytes such as β-alumina or organic polyfluorinated polymers with side chains including acid groups. Use of membranes (e.g., NAFION) allows the construction of very thin cells, but they suffer from the drawback of relatively high resistance.

Ir. *See* IRIDIUM.

iridium (Ir). Third row TRANSITION ELEMENT. Extraction from other PLATINUM GROUP METALS relies on its insolubility in molten lead, or the insolubility of ammonium chloroiridate in water. Iridium is the most chemically inert metal and as such is used in alloys in equipment requiring exceptional corrosion resistance. $Ir_{0.3}Pt_{0.7}$ is practically insoluble in AQUA REGIA.
$A_r = 192.22$, $Z = 77$, ES [Xe] $5d^7 6s^2$, mp 2410°C, bp 4130°C, $\rho = 22.65 \times 10^3$ kg m^{-3}. Isotopes: ^{191}Ir, 38.5 percent; ^{193}Ir, 61.5 percent.

iridium compounds. Iridium shows oxidation states +6 (IrF_6) to −1 $[Ir(CO)_4]^-$. Iridium(III) is the most stable state. Typical compounds are $IrCl_3$ and Na_3IrCl_6. Higher oxidation states are oxidizing. Iridium(IV) is stabilized by complexing, usually with nitrogen ligands. Iridium(I) is more common than iridium(II) (e.g., Vaska's compound *trans*-$[Ir(CO)(Ph_3P)_2Cl]$).

IRMS (isotope ratio mass spectrometry). *See* MASS SPECTROMETRY.

iron (Fe). First row TRANSITION ELEMENT. It is the fourth most abundant element in the earth's crust (5 percent), present as the ores haematite (Fe_2O_3), limonite (hydrated Fe_2O_3), magnetite (Fe_3O_4), siderite ($FeCO_3$), iron pyrites (FeS_2), chromite ($FeCr_2O_4$) and chalcopyrite ($CuFeS_2$). Ores are roasted to remove impurities, then reduced to the metal (CAST IRON) by carbon in a blast furnace. Iron is converted tó STEEL by the BASIC OXYGEN PROCESS. Iron is used extensively as the metal, and alloyed in steel. The metal readily oxidizes in air, when moisture is present the process known as rusting occurs (*see* RUST). Iron displays strong FERROMAGNETISM.
A_r = 55.847, Z = 26, ES $[Ar]3d^6 4s^2$, mp 1535°C, bp 2750°C, ρ = 7.87 × 10^3 kg m^{-3}. Isotopes: ^{54}Fe, 5.84 percent; ^{56}Fe, 91.68 percent; ^{57}Fe, 2.17 percent; ^{58}Fe, 0.31 percent.

iron compounds. Oxidation states −2, +1, +2, +3, +4 and +6 are known. Ferrous (+2) and ferric (+3) compounds are most common. Iron(II) compounds are oxidized by air, especially in basic solution ($E^{\ominus}_{Fe^{3+},Fe^{2+}}$ = 0.77 V). Mohr's salt (iron(II) ammonium sulphate $(NH_4)_2Fe(SO_4)_2.6H_2O$) is used as a reducing agent in volumetric analysis. Fe^{3+}, having high charge and small radius (53 pm), strongly attracts anions. Complexes of iron(II) and iron(III) are octahedral (e.g., $[Fe(CN)_6]^{3-}$) (*see* CYANO-FERRATES). Oxides, important as dyes, pigments and magnetic materials, are FeO, Fe_2O_3 and a mixed oxide with inverse spinel structure (Fe_3O_4) (*see* SPINELS). With other metal oxides, ferrates are formed (e.g., Na_2FeO_4, Sr_2FeO_4, $NaFeO_2$, $MgFe_2O_4$, $Na_4Fe(OH)_6$). A series of iron carbonyls are formed. $Fe(CO)_5$ is trigonal bipyramidal (Fe + CO, 200°C, 300 atm). Action of light on $Fe(CO)_5$ gives orange $Fe_2(CO)_9$ and heat gives dark green $Fe_3(CO)_{12}$. CO is readily displaced by a number of ligands as a route to synthesis of organometallic compounds of iron. Ferrocene (h^5–$C_5H_5)_2$Fe, first prepared in 1951, is the prototype METALLOCENE and an example of the SANDWICH COMPOUNDS.

iron pyrites. *See* IRON, SULPHUR.

irradiation. Exposure to any form of ELECTROMAGNETIC RADIATION or IONIZING RADIATION.

irreducible representation. *See* GROUP THEORY.

irreversible process. Process for which reversal cannot be achieved simply by changing a variable by an infinitesimal amount (*compare* REVERSIBLE PROCESS). All naturally occurring or spontaneous processes are irreversible because they cause a degradation of the quality of the ENERGY and hence an increase in the ENTROPY of the universe. WORK obtained in such a process is always less than the maximum obtained when the process is carried out reversibly.

IRS (internal reflectance spectroscopy). *See* ATTENUATED TOTAL REFLECTANCE.

IR spectrophotometry. *See* INFRARED SPECTROSCOPY.

isa. *See* IONIC STRENGTH ADJUSTOR.

isenthalpic process. Process carried out under conditions such that there is no change in ENTHALPY (i.e. dH = 0); for example, throttled expansion of gas under adiabatic conditions (*see* ADIABATIC PROCESS) when temperature of the gas decreases.

isentropic process. *See* ADIABATIC PROCESS.

iso-. Prefix indicating a single branching at the end of a carbon chain, thus iso-butanol $((CH_3)_2CHCH_2OH)$ is an isomer of butanol $(CH_3(CH_2)_3OH)$.

isobaric process. Process carried out at constant pressure (i.e. dP = 0). For a reversible isobaric process involving an ideal gas (*see* Figure)

$$q = \Delta H = \Delta U + P\Delta V = C_P \Delta T$$

See also ISOCHORIC PROCESS, ISOTHERMAL PROCESS.

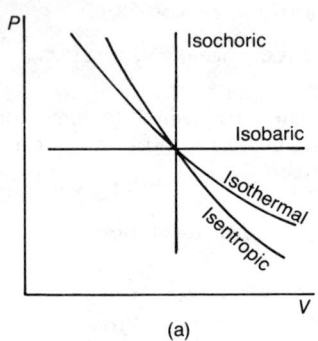

(a)

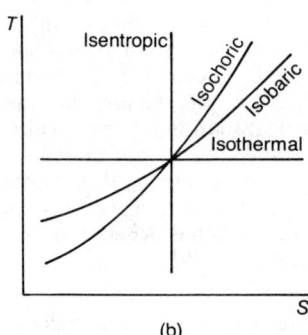

(b)

isobaric process
Plots for isobaric, isochoric, isentropic and isothermal process: (a) P–V; (b) T–S.

isobestic point. Wavelength at which the absorption of radiation is constant for all possible concentration ratios of a binary mixture. It is used in INFRARED SPECTROSCOPY and ELECTRONIC SPECTROSCOPY as a reference point.

isoborneol. *See* TERPENES.

isobutyl alcohol (iso-butanol). *See* BUTANOLS.

isobutyric acid. *See* 2-METHYLPROPANOIC ACID.

isochore. *See* VAN'T HOFF ISOCHORE.

isochoric process. Process carried out at constant volume (i.e. dV = 0, w = 0). For an isochoric reversible process involving an ideal gas (*see* ISOBARIC PROCESS, Figure)

$$q = \Delta U = C_V\Delta T = T\Delta S$$

See also ADIABATIC PROCESS, ISOTHERMAL PROCESS.

isocrotonic acid. *See* 2-BUTENOIC ACIDS.

isocyanates. Compounds containing the group –NCO. Alkyl isocyanates are prepared from dialkyl sulphates and potassium cyanate. Alkyl and aryl isocyanates are prepared from amines and carbonyl chloride

$$RNH_2 + COCl_2 \rightarrow RNHCOCl + HCl$$
$$\rightarrow RNCO + 2HCl$$

They are also formed as intermediates in the HOFMANN CONVERSION OF AMIDES and the CURTIUS REACTION. Isocyanates readily polymerize to isocyanuric esters, and hydrolyze in air to unstable carbamic acids, and react with alcohols and phenols to URETHANES, ammonia and anilines to substituted UREAS. They are used as precursors for polyurethane resins (*see* POLYMERS).

isocyanides. *See* CARBYLAMINES.

isoelectric focusing. Migration of AMPHOLYTES (e.g., proteins) through a pH gradient under an applied electric field. Molecules possessing an electric charge migrate towards a region in which they are isoelectric (*see* ISOELECTRIC POINT). The method is used to separate mixtures containing proteins of different pI, its success relying on the maintenance of a near-linear pH gradient.

isoelectric point (pI). pH at which a species (AMINO ACID, PROTEIN or COLLOID) does not move in an electric field. It is necessary to specify the nature of the buffer solution and its ionic strength. ISOELECTRIC FOCUSING is a method of separation that relies on species having different pI values.

isoelectronic. Having the same number of electrons (e.g., CO and N_2).

isoenzymes. Physicochemical variants of an enzyme that have the same function, but which can be distinguished and separated by electrophoresis.

isoindole. *See* HETEROCYCLIC COMPOUNDS (structure XXVIII).

isoionic. Referring to a solution of a protein in which there are no foreign ions present, except those arising from the dissociation of the solvent.

isoleucine (Ile). *See* AMINO ACIDS.

isomerases. *See* ENZYMES.

isomeric transition. *See* RADIOACTIVITY.

isomerism. Isomeric compounds have the same composition and molecular mass, but differ in a physical or a chemical property. Isomerism may be of several types.
structural isomerism. Due to different orders in which the atoms are joined.
1) Chain isomerism. Arises from the possibility of linking the same number of carbon atoms to produce either straight or branched chains. Butane is the first alkane to show chain isomerism (e.g., CH_3CH_2-CH_2CH_3 or $CH_3CH(CH_3)_2$). The number of isomers increases with increase in molar mass.
2) Position isomerism. Exhibited by compounds having the same carbon skeleton, but differing in the position occupied by a substituent group. Different monosubstituted derivatives are possible from C_3 (e.g., $CH_3CH_2CH_2X$ and CH_3CHXCH_3). Increasing the number of substituents increases the number of possible isomers. In unsaturated compounds (i.e. alkenes and alkynes) the number of unique locations for a double or triple bond determines the number of isomers (e.g., $CH_3CH_2CH=CH_2$ and $CH_3CH_2C\equiv CH$ or $CH_3CH=CHCH_3$ and $CH_3C\equiv CCH_3$). In symmetrical cyclic systems (i.e. cycloalkanes, benzene) all hydrogen atoms are equivalent and so at least two substituents must be present for the existence of positional isomers. For benzene there are three disubstituted isomers (1,2-, *ortho*-; 1,3-, *meta*-; 1,4-, *para*-). For non-symmetrical cyclic systems (i.e. cycloalkenes, heterocycles, polycyclic systems) positional

isomerism is possible with one substituent; polysubstitution leads to a greater number of isomers than in symmetrical systems.
3) Functional group isomerism. Exhibited by compounds having different functional groups, that is compounds with the same molecular formula belonging to different homologous series (e.g., CH_3CH_2OH and CH_3OCH_3).
4) Tautomerism. Special case of structural isomerism when the two isomers are directly interconvertible. A state of dynamic equilibrium exists between the two isomers in which one form may predominate or both forms are present to about the same extent. The rate of interconversion (which varies widely in different cases) may be slow enough to permit isolation of the two forms. The position of the equilibrium may be determined by NMR studies. Typical examples are the KETO-ENOL TAUTOMERISM (as in ethyl 3-oxobutanoate).
stereoisomerism. Due to differences in the spatial arrangement of atoms in the molecule.
1) Optical isomerism. Due to asymmetry (*see* OPTICAL ACTIVITY).
2) Rotational isomerism. (*See* CONFORMERS).
3) Geometrical or *cis–trans* isomerism. Characterized by compounds having the same structure but different configurations of dissimilar atoms or groups attached to two atoms joined either by a double bond or which form part of a ring structure. The double bond restricts free rotation of the bond between the atoms and permits the existence of two forms. This isomerism is not possible if one of the atoms forming part of the double bond carries identical groups; thus compounds of the type $Ca_2=Cb_2$ and $Ca_2=Cbd$ cannot form isomers, whereas those of the type $Cab=Cab$ and $Cab=Cad$ can

cis-form   trans-form

The prefixes *cis-* and *trans-* are used to distinguish between molecules in which the given groups lie on the same side or opposite sides of the plane of a double bond for example

HOOC   COOH
\   /
C = C
/   \
H   H

cis-form (maleic acid)

HOOC   H
\   /
C = C
/   \
H   COOH

trans-form (fumaric acid)

Cl   I
\   /
C = C
/   \
H   Br

(*Z*)-1-bromo- 2-chloro-1-
iodoethene

H₃C   H
\   /
C = C
/   \
H   CH₂CH₃
        \
         C = C
        /   \
       H   H

(*E,Z*)-2,4-heptadiene

or a ring (e.g., 1,2-dibromocyclopropanes) (*see* DIASTEREOISOMERS). For other double bonds (>C=N– or –N=N–) the prefixes *syn*- and *anti*- are used

R
\
C = N
/   \
H   COOMe

syn - or (*Z*)- form

R   COOMe
\   /
C = N
/
H

anti - or (*E*)- form

To overcome the difficulty of assigning *cis*- and *trans*- when there is no pair of identical or similar groups, a new pair of descriptors has been introduced which allows unambiguous designation of the configuration about all double bonds. For a pair of doubly bonded atoms A and B to which four different substituents *a,b,d,e* are attached, two isomers (I and II) are possible. All the atoms lie in a single plane; if a plane Q is drawn perpendicular to this plane and along the line of the double bond, then

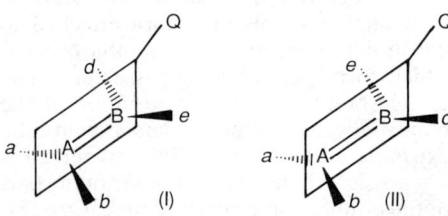

(I)   (II)

isomer I has atoms *a* and *d* on the same side of Q, whereas in II, they are on opposite sides. The priority of groups attached to each doubly bonded atom is established using the SEQUENCE RULES. The configuration in which the two groups of higher priority are on the same side of the reference plane is assigned the stereochemical descriptor Z (*zusammen* = together) and if on opposite sides the descriptor is E (*entgegen* = across). Thus maleic acid is the (*Z*)-isomer and fumaric acid the (*E*)-isomer. Using the rules the absolute configuration of the following compounds can be described

In many, but not all, cases *cis*-isomers correspond to (*Z*) and *trans*-isomers to (*E*).

isomerism in inorganic complexes. Possibilities for isomerism in inorganic derivatives are much greater than for organic compounds because of the widely varying coordination numbers.

1) Ionization isomerism. Occurs when there is an interchange of groups between the coordination sphere of the nuclear atoms and ions outside this sphere (e.g., $[PtCl_2(NH_3)_4]Br_2$ and $[PtBr_2(NH_3)_4]Cl_2$).

2) Hydration isomerization. Examples include $[Cr(H_2O)_6]Cl_3$, $[CrCl_2(H_2O)_4]Cl.2H_2O$ and $[CrCl(H_2O)_5]Cl.H_2O$.

3) Coordination isomerism. Occurs when both the anion and cation are complex (e.g., $[Pt(NH_3)_4][PtCl_4]$ and $[PtCl(NH_3)_3][PtCl_3(NH_3)]$) and also when the nuclear element is present in two different oxidation states.

4) Linkage isomerism. For example, the $-NO_2$ group may be coordinated to a metal ion either through the nitrogen or oxygen atom

$[CoNO_2(NH_3)_5]^{2+}$
nitropenta-amminecobalt(III) ion

$[Co(ONO)(NH_3)_5]^{2+}$
nitritopenta-amminecobalt(III) ion

the CNS^- can coordinate through nitrogen or sulphur.

5) Geometrical isomerism. In planar complexes of metals showing coordination number four

H₃N   Cl
\   /
Pt
/   \
H₃N   Cl

cis-form

H₃N   Cl
\   /
Pt
/   \
Cl   NH₃

trans-form

and in six-coordinate octahedral complexes

Cl   +
NH₃ —|— Cl
|
Co
/   \
NH₃ —|— NH₃
|
NH₃

cis-form (blue-violet)

Cl   +
NH₃ —|— NH₃
|
Co
/   \
NH₃ —|— NH₃
|
Cl

trans-form (green)

6) Optical isomerism. Occurs (as in organic compounds) when one form is the mirror image of the other: for example six-coordinate complexes containing bidentate ligands (*aa*) (e.g., $[CoCl_2en_2]^+$)

trans-form
(symmetrical),
not resolvable

two *cis*-forms
(unsymmetrical),
resolvable

isomers. *See* ISOMERISM.

isometric particles. Denoting a crystallo-graphic system in which the axes are perpendicular to each other as in a cubic crystal.

isomorphism. Denoting substances with the same crystal structure (irrespective of their chemical nature). Isomorphs can induce crystallization of each other and may crystallize on each other. They are able to form SOLID SOLUTIONS.

isonitriles. *See* CARBYLAMINES.

isopiestic. Describing solutions that have the same VAPOUR PRESSURE.

isopleth. Line of constant composition drawn on a PHASE DIAGRAM, to illustrate the effect of change of temperature on the various phases present. *See also* TWO-COMPONENT CONDENSED SYSTEM.

isopotential point. Point at which a surface (e.g., a colloidal particle) is completely dis-charged, the ELECTRICAL DOUBLE LAYER has collapsed, arid the ZETA-POTENTIAL and surface charge are zero. With increasing concentration of electrolyte, the zeta-potential at an interface approaches zero. The concentration of electrolyte at which the isopotential point is attained depends on the valence type of the electrolyte. *See also* POTENTIAL OF ZERO CHARGE.

isoprene. *See* 2-METHYL-1,3-BUTADIENE.

isoprene rule. Carbon skeletons of the TERPENES are built up of isoprene units, so that C-4 of one unit is attached to C-1 of the

other (*see* Figure); this arrangement is more common than 1,1- or 4,4-linkages. Although isoprene is produced on the pyrolysis of terpenes, it is not their actual precursor.

isoprenoid compounds. *See* TERPENES.

isopropyl alcohol. *See* PROPANOLS.

isoquinoline (2-benzazine, C_9H_7N). Nitrogen HETEROCYCLIC COMPOUND (struc-ture XXXIII), that occurs with QUINOLINE in coal tar; crystalline solid (mp 24°C) with a disagreeable odour, feebly basic ($pK_a = 8.6$), insoluble in water, but soluble in acids. It is synthesized by reacting benzaldehyde with aminoacetal to give a Schiff's base which undergoes cyclization with sulphuric acid (Pomeranz–Fritsch synthesis). In electrophilic substitution reactions, C-5 is the most reactive: nitration ($HNO_3 + H_2SO_4$, 0°C) gives predominantly 5-nitro-isoquinoline. With sodium amide 1-amino-isoquinoline is formed. The isoquinoline ring system constitutes the common struc-tural feature of several hundred alkaloids (e.g., morphine, heroin, codeine).

isosbestic point. *See* ISOBESTIC POINT.

isostere. Thermodynamic equation relat-ing properties of a system at constant composition. *See also* HEAT OF ADSORP-TION.

iso-stilbene. *See* STILBENE.

isostructural. Denoting the same lattice type and crystal structure, (e.g., caesium chloride and caesium iodide).

isotachophoresis. Electrophoretic separa-tion method (*see* ELECTROPHORESIS) in which a steady-state configuration is obtained in a moving boundary experiment. Separation relies on variations in IONIC MOBILITY. A leading electrolyte with a higher and a terminating electrolyte with a lower mobility than the ions to be separated sandwich the separated bands.

Capillary isotachophoresis is used as a preparative method.

isotactic polymer. *See* TACTICITY.

isothermal compressibility (κ). Measure of the change of volume of a substance with applied pressure at constant temperature, defined as

$$\kappa = -(1/V)(\partial V/\partial P)_T$$

isothermal process. Process in which the temperature remains constant (i.e. $dT = 0$). The process is carried out in a constant temperature bath or using a thermostat. For reversible isothermal processes involving an ideal gas

$$w = -RT \ln P_1/P_2$$

$$q = nRT \ln P_1/P_2$$

$$\Delta U = 0$$

isothiocyanates. Derivatives of the type RNCS. *See also* THIOCYANATES.

isotones. Atomic nuclei having the same number of neutrons but different numbers of protons (and mass numbers) (e.g., $^{38}_{50}Sr$ and $^{39}_{50}Y$).

isotonic. Describing solutions that have the same OSMOTIC PRESSURE. This is important from the physiological viewpoint, since biological membranes behave like SEMI-PERMEABLE MEMBRANES. Thus red blood cells in a solution of greater (lower) osmotic pressure will shrink (swell) as water passes out (in) across the cell membrane. Solutions with a higher (lower) osmotic pressure are termed hypertonic (hypotonic). Since osmotic pressure is not a readily measurable quantity, it is usual to calculate this from the depression of FREEZING POINT of the solution in question. A 0.9 percent (w/v) solution of sodium chloride is isotonic with blood.

isotope dilution analysis (IDA). Procedure for determining the concentration of X in a mixture when quantitative separation is extremely difficult (e.g., amino acid in protein hydrolyzate). A measured aliquot of an isotope of X of known specific activity is added to a known amount of the mixture.

The extent to which the specific activity is reduced is a measure of the amount of X originally present.

isotope effects (kinetics). Changes in the rates of reactions of molecules caused by the substitution of an isotope of an atom. The form of the POTENTIAL ENERGY SURFACE remains the same, but the vibrational energy of reactants and ACTIVATED COMPLEX may change because of changes in the ZERO-POINT ENERGY. The greater the mass difference between the isotopes the greater the isotope effect. Thus deuterium and tritium substituted for hydrogen result in the greatest effects (k_D/k_H = max 18, k_T/k_H = max 60).
primary isotope effect. One in which a bond to the isotopically substituted atom is made (heavier isotopes react faster) or broken (heavier isotopes react more slowly).
secondary isotope effect. In a secondary effect of the first kind, the bond to the isotopically labelled atom is not broken, but does suffer a spatial change (e.g., a change in hybridization of an atom to which it is attached). In a secondary isotope effect of the second kind, no obvious change occurs in the vicinity of the labelled atom.
    Kinetic isotope effects are used to determine mechanisms in chemistry.

isotope number (neutron excess). Difference between the number of neutrons and the number of protons in an isotope.

isotope ratio mass spectrometry (IRMS). *See* MASS SPECTROMETRY.

isotopes. Atoms which have the same atomic (proton) number (Z) but different numbers of neutrons (N) and hence different atomic mass numbers (A). The most precise way of denoting the isotopes is as $^A_Z X$ (e.g., ^1_1H, ^2_1H, ^3_1H). The atomic mass of an element is the average mass percentage of all its isotopes. Isotopes have similar chemical properties, but the slight difference in mass causes slight differences in physical properties. There are three kinds of isotopes: natural non-radioactive, natural radioactive and artificially radioactive (prepared by neutron bombardment). As a broad generalization, neutron-rich isotopes are radioactive (with very few exceptions) if the $N:Z$ ratio is greater than 1.25 for $Z = 1$–30, is greater than 1.40 for $Z = 31$–47 and is greater than 1.50 for $Z = 48$–76. *See also* ISOTOPE EFFECTS, NUCLIDE.

isotope separation. Methods based on slightly different physical properties include mass spectrometry, gaseous diffusion (separation of isotopes of uranium using UF_6), fractional distillation (D_2O), electrolysis, thermal diffusion, centrifugation and laser methods (excitation of one isotope and subsequent separation by electromagnetic methods).

isotropic. Denoting a medium whose physical properties are the same in all directions (e.g., crystals of the cubic system). *See also* ANISOTROPIC.

IUPAC. International Union of Pure and Applied Chemistry; international body that sets standards of procedure and nomenclature in chemistry.

J

J. *See* JOULE.

J. Rotational quantum number (*see* QUANTUM NUMBER); Massieu function.

j_0. *See* EXCHANGE CURRENT DENSITY.

Jahn–Teller effect. Distortion of octahedral d^4-high spin, d^7-low spin and d^9-COORDINATION COMPOUNDS to D_{4h} symmetry (*see* POINT GROUP). The Jahn–Teller theorem states that any orbitally degenerate electronic structure of a non-linear molecule is intrinsically unstable. There must be another structure that eliminates the degeneracy and has a lower energy. Copper(II) complexes (d^9) show Jahn–Teller distortions (*see* Figure).

$\bar{J}_i$. *See* PARTIAL MOLAR QUANTITY.

j–j coupling. Coupling between spin and orbital momentum of an electron in a heavy atom. This causes a failure of Russell–Saunders coupling (*see* TERM SYMBOLS) and the attendant SELECTION RULES for transitions between electronic levels. Thus transitions between singlet and triplet states become allowed for heavier atoms of the PERIODIC TABLE.

Jones reductor. Zinc AMALGAM, usually contained in a tube, used for the reduction of a solution (e.g., iron(III) to iron(II)) prior to estimation or titration.

joule (J; dimensions: $m\ l^2\ t^{-2}$; units: $N m = kg\,m^2\,s^{-2}$). SI unit of energy; defined as the WORK done when the point of application of a force of 1 NEWTON is displaced through a distance of 1 metre in the direction of the force (1 cal = 4.184 J).

Joule–Thomson effect (Joule–Kelvin effect). Change in temperature that occurs when a gas expands through a throttle from a high- to a low-pressure

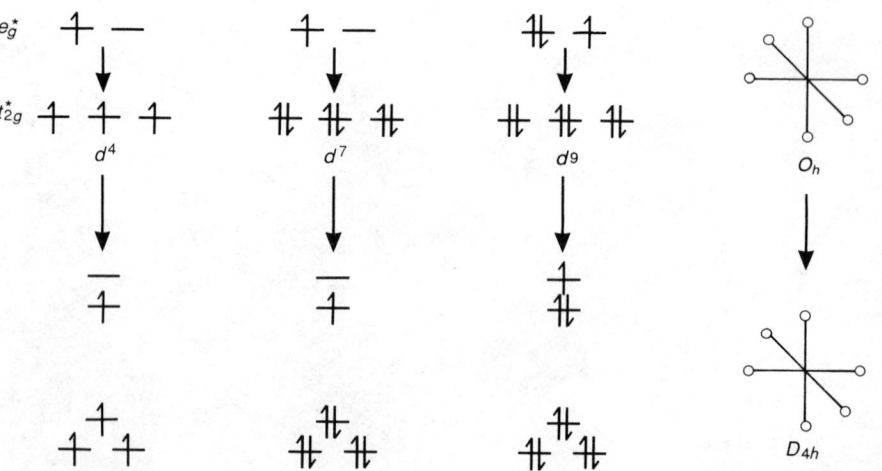

Jahn–Teller effect
Jahn–Teller distortion of octahedral transition metal complexes.

268

region under adiabatic conditions ($q = 0$). For most real gases, the temperature decreases because the attraction between the molecules retards their escape, thereby diminishing the kinetic energy and hence the temperature. The Joule–Thomson coefficient (μ_{JT}) depends on the temperature and pressure in general decreasing with increasing pressure, becoming negative at high pressures. For an IDEAL GAS, $\mu_{JT} = 0$. From the THERMODYNAMIC EQUATIONS OF STATE

$$\mu_{JT} = (\partial T/\partial P)_H = (\partial H/\partial P)_T/(\partial H/\partial T)_P$$
$$= [T(\partial V/\partial T)_P - V]/C_P$$

For a gas that obeys VAN DER WAALS EQUATION

$$\mu_{JT} = (2a/RT - b)/C_P$$

if $b < 2a/RT$, $\mu_{JT} > 0$; $b = 2a/RT$, $\mu_{JT} = 0$; $b > 2a/RT$, $\mu_{JT} < 0$. At a given pressure there is a particular temperature, the inversion temperature, at which μ_{JT} is zero; above this temperature the gas is heated by expansion and below it the gas is cooled. For hydrogen above 193 K and helium above 53 K, the effect of internal attraction is small compared with that of molecular size; this then acts as a repulsion and gives a negative Joule–Thomson effect.

junction potential. *See* LIQUID JUNCTION POTENTIAL.

K

K. Kelvin (*see* TEMPERATURE SCALES); POTASSIUM.

k. *See* BOLTZMANN CONSTANT, FORCE CONSTANT, RATE CONSTANT.

κ. *See* CONDUCTIVITY.

K_a. *See* ACID DISSOCIATION CONSTANT.

Kammerlingh Onnes equation. *See* EQUATION OF STATE.

kaolin. *See* ALUMINIUM.

Karl Fischer reagent. Solution of iodine and sulphur dioxide in a pyridine/methanol mixture used to titrate water. Hydrogen iodide is liberated and the course of the titration followed by the change in pH of the solution.

katharometer. *See* DETECTION SYSTEMS.

K_b. *See* BASE DISSOCIATION CONSTANT.

kcal. *See* CALORIE.

Kekulé structure. *See* BENZENE.

kelvin (K). SI unit of temperature; the fraction 1/273.15 of the thermodynamic temperature of the TRIPLE POINT of water. K is used both for the thermodynamic temperature and for the thermodynamic temperature interval.

Kelvin equation. Vapour pressure above a drop of radius *r* compared with the vapour pressure above a flat plane is related to the SURFACE TENSION by

$$\ln (P_{drop}/P_{flat}) = -2\gamma V_m/rRT$$

keratin. *See* PROTEINS.

kerosene. Fraction of PETROLEUM used as a fuel.

ketals. Diethers formed from ketones and trialkoxymethanes (orthoformates)

$$R_2CO + HC(OR')_3 \rightarrow R_2C(OR')_2$$

See also ACETALS.

ketens (ketenes). Compounds containing the group >C=C=O. Aldoketens are RCH=C=O and ketoketens are RR'C=C=O. Ketoketens are generally prepared by debrominating a α-bromoaryl bromide with zinc. Aldoketens are prepared by refluxing an acid chloride in pyridine. The simplest member is keten, ethenone ($H_2C=C=O$), prepared by treating bromoethanoyl bromide with zinc. It is a poisonous pungent gas (bp –41°C), which readily oxidizes to an unstable peroxide and dimerizes to diethenone. Reactions are typically of addition across >C=C< (e.g., ethenone reacts with water to give ethanoic acid). Photolysis of ethenone generates CARBENE and carbon monoxide.

ketimines. *See* AMINES.

10-keto-9,10-dihydroanthracene. *See* ANTHRONE.

keto–enol tautomerism. Form of tautomerism (*see* ISOMERISM) catalyzed by acids and bases. The keto and enol forms (*see* Figure), which are distinct molecules, are in dynamic equilibrium the position of which depends on the conditions (solvent, temperature, concentration, etc.). For simple aldehydes and ketones the equilibrium is far on the keto side (propanone <0.1 percent enol), whereas 1,3-dicarbonyl compounds contain much more enol (ethyl 3-oxobutanoate 7.5 percent enol; 2,4-pentanedione 80 percent enol). The enol form of most 1,3-dicarbonyl compounds exists as intra-molecularly hydrogen-bonded ring structures, chelate rings.

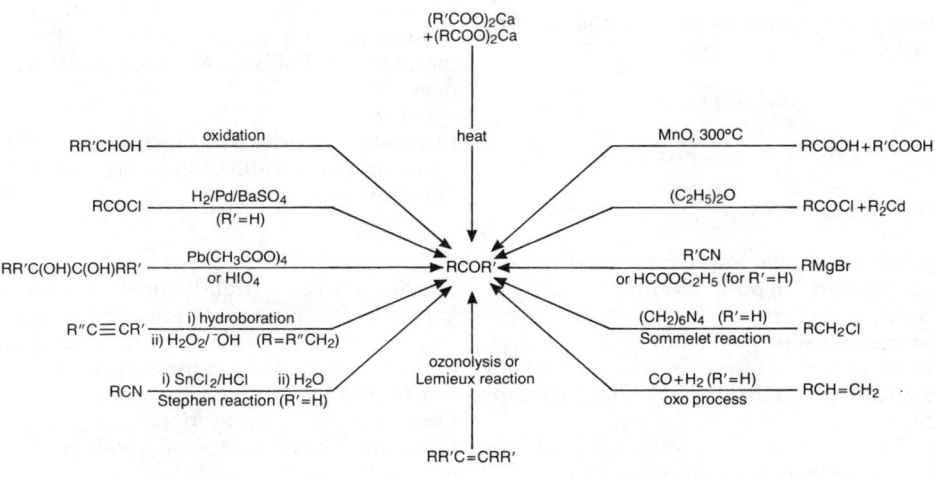

keto-enol isomerism

β-ketoglutaric acid (acetonedicarboxylic acid, oxoglutaric acid). *See* 3-OXO-PENTANEDIOIC ACID.

keto group. Compound or radical that has an OXO GROUP attached to two carbon atoms (C–CO–C). *See also* KETO–ENOL TAUTOMERISM, KETONES.

ketoketens. *See* KETENS.

ketols. Organic compounds containing both a keto group and a HYDROXYL GROUP (e.g., 4-HYDROXY-4-METHYL-2-PENTAN-ONE). *See also* ALDOL CONDENSATION.

ketones ($C_nH_{2n}O$). Compounds containing the OXO GROUP bonded to two carbon atoms. Ketones are named by the suffix –one appended to the longest alkyl chain such that the oxo group has the lowest number. Many of the preparations (Figure 1) and properties (Figure 2) of ketones are shared with ALDEHYDES. Ketones, however, differ from aldehydes in their lack of reducing properties. Esters of aliphatic ketones are prepared by BAEYER–VILLI-GER OXIDATION. Ketones containing the ETHANOYL GROUP undergo the haloform reaction (*see* TRIHALOMETHANES). KETALS are formed on treating a ketone with trialkoxymethane. Reduction by dissolving metals gives alcohols in alkaline solution and 1,2-DIOLS in acid, ketones being less reactive towards addition. *See also* DI-KETONES, PROPANONE.

5-ketopyrazoline. *See* PYRAZOLONE.

ketoses. Sugars containing a potential keto ($>C=O$) group; its presence may be masked by inclusion in a ring structure. *See also* HEXOSES, PENTOSES.

ketoximes (RR'C=NOH). Organic compounds, prepared by the reaction between ketones and hydroxylamine. If R and R' are

ketones
Figure 1. Preparation of aldehydes (R' = H) and ketones.

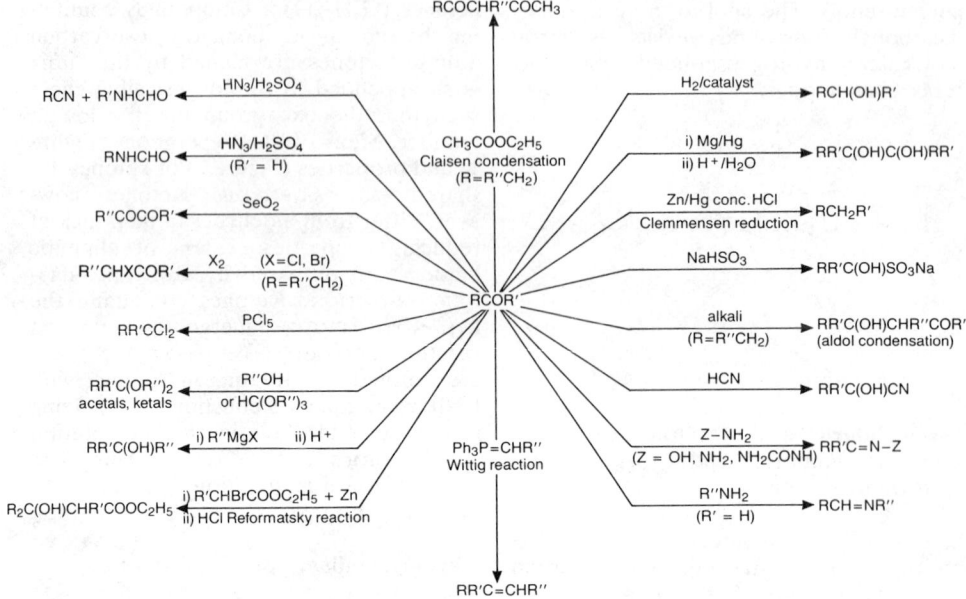

ketones
Figure 2. Reactions of aldehydes (R' = H) and ketones.

different, they exist in the *syn-* or *anti-* conformation (*see* ISOMERISM). They undergo the BECKMANN REARRANGEMENT and form ethanoyl derivatives with ethanoic anhydride

$$R_2C=NOH + (CH_3CO)_2O \rightarrow$$
$$R_2C=NOCOCH_3$$

They are reduced to primary amines. *See also* ALDOXIMES.

ketyl. *See* BENZOPHENONE.

K_f. *See* FORMATION CONSTANT.

kg. *See* KILOGRAM.

kieselguhr. Diatomaceous earth, soft fine-grained deposit consisting of the skeletons of diatoms formed in lakes; very porous and absorbent material. It is used as an absorbent filtering material, filler and insulator and in the manufacture of dynamite.

$k_{i,j}$. *See* SELECTIVITY COEFFICIENT.

Kiliani reaction. Method of ascending the sugar series in which the aldehyde group is

converted to the cyanohydrin, followed by the hydrolysis of the –CN to –COOH, the formation of a lactone between the –COOH and one of the –OH groups and finally reduction to the next higher aldose.

kilogram (kg). SI unit of MASS; it is equal to the mass of the international prototype kilogram (a cylindrical piece of platinum–iridium alloy) kept at the Bureau International des Poids et Mesures at Sèvres, near Paris.

kinematic viscosity (ν; dimensions: $l^2 t^{-1}$; units: $m^2 s^{-1}$). Ratio of the VISCOSITY of a liquid to its DENSITY

$$\nu = \eta/\rho$$

kinetic energy (translational energy). ENERGY of motion; WORK done by the body (system) possessing the energy when it is brought to rest. The kinetic energy ($mv^2/2$) can be resolved into components in three directions mutually at right angles. The rotational kinetic energy of a body having an angular velocity of ω is $I\omega^2/2$.

kinetic isotope effect. *See* ISOTOPE EFFECTS.

kinetics. *See* COLLISION THEORY, ORDER OF REACTION, RATE EQUATION, RATE OF REACTION, REACTION COORDINATE.

kinetic salt effect. Effect of the IONIC STRENGTH of a solution on the RATE CONSTANT of a reaction. Assuming the Debye–Hückel theory of ion activities (*see* DEBYE–HÜCKEL ACTIVITY COEFFICIENT EQUATION) the rate constant for a reaction between two ions of charge z_A and z_B is

$$\lg k = \lg k^0 + 2A z_A z_B I^{1/2}$$

where k^0 is the rate constant extrapolated to zero ionic strength ($I = 0$) and A the Debye–Hückel constant.

kinetic theory of gases. Assuming that the gas consists of elastic molecules of negligible size, which exert negligible forces on each other, moving with random motion, the pressure exerted by n molecules each of mass m in a volume V is

$$P = nm\bar{c}^2/3V$$

where $\bar{c}^2$ is the mean square speed of the molecules. At constant temperature, n, m and $\bar{c}^2$ are constant, hence PV = constant (i.e. BOYLE'S LAW). The kinetic (transitional) energy of the molecules is given by

$$nm\bar{c}^2/2 = 3PV/2 = 3RT/2$$

for 1 mole of gas. The root mean square speed is given by

$$(\bar{c}^2)^{1/2} = (3PV/mN_A)^{1/2} = (3P/\rho)^{1/2}$$
$$= (3RT/M_r)^{1/2}$$

for 1 mole of an ideal gas (*see* GRAHAM'S LAW OF EFFUSION). *See also* MAXWELL–BOLTZMANN DISTRIBUTION LAW.

Kirchhoff's equation. Expression of the variation of the ENTHALPY of a reaction with temperature,

$$[\partial(\Delta H)/\partial T]_P = \Delta C_P$$

Assuming the HEAT CAPACITY to be constant over a limited range, the above equation on integration becomes

$$\Delta H_2 = \Delta H_1 + \Delta C_P(T_2 - T_1)$$

Allowing for the variation of C_P with temperature (*see* HEAT CAPACITY)

$$\Delta H_2 = \Delta H_1 + \Delta a(T_2 - T_1) +$$
$$\Delta b(T_2^2 - T_1^2)/2 + \ldots$$

Analogous equations can be used to calculate changes in the INTERNAL ENERGY at different temperatures.

K_m (Michaelis constant). *See* ENZYME KINETICS.

Knoevenagel reaction. Base-catalyzed reaction between an ALDEHYDE and a compound containing an active methene group to give an unsaturated compound

$$R'CHO + H_2CR_2 \rightarrow R'CH{=}CR_2 + H_2O$$

where R is an activating group such as $-COOC_2H_5$. Thus ethanal reacts with propanedioic acid in the presence of pyridine to give butenoic acid.

Knudsen's method. *See* EFFUSION.

Kohlrausch's equation. *See* CONDUCTANCE EQUATIONS.

Kolbe–Schmitt reaction. *See* 2-HYDROXYBENZOIC ACID.

Kolbe synthesis. Electrolysis of a solution of an alkali metal salt of a carboxylic acid to yield carbon dioxide and an alkane

$$RCOO^- + R'COO^- \rightarrow R{-}R' +$$
$$2CO_2 + 2e$$

The reaction is favoured by high salt concentration, high current density and a platinum anode. Under non-ideal conditions side products (alkenes, esters and rearranged alkanes) are formed. It is used to initiate POLYMERIZATION and to manufacture dimethyl decanedioate from monomethyl hexanedioate.

Konig structure. *See* EXTENDED X-RAY ABSORPTION FINE STRUCTURE.

Koopman's theorem. The exact Hartree–Fock (*see* SELF-CONSISTENT FIELD) energy of a given MOLECULAR ORBITAL is the negative of the IONIZATION POTENTIAL for the removal of an electron from the orbital. This is strictly true only for closed shell molecules, but is valid for most stable molecules.

Kr. *See* KRYPTON.

Krafft temperature. Temperature above which the solubility of an association COLLOID (*see* MICELLE) increases rapidly. Below the Krafft temperature the unassociated compound has limited solubility. At the Krafft temperature the critical micelle concentration is reached with a concomitant increase in solubility. For a series of sodium alkyl sulphates the Krafft temperature rises from 8°C for the decyl sulphate to 56°C for the octadecyl sulphate.

Kroll process. Reduction of titanium tetrachloride with magnesium at red heat in the presence of argon or helium to give titanium metal.

krypton (Kr).   Noble gas (GROUP O ELEMENT), obtained by the fractionation of liquid air (1.1×10^{-3} percent of air). Krypton has six natural and 15 unstable isotopes. CLATHRATE COMPOUNDS have been prepared (under pressure) with hydroquinone and phenol. The difluoride (KrF_2), a white volatile solid (Kr + F_2 in electric discharge), decomposes slowly at room temperature and forms adducts with Lewis acids (e.g., $KrF_2.2SbF_5$). The orange–red spectral line of krypton-86 forms the basis of the SI unit of length, 1 metre is $1\,650\,763.73$ wavelengths of krypton-86. It is used with argon in fluorescent tubes and in electronic flash tubes for high-speed photography.
$A_r = 83.80$, $Z = 36$, ES $[Ar]3d^{10}4s^24p^6$, mp $-156.6°C$, bp $-152.3°C$, $\rho = 3.733 \text{ kg m}^{-3}$. Isotopes: ^{78}Kr, 0.35 percent; ^{80}Kr, 2.27 percent; ^{82}Kr, 11.56 percent; ^{83}Kr, 11.55 percent; ^{84}Kr, 56.90 percent; ^{86}Kr, 17.37 percent.

K_s.   *See* SOLUBILITY PRODUCT.

K-shell.   *See* ELECTRONIC CONFIGURATION, QUANTUM NUMBER.

kurchatovium.   *See* TRANSACTINIDE ELEMENTS.

K_w.   *See* AUTOPROTOLYSIS.

L

L. *See* LANGMUIR.

L. *See* ENTHALPY (fusion, etc.), LOSCH-MIDT'S CONSTANT, MOMENTUM (angular).

L- (configurational symbol). *See* OPTICAL ACTIVITY.

l. Azimuthal quantum number (*see* QUANTUM NUMBER).

Λ. *See* MOLAR CONDUCTIVITY.

Λ$_E$. *See* EQUIVALENT CONDUCTIVITY.

Λ$_i$. *See* MOLAR IONIC CONDUCTIVITY.

λ. *See* DECAY CONSTANT, LAMBDA POINT, MEAN FREE PATH, WAVELENGTH.

La. Lanthanum (*see* LANTHANIDE ELEMENTS).

label. ISOTOPE (radioactive or stable) that replaces a stable atom in a compound. The course of a chemical or biochemical reaction or physical process can be followed by tracing the radioactivity using a counter or the stable isotope in a mass spectrometer. A knowledge of the exact position of the isotope is essential for mechanistic studies; thus ethanoic acid can be labelled with carbon-14 in three different ways: $^{14}CH_3COOH$, $CH_3^{14}COOH$ and $^{14}CH_3^{14}COOH$. TRITIUM (^3_1H) is widely used in following biochemical reactions. *See also* ISOTOPE EFFECTS.

labile. Describing a compound in which certain attached groups (LIGANDS) can be readily replaced by other groups. The term is usually applied to coordination complexes in which ligands can be replaced by other ligands in an equilibrium reaction.

lachrymator. Gas that is strongly irritant to the eyes (i.e. tear gas).

β-lactam antibiotics. *See* ANTIBIOTICS.

lactams. Heterocyclic organic compounds

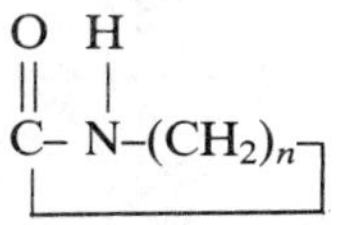

$$\begin{matrix} O & H \\ \| & | \\ C{-} & N{-}(CH_2)_n \end{matrix}$$

(e.g., $n = 3$ is 2-pyrrolidone, $n = 5$ is CAPROLACTAM). Lactams are formed from AMINO ACIDS which on heating eliminate water or by reduction of IMIDES of carboxylic acids. Lactams are poisonous solids. *See also* LACTONES.

lactic acid. *See* 2-HYDROXYPROPANOIC ACID.

lactides. Compounds formed by the elimination of water on the condensation of two molecules of a α-hydroxyacid (e.g., HYDROXYETHANOIC ACID). Thus 3,6-dimethyl-1,4-dioxan-2,5-dione (I) is formed from 2-HYDROXYPROPANOIC ACID.

$$2CH_3CH(OH)COOH \longrightarrow \quad + 2H_2O$$

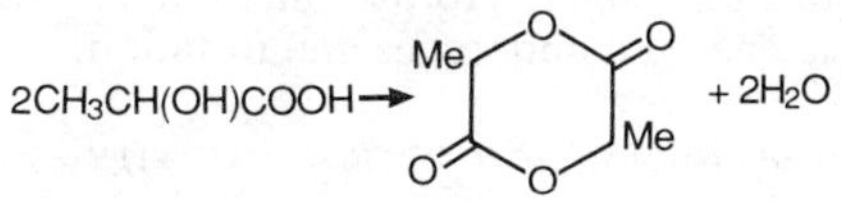

(I)

lactogenic hormone (prolactin). *See* HORMONES.

lactones. Anhydrides formed by an internal esterification of γ- or δ-hydroxyacids. This occurs readily on acidification of the sodium salt. β-Lactones are not usually derived from β-hydroxyacids (they tend to give α,β-unsaturated acids), but may be prepared by the reaction between ethenone

275

and an oxo compound. Lactones are named by the suffix –olide, with numbering of the ring atoms bonded to oxygen: for example, the lactones illustrated (*see* Figure) are 4-alkyl-1,4-butolide (I) and 5-alkyl-1,5-pentolide (II). Large-ring lactones are prepared by the oxidation of cyclic ketones. Salts are formed on refluxing with alkalis. Lactones are reduced to acids by sodium amalgam, and to diols by lithium tetrahydridoaluminate and GRIGNARD REAGENTS. LACTAMS can be formed with ammonia.

(I)

(II)

lactose (milk sugar). DISACCHARIDE (structure IV) of GLUCOSE and GALACTOSE, occurring in the milk of all mammals (about 5 percent). It is prepared commercially by the evaporation of whey (a byproduct in the manufacture of cheese) to crystallization. (mp 203°C (decomp.)). It is hydrolyzed by acid or a β-glucosidase to an equimolar mixture of glucose and galactose. It is a reducing sugar, forms an oxime and osazone, and undergoes mutarotation.

laevorotatory. *See* OPTICAL ACTIVITY.

laevulic acid. *See* OXOCARBOXYLIC ACIDS.

laevulose. *See* FRUCTOSE.

LAH (lithium aluminium hydride, lithium tetrahydridoaluminate). *See* ALUMINIUM INORGANIC COMPOUNDS.

lambda point (λ). Transition point between He_I and He_{II} (2.186 K), below which HELIUM becomes a superfluid.

Lambert's law. *See* BEER–LAMBERT LAW.

laminar flow. Streamline flow of a fluid in which there are no fluctuations or turbulence, successive particles passing the same point have the same velocity. It occurs at low REYNOLD'S NUMBERS (i.e. low velocity, high VISCOSITY). *See also* NEWTONIAN FLUID.

Landé g-factor (spectroscopic splitting factor). Factor appearing in equations for the energies of transitions involving electron spins (*see* ELECTRON SPIN RESONANCE SPECTROSCOPY).

$$g = 1 + [J(J+1) + S(S+1) - L(L+1)]/2J(J+1)$$

L and S are orbital and spin quantum numbers, respectively, for many electron atoms (*see* TERM SYMBOLS) and J is the quantum number for the coupling between L and S. For an uncoupled electron $g = 2$. Correction for relativistic effects increases this to 2.002 32.

langmuir (L). Unit used in adsorption and catalysis studies: the amount of material in one monolayer. *See also* ADSORPTION ISOTHERM.

Langmuir–Blodgett film. Series of monolayers built up on a solid surface by repeatedly dipping the surface through a surfactant (*see* SURFACE-ACTIVE AGENT) layer on a liquid (*see* Figure).

Langmuir–Hinshelwood mechanism. Kinetic description of a surface reaction in which the rate of reaction is proportional to the coverage of reactant(s), which in turn is given by the Langmuir isotherm (*see* ADSORPTION ISOTHERM). *See also* HETEROGENEOUS CATALYSIS.

Langmuir isotherm. *See* ADSORPTION ISOTHERM.

lanolin. Crude preparation of cholesterol (*see* STEROLS) and its esters obtained from wool fat. It readily emulsifies with water and is used, either alone or mixed with soft paraffin, as a base for many skin ointments.

lanosterol. *See* STEROLS.

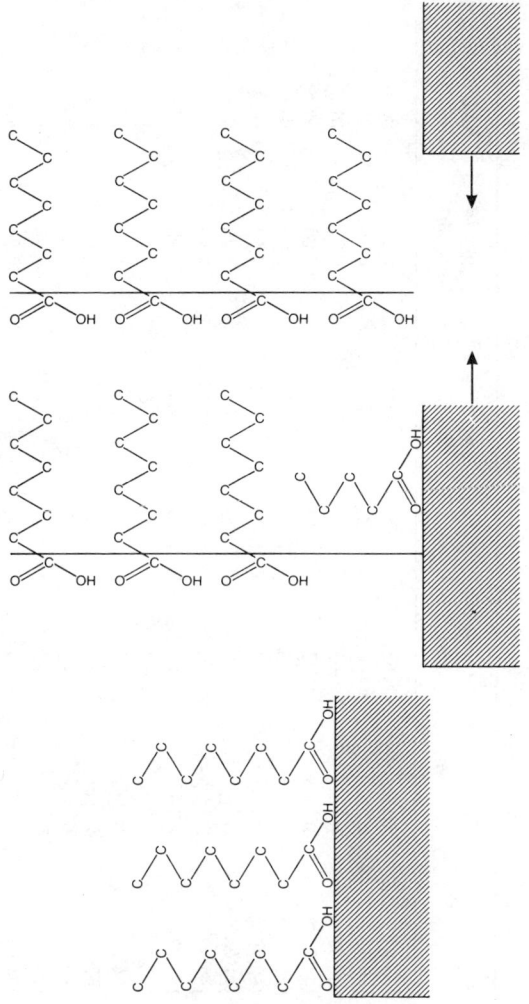

Langmuir–Blodgett film

lanthanide compounds.   Oxidation state +3 is predominant, although Eu^{2+}, Sm^{2+} and Ce^{4+} are known. The hydroxides $M(OH)_3$ are basic, and all elements form oxides M_2O_3, which resemble oxides of the ALKALINE EARTH ELEMENTS. Lanthanide fluorides are insoluble. Anhydrous chlorides (MCl_3) are formed by the action of ammonium chloride on the metal oxide. Salts of most mineral acids are known. The carbonates, phosphates and ethanedioates are insoluble, the ethanedioates being used to determine lanthanides gravimetrically. Double salts with, in particular, nitrates and sulphates are common. At high temperatures, lanthanides combine with non-metals directly to give 1:1 salts (e.g., MN, MP, MAs, MSb, MBi). Complexes are less common than with d-block elements due to the greater size of lanthanide ions and the unavailability of f-orbitals for bonding. Chelate compounds (e.g., with EDTA) are known.

lanthanide elements
Properties of the lanthanide elements.

Element	Symbol	Z	A_r	ES [Xe]+	Oxidation state	mp /°C	bp /°C	$\rho \times 10^{-3}$ /kg m^{-3}	Radius of M^{2+}/pm
Lanthanum	La	57	138.91	$5d^1\,6s^2$	+3	920	3454	6.14	106.1
Cerium	Ce	58	140.12	$4f^2\,6s^2$	+4, 3	798	3257	6.66	103.4
Praseodymium	Pr	59	140.91	$4f^3\,6s^2$	+3	931	3212	6.77	101.3
Neodymium	Nd	60	144.2	$4f^4\,6s^2$	+2, 3	1010	3127	7.00	99.5
Promethium	Pm	61	(145)	$4f^5\,6s^2$	+3	1168	2700	–	97.9
Samarium	Sm	62	150.36	$4f^6\,6s^2$	+2, 3	1077	1791	7.52	96.4
Europium	Eu	63	151.96	$4f^7\,6s^2$	+2, 3	822	1597	5.26	95.0
Gadolinium	Gd	64	157.5	$4f^7\,5d^1\,6s^2$	+3	1311	3233	7.89	93.8
Terbium	Tb	65	158.93	$4f^9\,6s^2$	+4, 3	1360	3041	8.28	92.3
Dysprosium	Dy	66	162.50	$4f^{10}\,6s^2$	+3	1409	2335	8.5	90.8
Holmium	Ho	67	164.93	$4f^{11}\,6s^2$	+3	1470	2720	8.80	89.4
Erbium	Er	68	167.26	$4f^{12}\,6s^2$	+3	1522	2510	9.01	88.1
Thulium	Tm	69	168.93	$4f^{13}\,6s^2$	+3	1545	1727	9.33	86.5
Ytterbium	Yb	70	173.04	$4f^{14}\,6s^2$	+2, 3	824	1193	6.98	85.8
Lutetium	Lu	71	174.97	$4f^{14}\,5d^1\,6s^2$	+3	1656	3315	9.84	84.8

lanthanide contraction. Decrease in the size of atoms and ions across the LANTHANIDE ELEMENTS (*see* Table). Electrons fill the 4*f*-shell from lanthanum (no 4*f*-electrons) to ytterbium and lutetium (14 4*f*-electrons), but the effect of increasing nuclear charge is to hold these electrons closer to the nucleus and thus decrease the size of the atom.

lanthanide elements. TRANSITION ELEMENTS having unfilled 4*f*-orbitals. Lanthanum itself (ES [Xe] $5d^1 6s^2$) is strictly a *d*-block transition element, but is usually included with the 14 true lanthanides ($Z = 58$–71) (*see* Table). The lanthanide elements together with SCANDIUM and YTTRIUM were known by the misnomer 'rare earth' elements. There are several ores containing lanthanides, principally monazite, the elements lanthanum, CERIUM, praseodymium and neodymium being in greatest abundance. The elements are separated by ION EXCHANGE CHROMATOGRAPHY, and the metals extracted by reduction of the trichlorides or trifluorides by calcium. The oxides of samarium, europium and ytterbium are reacted with lanthanum when the volatile metals may be separated. All lanthanides are reactive metals, being attacked by water, acids and air (to give mostly M_2O_3), and they react with hydrogen to give MH_2 and MH_3.

lanthanum (La). *See* LANTHANIDE ELEMENTS.

Laplace equation. Pressure inside a bubble or drop of radius *r* is related to the external pressure by

$$P_{int} = P_{ext} + 2\gamma/r.$$

Laporte selection rule. *See* SELECTION RULES.

large-ring compounds. Many organic compounds containing large numbers of carbon atoms in rings have been synthesized or occur naturally. For example, civitone and muskone are cyclic ketones with 17- and 16-membered rings, respec-

tively. They, and other cyclic ketones, have characteristic odours and are used in perfumery. Large rings are known containing double and triple bonds (*see* CYCLOALKENES). They may be prepared by linking α,ω-alkynes. The resulting cyclic tetrayne may be hydrogenated to a saturated compound (*see* CYCLOALKANES). Condensation reactions, either between molecules or by intermolecular coupling, can yield cyclic products (e.g., acyloin condensation of α,ω-dicarboxylic esters). *See also* CYCLOPHANES.

laser. Acronym for light amplification by stimulated emission of radiation. The device called a laser also includes some method of building up the signal such as multiple reflection by opposing mirrors. The lasing material is excited to a metastable state (*see* FLUORESCENCE, Figure) that is sufficiently long-lived to allow a build up in the number of molecules in the state. Population inversion occurs when there are more molecules in the excited state than the ground state, and this is achieved by the process called optical pumping. A photon emitted spontaneously has a high probability of inducing a similar transition in a second atom by stimulated emission. Each of these photons, which have the same energy, may in turn stimulate two more atoms and so on. The light is further amplified by travelling back and forth between two mirrors, one partially silvered and the other fully reflecting, leading to an intense, coherent beam of light finally passing through the partially silvered mirror (or Q-switch). Lasers may be solid, liquid or gas (*see* Table), emit continuously or may be pulsed, and operate over a wide range of wavelengths. The general properties of lasers are high-intensity, coherence and monochromaticity. Pulsed solid-state lasers (e.g., ruby) are highly coherent and intense (e.g., neodymium pulses at 10^{15} kW m^{-2}), but are less monochromatic because of imperfections in the crystal. Continuous wave gas lasers are more stable and directional, although usually (with the exception of carbon dioxide) of less power. Semiconductor lasers are tunable by varying the applied electric field. Dye lasers may also be tuned by non-linear frequency-doubling methods. In chemistry, the laser is used in spectroscopic techniques that require high-intensity monochromatic light (e.g., RAMAN SPECTROSCOPY, FLUORESCENCE). In kinetics, a fast pulse of laser light is used in the study of picosecond kinetics. Isotopes may be separated by a laser tuned to a particular absorption frequency of one isotope only which will cause reaction of that isotope.

laser spectroscopy. In FLUORESCENCE and RAMAN SPECTROSCOPY, the use of LASER light, because of its greater power and coherence, can increase the resolution and sensitivity. New effects that are unique to lasers include:

multiphoton absorption. Probability of a *N*-photon absorption increases as the *N*th power of the light. Two photon spectroscopy allows study of states of the same multiplicity as the ground state.

intercavity absorption. Sample placed in the cavity of a broad band dye laser creates missing lines at the absorption frequencies of the sample. The sensitivity is enhanced over conventional absorption spectroscopy by a factor equal to the number of times the beam is reflected across the cavity.

frequency mixing. Investigation of the non-linear response of dipoles on irradiation by two lasers of different frequencies.

Doppler-free spectroscopy. Atomic spectrum of gases is broadened by the Doppler effect. This is overcome by saturating the sample with a beam travelling in nearly the opposite direction to the analysis beam.

time-resolved spectroscopy. Short (less than 1 ps) pulses permit investigation of relaxation processes and the life-times of excited species.

Lassaigne's test. Test for the presence of nitrogen, halogens or sulphur in an organic compound. The material is fused with sodium and the product dissolved in water. Iron(II) sulphate with a trace of iron(III) and hydrochloric acid gives Prussian blue as an indication of nitrogen. Sodium halides, if formed, are detected by the addition of silver nitrate, and sulphur is tested for by the addition of sodium nitroprusside.

latent heat. ENTHALPY change during a CHANGE OF STATE.

latex. Aqueous dispersion of a POLYMER (*see* COLLOID). Natural latex is the raw material exuded from rubber trees. Latexes are used as a convenient way of handling resins and polymers which may be spread, moulded or electrodeposited on surfaces.

laser
Properties and uses of some common lasers.

Type	Examples	Excitation	Wavelength range/μm	Uses
Solid state	Ruby Nd−YAG[a] F-centre	Optical	0.5−3.1	Range finding, pollution monitoring
Gas	Carbon dioxide	Glow/arc discharge	0.15−100	Precision measurement, welding, cutting, medical, microcircuit fabrication
Semiconductor	GaAs GaAlAs	Electric field	0.3−40	Communications, pollution monitoring
Chemical		Chemical reaction	1.5−100	
Liquid organic dye		Flash tube, laser	0.3−1.5	Pollution monitoring

[a] YAG = yttrium aluminium garnet.

lattice. *See* CRYSTAL STRUCTURE.

lattice energy (U).   Energy change accompanying the formation of 1 mole of solid from the constituent gaseous ions

$$M^+(g) + A^-(g) \rightarrow M^+A^-(s)$$

$$\Delta H = U - 2RT$$

Energy is usually released in the process, thus values of U are negative. It is numerically equal to, but opposite in sign to, the energy of dissociation of the crystal. Lattice energies are obtained from thermodynamic data using the BORN–HABER CYCLE. The Born–Landé expression for the calculation of lattice energy is obtained by evaluating the energy of attractive coulombic forces between oppositely charged ions and the energy of repulsive Born forces

$$U = U_C + U_B$$

$$= -N_A A z^2 e^2 (1 - 1/n)/4\pi\varepsilon_0 r_0$$

where A is the MADELUNG CONSTANT, n the Born 'exponent' determined experimentally from the compressibility of the crystal (range: 5–12), and r_0 (nm) the interionic distance obtained from X-ray measurements.

lattice planes.   Series of parallel and equidistant planes in which the points in a CRYSTAL LATTICE can be arranged in an indefinite number of ways. The faces of the complete crystal are parallel to these planes, the most common faces correspond to the planes with the largest number of points.

Laue's method.   *See* X-RAY DIFFRACTION.

laughing gas (dinitrogen oxide). *See* NITROGEN OXIDES.

lauryl alcohol.   *See* 1-HYDROXYDODE-CANE.

lawrencium (Lr).   *See* ACTINIDE ELEMENTS.

layer lattice.   Crystal structure in which the atoms are chemically bonded in plane layers with weak, van der Waals, forces between adjacent layers. Cleavage occurs parallel to the layers leading to flaking. The lubricant properties of graphite and molybdenum sulphide depend on the ability of one plane of atoms to slide easily over another.

LCAO.   *See* LINEAR COMBINATION OF ATOMIC ORBITALS.

leaching.   Extraction of soluble components of a solid mixture by percolating a solvent through it.

lead (Pb).   GROUP IV ELEMENT; METAL. It occurs naturally as galena (PbS) and cerussite ($PbCO_3$), from which it is extracted by roasting to lead monoxide (PbO) and smelting with carbon. Lead is the end product of the uranium (^{206}Pb), actinium (^{207}Pb) and thorium (^{208}Pb) radioactive series (*see* RADIOACTIVE DECAY SERIES). It is a soft, dense, malleable metal which, although ELECTROPOSITIVE ($E^{\ominus}_{Pb^{2+},Pb} = -0.13$ V), is very unreactive due to a passivating layer of oxide and high OVERPOTENTIAL for hydrogen evolution. The major uses of lead are the LEAD–ACID BATTERY and in tetraethyllead (*see* ANTIKNOCK ADDITIVE, LEAD ORGANIC DERIVATIVES). Lead is also used in noise and weatherproofing of buildings, the manufacture of paints, lead shot, high-quality glass (lead crystal), as shielding for X- and γ-radiation and formerly for water and gas pipes. Lead compounds are highly toxic, which has resulted in a reduction in use of lead for situations in which ingestion is possible (e.g., in paints).
$A_r = 207.2$, $Z = 82$, ES [Xe] $4f^{14}\, 5d^{10}\, 6s^2$ $6p^2$, mp 327°C, bp 1744°C, $\rho = 11.3 \times 10^3$ kg m^{-3}. Isotopes: ^{204}Pb, 1.5 percent; ^{206}Pb, 23.6 percent; ^{207}Pb, 22.6 percent; ^{208}Pb, 52.3 percent.

lead–acid battery.   Rechargeable electrochemical storage device (*see* Figure). Electrodes are made of a grid of lead–antimony alloy filled with a paste of litharge and red lead in sulphuric acid. When these plates are charged, the anode material is con-

Cathode	$PbO_2 + 3H^+ + HSO_4^- + 2e$	$\underset{\text{charge}}{\overset{\text{discharge}}{\rightleftharpoons}}$	$PbSO_4 + 2H_2O$

Anode	$Pb + HSO_4^-$	$\underset{\text{charge}}{\overset{\text{discharge}}{\rightleftharpoons}}$	$PbSO_4 + H^+ + 2e$

Overall	$Pb + PbO_2 + 2H^+ + 2HSO_4^-$	$\underset{\text{charge}}{\overset{\text{discharge}}{\rightleftharpoons}}$	$2PbSO_4 + 2H_2O$

lead–acid battery
Electrode reactions of lead–acid battery.

verted into porous lead dioxide (PbO_2) and the cathode material into spongy lead. This battery is robust, can provide high discharge currents and has a long life. The main drawback is the weight of lead which results in a low energy density. The operating voltage is near to the theoretical value of 2.02 V.

lead inorganic compounds. Lead, unlike the earlier GROUP IV ELEMENTS, forms stable lead(II) salts (*see* INERT PAIR EFFECT), as well as the expected lead(IV) compounds. *lead(II) compounds.* Most lead(II) salts, with the exception of the nitrate and acetate, are insoluble in water. Lead oxide (PbO, orange-red litharge and yellow massicot) is prepared by the oxidation of lead. It is used in the LEAD–ACID BATTERY, as a pigment and an insecticide. It is hydrolyzed in water and dissolves in excess alkali to give plumbates(II). The halides are anhydrous and form complexes $[PbX_n]^{2-n}$ (where $n = 1, 2, 3, 4$) in water. Many lead(II) salts have basic forms (e.g., the pigment white lead is $2PbCO_3.Pb(OH)_2$). The nitride ($Pb(N_3)_2$) is a detonator for explosives. *lead(IV) compounds.* These are covalent and oxidizing ($E^\ominus_{Pb^{4+},Pb^{2+}} = 1.69$ V). PbF_4 ($PbF_2 + F_2$) is a strong fluorinating agent. $PbCl_4$ is unstable and decomposes to $PbCl_2$ and Cl_2. $PbBr_4$ and PbI_4 are not known. The complex ions $[PbCl_6]^{2-}$ and $[PbF_6]^{2-}$ are found. The dioxide (PbO_2) is an oxidizing agent which gives on heating red lead (Pb_3O_4) and other intermediate phases (e.g., Pb_2O_3, Pb_7O_{11}). The unstable hydride (PbH_4) is known.

lead organic derivatives. Organic derivatives of lead(II) are not well defined. The most extensive chemistry is of alkyl derivatives of lead(IV), R_4Pb and R_3PbX (where X = halogen, H). Tetramethyllead and tetraethyllead are added to motor fuel as ANTIKNOCK ADDITIVES. They are prepared electrochemically (*see* ORGANIC ELECTROCHEMICAL SYNTHESES) or by the reaction of chloroethane with a sodium–lead alloy. Tetraethyllead can also alkylate mercury and is used in the preparation of fungicidal organomercury compounds. All organic lead compounds are highly toxic.

lead tetraethyl. *See* LEAD ORGANIC DERIVATIVES.

leaving group. Group or ion that departs during a substitution or displacement reaction (*see* S_N2 MECHANISM). Good leaving groups include halide, tosylate and methylsulphonate ions, H_2O, N_2 and CO_2.

Le Chatelier's principle. Constraint applied to a system in equilibrium tends to shift the equilibrium in such a way as to nullify the effect of the constraint. For the exothermic reaction $2SO_2 + O_2 \rightleftharpoons 2SO_3$ an increase in pressure displaces the equilibrium to the right, whereas an increase in temperature displaces it to the left. *See also* VAN'T HOFF ISOCHORE.

lecithins. *See* PHOSPHOLIPIDS.

Leclanché cell (dry cell). Primary CELL, the anode of which is zinc, which forms the container, in a paste of ammonium chloride with zinc and mercury chlorides. The cathode is a carbon rod packed round with powdered manganese dioxide and carbon (*see* Figure). The Leclanché cell is cheap to make and has a good energy density (55–77 W h kg^{-1}), but its discharge performance is inferior to that of an ALKALINE CELL. The voltage given by this cell is about 1.6 V.

Cathode	$2MnO_2 + 2H_2O + 2e$	$\longrightarrow$ $2MnO(OH) + 2OH^-$
Anode	$Zn + 2OH^- + 2NH_4^+$	$\longrightarrow$ $Zn(NH_3)_2^{2+} + 2H_2O + 2e$
Overall	$Zn + 2MnO_2 + 2NH_4^+$	$\longrightarrow$ $Zn(NH_3)_2^{2+} + 2MnO(OH)$

Leclanché cell
Electrode reactions of Leclanché cell.

LEED. *See* LOW-ENERGY ELECTRON DIFFRACTION.

Lemieux reagent. Aqueous solution of sodium periodate and a trace of potassium permanganate (or osmium tetroxide). ALKENES are converted to KETONES and CARBOXYLIC ACIDS by this reagent.

Lennard–Jones (12,6) potential. Two types of force between molecules can be distinguished: (1) repulsive forces due to electrostatic repulsion between the outer electron clouds of the molecules; (2) attractive forces due to a correlation of the position of the electrons in one molecule with those in another in such a way that net electrical attraction occurs (*see* INTERMOLECULAR FORCES). Repulsive forces increase steeply with decreasing separation of the molecules and so the total intermolecular energy (i.e. algebraic sum of the repulsive and attractive energy) follows a steeply rising curve (*see* Figure). Although the behaviour at short distances is complicated, a good approximation to the curve is given by the Lennard–Jones equation

$$E = 4\varepsilon[(\sigma/r)^{12} - (\sigma/r)^6]$$

At small values of r the repulsive term $(\sigma/r)^{12}$ dominates, whereas at larger separations the negative attractive term is dominant. The parameter ε is the depth of the minimum of the curve which occurs at $r_{min} = 2^{1/6}\sigma$.

lepidolite. *See* LITHIUM, RUBIDIUM.

lepton. *See* ELEMENTARY PARTICLES.

Leu (leucine). *See* AMINO ACIDS.

leucine (Leu). *See* AMINO ACIDS.

leuco compounds. Colourless, soluble derivatives of DYES. For example the reduction of a VAT DYE containing an OXO GROUP gives the leuco compound containing C–ONa. *See also* INDIGOTIN.

Levich equation. *See* ROTATING DISC ELECTRODE.

Lewis acids and bases. According to G.N. Lewis an acid is a substance that can accept an electron pair from a base and a base is a substance that can donate an electron pair to an acid. Bases are similar to those defined by the Brønsted classification (*see* BRØNSTED–LOWRY ACIDS AND BASES), but Lewis acids comprise most cations and many other species (e.g., BF_3, $AlCl_3$, SO_3). Neutralization is the reaction between an acid and a base to form an adduct (*see* ADDITION COMPOUND). The theory encompasses ACID–BASE CATALYSIS and reactions between metals and ligands (*see* COORDINATION COMPOUND). However, obvious acids such as sulphuric acid and hydrochloric acid do not easily fit the theory, and the strengths of acids depends very much on the base. *See also* HARD AND SOFT ACIDS AND BASES.

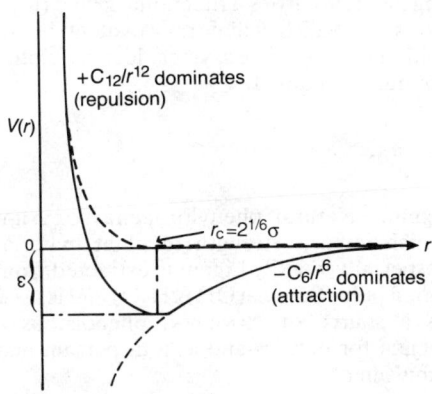

Lennard–Jones (12,6) potential
Repulsive and attractive forces.

ligand field stabilization energy
Ligand field stabilization energy for an octahedral complex in units of D_q ($\Delta_0/10$).

d	0	1	2	3	4	5	6	7	8	9	10
Low spin	0	−4	−8	−12	−16	−20	−24	−18	−12	−6	0
High spin	0	−4	−8	−12	−6	0	−4	−8	−12	−6	0

lewisite   (2-chloroethenyldichloroarsine, ClCH=CHAsCl$_2$). Pale yellow liquid (bp 190°C, with decomp.), prepared by passing ethyne through a mixture of anhydrous arsenic trichloride and aluminium trichloride. It is hydrolyzed by water and destroyed by alkalis and by oxidizing agents. It is a systemic poison and has been used as a war gas. The dithiol HSCH$_2$CH(SH)CH$_2$-OH (British anti-lewisite) is an effective antidote for poisoning by lewisite and by other heavy metals.

LFSE.   *See* LIGAND FIELD STABILIZATION ENERGY.

LH (luteinizing hormone).   *See* HORMONES.

Li.   *See* LITHIUM.

Librium.   Trade name for hydrochloride of CHLORDIAZEPOXIDE.

Liebermann's nitroso reaction. *See* NITROSAMINES.

ligand.   Group or atom that coordinates a metal ion. *See also* CHELATE LIGAND, COMPLEX COMPOUND, COORDINATE BOND, COORDINATION COMPOUND.

ligand field stabilization energy (LFSE). Difference in energy between a given electronic configuration of a transition metal complex and that of a reference state in which each *d*-electron has an average energy. Values for an octahedral complex are tabulated (*see* Table) showing the differences between low-spin and high-spin states (*see* HIGH-SPIN COMPLEX). *See also* LIGAND FIELD THEORY.

ligand field theory.   Use of MOLECULAR ORBITAL theory in describing the interactions between a transition metal (only the *d*-orbitals are considered) and a set of ligand orbitals. It is an extension of CRYSTAL FIELD THEORY in that ligands are no longer treated as simple electrostatic point charges. Crystal field splitting ($10D_q$, Δ_0) then arises as the energy between the nonbonding and anti-bonding molecular orbitals of the complex. *See also* HIGH-SPIN COMPLEX, LIGAND FIELD STABILIZATION ENERGY.

ligand immobilization.   DERIVATIZATION of a solid support (e.g., silica, cellulose) for use in AFFINITY CHROMATOGRAPHY, HIGH-PERFORMANCE LIQUID AFFINITY CHROMATOGRAPHY and HIGH-PERFORMANCE LIQUID CHROMATOGRAPHY. The support is derivatized to minimize general or nonspecific adsorption and to provide a suitable terminal functional group on the surface for the covalent attachment of ligands. The hydrophilic spacer attached to the supporting medium carries a functional group (e.g., epoxide) which will form covalent bonds with ligands (e.g., enzymes, lectins, biotin, immunoglobulins).

ligases.   *See* ENZYMES.

lignin.   Natural phenylpropene POLYMER which imparts strength to plant material (especially wood). Lignin is extracted from wood pulp (SO$_2$/Ca(OH)$_2$/H$_2$O) and is used as a source of VANILLIN, phenols, as a FILLER for plastics and as a dispersant and emulsifier.

lignite.   *See* COAL.

lignoceric acid. *See* TETRACOSANOIC ACID.

ligroin. *See* PETROLEUM.

lime. *See* CALCIUM COMPOUNDS.

limestone. *See* CALCIUM.

limiting conductivity. *See* INFINITE DILUTION.

limiting current density (i_L). Maximum possible current density at an electrode when the rate of the reaction is governed by transport of reactants to the electrode. This may be determined from solutions to Fick's equations (*see* FICK'S LAWS OF DIFFUSION). Using a simple model of diffusion near an electrode $i_L = nFDc_{bulk}/\delta$, where D is the diffusion coefficient and δ the thickness of the (Nernst) diffusion layer; δ is of the order of 0.5 mm in unstirred solutions.

limiting density. *See* DENSITY.

limonene. *See* TERPENES.

limonite. *See* IRON.

linalool. *See* TERPENES.

lindane. *See* 1,2,3,4,5,6-HEXACHLOROCYCLOHEXANE.

Lindemann theory. Explanation of the kinetics of unimolecular gas-phase reactions that show first-order kinetics at high pressures, but second-order kinetics at low pressures (*see* ORDER OF REACTION). The model is of a bimolecular collision (*see* MOLECULARITY OF REACTION) which produces an excited molecule (A*). A* may be deactivated by collision with another molecule of A or suffer a unimolecular decomposition to products (P).

$$A + A \underset{k_{-2}}{\overset{k_2}{\rightleftharpoons}} A^* + A; \quad A^* \overset{k_1}{\to} P$$

By the STEADY-STATE APPROXIMATION, $d[P]/dt = k_1k_2[A]^2/(k_1 + k_{-2}[A])$. At low pressures, $k_1 \gg k_{-2}[A]$ and so $d[P]/dt = k_2[A]^2$. At higher pressures, $k_{-2}[A] \gg k_{-1}$, therefore $d[P]/dt = k_1k_2/k_{-2}[A]$. Limitations of the theory arise from the assumption that all A* may react to give product. In complex molecules, the energy may be contained in vibrations that do not lead to product (i.e. A* is not necessarily the ACTIVATED COMPLEX).

Linde process. Method for LIQUEFACTION OF GASES based on the JOULE–THOMSON EFFECT. For liquefaction of air, the air is first freed of carbon dioxide and water, and compressed to 150 atm. The compressed gas passes through a copper coil (part of a heat exchanger) to an expansion nozzle in a Dewar flask. The emerging cooled air passes back through a second coil which surrounds the first. The expanded air cools the incoming air and eventually when it is below its critical temperature, at the high pressure (in excess of P_c) the air liquefies. For hydrogen and helium, the gases must be cooled (using liquid air or liquid hydrogen, respectively) below their inversion temperatures before they are expanded through the nozzle.

linear. *See* SHAPES OF MOLECULES.

linear combination of atomic orbitals (LCAO). Approximation to a molecular orbital as a sum of relevant atomic orbitals. Thus for a simple diatomic molecule A–B, a suitable molecular orbital would be $\psi = c_A\phi_A + c_B\phi_B$, where c_A and c_B are coefficients which may be determined by the variational method (*see* VARIATION PRINCIPLE).

linear free energy relationships. Linear relationship between free energies of activation for one homologous series of reactions and those of another. An equivalent relationship applies to free energies for the overall reactions. The HAMMETT EQUATION is equivalent to the existence of such a relationship.

$$\Delta G^{\ddagger} - (\rho/\rho')\Delta G'^{\ddagger} = \text{constant}$$

where $\Delta G^{\ddagger}$ and $\Delta G'^{\ddagger}$ and ρ and ρ' are the free energies of activation and reaction constants, respectively, for one homologous series of reactions and those for another, respectively (e.g., the plot of the free energy of activation for the benzoylations of substituted anilines against the free energy of dissociation of the aniline for a range of *meta-* and *para-*substituents is linear). The applicability of the equation is usually quite poor for *ortho-*substituents, the reason being that these exert not only an electron-attracting or repelling effect, but also a steric effect which leads to a decrease in the rate of reaction.

linear momentum. *See* MOMENTUM.

linear sweep voltammetry. Electrochemical technique in which the voltage at an electrode is scanned at a constant rate between limits while the current is displayed on a recorder or oscilloscope. In cyclic voltammetry, the voltage is scanned backwards and forwards between limits. Information is obtained about adsorption kinetics and intermediate species formed on surfaces.

line defect. *See* DEFECT STRUCTURE.

line spectra. *See* ATOMIC SPECTROSCOPY.

Lineweaver–Burk plot. *See* ENZYME KINETICS.

linkage isomerism. *See* ISOMERISM.

linoleic acid. *See* cis,cis-9,12-OCTADECADIENOIC ACID.

linolenic acid. *See* cis,cis,cis-9,12,15-OCTADECATRIENOIC ACID.

lipases. *See* ENZYMES.

lipids. Naturally occurring substances of a fat-like nature, characterized by their solubility in non-polar solvents. Strictly they are fatty acids or their derivatives. In practice, they vary from simple FATS and waxes to more complex PHOSPHOLIPIDS, glycolipids, TERPENES and STEROIDS.

lipoic acid (vitamin B complex). *See* VITAMINS.

lipophilic. 'Oil-loving'; applied to, for example, the hydrocarbon chain of a SURFACE-ACTIVE AGENT.

lipopolysaccharides. One of the main constituents, with lipoprotein, of the cell wall of gram-negative BACTERIA. They are responsible for the antigenic specificity of these organisms.

lipoproteins. *See* PROTEINS.

liposome. Spherical, microscopic, membrane-enclosed vesicle (20–30 nm diameter); prepared by the addition of an aqueous solution to a phospholipid gel. It is similar to a cell organelle, and the membrane resembles a cell membrane. Liposomes can be incorporated into living cells and are used to transport relatively toxic drugs into diseased cells. They are also used as models to investigate the behaviour of membranes (e.g., during anaesthesia) with respect to permeability changes.

liquefaction of gases. Conversion of a gaseous substance into the liquid form. The following methods are available (often a combination of methods is used): (1) vapour compression, provided the temperature is below the critical temperature; (2) cascade process (i.e. refrigeration at constant pressure by cooling the gas with successively colder fluids in a counter-current heat exchange system); (3) adiabatic expansion against the atmosphere in a reversible cycle; (4) JOULE–THOMSON EFFECT (*see* LINDE PROCESS). Large quantities of liquid gases are used industrially as coolants and as convenient sources of the pure gas.

liquid air. *See* LINDE PROCESS, LIQUEFACTION OF GASES.

liquid crystals. Certain organic compounds in a transition state between the solid and liquid forms. On heating the solid a cloudy liquid, with an ordered structure, is first formed; this changes to a clear liquid at higher temperatures. Both transitions occur at fixed temperatures. They are anisotropic, viscous, jelly-like materials resembling liquids in certain respects (e.g., viscosity) and crystals in other properties (e.g., birefringence and reflection). This behaviour results from the orientation of molecules parallel to each other in layers. Compounds that form liquid crystals possess a high degree of asymmetry which allows little rotation in the liquid state at low temperatures. The long thin crystals (e.g., ammonium oleate) showing only one-dimensional order are nematic crystals; the planar molecules (e.g., esters of cholesterol) exhibiting two-dimensional order arranged in distinct layers are cholesteric (axes of molecules parallel to plane of layer) or smetic (axes of molecules perpendicular to plane of layers) crystals. The application of a voltage causes disorder; hence their use in electronic display units. Cholesteric-type liquid crystals which scatter coloured polarized light are used in devices dependent on temperature determination.

liquid–gas equilibrium. *See* ONE-COMPONENT SYSTEM.

liquid junction potential. Potential difference established across the interface between two dissimilar electrolyte solutions due to diffusion of ions of different TRANSPORT NUMBER across the interface. For solutions of activities a_1 and a_2 between electrodes reversible to the cation of a 1:1 salt, the junction potential is

$$E_j = (2t_- - 1)(RT/nF) \ln (a_1/a_2)$$

Liquid junction potentials may be minimized by the use of a salt bridge consisting of a concentrated solution of potassium chloride, potassium nitrate, ammonium chloride, etc. in agar for which $t_+ = t_-$.

liquid–liquid chromatography. *See* CHROMATOGRAPHY.

liquid–liquid extraction. Extraction technique in which a component from a liquid mixture is partitioned (distributed) into a second immiscible solvent in which the component is more soluble (e.g., extraction of benzoic acid from an aqueous salt solution by shaking with toluene). The solution depleted in that component is known as the raffinate. *See also* COUNTER-CURRENT DISTRIBUTION.

liquid natural gas (LNG). Natural gas (i.e. METHANE) liquefied at –160°C at atmospheric pressure.

liquid oxygen explosives (LOX explosives). EXPLOSIVES based on liquid oxygen and a fuel, generally carbon black.

liquid petroleum gas (LPG). *See* PETROLEUM GAS.

liquid–solid equilibrium. *See* ONE-COMPONENT SYSTEM.

liquidus curve. *See* COMPLETELY MISCIBLE LIQUID SYSTEMS, TWO-COMPONENT CONDENSED SYSTEM.

liquid–vapour equilibrium. *See* ONE-COMPONENT SYSTEM.

lithium (Li). Group Ia ALKALI METAL. It is found naturally in the aluminosilicate ores lepidolite $(Li_2(F,OH)_2Al_2.(SiO_3)_3)$ and spodumene $(LiAl(SiO_3)_2)$. The ores are concentrated and converted to the chloride by an acid process—(1) H_2SO_4, (2) Na_2CO_3, (3) HCl—or an alkali process—(1) $CaCO_3$, (2) HCl. Lithium is extracted by electrolysis from a lithium chloride/potassium chloride eutectic. Lithium is a light, soft, silvery metal with the highest melting and boiling points of the alkali metals. It has a high specific heat ($34 \ kJ \ kg^{-1}K^{-1}$ at 0°C), but its corrosive nature precludes its extensive use in heat exchangers. Lithium dissolves in liquid ammonia giving a blue, electrically conducting solution. Uses of lithium are primarily in the preparation of lithium octadecanoate, a thickener for greases, and of lithium tetrahydridoaluminate (*see* ALUMINIUM INORGANIC COMPOUNDS) and as a flux in ceramics. Lithium reacts with oxygen to form the oxide, with water to give the hydroxide, with nitrogen (Li_3N) and with hydrogen (LiH).

$A_r = 6.941$, $Z = 3$, ES $[He]2s^1$, mp 180°C, bp 1330°C, $\rho = 0.531 \times 10^3 \ kg m^{-3}$. Isotopes: ^6Li, 7.3 percent; ^7Li, 92.7 percent.

lithium inorganic compounds. Lithium readily loses its 2s-electron to give Li^+ ($E^\ominus_{Li^+,Li} = -3.01 \ V$). Salts with the common anions are formed, all of which are ionic. Li^+ is small with a high charge density and so is strongly hydrated in water. The salts often crystallize with water of crystallization. Solutions of Li^+ show deviations from ideal behaviour in properties that depend on the ionic size. The most important compound is lithium hydroxide $(LiOH.H_2O)$ which is used industrially to prepare the octadecanoate and as a starting point for other lithium compounds. LiOH is a strong base, thus lithium salts of weak acids are hydrolyzed to give alkaline solutions. Lithium fluoride (LiF) is used as a flux and in IR SPECTROSCOPY because of its excellent dispersion in the region 1600–4000 cm^{-1}.

Compounds are formed directly with non-metallic elements (e.g., Li_3N, Li_2O (lithia), LiH, LiC, Li_2S). Lithium also reacts with ammonia to give the amide ($LiNH_2$), which on heating gives the imide (Li_2NH). Lithium carbonyl is unstable. The tetrahydridoborate ($LiBH_4$, *see* BORON INORGANIC COMPOUNDS) and tetrahydridoaluminate ($LiAlH_4$, *see* ALUMINIUM INORGANIC COMPOUNDS) are reducing agents, providing sources of H^- in organic reactions.

lithium organic derivatives. Organolithium compounds are more reactive than GRIGNARD REAGENTS. Alkyl- and aryllithium compounds are prepared from the chloride and lithium metal in benzene or hydrocarbon solvent

$$RCl + 2Li \rightarrow RLi + LiCl$$

Organolithium compounds are covalent. The structure of methyllithium has a methyl group sitting above each face of a tetrahedron of lithium. Each methyl coordinates to three lithium atoms. The bonding is similar to the multicentre bonds found in organoberyllium (*see* BERYLLIUM COMPOUNDS) and organoaluminium compounds. Further organolithium compounds are formed by exchange reactions with hydrogen, other metals or halogens in organic molecules (e.g., 3-bromopyridine gives 3-lithopyridine). Lithium reacts with ethynes to give $LiC\equiv CR$, used in the synthesis of vitamin A. All organolithium compounds react violently with air and water. Polymerization of 1-methyl-1,3-butadiene to a synthetic rubber is catalyzed by lithium alkyls with a high degree of stereospecificity.

lithocholic acid. *See* BILE ACIDS.

lithopone (Charlton white, Orr's white). White pigment 70 percent barium sulphate, 30 percent zinc sulphide.

litmus. INDICATOR not widely used.

litre (l). Unit of volume, defined as the volume of 1 kilogram of pure water at 4°C, equivalent to 1.000028 dm^3. It is not coherent with SI units.

lixivation. Separation of solid mixtures by LEACHING out soluble material.

LNG. *See* LIQUID NATURAL GAS.

London forces. *See* INTERMOLECULAR FORCES.

lone-pair electrons. ELECTRON PAIR in a molecule that does not take part in bonding. It may determine the shape of the molecule (*see* VALENCE SHELL ELECTRON PAIR REPULSION THEORY) or be donated to become a COORDINATE BOND.

lone-pair orbital. Non-bonding molecular orbital containing two electrons.

long-range forces. *See* INTERMOLECULAR FORCES, LENNARD–JONES (12,6) POTENTIAL.

Loschmidt's constant (L). Number of molecules per unit volume of an ideal gas at stp. $L = 2.686754 \times 10^{25}$ m^{-3}. $L = N_A/V_m$.

low-energy electron diffraction (LEED). Thermal electrons generated from a hot filament are allowed to diffract from the surface of a crystalline solid. Diffraction patterns may be recorded on a photographic plate and give information of the surface structure in a manner analogous to X-RAY DIFFRACTION for the entire crystal. The intensity of the refracted beam may also be measured as a function of electron energy. Surface faces are usually written (h, k, l) where h, k, l are the MILLER INDICES of the plane to which the surface is parallel. If the surface layer is significantly different from the underlying structure (e.g., in the case of adsorption), the structure is written ($n \times m$) (e.g., Si(100) (7 × 7)), where the unit cell of the new structure is n and m times that of the underlying one along each axis.

lowest unoccupied molecular orbital (LUMO). Empty molecular orbital of lowest energy. The highest occupied molecular orbital (HOMO) is the occupied molecular orbital of highest energy. When considering the possible formation of a bond between two molecules, the relative energies and shapes of the HOMO (nucleophile) and LUMO (electrophile) are of importance.

Lowry–Brønsted theory. *See* BRØNSTED–LOWRY ACIDS AND BASES.

LOX explosives. *See* LIQUID OXYGEN EXPLOSIVES.

LPG (liquid petroleum gas). *See* PETROLEUM GAS.

Lr. Lawrencium (*see* ACTINIDE ELEMENTS).

LSD (Lysergsäure Diethylamid, lysergic acid diethylamide). *See* ALKALOIDS.

***L*-shell.** *See* ELECTRONIC CONFIGURATION, QUANTUM NUMBER.

LTH (luteotrophic hormone). *See* HORMONES.

Lu. Lutetium (*see* LANTHANIDE ELEMENTS).

luciferins. Albumins present in fireflies which, under the influence of the enzyme luciferase and in the presence of ATP, are responsible for BIOLUMINESCENCE.

luminal. *See* BARBITURATES.

luminescence. Term covering all types of emission of light other than those arising from elevated temperatures. It includes PHOSPHORESCENCE and FLUORESCENCE. *See also* CHEMILUMINESCENCE, PHOTOLUMINESCENCE.

LUMO. *See* LOWEST UNOCCUPIED MOLECULAR ORBITAL.

lunar caustic (silver nitrate). *See* SILVER COMPOUNDS.

Lurgi process. *See* GASIFICATION.

lutein. *See* CAROTENOIDS.

luteinizing hormone (LH). *See* HORMONES.

luteotrophic hormone. *See* HORMONES.

lutetium (Lu). *See* LANTHANIDE ELEMENTS.

lutidines. *See* METHYLPYRIDINES.

lyases. *See* ENZYMES.

lycopene. *See* TERPENES.

Lyman series. *See* BALMER SERIES.

lyophilic colloid. COLLOID that is solvated by the dispersion medium (a hydrophilic colloid if the solvent is water). Such colloids are stable, often viscous and may set to a GEL on cooling. They are soluble macromolecular substances such as proteins or polymers or MICELLES (e.g., soaps and dyes). Lyophilic colloids, unlike LYOPHOBIC COLLOIDS, are not coagulated by added electrolyte or at the ISOELECTRIC POINT.

lyophobic colloid. COLLOID that is not solvated (a hydrophobic colloid if the solvent is water), but remains stable by electrostatic repulsion of the charged particles. Addition of electrolyte neutralizes the particles causing COAGULATION. They are also coagulated, for the same reason, at the ISOPOTENTIAL POINT. Lyophobic colloids may be stabilized by the addition of a LYOPHILIC COLLOID which adsorbs on the colloid surface (*see* PROTECTIVE COLLOID). Typical hydrophobic colloids are sols of gold, arsenic, alumina or silica. The term hydrophobic is used for any molecules or parts of molecules that are not solvated by water, for example, the hydrocarbon chain of a SURFACE-ACTIVE AGENT. *See also* DLVO THEORY, SCHULZE–HARDY RULE.

lyotropic series. LYOPHILIC COLLOIDS are salted out (*see* SALTING OUT) by high concentrations of ions which compete for water of crystallization. These may be arranged in a series of increasing power to salt out a colloid: Cs^+, Rb^+, NH_4^+, K^+, Na^+, Li^+, Ba^{2+}, Sr^{2+}, Ca^{2+}, Mg^{2+}, Al^{3+}, and CNS^-, I^-, NO_3^-, Cl^-, SO_4^{2-}, citrate^{3-}.

Lys (lysine). *See* AMINO ACIDS.

lysergic acid. *See* ALKALOIDS.

lysergic acid diethylamide (LSD). *See* ALKALOIDS.

lysine (Lys). *See* AMINO ACIDS.

lysol. Solution of isomeric cresols (*see* HYDROXYTOLUENES) in soapy water which is used as a disinfectant.

lysosome. Particle present in cells that contains hydrolytic enzymes. These enzymes, if not kept separate, could kill the cell; damage to the lysosome membrane leads to enzyme release, autolysis and cell death.

lysozyme. *See* ENZYMES.

lyxose. *See* PENTOSES.

M

M. *See* MOLARITY.

m. *See* METRE.

m. Magnetic quantum number (*see* QUANTUM NUMBER); MASS, MOLALITY.

m-. *See* META-.

μ. *See* CHEMICAL POTENTIAL, MAGNETIC MOMENT, REDUCED MASS.

μ$_s$. *See* SPIN MOMENT.

macrolides. *See* ANTIBIOTICS.

macromolecular crystal. Crystalline solid in which the atoms are all linked by covalent bonds; examples include diamond, boron nitride and silicon carbide. All very hard materials with high melting points. *See also* CRYSTAL FORMS.

macromolecule. Molecule of high molecular mass (usually > 10 000). Examples include proteins, haemoglobin, and natural and synthetic polymers.

Madelung constant (*A*). Constant that depends on the arrangement of the positive and negative ions in the crystal and not on the dimensions of the individual lattice. Its value is calculated by summing the mutual potential energies of all the ions in the lattice. Values for various lattices are: rock salt 1.747 56; zinc blende 1.638 05; fluorite 2.519 39; rutile 2.408. *See also* CRYSTAL FORMS, LATTICE ENERGY.

magic acid (anhydrous $HSO_3F.SbF_5$). SUPERACID that ionizes by $SbF_5 + 2HSO_3F \rightarrow H_2SO_3F^+ + SbF_5(SO_3PF)^-$. The addition of SbF_5 increases the acidity of HSO_3F from $H_0 = -15.1$ to about -20 (*see* HAMMETT ACIDITY FUNCTION). Many organic compounds protonate in magic acid; for example 2-methyl-2-propanol protonates then dehydrates to the 2-methylpropyl cation.

magnesia. *See* MAGNESIUM, MAGNESIUM COMPOUNDS.

magnesite. *See* MAGNESIUM.

magnesium (Mg). Group IIa ALKALINE EARTH METAL (*see* GROUP II ELEMENTS). It is found naturally in dolomite (*see* CALCIUM), carnallite (*see* POTASSIUM) and magnesite ($MgCO_3$), and in seawater (0.13 percent magnesium chloride). It is extracted from seawater by precipitation of magnesium hydroxide ($Mg(OH)_2$), conversion to magnesium chloride ($MgCl_2$) and electrolysis, or from dolomite by the thermal ferrosilicon process. The metal is light and silvery. It burns in air with a brilliant light to the oxide (magnesia), reacts with nitrogen to the nitride (Mg_3N_2) and carbon to the carbide (MgC_2) and then to Mg_2C_3. It is protected from extensive corrosion by a thin film of oxide. Mangesium is used in lightweight alloys, as a sacrificial anode (*see* CORROSION), oxygen scavenger and as a battery anode in metal–air cells. Compounds are used in ceramics, medicines (e.g., milk of magnesia), sugar-refining, refractories and cements, paper-making and tanning. Magnesium is an essential element for life.
A_r = 24.305, Z = 12, ES [Ne] $3s^2$, mp 650°C, bp 1110°C, ρ = 1.74 × 10^3 kg m^{-3}. Isotopes: ^{24}Mg, 78.6 percent; ^{25}Mg, 10.1 percent; ^{26}Mg 11.3 percent.

magnesium amalgam. *See* AMALGAMS.

magnesium compounds. Magnesium is a reactive ELECTROPOSITIVE metal ($E^{\ominus}_{Mg^{2+}.Mg}$ = -0.238 V), forming Mg^{2+} in its ionic compounds. Covalent character is seen in its organometallic chemistry (*see* GRIG-NARD REAGENTS). The hexaaquo ion is not hydrolyzed, and many salts crystallize with six molecules of water. The affinity of Mg^{2+} for water leads to the use of anhydrous salts as drying agents (e.g., $Mg(ClO_4)_2$). Magnesium compounds are used extensively by industry and agriculture. All have medicinal uses particularly milk of magnesia $(Mg(OH)_2)$, and Epsom salts $(MgSO_4)$. Magnesium chloride $(MgCl_2)$ is the starting point for magnesium metal and is also used to fireproof wood. Magnesia (MgO) is a refractory (mp 2800°C) and is used in fertilizers.

magnetic moment (μ; units: A m^{-2}). For a paramagnetic substance (*see* PARAMAGNETISM)

$$\mu = (3k/N_A)^{1/2}[\chi_M(T-\theta)]^{1/2}$$

where χ_M is the molar magnetic susceptibility and θ is the Curie temperature. *See also* CURIE'S LAW.

magnetic quantum number. *See* QUANTUM NUMBER.

magnetic resonance. *See* NUCLEAR MAGNETIC RESONANCE SPECTROSCOPY.

magnetic susceptibility (χ—per unit mass; χ_M—per mole; κ—per unit volume). For a substance placed in a magnetic field of strength H the magnetic induction B is given by $B = H + 4\pi I$, where I is the intensity of magnetization and $I/H = \kappa$, $\chi = \kappa/\rho$ (ρ is the density). The molar magnetic susceptibility $\chi_M = \chi M_r$. *See also* DIAMAGNETISM, FERROMAGNETISM, PARAMAGNETISM.

magnetite. *See* IRON, SPINELS.

malachite. *See* COPPER.

malachite green. *See* DYES.

maleic acid. *See* BUTENEDIOIC ACIDS.

maleic anhydride. *See* BUTENEDIOIC ACIDS.

malic acid. *See* HYDROXYBUTANEDIOIC ACID.

malonic acid. *See* PROPANEDIOIC ACID.

malonic ester. *See* PROPANEDIOIC ACID.

malononitrile. *See* DICYANOMETHANE.

malonyl urea. *See* BARBITURATES.

maltase (α-glucosidase). ENZYME, specific for splitting the α-glucoside linkage of maltose, found in the digestive systems of animals and in yeasts.

maltose. DISACCHARIDE (structure II); obtained by the partial acid hydrolysis of starch (mp 160–65°C). It is dextrorotatory. It is composed of two D-(+)-glucose units, differing from sucrose in that it is a reducing sugar; it forms an oxime and an osazone, and undergoes mutarotation. Hydrolysis by acids or the enzyme MALTASE yields two molecules of glucose. The hydroxyl group on the C-4 of one unit is involved in the glycoside linkage. It is used in brewing, soft drinks and food industry.

manganates(III), (IV), (V), (VI), (VII). *See* MANGANESE COMPOUNDS.

manganese (Mn). Group VIIb, first row TRANSITION ELEMENT. It is extracted by reducing the roasted ores (principally pyrolusite, MnO_2) with aluminum or carbon. Nodules of manganese are also found on the sea bed. Pure manganese is electroplated (*see* ELECTRODEPOSITION OF METALS) from solutions of manganese sulphate ($MnSO_4$) buffered with ammonium sulphate. The pure metal has few uses, being very easily oxidized in air, but is a constituent of ferrous STEEL, to which it is added as a deoxidant and desulphurizing agent. Manganese compounds are used in batteries (*see* LEAD–ACID BATTERY), dyes, paints and in chemical processes. Four crystal phases of the metal are known α, β (cubic), γ (fcc) and δ, the last being stable only at high temperatures.
$A_r = 54.9380$, Z = 25, ES [Ar] $3d^5 4s^2$, mp 1244°C, bp 1962°C, $\rho = 7.20 \times 10^3$ kg m^{-3}. Isotope: ^{55}Mn, 100 percent.

manganese compounds. Oxidation states –1 to +7 are found, of major importance are manganese(II) (manganous), manganese(III) (manganic), manganese(IV) (as

MnO_2) and manganese(VII) (as MnO_4^-, permangante). Manganese(II) ($E_{Mn^{2+},Mn}^{\ominus} = -1.03$ V) has a range of salts with anions (e.g., $MnCl_2$, $MnSO_4$). Their pale pink colour arises from the high-spin electronic structure of the d^5-ion (*see* HIGH-SPIN COMPLEX), which has no electronic transition which preserves spin multiplicity (*see* MULTIPLICITY, SELECTION RULES). Complex salts are formed by manganese(II) and manganese(III) (e.g., the perovskite $KMnO_3$) (*see* PEROVSKITE STRUCTURE) and K_4MnCl_6. Manganese(III) complexes show a JAHN–TELLER EFFECT and are oxidizing ($E_{Mn^{3+},Mn^{2+}}^{\ominus} = 1.51$ V). MnO_2 is used extensively as an oxidizing agent and as a battery anode (*see* LEAD–ACID BATTERY). Manganates are OXYANIONS of manganese. Potassium permanganate ($KMnO_4$) is a strong oxidizing agent used in analysis (*see* TITRATIONS) ($E_{MnO_4^-,Mn^{2+}}^{\ominus} = 1.49$ V; $E_{MnO_4^-,MnO_2}^{\ominus} = 1.68$ V). MnO_2^{2-} forms deep green salts which are stable in basic solutions. Hypomanganates (manganates(V), $[MnO_4]^{3-}$) are unstable in water. Manganates(IV) and (III) are found as mixed oxides: for example, $Mn^{II}Mn_2^{III}O_4$ has the spinel structure (*see* SPINELS).

mannans. *See* POLYSACCHARIDES.

Mannich reaction. CONDENSATION REACTION of an aldehyde, $[R_2NH_2]^+ Cl^-$ (R=H or alkyl) and a compound containing an active hydrogen atom (e.g., ketone, β-ketoester, nitroalkane). The reaction is catalyzed by acids or bases. For example

$$R_2NH + R'CHO \rightarrow R_2\overset{+}{N}=CHR' + OH^-$$

$$R_2\overset{+}{N}=CHR' + R''COCH_3 \rightarrow$$

$$R''COCH_2\overset{\cdot}{C}HR'\overset{+}{N}HR_2 \xrightarrow{-H^+}$$

$$R''COCH_2CHR'NR_2$$

The products are known as Mannich bases and are useful in syntheses.

mannitol. *See* HEXAHYDRIC ALCOHOLS.

mannose. HEXOSE. D-(+)-Mannose (mp 132°C) is obtained by carefully oxidizing mannitol with nitric acid. Commercially it is obtained by the hydrolysis of the mannan seminine, which occurs in the shell of the ivory nut.

mannuronic acid. *See* URONIC ACIDS.

marble. *See* CALCIUM.

marijuana. *See* TETRAHYDROCANNABINOLS.

Markownikoff's rules. During addition to a double bond the more negative part of the addendum adds to the carbon atom that is joined to the least number of hydrogen atoms. Thus HCl adds to $CH_3CH=CH_2$ to give $CH_3CHCl–CH_3$ and not $CH_3CH_2–CH_2Cl$. *See also* ALKENES.

marsh gas. *See* METHANE.

Marsh's test. Sensitive test for ARSENIC. The volatile arsine (AsH_3) (Zn + As compound + HCl) is passed through a heated glass tube where it decomposes to give a brown deposit of arsenic. ANTIMONY reacts in a similar manner, but the antimony stain is insoluble in sodium hypochlorite (NaOCl) solution.

mass (*m*; dimensions: *m*; units: kg). Measure of a body's tendency to resist a change of motion. Mass and WEIGHT are often used synonymously; weight depends on where it is measured because of the variation of g with location, but mass is constant wherever it is measured. According to Einstein's theory, if the energy of a body increases then its mass increases.

$$\Delta m = \Delta U/c^2$$

The increase in mass is only significant for extremely high energies.

mass action, law of. Rate at which a chemical reaction occurs at a given temperature is proportional to the active masses (usually taken as concentrations or activities) of the reactants. For the reaction A + B → C

$$\text{rate of reaction} = v = k[A][B]$$

where *k* is the rate or velocity constant.

mass average molecular mass. *See* NUMBER AVERAGE MOLECULAR MASS.

mass defect. Difference between the mass of the nucleus (*M*) and the total mass of its individual NUCLEONS. It is the mass equiva-

mass spectrometry
Ionization methods in mass spectrometry.

Method	Ionization process	Comments
Electron impact (EIMS)	Beam of electrons (6–100 V) $$M + e \rightarrow M^+ + 2e$$	Most common method. Produces many fragment ions especially with high-energy electrons
Chemical ionization (CIMS)	Ionized reagent gas (e.g., CH_4^+, NH_3^+, SF_6^-) used to effect ionization. $$MH + NH_3^+ \rightarrow MNH_4^+$$ $$M + CH_4^+ \rightarrow M^+ + CH_4$$	Reagent gas chosen to ionize specific functional groups. Gives fewer fragment ions
Field ionization (FIMS)	Molecule ionized directly on a high-voltage wire or blade by the TUNNEL EFFECT	Lower sensitivity than EIMS, but little fragmentation. Used for high-molecular-mass compounds that are not easily vaporized
Field desorption (FDMS)	Wire is heated or a laser is used to desorb a species	
Secondary ion (SIMS)	Surface bombardment by ions (e.g., 5–20 keV) A^+ and ejected ions are analyzed	Positive and negative ions ejected. Very sensitive (10^{-15}–10^{-19} g). Beam may be scanned across a surface to produce an image (ion probe). Used in conjunction with Auger spectroscopy (*see* AUGER EFFECT)

lent of the binding energy ($E = mc^2$). For ^4_2He the measured mass is 4.002 604 amu, the mass of the nucleons is 4.032 980 amu, giving a mass defect of 0.030 376 amu, equivalent to a binding energy of 28.3 MeV. The mass defect accounts for the high energy released in NUCLEAR FISSION.

massicot. *See* LEAD INORGANIC COMPOUNDS.

Massieu function (J; units: J K^{-1}). Possible thermodynamic function, superseded by the HELMHOLTZ FUNCTION, defined as

$$J = -A/T = -U/T + S$$

mass number. *See* NUCLEON NUMBER.

mass spectrometry. Method of analysis in which the unknown sample is ionized (usually as the positive ion) in a vacuum, and the abundance of ions of different mass-to-charge ratios (m/e) is measured. The resulting mass spectrum is a unique record of the compound, being composed of a parent peak of the molecular ion (this is not always present) and those of fragment ions. The separation and detection of ions of different m/e may be accomplished by electric and magnetic fields in a double-focusing spectrometer, by a quadrupole (*see* NUCLEAR SPIN) mass analyzer, in a TIME-OF-FLIGHT SPECTROMETER or by ION CYCLOTRON RESONANCE. The means of ionization and the uses of the techniques derived from them are listed (*see* Table). The resolution of a mass spectrometer may be 50 000 with a sensitivity of 1 part in 10^6. Gas chromatography mass spectrometry (GCMS) is a powerful analytical tool in which the mass spectrum of each component of a gas chromatogram is recorded. Isotope ratio mass spectrometry (IRMS) is a high-resolution method of determining the ratios of isotopes in a sample, often after artificial enrichment. Organic and biological compounds may be converted to simpler molecules (e.g., CO_2, H_2O, N_2) to follow isotopes of carbon, hydrogen and nitrogen. Metastable peaks, seen as broadened peaks of non-integral mass, arise from unstable ions which decompose in the region between the ionization chamber and the analyzer. Other methods include fast atom bombardment (FAB), photoionization, fission fragment ionization, spark source ionization, thermal ionization and thermospraying.

mass-to-charge ratio. *See* MASS SPECTRO-METRY.

matrix isolation. Technique for trapping reactive atoms, radicals or molecules at low temperature for spectroscopic study. The molecules and a matrix gas (usually argon, but also krypton, neon, xenon and nitrogen) are sprayed on to a surface cooled by liquid helium. The trapped species may be then studied by absorption spectroscopy, ELECTRON SPIN RESONANCE SPECTROSCOPY or laser excitation spectroscopy (*see* LASER SPECTROSCOPY).

matte. Residual mixture of sulphides formed during SMELTING.

maximum work function. *See* HELM-HOLTZ FUNCTION.

Maxwell–Boltzmann distribution law. Out of a total number of molecules (N), the probability of finding molecules in the speed range c and $(c + dc)$ is

$$p = dN/N$$
$$= 4\pi(m/2\pi kT)^{3/2}c^2 \exp(-mc^2/2kT)\, dc$$

or in terms of the distribution of kinetic energy ($E = mc^2/2$)

$$dN/N = 2\pi(1/\pi kT)^{3/2} E^{1/2} \times$$
$$\exp(-E/kT)\, dE$$

The plot of $(dN/N)/dc$ or $(dN/N)/dE$ against c or E, respectively, becomes broader and less peaked at higher temperatures as the average speed or energy increases and the distribution about the peak becomes wider (*see* Figure). When the energy of a gas can be expressed as the sum of two square terms, the total number of molecules (n) in a plane having an energy in excess of E is given by $n/N = \exp(-E/kT)$. This equation finds extensive application in reaction kinetics. *See also* MOLECULAR SPEED.

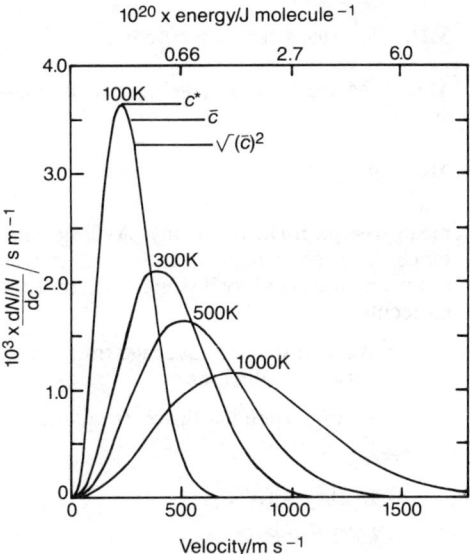

Maxwell–Boltzmann distribution
Distribution of molecular speeds (oxygen).

Maxwell's equations. Set of equations following from the fact that U, H, G and A are STATE FUNCTIONS.

$$(\partial T/\partial V)_S = -(\partial P/\partial S)_V$$
$$(\partial V/\partial T)_P = -(\partial S/\partial P)_T$$
$$(\partial T/\partial P)_S = (\partial V/\partial S)_P$$
$$(\partial P/\partial T)_V = (\partial S/\partial V)_T$$

These relationships are used to derive other thermodynamic equations (e.g., the CLAPEYRON–CLAUSIUS EQUATION from the last one).

McConnell equation. *See* ELECTRON DEN-SITY.

McLafferty rearrangement. Fragmentation process in MASS SPECTROMETRY which involves the six-centre transfer of a γ-hydrogen to an atom at a double bond followed by cleavage of the β-bond: for example, ionized 2-pentanone fragments to the 2-hydroxypropenyl cation and ethene.

MCPA. (2-methyl-4-chlorophenoxyacetic acid). *See* AUXINS.

MCPB (4-(4-chloro-2-methylphenoxy)-butyric acid). *See* AUXINS.

MD. *See* MOLECULAR DYNAMICS.

Md. Mendelevium (*see* ACTINIDE ELEMENTS).

Me. *See* METHYL GROUP.

mean free path (λ; units: m). Average distance through which a molecule moves between successive collisions with another molecule.

$$\lambda = \text{average speed} \times \text{average time between collisions}$$

$$= \text{average speed/collision frequency}$$

$$= \bar{c}/Z$$

$$= 1/(2^{1/2}\pi d^2(N/V))$$

$$= kT/2^{1/2}\sigma^2 P$$

where d is the molecular diameter, σ the COLLISION CROSS-SECTION and N the number of molecules in volume V. In air at atmospheric pressure, $\lambda = 10^{-7}$ m. Experimental values for λ can be obtained from measurements of gas viscosities.

mean life (units: day, min, s). Reciprocal of the DECAY CONSTANT (λ) for a RADIONUCLIDE.

mechanical equivalent of heat. Amount of mechanical energy equivalent to unit amount of thermal energy (1 cal = 4.184 J). The concept loses its significance in SI units, where all forms of energy are expressed in joules.

Meerwein reaction. Reaction of a diazonium salt with a compound containing an activated ethenic bond to give arylated products in the presence of sodium ethanoate and a copper(II) salt (e.g., formation of STILBENE).

$$PhN_2Cl + PhCH=CHCOOH \rightarrow$$
$$PhCH=CHPh + N_2 + CO_2 + HCl$$

MEK (methyl ethyl ketone). *See* BUTANONE.

melamine. *See* 2,4,6-TRIAMINO-1,3,5-TRIAZINE.

melanin. One of the pigments occurring in hair, skin and eyes of animals; a complex polymer of tyrosine (*see* AMINO ACIDS) derivatives and proteins.

melibiose. *See* DISACCHARIDES.

mellitic acid. *See* BENZENEHEXACARBOXYLIC ACID.

melting point. Temperature at which liquid first appears when a solid is heated at atmospheric pressure; during the cooling of a liquid the temperature remains constant at the FREEZING POINT owing to the evolution of the enthalpy of fusion. The melting point of a solid or the freezing point of a liquid is an invariant point ($f' = 0$) (*see* INVARIANT SYSTEM).

In a TWO-COMPONENT CONDENSED SYSTEM in which compound formation occurs, if the compound is in equilibrium with liquid of the same composition, it is said to have a congruent melting point. On the other hand, if the compound is not very stable and decomposes at temperatures below its melting point it is said to have an incongruent melting point. The latter case is known as meritectic melting, the solid phase establishing equilibrium with a second solid phase and a liquid of different composition (i.e. a chemical reaction).

$$\text{solid AB} \rightleftharpoons \text{solid B} + \text{liquid}$$

Peritectic melting involves no bond rupture and may be represented as

$$\text{solid solution } \alpha \rightleftharpoons \text{solid solution } \beta + \text{liquid}$$

Since solid solutions are not compounds, this is not a chemical reaction but a phase change.

membrane electrode. Probe used as a chemical sensor in which the electrode is separated from the analyte by a membrane. Membranes are constructed from plastics such as POLYTETRAFLUOROETHENE, polypropene, and CELLOPHANE. In biosensors, the active enzymes may be immobilized in the membrane. *See also* GAS-SENSING MEMBRANE PROBE, ION-SELECTIVE ELECTRODE.

membrane equilibrium. *See* DONNAN MEMBRANE EQUILIBRIUM.

mendelevium (Md). *See* ACTINIDE ELEMENTS.

***p*-menthadiene** (α-terpene). *See* TERPENES.

menthol. *See* TERPENES.

menthone. *See* TERPENES.

mercaptals $(RR'C(SR'')_2)$. ACETALS in which sulphur replaces oxygen. Oily liquids (aldehyde + thiol), which are not decomposed by dilute acids or alkalis.

mercaptans. *See* THIOLS.

mercapto group. The group –SH (or thiol).

mercuration. Formation of mercury(II) derivatives of aromatic compounds (mono-, di- and polymercurated compounds are known) by: (1) direct replacement of a hydrogen atom in benzene, polynuclear hydrocarbon or heterocyclic compounds with the –HgOCOCH$_3$ group; for example, phenylmercury(II) ethanoate (PhHgOCOCH$_3$) is obtained by heating benzene with mercury(II) ethanoate and (2) replacement of a functional group (as in diazonium salts, sulphinic esters) with –HgCl. With substituted benzene derivatives the orientation may be unusual. These organomercurials are useful synthetic intermediates, such as in the preparation of aromatic halides. *See also* METALATION.

mercury (Hg). Group IIb metal (*see* GROUP II ELEMENTS). It is the only metal that is a liquid at room temperature. It is found naturally as cinnabar (HgS) which is roasted to give mercury vapour. Mercury is not attacked by air and dilute acids, but reacts with hot oxidizing acids, halogens and sulphur. Mercury is used in thermometers, vacuum gauges (McLeod gauge), diffusion pumps and in electrical contacts and relays. Mercury AMALGAMS are used in dentistry and in cells (*see* MERCURY CELL, STANDARD CELL). Mercury vapour and soluble mercury(II) compounds are highly toxic.

A_r = 200.59, Z = 80, ES [Xe] $5d^{10} 6s^2$, mp –38.9°C, bp 357°C, ρ = 13.6 × 10^3 kg m^{-3}. Isotopes: ^{196}Hg, 0.15 percent; ^{198}Hg, 10.0 percent; ^{199}Hg, 16.8 percent; ^{200}Hg, 23.1 percent; ^{201}Hg, 13.2 percent; ^{202}Hg, 29.8 percent; ^{204}Hg, 6.8 percent.

mercury cell. Overall reaction of the mercury dry cell is M + HgO → MO + Hg, where M is zinc or cadmium. The electrolyte is concentrated potassium hydroxide solution, and the metal ions dissolving from the anode are precipitated as the oxide and hydroxide. The cathode is a compressed mixture of mercury(II) oxide and graphite, and the cathode reaction is HgO + H$_2$O + 2e → Hg + 2OH$^-$. The electrolyte is not consumed in the reaction, so that little is necessary. The energy density of the cell is therefore high, and the cell can be made in very small sizes for calculators, etc. The cell has other advantages: the voltage (1.35 V for zinc, 0.9 V for cadmium) is stable, and it can give high currents without loss in performance; its shelf life is also high.

mercury inorganic compounds, Mercury is an unusual member of group IIb in that it forms mercury(I) compounds containing Hg_2^{2+}, as well as mercury(II) compounds (Hg^{2+}). It is not so ELECTROPOSITIVE as zinc and cadmium ($E_{Hg^{2+},Hg}^{\ominus}$ = 0.854 V, $E_{Hg_2^{2+},Hg}^{\ominus}$ = 0.789 V) and because of the large size of Hg^{2+} many compounds have covalent character.

mercury(I) compounds. The number of stable mercury(I) compounds is restricted by their tendency to disproportionate to mercury and mercury(II). For example the oxide, hydroxide and sulphide are only known for mercury(II). In compounds of mercury(I), the mercury–mercury bond length is about 250 pm. Calomel (Hg$_2$Cl$_2$) is important for its use in the calomel REFERENCE ELECTRODE, as an insecticide and as a purgative. Other mercury(I) halides are known. Ionic bonds are formed in complexes with phosphates.

mercury(II) compounds. Mercury(II) oxide (HgO) is formed by the action of oxygen on mercury at 350°C (red) or of alkalis on soluble mercury(II) salts (yellow). HgO is polymeric with zig-zag chains of atoms. Mercury(II) sulphide (so-called vermilion in its red form) is prepared by bubbling hydrogen sulphide into mercury(II) nitrate ($Hg(NO_3)_2$). Mercury(II) fulminate ($Hg(OCN)_2$) is explosive when dry and is used as a detonator.

mercury organic derivatives.   Many mercury(II) organic derivatives (R_2Hg and RHgX) are known. They may be prepared from mercury(II) chloride and a GRIGNARD REAGENT in the appropriate mole ratios. Arylmercury compounds may be prepared by reaction with mercury(II) ethanoate (e.g., $ArH + Hg(CH_3COO)_2 \rightarrow ArHg(CH_3COO) + CH_3COOH$ (*see* BENZENEDICARBOXYLIC ACIDS). RHgX are covalent for X = halogen, CN, SCN, OH, and so are non-polar, soluble in organic solvents, etc. For $X = SO_4^{2-}$, NO_3^-, the compound is ionic (e.g., $(RHg)^+ NO_3^-$). The equilibrium $R_2Hg + HgX_2 \rightleftharpoons 2RHgX$ has values for the equilibrium constant in the range 10^5–10^{11}. R_2Hg (e.g., diethylmercury) are non-polar, volatile and toxic. They are used to prepare organometallics by exchange. Complexes with compounds containing multiple bonds are thought to be intermediates in reactions in which mercury salts are catalysts (e.g., the conversion of ethyne to ethanal).

mercury thermometer.   *See* THERMOMETER.

meritectic melting.   *See* MELTING POINT.

mescaline.   *See* ALKALOIDS.

mesitylene (1,3,5-trimethylbenzene, $C_6H_3(CH_3)_3$). Liquid (bp 164.7°C), occurs in small amounts in coal tar and crude petroleum; prepared by the distillation of propanone with concentrated sulphuric acid.

meso-**isomer.**   Isomer (*see* ISOMERISM) which is optically inactive by internal compensation. In the molecule *Cabd–Cabd* with two asymmetric carbon atoms, if the upper and lower halves reinforce each other then the molecule will be dextro- or laevorotatory, but if they are in opposition the molecule as a whole will not show optical activity.

mesons.   Subatomic particles (including kaon, pion and psi-particles) existing with positive, zero or negative charge; when charged they carry the electronic charge; mean life of about 10^{-7} s. They are found in cosmic radiation and result from high-energy nuclear collisions and are believed to participate in the forces holding the atomic nucleus together.

mesoperiodic acid.   *See* IODATES.

messenger RNA (mRNA).   *See* NUCLEIC ACIDS.

mestranol (3-methoxy-17α-ethynyloestradiol). *See* HORMONES.

Met (methionine).   *See* AMINO ACIDS.

meta- (*m*-).   Prefix for disubstituted BENZENE derivatives designating the position of groups which are separated by an unsaturated carbon atom in the ring (1,3-position). *See also* ORIENTATION IN THE BENZENE RING.

meta-.   Prefix (written in full) denoting the least hydrated acid of a series formed from the same anhydride (e.g., metaphosphoric acid—*see* PHOSPHORUS OXOACIDS).

metabolism.   Physical and chemical changes that occur in living organisms. These include the synthetic (anabolic) reactions, such as the production of proteins and fat and the destructive (catabolic) reactions such as the breakdown of glucose to carbon dioxide and water which provide the energy for the anabolic reactions.

meta-**directing groups.**   *See* ORIENTATION IN THE BENZENE RING.

metaformaldehyde.   *See* METHANAL.

metal.   Element having a crystalline form with an incompletely filled conduction band which leads to facile electrical and thermal conduction (*see* BAND THEORY OF SOLIDS,

metal carbonyls
Binary metal carbonyls ($M_n(CO)_m$).

Metal:	V	Cr	Mo	W	Mn	Tc	Fe
n	1	1	1	1	2	2, 3	1, 2, 3
m	6	6	6	6	10	10, 12	5, 9, 12
Metal:	Ru	Os	Co	Rh	Ir	Ni	
n	1, 3, 6	1, 3, 6	2, 4	2, 4, 6	2, 4	1	
m	5, 12, 18	5, 12, 18	8, 12	8, 12, 16	8, 12	4	

Figure). A characteristic of metallic conduction (a term applied to materials other than metals) is a negative temperature coefficient. Metals are lustrous, ductile and are generally ELECTROPOSITIVE (i.e. they form compounds in positive OXIDATION STATES). The boundary between metals and non-metals is not sharp (*see* METALLOID, PERIODIC TABLE). Metals usually crystallize in bcc, fcc or hcp lattices (*see* CLOSE-PACKED STRUCTURES). *See also* FERMI ENERGY, INSULATOR, SEMICONDUCTOR.

metalation. Replacement of a relatively acidic proton in an aromatic compound by a metal atom (lithium, sodium, potassium) (e.g., PhNa—heat benzene with alkylsodium—some *p*-disodium compound is also formed; PhLi—PhBr + Li). Treatment of these compounds with carbon dioxide gives benzoate and some 1,4-benzenedicarboxylate (terephthalate). The lithium compounds, which undergo the general reactions of GRIGNARD REAGENTS, are used to prepare a variety of compounds. *See also* MERCURATION.

metal carbonyls. Complexes of TRANSITION ELEMENTS and carbon monoxide (CO) (*see* CARBON OXIDES). (For known binary carbonyls (metal and CO only)—*see* Table.) Other transition metals form anionic carbonyls (e.g., $[Pt_6(CO)_{12}]^{2-}$) or compounds with additional ligands. They may be prepared by direct reaction between the metal (iron, nickel) and CO under pressure, but more conveniently by the reduction of a salt of the metal in the presence of CO. Metal carbonyls are of importance because of the ease of replacement of CO by other ligands. They are thus catalysts in some organic syntheses and are intermediates in reactions involving CO

(e.g., FISCHER–TROPSCH PROCESS, OXO PROCESS). INFRARED SPECTROSCOPY may be used to probe the electronic and molecular structure of the complex. The stability of metal carbonyls arises from BACK-BONDING into π^*-orbitals of CO in addition to weak σ-donation (*see* CARBON-TO-METAL BOND, IR SPECTROSCOPY, MOLECULAR ORBITAL).

metal cluster compound. Compound containing covalent or coordinate metal-to-metal bonds. There are bridged and unbridged compounds with simple metal–metal bonds with BOND ORDERS of up to four (e.g., in $[Re_2Cl_8]^{2-}$), and triangular (e.g., $Ru_3(CO)_{12}$), tetrahedral ($[cpFe-(CO)]_4$), octahedral ($Rh_6(CO)_{16}$), cubic ($Ni_8(CO)_8(h^4-PPh)_6$) clusters. Clusters are found extensively in METAL CARBONYLS (e.g., $Os_6(CO)_{18}$, $Rh_4(CO)_{12}$). Unlike the boron hydrides (*see* BORON INORGANIC COMPOUNDS), metal cluster compounds are usually not electron-deficient (*see* ELECTRON-DEFICIENT COMPOUND). Metal cluster compounds have potential uses as heterogeneous catalysts (*see* HETEROGENEOUS CATALYSIS).

metaldehyde. *See* ETHANAL.

metal electrode. *See* HALF-CELL.

metallic conduction. *See* BAND THEORY OF SOLIDS, METAL.

metallocenes. Organometallic compounds bis(h^5-cyclopentadienyl) metal (*see* Figure) named after the first compound of this type made—ferrocene. Many metallocenes are SANDWICH COMPOUNDS, but the important feature is the pentahapto bonding between the delocalized π-electrons of

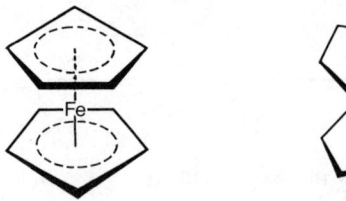

metallocenes

the cyclopentadienyl ring (often abbreviated to cp) and the metal. Metallocenes are prepared by treating the sodium derivative of cyclopentadiene (*see* CYCLOALKENES) with an anhydrous salt of the metal in THF (tetrahydrofuran). They are known for the following metals: titanium, vanadium, chromium, molybdenum, tungsten, manganese, iron, cobalt, nickel, osmium and ruthenium. Ferrocene is the most stable (mp 174°C, bp 249°C). It is diamagnetic and shows AROMATIC properties (e.g., the ring protons may be substituted) and it undergoes the FRIEDEL–CRAFTS REACTION. Ferrocene is reversibly oxidized to the ferrocinium cation ($E^{\ominus} = +0.56$ V versus standard hydrogen electrode). Metallocenes of the iron group are the only followers of the EIGHTEEN-ELECTRON RULE. Many other metallocenes are unstable and easily oxidized.

metallochromic indicator. *See* INDICATORS.

metalloid. Element that has the physical appearance and properties of a METAL, but behaves chemically in some respects as a non-metal. Metalloids are often SEMICONDUCTORS, and their compounds are amphoteric. The definition is not exact, but it usually includes the elements arsenic, antimony, silicon, germanium, bismuth, tellurium and polonium. In the PERIODIC TABLE, the metalloids are found in a diagonal block which separates the obvious metals from the non-metals.

metal–metal bond. *See* METAL CLUSTER COMPOUND.

metal–metal oxide electrode. *See* ELECTRODE.

metaperiodic acid. *See* IODATES.

metaphosphoric acid. *See* PHOSPHORUS OXOACIDS.

metasilicates. *See* SILICATES.

metastable. *See* MASS SPECTROMETRY.

metastyrene. *See* STYRENE.

metathesis. *See* DOUBLE DECOMPOSITION.

metavanadates. *See* VANADIUM COMPOUNDS.

methacrylic acid. *See* 2-METHYLPROPENOIC ACID.

methadone (amidone, 6-dimethylamino-4,4-diphenyl-3-heptanone, $CH_3CH_2COC(C_6H_5)_2CH_2CH(CH_3)N(CH_3)_2$). Prepared by the reaction of 1-dimethylamino-2-chloropropane with diphenylmethylcyanide in the presence of sodamide, followed by conversion to a mixture of ketones from which methadone is separated. The hydrochloride is soluble in water; the free base is precipitated from solution at pH > 6. It has a bitter taste, and is an addictive narcotic analgesic, which is less sedating than morphine (*see* ALKALOIDS), but acts for more prolonged periods.

methanal (formaldehyde, HCHO). Colourless pungent gas (bp –21°C). Simplest ALDEHYDE, but having only hydrogen atoms bonded to the OXO GROUP methanal has properties atypical of aldehydes in general. Industrial preparation is by: (1) dehydrogenation or oxidation of methanol over a silver catalyst, the product from oxidation after distillation is formalin (40 percent methanal, 8 percent methanol, 52 percent water); (2) oxidation of NATURAL GAS or PETROLEUM GAS over a metal oxide catalyst. It is a powerful disinfectant and antiseptic, and is used in preserving anatomical specimens. Its main use is in the manufacture of POLYMERS, DYES and for hardening CASEIN and GELATINE. Reactions peculiar to methanal include the reaction with ammonia to give hexamethenetriamine (I, $(CH_2)_6N_4$), the methylation of primary and secondary AMINES and polymerization in aqueous solutions to dihydroxymethane. Evaporation produces a white solid, paraformaldehyde $(HCHO)_n.H_2O$ (where $n = 6-50$). Higher polyoxymethenes are formed with concentrated sulphuric acid. On standing methanal gas polymerizes to trioxymethene (II, 1,3,5-trioxane, metaformaldehyde) which is a useful source of pure methanal.

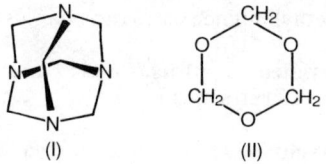

(I)　　　　　　　(II)

methanamide (formamide, $HCONH_2$). Colourless, viscous liquid (decomp. 210°C); manufactured by the reaction between carbon monoxide and ammonia. It is converted to METHANOIC ACID by concentrated acids and to methanoates by alkalis. Methanamide forms compounds with metals, and is used as a solvent as is its derivative N,N-dimethylmethanamide.

methane (marsh gas, CH_4). ALKANE; major constituent of NATURAL GAS (mp −182.6°C, bp −161.6°C). It is formed naturally by the decay of organic matter. Its uses are as a feedstock for PETROCHEMICALS (*see* Figure 1) and as a fuel. It is manufactured from carbon monoxide and hydrogen over a nickel catalyst at 230–50°C.

methanethiol (CH_3SH). Sulphur analogue of METHANOL. *See also* THIOLS.

methanides. *See* CARBIDES.

methanoic acid (formic acid, HCOOH). CARBOXYLIC ACID; colourless pungent liquid (mp 8.4°C, bp 100.5°C) found naturally in stinging nettles, ants, etc. It is manufactured as the sodium salt by heating sodium hydroxide with carbon monoxide at 210°C and 6–10 atm. The acid is obtained by distillation with sulphuric acid. It is prepared by heating glycerol with excess ethanedioic acid at 100–110°C. It is the strongest of the alkyl monocarboxylic acids ($pK_a = 3.75$) and is a powerful reducing agent. Methanoic acid decomposes on heating or at room temperature over an iridium or rhodium catalyst to carbon monoxide and hydrogen. On heating the sodium salt is converted to the ethanedioate and hydrogen or with alkali to sodium carbonate and hydrogen. METHANAL is produced on heating calcium or zinc methanoate. Methanoic acid is used in textile dyeing and finishing, in leather tanning and as an intermediate in synthesis.

methanol (methyl alcohol, CH_3OH). Simplest primary ALCOHOL; colourless liquid (bp 64.5°C). It is poisonous, small doses causing blindness. It occurs as esters in plant oils. It is manufactured from SYNTHESIS GAS using catalysts at 300°C and 300 atm. It shows typical properties of a primary MONOHYDRIC ALCOHOL: readily oxidized to METHANAL, reacts with sulphuric acid to methyl hydrogen sulphate, dimethyl sulphate and dimethyl ether. Heating with soda lime gives sodium methanoate and hydrogen. It is used extensively as a solvent for many inorganic and organic compounds, and in the manufacture of methanal, methanoic acid, methyl chloride and methyl esters of organic acids.

methanoylation (formylation). Introduction of a methanoyl (formyl) group into an aromatic compound to give an aldehyde (essentially a type of FRIEDEL–CRAFTS REACTION). *See also* GATTERMANN–KOCH ALDEHYDE SYNTHESIS, REIMER–TIEMANN REACTION.

methanoyl group (formyl group). The group HCO–.

methene group (methylene group). The group –CH_2–.

methicillin. *See* ANTIBIOTICS.

methionine (Met). *See* AMINO ACIDS.

methoxybenzene. *See* AROMATIC ETHERS.

methoxy group. The group CH_3O– (MeO–).

1-methoxy-4-(1-propenyl)benzene (anethole). *See* PHENOLS.

6-methylacetopyranone (dehydroacetic acid). Colourless needles (mp 109°C), prepared by refluxing ethyl 3-oxobutanoate with a trace of sodium hydrogen carbonate. It is also formed by the POLYMERIZATION of diketen (*see* KETENS). It exists as a tautomer (*see* ISOMERISM) with the more stable lactone form predominating.

methyl alcohol. *See* METHANOL.

methylamines ($(CH_3)_nNH_{3-n}$, where $n = 1$, 2, 3). AMINES; manufactured by the reaction between methanol and ammonia at 450°C over an alumina catalyst and the fractional distillation of the mixture produced. Solutions of methylamines are alkaline.
monomethylamine. Flammable gas (bp −7.6°C) which has an ammonia-like smell and is soluble in water. It it used as a refrigerant and in the manufacture of herbicides and surfactants.
dimethylamine. Pungent gas (bp 7°C) used to make herbicides, surfactants, N,N-DIMETHYLMETHANAMIDE and dimethylacetamide.
trimethylamine. Gas (bp 3.5°C) with a fishy smell. It occurs in sugar residues and reacts with hydrochloric acid to give chloromethane. It reacts with ethene oxide to give choline (*see* VITAMINS).

4-methylaminophenol. *See* AMINOPHENOLS.

N-methylanilines. Mono- ($C_6H_5NHCH_3$) and dimethylanilines ($C_6H_5N(CH_3)_2$) are prepared by heating aniline with iodomethane

$$PhNH_2 + MeI \rightarrow PhNHMe.HI$$

$$PhNHMe + MeI \rightarrow PhNMe_2.HI$$

They are liquids (bp 196°C and 193°C, respectively) which may be separated by ethanoylating monomethylaniline. They are stronger bases than aniline (pK_a 4.85 and 5.06, respectively). They resemble secondary and tertiary aliphatic amines, but they may also be substituted in the *ortho-* and *para*-positions of the ring. With nitrous acid, monomethylaniline gives N-nitroso-N-methylaniline (*see* NITROSAMINES).

methylated spirits. *See* INDUSTRIAL METHYLATED SPIRITS.

methylation. Substitution of a METHYL GROUP for a hydrogen atom in an organic compound. Methylation of a hydroxyl group gives a methyl ether, of an amine gives a secondary or tertiary amine. Acidic hydroxyl groups are methylated by DIAZOMETHANE. *See also* ALKYLATION, FRIEDEL–CRAFTS REACTION.

methylbenzene. *See* TOLUENE.

methylbenzenols. *See* HYDROXYTOLUENES.

methyl bromide. *See* BROMOMETHANE.

2-methyl-1,3-butadiene (isoprene, $CH_2=C(CH_3)CH=CH_2$). Colourless liquid with penetrating odour (bp 34°C), obtained (with other products) during the pyrolysis of rubber or by passing the vapour of turpentine oil through a red-hot metallic tube. Commercially it is prepared: (1) by the dehydrogenation of C_5 alkenes and (2) from propene (*see* Figure).

It can be purified by formation of an adduct with sulphur dioxide (cyclic sulphone) at room temperature, followed by its thermal decomposition at 115°C. It undergoes the normal reactions of a conjugated system, for example, the addition of bromine to give

$$2CH_3CH=CH_2 \xrightarrow{AlPr_3} CH_2=C(CH_3)C_3H_7 \xrightarrow{\text{isomerization}} (CH_3)_2C=CHC_2H_5$$

$$\xrightarrow[\text{[heat + HBr + steam]}]{\text{demethylation}}$$

$$CH_2=C(CH_3)CH=CH_2$$

the 1,4-dibromide and the tetrabromide, and polymerization to TERPENES and rubber.

methylcelluloses. CELLULOSE ethers in which some of the hydroxyl groups of cellulose are replaced by METHOXY GROUPS on treatment with alkali and chloromethane. Those of low methoxy content are soluble in cold water and precipitate out when heated. They are used chiefly as thickeners, in textile printing and in the food industry (E461).

methyl chloride. *See* CHLOROMETHANE.

2-methyl-4-chlorophenoxyacetic acid. *See* AUXINS.

methyl cyanide. *See* ETHANENITRILE.

methylene blue (methylthionine chloride, $C_{16}H_{18}N_3SCl.3H_2O$ (medicinal), $(C_{16}H_{18}N_3SCl)_2ZnCl_2.H_2O$ (dye)). Dark green crystals with a bronze lustre, soluble in water to give a deep blue solution; prepared by the oxidation of 4-aminodimethylaniline with iron(III) chloride in the presence of hydrogen sulphide. It is used medicinally as an antiseptic, in dyeing cotton and wool, as a stain and a redox indicator for biological systems.

$(CH_3)_2N$ —[structure] $N(CH_3)_2$ Cl^-

methylene blue

methylene chloride. *See* DICHLOROMETHANE.

methylene group. *See* METHENE GROUP.

methyl ether. *See* DIMETHYL ETHER.

(1-methylethyl)benzene. *See* CUMENE.

4-methyl-(1-methylethyl)benzene. *See* p-CYMENE.

methylethylcarbinol. *See* BUTANOLS.

1-methylethyl group (isopropyl group). The group $(CH_3)_2CH–$.

methyl ethyl ketone (MEK). *See* BUTANONE.

methyl group The ALKYL GROUP $CH_3–$ (Me–).

methyl 2-hydroxybenzoate (oil of wintergreen, methyl salicylate, $HOC_6H_4COOCH_3$). Natural ESSENTIAL OIL (bp 223°C), prepared by the esterification of 2-hydroxybenzoic acid. It is used extensively for its medicinal properties as an antiseptic and analgesic.

3-methylindole. *See* SKATOLE.

methyl iodide. *See* IODOMETHANE.

methyllithium. *See* LITHIUM ORGANIC DERIVATIVES.

methyl methacrylate. *See* 2-METHYLPROPENOIC ACID, POLYMERS.

5-methyl-2-(1-methylethyl)phenol (thymol). *See* PHENOLS.

methylnaphthalenes. Occur extensively in crude oil and reformed petroleum; prepared by the FRIEDEL–CRAFTS REACTION (MeI + naphthalene at low temperatures) to give 1- and 2-methylnaphthalenes; with alcohols and aluminium trichloride 2,6-dialkylnaphthalenes are formed. Oxidation gives the corresponding naphthalenecarboxylic acid.

methyl phenyl ketone. *See* ACETOPHENONE.

2-methylpropanoic acid (isobutyric acid). Liquid (bp 152–5°C), rancid-smelling saturated CARBOXYLIC ACID.

2-methyl-1-propanol. *See* BUTANOLS.

2-methyl-2-propanol. *See* BUTANOLS.

2-methylpropene. *See* BUTENES.

2-methylpropenoic acid (methacrylic acid, $CH_2=C(CH_3)COOH$). Solid α,β-unsaturated CARBOXYLIC ACID (mp 16°C), prepared by removing HBr (ethanolic KOH) from 2-bromo-2-methylpropenoic acid. It readily polymerizes to an acrylic resin.

Methyl 2-methylpropenoate (methyl methacrylate) is a precursor for Perspex acrylate resins (*see* POLYMERS). This ester is prepared by treating propanone with hydrogen cyanide and the resulting cyanohydrin with methanol in sulphuric acid.

methylpyridines. Present in coal tar.
picolines (C_6H_7N). Three isomers: 2-methylpyridine is used for the manufacture of the monomer 2-ethenylpyridine; 3-methylpyridine for nicotinic acid; 4-methylpyridine for isoniazid.
lutidines (dimethylpyridines, C_7H_9N). Best known is 2,6-lutidine, an oily liquid isolated from the basic fraction of coal tar.
collidines (trimethylpyridines, $C_8H_{11}N$). Best known is *s*-2,4,6-trimethylpyridine; a base that is superior to pyridine as a solvent for elimination reactions.

methyl red. *See* INDICATORS.

methyl salicylate. *See* METHYL 2-HYDROXYBENZOATE.

methylthionine chloride. *See* METHYLENE BLUE.

3-methyluracil (thymine). *See* PYRIMIDINES.

metol. *See* AMINOPHENOLS.

metre (m). Basic SI unit of length; defined as the length equal to 1 650 763.73 wavelengths in vacuum of the radiation corresponding to the transition between the levels $2p^{10}$ and $5d^5$ of the krypton-86 atom.

metric system. *See* CGS UNITS, MKS UNITS, SI UNITS.

Mg. *See* MAGNESIUM.

mho (reciprocal ohm, Ω^{-1}). Archaic unit of conductance equal to the SIEMENS.

micas. *See* ALUMINIUM.

micelle. Particle of an aggregation COLLOID (*see* COLLOIDAL ELECTROLYTE) formed from molecules containing lyophilic (*see* LYOPHILIC COLLOID) and lyophobic (*see* LYOPHOBIC COLLOID) parts (e.g., soaps with LIPOPHILIC (hydrophobic) hydrocarbon chains and ionic head groups). At low concentrations, the molecules are dissolved and not associated, but at the critical micelle concentration (cmc) molecules come together into spherical micelles with the head groups pointing outwards into the solvent and the lyophobic chains inside where the solvent is excluded. At higher concentrations, a lamellar structure may be adopted. At the cmc, the surface tension and the osmotic pressure change less rapidly with concentration, the TURBIDITY rises sharply and the molar conductivity no longer follows the usual relations of strong electrolytes. The solubility of micelles is also higher (*see* KRAFFT TEMPERATURE).

Michael condensation. Base-catalyzed addition of an α,β-unsaturated KETONE to a compound containing an active METHENE GROUP (e.g., diethyl propanedioate—*see* PROPANEDIOIC ACID, ETHYL 3-OXOBUTANOATE). The term is now used for the wider class of additions of NUCLEOPHILES across double bonds conjugated with electron-withdrawing groups. Michael condensations are reversible, the reverse reaction being known as Michael retrogression. *See also* CONDENSATION REACTION, CYANOETHYLATION.

Michaelis–Menten equation. *See* ENZYME KINETICS.

Michael retrogression. *See* MICHAEL CONDENSATION.

Michler's ketone $(CH_3)_2NC_6H_4COC_6H_4N(CH_3)_2$. Precursor of triphenylmethane DYES, prepared by treating *N*-dimethylaniline with carbonyl chloride in the presence of zinc chloride.

microwave radiation. *See* ELECTROMAGNETIC SPECTRUM.

microwave spectra. *See* ROTATIONAL SPECTRA.

milk of magnesia. *See* MAGNESIUM COMPOUNDS.

milk sugar. *See* LACTOSE.

Miller indices. Convenient notation for the simple description of a crystal face or plane. If *a*, *b* and *c* represent the periodicities along the crystallographic axes *x*, *y* and *z*, respectively, then the plane ABC drawn to intercept these axes at *a*, *b* and *c* is known as the parametral plane and is designated by the Miller indices (111) (*see* Figure). For any other plane LMN, making intercepts *a/h*, *b/k* and *c/l* along the axes, the Miller indices are expressed as the ratio of the intercepts of the parametral plane to those of the plane (i.e. (*hkl*) where *h*, *k* and *l* are intercepts). If, as in the example, *a/h* = *a*/2, *b/k* = *b*/4 and *c/l* = *c*/3 then the plane LMN is designated (243). The Miller indices of a face are thus inversely proportional to the intercepts of that face on the chosen axes. If the plane had cut the *y* axis at −*b*/4, the other intercepts being unchanged, the Miller indices would be written (2$\bar{4}$3). For the plane ABDE (parallel to the *z* axis) the indices are (110) and for BDFG (parallel to both the *x* and *z* axes) the indices are (010). For a perfect cube, with crystallographic axes chosen parallel to the three edges at right angles and meeting at the centre of the cube, the symbol for the cubic form is (100), the various faces being represented by (100), (010), (001), ($\bar{1}$00), (0$\bar{1}$0) and (00$\bar{1}$).

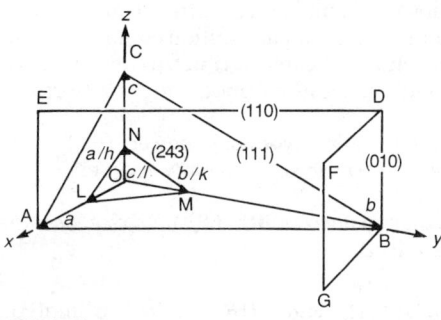

Miller indices

millerite. *See* NICKEL.

mineralocorticoids. *See* CORTICOSTEROIDS.

mirror plane. *See* SYMMETRY ELEMENTS.

Misch metal. Alloy of cerium (about 25 percent) and other LANTHANIDE ELEMENTS. It is PYROPHORIC and used in lighter flints.

miscible liquid mixtures. *See* BINARY LIQUID MIXTURE.

mispickle. *See* ARSENIC.

Mitcherlich's law of isomorphism. Compounds having analogous formulae with related atoms in the same valence state often crystallize in the same form, or are isomorphous (*see* ISOMORPHISM). Examples include the alums formed by trivalent iron, chromium and aluminium.

mixed crystals. Homogeneous crystals deposited together from a solution containing two (normally isomorphous) substances, which contain both substances.

mks units. Metric system of units based on the METRE, KILOGRAM and SECOND which grew from the earlier CGS UNITS. The electrical units used to augment the basic units was the AMPERE and the permeability of free space (magnetic constant) was taken as 10^{-7} H m^{-1}. The mks system has now been superseded by SI UNITS.

mmHg. Unit of pressure, equal to the pressure exerted under standard gravity by a height of 1 millimetre of mercury (1 mmHg = 133.322 Pa or N m^{-2}).

Mn. *See* MANGANESE.

MNDO (modified neglect of differential overlap). *See* MOLECULAR ORBITAL CALCULATIONS, SEMIEMPIRICAL MOLECULAR ORBITAL THEORY.

Mo. *See* MOLYBDENUM.

mo. *See* MOLECULAR ORBITAL.

mobile phase. Phase in chromatographic systems that carries the solutes along and over the column material; in GAS–LIQUID CHROMATOGRAPHY it is an inert gas, in liquid chromatography (*see* CHROMATOGRAPHY) it is a solvent. *See also* STATIONARY PHASE.

mobility. *See* IONIC MOBILITY.

moderator. Substance of low atomic mass (e.g., beryllium, graphite, deuterium in heavy water) which is capable of reducing the speed of neutrons but which has little tendency for neutron absorption.

Mohr method. Titration of a chloride solution with silver nitrate in the presence of sodium chromate ($NaCrO_4$). Formation of a red precipitate of silver chromate indicates the end-point.

Mohr's salt. *See* IRON COMPOUNDS.

Moh's scale. Measure of hardness, in which a substance is compared with 10 reference minerals of increasing hardness: 1, talc; 2, gypsum; 3, calcite; 4, fluorite; 5, apatite; 6, orthoclase; 7, quartz; 8, topaz; 9, corundum; 10, diamond. A member higher in the series can scratch a lower member.

molality (m; units: mol kg^{-1}). A molal solution is prepared by dissolving one-thousandth of the formula weight in kilograms (i.e. formula weight in gram) of solute in 1 kilogram of solvent. The mean ionic molality ($m_\pm$) is

$$m_\pm = (m_+^{\nu_-} m_-^{\nu_-})^{1/\nu}$$

See also CONCENTRATION UNITS.

molar (1). Denoting that an EXTENSIVE PROPERTY is being expressed per amount of substance, usually per mole (e.g., the molar heat capacity is the HEAT CAPACITY per mole, $J K^{-1} mol^{-1}$). (2) Denoting a solution that contains 1 mole of the solute per cubic decimeter of solution. *See also* MOLARITY.

molar absorption coefficient (ε). *See* BEER–LAMBERT LAW.

molar conductivity (Λ) (units: S m^2 mol^{-1}). CONDUCTIVITY per unit molar concentration ($\Lambda = \kappa/c$). κ is not directly dependent on the concentration, but may be affected by the degree of dissociation (*see* ARRHENIUS DISSOCIATION THEORY) of a weak electrolyte or the variation of the MOLAR IONIC CONDUCTIVITY of a strong electrolyte with concentration. The molar conductivity of an electrolyte is related to the molar conductivities of the individual ions by Kohlrausch's law. Thus

$$\Lambda_{\text{electrolyte}} = z_+\nu_+\lambda_+ + z_-\nu_-\lambda_-$$

The limiting conductivity is the value of the conductivity extrapolated to zero concentration. This may be obtained from a plot of Λ against $c^{1/2}$ (*see* CONDUCTANCE EQUATIONS) for strong electrolytes, or from the combination of single molar ion conductivities (e.g., Λ of ethanoic acid is the sum of Λ of sodium ethanoate and hydrochloric acid minus Λ of sodium chloride).

molar enthalpy. *See* ENTHALPY.

molar entropy. *See* ENTROPY.

molar extinction coefficient (ε). *See* BEER–LAMBERT LAW.

molar free energy. *See* GIBBS FUNCTION.

molar heat capacity. *See* HEAT CAPACITY.

molar ionic conductivity (λ_i; units: S m^2 mol^{-1}). Contribution of the ith ion to total MOLAR CONDUCTIVITY. It is obtained by combining experimental values for the molar conductivity of the electrolyte with the TRANSPORT NUMBER of each ion, thus $\lambda_+ = t_+\Lambda$ and $\lambda_- = t_-\Lambda$.

molarity (M; units: mol dm^{-3}, mol l^{-1}). Obsolete unit; its use with SI units is discouraged. A molar solution contains 1 mole of solute in 1 cubic decimetre (i.e. 1 litre) of solution. *See also* CONCENTRATION UNITS.

molar optical rotatory power (α_m). *See* SPECIFIC OPTICAL ROTATORY POWER.

molar polarizability (P_m). *See* POLARIZATION.

molar refraction ([R]; units: m^3 mol^{-1}). Defined by

$$[R] = (M_r/\rho)[(n^2 - 1)/(n^2 + 2)]$$

The value of [R] depends only on the wavelength of the light used to measure the refractive index (n). As a molar volume, it is an additive and a constitutive property; refraction equivalents for various atoms, radicals and bonds have been determined. For a liquid mixture

$$[R]_{A,B} = x_A[R]_A + x_B[R]_B$$
$$= [(x_A M_A + x_B M_B)/\rho_{A,B}]$$
$$\times [(n_{A,B}^2 - 1)/(n_{A,B}^2 + 2)]$$

where $n_{A,B}$ and $\rho_{A,B}$ are the refractive index and density of the mixture, respectively. Refractive index measurements thus provide a useful method for the analysis of mixtures.

molar rotation. *See* SPECIFIC OPTICAL ROTATORY POWER, SPECIFIC ROTATION.

molar volume (V_m). Volume occupied by 1 mole of a substance under specified conditions ($V_m = M_r/\rho$). *See also* MOLAR REFRACTION, PARACHOR.

molasses. *See* SUCROSE.

mole (n_i). Amount of substance that contains as many elementary units (atoms, molecules, ions, radicals) as there are atoms in 0.012 kilogram of carbon-12.

$$n_i = w_i/M_r(i)$$

The total number of molecules is $n_i N_A$.

molecular beam. Narrow stream of atoms, molecules or ions produced at low pressure having defined velocity, orientation, etc. Molecular beans may be allowed to intersect in order to study the molecular dynamics of reactions and to measure the COLLISION CROSS-SECTION of molecules. They are also used to study surfaces and reactions at surfaces.

molecular crystal. Aggregate of covalent molecules (e.g., I_2) which form periodic, symmetrical arrays. On account of the intermolecular bonds such aggregates have low melting and boiling points.

molecular diameter (d; units: m). When two molecules, assumed to be solid elastic spheres, come within a distance of their diameter (d) a collision is recorded. Typical values are: hydrogen, 0.29 nm; benzene, 0.53 nm. *See also* COLLISION CROSS-SECTION, MEAN FREE PATH.

molecular dynamics (MD). Method of numerical simulation of the movements of molecules and atoms. From the initial positions and velocities of a number of atoms, the movements of each atom are calculated knowing the INTERMOLECULAR FORCES between them. After many small-time steps different thermodynamic functions may be calculated (e.g., energy of the system and pressure if a gas).

molecular ellipticity. *See* CIRCULAR DICHROISM.

molecular flow. Flow of a gas in which the MEAN FREE PATH of the molecules is large compared with the dimensions of the pipe. This type of flow occurs at low pressures where most collisions are with the walls of the vessel rather than with other molecules. Flow characteristics of the gas then depend on the molar mass of the gas rather than on its VISCOSITY. *See also* NEWTONIAN FLUID.

molecular formula. *See* FORMULA.

molecular ion. *See* MASS SPECTROMETRY.

molecularity of reaction. Number of molecules that form the ACTIVATED COMPLEX in a given step of a chemical reaction. Thus unimolecular reactions involve a single molecule, bimolecular reactions two molecules, etc. Molecularity refers to the mechanism of a reaction and is distinct from the experimentally determined ORDER OF REACTION.

molecular mass. *See* RELATIVE MOLECULAR MASS.

molecular mechanics. *See* FORCE FIELD CALCULATION.

molecular orbital (mo). WAVEFUNCTION of an electron that extends over the atoms of a molecule. Molecular orbitals may be constructed by LINEAR COMBINATIONS of ATOMIC ORBITALS. Those formed by the overlap of ATOMIC ORBITALS along the internuclear axis are known as σ-orbitals if the overlap is constructive and leads to bonding, or σ*-orbitals if the overlap is destructive or antibonding. Bonding and antibonding orbitals in which the overlap occurs away from the internuclear axis are called π and π*, respectively. An alternative nomenclature for molecular orbitals uses the subscripts g (gerade) and u (ungerade) for bonding and antibonding orbitals, respectively. An example of π-orbitals arises in the bonding of BENZENE where the overlap of six $2p_z$-orbitals of carbon creates electron density above and

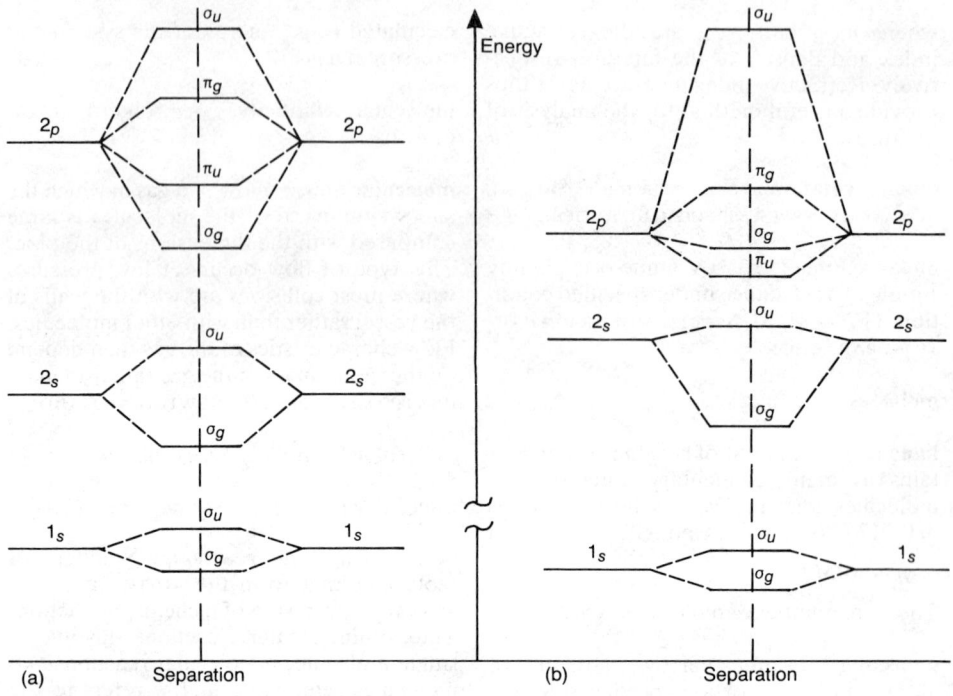

molecular orbital
Energies of the molecular orbitals of first-row diatomic molecules relative to the atomic orbitals. (a) O_2 and F_2, (b) Li_2 to N_2.

below the plane of the molecule (*see* AROM-ATICITY). σ molecular orbitals are formed in the hydrogen molecule from overlap of the $1s$ atomic orbitals of each atom. These may be written as

$$\psi_{1s\sigma^*} = 2^{-1/2}(\phi_{1sA} + \phi_{1sB})$$

and

$$\psi_{1s\sigma^*} = 2^{-1/2}(\phi_{1sA} - \phi_{1sB})$$

where A and B refer to each atom. Bonding molecular orbitals have energies less than the atomic orbitals from which they are formed, the energies of antibonding orbitals are greater, whereas nonbonding molecular orbitals have the same energy as their generating atomic orbitals. (For molecular orbitals of homonuclear diatomics—*see* Figure.) To determine the structure of a diatomic molecule two electrons are fed into each molecular orbital (*see* AUFBAU PRINCIPLE, PAULI EXCLUSION PRINCIPLE) taking due account of HUND'S RULES.

molecular orbital calculations. Methods of calculating energies, geometries and electron distributions in molecules. They divide into SEMIEMPIRICAL MOLECULAR ORBITAL THEORY and *AB-INITIO* CALCULATIONS. *See also* MOLECULAR ORBITAL, SELF-CONSISTANT FIELD.

molecular partition function. *See* PARTITION FUNCTION.

molecular sieve. Fine, porous material used for separations based on differences of molecular dimensions. Synthetic alkali metal aluminosilicates, with cations in their crystal structure which attract polar molecules and with varying pore size, are particularly useful. It is used to dry gases and to remove water from organic solvents and sulphur from petroleum. *See also* ZEOLITE.

molecular speed (units: m s^{-1}). Molecular speeds can be measured directly using spinning discs with slots in them which permit the passage of only those molecules moving through them at the appropriate speed. The number of molecules passing through are collected in a detector for counting. A full distribution of speeds is obtained by using different rates of rotation of the disc. There are several ways of quoting the speed of molecules:

(1) mean speed $(\bar{c})$

$$\bar{c} = (8kT/\pi m)^{1/2}$$

$$= 145.51 \, (T/M_r)^{1/2}$$

(2) root mean square speed $((\bar{c^2})^{1/2})$

$$(\bar{c^2})^{1/2} = (3kT/m)^{1/2}$$

$$= 157.93 \, (T/M_r)^{1/2}$$

(3) most probable speed (c^*)

$$c^* = (2kT/m)^{1/2}$$

$$= 128.95 \, (T/M_r)^{1/2}$$

(4) average relative speed of two molecules $(2^{1/2}\bar{c})$.

See also MAXWELL–BOLTZMANN DISTRIBUTION LAW.

molecular weight. See RELATIVE MOLECULAR MASS.

molecule. Smallest amount of a chemical compound that can have a separate existence and shows the properties of that compound. Molecules are usually held together by COVALENT BONDS or COORDINATE BONDS. In the case of ionic substances, the smallest neutral entity (e.g., NaCl in sodium chloride) is considered to be the molecule. In a similar way, covalently bonded crystals (e.g., BN) are considered to be made up of molecules of BN although they do not have an existence separate from the crystal. See also ATOM, ELEMENT, MOLECULAR CRYSTAL.

mole fraction (x_i). Dimensionless quantity; the mole fraction of the ith component in a mixture is given by

$$x_i = \frac{\text{no. of molecules of } i\text{th component}}{\text{total no. of moles of all components}}$$

$$= \frac{n_i}{\Sigma \, n_i}$$

$$x_A + x_B + x_C + \ldots = 1$$

See also CONCENTRATION UNITS.

Molisch's test. General test for carbohydrates. A small amount of an ethanolic solution of 1-hydroxynaphthalene is mixed with the test solution and concentrated sulphuric acid is poured slowly down the side of the tube. A violet ring at the junction of the two liquids indicates the presence of a carbohydrate.

molten salts. See FUSED SALTS.

molybdates. See MOLYBDENUM COMPOUNDS.

molybdenite. See MOLYBDENUM.

molybdenum (Mo). Second row, group VIb TRANSITION ELEMENT. It is extracted from the ore molybdenite (MoS_2). Ore is roasted to MoO_3, which is purified by further heating at 1000°C or via ammonium molybdenate (($NH_4)_2Mo_2O_7$). The metal is produced by reduction in hydrogen at 1090°C. Principal use is in STEEL. Molybdenum is an important element in biological systems (e.g., in PHOTOSYNTHESIS).
$A_r = 95.94$, $Z = 42$, ES [Kr] $4d^5 \, 5s^1$, mp 2617°C, bp 4612 °C, $\rho = 10.22 \times 10^3 \, kg \, m^{-3}$. Isotopes: ^{92}Mo, 15.8 percent; ^{94}Mo, 9.12 percent; ^{95}Mo, 15.70 percent; ^{96}Mo, 16.50 percent; ^{97}Mo, 9.45 percent; ^{98}Mo, 23.75 percent; ^{100}Mo, 9.6 percent.

molybdenum compounds. Oxidation states –2 to +6 are found, although no simple molybdenum ions are known. The molybdenyl cation MoO_2^{2+} exists, possibly in MoO_2Cl_2. Nine oxides with stoichiometries from MoO_2 to MoO_3 are known, some with molybdenum–molybdenum bonds (e.g., red Mo_4O_{11}, red–blue $Mo_{17}O_{47}$). Molybdic acid (H_2MoO_4 ($MoO_3.H_2O$)) forms a series of molybdates of the ion MoO_4^{2-}. Acidification of molybdates yields polymeric molybdates $Mo_7O_{24}^{6-}$ then $Mo_8O_{26}^{4-}$. Compounds with sulphur, selenium and tellurium are similar to the oxides. MoS_2 is a solid lubricant. Molybdenum hexacarbonyl ($Mo(CO)_6$) (Mo + CO under pressure) is a starting point for the preparation of organomolybdenum compounds. Molybdenum compounds are used as pigments.

moment of inertia (I; units: kg m^2). Moment of inertia of a massive body about an axis is the sum of all the products of the magnitude of each element ($\delta\, m$) and the square of the distance from the axis (r)

$$I = \Sigma\, \delta m r^2$$

For a diatomic molecule

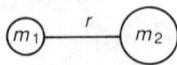

$$I = m_1 m_2 r^2 / (m_1 + m_2)$$

and for a triatomic linear molecule

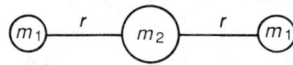

$$I = 2m_1 r^2$$

See also PARTITION FUNCTION.

momentum.
linear momentum (p). Product of a particle's MASS and its VELOCITY ($p = mv$).
angular momentum (L). Product of the angular velocity of a particle and its MOMENT OF INERTIA about the axis of rotation ($L = I\omega$).

Monastral blue 5025. *See* PHTHALOCYANINS.

monazite. *See* ACTINIDE ELEMENTS (thorium), CERIUM.

Mond process. *See* NICKEL.

monochromatic. Having a single wavelength (or, in practice, a narrow range of wavelengths); also used for monoenergetic electrons. Monochromatic radiation may be produced using a prism or grating, or for X-rays by diffraction through quartz crystals. *See also* POLYCHROMATIC.

monoclinic system. *See* CRYSTAL SYSTEMS.

monodentate ligand. *See* LIGAND.

monodisperse. Describing a COLLOID consisting of particles of similar size. A polydisperse colloid has particles of a range of sizes.

monoethanolamines.   *See*   ETHANOLAMINES.

monohydric alcohols. ALCOHOLS with one hydroxyl group. The three classes (primary, secondary and tertiary) have similar properties owing to the presence of the same functional group; their relative acidity is *tert-*<*sec-*<methanol, thus *tert-*butoxide is a stronger base than is methoxide. The lower alcohols are volatile liquids; the higher ones (above C$_{12}$) are solids. The boiling point of an alcohol is higher than that of the alkane of approximately the same relative molecular mass because of HYDROGEN BOND formation. All are strongly polar; their solubility in water decreases with increasing chain length.
general methods of preparation (primary, secondary, tertiary): (1) acid-catalyzed hydration of an alkene (RCH=CH$_2$, RCH=CHR′, RR′C=CHR″); (2) acid hydrolysis of the corresponding ester or halide; (3) from the GRIGNARD REAGENT with methanal (HCHO), R′CHO or R′R″CO, respectively; (4) reduction of the aldehyde for primary or ketone for secondary alcohol.
general reactions: (1) oxidation (K$_2$Cr$_2$O$_7$ + H$^+$): RCH$_2$OH gives RCHO and then RCOOH; RR′CHOH gives RR′CO whereas RR′R″COH breaks down to compounds containing fewer carbons than the original alcohol; (2) with organic acids, acid chlorides or anhydrides all classes form an ESTER; (3) with hydrogen halides all classes form alkyl halides (tertiary more rapidly than secondary or primary); (4) with PX$_3$ alkyl halides are formed (trialkyl phosphites can also be formed); (5) sulphuric acid dehydrates the alcohol to the corresponding alkene; (6) with sodium and alkyl bromide primary and secondary alcohols form ethers. *See also* BENZYL ALCOHOL, BUTANOLS, CYCLOHEXANOL, ETHANOL, 1-HYDROXYDODECANE, METHANOL, PROPANOLS.

monolayer. *See* LANGMUIR–BLODGETT FILM, SURFACE FILM.

monomer. *See* DIMER, OLIGOMER, POLYMERS.

monomethylamine. *See* METHYLAMINES.

monosaccharides. Smallest carbohydrate molecules, including the four-, five- and six-carbon sugars (*see* TETROSES, PENTOSES, HEXOSES).

monosodium glutamate. *See* AMINO ACIDS.

monoterpene. *See* TERPENES.

monotropy. When the transition temperature between two allotropic or polymorphic forms lies above the melting point of the solid, the system exhibits monotropy. Only one form is stable at all temperatures up to the melting point, the other form is metastable. The transformation is irreversible (*see* ONE-COMPONENT SYSTEM, Figure). Examples include phosphorus (violet stable, yellow metastable), iodine monochloride, carbon. *See also* ENANTIOTROPY, PHASE RULE.

Monte Carlo method. Method of numerical calculation that relies on a random poll of the initial conditions of a model chosen to be characteristic of the real system. For example, given a POTENTIAL ENERGY SURFACE it is possible to calculate the trajectories of reactions as they pass over it from initial conditions (molecular velocities, orientation, etc.) that are randomly assigned to mimic a system in thermal equilibrium at a particular temperature. From a large number of these calculations, parameters of the system such as the rate constant may be determined.

montmorillonite. Aluminosilicate clay mineral of general formula $Al_2O_3.4SiO_2.H_2O$, but of variable composition due to isomorphous substitution. Two varieties are known: one which swells and gives gels, the other non-swelling form has marked absorptive properties. *See also* BENTONITE, SILICATES.

mordant. Chemical that is applied to a material before dyeing to facilitate the fixing of the dye. Aluminium sulphate or an aluminium ALUM is a mordant for ALIZARIN, and chromium salts are used extensively for ACID DYES. *See also* DYES.

morphine. *See* ALKALOIDS.

Morse potential. Potential energy (U) of a HARMONIC OSCILLATOR that dissociates at greater internuclear separation as a function of internuclear distance (r)

$$U = D_e\{1 - \exp[-\beta(r - r_e)^2]\}$$

D_e is the spectroscopic DISSOCIATION ENERGY, r_e the equilibrium separation and β a constant (*see* Figure).

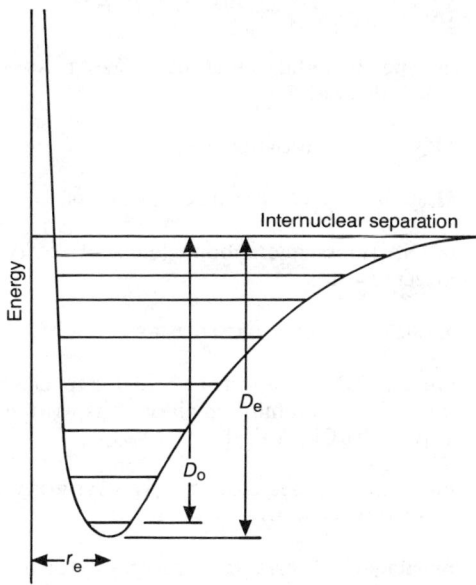

morse potential

Moseley's law. Linear relationship between the square roots of the frequencies of the corresponding lines in the X-ray spectra of a set of elements and their atomic numbers.

Mössbauer spectroscopy (nuclear gamma resonance fluorescence). Method of analysis that depends on recoil-free gamma-ray resonance absorption. Absorption of a gamma-photon causes a transition to an excited nuclear state. The decay process emits a photon of an equivalent energy if the atom is not able to recoil: for example, if it is in a crystal, which may then be readsorbed and so on in a resonant process. A small CHEMICAL SHIFT is seen due to a change in oxidation state or nuclear quadrupole coupling, and the modulation of the gamma-source required is provided by moving the sample relative to the source with a velocity of about 1 mm s^{-1}. The difference in energy is given by the Doppler effect. Nuclei that exhibit a Mössbauer effect are ^{57}Fe, ^{83}Kr, ^{119}Sn, ^{129}I, ^{129}Xe, ^{133}Cs, ^{151}Eu and ^{197}Au.

mother liquor. Solution that remains after a compound has crystallized. *See also* CRYSTALLIZATION.

moving boundary method. *See* TRANSPORT NUMBER.

MQ. *See* AMINOPHENOLS.

M_r. *See* RELATIVE MOLECULAR MASS.

mRNA (messenger ribonucleic acid). *See* NUCLEIC ACIDS.

M-shell. *See* QUANTUM NUMBER.

mu (μ). Designation of a bridging species, thus gaseous aluminium chloride is written as $[Cl_2Al(\mu Cl)_2AlCl_2]$.

mucic acid. *See* 2,3,4,5-TETRAHYDROXYHEXANEDIOIC ACID.

mucilages. POLYSACCHARIDES, extracted from plants, which swell in water. They contain GALACTOSE, arabinose (*see* PENTOSES) and glucuronic acid (*see* URONIC ACIDS) residues.

mucins. Substances that confer the viscid, slippery properties to fish eggs, slugs, saliva and various other tissue and glandular secretions. Substances responsible for blood group specificity are glycoproteins that appear to be related closely in structure to the mucins. Hydrolysis of the carbohydrate fraction gives HEXOSES, hexosamines (*see* AMINO SUGARS) and SIALIC ACIDS.

mucopeptides. *See* PEPTIDES, PEPTIDOGLYCANS.

mucopolysaccharides. Group of polysaccharides, present in animal connective tissue, that consist of DISACCHARIDE repeating units, one of which is always an AMINO SUGAR, such as glucosamine. The most abundant are hyaluronic acid, chrondroitin and heparin.

mucoproteins. *See* PROTEINS.

mull. Dispersion of a solid in a liquid frequently used in INFRARED SPECTROSCOPY to obtain a vibrational spectrum of the solid. The mulling agent is chosen to have no vibrational modes that overlap with those of the solid. In practice two different agents are used in successive experiments: Nujol, a high-boiling petroleum fraction, has only carbon–hydrogen and carbon–carbon modes and HEXACHLORO-1,3-BUTADIENE (Fluorolube) which contains no carbon–hydrogen bonds.

Mulliken electronegativity scale. *See* ELECTRONEGATIVITY.

Mulliken symbols. Denote the symmetry of electronic states of molecules. A and B are singly degenerate (symmetrical and unsymmetrical, respectively, with respect to rotation about the principal axis), E doubly degenerate and T triply degenerate. Subscripts may be 1 or 2 (symmetrical and unsymmetrical, respectively, with respect to rotation about a C_2-axis perpendicular to the principal axis, or with respect to reflection in a σ_v plane), and *g* and *u* (symmetrical or unsymmetrical with respect to inversion). If the molecule has a σ_h-plane (*see* SYMMETRY ELEMENTS), superscripts ' and " denote symmetry or antisymmetry with respect to reflection. *See also* GROUP THEORY, TERM SYMBOLS.

multidentate ligand. *See* CHELATE LIGAND, LIGAND.

multiple proportions, law of. When two elements A and B combine to form more than one compound, the weights of B that combine with a fixed weight of A are in the ratio of small whole numbers. *See also* CHEMICAL COMBINATION, LAWS OF.

multiplet. Group of lines (doublet, triplet, quartet, etc.) in the fine structure of ELECTRON SPIN RESONANCE SPECTROSCOPY or NUCLEAR MAGNETIC RESONANCE SPECTROSCOPY (NMR) caused by spin–spin coupling between electrons and nuclei. The number and intensity of lines may be calculated from the different possible ways of combining the spins of each atom. Thus in the proton NMR of bromoethane, two bands due to methyl and methene groups are seen at medium resolution which splits at high resolution into a triplet (intensity 1:2:1) and a quartet intensity (1:3:3:1), respectively (*see* Figure).

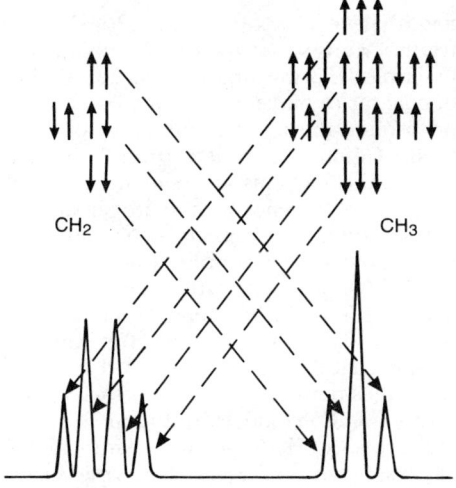

multiplet
Multiplet splitting in the nuclear magnetic resonance spectrum of ethyl bromide.

multiplicity. Number of possible orientations of the total spin of an atom. *See also* MULTIPLET.

Mumetal. Ferromagnetic (*see* FERRO-MAGNETISM) ALLOY (78 percent nickel, 17 percent iron, 5 percent copper) that has a high permeability and is used to shield devices from external magnetic fields. Modern formulations contain chromium and molybdenum.

muon. *See* ELEMENTARY PARTICLES.

muramic acid. *See* AMINO SUGARS.

mureins. *See* PEPTIDOGLYCANS.

muscovite. *See* SILICATES.

mustard gas (2,2'-dichlorodiethylsulphide, $(ClCH_2CH_2)_2S$). Colourless, heavy, oily liquid (mp 13°C, bp 217°C), manufactured by treating sulphur monochloride with ethene; decomposed violently with bleaching powder. It is a powerful vesicant and poison causing conjunctivitis and temporary blindness; it is an important agent in chemical warfare.

mutagen. Agent capable of causing a mutation or increasing the rate of mutation. Chemical mutagens may cause changes in the DNA molecules; thus nitrous acid converts cytosine to uracil, alkylating agents (e.g., NITROGEN MUSTARDS) produce changes in the base pairs and acridines cause additions or deletions of base pairs. Electromagnetic radiation can also induce mutation.

mutarotation. Change in the optical activity that occurs when an optically active substance (e.g., glucose) is dissolved in water. It is frequently catalyzed by acids and/or bases.

myoglobin. *See* PROTEINS.

myrcene. *See* TERPENES.

N

N. *See* NEWTON, NITROGEN.

N. *See* NEUTRON NUMBER.

n. Number of electrons transferred in a cell reaction; principal quantum number (*see* QUANTUM NUMBER).

ν. *See* FREQUENCY, KINEMATIC VISCOSITY.

ν̄. *See* WAVENUMBER.

N_A. *See* AVOGADRO CONSTANT.

Na. *See* SODIUM.

NAA (1-naphthylacetic acid). *See* AUXINS.

NAD^+ (nicotinamide adenine dinucleotide). *See* COENZYMES.

$NADP^+$ (nicotinamide adenine dinucleotide phosphate). *See* COENZYMES.

Nafion. Organic polyfluorinated polymer containing sulphonic acid groups that can act as an ION-SELECTIVE MEMBRANE.

$$(CF_2CF_2CFCF_2)_n$$
$$|$$
$$(OCF_2CF)_yOCF_2CF_2SO_3H$$
$$|$$
$$CF_3$$

Nafion

Napalm. *See* SOAP.

naphtha. Light-boiling fraction of PETROLEUM in the range 40–150°C. Particular boiling fractions are known as special boiling point spirits and are used as industrial solvents. Naphthas are the basic ingredients of GASOLINE.

naphthacene ring system. *See* POLYCYCLIC HYDROCARBONS.

naphthalene ($C_{10}H_8$). POLYCYCLIC HYDROCARBON (structure I) containing two rings. It is the largest single constituent of coal tar (6 percent); obtained as crystals on cooling the middle and heavy oil fractions. Commercially it is prepared from petroleum fractions by demethylation of methylnaphthalenes with hydrogen at high temperatures and pressures. Naphthalene is a white, lustrous solid (mp 80°C, bp 218°C), with a penetrating tar-like smell, insoluble in water but steam-volatile. It is a typical aromatic hydrocarbon (*see* AROMATICITY, HÜCKEL'S RULE). Although it resembles benzene, it is more reactive and forms addition and substitution products more readily. Hydrogenation with a nickel catalyst gives 1,2,3,4-TETRAHYDRONAPHTHALENE; more vigorous treatment gives BICYCLO[4.4.0]DECANE, dry chlorine on the solid gives naphthalene tetrachloride ($C_{10}H_8Cl_4$). Oxidation with concentrated sulphuric acid and mercury(II) sulphate gives PHTHALIC ANHYDRIDE, with acid potassium permanganate phthalic acid (*see* BENZENEDICARBOXYLIC ACIDS) and with chromium(III) oxide 1,4-naphthaquinone (*see* QUINONES). There are two possible monosubstituted products and 10 disubstituted products of the type $C_{10}H_6X_2$; the 1, 4, 5 and 8 (or α-) positions are equivalent, as are the 2, 3, 6 and 7 (or β-) positions. In ELECTROPHILIC SUBSTITUTION reactions, the first group enters the 1-position except in sulphonation reactions at high temperatures (giving the 2-substituent) and in some FRIEDEL–CRAFTS REACTIONS (acylation in carbon disulphide as solvent substitution occurs in the 1-position, in nitrobenzene in the 2-position). If one ring is substituted with a deactivating group (e.g., –NO₂) then further substitution occurs in the unsubstituted ring (e.g., formation of 1,8- and some 1,5-dinitronaphthalenes on reaction with the NITRONIUM ION). Naphthalene reacts with metallic sodium in tetrahydrofuran to give a deep green solution containing $[C_{10}H_8]^- Na^+$, a useful reagent in METALATION reactions. Naphthalene is used as an insecticide and in the industrial synthesis of phthalic anhydride, 2-hydroxynaphthalene and many dyes.

naphthalenecarboxylic acids (naphthoic acids). Prepared by the hydrolysis of the cyanides, by the oxidation (NaOCl) of ethanoylnaphthalenes or by a GRIGNARD REAGENT. Naphthalene-1-carboxylic acid (mp 161°C) and naphthalene-2-carboxylic acid (mp 185°C) both eliminate carbon dioxide on heating with soda line, and show the general reactions of CARBOXYLIC ACIDS.

naphthalenesulphonic acids. When naphthalene is treated with concentrated sulphuric acid at 40°C the main product is naphthalene-1-sulphonic acid (decomposed by superheated steam), at 160°C the main product is the 2-isomer (stable to steam). Fusion with sodium hydroxide gives the corresponding hydroxynaphthalene. Drastic sulphonation of naphthalene yields di- and trisulphonic acids, which are important dyestuff intermediates.

naphthaquinones. See QUINONES.

naphthenes. Any of the five- and six-membered CYCLOALKENES found in crude petroleum.

naphthoic acids. See NAPHTHALENECARBOXYLIC ACIDS.

naphthols. See HYDROXYNAPHTHALENES.

1-naphthylacetic acid. See AUXINS.

naphthylamines. See AMINONAPHTHALENES.

nascent hydrogen. Highly active hydrogen prepared in situ by electrolysis or by the dissolution of an electropositive metal in acid. It is used for reduction; the active species are hydrogen atoms or hydride ions adsorbed on the surface of the metal which thus acts as a CATALYST.

natural gas. Gas (95 percent METHANE + other ALKANES + nitrogen and carbon dioxide) found in underground deposits which may be associated with crude oil. 'Dry gas' contains methane as the main hydrocarbon. 'Wet gas', found with oil, also contains higher hydrocarbons (e.g., pentane, hexane). It is used as a fuel and raw material (see PETROCHEMICALS).

natural products. Compounds produced by living organisms. This definition encompasses a wide range of compounds such as CARBOHYDRATES, LIPIDS, NUCLEIC ACIDS and PROTEINS, all of which play an important and primary role in metabolic processes. There are other organic compounds, some of extraordinary complexity, produced naturally that are not primary metabolites. Many of these (e.g., ALKALOIDS, fungal metabolites and ANTIBIOTICS) do not seem to have any obvious metabolic or evolutionary function. Regardless of their utility to the parent organism their value to man as drugs, herbs, flavourings, poisons, dyes, etc. is undisputed. See also STEROLS, TERPENES.

Nb. See NIOBIUM.

NBS (N-bromosuccinimide). See N-BROMOBUTANIMIDE.

Nd. Neodymium (see LANTHANIDE ELEMENTS).

Ne. See NEON.

Néel temperature. See ANTIFERROMAGNETISM.

negative adsorption. See ADSORPTION, ADSORPTION ISOTHERM.

negatron (β^-). See BETA-DECAY.

neighbouring group participation. Intramolecular NUCLEOPHILIC SUBSTITUTION which may occur if a nucleophilic group is contained in the same molecule as a good leaving group; for example, the formation of tetrahydrofuran from 4-chlorobutanol (see Figure). The rate is first-order in 4-chorobutanol, although the reaction is more like a simple S_N2 reaction than a S_N1 reaction (i.e. a displacement, not an ionization followed by an addition). The chain length in such a reaction is critical. The reaction is even faster in the presence of a

4-chlorobutanol   tetrahydrofuran

neighbouring group participation
Formation of tetrahydrofuran.

base which converts the –OH (nucleophile) into an alkoxide, an even better nucleophile. When neighbouring group participation occurs during the rate-determining step the reaction may occur faster than predicted (i.e. the neighbouring group has anchimerically assisted the displacement).

nematic crystals. *See* LIQUID CRYSTALS.

nembutal. *See* BARBITURATES.

neo-. Prefix indicating a hydrocarbon in which at least one carbon atom is connected directly to four other carbon atoms (e.g., neopentane, $C(CH_3)_4$; neopentyl $(CH_3)_3CCH_2$–).

neodymium (Nd). *See* LANTHANIDE ELEMENTS.

neon (Ne). Noble gas (GROUP O ELEMENT) obtained by the fractionation of liquid air (1.82×10^{-3} percent of air). It is completely inert and has no normal chemistry. Neon is used in low-pressure discharge tubes for coloured signs and in gas lasers. Liquid neon is an economic cryogenic refrigerant. $A_r = 20.179$, Z = 10, ES [He] $2s^2\, 2p^6$, mp –248.67°C, bp –246.05°C, $\rho = 0.9$ kg m^{-3}. Isotopes: ^{20}Ne, 90.92 percent; ^{21}Ne, 0.26 percent; ^{22}Ne, 8.82 percent.

neoprene (poly-(2-chloro-1,3-butadiene)). Synthetic rubbers whose molecular mass is controlled by the addition of sulphur. Neoprene shows high tensile strength and is resistant to common solvents. Its uses are as wire and cable covering and in paper-making.

nephelauxetic effect. Expansion in the d-electron cloud of a transition element ion due to the presence of a ligand. The effect of different ligands is largely independent of the metal ion and is demonstrated by changes in the electronic spectrum. The nephelauxetic series of ligands is arranged in increasing order of the effect: F$^-$, H$_2$O, NH$_3$, (COO$^-$)$_2$, en, NCS$^-$, Cl$^-$, CN$^-$, Br$^-$, I$^-$.

nephelometry. Quantitative analysis by measurement of the scattered light (*see* SCATTERING) from a precipitate.

neptunium (Np). *See* ACTINIDE ELEMENTS.

neral. *See* TERPENES.

Nernst equation. Equation relating the ELECTRODE POTENTIAL of a reversible HALF-CELL to the activities of the ions in solution. In general for the reaction ox + $ne \rightleftharpoons$ red

$$E = E^{\ominus} + (RT/nF) \ln (a_{ox}/a_{red})$$

$E^{\ominus}$ is the standard electrode potential of the half-cell.

Nernst heat theorem. ENTROPY change accompanying transformations between condensed phases in chemical equilibrium approaches zero as the temperature approaches 0 K. From the GIBBS–HELMHOLTZ EQUATION, $\partial(\Delta G)/\partial T$ approaches zero asymptotically as the temperature approaches 0 K (i.e. ΔG approaches ΔH asymptotically in the vicinity of 0 K). Mathematically the heat theorem may be expressed as

$$\lim_{T \to 0} (\Delta S) = \lim_{T \to 0} \partial(\Delta G)/\partial T$$
$$= \lim_{T \to 0} \partial(\Delta H)/\partial T = 0$$

The postulate has been confirmed, provided it is limited to a perfectly crystalline substance. *See also* THIRD LAW OF THERMODYNAMICS.

nerol.   *See* TERPENES.

nerolins.   *See* HYDROXYNAPHTHALENES.

nerve gas.   *See* FLUOROPHOSPHATES.

Nessler's reagent.   Alkaline solution of mercury(II) iodide in potassium iodide, used for detecting and estimating ammonia with which it forms a brown coloration or precipitate.

neuraminic acid.   *See* AMINO SUGARS.

neutralization.   *See* ACIDS AND BASES.

neutrino (ν).   Elementary particle (zero mass, zero charge, spin 1/2) emitted spontaneously with a beta-ray in positron decay to conserve energy and momentum for each decaying nucleus. The sum of the kinetic energies of the neutrino and the beta-particle always equals E_{max} for a particular transition. Antineutrinos are emitted in negatron BETA-DECAY.

neutron (^1_0n; mass: $1.674\ 95 \times 10^{-27}$ kg). Fundamental particle of matter having almost the same mass as a PROTON, but with no electric charge being a constituent of the nuclei of all atoms except protium. Neutrons are liberated from the nucleus by fission of uranium-235 or plutonium, each nucleus yielding an average of 2.5 neutrons, and are also produced by bombardment of other atoms. They have a magnetic moment of 1.91 B.M and interact with atoms with unpaired electrons. Free neutrons (mean life: 12 min) are radioactive with tremendous penetrating power, decaying into a proton, an electron and an antineutrino. They have a highly damaging effect on living tissue requiring shielding on all equipment in which they are used. 'Fast' neutrons bring about the chain reaction required in the atom bomb. In nuclear reactors, their energy is partially absorbed by a MODERATOR, resulting in 'slow' or thermal neutrons. Neutrons are used for making radioactive isotopes and in NEU-TRON ACTIVATION ANALYSIS.

neutron activation analysis.   Non-destructive test in which the sample to be analyzed is bombarded by NEUTRONS, thereby activating the elements composing it. The subsequent measurement of their radio-activity and its various energy levels provide an index of its chemical composition. *See also* ACTIVATION ANALYSIS.

neutron diffraction.   Scattering of thermal neutrons (kinetic energy ~ 0.025 eV, $\lambda = 0.1$ nm). There are two types of interaction: (1) between neutrons and the atomic nucleus, giving rise to diffraction patterns complementing those from X-ray diffraction, but which are particularly suitable for investigating light atoms; (2) between the magnetic moments of the neutrons and the spin and orbital magnetic moments of the atom, providing information on anti-ferromagnetic and ferrimagnetic material.

neutron excess.   *See* ISOTOPE NUMBER.

neutron flux (ϕ; units: neutron s^{-1} m^{-2}). Intensity of neutron radiation, expressed as the number of neutrons passing through an area of 1 m^2 in 1 s. A typical value for a reactor is 10^{16} neutron s^{-1} m^{-2}.

neutron number (N).   Number of neutrons in a nucleus.

Newman projection formula.   Structural formula obtained by viewing a molecule along the bond of two carbon atoms, in which the carbon atom nearer the eye is designated by three equally spaced radii and the far carbon atom by three equally spaced radial extensions. Thus for ethane two extremes of the arrangement of one methyl group with respect to the other may

Newman projection formula

be obtained by rotation about the carbon–carbon bond; eclipsed and staggered arrangement (*see* Figure). The DIHEDRAL ANGLE (ω) (i.e. the angle in the plane of the paper between the front carbon–hydrogen bond and the rear carbon–hydrogen bond) varies from 0 to 360°C as one methyl group rotates. This angle is 0° when the hydrogens on the rear carbon are directly behind those on the front carbon (in the diagram they are offset for clarity); this is the eclipsed conformation; 60° corresponds to the staggered, 120° to eclipsed, etc. *See also* CONFORMERS.

newton (N; dimensions: $m\,l\,t^{-2}$; units: N = kg m s^{-2} = J m^{-1}). SI unit of force; defined as that force which when applied to a body having a mass of 1 kg gives it an acceleration of 1 m s^{-2}.

newtonian fluid. Fluid in which the velocity gradient or shear rate ($S = \mathrm{d}v/\mathrm{d}x$) is directly proportional to the shear stress (τ); that is the external force required to overcome the internal friction between adjacent layers of fluid

$$\eta = \text{shear stress/shear rate} = \tau/S$$

where η is the coefficient of VISCOSITY of the fluid. Many liquids and gases exhibit newtonian behaviour over a wide range of temperatures and pressures. However, solutions of polymers, emulsions and suspensions exhibit non-newtonian behaviour (i.e. η varies with the shear rate). This anomalous behaviour arises from the orientation of the particles or the disruption of the internal structure. Time-independent non-newtonian liquids (*see* Figure) are those whose viscosity is not dependent on the duration of flow. Time-dependent non-newtonian liquids are those whose viscosity depends on the duration of flow. There is no unique relation between η and S so consistency curves cannot be drawn. Thixotropic liquids become runny when stirred or shaken (i.e. the apparent viscosity decreases with time) and when left to stand they reset forming a gel. This reversible gel–liquid–gel sequence is known as thixotropy. Examples include ketchup and non-drip paints. For some liquids, the rebuilding of structure occurs more rapidly when they are stirred, such systems exhibit rheopexy. In viscoplastic systems, there is no fluid flow until a critical yield stress is applied and exceeded. Data from stress/rate relationships are important in the production of paints, greases, putties and jellies.

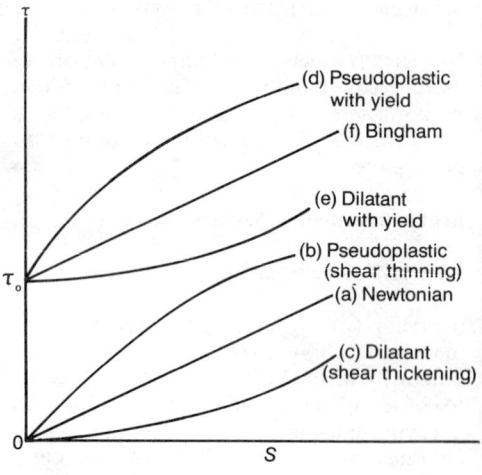

newtonian fluid
Plots of shear stress, τ' (N m^{-2}) against shear rate S (s^{-1}), for time-independent non-newtonian liquids (consistency curves). (a) Newtonian liquid, η constant and independent of shear rate. (b) Liquid exhibiting shear thinning or *pseudoplasticity*, η decreases as shear rate increases, the slope at any point gives the apparent viscosity (the shear rate must be quoted). (c) Liquid exhibiting shear thickening or *dilatancy*, η increases as shear rate increases. (d) Plastic substances with a yield value, τ_0, i.e. the substance only becomes fluid above a critical shear stress. (e) Plastic substance with a yield value exhibiting dilant behaviour in liquid state. (f) Bingham plastic, a material which deforms elastically until the yield stress is exceeded when it behaves as a newtonian liquid (e.g. fresh cement).

Newton's law of cooling. Rate at which a body loses heat is proportional to the difference in temperature between the body and its surroundings. It is an empirical law.

Newton's laws of motion. (1) A body continues in a state of rest or uniform motion in a straight line unless it is acted upon by external forces. (2) Rate of change of momentum of a moving body is proportional to and in the same direction as the force acting on it

$$F = ma$$

where a is the acceleration. (3) If one body exerts a force on another, there is an equal and opposite force, called a reaction, exerted on the first body by the second.

Ni. *See* NICKEL.

n_i.   See TRANSFERENCE NUMBER.

niacin.   See NICOTINIC ACID, VITAMINS.

niccolite.   See NICKEL.

nichrome.   ALLOY of nickel (about 80 percent) and chromium (about 20 percent) having high electrical resistance and good thermal stability. The alloy is used in heating elements.

nickel (Ni).   First row TRANSITION ELEMENT. It occurs naturally as sulphide ores—pentlandite and pyrrholite $((Ni,Fe)_xS_y)$, millerite (NiS)—and also as arsenides and antimonides—niccolite (NiAs), breithamptite (NiSb). Metal is extracted from roasted ores by reduction by carbon. Nickel may be electrolytically purified or converted to the carbonyl $Ni(CO)_4$ which decomposes to the pure metal at 180°C (Mond process). Uses are in alloys (e.g., INVAR, Monel), as a green colouring in glass and ceramics, and as a HYDROGENATION catalyst (see RANEY NICKEL). Nickel shows weak FERROMAGNETISM, is stable in alkaline solution, but dissolves in acids ($E^{\ominus}_{Ni^{2+},Ni} = -0.24$ V). PASSIVITY is shown in concentrated nitric acid.
$A_r = 58.69$, $Z = 28$, ES [Ar] $3d^8 4s^2$, mp 1453°C, bp 2732°C, $\rho = 8.90 \times 10^3$ kg m^{-3}. Isotopes: ^{58}Ni, 67.76 percent; ^{60}Ni, 26.16 percent; ^{61}Ni, 1.25 percent; ^{62}Ni, 3.66 percent; ^{64}Ni, 1.16 percent.

nickel arsenide lattice.   See CRYSTAL FORMS.

nickel compounds.   Nickel chemistry is almost exclusively that of nickel(II). Salts of all the common anions are known. Complexes are common, the green colour of most salts arising from octahedral $[Ni(H_2O)_6]^{2+}$. Tetrahedral complexes are blue (e.g., $[NiCl_4]^{2-}$ in ethanol) and planar complexes are red (e.g., bis(dimethylglyoximato)nickel(II)). Nickel (IV) is found in complex salts, such as $KNiF_6$ and $BaNiO_3$. An intermediate oxide NiO(OH) is formed electrochemically from the metal at anodic potentials. Nickel carbonyl $Ni(CO)_4$ (Ni + CO, 100°C) is highly toxic. Many organic derivatives are known. The bi-π propenyls are formed from GRIGNARD REAGENTS and Ni^{2+} (e.g., $(h^3-C_3H_5)_2Ni$). The SANDWICH COMPOUND nickelocene $(h^5-C_5H_5)_2Ni$ is known. The CLATHRATE COMPOUND $Ni(CN)_2.NH_3.C_6H_6$ (see METALLOCENES) has trapped benzene in a cage structure which may be released when heated or dissolved in water.

nicotinamide.   See COENZYMES, NICOTINIC ACID, VITAMINS.

nicotinamide   adenine   dinucleotide (NAD^+).   See COENZYMES.

nicotinamide adenine dinucleotide phosphate $(NADP^+)$.   See COENZYMES.

nicotine.   See ALKALOIDS.

nicotinic acid (niacin, 3-pyridinecarboxylic acid, $C_6H_5NO_2$).   White solid (mp 237°C), prepared by oxidizing nicotine (see ALKALOIDS) (HNO_3) or 3-methylpyridine ($KMnO_4$). It is formed during the biochemical degradation of tryptophan. As the pellagra-preventing component of vitamin B complex, it is an essential constituent of mammalian diets. The amide, nicotinamide, is a component of the COENZYMES NAD^+ and $NADP^+$.

nido.   See CARBORANES.

ninhydrin.   See 1,2,3-TRIKETOHYDRINDENE HYDRATE.

niobium (Nb).   Second row, group Vb TRANSITION ELEMENT, once known as columbium (Cb). It is mostly found with TANTALUM, the principal ore being columbite–tantalite (Fe,Mn) $(Nb,Ta)_2O_6$. Extraction from aqueous hydrofluoric acid with 2-methyl-2-pentanone leaves niobium in the aqueous phase. Complexing with boric acid and treatment with ammonia precipitates Nb_2O_5. Metal is obtained by reduction with carbon or ELECTROLYSIS of fused K_2NbF_7. Niobium is inert to all mineral acids except hydrofluoric acid, reacts with oxygen (Nb_2O_5), halogens (NbX_5, NbX_4) and other non-metals (e.g., NbN, NbC). The metal is alloyed in special STEELS.
$A_r = 92.9064$, $Z = 41$, ES [Kr] $4d^4 5s^1$, mp 2468°C, bp 4742°C, $\rho = 8.57 \times 10^3$ kg m^{-3}. Isotope: ^{93}Nb, 100 percent.

niobium compounds. Oxidation state +5 is most common. Reduction of niobium(v) compounds (e.g., $NbCl_5$ + Nb) gives niobium(IV) in which metal clusters are found (e.g., Nb_3Cl_8, Nb_6Cl_{14}). Niobates (NbO_4^{3-}) are insoluble in water. Dissolution in hydrofluoric acid gives species such as $[NbOF_5]^{2-}$ and $[NbOF_6]^{3-}$. $[NbF_7]^{2-}$ is known in the crystal phase, $[NbF_6]^-$ is the highest coordinate species in solution. Niobium does not have an extensive organometallic chemistry; known compounds are $RNbCl_4$ (R = alkyl or aryl), and cyclopentadienyl and carbonyl derivatives (e.g., $(h^5-C_5H_5)_2NbCl_3$, $(h^5-C_5H_5)Nb(CO)_4$, $(h^5-C_5H_5)Nb(CO)_2(Ph_2C_2)$).

nitramide. See NITROGEN OXOACIDS.

nitrates. Salts of nitric acid, formed by the action of nitric acid on the metal, oxide, hydroxide or carbonate. All metallic nitrates are soluble in water; some (e.g., $UO_2(NO_3)_2$) are readily extracted into organic solvents. Nitrates are generally ionic in character; heavy metal and anhydrous nitrates have covalently bonded nitrate groups. On heating most nitrates decompose to the metal oxide, dinitrogen tetroxide (N_2O_4) and oxygen; sodium and potassium nitrates give nitrites and oxygen; ammonium nitrate gives dinitrogen oxide and water. Nitrates are strong oxidizing agents. With concentrated sulphuric acid nitrates give nitric acid and dinitrogen tetroxide (N_2O_4). Devarda's alloy and sodium hydroxide reduce nitrates to ammonia. Nitrates are used in the manufacture of explosives and fertilizers. See also NITROGEN OXOACIDS.

nitration. Replacement of a hydrogen atom in an organic compound with a NITRO GROUP. Aromatic nitration is effected by concentrated nitric acid, mixed or nitrating acid (conc. H_2SO_4 + conc. HNO_3), nitric acid in an organic solvent such as ethanoic acid, or nitromethane, ethanoyl nitrate or nitronium salts. The active nitrating agent is the nitronium cation (NO_2^+) which is an electrophile (see ELECTROPHILIC SUBSTITUTION). See also NITROALKANES, NITROBENZENE.

nitric acid. See NITROGEN OXOACIDS.

nitric oxide. See NITROGEN OXIDES.

nitrides. Compounds of nitrogen with other elements prepared by the action of nitrogen gas or ammonia on the element. Three general classes exist: (1) ionic nitrides formed by electropositive elements formally containing the N^{3-} ion (e.g., Li_3N) which are readily hydrolyzed to ammonia; (2) covalent nitrides formed by the' less electropositive elements (e.g., BN, S_4N_4, P_3N_5) whose properties depend on the element with which nitrogen is combined (ring compounds are often formed); (3) transition metal nitrides (analogous to the borides and carbides in constitution and properties), in which nitrogen atoms occupy interstices in close-packed metal lattices. These, together with boron nitride (BN), silicon nitride (Si_3N_4) are metallic in appearance, hardness and electrical conductivity.

nitrification. Process, brought about by bacteria (e.g., *Nitrosomonas* and *Nitrobacter*), in which nitrogen (usually in the form of ammonium compounds) in plant and animal waste and dead remains is oxidized to nitrites and then nitrates, which (unlike ammonia) can be taken up by plants. See also NITROGEN CYCLE.

nitriles. Organic compounds containing the cyanide group ($-C\equiv N$). They are prepared by: the dehydration (P_2O_5) of amides; the catalyzed (Al_2O_3) reaction between ammonia and a carboxylic acid; the reaction between an alkyl halide and potassium cyanide; and from a GRIGNARD REAGENT and cyanogen chloride. They are hydrolyzed to the carboxylic acid, give ALDEHYDES ($SnCl_2$) or primary AMINES ($LiAlH_4$, $NaBH_4$ or Na/NH_3) on reduction, AMIDINES with ammonia, ESTERS with alcohols in concentrated sulphuric acid, and KETONES with Grignard reagents. Nitriles (e.g., BENZONITRILE, ETHANENITRILE) are used as solvents, chemical intermediates, fuel additives, insecticides, herbicides and additives to electroplating baths.

nitrites (nitrates(III)). Salts of the fairly strong nitrous acid (HNO_2) (see NITROGEN OXOACIDS). They are formed by the reduction of nitrates with carbon or lead. Nitrites can act as oxidizing agents (e.g., against I^-

and Fe^{2+}), but are more usually reducing agents.

nitrito group. The group –ONO. It is a tautomer (*see* ISOMERISM) of the nitro group (*see* NITROALKANES).

nitroalkanes (nitroparaffins; RNO_2, where R = alkyl). Toxic liquids, prepared by heating an alkyl halide with silver nitrite. Nitritoalkanes (*see* NITRITO GROUP) are also formed in the reaction. Industrially alkanes are directly nitrated with nitric acid at 400°C to give a mixture of products. Nitro compounds containing α-hydrogen atoms show tautomerism (*see* ISOMERISM) between the nitro (RCH_2NO_2) and acinitro ($RCH=NO(OH)$) forms. In alkali a nitro compound gives the salt of the aci-form which may be hydrolyzed to an aldehyde or ketone. Nitroalkanes are reduced to primary amines ($LiAlH_4$ or Zn/HCl) to N-alkylhydroxylamines (Zn/NH_4Cl) or to oximes ($SnCl_2$/HCl). The lower nitroalkanes are used as propellants, solvents for polar compounds and as chemical intermediates.

nitrobenzene (oil of mirbane, $C_6H_5NO_2$). Colourless, refractive liquid (bp 211°C) with distinctive smell, prepared by the action of a cold mixture of fuming nitric acid and concentrated sulphuric acid in which the nitrating agent is the NITRONIUM ION. The nitro group is *meta*-directing and so further nitration yields 1,3-dinitrobenzene; sulphonation gives 3-nitrobenzenesulphonic acid. Reduction gives nitrosobenzene, N-phenylhydroxylamine and aniline. Azoxybenzene ($C_6H_5NO=NC_6H_5$), AZOBENZENE and hydrazobenzene may also be derived from nitrobenzene by reduction. The major use of nitrobenzene is in the manufacture of aniline for use in the dye industry.

nitrocellulose. See CELLULOSE NITRATE.

nitrogen (N). Colourless, odourless gas, GROUP V ELEMENT. It forms 78.1 percent by volume of the earth's atmosphere from which it can be obtained by liquefaction and distillation. It also occurs as nitrates and in proteins (16 percent by weight). Nitrogen is prepared in the laboratory by heating ammonium nitrite, or barium or sodium azide. The common form (dinitrogen, N_2) has a high dissociation energy and is fairly inert. At room temperature, it reacts with metallic lithium to give lithium nitride (Li_3N) and with certain transition metals complexes (e.g., $[Ru(NH_3)_5.H_2O]^{2+}$ gives $[RuN_2(NH_3)_5]^{2+}$). At higher temperatures nitrogen becomes more reactive; at red heat calcium gives calcium nitride (Ca_3N_2) and at bright red heat boron and aluminium form the nitrides (BN and AlN, respectively). In the presence of catalyst and at high temperatures, it combines with oxygen (BIRKELAND–EYDE PROCESS) and hydrogen (HABER PROCESS). Its octet of electrons may be completed by: (1) electron gain from the nitride ion (N^{3-}); (2) formation of electron pair bonds (e.g., ammonia, azo compounds, RNO_2); (3) formation of electron pair bonds with electron gain (e.g., amide ion, NH_2^-); (4) formation of electron pair bonds with electron loss (e.g., R_4N^+). Since it is such an electronegative element it enters extensively into hydrogen bond formation both as a proton donor –NH–X and as an acceptor =N–HX.
A_r = 14.0067, Z = 7, ES [He] $2s^2 2p^3$, mp –209.86°C, bp –195.8°C; ρ = 1.2506 kg m^{-3}. Isotopes: ^{14}N, 99.63 percent; ^{15}N, 0.37 percent.

nitrogen cycle. Major cycle for the circulation of nitrogen in the environment (*see* Figure). Although the element is essential to all forms of life, atmospheric nitrogen is unavailable to most organisms (*compare* CARBON CYCLE), but can be assimilated by certain bacterial species (*see* NITROGEN FIXATION) and thus made available to other organisms. Some nitrogen is returned to the atmosphere by DENITRIFICATION.

nitrogen fixation.
chemical fixation. Direct combination of atmospheric nitrogen with oxygen to give nitrogen oxides (BIRKELAND–EYDE PROCESS) and with hydrogen to give ammonia (HABER PROCESS).
biological fixation. Ability to fix free nitrogen is limited to certain bacteria (e.g., *Azotobacter*) and blue–green algae (Cyanobacteria). Some bacteria in symbiotic association with cells in the roots of leguminous plants (peas and beans) are able to fix nitrogen. Cultivation of legumes is one way of increasing soil nitrogen without the addition of chemical fertilizers.

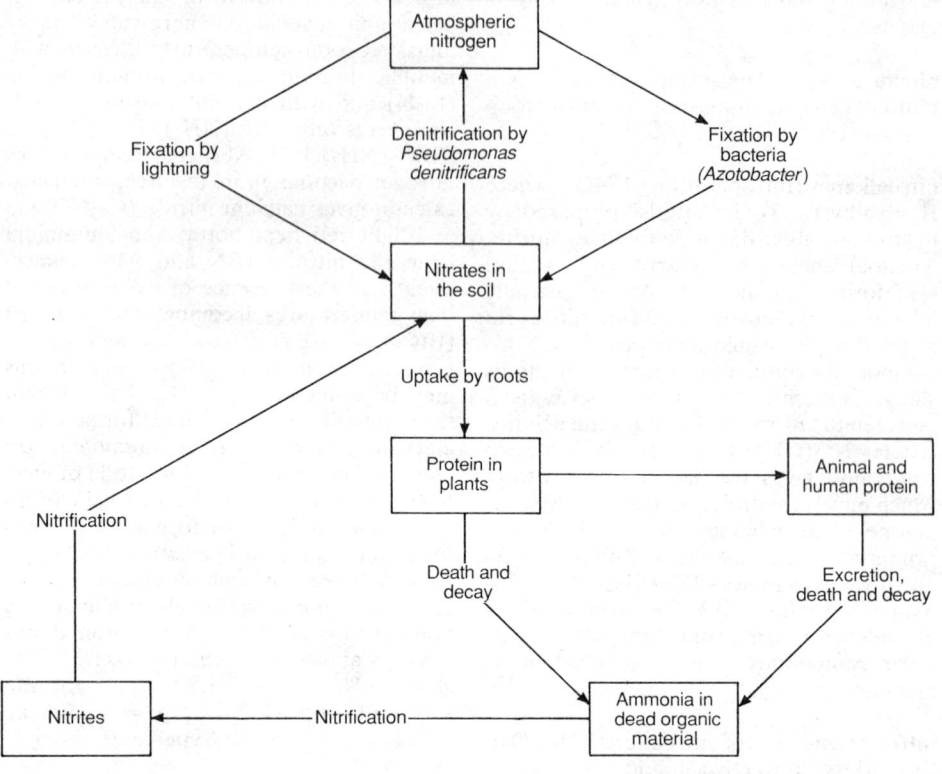

nitrogen cycle

nitrogen halides. Trihalides of nitrogen with fluorine and chlorine have been isolated; for bromine and iodine only the ammonia complexes $NBr_3(NH_3)_6$ and $NI_3(NH_3)$, respectively, are known. With the exception of nitrogen trifluoride, the halides and halogen azides (XN_3) are reactive and potentially hazardous substances.

nitrogen trifluoride (NF_3). Very stable gas (electrolysis of NH_4F in anhydrous HF), only reactive at temperatures in excess of 250°C. It is unaffected by water at room temperature, does not decompose on heating and in the presence of copper forms copper(II) fluoride (CuF_2) and tetrafluorohydrazine. The molecule has a pyramidal structure, low dipole moment and is devoid of donor properties.

tetrafluorohydrazine (N_2F_4). Highly reactive gas (NF_3 + Cu) which dissociates in the gas and liquid phase to nitrogen difluoride (NF_2). It shows the characteristics of a free radical, abstracting H from thiols (RSH) giving RSSR, and reacting violently with H_2 in a chain reaction. With Cl_2 it forms $NClF_3$.

nitrogen difluoride (N_2F_2, FN=NF). Gas that exists as *cis*- and *trans*-isomers. It is obtained by the reaction of aluminium trichloride with N_2F_4.

nitrogen hydrides.

ammonia (NH_3).   Colourless gas (mp −77.7°C, bp −33.35°C; ρ = 0.7710 kg m^{-3} (760 mmHg)) with pungent smell, occurring naturally in gases from hot springs. It is manufactured from the elements in the HABER PROCESS and is generated in the laboratory by the treatment of an ammonium salt with a base. The molecule is pyramidal (HN̂H = 107°C). It burns in oxygen to give nitrogen, but in the presence of platinum forms nitrogen monoxide and nitrogen dioxide (*see* NITROGEN OXIDES). The gas is very soluble in water forming hydrates in which NH_3 and water are linked by hydrogen bonds; the formation of ammonium hydroxide is very improbable, the solution is a weak base ($pK_b = 4.75$). A saturated solution of NH_3 has a density of 0.880 g cm^{-3} ('880 ammonia'). NH_3 reacts with acids forming AMMONIUM SALTS. The gas is readily liquefied; the liquid resembles water in its physical properties. It is a good ionizing solvent ($\varepsilon_r = 22$ at $-33°C$), with an autoprotolysis constant of 32 and a dipole moment of 1.47 D. Through its lone pair of electrons, the molecule is a strong proton acceptor facilitating the dissociation of weak acids. It is a better solvent than water for organic compounds, but a poorer one for inorganic compounds (unless complexes are formed, e.g., silver iodide). Many active metals (sodium, calcium) dissolve to give blue solutions in which there are solvated metal ions and solvated electrons. It is extensively used in the manufacture of nitric acid, urea, nylon, fertilizers, explosives and plastics.

hydrazine (N_2H_4). Fuming, colourless liquid (bp $113.5°C$) with a high dielectric constant ($\varepsilon_r = 52$). It is manufactured by the reaction of NH_3 with sodium chlorate(I) (NaOCl) in the presence of gelatin (Raschig synthesis) or by the reaction of Cl_2 and NH_3 in the presence of a ketone. It forms an azeotrope with water (*see* AZEOTROPIC MIXTURE). Anhydrous N_2H_4 is obtained by precipitation of the sulphate ($H_2SO_4 +$ EtOH), followed by distillation of the sulphate with concentrated potassium hydroxide. It is a bifunctional base ($pK_1 = 6.07$, $pK_2 = 15.05$) forming two series of hydrazinium salts; those of $N_2H_5^+$ are stable in water, whereas those of $N_2H_6^{2+}$ are extensively hydrolyzed. The molecule is similar in properties to HYDROGEN PEROXIDE. Anhydrous N_2H_4 burns in oxygen, reacts readily with the halogens and decomposes on heating to nitrogen and ammonia. The liquid is a good solvent for sulphur, selenium, phosphorus and arsenic. Although showing oxidizing properties, it is a powerful reducing agent in which nitrogen is the most common product; under specific conditions ammonia and hydrazoic acid can be obtained. It exists in the *gauche* form (*see* CONFORMERS). Hydrazine undergoes mono- and di-condensation with aldehydes and ketones to give HYDRAZONES and

azines, respectively. The substituted hydrazine, phenylhydrazine, forms phenylhydrazones with ketones. It is used as a deoxidant for boiler water, and organic derivatives are used as high-energy fuels, antioxidants and herbicides.

diimine (diimide, HN=NH). Parent compound of AZO COMPOUNDS. It cannot be isolated, but there is evidence of its transient existence in the gaseous phase and in solution. It is commonly obtained in the oxidation of hydrazine by two-electron oxidants (e.g., O_2, peroxides); the *in situ* preparation is used for reduction of symmetrical multiple bonds by stereospecific *cis*-addition of hydrogen. In the absence of a substrate for reduction, diimine either decomposes to nitrogen and hydrogen or to nitrogen and hydrazine.

hydrazoic acid (HN_3). Although formally a hydride of nitrogen it has no essential relationship to ammonia and hydrazine. It is present in the aqueous distillate from the action of sulphuric acid on sodium azide ($NaNH_2 + NaNO_3$ at $175°C$). Anhydrous HN_3 ($pK_a = 4.75$), obtained by distillation of the solution, is a colourless liquid (bp $37°C$) that is dangerously explosive. Metals dissolve with the formation of the azide, ammonia and nitrogen. The covalent azides (e.g., organic azides and those of the heavy metals silver, copper, lead, mercury) explode when struck. Azides of electropositive metals are not explosive and decompose smoothly and quantitatively on heating to the metals and nitrogen; lithium azide (LiN_3) is converted to the nitride. The linear azide ion (N_3^-) functions as a ligand in complexes of the transition metals; in general, it behaves rather like a halide ion. Azides are used synthetically in organic chemistry.

hydroxylamine (NH_2OH). Ammonia with one hydrogen atom replaced by hydroxyl. It is a colourless solid which is very soluble in water to give a basic solution ($pK_b = 8.2$). It is prepared by the reduction of nitrates or nitrites electrolytically or with sulphur dioxide or by the reduction of nitrogen dioxide with hydrogen in hydrochloric acid (platinized charcoal as catalyst). Free NH_2OH must be stored at $0°C$ to avoid decomposition. Its salts (e.g., the hydrochloride) are stable, water-soluble solids. Although it can serve as either an oxidizing or a reducing agent, it is usually used as the latter. It forms oximes with aldehydes and ketones. It is used in the manufacture of CAPROLACTAM and as a free-radical scavenger.

nitrogen mustards $(RN(CH_2CH_2Cl)_2$, where R is an alkyl, alkylamine or alkyl chloride group). *N*-Analogues of MUS-TARD GAS. Methyl bis-(2-chloroethyl)-amine hydrochloride is prepared by treating methyldiethanolamine with phosphorus trichloride or thionyl chloride. These compounds have vesicant and other properties similar to those of mustard gas. Their mutagenic activity (*see* MUTAGEN) is believed to be due to the alkylation of cellular material.

nitrogen oxides.
dinitrogen oxide (nitrous oxide, N_2O). Unreactive colourless gas (heat NH_4NO_3), decomposing into oxygen and nitrogen at 500°C and supporting the combustion of many compounds. It is slightly soluble in water giving a neutral solution. It is a linear molecule (N–N–O). It is used as a mild anaesthetic (laughing gas).
nitrogen monoxide (nitric oxide, NO). Colourless gas, soluble in water, ethanol and ether, that condenses to a dark blue paramagnetic liquid. It has a slight tendency to dimerize. NO is obtained by the reduction of nitric acid, nitrites or nitrates (e.g., $Cu + 8$ mol dm^{-3} HNO_3) and industrially by the catalytic oxidation of ammonia. Direct combination of the elements only occurs to a slight extent at very high temperatures (*see* BIRKELAND–EYDE PROCESS). NO reacts readily with oxygen to form nitrogen dioxide and the halogens to give nitrosyl halides (NOX) (*see* NITROGEN OXOHALIDES). Removal of an electron in the π^*-orbital gives the NITROSONIUM ION (NO^+), and gain of an electron gives NO^- as found in the NITROSYLS.
dinitrogen trioxide (N_2O_3). Exists pure only in the solid state at low temperatures ($< -100°C$). It is prepared from stoichiometric quantities of nitrogen monoxide (NO) and oxygen or NO and dinitrogen tetroxide (N_2O_4) at low temperatures. It is the anhydride of nitrous acid. The gas contains $ONNO_2$ molecules, which persist into the liquid state. The vapour is largely dissociated into NO and N_2O_4.
nitrogen dioxide (NO_2), *dinitrogen tetroxide* (N_2O_4). Prepared by heating lead nitrate

and condensing the gas evolved, or by the action of concentrated nitric acid on copper. It is colourless and diamagnetic in the dimeric form (N_2O_4), which is found pure only in the solid state. The yellow liquid form contains about 1 percent of the monomer, whereas in the vapour at 100°C 90 percent exists as the brown paramagnetic monomer. NO_2 is a bent molecule (ONO = 134°), whereas the dimeric form is largely planar O_2NNO_2. The oxides react with water to give a mixture of nitrous (HNO_2) and nitric acid (HNO_3), a fairly strong oxidizing agent. The monomeric form, an odd-electron molecule, has many of the characteristics of a free radical: (1) it associates with other radicals; (2) it abstracts H from saturated hydrocarbons ($RH + NO_2 \rightarrow R^. + HONO$); (3) it adds to unsaturated hydrocarbons. In anhydrous acids (e.g., H_2SO_4 and HNO_3), N_2O_4 dissociates into the nitrosonium (NO^+) and nitronium (NO_2^+) ions. Liquid N_2O_4, which self-ionizes (NO^+ and NO_3^-), forms molecular addition compounds with a variety of donor compounds (e.g., $[base.NO^+] NO_3^-$). It also reacts with metals (sodium, potassium, zinc, silver, lead, mercury) giving the nitrate and NO, and with salts giving nitrates; hydroxylic and secondary amino solvents are usually nitrosated to RONO and R_2NNO. Mixtures of N_2O_4 with organic solvents are very reactive (e.g., copper dissolves in N_2O_4/ethyl ethanoate to give $Cu(NO_3)_2.N_2O_4$, which on heating gives the anhydrous nitrate).
dinitrogen pentoxide (N_2O_5). White solid obtained by the dehydration of nitric acid with phosphorus pentoxide. The gaseous compound has the structure $O_2N-O-NO_2$, whereas the solid in its stable form is nitronium nitrate (NO_2^+ NO_3^-). It is a strong oxidizing agent which readily decomposes to NO_2 and O_2. In concentrated sulphuric acid the NITRONIUM ION is formed.
nitrogen trioxide (NO_3). Unstable white solid formed in the catalyzed reaction of N_2O_5 with ozone.

nitrogen oxoacids.
hyponitrous acid ($H_2N_2O_2$). Solution of the sodium salt is formed from sodium nitrite ($NaNO_2$) and sodium amalgan, from which the insoluble silver salt is precipitated. This on treatment with hydrogen chloride in dry

ether gives the free acid. The free acid is stable in solution, but the solid decomposes spontaneously. The salts of the weak acid behave as reducing agents. The hyponitrite ion has the *trans*-configuration.

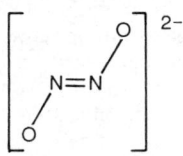

nitrogen oxoacids
Configuration of hyponitrite ion.

nitramide (H_2NNO_2). Weak acid, isomeric with hyponitrous acid.

nitroxylic acid (H_2NO_2). Free acid is unknown, the yellow sodium salt is formed by the electrolysis of sodium nitrite in liquid ammonia. The salt decomposes on heating.

nitrous acid (HNO_2). Solution of this unstable, fairly strong acid ($pK_a = 3.37$) can be prepared free of salts by treating barium nitrite with sulphuric acid. It can act as an oxidizing agent (e.g., towards I^-, Fe^{2+}) but is more normally a reducing agent being oxidized to nitric acid. The thermally stable group I and II metal nitrites are prepared by heating the nitrate with a reducing agent. HNO_2 (prepared *in situ*) is used in the preparation of DIAZONIUM COMPOUNDS. Numerous organic derivatives of the NO_2 group are known: the nitrito compounds (R–ONO) and the nitro compounds (RNO_2).

hyponitric acid ($H_2N_2O_3$). Free acid is unstable, but the sodium salt is formed by the reaction of hydroxylamine with ethyl nitrite in methanol in the presence of sodium methoxide.

peroxonitrous acid (HOONO). Formed as an intermediate in the oxidation (H_2O_2) of nitrous to nitric acid. The acid is unstable, but the anion is stable in alkaline solution. It is isomeric with nitric acid.

nitric acid (HNO_3). Produced industrially by the catalytic oxidation of ammonia, the resulting nitrogen monoxide is mixed with oxygen and absorbed in water; the commercial acid, a constant boiling mixture, is about 67 percent HNO_3. Pure HNO_3, a colourless liquid or white crystalline solid is obtained by treating potassium nitrate with 100 percent sulphuric acid and distilling *in vacuo*. The anhydrous acid is a good ionizing solvent with a high self-ionization constant ($NO_2^+ + NO_3^-$). With water it

forms mono- and tri-hydrates. Red 'fuming HNO_3' contains dissolved nitrogen dioxide. It is a strong acid forming nitrates with bases, carbonates, etc. The concentrated acid is a powerful oxidizing agent being reduced to nitrogen oxides by metals, of the metals only gold, platinum, rhodium and iridium are unattacked; aluminium and iron are rendered 'passive'. Non-metals are oxidized to oxoacids or oxides; fluorine and chlorine react to give the corresponding nitryl. The ability of concentrated HNO_3 in the presence of concentrated sulphuric acid to nitrate many organic compounds is due to the presence of the NITRONIUM ION. Gaseous HNO_3 is a planar molecule. HNO_3 is used extensively in the production of fertilizers, explosives and dyestuffs.

nitrogen oxohalides.

nitrosyl halides (NOX, where X = F, Cl, Br). Series of covalent compounds prepared by the direct union of the halogen with nitrogen monoxide (NO). They are unstable, decomposing into the halogen and NO. They are very reactive and powerful oxidizing agents, decomposing in water to nitric acid, nitrous acid, NO and HX.

nitryl halides (XNO_2). Known only for fluorine ($N_2O_4 + CoF_3$, 300°C) and chlorine ($ClSO_3H + HNO_3$, 0°C). They are reactive compounds, decomposed by water to nitric acid and HX.

nitroglycerin ($O_2NOCH_2(ONO_2)CH_2ONO_2$). Oily liquid prepared by the nitration of GLYCEROL with nitric acid and sulphuric acid. It is a very powerful and dangerous EXPLOSIVE used in the manufacture of dynamite. Medicinally it is used to give relief from pain in angina pectoris.

nitro group. The group $-NO_2$. *See also* NITROALKANES, NITROBENZENE.

nitrones. Unsaturated analogues (*see* Figure) of tertiary AMINE OXIDES, formed

$$ArC=\overset{+}{N}-R \quad \longleftrightarrow \quad Ar-\overset{O^-}{\underset{H}{\overset{+}{C}-N}}-R$$

by the reaction of an aromatic aldehyde or ketone with a *N*-substituted hydroxylamine. They add to carbon–carbon double bonds to give isoxazolidines, and with UV irradiation give oxaziranes which at elevated temperatures are isomerized to the nitrone.

nitronium ion (nitryl ion, NO_2^+). Electrophilic ion found in mixtures of concentrated nitric and sulphuric acids and in solutions of nitrogen oxides in nitric acid. Nitronium salts (e.g., NO_2^+ ClO_4^-) can be isolated, but are extremely reactive towards water and organic matter. In the nitration of aromatic hydrocarbons with concentrated nitric acid/sulphuric acid mixtures, it is the NO_2^+ ion that attacks the benzene ring.

nitroparaffins.   *See* NITROALKANES.

nitrophenols.   Phenol nitrates so rapidly that for mononitration dilute nitric acid at room temperature is required. The *ortho*- and *para*-isomers so formed can be separated by steam distillation. Reduction gives the corresponding AMINOPHENOL. They are all pale yellow or colourless solids; the salts are deep yellow. The nitro group increases the acidity of phenol.
2-nitrophenol (mp 45°C).
3-nitrophenol (mp 57°C, pK_a 8.35). Formed by diazotization of 3-nitroaniline and hydrolysis of the diazo compound.
4-nitrophenol (mp 114°C, pK_a 7.14).
2,4,6-trinitrophenol (picric acid). Yellow solid (mp 122°C), insoluble in cold, but soluble in hot water, strong acid (pK_a 0.38); prepared by nitration of phenolsulphonic acid (PhOH + HNO_3 causes excessive oxidation). It forms charge transfer complexes, 'picrates', with aromatic hydrocarbons, amines and phenols which are used for identification. Aromatic ring is oxidized by alkaline hypochlorite (NaOCl) to trichloronitromethane (Cl_3CNO_2). Picric acid reacts with phosphorus pentachloride to 2,4,6-trinitrochlorobenzene and with metals forming picrates which are sensitive to shock. Picric acid is used in the dyeing industry and was once used as an explosive.

nitroprussides.   *See* CYANOFERRATES.

nitrosamines   (R_2NNO). Yellow oils, prepared by the reaction between second-

ary amines and nitrous acid (a source of NO^+)

$$R_2NH + NO^+ \rightarrow R_2NNO + H^+$$

Nitrosamines are used to detect secondary amines by Liebermann's nitroso reaction. In this the nitrosamine is warmed with phenol and concentrated sulphuric acid, and the resulting solution is treated with alkali to give a deep-blue solution.

nitroso compounds.   Organic compounds containing the nitroso group –N=O. C-Nitroso compounds are prepared by the addition of nitrosyl chloride to alkenes, by the action of nitrous acid on secondary nitroalkanes, or by the oxidation (Caro's acid, *see* SULPHUR OXOACIDS) of primary amines with a α-tertiary alkyl group. Nitroso compounds have an unpaired electron and thus are blue or green. Tertiary nitroso compounds dimerize to colourless compounds or if they have α-hydrogen atoms tautomerize to the OXIME

$$R_2CHNO \rightleftharpoons R_2C=NOH$$

They are oxidized by nitric acid to nitro compounds and reduced (Sn/HCl) to primary amines. Aromatic nitroso compounds are prepared by oxidation of the arylhydroxylamine ($K_2Cr_2O_7/H_2SO_4$) or amine (Caro's acid). Nitrosobenzene exists as colourless crystals of the dimer (mp 68°C) and is reduced to azobenzene (Pd/H_2), phenylhydroxylamine (C_6H_5NHOH) (P/HI), azoxybenzene ($C_6H_5N=NOC_6H_5$) (Ph_3P) and aniline (Fe/HCl). It is oxidized to nitrobenzene by hydrogen peroxide or nitric acid and undergoes CONDENSATION REACTIONS in which the nitroso group acts like an OXO GROUP. *See also* NITROSAMINES.

nitrosonium ion (NO^+).   Ion, intermediate in size between H_3O^+ and NH_4^+, formed when dinitrogen trioxide or dinitrogen tetroxide is dissolved in concentrated sulphuric acid. Nitrosonium hydrogen sulphate (NO^+ HSO_4^-), a white solid (SO_2 + fuming HNO_3) is an intermediate in the lead chamber process for the manufacture of sulphuric acid. The compounds NO^+ ClO_4^- and NO^+ BF_3^- (NOBr + BF_3) are prepared in non-hydroxylic solvents; they are isomorphous with the corresponding hydroxonium and ammonium salts. The salts are readily hydrolyzed to nitrous acid and must be handled under anhydrous conditions. $NOBF_3$ reacts with aromatic primary amines to give diazonium tetrafluoroborates, which on pyrolysis give aromatic fluoro derivatives.

nitrosyl halides. *See* NITROGEN OXOHALIDES.

nitrosyls. Compounds containing NO bonded to metals through the nitrogen atom. In general they are similar to the CARBONYLS, but the metal–nitrogen bonds are more stable to substitution than are the metal–carbon bonds. Sodium nitrosyl (NaNO) is prepared by the action of nitrogen monoxide on the metal dissolved in liquid ammonia. Unstable nitrosyls are formed by iron, ruthenium and nickel; iron nitrosyl (Fe(NO)) ($Fe(CO)_4$ + NO under pressure at 50°C) is the most stable. The nitrosyl carbonyls (e.g., $Co(NO)(CO)_3$, $Fe(NO)_2(CO)_2$) are more stable than the nitrosyls. Nitrosyl halides ($Fe(NO)_2X$) and $Co(NO)_2X$ are known.

nitrous acid. *See* NITROGEN OXOACIDS.

nitrous oxide. *See* NITROGEN OXIDES.

nitroxylic acid. *See* NITROGEN OXOACIDS.

nitryl halides. *See* NITROGEN OXOHALIDES.

nitryl ion. *See* NITRONIUM ION.

NMR. *See* NUCLEAR MAGNETIC RESONANCE SPECTROSCOPY.

No. Nobelium (*see* ACTINIDE ELEMENTS).

nobelium (No). *See* ACTINIDE ELEMENTS.

noble gas electron structure. Having a filled outer shell of electrons (*see* ELECTRONIC CONFIGURATION) and the consequent stability it imparts. Thus in sodium chloride, Na^+ Cl^-, the sodium atom loses an electron to achieve the electronic structure of neon, and chlorine gains an electron to give that of argon. *See also* GROUP O ELEMENTS.

noble gases. *See* GROUP O ELEMENTS.

noble metals. *See* PLATINUM GROUP METALS.

nodal plane. Plane through a molecule in which a molecular orbital has zero electron density. For example, the π^*-orbital in ethene has a nodal plane bisecting the carbon–carbon bond.

non-aqueous solvents. Solvents other than water that include FUSED SALTS (e.g., LiF), organic solvents (e.g., dimethylsulphoxide) and other liquids (e.g., H_2SO_4). The ability of solvents to undergo acid/base types of interaction is important (*see* LEWIS ACIDS AND BASES). They are classed as acidic (e.g., ethanoic acid), basic (e.g., ethers), aprotic (e.g., tetrachloromethane) or amphiprotic (e.g., liquid ammonia) (*see* PROTOPHILIC SOLVENT). The ability to solvate an ionic solute may be related to the relative PERMITTIVITY or the DONICITY of the solvent. Organic solvents are used extensively in industry for the manufacture of paints, inks, etc., synthetic fibres and polymers, and for cleaning and degreasing.

non-benzenoid. Describing a monocyclic conjugated system in which there are ($4n$ + 2) π-electrons, not containing a benzene ring (e.g., ANNULENES). There are also non-benzenoid aromatics which contain more than one ring (e.g., AZULENE). *See also* HÜCKEL'S RULE.

non-bonding orbital. *See* MOLECULAR ORBITAL.

non-cumulative double bonds. *See* CUMULATIVE DOUBLE BONDS.

non-ionic detergent. *See* SURFACE-ACTIVE AGENT.

non-newtonian fluid. *See* NEWTONIAN FLUID.

non-stoichiometric compound. Solid (BERTHOLLIDE COMPOUND) in which the atoms combine in ratios that are not small whole numbers. Binary compounds are known having an excess or deficit of either atom (e.g., $Zn_{1+\delta}O$, $Fe_{1-\delta}O$, $UH_{3-\delta}$, $UO_{2+\delta}$) or some show deviations either side of stoichiometric (e.g., $NiO_{1\pm\delta}$) depending on the method of preparation. INTERCALA-

TION COMPOUNDS and substitutional SOLID SOLUTIONS may also be non-stoichiometric. Non-stoichiometric compounds are generally metallic or semiconducting, coloured and show unusual reactivity. They do not follow the law of equivalent proportions (*see* EQUIVALENT PROPORTIONS, LAW OF).

nor-. Prefix to designate a homologue with one less methene group (e.g., adrenaline and noradrenaline). In TERPENE nomenclature, nor- indicates the loss of all methyl groups from the parent compound.

noradrenaline. *See* HORMONES.

norepinephrine (noradrenaline). *See* HORMONES.

normality. Solution that contains 1 gram-equivalent of the solute in 1 litre of solution (e.g., 1 N hydrochloric acid). It is not coherent with SI units, and its use is discouraged.

normalization. WAVEFUNCTION (ψ) is normalized if it satisfies the condition

$$\int_{-\infty}^{\infty} \psi^* \, \psi \, d\tau = 1$$

where τ is $dx \, dy \, dz$. In terms of the probability of finding an electron, the normalization condition is an assertion that the electron must be somewhere.

normal mode. Independent vibration of a molecule. There are $(3N–5)$ normal modes in a linear N-atom molecule, and $(3N–6)$ in a non-linear molecule. They may be classed as stretching (labelled N) or bending (labelled ν or δ), depending on whether a bond length or bond angle is altered. Other terms, such as scissor, rock and wag, are also used to describe bending modes. If the vibration is in phase or out of phase with the molecule turned through 180° about its axis of symmetry, the mode is classed as symmetric or antisymmetric, respectively. Vibrations are also classified as parallel ($\parallel$) or perpendicular ($\perp$) according to the change of dipole with respect to the principal axis of the molecule. (For normal modes of water—*see* Figure.) *See also* INFRARED SPECTROSCOPY.

normal solution. *See* NORMALITY.

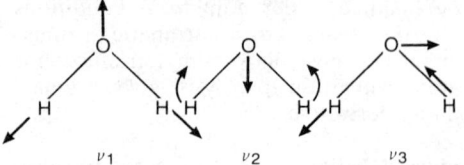

normal mode
Normal modes of water.

normal temperature and pressure. *See* STANDARD TEMPERATURE AND PRESSURE.

novobiocin. *See* ANTIBIOTICS.

Np. Neptunium (*see* ACTINIDE ELEMENTS).

NQR. *See* NUCLEAR QUADRUPOLE RESONANCE SPECTROSCOPY.

ntp. *See* STANDARD TEMPERATURE AND PRESSURE.

nuclear energy. Energy liberated by: (1) the splitting or fission of an atomic nucleus; (2) the union or fusion of two atomic nuclei; and (3) the radioactive decay of a nucleus. The energy emitted in (1) is equivalent to the slight difference in mass between the nucleus as a whole and its constituents (*see* MASS DEFECT). The release of energy is virtually instantaneous; it may involve a large number of nuclei (atom bomb) or comparatively few nuclei (nuclear reactor) where the rate of release is under control. In the fusion process (2) the reaction is instantaneous and the energy released much greater than in fission, but no means of controlling it has so far been found. In radioactive decay (3) the rate of release of energy is extremely slow.

nuclear fission. Splitting of an atomic nucleus (e.g., uranium) into two parts (fission products) with the emission of two or three neutrons and the release of energy equivalent to the difference between the mass of the neutrons and fission products and that of the original nucleus. These neutrons, in turn, split other atoms so starting a chain reaction. Fission may be spontaneous or as a result of irradiation by neutrons. For the reaction

$$^{235}_{92}U + ^{1}_{0}n \longrightarrow ^{148}_{57}La + ^{85}_{35}Br + 3^{1}_{0}n$$

the energy released is 3×10^{-11} J per uranium-235 nucleus; for 1 kg of uranium-235 this is equivalent to 20 000 MW h.

nuclear fusion. Nuclear reaction in which atomic nuclei of low atomic number fuse to form heavier nuclei, with the release of large amounts of energy. As both nuclei are positively charged, there are strong repulsive forces which can only be overcome if the nuclei have very high kinetic energies. This is achieved at very high temperatures (about 10^8 K), when fusion is self-sustaining, the reactants being in the form of a plasma (i.e. nuclei and free electrons). The fusion (hydrogen) bomb, the sun and stars generate energy in this way. A typical fusion reaction is

$$^6_3Li + \,^2_1H \longrightarrow 2\,^4_2He$$

$$\Delta U = -35.8 \times 10^{-13} \, J$$

By comparison 4×10^{-19} J are released for the formation of one molecule of water from its elements. If controlled fusion can be achieved, it will have the advantages of a cheap and unlimited source of energy with no radioactive byproducts such as are produced in a fission reactor.

nuclear gamma resonance fluorescence. *See* MÖSSBAUER SPECTROSCOPY.

nuclear gyromagnetic ratio (γ; units: s^{-1} T^{-1}). Ratio between the MAGNETIC MOMENT (μ) and the angular MOMENTUM (L).

$$\gamma = 2\mu\pi/hL$$

For a proton $\gamma_p = 2.6752 \times 10^8 \, s^{-1} \, T^{-1}$. *See also* NUCLEAR SPIN.

nuclear magnetic resonance spectroscopy (NMR spectroscopy). Nuclear energy levels of elements having NUCLEAR SPIN (e.g., ^1H, ^{13}C, ^{15}N) are split by a magnetic field. Transitions between these levels may be induced by radiofrequency ELECTROMAGNETIC RADIATION. A plot of the intensity of absorption against radiofrequency (at fixed magnetic field) or field (fixed radiofrequency) is the NMR spectrum. Typical frequencies used are between 60 and 400 MHz, which correspond to magnetic fields of 1.5–8 T. FOURIER TRANSFORM ANALYSIS has been used successfully in NMR. The position of lines characteristic to a nucleus

in a given environment (CHEMICAL SHIFT) aids identification of the molecular structure, and the integrated intensities give the relative numbers of the particular nuclei. Extent of the chemical shift (δ) is expressed on a scale in parts per million (ppm) relative to a standard which for protons is tetramethyl silane (TMS)

$$\delta = (\nu - \nu_{TMS})/\nu_{TMS} \times 10^6 \, ppm$$

Proton chemical shifts are given on a tau scale where $\tau = 10 - \delta$.

At high resolution, NMR lines split into multiplets due to spin–spin coupling between groups of equivalent nuclei (*see* MULTIPLET, Figure). The magnitude of the splitting is a few hertz and is quoted as a coupling constant (J) which is the energy difference (expressed in hertz) between two nuclei having parallel spins and two of opposite spins. In complex spectra where the extent of coupling can make interpretation impossible, irradiation of the sample at the resonance frequency of one nucleus wipes out (decouples) the coupling due to that nucleus and thus leads to a simpler spectrum. A side effect of spin decoupling is the nuclear Overhauser effect (NOE) by which the intensity of the signal from nuclei close to the decoupled nucleus is perturbed. The effect is proportional to $1/r^6$ (r is the internuclear separation), and so NOE is a sensitive probe of the immediate environment of a given nucleus. The use of lanthanide shift reagents also aids determination of molecular structure. A large ion (e.g., complexed Eu^{3+}) coordinates with electron-rich sites of a molecule and deshields (i.e. shifts lines downfield) other nuclei by an extent that depends on the distance from the ion. Analysis of line shapes gives information of rapid molecular processes (e.g., proton exchange, interconversion of isomers, see ISOMERISM). Other kinetic measurements are of T_1, the spin-lattice relaxation time (of the order of seconds in liquids and minutes in solids) and of T_2, the spin–spin relaxation time (about T_1 in liquids and much shorter in solids).
CIDNP (chemically induced dynamic nuclear polarization). Phenomenon of emission or enhanced absorption during FREE RADICAL reactions.
INDOR (internuclear double resonance). Performed by monitoring the signal at a fixed frequency while scanning the decoupl-

ing frequency. This gives a spectrum of all the nuclei coupled to a particular nucleus which may be more simple to interpret than that of the molecule as a whole.

nuclear paramagnetism. *See* NUCLEAR SPIN.

nuclear quadrupole resonance spectroscopy (NQR spectroscopy). Atomic nuclei possess an electric quadrupole moment if the NUCLEAR SPIN quantum number $I \geqslant 1$. This is expressed as eQ, where Q has the units of length squared and typical values of 3×10^{-32} m^2 (^{2}H) to 8×10^{-29} m^2 (^{127}I). Transition between nuclear quadrupole levels of a solid may be observed in the radiofrequency range of the ELECTROMAGNETIC SPECTRUM. The absorption frequency measures eQq, where q is the electric field gradient (V m^{-2}) at the nucleus which may therefore give an indication of the degree of covalent bonding. A larger value of q suggests a removal of an electron away from the nucleus creating an asymmetrical charge distribution.

nuclear spin. Property that imparts angular momentum (L) to a nucleus. L is quantized

$$L = h/2\pi \, [I(I + 1)]^{1/2}$$

where I is the nuclear spin QUANTUM NUMBER. The magnetic moment of the nucleus μ is

$$\mu = g_N \mu_N \, [I(I + 1)]^{1/2}$$

g_N is the nuclear g-factor (*compare* LANDÉ g-FACTOR for ELECTRON SPIN) and μ_N is the nuclear magneton ($\mu_N = eh/4\pi m_p$); m_p is the proton mass (*compare* BOHR MAGNETON) and has a value of $5.050\,824 \times 10^{-27}$ A m^2. The product $g_N \mu_N$ is called the magnetogyric or gyromagnetic ratio (γ). The spin of a nucleus may couple with rotational levels causing them to split. The effect is usually small except in molecules having an odd number of electrons. Nuclei may also have a quadrupole moment if $I \geqslant 1$ (*see* NUCLEAR QUADRUPOLE RESONANCE SPECTROSCOPY) Nuclear properties of some atoms are listed (*see* Table). *See also* NUCLEAR GYROMAGNETIC RATIO.

nucleation. Formation of a nucleus prior to crystallization.

nuclear spin
Nuclear properties of some atoms.

Nucleus	Spin	Magnetic moment (units of μ_N)	Quadrupole moment $/10^{-30}$ m^2
^{1}H	1/2	2.79270	0.0
^{2}H	1	0.85738	+0.28
^{12}C	0	0.0	0.0
^{13}C	1/2	0.70216	0.0
^{14}N	1	0.40357	+2.00
^{16}O	0	0.0	0.0
^{17}O	5/2	−1.8930	−3.01
^{19}F	1/2	2.6273	0.0
^{35}Cl	3/2	0.82089	−7.90
^{79}Br	3/2	2.0990	+33.20
^{127}I	5/2	2.7939	−78.50

nucleic acids. Polynucleotides that are essential components of all living cells. They consist of a long, unbranched chain with a backbone of sugar (ribose or deoxyribose, *see* PENTOSES) and phosphate units with heterocyclic bases protruding from the chain at regular intervals. There are two main types: deoxyribonucleic acid (DNA), the genetic material of most living cells and a major constituent of chromosomes within the cell nucleus, and ribonucleic acid (RNA) concerned with protein synthesis found mainly in the cytoplasm (although synthesized in the nucleus). On hydrolysis nucleic acids are broken down first to NUCLEOTIDES, then NUCLEOSIDES and finally PURINES, PYRIMIDINES, sugar and phosphate.

The DNA molecule consists of two helical chains coiled round the same axis in which alternate deoxyribose molecules are connected through C-3 and C-5 by phosphate groups, with bases attached to the sugars (*see* Figure). HYDROGEN BONDS between the bases hold the chains together. Spatial considerations demand that one of the bases in a pair be a purine and the other a pyrimidine (purine–purine pairs would be too large and pyrimidine–pyrimidine pairs too small). This is satisfied by the pairing of thymine with adenine and cytosine with guanine, consequently a given sequence of bases in one chain determines that in the other. DNA replicates on cell division so that the two daughter molecules

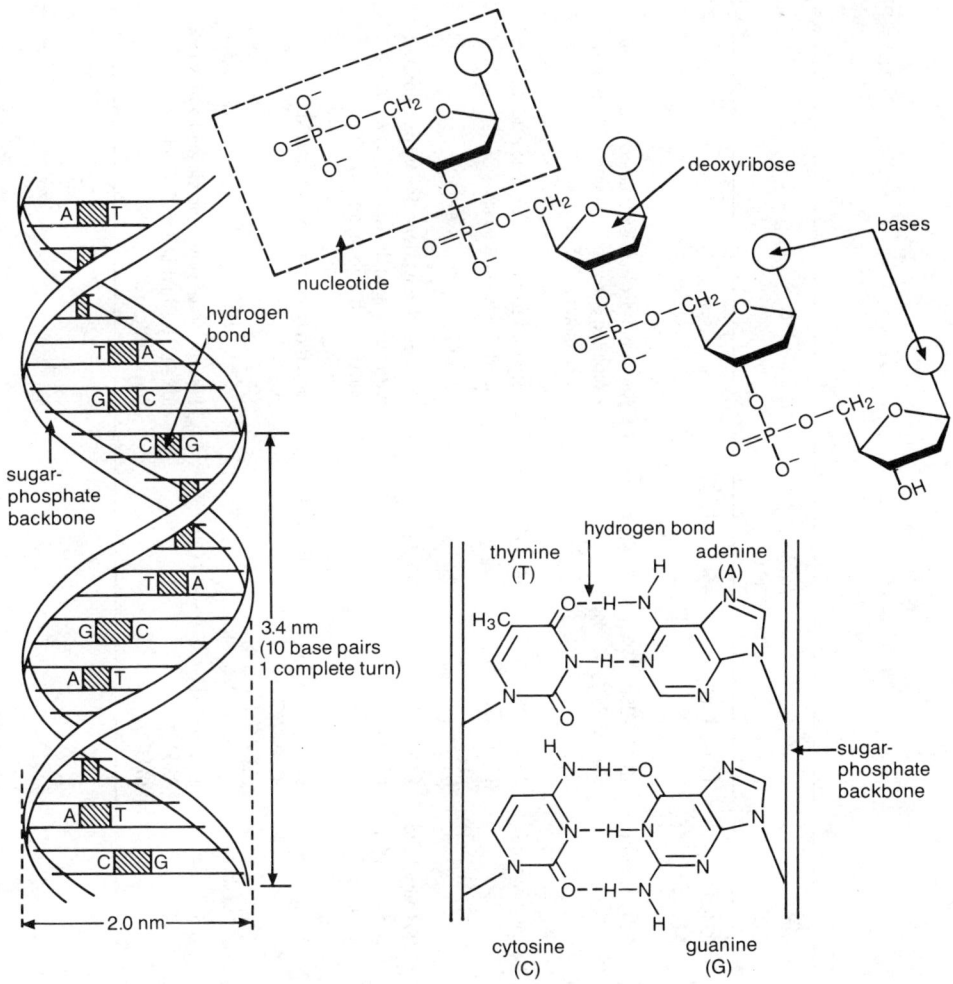

nucleic acids
Molecular structure of DNA (adapted from *Concise Dictionary of Chemistry*, Oxford Science Publications).

are identical to the parent molecule. In the presence of polymerases the hydrogen bonds break, the strands unwind and, from nucleotides in the nucleus, each strand directs the synthesis of a complementary strand.

In the RNA molecule, the sugar is ribose and uracil replaces thymine. Messenger RNA (mRNA) conveys information in the form of base sequence from DNA to RNA on RIBOSOMES where protein synthesis occurs from amino acids associated with the RNA template. A specific sequence of three bases—a codon—codes for each amino acid residue; thus the amino acid sequence is ultimately based on the order of the nucleotides in the DNA molecule (*see* GENETIC CODE). Ribosomal RNA (rRNA) together with ribosomal protein form the ribosomes. Transfer RNA (tRNA) is responsible for transporting specific amino acid molecules to mRNA or the ribosome during protein synthesis.

nucleon.   Collective name for particles of mass number 1, that is PROTONS and NEUTRONS, the essential constituents of atomic nuclei.

nucleosides
Properties of nucleosides.

Compound	Formula	mp /°C	ε (λ/nm)	$[\alpha]_D^t$ (t/°C)	Comments
Adenosine (adenine-9-β-D-ribofuranoside)	$C_{10}H_{13}N_5O_4$ $M_r = 267.24$	234–5	15 100 (260)	−61.7 (11)	
Cytidine (cytosine-1-β-D-ribofuranoside)	$C_9H_{13}N_3O_5$ $M_r = 243.22$	220 (decomp.)	13 400 (280) pH 2.2	+31 (25)	Separated from other nucleosides by ion exchange chromatography. pK (amino) 4.22; (sugar) 12.5
Guanosine (guanidine-9-β-D-ribofuranoside)	$C_{10}H_{13}N_5O_5$ $M_r = 283.24$	240 (decomp.)	26 800 (188.3) 13 700 (252.5) pH 5.5	−60.5 (20)	Prepared from yeast nucleic acid. Sol. in mineral acids and bases
Inosine (hypoxanthine-9-β-D-ribofuranoside)	$C_{10}H_{12}N_4O_5$ $M_r = 268.23$	90 (dihydrate) 218 (anhyd., decomp.)	12 200 (248.5) pH 6.0	−49.2 (18)	Prepared from adenosine by deamination with adenosine deaminase; or action of $NaNO_2$ and acid on adenosine
Thymidine (thymine-2-deoxy-β-D-ribofuranoside)	$C_{10}H_{14}N_2O_5$ $M_r = 242.23$	185	9800 (206.5) 9700 (267) pH 7.0	+30.6 (25)	Isolated from thymonucleic acid
Uridine (uracil-1-β-D-ribofuranoside)	$C_9H_{12}N_2O_6$ $M_r = 244.2$	165	10 100 (261) 9800 pH 7.3	+ 4 (20)	Hydrolysis of yeast nucleic acid with weak alkalis. Refluxing with HCl gives furfuraldehyde

nucleon number (mass number, A). Number of NUCLEONS in the nucleus of a particular NUCLIDE ($A = N + Z$).

nucleophile (nucleophilic reagent). Compound or group capable of donating electrons. Examples include oxidizing agents, Lewis bases, negative ions (e.g., $-OH^-$, $-OR^-$, Cl^-) and molecules with electron pairs (e.g., NH_3, R_3N). In organic reactions they tend to attack positively charged parts of a molecule. *See also* ELECTROPHILE.

nucleophilic addition. Reaction in which the first step is the attachment of a NUCLEOPHILE to a positive (electron-deficient) part of a molecule, for example

$$-O = C - + Y^- \longrightarrow {}^-O - C - Y$$

nucleophilicity. Ability to form bonds to carbon atoms; a measure of the rate of reaction of a NUCLEOPHILE with a substrate. Factors affecting the relative nucleophilicities of reagents include the BASICITY, the polarizability of the attacking electron pair and the solvation shell surrounding the nucleophile. Nucleophilicity (kinetic) and basicity (thermodynamic) are not synonymous, but are related. Deviations do occur, thus RS^- is more reactive as a nucleophile but less basic than RO^-; I^- is a weak base but powerful nucleophile, but F^- is a poor nucleophile.

nucleophilic reagent. *See* NUCLEOPHILE.

nucleophilic substitution. Reaction in which a NUCLEOPHILE (Z^-) displaces another group from a compound

$$R-X + Z^- \rightarrow R-Z + X^-$$

where R is alkyl (some aryl) or metal group, and Z a wide variety of organic or inorganic anions, X^- is the leaving group. The reaction

$$R-Cl + OH^- \rightarrow R-OH + Cl^-$$

may proceed by a S_N1 MECHANISM, favoured by tertiary aliphatic halides or by a S_N2 MECHANISM favoured by primary aliphatic halides. Aryl halides undergo nucleophilic substitution with difficulty.

nucleoproteins. *See* NUCLEIC ACIDS, PROTEINS.

nucleosides. Glycosides of heterocyclic bases, particularly PURINES and PYRIMIDINES (i.e. base + sugar); partial hydrolysis products of NUCLEIC ACIDS. They are crystalline substances, sparingly soluble in water and less soluble in ethanol. The nucleosides (*see* Table) forming part of the molecule of ribonucleic acid are the 9-β-D-ribofuranosides of adenine and guanine and the 1-β-D-ribofuranosides of cytosine and uracil. In deoxyribonucleic acid the sugar is deoxyribose and the base thymine replaces uracil.

nucleotides. Compounds containing one purine or pyrimidine base, one pentose unit and one phosphate unit. They are obtained by the hydrolysis of NUCLEIC ACIDS. 5'-Adenylic acid is a typical nucleotide. Some of the properties of the nucleotides are listed (*see* Table). *See also* ADENOSINE TRIPHOSPHATE, COENZYMES.

nucleotides
5'-Adenylic acid.

nucleus. (1) Positively charged, high-density central mass of an atom consisting of protons and neutrons. The mass and radioactive properties are associated with the nucleus, whereas the chemical properties and spectra are associated with the orbiting electrons. The atomic nucleus can be split by bombardment with neutrons from an external source. (2) Central portion of a plant or animal cell containing the genetic material and separated from the surrounding cytoplasm by a membrane. (3)

nucleotides

Nucleotide/base/sugar	mp/°C	ε (λ/nm)	$[\alpha]_D^t$ (t/°C)	pK	Comments
5′-Adenylic acid/ adenine/D-ribose (adenosine monophosphate, AMP) $C_{10}H_{14}N_5O_7P$ $M_r = 347.23$	196–200	15 400 (259) (pH 7.0)	–47.5 (20) (in NaOH)	3.8; 6.2	Prepared by hydrolysis of ATP with $Ba(OH)_2$ or enzymic phosphorylation of adenosine
3′-Cytidylic acid/ cytosine/D-ribose (cytosine monophosphate, CMP) $C_9H_{14}N_3O_8P$ $M_r = 323.21$	232 (decomp.)	13.0 (279)	+49.4 (20)	0.8; 4.28; 6.0	Prepared from yeast RNA; mod. sol. in H_2O. Ribonuclease inhibitor
5′-Guanylic acid/ guanine/D-ribose (guanosine monophosphate, GMP) $C_{10}H_{14}N_5O_8P$ $M_r = 363.23$	190–200 (decomp.)	13 700 (252) (of di-Na salt)			Isolated from sardines or yeast extract; direct biosynthesis using bacteria or enzyme
5′-Inosinic acid/ hypoxanthine/D-ribose (inosine monophosphate, IMP) $C_{10}H_{13}N_4O_8P$ $M_r = 348.22$	Syrup		–18.5 (20) (in HCl)	2.4; 6.4	Isolated from meat extract, dried sardines; prepared by deamination of muscle adenylic acid. Used as flavour intensifier
5′-Uridylic acid/uridine/ D-ribose (uridine monophosphate, UMP) $C_9H_{13}N_2O_9P$ $M_r = 324.19$		10 000 (262) (pH 7.0)		6.4; 9.5	Widely distributed in nature

Characteristic structure of a broad group of chemical compounds (e.g., the benzene nucleus). (4) Any small particle that can serve as the basis for crystal growth (*see* NUCLEATION).

nuclide. Atomic nucleus characterized by its ATOMIC NUMBER (Z) and NUCLEON NUMBER (A) (e.g., ^A_ZX). *See also* ISOTOPES.

nujol. *See* MULL.

number average molecular mass. Average measure of a POLYMER may be made in terms of the average number of units or the average weight. The number average molecular mass is $\Sigma C_i M_i / \Sigma C_i$ where C_i is the concentration of polymer length i and molecular mass M_i. The weight average molecular mass is $\Sigma C_i M_i^2 / \Sigma C_i M_i$.

Nutrasweet. Trade name for a SWEETENING AGENT.

Nylon. Synthetic polyamides with fibre-like structures. They are manufactured by the CONDENSATION of a DICARBOXYLIC ACID with a diamine, or by the self-condensation of an amino acid. Nylons $H(NH(CH_2)_xNH-CO(CH_2)_yCO)_nOH$ are named Nylon-$x,y+2$ or $H(NH(CH_2)_xCO)_nOH$ are named Nylon-$x+1$. Nylon-6 is made by the polymerization of N-aminocaprolactam (*see* CAPROLACTAM) with a trace of water at 240–80°C. Nylon-6,6 is produced by the condensation of HEXANEDIOIC ACID and 1,6-diaminohexane. Nylons are important engineering thermoplastics as well as synthetic fibres. They have high melting points, are insoluble in many solvents, and are tough with good impact resistance and low friction. *See also* POLYMERIZATION, POLYMERS.

nystatin. *See* ANTIBIOTICS.

O

O. *See* OXYGEN.

o-. *See* ORTHO-.

Ω. *See* OHM.

ω. Angular velocity (*see* VELOCITY); DI-HEDRAL ANGLE.

O-branch. *See* VIBRATION–ROTATION SPECTRA.

obsidian. *See* SILICA.

occlusion. (1) Trapping of small pockets of liquid in a crystal during crystallization. (2) Retention of a gas by a solid such that the atoms or molecules of the gas occupy interstitial positions in the solid lattice (e.g., palladium can occlude hydrogen).

ocimene. *See* TERPENES.

***cis,cis*-9,12-octadecadienoic acid** (linoleic acid, $C_{17}H_{31}COOH$). Unsaturated CARB-OXYLIC ACID (bp 228°C, 14 mmHg), occur-ring naturally as the glyceryl ester in lin-seed oil, hemp oil and in mammalian lipids. It is an ESSENTIAL FATTY ACID.

n-octadecanoic acid (stearic acid, $CH_3(CH_2)_{16}COOH$). CARBOXYLIC ACID (mp 70°C), found naturally as glycerides in animal and vegetable fats. The sodium and potassium salts (stearates) are the constit-uents of soap. *See also* HEXADECANOIC ACID, PALMITIN, STEARIN.

***cis,cis,cis*-9,12,15-octadecatrienoic acid** (linolenic acid, $C_{17}H_{29}COOH$). Unsatur-ated CARBOXYLIC ACID (bp 230–32°C, 1 mmHg); the most abundant natural tri-ene. It occurs as the glyceryl ester in poppy seed oil, linseed oil, etc. It is an ESSENTIAL FATTY ACID.

***cis*-9-octadecenoic acid** (oleic acid, $C_{17}H_{33}$-COOH). Unsaturated CARBOXYLIC ACID (mp 16°C). It occurs widely as the glyceryl ester in oils and fats; it forms one-third of the fatty acids of cow's milk.

octahedral coordination. *See* SHAPES OF MOLECULES.

octane number. Measure of the tendency of a petrol formulation to prevent knock-ing in an internal combustion engine on a scale in which 2,2,4-trimethylpentane (iso-octane) is given a value of 100 and n-heptane zero. Addition of tetraethyllead (*see* LEAD ORGANIC DERIVATIVES) im-proves the octane number of petrol.

octanoic acid (caprylic acid). Oily liquid (bp 239.7°C) CARBOXYLIC ACID, with un-pleasant smell.

2-octanol (capryl alcohol, $CH_3(CH_2)_5CH$-$(CH_3)OH$). Liquid (bp 178.5°C) with a strong odour obtained by the action of concentrated sodium hydroxide on castor oil. It undergoes polymerization when heated with bases and is used as a foam-reducing agent.

octet. Group of eight valence electrons associated with stability in atoms and com-pounds of the first two periods (*see* PERI-ODIC TABLE) typified by the noble gases (*see* NOBLE GAS ELECTRONIC STRUCTURE). TRANSITION ELEMENTS, atoms in ELEC-TRON-DEFICIENT COMPOUNDS and radicals may not have an octet of electrons (e.g., NO, BF_3 and SF_6). *See also* EIGHTEEN-ELECTRON RULE.

oestradiol-17β. *See* HORMONES.

oestrogens (estrogens). Group of C_{18} steroid HORMONES produced from ethano-ate and cholesterol chiefly by the ovaries and placenta, but also by testes and adrenal cortex in all vertebrates. They stimulate the growth and maintenance of the reproductive organs in female mammals and are responsible for female secondary sexual characteristics. The predominant natural oestrogens are oestradiol-17β and oestrone.

oestrone. *See* HORMONES.

ohm (Ω; dimensions: $\varepsilon^{-1} l^{-1} t$; kg m^2 s^{-3} A^{-2} = VA^{-1}). Electrical resistance between two points on a conductor when a constant voltage of 1 volt applied between those points produces a current of 1 ampere.

Ohm's law. A conductor obeys Ohm's law if the voltage (V) across a conductor is proportional to the current (I) flowing through it. The constant of proportionality is the RESISTANCE (R) of the sample and has the units of ohms. Thus $V = IR$.

OHP. *See* OUTER HELMHOLTZ PLANE.

oil of bitter almonds. *See* BENZALDEHYDE.

oil of mirbane. *See* NITROBENZENE.

oil of wintergreen. *See* METHYL 2-HYDROXYBENZOATE.

oleandomycin. *See* ANTIBIOTICS.

olefin complex. COORDINATION COMPOUND in which the ligand is an olefin π-bonded to a metal. Many olefin complexes are known with TRANSITION ELEMENTS, and it is proposed that they are intermediates in catalyzed reactions of olefins. Olefin complexes are prepared by: (1) direct reaction (e.g., Vaska's compound, *see* IRIDIUM COMPOUNDS, reacts with alkenes); (2) displacement of a weaker ligand (e.g., CO); and (3) reductive addition (e.g., in aqueous ethanol)

$$(Ph_3P)_2PtCl_2 + N_2H_4 + C_2H_4 \rightarrow$$

$$(Ph_3P)_2Pt(C_2H_4)$$

See also π-COMPLEXES.

olefins. *See* ALKENES.

oleic acid. *See* cis-9-OCTADECENOIC ACID.

olein (triolein, glycerol trioleate, $(C_{17}H_{33}COO)_3C_3H_5$). Triglyceride of oleic acid, occurring in most fats and oils. It constitutes 70–80 percent of olive oil.

oleoresin. Semi-solid mixture of the resin and the essential oil of the plant from which they are derived. Oleoresins have a pungent taste and peculiar odour; they are sometimes referred to as balsams.

oleum. *See* SULPHUR OXOACIDS.

oligomer. POLYMER consisting of a known, small number of MONOMER units named dimer, trimer, tetramer, etc.

oligosaccharides. *See* SUGARS.

one-component system. System in equilibrium which, according to the PHASE RULE, can exist in no more than three phases (*see* Figure). In any area $p = 1$; $f = 2$ (bivariant), on a line where two phases are in equilibrium $p = 2$; $f = 1$ (univariant) and at a point (TRIPLE POINT) where three phases are in equilibrium $f = 0$. G (on the extension of EB) is a position of metastable equilibrium between supercooled liquid and vapour. The slope of lines BC and EF is determined by the relative molar volumes of the states in equilibrium. For the WATER system, BC has a negative slope. *See also* CLAPEYRON–CLAUSIUS EQUATION.

onium ion. Cations R_xA^+ (e.g., ammonium, R_4N^+; phosphonium, R_4P^+; oxonium, R_3O^+; sulphonium, R_3S^+; iodonium, R_2I^+).

opacity. *See* ABSORBANCE.

open shell molecule. *See* CLOSED SHELL MOLECULE.

open system. *See* SYSTEM.

opium alkaloids. *See* ALKALOIDS.

OPLC. *See* OVERPRESSURE LAYER CHROMATOGRAPHY.

optical activity. Phenomenon exhibited by some compounds (solids, liquids and solutions) which when placed in a beam of plane-polarized light rotate the plane of polarization to the right (dextrorotatory) or to the left (laevorotatory). Optical activity

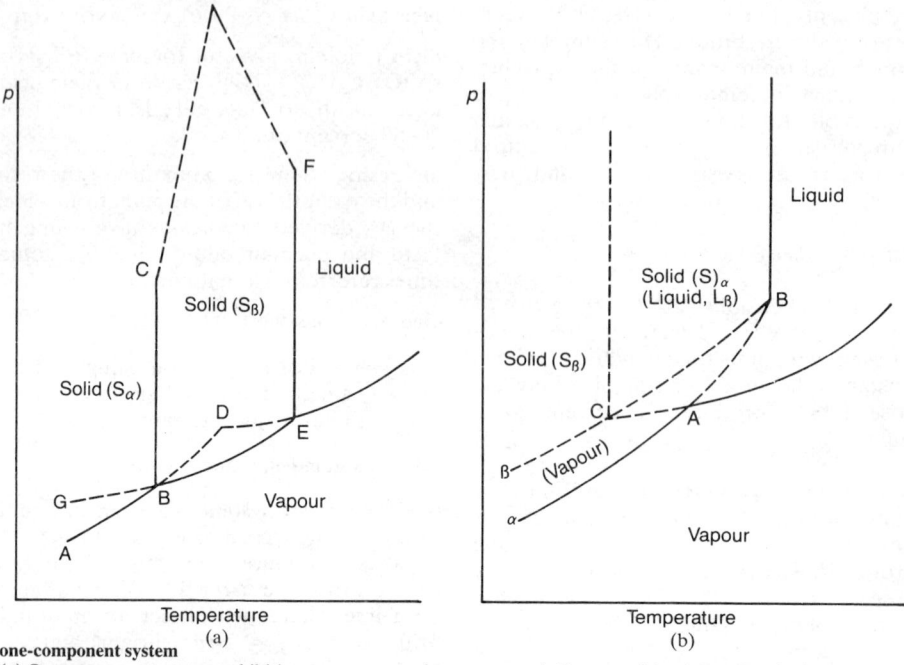

one-component system
(a) One-component system exhibiting ENANTITROPY (e.g., SULPHUR). Four possible triple points (B (S_α, S_β, vapour), E (S_β, liquid, vapour), C/F (S_α, S_β, liquid), D (S_α, liquid, vapour). (b) One-component system exhibiting MONOTROPY (e.g., PHOSPHORUS). B is the transition temperature, A (triple point) is the melting point of S_α and C is the melting point of metastable S_β.

is associated with asymmetry of a molecule, an ion or a crystal lattice. Molecules may be asymmetric due to the presence of a chiral centre or from restricted rotation about a central bond (*see* CHIRALITY). The prefixes *d*- and *l*- were once used to indicate the particular isomer, but the correct prefixes are now (+)- for dextro, (−)- for laevo- and (±)- for racemic compounds. The prefixes are arbitrary since the rotation can change with the wavelength of the light (the Cotton effect, *see* OPTICAL ROTATORY DISPERSION). With carbohydrates and amino acids the prefixes D- and L- are used to indicate configuration and not direction of rotation. Stereochemical relationships are more important than the direction of rotation, and the symbols D- and L- are now used to indicate relationships between configurations. The convention adopted for the establishment of stereochemical relationships is that the formula of D-(+)-glyceraldehyde (2,3-dihydroxypropanal) is represented by I with the hydrogen atom at the left, the hydroxyl group at the right and the aldehydic group at the top corner; II is the L-(−)- isomer.

Thus any compound that can be prepared from or converted into D-(+)-glyceraldehyde will belong to the D-series; similarly for the L-series. When representing the relative configurational relationships of molecules containing more than one asymmetric carbon, the asymmetric carbon of glyceraldehyde is always drawn at the bottom; the rest of the molecule being built up from this unit.

The same nomenclature applies to the amino acids, the configurational family to which the α-carbon belongs is denoted by D- or L-. On account of the ambiguous definitions of D- and L-, this system has been superseded by a convention which allows unequivocal designation of the absolute spatial configuration of groups attached to a chiral centre. The symbols used for the two enantiomers are (R) (rectus: right) and (S) (sinster: left); what was D in the old system, in most cases, becomes (R); similarly L becomes (S). The definitions of (R) and (S) are quite different from those of D- and L-. The order of priority of groups is assigned according to the SEQUENCE RULES. If the priority for the groups in the molecule Cabde is a, b, d, e, the spatial pattern can be described as righthanded (R) or left-handed (S) according to whether the sequence a→b→d→e is clockwise or anti-clockwise when viewed from an external point on the side remote from e (the atom or group of lowest priority). It is analogous to a steering wheel facing the driver with the group of lowest priority down the steering column. Since the sequence is clockwise this is the (R)-form. For D-(+)-glyceraldehyde I, the priority sequence is –OH, –CHO, –CH₂OH, H corresponding to a, b, d in a clockwise fashion, thus D-(+)-glyceralde-hyde is (R)-glyceraldehyde and II is (S)-glyceraldehyde. For a molecule containing two or more asymmetric carbon atoms, each chiral centre is assigned a configuration according to the rules. Thus (+)-tartaric acid is (R,R)-tartaric acid.

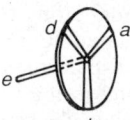 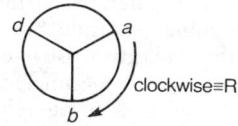

optical anisotropy. *See* ANISOTROPIC.

optical density. *See* ABSORBANCE.

optical exultation. Abnormal increase in the refractive index of a compound due to the presence of a system of conjugated bonds (*see* CONJUGATION).

optical rotatory dispersion (ORD). Variation in the rotation of the plane of POLARIZED RADIATION with wavelength. The molecular rotation goes through a minimum then zero at the wavelength of maximum absorption, then a maximum

(positive Cotton effect), and vice versa (negative Cotton effect) (*see* Figure). ORD is measured directly as the difference in absorbance of left and right circularly polarized light produced by passing plane-polarized light through a Pockel cell (crystal of ammonium hydrogen phosphate) across which an alternating voltage is applied. The sign of the Cotton effect may be used to determine absolute CONFIGURATION. Optically active organic molecules and transition metal complexes are studied using ORD. *See also* OPTICAL ACTIVITY.

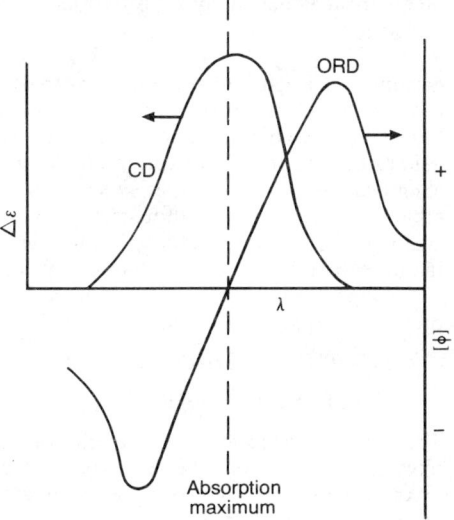

optical rotatory dispersion
Optical rotatory dispersion (ORD) showing the Cotton effect.

optoacoustic spectroscopy. *See* PHOTO-ACOUSTIC SPECTROSCOPY.

orbital. WAVEFUNCTION of an atom or molecule. A spin orbital is an orbital coupled with a specification of the spin of an occupying electron. *See also* ATOMIC ORBI-TAL, MOLECULAR ORBITAL.

ORD. *See* OPTICAL ROTATORY DISPER-SION.

order of reaction. Experimentally determined quantity, which is the power to which the concentration of a species is raised in the RATE EQUATION. The overall order is the sum of the orders with respect to the individual species. Orders may be simple whole numbers (i.e. zero, first, second, etc.), but may also be fractional (e.g., the thermal decomposition of acetal is of three-halves order) or negative. In complex reaction schemes (e.g., CHAIN REACTION), it may not be possible to determine orders as defined above. The order of reaction is not necessarily given by the STOICHIOMETRY of a reaction nor be the same as the MOLECULARITY OF REACTION. See also PSEUDO-ORDER OF REACTION, RATE CONSTANT.

ore. Naturally occurring compound of a metal from which it may be extracted by REFINING.

organic   electrochemical   syntheses. Organic syntheses which rely on reactive intermediates may be carried out by generating the species in situ in an electrochemical cell. The KOLBE SYNTHESIS of alkanes is an important industrial process. Adiponitrile, an intermediate in the manufacture of NYLON, is prepared by the electrodimerization of cyanoethene in the presence of quaternary ammonium salts.

$$2CH_2=CHCN + 2H_2O + 2e \rightarrow$$
$$CN(CH_2)_4CN + 2OH^-$$

The anodic reaction is the evolution of oxygen from water. Tetraethyllead (see LEAD ORGANIC DERIVATIVES) is generated at a sacrificial lead anode

$$Pb + 4C_2H_5^- \rightarrow (C_2H_5)_4Pb + 4e$$

The GRIGNARD REAGENT tetraethylmagnesium is formed via the cathodic reaction of $MgCl^+$ to Mg. HYDROCARBONS may be fluorinated electrochemically by electrolysis of the hydrocarbon in liquid hydrogen fluoride at a nickel anode. POLYMERIZATION may be induced electrochemically in a number of ways. Discharge of an ion at an electrode may produce an atom, radical or radical anion which can initiate polymerization. A stable initiator may be produced or an inhibitor removed by electrolysis. Electropolymerization may be classed as cationic, anionic, free radical or condensation depending on the intermediate involved.

organometallic compounds. Compounds containing a direct link between a metal atom and a carbon atom in an organic compound (e.g., organic derivatives of particular metallic elements). See also CARBON-TO-METAL BOND, GRIGNARD REAGENTS.

organosol. Colloidal dispersion of any insoluble material in an organic liquid (e.g., rubber in toluene).

orientation in the benzene ring. Only one compound is formed when one group is introduced into a benzene molecule. When a second group is introduced, three isomers are possible ortho-, meta- and para- (o-, m- and p-). Pure o-, m-, p-substitution is rare, the three isomers being obtained simultaneously, but not all at the same rate. Experimentally the main product is either a mixture of the o- and p-isomers, associated with activation of the benzene nucleus (i.e. reaction faster than in benzene) or the m-isomer associated with deactivation of the nucleus (i.e. reaction slower than in benzene). The nature of group A (already present in the ring) determines the position taken by the incoming group.
class I.   o-, p-Directing groups in which the atom adjacent to the nucleus has at least one lone pair of electrons; this give rise to increased electron densities in the o- and p-positions. Substitution with an ELECTROPHILE takes place in these positions. A may be R, OH, OR, NH_2, NHR, NR_2, $NHCOCH_3$, Cl, Br, I, F, SH or Ph.
class II.   m-Directing groups with at least one strongly electron-attracting atom or double or triple bond conjugated to the ring; this gives rise to electron displacement away from the nucleus resulting in the m-position having a relatively high electron density compared with the o- and p-positions. Substitution with an electrophile occurs at this position. A may be NO_2, CHO, COOH, COOR, SO_3H, SO_2Cl, $COCH_3$, CN, NH_3, NR_3.
The position of a third group entering the ring depends on the nature and relative directing powers of the two groups already present.

ornithine.   See AMINO ACIDS.

orpiment.   See ARSENIC.

Orsat gas analysis.   See GAS ANALYSIS.

ortho- (*o-*). (1) Prefix for disubstituted BENZENE derivatives designating the position of groups which are adjacent in the ring (1,2-position). *See also* ORIENTATION IN THE BENZENE RING. (2) *See ORTHO*-HYDROGEN.

ortho-. Prefix (written in full) denoting the most hydrated acid of a series formed from the same anhydride (e.g., orthophosphoric acid, *see* PHOSPHORUS OXOACIDS).

orthoclase. *See* SILICATES.

ortho-hydrogen. H_2 may exist as one of two spin isomers, with the NUCLEAR SPIN on each atom parallel (*ortho-*) or antiparallel (*para-*). At room temperature $^1H_2$ is 75 percent *ortho-* and 25 percent *para-*. Interconversion takes place only over a catalyst such as activated charcoal (*see* CARBON) when pure *para-*hydrogen may be formed at liquid hydrogen temperatures. Chemically the isomers are identical. There are small differences in physical properties, *para-*hydrogen has a lower melting point and boiling point by 0.13°C and has a greater thermal conductivity.

ortho-, para-directing group. *See* ORIENTATION IN THE BENZENE RING.

orthophosphates. *See* PHOSPHATES.

orthophosphites. *See* PHOSPHATES.

orthophosphoric acid. *See* PHOSPHORUS OXOACIDS.

orthorhombic system. *See* CRYSTAL SYSTEMS.

orthosilicates. *See* SILICATES.

orthovanadates. *See* VANADIUM COMPOUNDS.

Os. *See* OSMIUM.

osazones. Yellow crystalline compounds formed by the reaction of an ALDOSE or a KETOSE with excess phenylhydrazine

Although principally used for the identification of sugars, the serious disadvantage is that the asymmetric centre at C-2 is lost; thus glucose and mannose (*see* HEXOSES) give the same osazone which is identical to that of the ketose fructose. On hydrolysis with hydrochloric acid, both phenylhydrazino groups are eliminated forming a dicarbonyl compound known as an osone: for example for glucose

$$
\begin{array}{c}
CH=NNHPh \\
| \\
C=NNHPh \\
| \\
(CHOH)_3 \\
| \\
CH_2OH
\end{array}
\xrightarrow{\ HCl\ }
\begin{array}{c}
CHO \\
| \\
CO \\
| \\
(CHOH)_3 \\
| \\
CH_2OH
\end{array}
$$

oscillating reaction. Chemical reaction in which a concentration varies regularly in time or space. An architypal redox reaction which shows oscillatory behaviour is the Belusov–Zhabotinski reaction of a mixture of bromate(III), bromide, propanedioic acid and cerium(III) or cerium(IV).

oscillator strength (*f*). Dimensionless quantity ($0 \leqslant f \leqslant 1$) that is a measure of the coupling between radiation and an electronic transition over a range of frequencies (v_1, v_2). It is defined as

$$ f = (4m_ec\varepsilon_0/N_Ae^2) \int_{v_2}^{v_2} 2.303\ \varepsilon\ \mathrm{d}v $$

where ε is the molar extinction coefficient (*see* BEER–LAMBERT LAW).

osmiridium. Naturally occurring or synthetic alloy of osmium (15–40 percent), iridium (50–80 percent) and smaller amounts of other PLATINUM GROUP METALS. It is a hard alloy used in the manufacture of sparking points and fountain pen nibs.

osmium (Os). Third row TRANSITION ELEMENT of the PLATINUM METALS GROUP. It is rare and occurs naturally as the sulphide and as OSMIRIDIUM. Osmium is extracted by distillation of osmium tetroxide (OsO_4) from solutions of osmium

$$
\begin{array}{c}
CHO \\
| \\
CHOH \\
| \\
CHOH \\
|
\end{array}
\xrightarrow{3PhNHNH_2}
\begin{array}{c}
CH=NNHPh \\
| \\
C=NNHPh \;+\; PhNH_2 + NH_3 \\
| \\
CHOH \\
|
\end{array}
\xleftarrow{3PhNHNH_2}
\begin{array}{c}
CH_2OH \\
| \\
C=O \\
| \\
CHOH \\
|
\end{array}
$$

compounds and nitric acid. Osmium is used for hardening platinum and has catalytic properties. The metal tarnishes in air to OsO_2, dissolves in nitric acid and reacts with chlorates(I) to give osmates(VI) ($[OsO_4]^{2-}$).

$A_r = 190.2$, $Z = 76$, ES [Xe] $5d^6 6s^2$, mp $3045°C$, bp $5027°C$, $\rho = 22.6 \times 10^3$ kg m^{-3}, Isotopes: ^{188}Os, 13.3 percent; ^{189}Os, 16.1 percent; ^{190}Os, 26.4 percent; ^{192}Os, 41.0 percent.

osmium compounds. Oxidation states 0 to +8 are found. Red osmates(VIII) ($[OsO_4(OH)_2]^{2-}$) are produced when OsO_4 is dissolved in strong alkali. Reduction then gives osmates(VI). Osmates(VII) are found in mixed oxides. Several halides and oxyhalides are known, principally black $OsCl_4$.

osmosis. Tendency of solvent molecules to pass through a semipermeable membrane from a more dilute to a more concentrated solution. See also OSMOTIC PRESSURE.

osmotic coefficient (g). Simple method of including in one coefficient all the factors that cause departure from ideal behaviour in solution. It is defined by

$$\mu_i = \mu_i^{\ominus} + g_i RT \ln x_i$$

Comparing this equation with that for CHEMICAL POTENTIAL gives

$$\ln \gamma_i = (g_i - 1) \ln x_i$$

The limiting value of g_i is 1 for ideal behaviour. It is the factor by which the OSMOTIC PRESSURE, elevation of BOILING POINT or depression of FREEZING POINT of an ideal solution must be multiplied to give these quantities for an actual solution of the same concentration.

osmotic pressure (Π; dimensions: $m\, l^{-1}t^{-2}$; units N m^{-2}). Minimum excess pressure necessary to prevent the transfer of solvent molecules A through the membrane. Osmotic pressure is a COLLIGATIVE PROPERTY, depending only on the concentration of the solution. Assuming ideal behaviour

$$\Pi \bar{V}_A = RT \ln p^{\ominus}/p = -RT \ln x_A$$

$$\Pi = n_B RT/V = c_B RT = w_B RT/M_{r,B}$$

This equation is only valid at infinite dilution, Π for a non-electrolyte can be written as

$$\Pi/w_B = RT/M_{r,B} + Cw_B + Dw_B^2 + \ldots$$

The plot of Π/w_B against w_B is linear, $M_{r,B}$ can be obtained from the intercept. Osmotic pressure measurements are more useful for determining the molecular masses of macromolecules than are the elevation of the BOILING POINT or depression of the FREEZING POINT.

osones. *See* OSAZONES.

Ostwald's dilution law. Relationship between the ACID DISSOCIATION CONSTANT of a weak acid and the MOLAR CONDUCTIVITY. If the degree of dissociation (α) is given by Λ/Λ^0(Λ^0 is the conductivity at infinite dilution), then for a monobasic acid HA $\rightleftharpoons$ H$^+$ + A$^-$, $K_a = \Lambda^2 c/\Lambda^0(\Lambda^0 - \Lambda)$, where c is the concentration of the acid.

outer Helmholtz plane (OHP). Locus of the centres of the plane of ions attracted to a charged surface forming an ELECTRICAL DOUBLE LAYER (*see* Figure).

outer potential (Volta potential, psi-potential, ψ). Between approximately 10^{-3} and 10^{-5} cm from a metal surface a region exists in which the potential due to the metal is independent of distance. The outer potential is that potential near an electrified interface just outside the range of image forces. An outer potential may also be identified for an electrolyte and thus for an electrode immersed in an electrolyte, the outer potential difference is given by

$$\Delta\psi = \psi_{metal} - \psi_{solution}$$

$\Delta\psi$ may be measured or calculated from simple electrostatics and a reasonable model of the structure of the metal/solution interface.

outer sphere. Refers to ions outside the shell of metal ion and ligand (the inner sphere) of a complex compound.

outer sphere mechanism. One which involves electron transfer without disruption of the inner sphere. For example

$$[Fe(CN)_6]^{4-} + [IrCl_6]^{2-} \rightarrow$$

$$[Fe(CN)_6]^{3-} + [IrCl_6]^{3-}$$

inner sphere mechanism.   One in which the reactants share a ligand in the intermediate complex. For example NCS in the reaction

$$[Co(NH_3)_5NCS]^{2+} + [Cr(H_2O)_6]^{2+} +$$

$$5H_3O^+ \longrightarrow [Cr(H_2O)_5SCN]^{2+} +$$

$$[Co(H_2O)_6]^{2+} + 5NH_4^+$$

Overhauser effect.   *See* NUCLEAR MAGNETIC RESONANCE SPECTROSCOPY.

overpotential (η).   Difference between the value of the ELECTRODE POTENTIAL when current flows and that at equilibrium. Factors contributing to the overpotential are as follows. (1) Charge transfer or activation overpotential. This occurs by virtue of the ACTIVATION ENERGY for the transfer of an electron (*see* BUTLER–VOLMER EQUATION). (2) Diffusion or concentration overpotential. This arises from rate-limiting transport of the reactants to the OUTER HELMHOLTZ PLANE (*see* DIFFUSION, LIMITING CURRENT DENSITY). (3) Reaction overpotential. In a multistep reaction, a rate-limiting chemical step will contribute to the overall overpotential. (4) Crystallization overpotential. In the ELECTRODEPOSITION OF METALS, the activation barrier to the incorporation of an adsorbed metal atom into the lattice gives rise to an overpotential.

overpressure   layer   chromatography (OPLC).   Extension of the theory and practice of HIGH-PERFORMANCE LIQUID CHROMATOGRAPHY (HPLC); a planar HPLC technique. The sorbent layer (in the form of a thin, wide cross-section chromatography plate) is completely covered with a flexible membrane under external pressure in a vapour-free state and the solvent is introduced by a pumping system. Advantages are smaller plate height over longer migration distances, ultrarapid separation with excellent resolution, closed system (no air interactions), small consumption of solvents and parallel analysis of many samples.

oxalic acid.   *See* ETHANEDIOIC ACID.

oxane (tetrahydropyran).   *See* HETEROCYCLIC COMPOUNDS (structure XXII).

oxazole   (C_3H_3NO).   Five-membered HETEROCYCLIC COMPOUND (structure IX), only known in certain derivatives containing an aromatic group. Fischer oxazole synthesis involves the condensation of equimolar amounts of an aldehyde cyanohydrin and an aromatic aldehyde in dry ether with hydrogen chloride.

oxidant.   *See* OXIDIZING AGENT.

oxidases.   *See* ENZYMES.

oxidation.   Originally   oxidation   was simply regarded as a chemical reaction with oxygen; the reverse process, the loss of oxygen was called reduction. A more general definition, which combines both oxidation and reduction, is that oxidation involves a loss of electrons and reduction a gain of electrons

$$\text{oxidized state} + ne \underset{\text{oxidation}}{\overset{\text{reduction}}{\rightleftharpoons}} \text{reduced state}$$

Oxidation and reduction occur simultaneously, thus two redox systems are always involved

$$red_1 + ox_2 \rightleftharpoons red_2 + ox_1$$

Every electrode is based on this reaction involving the transfer of electrons, so that when two electrodes are joined in a cell, oxidation occurs at the anode and reduction at the cathode. This definition, which applies only to reactions in which electron transfer occurs, can be extended to reactions between covalent molecules using the concept of OXIDATION STATE, in which oxidation (reduction) is a process in which there is an increase (decrease) in the oxidation state. Thus in

$$2H_2 + O_2 \rightarrow 2H_2O$$

the hydrogen in water is in the $+1$ oxidation state and the oxygen -2; the hydrogen is oxidized and the oxygen reduced. In organic chemistry the most obvious effect of oxidation is the increase in the proportion of oxygen in the molecule (e.g., $RCH_2OH + O \rightarrow RCOOH$); selective oxidizing agents include selenium dioxide, chromium trioxide and aluminium *tert*-butoxide.

oxidation–reduction indicator.   *See* INDICATORS.

oxidation–reduction potential. *See* REDOX ELECTRODE SYSTEM.

oxidation–reduction system. *See* REDOX ELECTRODE SYSTEM.

oxidation state. Formal number of electrons associated with an element in a compound on a scale in which the oxidation state of a non-combined element is zero. The oxidation state of electronegative atoms is the charge on the atom that would be required to create an inert gas electronic structure. Thus oxygen is usually –2 and halogens –1. The oxidation state of a combined electropositive atom is then calculated from the requirement that the sum of all oxidation states of elements is the molecular charge. For example, the oxidation state of sulphur in SO_4^{2-} is +6 and in SO_3^{2-} is +4. Coordinate bonds between ligands and metals do not affect the oxidation state, so in $[Co(CN)_6]^{3-}$ the oxidation state of cobalt is +3. In systematic nomenclature, the oxidation state is written in roman numerals (e.g., copper(II) sulphate, manganate(VII) (the permanganate ion, MnO_4^-)).

oxidative addition. Reaction in which there is direct addition to an element or multiple bond resulting in an increase in the oxidation state (e.g., addition of Cl_2 to PCl_3 giving PCl_5).

oxidative phosphorylation. Process whereby the reaction

$$ADP^{3-} + P_i + H^+ \longrightarrow ATP^{4-} + H_2O$$

$$\Delta G^{\ominus\prime} = 29.3 \text{ kJ mol}^{-1}$$

is coupled to the ELECTRON TRANSPORT CHAIN resulting in the generation of ADENOSINE TRIPHOSPHATE and the storage of chemical energy.

oxides. Binary compounds formed between elements and oxygen. General classification is as follows: (1) acidic (e.g., SO_2, P_4O_{10}, CrO_3); (2) basic (e.g., CrO, CuO, CaO); (3) amphoteric (e.g., ZnO, Cr_2O_3, Al_2O_3); (4) neutral (e.g., NO, CO). Many oxides (e.g., FeO) are NON-STOICHIOMETRIC COMPOUNDS, containing metal atoms in two or more oxidation states. Metal oxides are predominantly ionic, MO-type with O^{2-} and M^{2+} ions arranged in 4:4 (BeO) or 6:6 (MgO, rock salt lattice) coordination; MO_2-type for large metal ions (e.g., ThO_2) have 8:4 coordination (fluorite unit cell) and for small metal ions (e.g., SnO_2) 6:3 coordination (rutile lattice); M_2O_3-type oxides have corundum-type lattice (*see* CRYSTAL FORMS). *See also* PEROXIDES, SUPEROXIDES.

oxidizing acid. Acid that can act as a strong oxidizing agent as well as an acid (e.g., nitric acid).

oxidizing agent (oxidant). Substance that brings about OXIDATION of another substance, itself being reduced (e.g., MnO_4^- oxidizes Fe^{2+} to Fe^{3+} and is reduced to Mn^{2+}). Oxidizing agents contain atoms with a high OXIDATION STATE, gaining electrons during the process of oxidation.

oxidoreductases. *See* ENZYMES.

oximes ($RR'C=NOH$). Organic compounds (i.e. ALDOXIMES ($R' = H$) and KETOXIMES ($R' = $ alkyl or aryl)). *See also* HEXOSES.

oxine. *See* 8-HYDROXYQUINOLINE.

oxoacid. Acid in which the acidic hydrogen atom(s) are bound to oxygen atoms. In sulphuric acid (*see* SULPHUR OXOACIDS), the two acidic hydrogens are on the –OH groups bound to sulphur.

oxocarboxylic acids. Compounds containing both an OXO GROUP and a CARBOXYL GROUP. They are named α-, β-, γ-keto- or aldehydic acids, depending on the position of the oxo group relative to the carboxyl group (i.e. 1,2; 1,3; 1,4, respectively). The properties of oxocarboxylic acids are governed by the separation of the groups. If they are far apart the molecule acts independently as a KETONE or ALDEHYDE, and as a CARBOXYLIC ACID. OXOETHANOIC ACID (glyoxylic acid) is the simplest aldehydic acid, and 2-OXOPROPANOIC ACID (pyruvic acid) is the simplest ketonic acid. ETHYL 3-OXOBUTANOATE, an important synthetic reagent, is the ethyl salt of the β-ketoacid 3-oxobutanoic acid. The simplest γ-ketoacid is 4-oxopropanoic acid (laevulic acid). It is prepared by the reaction between ethyl 3-

oxobutanoate and ethyl chloroethanoic acid. It exists in two tautomeric forms (*see* ISOMERISM), in a chain or a ring

CH₂COCH₃ — CH₂COOH ⇌ H₂C—C(OH)CH₃ / H₂C—C structure

$$CH_2COCH_3 \atop CH_2COOH \quad \rightleftharpoons \quad {H_2C-C(OH)CH_3 \atop H_2C-C}$$

3-oxobutanoic acid. *See* OXOCARBOXYLIC ACIDS.

oxoethanoic acid (glyoxylic acid, CHO-COOH). The simplest aldehydic acid. A viscous liquid (bp 98°C), prepared by the hydrolysis of dichloroethanoic acid. It crystallizes with one molecule of water. Oxoethanoic acid shows all the reactions of an ALDEHYDE and a CARBOXYLIC ACID. It undergoes the CANNIZZARO REACTION to hydroxyethanoic and ethanedioic acids. Reduction with nascent hydrogen gives 2,3-DIHYDROXYBUTANEDIOIC ACID.

3-oxoglutaric acid (acetonedicarboxylic, β-ketoglutaric acid). *See* 3-OXOPENTANE-DIOIC ACID.

oxo group (carbonyl group). The group >C=O found in ALDEHYDES, KETONES and CARBOXYLIC ACIDS.

oxolane. *See* TETRAHYDROFURAN.

oxonium ion. Ion of the type R₃O⁺, in which R represents hydrogen or an organic group. The HYDROXONIUM ION (hydronium ion) (H₃O⁺) results from the ionization of water.

3-oxopentanedioic acid (acetonedicarboxylic acid, CO(CH₂COOH)₂). Needles (mp 138°C (decomp.)), prepared by the action of fuming sulphuric acid on citric acid (*see* 2-HYDROXYPROPANONE-1,2,3-TRICARB-OXYLIC ACID). It is readily hydrolyzed to propanone and carbon dioxide. The acid or its diethyl ester, which is used in synthesis, reacts with sodium in a similar way to ETHYL 3-OXOBUTANOATE.

oxo process. Industrial process for the production of ALCOHOLS. ALDEHYDES from a HYDROFORMYLATION process are hydrogenated over a copper–zinc catalyst in an integrated process. 'Oxo' may also be used to describe the hydroformylation step itself. *See also* FISCHER–TROPSCH PROCESS.

2-oxopropanoic acid (pyruvic acid, CH₃CO-COOH). The most simple ketoacid; liquid (bp 165°C), prepared by heating 2,3-DIHYDROXYBUTANEDIOIC ACID with potassium hydrogen sulphate at 220°C. It acts as both a KETONE and a CARBOXYLIC ACID. In common with other α-ketoacids, 2-oxopropanoic acid decarboxylates (*see* DE-CARBOXYLATION) in warm sulphuric acid to an aldehyde (ethanal). In concentrated sulphuric acid, it decarbonylates (*see* DECARBONYLATION) to ethanoic acid. 2-Oxopropanoic acid is an important biological acid, being an intermediate in the metabolism of sugars.

6-oxy-2-aminopurine (guanine). *See* PURINES.

2-oxy-4-aminopyrimidine (cytosine). *See* PYRIMIDINES.

oxyanion. Anion containing oxygen in oxidation state –2. Examples are OXIDES, SULPHATES, NITRATES and PHOSPHATES.

oxygen (O). Colourless, odourless gas belonging to group VI (*see* GROUP VI ELEMENTS). It is the most abundant element in the earth's crust (47.2 percent by weight) and forms 21 percent by volume of the atmosphere from which it can be obtained by liquefaction and distillation. Atmospheric oxygen is vital for aerobic respiration. The common form is diatomic (dioxygen, O₂), the liquid and solid forms of which are pale blue in colour and are strongly paramagnetic. The gas is sparingly soluble in water and readily soluble in some organic solvents. It is a reactive gas and a strong oxidizing agent, combining with most other elements (usually only at elevated temperatures) to form OXIDES (*see* Figure). Its predominant chemistry is in the –2 oxidation state in oxides, ALKOXIDES and ETHERS. It is used in steel-making, synthetic processes (nitric acid, sulphuric acid, 1,2-epoxyethane), welding (oxyhydrogen, oxyethyne), mining, rocket

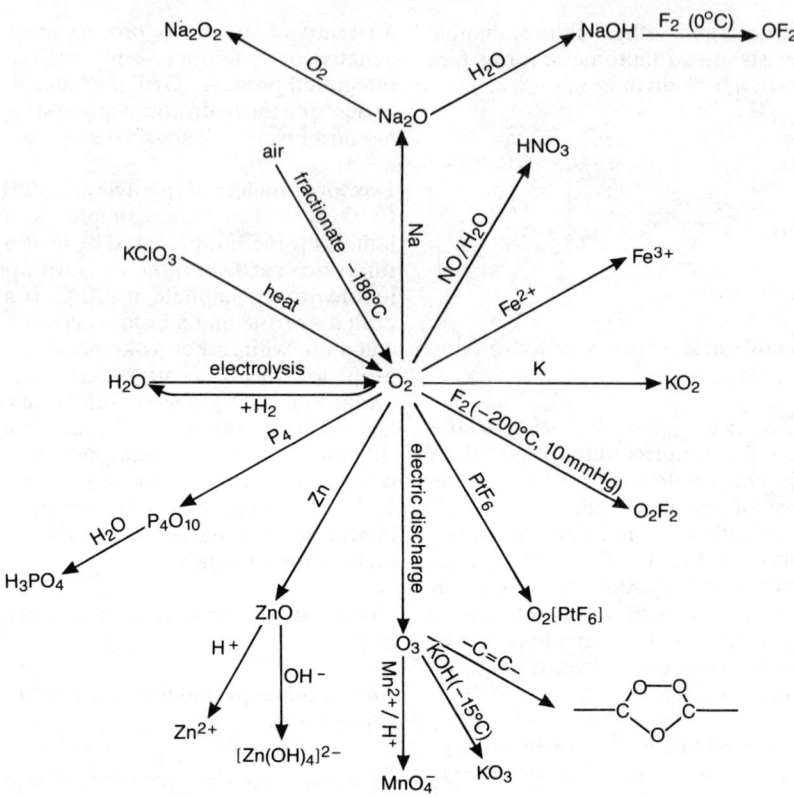

oxygen
Some reactions of oxygen.

fuels, sewage digestion and in breathing apparatus.

A_r = 15.9994, Z = 8, ES [He] $2s^2 2p^4$, mp −218.4°C, bp −182.962°C, ρ = 1.429 kg m^{-3}. Isotopes: ^{16}O, 99.75 percent; ^{17}O, 0.0374 percent; ^{18}O, 0.2039 percent.

oxygen consumed (OC, chemical oxygen demand). Measure of the oxidizable components present in water; its value is dependent on the structure of organic compounds present and the manipulative procedure (e.g., determination of dichromate oxygen consumption, DOC). It does not differentiate stable from unstable organic matter and so does not necessarily correlate with BIOCHEMICAL OXYGEN DEMAND.

oxygen fluorides. The colourless gas oxygen difluoride (OF$_2$) is made by the action of fluorine on sodium hydroxide

solution. It is not an acid anhydride, reacting with water to give fluoride ions and free oxygen. The bonding is essentially covalent. Other unstable fluorides (O$_2$F$_2$, O$_3$F$_2$ and O$_4$F$_2$) are obtained when oxygen/fluorine mixtures are subjected to an electric discharge at low pressures and temperatures.

oxygen halides. *See* BROMINE OXIDES, CHLORINE OXIDES, IODINE OXIDES, OXYGEN FLUORIDES.

oxygen probe. HALF-CELL; in acid O$_2$ + 4H$^+$ + 4e $\rightleftharpoons$ 2H$_2$O ($E^\ominus$ = 1.229 V), in alkali O$_2$ + 2H$_2$O + 4e $\rightleftharpoons$ 4OH$^-$ ($E^\ominus$ = 0.403 V). Unlike the HYDROGEN ELECTRODE, the achievement of the true equilibrium potential is difficult as on metal electrodes a series of oxides are formed which leads to mixed potentials. The oxygen electrode reaction is of importance in water ELECTROLYSIS, and as the cathode reaction in many FUEL CELLS. Oxygen is determined by a GAS-SENSING MEMBRANE PROBE in which the gas diffuses through a membrane and is reduced at a platinum or silver electrode. This is the basis of the Clark oxygen electrode.

oxyhaemoglobin.   *See* HAEMOGLOBIN.

6-oxypurine (hypoxanthine). *See* PUR-INES.

oxytetracycline.   *See* ANTIBIOTICS.

oxytocin.   *See* HORMONES.

ozone (trioxygen, O_3). Highly reactive allotropic form of OXYGEN, formed by the action of an electrical discharge or UV radiation on oxygen. The molecule (mp $-193°C$, bp $-112°C$) is bent with considerable double-bond character. It is a powerful oxidizing agent (in acid solution, $E^\ominus = +2.07$ V) oxidizing sulphur to sulphuric-(VI) acid, silver(I) compounds to the $+2$ state and converting olefinic compounds to OZONIDES. It reacts with potassium hydroxide to give potassium ozonide (KO_3).

ozonides. (1) Compounds formed by the reaction of OZONE with alkali metal hydroxides (potassium, rubidium, caesium). These contain the O_3^- ion which is paramagnetic with one unpaired electron. (2) Compounds formed by the action of ozone on various classes of unsaturated organic compounds (e.g., ALKENES)

$$RCH=CR'R'' + O_3 \longrightarrow$$

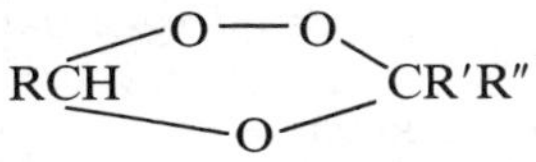

Ozonides are readily decomposed by water and on reduction yield the aldehyde RCHO and the ketone $R'R''CO$. Most are thick, explosive oils, but are seldom isolated.

ozonolysis. Addition of ozone to an unsaturated organic compound to give an OZONIDE. *See also* ALKENES.

P

P. *See* PHOSPHORUS, POISE.

P. Total pressure (*see* PRESSURE), POLARIZATION, POWER.

p. *See* MOMENTUM, PARTIAL PRESSURE.

II. *See* OSMOTIC PRESSURE.

π-bonding. *See* MOLECULAR ORBITAL.

π-complexes. COORDINATION COMPOUND in which there is a π-bond (*see* MOLECULAR ORBITAL) between metal and ligand. Ligands that have fully occupied π-orbitals are known as π-donors, the electrons being delocalized over the metal. If metal t_{2g} orbitals interact with the empty π*-ligand orbitals, the ligand is known as a π-acceptor and acts as a Lewis acid (*see* LEWIS ACIDS AND BASES). *See also* OLEFIN COMPLEX, SANDWICH COMPOUND.

π-donor complex. *See* CARBON-TO-METAL BOND.

π-orbital. *See* MOLECULAR ORBITAL.

Φ. *See* QUANTUM EFFICIENCY.

φ. *See* ATOMIC ORBITAL, FLUIDITY, INTERFACIAL POTENTIAL, NEUTRON FLUX.

Ψ. *See* MOLECULAR ORBITAL.

p-. *See* PARA-.

Pa. Protactinium (*see* ACTINIDE ELEMENTS), PASCAL.

packing fraction. MASS DEFECT divided by the mass number (i.e. $(M - A)/A$, where M is the actual mass). It is positive for the lightest and heaviest elements and negative for intermediate elements (e.g., chlorine -0.00042).

paint. Uniformly dispersed mixture with viscosity ranging from a thin liquid to a semisolid paste consisting of: (1) a drying oil, synthetic resin or film-forming component, the 'binder'; (2) a solvent or thinner; (3) an organic or inorganic PIGMENT (e.g., Pb_3O_4, TiO_2). The binder and solvent are collectively known as the 'vehicle'. Paints are used to protect a surface from corrosion, oxidation or other types of deterioration and to provide decorative effects. Emulsion paint is composed of two dispersions blended together: (1) dry powders (colouring material, fillers and extenders) milled with water; (2) a resin dispersion (a latex formed by emulsion polymerization or resin in emulsion form). Surfactants and PROTECTIVE COLLOIDS are added as stabilizers. In emulsion paints, the binder is in a water-dispersed form, whereas in solvent paint it is in solution form. Principal latex paints are styrene–butadiene, polyvinyl ethanoate and acrylic resins, 20–25 percent dry ingredients, 40 percent latex, 20–30 percent water plus stabilizers.

palladium (Pd). Second row TRANSITION ELEMENT. It is extracted from other PLATINUM GROUP METALS as the dichlorodiammine complex, which is readily reduced to the metal by hydrogen. Palladium is a white, ductile metal used in alloys with platinum or gold (white gold) or as an electrical contact. When supported on alumina or carbon, palladium is used as a HYDROGENATION and dehydrogenation catalyst. Palladium tarnishes in oxygen or air forming PdO and dissolves in concentrated nitric acid and hot sulphuric acid. It is attacked by moist halogens.
$A_r = 106.42$, $Z = 46$, ES [Kr]$4d^{10}$, mp 1552°C, bp 3140°C, $\rho = 12.0 \times 10^3$ kg m^{-3}. Isotopes: ^{102}Pd, 1.0 percent; ^{104}Pd, 11.0 percent; ^{105}Pd, 22.2 percent; ^{106}Pd, 27.3

percent; ^{108}Pd, 26.7 percent; ^{110}Pd, 11.8 percent.

palladium compounds. Oxidation states +2 (palladous) and +4 (palladic) comprise the known chemistry of palladium. A range of fluorides is known (e.g., $[PdF_6]^{2-}$, PdF_4, Pd_2F_6 ($Pd(II)Pd(IV)$), PdF_2), but only the chloride $PdCl_2$ is known; the complex ions $[PdCl_4]^{2-}$ and $[PdCl_6]^{2-}$ are formed in aqueous hydrochloric acid. Palladium(II) nitrate, nitrite, oxide and a range of amines are known. Organic derivatives of palladium are less stable than those of platinum. Alkenyl and allyl compounds (e.g., $(C_3H_5PdCl)_2$) have been characterized and are thought to be intermediates in the catalytic WACKER PROCESS. Hydrogen is adsorbed to $PdH_{0.6}$. Hydrido compounds having palladium–hydrogen bonds are formed by reduction of chlorides.

palmitic acid. See HEXADECANOIC ACID.

palmitin (tripalmitin, glycerol tripalmitate, $(C_{15}H_{31}COO)_3C_3H_5$). Triglyceride of palmitic acid (see HEXADECANOIC ACID); a constituent of most animal and vegetable fats. It is used in the manufacture of SOAP. See also GLYCERIDES.

pantothenic acid (vitamin B complex). See VITAMINS.

papain. See ENZYMES.

papaverine. See ALKALOIDS.

paper. Semisynthetic product made by chemically processing cellulosic fibres, mainly obtained from soft woods. The properties of the paper depend on the degree of hydration and subdivision of the particles. Sizing (with alums or colloidal mixtures) renders the paper less porous and water-repellant. Inorganic fillers (e.g., $BaSO_4$) are often added to improve the finish.

paper chromatography. Form of partition chromatography (although adsorption and ion exchange have small roles) in which the paper acts both as SUPPORT MEDIUM and, by virtue of the water held in its fibres, as the STATIONARY PHASE. The MOBILE PHASE, which depends on the mixture to be separated, is an organic solvent or mixture of solvents which may travel up or down the paper. The paper is dried, and the separated components located by suitable means (e.g., production of a coloured or fluorescent product) and identified by comparison with the chromatograms of reference compounds (see also R_f). Limits of detection are between 1 and 50 μg. The technique has been largely replaced by THIN LAYER CHROMATOGRAPHY which is more rapid.

para- (*p-*). Prefix for disubstituted BENZENE derivatives designating the position of groups at opposite ends of the ring (1,4-position).

paracetamol. See 4-ACETAMIDOPHENOL.

parachor ($[P]$; units: $m^3 mol^{-1}$). Empirical measure of molecular volume

$$[P] = \gamma^{1/4} M_r/(\rho_l - \rho_g) = \gamma^{1/4} M_r/\rho_l$$

where ρ_g, the density of the vapour, can be neglected in comparison with ρ_l, the density of the liquid; γ is the SURFACE TENSION. Although primarily an additive property, it is not without constitutive influence; parachor equivalents of atoms, groups and bonds were once used extensively in establishing molecular structures.

paracyanogen. See CYANOGEN.

paraffin. See KEROSENE.

paraffins. See ALKANES.

paraffin wax. Solid hydrocarbon obtained from PETROLEUM. It consists mainly of macrocrystalline n-alkanes (C_{20} and above). High-molecular-mass branched alkanes and cycloalkanes form so-called microcrystalline wax. It is used to manufacture WAXES and polishes, candles, and as a chemical feedstock for CRACKING.

paraformaldehyde. See METHANAL.

***para-*hydrogen.** See ORTHO-HYDROGEN.

paraldehyde. See ETHANAL.

parallel reactions. Reactions of a molecule by independent routes which may occur under the same reaction conditions.

The overall rate is governed by the fastest route (for first-order parallel reactions, *see* INTEGRATED RATE EQUATIONS).

paramagnetism. Alignment of magnetic dipoles in an external magnetic field. Free electrons in solids show a weak paramagnetic effect, but this is overwhelmed in ions of TRANSITION ELEMENTS by the contribution of unpaired electrons (*see* LIGAND FIELD THEORY). The magnetic susceptibility is inversely related to temperature (*see* CURIE'S LAW).

paraperiodic acid. *See* IODATES.

paraquat (1,1'-dimethyl-4,4'-bipyridylium dimethyl sulphate or chloride). White, highly poisonous contact HERBICIDE; prepared by the quaternization of 4,4'-dipyridyl with dimethyl sulphate or chloromethane.

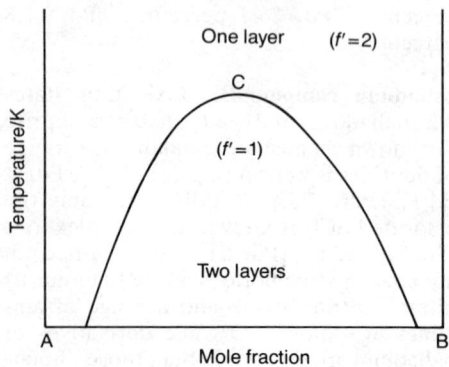

parathion. *See* INSECTICIDE.

parathyroid hormone (PTH). *See* HORMONES.

partially miscible liquid systems. Mixture of two dissimilar liquids that differ in polarity or extent of association. The forces A–B are considerably less than A–A and B–B and so the escaping tendency is increased, leading to positive deviations from RAOULT'S LAW and the separation of the liquid phase into two layers, saturated solutions of A in B and B in A, known as conjugate solutions. The range of miscibility is a function of temperature at constant pressure; an increase in temperature usually results in an increase in the mutual solubility of the two components (*see* Figure) until at temperatures above the critical solution temperature or consolute temperature (*C*) the system becomes homogeneous at all compositions (e.g., phenol/water). Some systems show decreasing mutual solubility with increase of temperature (e.g., triethylamine/water), whereas a few show a closed solubility curve with upper and lower consolute temperatures (e.g., nicotine/water).

partially miscible liquid system
Temperature–composition curve for partially miscible liquid system.

partial molar Gibbs function. *See* CHEMICAL POTENTIAL.

partial molar quantity ($\bar{J}_i$).

$$\bar{J}_i = (\partial J/\partial n_i)_{T,P,n_j}$$

Increase in any EXTENSIVE PROPERTY of a system when 1 mole of the *i*th component is added to such a large amount of the system that the composition and temperature and pressure do not change appreciably when the addition is made

$$J = f(T, P, n_A, n_B, \ldots, n_i)$$
$$dJ_{T,P} = \bar{J}_A dn_A + \bar{J}_B dn_B + \ldots = \sum_i \bar{J}_i \, dn_i$$

A change of composition (except for ideal binary liquid mixtures) results in changes in $\bar{J}_i$, as well as in n_i. Hence at constant T and P

$$n_A d\bar{J}_A + n_B d\bar{J}_B + \ldots = \sum_i n_i \, d\bar{J}_i = 0$$

For two-component systems

$$d\bar{J}_A = -n_B \, d\bar{J}_B/n_A$$

This is a form of the GIBBS–DUHEM EQUATION. Partial molar quantities can be defined for any extensive STATE FUNCTION (e.g., CHEMICAL POTENTIAL, PARTIAL MOLAR VOLUME).

partial molar volume ($\bar{V}_i$; units: $m^3 mol^{-1}$). PARTIAL MOLAR QUANTITY defined by

$$\bar{V}_i = (\partial V/\partial n_i)_{T,P,n_j}$$

The total volume of a solution is not, in general, equal to the sum of the volumes of the individual components owing to changes in the character of the molecules on mixing; it is given by

$$V = \bar{V}_A n_A + \bar{V}_B n_B + \ldots$$

whence

$$n_A d\bar{V}_A + n_B d\bar{V}_B + \ldots = 0$$

This is a form of the GIBBS–DUHEM EQUATION. Partial molar volumes may be determined from a knowledge of the density of solutions of different concentrations or from the measured volume of solutions of different concentrations.

partial pressure (p; units: Nm^{-2}, mmHg, atm). PRESSURE that the same amount of the component in a gas mixture would exert if it were present in a vessel of the same volume as that occupied by the mixture. *See also* DALTON'S LAW OF PARTIAL PRESSURES.

particle in a box. Quantum mechanical (*see* QUANTUM MECHANICS) model in which the potential energy in the vertically sided box is zero and that outside infinite. The SCHRÖDINGER EQUATION for a one-dimensional box length L is therefore

$$(h^2/8\pi^2 m)(\partial^2 \psi/\partial x^2) = E\psi$$

This has a solution $\psi = (2/L)^{1/2} \sin(n\pi x/L)$. n ($= 1, 2, \ldots$) is the QUANTUM NUMBER of the particular ENERGY LEVEL, which has energy $E = n^2 h^2/8mL^2$. The particle in a box provides a useful and simple model for electrons in molecules. The free electron molecular orbital model uses a series of square wells to represent atoms in a conjugated molecule (*see* CONJUGATION).

partition chromatography. *See* CHROMATOGRAPHY, PAPER CHROMATOGRAPHY.

partition coefficient (D). Ratio of the equilibrium activities (concentrations) of a solute distributed between two immiscible solvents (A and B) (i.e. $a_A/a_B = $ constant or, in dilute solution, $c_A/c_B = D$). The relationship is only valid when the solute does not react with either solvent and undergoes no association or dissociation. For the species common to both solvents, if the fraction associated or dissociated is α

$$D = (1 - \alpha_A) c_A / (1 - \alpha_B) c_B$$

Like any EQUILIBRIUM CONSTANT, D varies with temperature and with concentration.

partition function (Zustandssume) (Q). The canonical partition function defined by

$$Q = \sum_i g_i \exp(-\varepsilon_i/kT)$$

$$= g_0 \exp(-\varepsilon_0/kT) + g_1 \exp(-\varepsilon_1/kT) +$$

$$\ldots + g_i \exp(-\varepsilon_i/kT)$$

determines how the microstates (particles, molecules) in an assembly will distribute over the possible quantum states or energy levels (ε_0, ε_1, ε_2, . . ., ε_i); g_i is the DEGENERACY of the ith energy level. The potentially infinite series converge more rapidly the lower the temperature and the larger the energy spacing between successive quantum states. The series can be terminated as soon as the ratio $\varepsilon/kT \gg 1$; when the first energy step above the ground state is large (vibration or electronic energy) Q has a value of unity. However, for many quantum states $\varepsilon \ll kT$ (translational energy) $Q \gg 1$. The magnitude of Q reflects the extent to which the particles of the assembly can spread over the various quantum states characteristic of their species (*see* BOLTZMANN DISTRIBUTION). The partition function plays a central role in the calculation of molar values of thermodynamic parameters from spectroscopic data

$$U = N_A kT^2 (\partial \ln Q/\partial T)_V$$

$$S = N_A k \ln Q + N_A kT (\partial \ln Q/\partial T)_V$$

$$A = -N_A kT \ln Q$$

$$G = -N_A kT \ln Q + N_A kTV (\partial \ln Q/\partial V)_T$$

$$P = N_A kT (\partial \ln Q/\partial V)_T$$

If the assembly is a perfect gas where the molecules have no identifiable positions, Q is smaller by a factor of $N_A!$. Since the total energy of a molecule (U_{tot}) is the sum of energies of different kinds (translation, rotation, vibration and electronic)

$$U_{tot} = \varepsilon_t + \varepsilon_r + \varepsilon_v + \varepsilon_e$$

the total partition function is therefore factorizable

$$Q_{tot} = Q_t \times Q_r \times Q_v \times Q_e$$

Thus separate values of Q and hence contributions of thermodynamic parameters for the separate energy levels can be evaluated for quite complex molecules (*see* Table).

For most molecules the excited electronic energy levels lie so far above the ground

partition function
Partition functions and the thermodynamic functions for the different types of energy.[a]

	Translational	Rotational (a) linear (b) non-linear	Vibrational (for each characteristic frequency)
Energy (ε)	$mu_x^2/2 = p^2h^2/8ma^2$ (for 1 degree of freedom)	$J(J+1)h^2/8\pi^2 I$	$vh\nu$
Partition function (Q)	$(2\pi mkT)^{3/2}V/h^3$ $= (2\pi m/h^2)^{3/2}k^{5/2}T^{5/2}/p^{\ominus}$	(a) $8\pi^2 IkT/h^2\sigma = T/\theta_r\sigma$ (b) $8\pi^2(8\pi^3 I_A I_B I_C)^{1/2}(kT)^{3/2}/h^3\sigma$	$1/[1 - \exp(-h\nu/kT)]$ $= 1/[1 - \exp(-\theta_v/T)]$
Internal energy (U) Enthalpy (H)	$\left.\begin{matrix}3N_AkT/2 \\ 5N_AkT/2\end{matrix}\right\}$	(a) $\left.\begin{matrix}N_AkT \\ 3N_AkT/2\end{matrix}\right\}$ (b)	$N_Ak\theta_v/[\exp(\theta_v/T) - 1]$
Heat capacity (C_V) (C_P)	$\left.\begin{matrix}3N_Ak/2 \\ 5N_Ak/2\end{matrix}\right\}$	(a) $\left.\begin{matrix}N_Ak \\ 3N_Ak/2\end{matrix}\right\}$ (b)	$N_Ak\{[(\theta_v/T)^2 \exp(\theta_v/T)]/[\exp(\theta_v/T) - 1]^2\}$
Entropy (S)	$N_Ak(2.5 + \ln Q_t)$	(a) $N_Ak(1 + \ln Q_t)$ $= N_Ak(1 + \ln T/\theta_r\sigma)$ (b) $N_Ak\{3/2 + \ln[(\pi^2/\sigma)(T^3/\theta_A\theta_B\theta_C)^{1/2}]\}$	$N_Ak\{(\theta_v/T)/[\exp(\theta_v/T) - 1] - \ln[1 - \exp(-\theta_v/T)]\}$
Gibbs function (G)	$-N_AkT \ln Q_t$	$-N_AkT \ln Q_r$	$N_AkT \ln[1 - \exp(-\theta_v/T)]$

[a] J, the rotational quantum number; v, the vibrational quantum number; I, the MOMENT OF INERTIA; ν, the characteristic frequency; σ, the SYMMETRY NUMBER; θ_r ($= h^2/8\pi^2 Ik$) is the characteristic temperature of rotation; θ_v ($= h\nu/k$) the characteristic temperature of vibration is generally large compared to room temperature, most gas molecules are vibrationally unexcited.

state compared with kT that all the molecules can be assumed to be in the ground state; thus $Q_e = g_0$. The electronic contributions in a few diatomic molecules (e.g., nitrogen monoxide and oxygen) are due to the presence of two separate ground state electronic levels lying very close together.

pascal (Pa). SI unit of PRESSURE; equal to a pressure of $1\,N\,m^{-2}$.

Paschen series. *See* BALMER SERIES.

passivity. A metal surface is said to be passive when, although exposed to conditions in which it is thermodynamically unstable, it remains unattacked. Metals may be rendered passive by the action of an

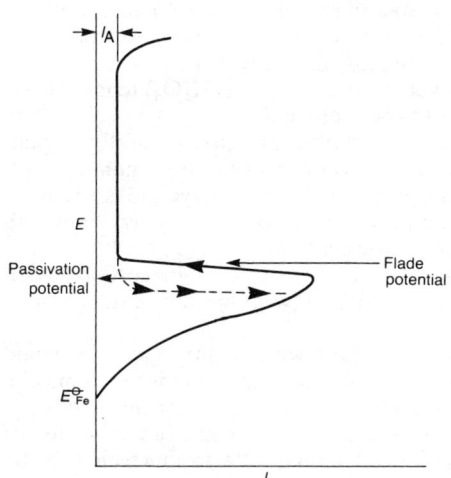

passivity
Passivation curve for iron where I_A = anodic protection current.

oxidizing acid (e.g., concentrated nitric acid) or an anodic potential: for example, when the anodic voltage on iron is increased (*see* Figure). The current initially increases as expected, but at the passivation potential

it falls to a low value. Some passivation potentials, against the standard hydrogen electrode, are Fe = 0.58 V, Ni = 0.36 V and Cr = –0.22 V. The Flade potential is at the point of inflexion in the current–voltage curve on the anodic side of the passivation potential.

path function. *See* STATE FUNCTION.

Pauli exclusion principle. In an atom, no two electrons may have the same quantum numbers. This means that the maximum occupancy of an orbital is two, the electrons having spin quantum numbers of +1/2 and –1/2. The Pauli exclusion principle is a consequence of the condition which requires electron wavefunctions to be anti-symmetric to exchange of electrons. *See also* SLATER DETERMINANT.

Pauling radius. *See* IONIC RADIUS.

Pauling's electronegativity scale. *See* ELECTRONEGATIVITY.

Pb. *See* LEAD.

p-band. *See* BAND THEORY OF SOLIDS.

P-branch. *See* INFRARED SPECTROSCOPY.

Pd. *See* PALLADIUM.

pectic acids. *See* PECTINS.

pectinic acids. *See* PECTINS.

pectins. Compounds formed on the partial acid hydrolysis of water-soluble protopectins which comprise the major pectic component of primary cell walls of fruits and berries. The pectins or, pectinic acids, contain both carboxyl and methylcarboxyl groups; hydrolysis of these yields the pectic acids, high-molecular-mass polymers. Enzymic hydrolysis of purified pectic acids gives up to 85 percent D-galacturonic acid

pectins

(see URONIC ACIDS). In the pectins, the carboxyl groups are partially esterified. Pectins, manufactured as white powders, have the properties of forming gels with sugars and acids and are used in the food industry (E440a), and as emulsifying and gelling agents assisting in the setting of jams and jellies.

penicillinase. *See* ENZYMES.

penicillins. *See* ANTIBIOTICS.

pentacene. *See* POLYCYCLIC HYDROCARBONS (structure IV).

pentaerythritol (tetra(hydroxymethyl)-methane, $C(CH_2OH)_4$). Colourless crystalline compound (mp 263°C). Condensation of methanal and ethanal in the presence of calcium hydroxide yields the intermediate tris(hydroxymethyl)ethanal (($HOCH_2)_3$-CCHO), which under the reaction conditions undergoes a 'crossed Cannizzaro' reaction with methanal to give pentaerythritol. Some di- and tripentaerythritols are formed during this ALDOL CONDENSATION. It is used in resins, plastics and drying oils and in the manufacture of PENTAERYTHRITOL TETRANITRATE.

pentaerythritol tetranitrate (PETN, $C(CH_2ONO_2)_4$). Very powerful and extremely violent explosive made by nitrating pentaerythritol. Mixed with trinitrotoluene (*see* EXPLOSIVES) it forms the explosive pentolite. It is used medicinally to relieve the pain of angina pectoris.

pentamethylene diamine (cadaverine). *See* POLYAMINES.

pentanedioic acid (glutaric acid, HOOC-$(CH_2)_3$COOH). Saturated DICARBOXYLIC ACID (mp 97°C, bp 303°C), prepared by treating 1,3-dichloropropane with concentrated hydrochloric acid. On heating with ethanoic anhydride or thionyl chloride pentanedioic anhydride is formed.

2,4-pentanedionato group (acetylacetonato group). *See* DIKETONES.

2,4-pentanedione (acetyl acetone). *See* DIKETONES.

pentanoic acid. *See* CARBOXYLIC ACIDS.

pentlandite. *See* NICKEL.

pentobarbital. *See* BARBITURATES.

pentolite. *See* PENTAERYTHRITOL TETRANITRATE.

pentosans. HEMICELLULOSES which, on hydrolysis, give PENTOSES. An example is XYLAN.

pentoses ($C_5H_{10}O_5$). CARBOHYDRATES with five carbon atoms. The eight optically active forms of the aldopentoses corresponding to the D- and L-isomers are shown in the Fischer and Haworth projection formulae (*see* Figure). The chemical properties of aldopentoses are similar to those of HEXOSES. Pentoses do not undergo fermentation and are converted quantitatively to 2-FURALDEHYDE when warmed with dilute acid. They are broken down quantitatively by periodic acid (HIO_4) to methanoic acid and methanal.

L-(+)-arabinose. Occurs naturally as pentosans in various gums (e.g., gum arabic), mucilages and some glycosides. Usually obtained by hydrolysis of cherry gum with sulphuric acid (mp 150°C).

D-(−)-arabinose. Occurs in certain glycosides and in polysaccharides of the tubercle bacillus.

D-(+)-xylose (wood sugar). Occurs in wood gums, bran and straw from which it may be obtained by acid hydrolysis (mp 145°C).

D-(+)-ribose. Occurs in NUCLEIC ACIDS of plant and animal cells, from which it can be extracted by hydrolysis (mp 95°C). It has the furanose structure.

The remaining aldopentoses are synthetic compounds.

2-deoxyribose (β-D-2-deoxyribofuranoside, $HOCH_2(CHOH)_2CH_2CHO$). Very important aldopentose, in which a hydroxyl group is replaced by hydrogen, is obtained by the hydrolysis of deoxyribonucleic acid (*see* NUCLEIC ACIDS).

Ketopentoses with the structure $HOCH_2CO(CHOH)_2CH_2OH$ have been prepared; with two asymmetric carbon atoms they can exist in four optically active forms, the D- and L-forms of ribulose and xylulose. The diphosphate of ribulose is a key intermediate in the fixation of atmospheric carbon dioxide.

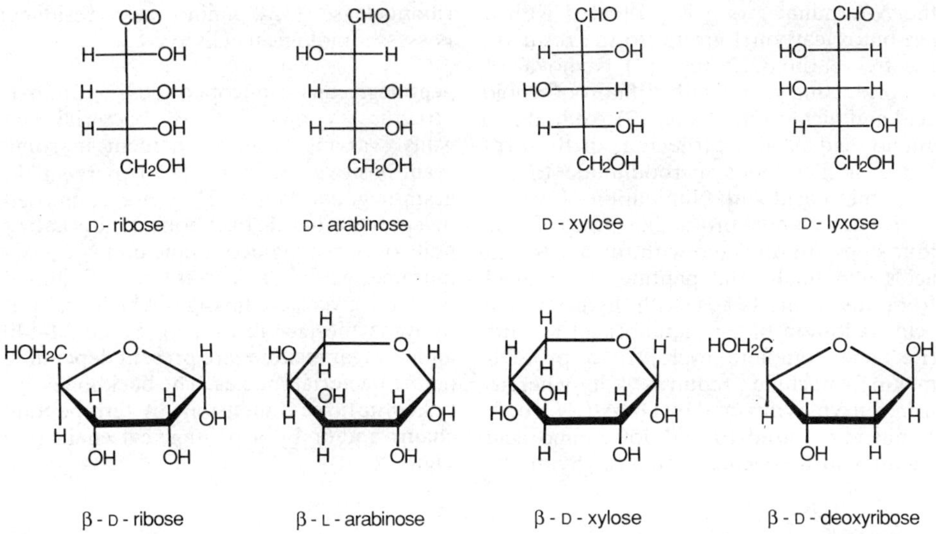

Fischer projections for D-ribose, D-arabinose, D-xylose, D-lyxose and Haworth projections for β-D-ribose, β-L-arabinose, β-D-xylose, β-D-deoxyribose.

pentoses
Fischer and Haworth projection formulae for aldopentoses.

pentyl group.   *See* AMYL GROUPS.

pepsin.   *See* ENZYMES.

peptide antibiotics.   *See* ANTIBIOTICS.

peptides.   Substances composed of two or more amino acids, designated as di-, tri-, oligo- or polypeptide according to the number of AMINO ACIDS linked together by an amide or peptide bond (-CONH-) (i.e. 2, 3, ~ 10 and 100–300, respectively). In naming peptides, the amino or *N*-terminal end is written to the left and the amino acids named, left to right, replacing '-ine' by '-yl' except for the *C*-terminal amino acid. For example leucylalanylmethionine.

Peptides are formed by the elimination of a water molecule between two amino acids, one with a protected amino and the other with a protected carboxyl group; the protecting groups are removed to leave the peptide.

$$H_2NCH(R)COOH +$$

$$HNHCH(R')COOH \longrightarrow$$

$$H_2NCH(R)CONHCH(R')COOH$$

N,N'-Dicyclohexylcarbodiimide in one step activates the carboxyl group and effects coupling between an amino group and an acid group with the removal of water. The amide group is essentially planar as a result of resonance (–CN̂C– is 123°), the carbon–nitrogen bond is about 40 percent double bond in character, consequently rotation about the bond is restricted. The *trans*-configuration is more stable than the *cis*-form. Hydrogen bonds play an essential role in stabilizing peptide chain conformation. Peptides arise in the partial hydrolysis of PROTEINS, the different peptide species formed being separated by electrophoresis. From the amino acid sequence (*see* EDMAN DEGRADATION), the primary structure of the protein can be established. Naturally occurring peptides possessing important physiological activity include: the tripeptide glutathione (L-glutamyl-L-cysteinylglycine), a constituent of practically all cells; HORMONES; and the vasodilator bradykinin (Arg–Pro–Pro–Gly–Phe–Ser–Pro–Phe–Arg). A number of ANTIBIOTICS (e.g., bacitracin and polymyxin) are peptides.

peptide synthesis (Merrified solid-phase synthesis).   Synthetic polypeptides (with up to 100 amino acid residues) have been prepared in a four-stage process using polystyrene beads with active –CH$_2$Cl groups. (1) Attachment of an amino acid 1, in which the *N*-terminal group is protected with a *tert*-butoxycarbonyl group, to the resin via the free carboxyl group. (2) Removal of the protecting group with trifluoroethanoic acid and dichloromethane. (3) Addition of amino acid 2, also protected, in the presence of dicyclohexylcarbodiimide to the first amino acid with elimination of water. (4) Removal of the protecting group. These four steps are repeated with further amino acids and finally the peptide is removed from the resin beads with hydrobromic acid. Addition of one amino acid residue (i.e. one complete cycle) in a protein-making 'machine' requires 4 h, whereas using enzymes directed by DNA/RNA only 1 min is required to add 150 amino acid residues in a specific sequence. Synthetic ribonuclease (124 amino acid residues) possesses biological activity.

peptidoglycans (mucopeptides, mureins). Structural components of bacterial cell walls (varying from 95 percent in some gram-positive cells to 5–10 percent in gram-negative cells). They are composed of a polysaccharide backbone of alternating units of *N*-acetylglucosamine and *N*-acetyl-muramic acid (*see* AMINO SUGARS) joined by β-1,4-glycosidic linkages which are split by lysozyme (*see* ENZYMES). Some 20–140 amino sugar residues are present depending on the bacterial species. The backbones are linked in three dimensions by tetrapeptide chains and/or by a pentaglycyl chain (*see* Figure).

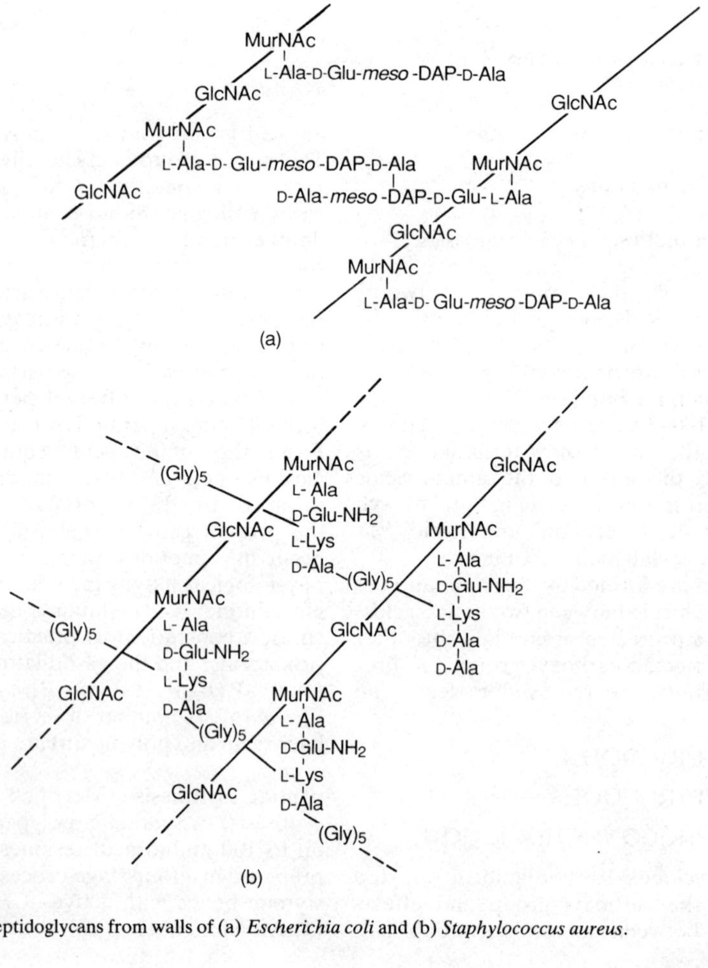

(a)

(b)

peptidoglycans
Structures of peptidoglycans from walls of (a) *Escherichia coli* and (b) *Staphylococcus aureus*.

peptization. Dispersion of a COLLOID by changing the composition of the dispersion medium (i.e. the reverse of COAGULATION). Peptization may be brought about by the addition of ions which will adsorb on the surface of the particles and thus stabilize the sol (*see* LYOPHOBIC COLLOID), dilution of the dispersion medium and by DIALYSIS. In peptization, V_R (*see* DLVO THEORY) is modified to create a maximum in the potential energy curve which acts as a barrier opposing coagulation.

peracetic acid. *See* PERETHANOIC ACID.

peracids. Oxoacids containing the group M–O–O– (e.g., perboric acid, *see* BORON INORGANIC COMPOUNDS; persulphuric acid, *see* SULPHUR OXOACIDS), prepared by the action of hydrogen peroxide on, or electrochemical oxidation of the oxoacid. Permanganates, perchlorates or periodates are not salts of peracids.

perboric acid. *See* BORON INORGANIC COMPOUNDS.

perbromates. *See* BROMATES.

perchlorates. *See* CHLORATES.

perchloroethane. *See* HEXACHLOROETHANE.

perchloryl fluoride. *See* CHLORINE OXIDES.

perethanoic acid (peracetic acid, CH_3CO-(OOH)). Organic PERACID, prepared by treating ETHANOIC ANHYDRIDE with concentrated hydrogen peroxide. It is an unpleasant-smelling liquid that explodes when heated above 110°C. It is a powerful oxidizing agent and is used in the preparation of ethene oxide (*see* ALKENE OXIDES). In solution it forms an intramolecular HYDROGEN BOND.

perfect gas. *See* IDEAL GAS.

peri-. Prefix used to denote groups occupying the 1,8-positions in the NAPHTHALENE ring. It is preferable to use numbers to indicate positions.

period. *See* PERIODIC TABLE.

periodates. *See* IODATES.

periodic table. Table of the elements written in sequence in increasing order of ATOMIC NUMBER (*see* Appendix, Table 4). It is arranged in rows called periods, and columns called groups, such that elements with similar ELECTRONIC CONFIGURATION lie beneath one another. In this way similarities and trends in properties may be highlighted. The original concept of D. Mendeleev was based on atomic weights, and although this gives mostly the correct arrangement it is now known that the electronic structure, and thus the atomic number, should be used. The lighter elements fit neatly into groups of eight as only *s*- and *p*-orbitals are filled (*see* ATOMIC ORBITAL). Space is then found for the *d*-block elements in three transition series which contain elements with partially filled *d*-orbitals. In turn the second and third transition series are punctuated by elements with incomplete *f*-shells —the LANTHANIDE ELEMENTS and ACTINIDE ELEMENTS. Taken together these elements are known as the TRANSITION ELEMENTS, and they are distinguished from the main group elements by incomplete inner shells. Properties that vary systematically across or down the table are OXIDATION STATES, whether the elements are metallic or non-metallic, atomic volume and any other properties that depend on electronic structure. A DIAGONAL RELATIONSHIP is found between some members of the main group elements. The properties of unknown elements have been predicted from their positions in the table. Thus element 106 should be a member of the group chromium, molybdenum and tungsten.

peritectic melting. *See* MELTING POINT.

Perkin reaction. CONDENSATION REACTION between an aromatic aldehyde (*see* BENZALDEHYDE, Figure) and the anhydride of an aliphatic acid containing two α-hydrogen atoms, in the presence of a sodium salt of the acid to give a β-arylpropenoic acid.

permanent hardness. *See* HARDNESS OF WATER.

permanganates. *See* MANGANESE COMPOUNDS.

permanganate titration. Potassium permanganate is an important volumetric oxidizing agent which requires no indicator. In the presence of sulphuric acid, it is reduced (e.g., oxalic acid, iron(II) ammonium sulphate) to manganese(II) — a five-electron reaction. *See also* TITRATIONS.

permangic acid. *See* MANGANESE COMPOUNDS.

permeability (magnetic permeability; μ; units: $H\ m^{-1} = kg\ m\ s^{-2}\ A^{-2}$). Ratio of the magnetic flux density (B) in a substance to the external field strength (H) (i.e. $\mu = B/H$). The permeability of free space, μ_0, is known as the magnetic constant and has the value of $4\pi \times 10^{-7}\ H\ m^{-1}$. The relative permeability of a substance $\mu_r = \mu/\mu_0$ is dimensionless.

permeation chromatography. *See* GEL FILTRATION CHROMATOGRPAHY.

permittivity (ε; units: $Fm^{-1} = A^2 s^4 kg^{-1} m^{-3}$). Constant of a dielectric medium that relates the electric field strength (E) to the charge density (σ), $E = 4\pi\sigma/\varepsilon$. Permittivity values are expressed relative to that of a vacuum ($\varepsilon_0 = 8.854 \times 10^{-12}\ Fm^{-1}$), $\varepsilon = \varepsilon_r\varepsilon_0$. The relative permittivity (ε_r) is also known as the dielectric constant. The potential between ions in a solution is $-z_i e/4\pi\varepsilon$. The effect of the dielectric is thus to lower the attractive forces between ions and so electrolytes are more likely to dissociate in solvents of high permittivity (e.g., water $\varepsilon_r = 80$) than in those of low permittivity (e.g., ethanoic acid $\varepsilon_r = 6$).

Permutit process. *See* HARDNESS OF WATER.

perovskite. *See* TITANIUM.

perovskite structure. Structure shown by many $MM'X_3$ compounds, in which M and M' are cations and X is F^- or O^{2-}. M' is at the centre of a cube with M at the corners and X at the face centres (M and X are ccp, *see* CLOSE-PACKED STRUCTURES). M is surrounded by 12 X neighbours and M' by six X (octahedral) neighbours. Examples include the mineral perovskite ($CaTiO_3$), $KNiF_3$, $BaTiO_3$.

peroxidases. *See* ENZYMES.

peroxides. Compounds containing the O_2^{2-} ion are known for the alkali and alkaline earth metals. The ion does not exist free in solution due to the rapid reaction with water or dilute acids to give HYDROGEN PEROXIDE. Peroxides are all powerful oxidizing agents. Magnesium, the lanthanide and the uranyl peroxides are intermediate in character between the ionic and the essentially covalent peroxides of zinc, cadmium and mercury. *See also* OXIDES, SUPEROXIDES.

peroxodisulphuric acid. *See* SULPHUR OXOACIDS.

peroxomonosulphuric acid. *See* SULPHUR OXOACIDS.

peroxonitrous acid. *See* NITROGEN OXOACIDS.

perspex (poly(2-methylpropenoate), poly(methylmethacylate)). *See* POLYMERS.

persulphuric acid. *See* SULPHUR OXOACIDS.

perturbation molecular orbital theory (PMO theory). Theory of the energetics of organic reactions in which the changes that take place are treated as small perturbations of the original structure. For example, in the protonation of pyridine to give the pyridinium ion

$$\Delta E = E_{N-H} + \delta E_\pi$$

where E_{N-H} is the nitrogen–hydrogen σ-bond strength and δE_π is the effect on the aromatic π-system (*see* AROMATICITY). δE_π is estimated in terms of the changes in the Coulomb and resonance integrals derived in the HÜCKEL MOLECULAR ORBITAL THEORY approach.

PES (photoelectron spectroscopy). *See* ELECTRON SPECTROSCOPY.

pesticide. Substance used to destroy or inhibit the action of plant or animal pests; the term includes INSECTICIDES, HERBICIDES, fungicides and RODENTICIDES. Virtually all are toxic to man to a greater or lesser extent.

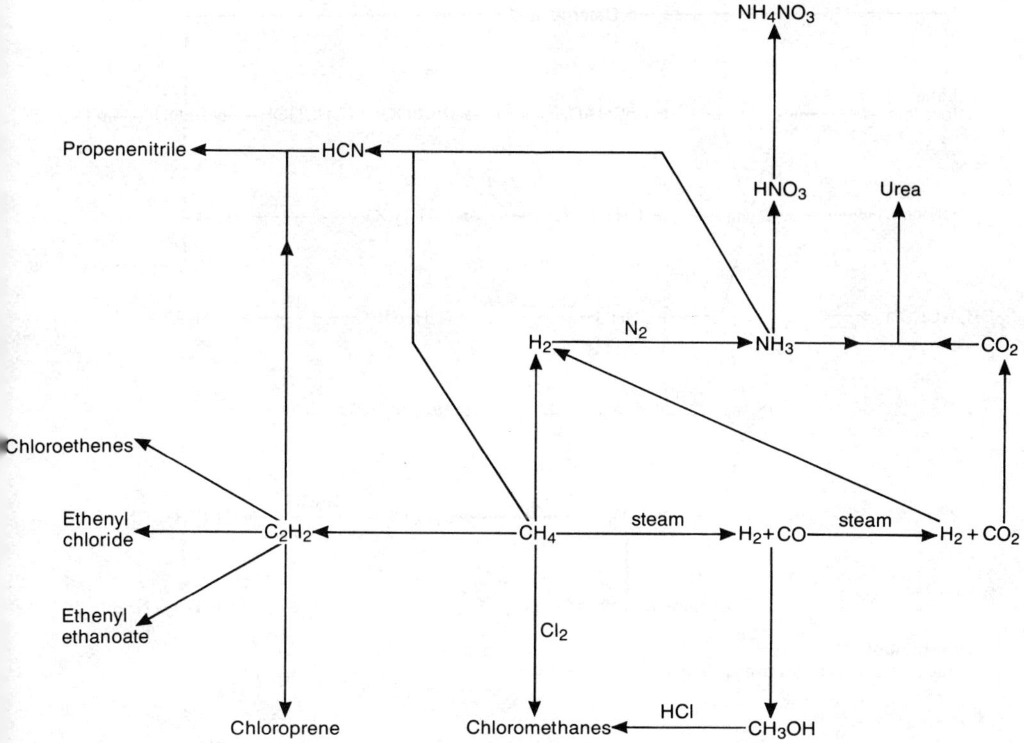

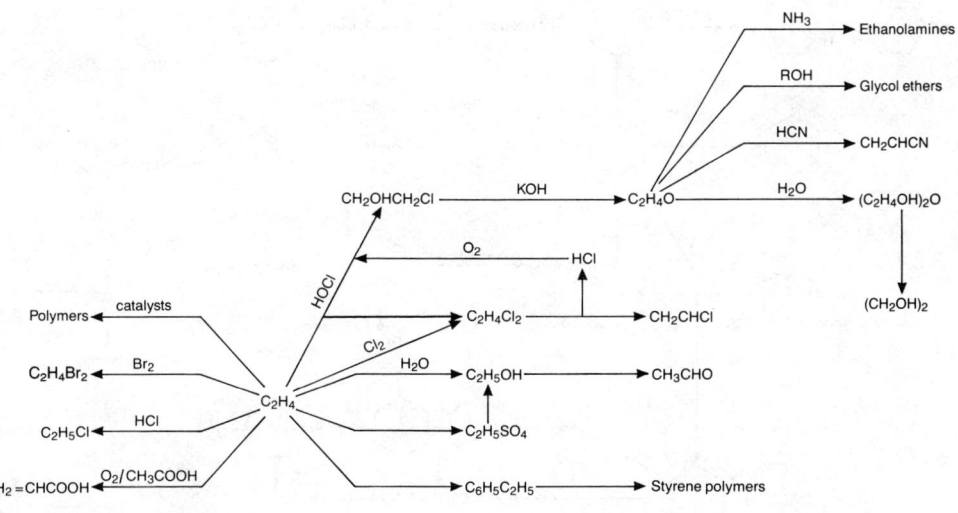

petrochemicals
Figure 1. Petrochemicals from methane.

petrochemicals
Figure 2. Petrochemicals from ethene.

continued overleaf

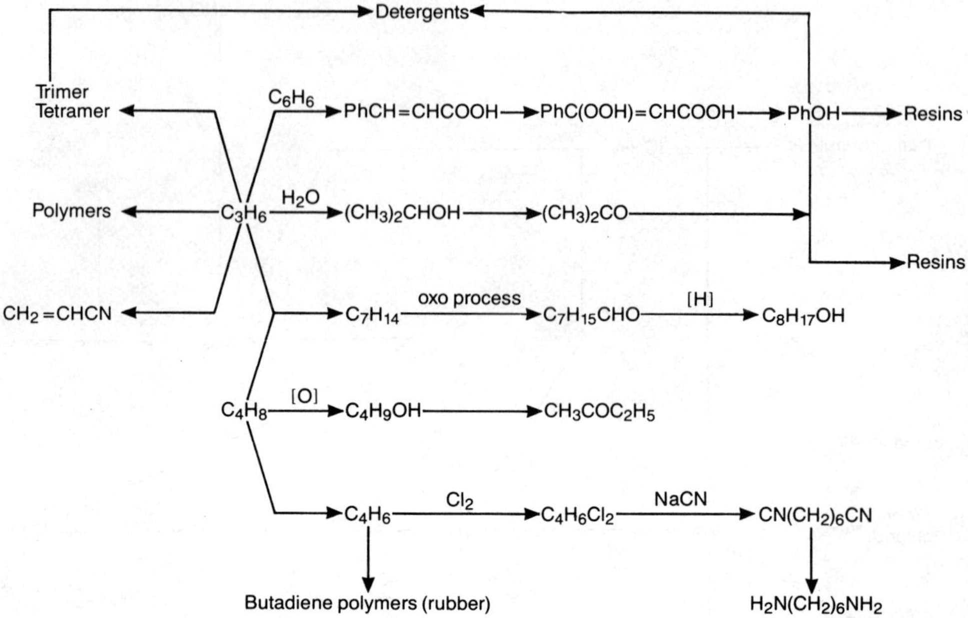

petrochemicals
Figure 3. Petrochemicals from propene and butene.

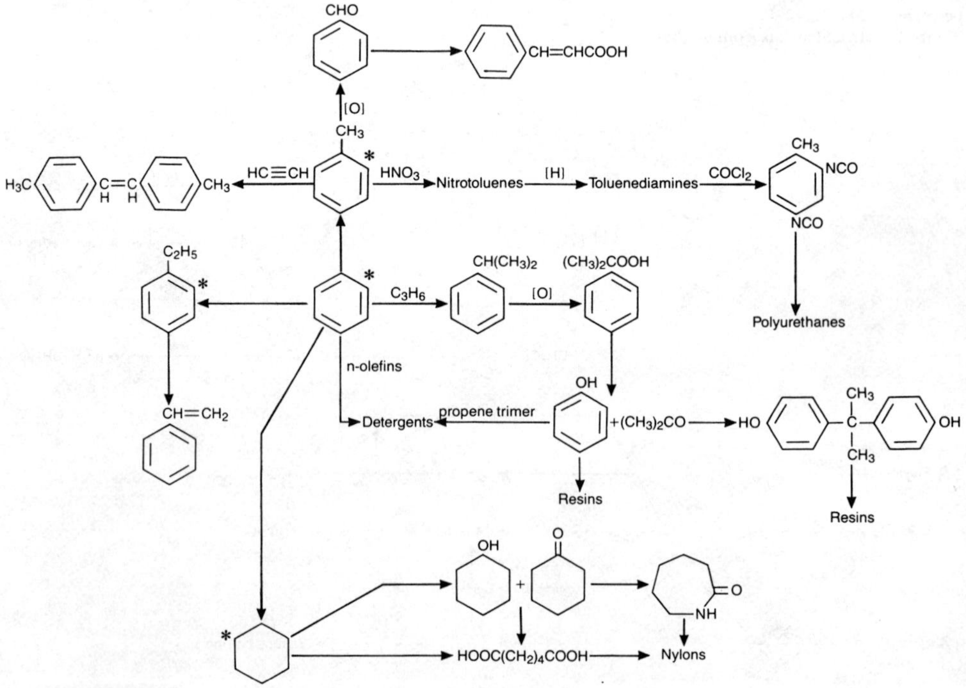

petrochemicals
Figure 4. Petrochemicals from cyclic and aromatic hydrocarbons. (* obtained directly from petroleum.)

PETN. *See* PENTAERYTHRITOL TETRA-NITRATE.

petrochemicals. Chemicals obtained from PETROLEUM. The most important starting compounds are METHANE (*see* Figure 1), ETHENE (*see* Figure 2), PROPENE and BUTENES (*see* Figure 3) and cyclic and aromatic hydrocarbons (*see* Figure 4).

petrol. United Kingdom name for motor and aviation fuel (*see* GASOLINE).

petrolatum. Hydrocarbon jelly, generally C_{18}–C_{22}, produced as a fraction of PETROLEUM. It is used for pharmaceutical preparations as Vaseline and petroleum jelly.

petroleum (crude oil). Naturally occurring mixture of HYDROCARBONS. The composition, especially that of the higher boiling fractions, varies with the source. The composition of a typical set of fractions obtained by DISTILLATION from petroleum are tabulated (*see* Table). CRACKING of higher oils leads to the more useful lower molecular mass fractions (naphthas).

petroleum ether. Volatile hydrocarbon distillate of PETROLEUM (C_5–C_6), characterized by its boiling range (e.g., 40–60°C, or 60–80°C).

petroleum gas. Mixture of propanes (commercial propane 92 percent) and butanes (commercial butane 85 percent) derived from natural gas, or from PETROLEUM by catalytic reforming (*see* HYDROFORMING). Under moderate pressure or refrigeration, petroleum gas liquefies to give LPG, which is used as a fuel with a high calorific value of about 10^5 kJ m^{-3}.

Pfund series. *See* BALMER SERIES.

PG. *See* PROSTAGLANDINS.

Ph. *See* PHENYL GROUP.

petroleum
Fractions obtained from petroleum.

Name	Boiling point /t°C	Composition (carbon number)	Uses
Natural gas	< room temperature	1–4	Fuel, starting material for many organic compounds
Light petrol (petroleum ether)	20–100	5–7	Solvents, special boiling point spirits
Benzene	70–90	6–7	Dry cleaning
Ligroin (light naphtha)	80–120	6–8	Solvent, white spirit
Gasoline (petrol)	70–200	6–11	Motor fuel
Kerosene (paraffin oil)	200–300	12–16	Lighting, heating
Gas oil (heavy oil)	> 300	13–18	Fuel oil
Lubricating oil	> 300[a]	16–20	Lubricants
Greases, petroleum jelly	> 300[a]	18–22	Pharmaceutical preparations
Paraffin wax	> 300[a]	20–30	Candles, greaseproof paper
Residua (asphaltic bitumen)	> 300[a]	30–40	Asphalt, tarmacadam, petroleum coke

[a] Not distilled at atmospheric pressure.

pH. Negative logarithm to base 10 of the hydrogen ion activity of a solution (i.e. pH = $-\lg a_{H^+}$). Although the potential of electrodes such as a GLASS ELECTRODE or HYDROGEN ELECTRODE depends on a_{H^+}, the impossibility of measuring a single electrode potential precludes the determination of the pH value of a solution as defined. The modern definition of pH is an operational one in which a suitable indicator electrode is combined with a REFERENCE ELECTRODE, and the emf of the cell so formed is determined. The pH values of two solutions A and B determined by this cell are related to the emf values by

$$pH_A = pH_B + (E_A - E_B) \, F/2.303RT$$

It is assumed that any LIQUID JUNCTION POTENTIALS in the cell cancel out. In this way, a scale of pH values may be drawn up given one or more standard solutions. If the IONIC STRENGTH of the solution is not too great, it is possible to replace the definition above with one based on concentrations (i.e. pH = $-\lg c_{H^+}$). pH values are measured electrochemically as described above or using INDICATORS.

phase. Homogeneous, physically distinct and mechanically separable part of a system. Each phase is separated from other phases by a physical boundary. Ice/water/water vapour are three phases; any number of gases mixing in all proportions constitute one phase. At any particular temperature and pressure, the stable form is the one in which the COMPONENT has the smallest CHEMICAL POTENTIAL.

phase diagram. Diagram showing the conditions of equilibrium between various phases. For a ONE-COMPONENT SYSTEM, it is a pressure–temperature diagram. For a TWO-COMPONENT CONDENSED SYSTEM, a three-dimensional diagram relating pressure, composition and temperature or a two-dimensional diagram relating composition with temperature (pressure) at constant pressure (temperature) is required. For a THREE-COMPONENT SYSTEM, a triangular diagram is used to show the variation of composition at constant pressure and temperature.

phase rule. Formulated by Gibbs to provide a general relationship between the number of degrees of freedom (f), the number of components (c) and the number of phases (p) of a system in equilibrium

$$p + f = c + 2$$

The law applies to all macroscopic systems in a state of heterogeneous equilibrium (chemical or physical) and which are influenced only by changes in temperature, pressure and concentration. Systems are classified according to the phase rule as a ONE-COMPONENT SYSTEM, a TWO-COMPONENT CONDENSED SYSTEM or a THREE-COMPONENT SYSTEM. See also COMPONENT, DEGREE OF FREEDOM, PHASE.

phase transition. Change of state between solid, liquid and vapour phases.

Phe (phenylalanine). See AMINO ACIDS.

phen. See 1,10-PHENANTHROLINE.

phenacetin (4-(N-acetyl)-phenetidine, 4-$C_2H_5OC_6H_4NHCOCH_3$). White solid (mp 135°C), prepared by the N-acetylation of a 4-aminophenetole with ethanoyl chloride. It possesses antipyretic and analgesic properties similar to those of ASPIRIN, but is toxic to the kidneys and has been reported to be carcinogenic.

phenacite. See SILICATES.

phenanthrene ($C_{14}H_{10}$). Angular POLYCYCLIC HYDROCARBON (structure IX), with aromatic character, structurally related to certain alkaloids (e.g., morphine) and steroids (e.g., cholesterol). It occurs in the anthracene oil fraction of COAL TAR. It is a white solid (mp 99°C), a solution in benzene shows blue fluorescence. It is nearly as reactive as its isomer ANTHRACENE, and is reduced to 9,10-dihydrophenanthrene, adds bromine to form 9,10-phenanthrene dibromide and undergoes the FRIEDEL–CRAFTS REACTION in the 3-position, is nitrated to three mono-derivatives (the 3-isomer predominating). Oxidation (CrO_3) yields phenanthraquinone. It is considered to be carcinogenic.

1,10-phenanthroline ($C_{12}H_8N_2$). HETEROCYCLIC COMPOUND (structure XXXVII) formed by heating 1,2-diaminobenzene with glycerol, nitrobenzene and sulphuric acid. It is slightly soluble in water, and forms a complex with iron(II) which is used as an indicator in redox titrations.

phenanthro[3,4c]phenanthrene. *See* HEXA-HELICENE.

phenazine ($C_{12}H_8N_2$). Pale yellow crystal-line HETEROCYCLIC COMPOUND (structure XXXVIII), prepared by boiling 2,2'-di-nitrodiphenylamine with tin(II) chloride in hydrochloric acid/ethanoic acid followed by oxidation with hydrogen peroxide. It is insoluble in water, slightly soluble in ethanol and soluble in acids to give yellow–red solutions.

phenetidines. Ethyl ethers of AMINO-PHENOLS. *See also* PHENACETIN.

phenetole. *See* AROMATIC ETHERS.

phenobarbitone. *See* BARBITURATES.

phenol (benzenol, carbolic acid, hydroxy-benzene. C_6H_5OH, PhOH). Colourless, crystalline compound (mp 43°C, bp 183°C), turning pink on exposure to air and light. It is soluble in water at room temperature, larger amounts of phenol give two-phase system (*see* PARTIALLY MISCIBLE LIQUID SYSTEMS), but is completely miscible with water at temperatures above 84°C. It is a component of COAL TAR, obtained by the alkali fusion of benzenesulphonic acid. Phenol is manufactured by (1) heating chlorobenzene with a 10 percent solution of sodium carbonate or sodium hydroxide under pressure at 300°C (10 percent di-phenyl ether is added to prevent excessive formation of this compound); (2) catalytic vapour-phase.reaction of steam and chloro-benzene at 425°C; (3) oxidation of CUMENE. It is a weak acid (pK_a 9.98), forming metal-lic salts, the phenates. It is readily halo-genated, nitrated and sulphonated to give *o*- and *p*-derivatives (–OH group is *o*,*p*-directing, *see* ORIENTATION IN THE BENZENE RING), and forms aniline with ammonia under pressure. Phenates react with alkyl halides to form phenolic ethers (e.g., $C_6H_5OCH_3$). Reduction (H_2/Ni) gives cyclohexanol, oxidation ($KMnO_4$) breaks the ring giving a variety of products and with hydrogen peroxide in the presence of iron(II) ions gives a mixture of 1,2- and 1,4-

dihydroxybenzenes. Phenol undergoes the REIMER–TIEMANN REACTION and the GATTERMANN ALDEHYDE SYNTHESIS with the formation of aldehydes, the FRIEDEL–CRAFTS REACTION to give mainly 4-alkyl-phenols and couples with diazonium salts to form hydroxyazo compounds (e.g., *p*-$HOC_6H_4N_2C_6H_5$). Phenol condenses with aliphatic and aromatic aldehydes in the *o*- and *p*-positions, the most important being the condensation with methanal under alkaline conditions to give mainly 4-hydr-oxybenzyl alcohol (some *o*-isomer is formed). When larger amounts of methanal are used bishydroxymethylphenol and 4,4'-dihydroxydiphenylmethane are formed, these condensations form the basis of phenol–formaldehyde resins (*see* POLY-MERS). Phenol is used in the production of phenolic and epoxy resins, capro-lactam, alkylphenols, dyes, pharmaceuti-cals (powerful antiseptic) and perfumes.

phenol–formaldehyde resins. *See* POLY-MERS.

phenolic ethers. *See* AROMATIC ETHERS.

phenolic resins. *See* POLYMERS.

phenolphthalein. *See* INDICATORS.

phenols (arenols). Aromatic compounds containing one or more hydroxyl groups attached directly to the ring. They are more polar and more acidic and form stronger hydrogen bonds than the corresponding saturated alcohols, resulting in higher boil-ing points, water solubility and increased ability to act as solvents for reasonably polar organic molecules. Many monohydric phenols occur in coal tar. General methods of preparation are by: (1) hydrolysis of aryldiazonium salts ($ArN_2Cl + H_2O \rightarrow ArOH$); (2) alkaline fusion of arylsulphonic acids; (3) industrially by heating aryl halides with sodium hydroxide at high tempera-tures and pressures. The electron-rich π-orbital system of phenols (especially the phenate ions) make then susceptible to ELECTROPHILIC SUBSTITUTION; with brom-ine water substitution occurs *o*- or *p*- to the –OH group. Phenols are soluble in alkalis to give alkali metal phenates (phenoxides) and are reduced (H_2/Ni) to the correspond-ing cyclohexanol; oxidation generally opens

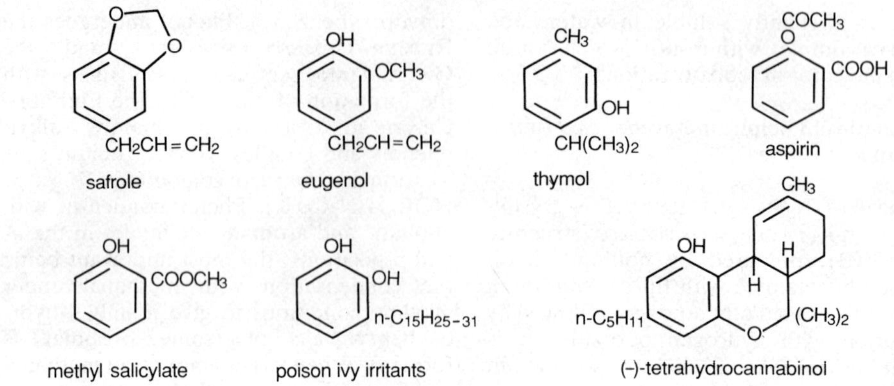

CH₂CH=CH₂
safrole

CH₂CH=CH₂
eugenol

thymol
CH(CH₃)₂

aspirin

methyl salicylate

poison ivy irritants

(–)-tetrahydrocannabinol

phenols

the ring. Several important aromatic compounds have more than one arene hydroxyl group (see DIHYDROXYBENZENES, TRIHYDROXYBENZENES). Many phenols and phenol ethers occur in nature (see Figure). For example: 2-HYDROXYBENZOIC ACID (salicylic acid, obtained from the willow tree).

thymol (5-methyl-2-(1-methylethyl)phenol). Flavouring constituent of thyme and other essential oils, used as an antiseptic ingredient in mouthwashes.

safrole (5-(2-propenyl)-1,3-benzodioxole). Acetal closely related to eugenol, the main constituent of oil of sassafras.

anethole (1-methoxy-4-(1-propenyl)benzene). Constituent of aniseed oil.

Certain phenols have powerful physiological effects (e.g., urushiol, a constituent of poison ivy and poison oak, and (–)-TETRAHYDROCANNABINOL). *See also* HYDROXYDIMETHYLBENZENE, HYDROXYNAPHTHALENES, HYDROXYTOLUENES, PHENOL.

phenothiazine ($C_{12}H_9NS$). HETEROCYCLIC COMPOUND (structure XXXIX), prepared by fusing diphenylamine with sulphur. It consists of pale yellow crystals (mp 185°C) which darken on exposure to light and air. It is insoluble in water. Although poisonous to man, it is used as an anthelmintic in veterinary practice. Substituted phenothiazines (e.g., CHLORPROMAZINE) form an important group of tranquillizers.

phenoxazine. *See* HETEROCYCLIC COMPOUNDS (structure XL).

phenoxy group. The group C_6H_5O- (PhO–).

phenylalanine (Phe). *See* AMINO ACIDS.

phenylamine. *See.* ANILINE.

phenylbenzene. *See* BIPHENYL.

***N*-phenylbenzeneamine.** *See* DIPHENYLAMINE.

phenyl cation ($C_6H_5^+$). Very unstable CARBONIUM ION.

2-phenylchromone. *See* FLAVONE.

phenyl cyanide. *See* BENZONITRILE.

phenylenediamines. *See* DIAMINOBENZENES.

phenylene group. The bivalent group $C_6H_4<$ with appropriate prefix (*o*-, *m*-, *p*-) (e.g., *m*-phenylenediamine).

phenylethyl ether. *See* AROMATIC ETHERS.

phenyl group. The group C_6H_5- (Ph–) present in benzene derivatives.

phenylhydrazine ($C_6H_5NHNH_2$). Colourless, refractive oil (bp 241°C), prepared by the reduction of benzenediazonium chloride ($SnCl_2/HCl$; commercially Na_2SO_3 then Zn/CH_3COOH). It oxidizes in air, but may be stabilized by the formation of the hydrochloride ($C_6H_5NHNH_2.HCl$). It is a powerful reducing agent, reducing FEHLING'S SOLUTION in the cold. It reacts with compounds containing the OXO GROUP to give phenylhydrazones (*see* HYDRAZONES) and with sugars to give OSAZONES. It is reduced (Zn/HCl) to aniline and ammonia, and is used in the FISCHER INDOLE SYNTHESIS.

N-phenylimides. *See* SCHIFF'S BASES.

phenylhydrazone. *See* HYDRAZONES.

phenylmagnesium halides. *See* GRIGNARD REAGENTS.

phenylmethane. *See* TOLUENE.

phenylmethanol. *See* BENZYL ALCOHOL.

phenylmethylamine. *See* AMINOTOLUENES.

3-phenylpropenoic acids (cinnamic acids, $C_6H_5CH=CHCOOH$). The *trans*-isomer (mp 133°C) occurs naturally in amber, balsams and resins and may be prepared by the PERKIN REACTION, the KNOEVENAGEL REACTION or by the CLAISEN CONDENSATION between benzaldehyde and ethyl ethanoate in the presence of sodium ethoxide. It gives STYRENE on distillation and photodimerizes (*see* PHOTODIMERIZATION) to 3,4-diphenyl-1,2-cyclobutanedioic acid and 2,4-diphenyl-1,3-cyclobutanedioic acid. It is used in perfumes, flavours and pharmaceuticals. The *cis*-isomer (mp 68°C) is unstable and readily converts to the *trans*-form.

phloroglucinol. *See* TRIHYDROXYBENZENES.

phosgene. *See* CARBONYL CHLORIDE.

phosphatases. *See* ENZYMES.

phosphates. Formally any salt of a PHOSPHORUS OXOACID; normally applied to the salts of tribasic phosphoric acid.

phosphatides. *See* PHOSPHOLIPIDS.

phosphatidylcholine (lecithin). *See* PHOSPHOLIPIDS.

phosphazines. *See* PHOSPHONITRILIC DERIVATIVES.

phosphides. Compounds of the elements with phosphorus. Non-metals (e.g., sulphur) form covalent materials; the more electropositive elements form ionic derivatives. Sodium phosphide (Na_3P) (Na + P) is used for sea-flares, as it is rapidly hydrolyzed to a mixture of flammable PHOSPHORUS HYDRIDES.

phosphine. *See* PHOSPHORUS HYDRIDES.

phosphites (($RO)_3P$). Organic derivatives of phosphorous (phosphonic) acid, prepared by the action of phosphorus trichloride on an alcohol or phenol. They are useful intermediates in organic syntheses and form complexes with transition metals. Phosphites are used as ANTIOXIDANTS in lubricating oils.

phosphocreatine. *See* CREATINE.

phospholipids. Group of fat-like substances, containing a phosphate group and one or more fatty acid residues, which are essential components of cell membranes. Hydrolysis yields fatty acids, phosphoric acid and a base. They are amphoteric with a polar and a non-polar region.
lecithins, cephalins and related compounds. Based on a glycerol backbone, with a phosphate group. In lecithin (phosphatidylcholine), R' and R" are fatty acid residues, usually one is saturated and the other unsaturated. In the cephalins, ethanolamine ($H_2NCH_2CH_2OH$) or serine (*see* AMINO ACIDS) replaces choline (*see* VITAMINS). Commercial lecithin is a mixture of phosphatides and glycerides obtained in the manufacture of soyabean oil. They are widely used in the food industry (E322) as surfactants, emulsifiers and antioxidants.

$$CH_2OCOR'$$
$$CHOCOR''$$
$$CH_2OPOCH_2CH_2\overset{+}{N}(CH_3)_3$$

lecithins

sphingomyelins. R is a fatty acid residue (usually TETRACOSANOIC ACID). They occur abundantly in brain tissues in association with CEREBROSIDES, which they resemble in many of their properties. On hydrolysis they split into choline, SPHINGOSINE, phosphoric acid and a fatty acid.

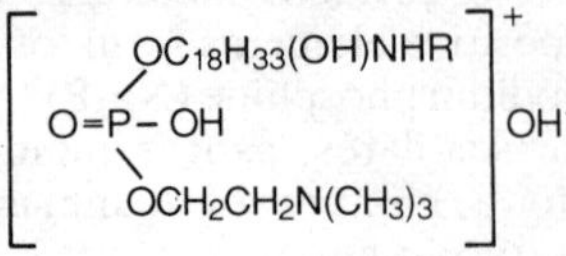

$$\left[\begin{array}{c} OC_{18}H_{33}(OH)NHR \\ / \\ O=P-OH \\ \backslash \\ OCH_2CH_2N(CH_3)_3 \end{array} \right]^{+} \quad OH^{-}$$

sphingomyelins

phosphomolybdates. *See* HETEROPOLY ACIDS AND ANIONS.

phosphonitrilic derivatives (phosphazines, $(PNR_2)_n$). Cyclic or chain compounds containing alternating phosphorus and nitrogen atoms with two substituents on the phosphorus atoms. The simplest, hexachlorocyclotriphosphazine, formed by heating phosphorus pentachloride with ammonium chloride in chlorobenzene, is a key intermediate in the synthesis of other phosphazines. Replacement of Cl by OH, OR, NR_2, NHR or R gives partially or fully substituted derivatives. When the chlorides are heated to 300°C, they are converted to rubber-like materials which are potentially useful as far as their physical and mechanical properties are concerned, but they have been generally useless because of their chemical (hydrolytic) instability.

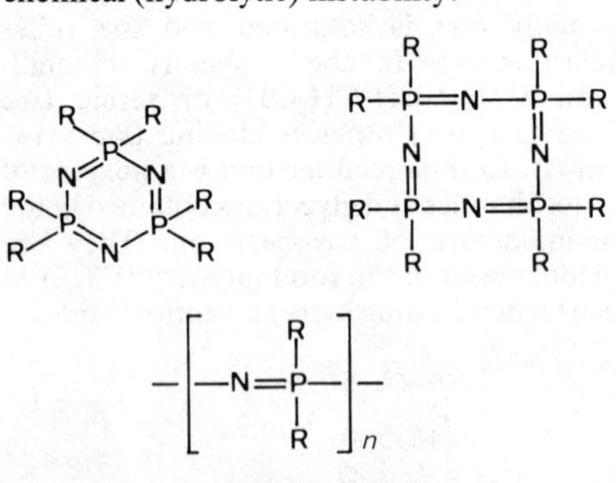

phosphonitrile derivatives

phosphonium salts. White crystalline salts analogous to ammonium halides such as phosphonium iodide (PH_4I) (dry HI +

PH_3). They readily dissociate into phosphine (PH_3) and the hydrogen halide (the chloride is not stable at normal temperatures and pressures). PH_3 dissolves in very strong acids (e.g., $BF_3.H_2O$) where it is protonated to PH_4^{+}. Organophosphonium salts (e.g., $(Ph_3PH)^{+}(BF_3)^{-}$) are more stable than unsubstituted salts. Triphenylalkylphosphonium salts are used in the preparation of phosphoranes (*see* YLIDES).

phosphoproteins. *See* PROTEINS.

phosphor. Substance that exhibits LUMINESCENCE, especially PHOSPHORESCENCE. Excitation of the phosphor may be by electromagnetic radiation (e.g., $Ca_5(F,Cl)(PO_4)_3(Sb,Mn)$ is used in fluorescent lights), electrons (e.g., $YVO_4.Eu$ is a red phosphor in television tubes) or energetic particles (e.g., ZnS in SCINTILLATION COUNTERS for alpha-particles). Organic phosphors are used in optical brighteners.

phosphoranes. Derivatives of five-covalent phosphorus (e.g., Ph_5P) and the Wittig intermediate (*see* PHOSPHORUS ORGANIC DERIVATIVES).

phosphor bronze. *See* BRONZE.

phosphorescence. Relaxation of a molecule in an excited triplet state to a singlet ground state by the emission of a photon. Population of the triplet state is by forbidden intersystem crossing from an excited singlet state. As the relaxation process itself is also forbidden, the triplet state is long-lived and so may be quenched by collisions with solvent molecules in solutions or other species (e.g., oxygen). A relationship exists between FLUORESCENCE (*see* Figure), ABSORBANCE and phosphorescence.

phosphorous acid. *See* PHOSPHORUS OXOACIDS.

phosphorus (P). Non-metal of Group V (*see* GROUP V ELEMENTS). It occurs naturally (0.11 percent of the lithosphere) as apatite and fluorapatite. The element is obtained by the reduction of phosphate rock with coke and silica in an electric furnace; it is distilled over and condensed under water as white phosphorus. There are many allotropes, but three main forms

occur: white phosphorus, the most reactive, contains tetrahedral (P_4) molecules; red phosphorus, a polymeric material made by heating the white form at 400°C for several hours, has a highly complex structure; black phosphorus, made by heating the white form under high pressure in the presence of mercury as a catalyst, is a flaky crystalline solid like graphite. There are considerable differences in the chemical reactivity of the allotropes: white phosphorus inflames in air (stored under water) and is soluble in carbon disulphide and benzene; the red and especially the black allotropes are stable in air. White phosphorus combines with oxygen (P_4O_6, P_4O_{10}), sulphur (range of sulphides), halogens (PX_3, PX_5) and metals (PHOSPHIDES), and reacts with sodium hydroxide to give phosphine (see PHOSPHORUS HYDRIDES) and hypophosphite. Differences between nitrogen and phosphorus in their chemistries are due to: (1) no known $p\pi$–$p\pi$ bonds in phosphorus (e.g., existence of $P(OR)_3$, but not $N(OR)_3$); (2) weak to moderate $d\pi$–$p\pi$ bonding in phosphorus; (3) the possibility of valence expansion (e.g., existence of Ph_5P, $[P(OR)_6]^-$). Stable oxidation states are +5 (e.g., PF_5, six-coordinate complexes $[PF_6]^-$, four-coordinate tetrahedral ions, PO_4^{3-}) and +3 (mainly three-coordinate pyramidal, e.g., PH_3). There is a significant tendency to catenation forming a series of cyclic compounds $((RP)_n)$. Phosphorus compounds are important as fertilizers, but are also widely used in matches, pesticides, alloys (e.g., phosphor bronze), detergents, electrical components (GaP), foods and drinks. Phosphates are an essential constituent of living organisms (e.g., NUCLEIC ACIDS and in bones).
A_r = 30.97376, Z = 15, ES [Ne]$3s^2 3p^3$, mp (white) 44.1°C, bp (white) 280°C, ρ = 1.82 × 10^3 kg m^{-3}. Isotope: ^{31}P, 100 percent.

phosphorus halides
phosphorus trihalides (PX_3). Prepared by direct halogenation with phosphorus in excess. All are volatile, the gaseous molecules having a pyramidal structure. They are all good Lewis bases (with lone pair of electrons), forming stable complexes (e.g., $Ni(PCl_3)_4$). Phosphorus trichloride is the most common trihalide. (For reactions of phosphorus trichloride — see Figure.)
phosphorus pentahalides (PX_5, where X = F, Cl or Br). Produced by addition of another mole of the halogen to PX_3. Thermal stability decreases as follows: F > Cl > Br. Phosphorus pentachloride has a trigonal bipyramidal structure in the gaseous state, but the solid contains the tetrahedral $[PCl_4]^+$ and octahedral $[PCl_6]^-$ ions forming salts with other chlorine acceptors (e.g., $[PCl_4]^+$ $[NbCl_6]^-$). Phosphorus pentabromide exists as $[PBr_4]^+$ and Br^-. Phosphorus pentachloride is hydrolyzed by water to phosphorus oxochloride and finally orthophosphoric acid; it reacts with ammonium chloride to form PHOSPHONITRILIC DERIVATIVES, with hydrogen sulphide at red heat to phosphorus pentasulphide and with sulphur dioxide to phosphorus oxochloride.
phosphorus oxohalides (POX_3). Formed from the partial hydrolysis of PX_3 or by the direct action of oxygen on PX_3. Hydrolysis gives orthophosphoric acid; the reactions are very similar to those of phosphorus trichloride.

phosphorus hydrides
phosphine (PH_3). Colourless, poisonous gas (bp –87.7°C) formed by the action of water on calcium phosphide or the reaction of yellow phosphorus with sodium hydroxide. The molecule is pyramidal (HP̂H = 93.7°). It is a weaker base than ammonia, but is a stronger reducing agent. When pure it is not spontaneously flammable; the presence of diphosphine causes it to burst into flame. It is only sparingly soluble in water, but reacts with some acids to give PHOSPHONIUM SALTS. Alkyl and aryl phosphines (similar in structure to AMINES) are more stable than phosphine, but are highly flammable.
diphosphine (P_2H_4). Minor product formed during the reaction of phosphides with water, separated from phosphine by freezing out to a yellow liquid. It is spontaneously flammable and decomposes on storage to polymeric solids. Unlike hydrazine (see NITROGEN HYDRIDES) it has no basic properties, but is readily oxidized and is a strong reducing agent.

phosphorus organic derivatives. Names and structures of representative organophosphorus compounds are as follows: trimethylphosphine oxide $((CH_3)_3P=O)$, methylenetriphenylphosphorane (triphenylphosphoniummethylide) $(Ph_3P=CH_2 \leftrightharpoons Ph_3P^+{}^-CH_2)$, pentaphenylphosphorane (Ph_5P).

The nomenclature of the organophosphorus acids is based on the parent acid (some of the parent compounds are unstable and cannot be isolated).

phosphites $((RO)_3P)$. Derivatives of phosphorous acid $(PCl_3 + ROH \text{ or } ArOH)$.

They are used as intermediates in organic syntheses and form complexes with transition metals, used as antioxidants in lubricating oils.

phosphonites $(RP(OR')_2)$. Derivatives of phosphonous acid $(HP(OH)_2)$.

phosphinites (R_2POR'). Derivatives of phosphinous acid $(H_2P(OH))$.

phosphonates $(RP(O)(OR')_2)$. Derivatives of phosphonic acid $(HP(O)(OH)_2)$.

phosphinates $(R_2P(O)OR')$. Derivatives of phosphinic acid $(H_2P(O)(OH))$.

phosphates $((RO)_3P=O)$. Derivatives of phosphoric acid $((OH)_3P=O)$.

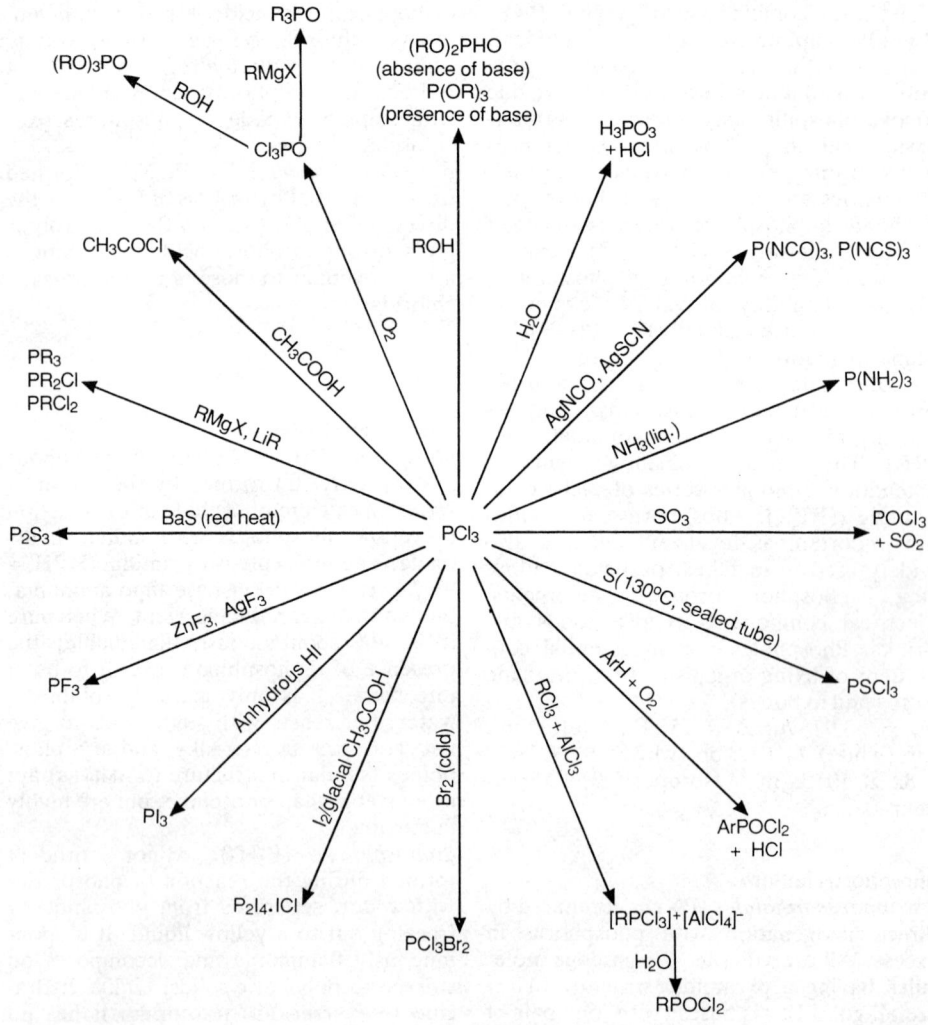

phosphorus halides
Some important reactions of phosphorus trichloride (many are typical for other trihalides).

Trivalent phosphorus compounds have a high nucleophilic reactivity (e.g., the reaction of R_3P with alkyl halide forming the quaternary salt $R_4P^+X^-$). The reaction of an alkyl halide with a trialkyl phosphite gives a phosphonate ester (Arbusov reaction). Halides, except fluoride, are readily displaced from trivalent phosphorus by a wide variety of nucleophiles (e.g., the reaction of R_2PCl with $R'MgX$, $R'O^-$ and $R'COO^-$ gives, respectively, R_2PR', R_2POR' and R_2POCOR'). Phosphonium salts can undergo a variety of reactions:

$$\overset{+}{R_3P}-CH_2CH_2R'$$

S_N2 on carbon $(X^-) \rightarrow$

$$R_3P + XCH_2CH_2R'$$

E2 elimination (B:) $\rightarrow$

$$R_3P + CH_2{=}CHR + BH$$

substitution on P $(OH^-) \rightarrow$

$$R_2P(O)CH_2CH_2R' + RH$$

YLIDE formation (B:) $\rightarrow$

$$R_3P^+\ {}^-CHCH_2R' + BH$$

Trivalent phosphorus compounds react with halogens to give adducts (R_3PX_2). In solution such compounds are largely in the ionized form, reacting with alcohols to give a phosphine oxide. Organic phosphate derivatives play important roles in biological systems, ranging from mobile energy sources (ATP) to structural functions in hereditary material (deoxyribonucleic acid).

phosphorus oxides. Phosphorus forms a series of oxides.

phosphorus(v) oxide (phosphorus pentoxide, P_4O_{10}). Main product of the combustion of phosphorus in excess oxygen. It is a white crystalline compound, which sublimes at 360°C. It is an acidic oxide with great affinity for water, and is used as a drying agent and as a dehydrating agent (e.g., $HNO_3 \rightarrow N_2O_5$, $H_2SO_4 \rightarrow SO_3$, organic amides $\rightarrow$ nitriles) being converted to PHOSPHORUS OXOACIDS. Fusion with basic oxides gives solid phosphates. The molecule (in solid and gaseous phase) is a P_4

tetrahedron with an oxygen atom on each edge and on each phosphorus atom.

phosphorus(III) oxide (phosphorus trioxide, P_4O_6). Colourless, volatile compound, formed when phosphorus is burnt in a deficiency of oxygen. It burns in excess oxygen to P_4O_{10}, reacts with chlorine to phosphorus oxochloride and dissolves in water to give a mixture of oxoacids. Its structure is similar to that of P_4O_{10}, but the four non-bridging oxygen atoms are missing.

phosphorus oxoacids. All oxoacids have POH groups in which the hydrogen is ionizable; hydrogen in phosphorus–hydrogen groups is not ionized. Because of the two oxidation states and the possibility of sharing the oxygen atoms, there are a vast number of oxoacids or ions. Some are of great technical importance, but with the exception of the simpler species their structures are not fully understood.

hypophosphorous acid (H_3PO_2, $H_2P(O)OH$). Monobasic acid. Salts are prepared by boiling white phosphorus with alkali. The acid, a white crystalline solid ($pK_a = 1.2$), and its salts are strong reducing agents, being oxidized to orthophosphoric acid.

phosphorous acid (H_2PO_3, $HP(O)(OH)_2$). Dibasic acid (PCl_3 or $P_4O_6 + H_2O$). It is a deliquescent solid with strong reducing activity.

hypophosphoric acid ($H_4P_2O_6$, $(HO)PH(O)OP(O)(OH)_2$). Sodium salt is made by the action of alkaline sodium chlorate(I) on red phosphorus. Commonest salts have only two hydrogen atoms replaced. Acids and salts are stable towards oxidizing and reducing agents.

orthophosphoric acid (H_3PO_4, $OP(OH)_3$). Prepared, usually as 85 percent syrupy acid, by direct action of sulphuric acid on phosphate rock or by hydration of phosphorus pentoxide. The pure acid is a colourless, crystalline solid, very stable with no oxidizing properties below 350°C. Tribasic acid gives rise to tetrahedral anions.

pyrophosphoric acid ($H_4P_2O_7$, $(HO)_2(O)POP(O)(OH)_2$). Colourless solid crystallized from 80 percent phosphorus pentoxide solution. All hydrogen atoms are replaceable.

metaphosphoric acids (empirical formula: HPO_3). Obtained by further dehydration of orthophosphoric acid by heating at 316°C. Their nature is in doubt; the existence of a monomer is unlikely. Many species are known: a cyclic $[P_3O_9]^{3-}$ ion and various chain phosphates. The so-called sodium hexametaphosphate (made by rapidly cooling molten metaphosphate) is a long-chain compound, soluble in water. It acts as a water softener removing calcium ions by chelation to the colloidal polyanion. *polyphosphoric acids.* Polyphosphate is usually applied to salts of the chain-like ions $[P_nO_{3n+1}]^{-n-2}$ (e.g., $[P_3O_{10}]^{5-}$).

phosphorus sulphides. Phosphorus and sulphur combine directly above 100°C giving a number of covalent sulphides (e.g., P_4S_3, P_4S_5, P_4S_7, P_4S_{10}). The structures are based on a P_4 tetrahedron with some bridging sulphur atoms; P_4S_{10} has a similar structure to P_4O_{10}. Phosphorus trisulphide, used in 'strike anywhere matches' is attacked only slowly by cold water; hot water gives phosphine, hydrogen sulphide and phosphorus oxoacids; reaction with potassium hydroxide produces phosphine, hypophosphite and sulphide ions. P_4S_{10} reacts with alcohols giving dialkyl and diaryl dithiophosphates $(RO)_2P(S)SH$, which form the basis of extreme pressure lubricants, of oil additives and of flotation agents.

phosphorylation. Esterification of organic compounds with phosphoric acid. Biological phosphorylations are catalyzed by phosphorylases, ATP often being the source of the phosphate group and the necessary energy.

phosphoryl chloride (phosphorus oxochloride). *See* PHOSPHORUS HALIDES.

photoacoustic spectroscopy (optoacoustic spectroscopy). Technique that measures absorption of ELECTROMAGNETIC RADIATION directly by the detection of sound generated in the material. It is a very sensitive method allowing measurement of values of ABSORBANCE of 10^{-10}. The incident beam, which may be a tunable laser, is modulated at acoustic frequencies; a sensitive microphone or piezoelectric transducer detects the emission. The absorbed radiation eventually appears as heat which, in gaseous samples, causes a periodic fluctuation in the temperature of the gas which is manifested as sound waves. Nitrogen monoxide has been detected at a concentration of about 10^7 molecules cm^{-3} (1 part per trillion).

photochemistry. Study of reactions initiated by light. The absorption of a PHOTON of sufficient energy to excite an electronic transition may lead to the dissociation of the molecule (photolysis). The necessity of absorption for reaction to occur is stated in the GROTTHUS–DRAPER LAW. The primary photochemical process requires only a single photon to be absorbed (Einstein–Stark law). Subsequent reactions, especially if CHAIN REACTIONS are involved, may lead to a QUANTUM EFFICIENCY of greater than unity. If continuous absorption of light is needed to sustain the reaction, a photostationary state is attained which is limited by the flux of photons. The transfer of energy from a photoexcited atom or molecule, which itself does not react to another molecule is known as photosensitization. For example, excited mercury atoms cause the dissociation of hydrogen, and in many biological processes the initial molecule which absorbs a photon (*see* CHLOROPHYLL) passes the energy on through a chain of reactions (*see* PHOTOSYNTHESIS).

photodimerization. Dimerization (*see* DIMER) of two molecules containing multiple bonds under the influence of light. Thermal dimerization of olefins in a concerted process is forbidden, but the photoreaction is allowed (*see* WOODWARD–HOFFMANN RULE). An example is the dimerization of the *trans*-isomer of 3-PHENYLPROPENOIC ACID.

photoelectron spectroscopy. *See* ELECTRON SPECTROSCOPY.

photographic developer. Agent that specifically reduces silver halide grains in an exposed photographic film. *See also* AMINOPHENOLS.

photoionization detector (PID). *See* DETECTION SYSTEMS.

photoluminescence. General term to describe LUMINESCENCE caused by the absorption of light. This may be FLUORES-CENCE or PHOSPHORESCENCE.

photolysis. Dissociation of a molecule by the absorption of light. *See also* PHOTO-CHEMISTRY.

photon. Quantum of radiation of energy $h\nu$. In corpuscular terms, the photon is the particle of radiation.

photophosphorylation. *See* PHOTOSYN-THESIS.

photosensitization. *See* PHOTOCHEMISTRY.

photosynthesis. Process by which plants convert atmospheric carbon dioxide to carbohydrates using light as the energy source. The energy absorbed by CHLORO-PHYLL is used to drive two separate processes:
1) reduction of $NADP^+$ to NADPH

$$NADP^+ + 2e + H^+ \rightarrow NADPH$$

2) oxidation of water to oxygen

$$H_2O \rightarrow 2H^+ + 1/2\,O_2 + 2e$$

These reactions occur at different locations within an ELECTRON TRANSPORT CHAIN; and in the process of transferring electrons, ADP is converted to ATP (*see* ADENOSINE TRIPHOSPHATE) (with the resultant storage of energy), a process called photophosphorylation. The end result of the photochemical part of photosynthesis is the formation of NADPH, ATP and oxygen, the oxygen being released to the atmosphere. The ATP and NADPH are utilized in a series of dark reactions in which atmospheric carbon dioxide is reduced to the level of a carbohydrate, for example

$$6CO_2 + 12NADPH + 12H^+ \rightarrow$$

$$C_6H_{12}O_6 + 12NADP^+ + 6H_2O$$

The oxygen formed comes from the water and not from the carbon dioxide.

phthalic acid. *See* BENZENEDICARBOXYLIC ACIDS.

phthalic anhydride ($C_6H_4(CO)_2O$). ANHYDRIDE of BENZENE-1,2-DICARBOXYLIC ACID (mp 28°C); prepared by the oxidation of naphthalene (H_2SO_4, HgSO$_4$ at 300°C or

O_2, V_2O_5 at 400°C) or *o*-XYLENE (KMnO$_4$ then heat). It condenses with benzene (AlCl$_3$) to anthraquinone, and with phenols to give dyes. It is used to manufacture plasticizers and resins.

phthalocyanins. Tetraazatetrabenzo derivatives of PORPHYRINS; class of insoluble organic colouring materials (blue to green in colour) used as PIGMENTS, but with limited use as dyestuffs. Copper phthalocyanin (Monastral blue 5025), a typical example, is obtained by the exothermic condensation of four molecules of 1,2-benzenedicarbonitrile (1,2-phthalonitrile) in the presence of copper metal at 200°C (*see* Figure). A cheaper method is to heat phthalic anhydride with urea in the presence of copper(I) chloride and a catalyst such as vanadium or aluminium oxide. The molecule is stable to heat at 500°C. Other phthalocyanins are obtained from the copper compound by replacement of the metal. The introduction of an increasing number of chlorine atoms in the benzene rings gives rise to greener compounds. As pigments they must be suitably dispersed in a binder. The water-soluble sulphonate derivatives are used as dyestuffs.

phthalocyanin
Synthesis of copper phthalocyanin.

phylloquinone (vitamin B complex). *See* VITAMINS.

physical adsorption. *See* CHEMICAL AD-SORPTION.

physical change. One that does not involve a chemical reaction, that is the breaking or making of bonds (e.g., the expansion of a gas, evaporation of a liquid).

physiological saline. *See* RINGER'S SOLUTION.

physisorption. *See* CHEMICAL ADSORPTION.

phytol. *See* TERPENES.

p*I*. *See* ISOELECTRIC POINT.

pi-bonding. π-Bonding (*see* MOLECULAR ORBITAL).

picolines. *See* METHYLPYRIDINES.

pi-complex. *See* π-COMPLEXES.

picric acid (2,4,6-trinitrophenol). *See* NITROPHENOLS.

PID (photoionization detector). *See* DETECTION SYSTEMS.

pi-donor complex. π-Donor complex (*see* CARBON-TO-METAL BOND).

pigments. Coloured particles which are dispersed in a liquid (paint), or mixed with a solid (e.g., a polymer). Common inorganic pigments are titanium dioxide (white), iron oxides (red, brown), chromates (yellow, red), cyanoferrates (blue) and carbon (black). Insoluble DYES are also used as pigments.

pimelic acid. *See* HEPTANEDIOIC ACID.

pinacol. *See* 2,3-DIMETHYL-2,3-BUTANEDIOL.

pinacolone. *See* 3,3-DIMETHYL-2-BUTANONE.

pinacol–pinacolone rearrangement. Conversion of a 1,2-DIOL to a KETONE in acid conditions by a 1,2-shift of an alkyl or aryl group. Water is eliminated and the mechanism is thought to be via a CARBOCATION.

$$R_2C-CR_2 \underset{H^+}{\rightleftharpoons} R_2C-CR_2 \underset{-H_2O}{\rightleftharpoons} R_2C-CR_2 \rightleftharpoons$$
$$\;\;| \;\;\; | \qquad\qquad | \;\;\; | \qquad\qquad |$$
$$HO\;\; OH \qquad\quad HO\;\; \overset{+}{O}H_2 \qquad\quad HO$$

$$\overset{+}{R}C-CR_3 \underset{-H^+}{\rightleftharpoons} RCOCR_3$$
$$\;\;|$$
$$HO$$

pinacols. 1,2-GLYCOLS (RR'COH–HOCR″R‴, where R–R‴ are alkyl or aryl groups; prepared by the reduction of 1,2-diketones. Pinacol is 2,3-DIMETHYL-2,3-BUTANEDIOL. *See also* PINACOL–PINACOLONE REARRANGEMENT.

α-pinene. *See* TERPENES.

pion. *See* ELEMENTARY PARTICLES.

pi-orbital. *See* MOLECULAR ORBITAL.

piperazine (hexahydropyrazone). Solid (mp 104°C (anhyd.)); prepared by the action of alcoholic ammonia on 1,2-dichloroethane. It is used in human and veterinary medicine as a deworming agent. It reacts with 1,2-dichloroethane to give 1,4-diazobicyclo[2.2.2]octane (triethylenediamine), a tricyclic compound with a cage structure, which is used as a catalyst in the production of polyurethanes. Substituted piperazines can be prepared from *N*-substituted bis-2-chloroethylamines and ammonia or primary amines.

piperidine (hexahydropyridine, $C_5H_{11}N$). HETEROCYCLIC COMPOUND (structure XV); colourless liquid (bp 106°C) with ammonialike smell, made by the electrolytic hydrogenation of PYRIDINE. It occurs in pepper as piperine (*see* ALKALOIDS) from which it can be extracted by treatment with alkali. It is a strong base giving the reactions of a secondary amine. Substituted piperidines are made by the reduction of the corresponding pyridine.

piperine. *See* ALKALOIDS.

pitch. Residual material after distillation of TAR.

pitchblende. *See* POLONIUM, RADIUM, URANIUM.

PIXE (proton-induced X-ray emission). *See* ACTIVATION ANALYSIS.

pK_a. *See* ACID DISSOCIATION CONSTANT.

pK_b. *See* BASE DISSOCIATION CONSTANT.

plait point. *See* THREE-COMPONENT SYSTEM.

Planck constant ($h = 6.626 \times 10^{-34}$ Js). Constant of proportionality between frequency v and the QUANTUM of energy E ($E = hv$). It appears in many equations of QUANTUM MECHANICS as $h/2\pi$ when it is given the symbol $\hbar$ ('crossed h') and has the value 1.055×10^{-34} Js. *See also* PARTICLE IN A BOX, SCHRÖDINGER EQUATION.

Planck function (Y; units: JK^{-1}). Possible thermodynamic function, superseded by the GIBBS FUNCTION.

$$Y = -G/T = S - H/T$$

plane of symmetry. *See* SYMMETRY ELEMENTS.

plasma. 'Gas' of ions, atoms and electrons formed in an electric discharge or in extremely hot nuclear processes (*see* NUCLEAR FUSION). *See also* INDUCTIVELY COUPLED PLASMA.

plaster of Paris. Semihydrated form of calcium sulphate ($CaSO_4.1/2H_2O$) formed by heating gypsum ($CaSO_4.2H_2O$). When water is added the material expands and sets, making it useful in moulds.

plastic flow. *See* VISCOSITY.

plasticizer. Additive to hard PLASTICS which makes them softer and workable. For example TRICRESYL PHOSPHATE added to PVC (*see* POLYMERS) makes it soft and rubber-like.

plastics. Synthetic POLYMERS of high molecular mass that may be formed or moulded at high temperatures. Thermoplastics are linear polymers softened by heat and which harden on cooling (e.g., PVC, Perspex). They are soluble in organic solvents. Thermosetting plastics (resins) are three-dimensional polymers formed by an irreversible heat treatment (e.g., urea–methanal resin).

plate column. Vertical cylindrical column, used for liquid–vapour contacting, containing a series of plates or trays spaced 1 m apart throughout the column. Liquid flows down the column from plate to plate while the vapour passes upwards through the holes in the plates. It is used in gas absorption and fractional distillation. The degree of absorption or separation depends on the number of plates in the column.

plate efficiency. In a distillation or absorption column, if a plate was 100 percent efficient the liquid and vapour leaving the plate would be in equilibrium. In practice, the efficiency is considerably less, thus the number of plates required to achieve a given degree of separation or absorption is greater than the theoretical number.

platinum (Pt). Third row TRANSITION ELEMENT. It is extracted from native mixtures of PLATINUM GROUP METALS as the soluble chloroplatinic acid or the hexanitro complex. The metal is then recovered by thermal decomposition or reduction in hydrogen. It is used in jewellery, laboratory ware, thermocouples and as a catalyst (e.g., for oxidation of pollutants from motor car exhausts). High-surface-area platinum black is precipitated from platinum(II) complexes by reduction or electrolysis. In this form, gases (especially hydrogen, oxygen and carbon monoxide) are adsorbed readily. Platinum is inert at room temperature to acids and alkalis, but is dissolved by AQUA REGIA. The metal remains bright in oxygen, even though a thin film of PtO_2 is formed.
$A_r = 195.08$, $Z = 78$, ES [Xe] $5d^9 6s^1$, mp 1772°C, bp 3827°C, $\rho = 21.45 \times 10^3$ kg m^{-3}. Isotopes: ^{194}Pt, 32.9 percent; ^{195}Pt, 33.8 percent; ^{196}Pt, 25.2 percent; ^{198}Pt, 7.2 percent.

platinum compounds. Oxidation states +1, +2, +4, +5 and +6 are found with platinum(II) (platinous) and platinum(IV) (platinic) being the most common. Complexes of platinum are square planar (e.g., $PtCl_2(NH_3)_2$) or octahedral (e.g., $[PtCl_6]^{2-}$). A series of oxides (PtO, PtO_2 and possibly PtO_3) is formed on anodizing the metal. The most important compound is chloroplatinic acid (H_2PtCl_6) produced by the dissolution of the metal in AQUA REGIA and evaporation from hydrochloric acid. Organic derivatives with alkyl and aryl ligands are synthesized from GRIGNARD REAGENTS and platinum halides. Some complexes of platinum have been shown to be effective anticancer drugs.

platinum group metals. TRANSITION ELEMENTS ruthenium, osmium, rhodium, iridium, palladium and platinum. They occur together as native alloys (*see* OSMIRIDIUM) and in sulphide ores of the group Ib metals. They are characterized by electronegativity and high catalytic activity.

platinum resistance thermometer. *See* THERMOMETER.

pleochromic crystal. Crystal that transmits different colours when viewed from different directions. Potassium cyanoplatinate ($K_2Pt(CN)_4$) is pleochromic.

Plexiglass. Commercial name for a clear poly(methyl 2-methylpropenoate) plastic.

plutonium (Pu). Radioactive, one of the ACTINIDE ELEMENTS. Its principal isotope is plutonium-239 formed by the capture of a neutron by uranium-238 and subsequent decay of uranium-239 (β^- 23.5 min) and neptunium-239 (β^- 2.33 days). Uses of plutonium are as a nuclear fuel, in the production of radioisotopes and as fissile material for nuclear weapons. Plutonium-238 is also used as a heat source in spacecraft. Plutonium metal is obtained by the reduction of PuF_4 with calcium. Plutonium is a silvery, dense, toxic metal which is very reactive. Oxidation states +3 to +7 are known. Plutonium(IV) is most stable (e.g., PuO_2), but in aqueous solutions complexes of states (III) to (VI) coexist.
$A_r = 239.13$, $Z = 94$, ES [Rn]$5f^67s^2$, mp 641°C, bp 3232°C, $\rho = 19.82 \times 10^3$ kg m^{-3}.

Pm. Promethium (*see* LANTHANIDE ELEMENTS).

PMO. *See* PERTURBATION MOLECULAR ORBITAL THEORY.

PMR (proton magnetic resonance). *See* NUCLEAR MAGNETIC RESONANCE SPECTROSCOPY.

Po. *See* POLONIUM.

Pockel cell. *See* OPTICAL ROTATORY DISPERSION.

point defeat. *See* DEFECT STRUCTURE.

point group. Classification of the symmetry of a molecule. Molecules with only C_n or C_n plus σ_v or σ_h are classed as C_n, C_{nv} or C_{nh} (*see* SYMMETRY ELEMENTS). If there are nC_2 axes perpendicular to the principal axis the molecules belong to D_n, D_{nv} or D_{nh} and if there is a σ_d in addition the molecule is D_{nd}. Molecules with S_n axes plus mirror planes can belong to S_n, S_{nv} and S_{nh}. More complex groups found in chemistry are the tetrahedral groups T, T_d and T_i (with a centre of inversion), the octahedral groups O and O_h and the icosahedral groups I and I_h. *See also* CRYSTAL SYSTEMS.

poise (P). CGS UNIT of VISCOSITY, not coherent with SI units; P = dyne cm^{-2} s = 10^{-1} N m^{-2} s.

Poiseuille's equation. Equation for the determination of the VISCOSITY of a fluid

$$\eta = \pi r^4 Pt/8Vl$$

where t is the time required for a volume V to flow through a tube of radius r and length l under a pressure P.

poison. (1) Any substance that is harmful to living tissues when applied in relatively small doses. The important factors are quantity or concentration, duration of exposure, particle size or physical state of the substance, its affinity for and solubility in tissue fluids, and the sensitivity of the tissues or organ. There is no sharp demarcation between poisons and non-poisons. (2) In nuclear technology, any material with high capture probability for neutrons which may divert neutrons from fission chain reactions. (3) Any substance that reduces or destroys the activity of a catalyst.

polar. Description of a molecule or radical that has or is capable of developing an electrical charge: for example, water, ethanol and sulphuric acid are polar, whereas most hydrocarbons are non-polar.

polar bond. Bond formed from two atoms of dissimilar electronegativity in which the electrons are held closer to the more electronegative atom. This creates an uneven charge distribution. *See also* COVALENT BOND, IONIC BOND.

polarizability (α; units: $J^{-1}C^2m^2$). When a molecule (polar or non-polar) is exposed to an electric field, the electronic distribution and nuclear configuration become distorted resulting in an induced dipole moment in the molecules. Its magnitude is related to the applied field strength (E)

$$P_{induced} = \alpha E$$

Usually polarizabilities are expressed as a volume by dividing by $4\pi\varepsilon_0$ (units: $J^{-1}C^2$ m^{-1}) usually referred to as polarizability volumes α' ($\alpha 4\pi\varepsilon_0$). Typical values of α' are 0.20×10^{-24} cm^3 for helium, 10.5×10^{-24} cm^3 for tetrachloromethane.

polarization. (1) Molecules of polar liquids exhibit two effects when subjected to an electric field: (a) an electronic polarization, due to the displacement of electrons within the molecule towards the positive pole of the field and (b) orientation polarization due to the alignment of the permanent dipoles in the electric field. In addition, nuclei with a positive charge are slightly displaced relative to one another; this is atomic polarization and is of very small magnitude. The total polarization is the sum of orientation and electronic polarization. The molar polarizability, given by

$$P_m = (M_r/\rho)[(\varepsilon_r - 1)/(\varepsilon_r + 2)]$$
$$= (N_A/3\varepsilon_0)(\alpha + p^2/3kT)$$

where p is the dipole moment, is valid for dilute gases and approximately so for slightly polar liquids, but not for liquids of high permittivity. The intercept of the graph of P_m against $1/T$ gives the POLARIZABILITY (α). *See also* DEBYE EQUATION. (2) In electrochemistry, deviation of an electrode potential from equilibrium. A polarizable electrode is therefore one which follows any applied potential. An ideal non-polarizable electrode can allow large currents to pass in either direction without changing its potential from equilibrium. *See also* DEPOLARIZATION.

polarized radiation. Plane-polarized radiation (usually light) has the electric vector oriented along one axis only (*see* ELECTROMAGNETIC RADIATION). Passage through optically active substances (*see* OPTICAL ACTIVITY) rotates the plane of the radiation. The polarized radiation may be considered to be the sum of two vectors of constant magnitude whose orientation is changing at constant angular velocity. The vectors are termed right and left circularly polarized, depending on the rotation of the vector looking towards the source. *See also* OPTICAL ROTATORY DISPERSION.

polar molecule. Compound that is ionic (e.g., sodium chloride) or that has a large permanent DIPOLE MOMENT (e.g., water). Lone pairs of electrons also make molecules strongly polar. In solution and in the fused state, polar molecules ionize and impart electrical conductivity.

polarography. Electrochemical technique in which ions or molecules in a sample solution are subjected to a ramped potential at a DROPPING MERCURY ELECTRODE. A polarogram is a plot of current against the applied potential and is of sigmoidal shape arising from an increasing then saturating current (*see* LIMITING CURRENT DENSITY). The potential at which dI/dV is a maximum (i.e. at which the current is one-half the diffusion-limited current I_d) is known as the half-wave potential ($E_{1/2}$) and is characteristic of the species reacting. The Ilkovic equation relates I_d (A) to the concentration of the reacting species c (mol dm^{-3})

$$I_d = 6.07 \times 10^{-5} D^{1/2} m^{2/3} t^{1/6} c$$

m (mg s^{-1}) is the rate of flow of the mercury drop, t (s) the drop life and D (m^{2}s^{-1}) the diffusion coefficient (*see* FICK'S LAW OF DIFFUSION). Polarography is used in the quantitative analysis of inorganic ions with detection limits of about 10^{-7}–10^{-8} mol dm^{-3} and to detect and identify organic species. *See also* STRIPPING VOLTAMMETRY.

polar solvent. Solvent with a high dipole moment and usually a good solvent for ions.

pollucite. *See* CAESIUM.

polonium (Po). Rare radioactive metallic element of group VI, which occurs in uranium ores (100 μg Mg^{-1}) as a product of the RADIOACTIVE DECAY SERIES. The most accessible isotope, polonium-210 (half-life: 138 days) is obtained by neutron irradiation of bismuth-209

$$^{209}_{83}Bi(n,\gamma)\,^{210}_{83}Bi \longrightarrow\,^{210}_{84}Po + \beta^-$$

The polonium is separated from the bismuth electrochemically. Polonium-210 is virtually a pure alpha-emitter; its high specific activity (4.5 Ci mg^{-1}) makes it a very dangerous material. The metal, obtained by thermal decomposition of the sulphide, is dimorphic; the low-temperature α-form has a simple cubic lattice and the high-temperature β-form a rhombohedral form. The metal reacts with the halogens giving the tetrahalide, with oxygen at 250°C forming the dioxide and with lead and mercury forming polonides containing the Po^{2-} ion. Intimately mixed with beryllium it forms a useful, weak neutron source.
A_r = 209.0, Z = 84, ES [Xe] $5d^{10}\,6s^2\,6p^4$, mp 254°C, bp 962°C, ρ = 5.51 × 10^3 kg m^{-3}. There are 34 isotopes.

polonium compounds. Study of polonium chemistry is difficult owing to the intense alpha-radiation which causes damage to solutions and solids, evolves much heat and necessitates special handling techniques. Its chemistry shows many similarities to that of TELLURIUM; the oxide and hydroxide are more basic than the corresponding TELLURIUM COMPOUNDS. The tetrahalides (Po + halogen) are readily hydolyzed, form complex halides (M$_2$PoX$_6$) and undergo thermal dissociation to PoX$_2$. Polonium is sufficiently metallic to form highly coloured oxoacid salts (e.g., Po$_2$O$_3$.SO$_4$, Po(SO$_4$)$_2$).

polyamides. See NYLON, POLYMERS.

polyamines. Compounds that contain two or more amino groups. At physiological pH these compounds are multivalent cations and are found associated with anionic sites, such as the anionic sites of DNA. A general method of preparation is

$$Br(CH_2)_nBr \xrightarrow{KCN} NC(CH_2)_nCN$$
$$\xrightarrow{Na/EtOH} H_2NCH_2(CH_2)_nCH_2NH_2$$

cadaverine (pentamethylene diamine, H$_2$N(CH$_2$)$_5$NH$_2$). Poisonous, syrupy fuming liquid (bp 178°C) with a disagreeable odour. It is associated with putrescine in decaying tissues, arising from the decarboxylation of lysine. Heating of the hydrochloride gives PIPERIDINE.
hexamethylene diamine (1,6-diaminohexane, H$_2$N(CH$_2$)$_6$NH$_2$). Colourless, crystalline solid (mp 39°C), not occurring naturally; manufactured by the catalytic hydrogenation of adiponitrile. Used in the manufacture of NYLON.
putrescine (tetramethylene diamine, 1,4-diaminobutane, H$_2$N(CH$_2$)$_4$NH$_2$). Poisonous solid (mp 27°C) with disagreeable odour. It is found in dead bodies, formed by the putrefaction of proteins arising from the decarboxylation of ornithine.
spermine (H$_2$N(CH$_2$)$_3$NH(CH$_2$)$_4$NH(CH$_2$)$_3$NH$_2$). Deliquescent crystalline solid isolated from human sperm.
spermidine (H$_2$N(CH$_2$)$_3$NH(CH$_2$)$_4$NH$_2$). Liquid isolated from sperm.

poly-(2-chloro-1,3-butadiene). See NEOPRENE, POLYMERS, RUBBER.

polychloroprene. See NEOPRENE, POLYMERS, RUBBER.

polychromatic. Radiation having a range of wavelengths. Many primary sources are polychromatic, in particular BLACK BODY RADIATION. See also MONOCHROMATIC.

polycyclic hydrocarbons. Polynuclear hydrocarbons in which two or more rings are fused together in the *ortho*-positions, as distinct from those in which the rings are isolated (e.g., BIPHENYL). The higher polycyclic aromatic compounds are conveniently divided into three classes: linear, angular, condensed (*see* Figure). The linear compounds become increasingly reactive as additional rings are added; they also become more intensely coloured. Hexacene (V) is stable in solution reacting instantly with maleic anhydride and with oxygen. Heptacene (VI) is so reactive that it has not been obtained pure; it is stabilized by angular benzo groups, thus dibenzoheptacene (VII) is stable (violet-red) in solution but is readily oxidized by air, whereas tetrabenzoheptacene (VIII) is a stable orange compound unaffected by air. After phenan-

Linear polycyclic hydrocarbons

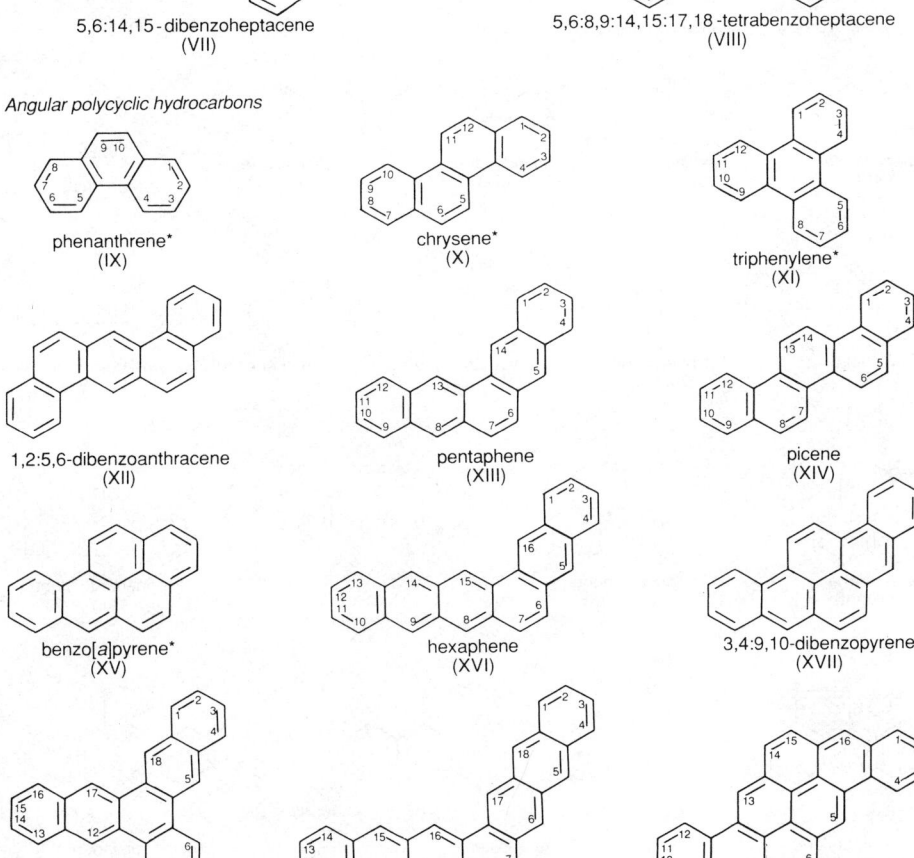

naphthalene*
(I)

anthracene*
(II)

naphthacene
(III)

pentacene
(IV)

hexacene
(V)

heptacene
(VI)

5,6:14,15-dibenzoheptacene
(VII)

5,6:8,9:14,15:17,18-tetrabenzoheptacene
(VIII)

Angular polycyclic hydrocarbons

phenanthrene*
(IX)

chrysene*
(X)

triphenylene*
(XI)

1,2:5,6-dibenzoanthracene
(XII)

pentaphene
(XIII)

picene
(XIV)

benzo[a]pyrene*
(XV)

hexaphene
(XVI)

3,4:9,10-dibenzopyrene
(XVII)

trinaphthylene
(XIX)

heptaphene
(XX)

pyranthrene
(XXII)

polycyclic hydrocarbons
Compounds asterisked (*) have a separate entry.

continued overleaf

Condensed polycyclic hydrocarbons

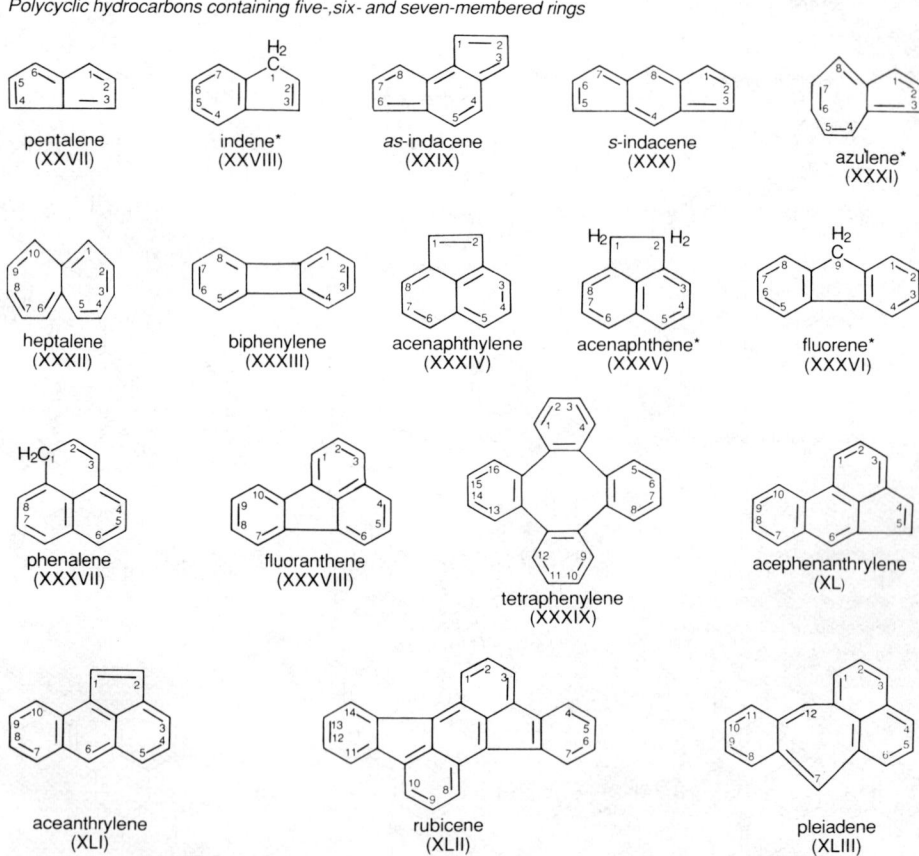

pyrene
(XXIII)

perylene
(XXIV)

(30) coronene
(XXV)

ovalene
(XXVI)

Polycyclic hydrocarbons containing five-,six- and seven-membered rings

pentalene
(XXVII)

indene*
(XXVIII)

as-indacene
(XXIX)

s-indacene
(XXX)

azulene*
(XXXI)

heptalene
(XXXII)

biphenylene
(XXXIII)

acenaphthylene
(XXXIV)

acenaphthene*
(XXXV)

fluorene*
(XXXVI)

phenalene
(XXXVII)

fluoranthene
(XXXVIII)

tetraphenylene
(XXXIX)

acephenanthrylene
(XL)

aceanthrylene
(XLI)

rubicene
(XLII)

pleiadene
(XLIII)

polycyclic hydrocarbons
Compounds asterisked (*) have a separate entry.

threne (IX), the next angular compounds chrysene (X) and dibenzoanthracene (XII) are colourless compounds which do not react with maleic anhydride or form photo-oxides. Carcinogenic compounds in this group include 1,2:5,6-dibenzoanthracene (XII), benzo[a]pyrene (XV) and 3,4:9,10-dibenzopyrene (XVII). Examples of condensed ring hydrocarbons include pyrene (XXIII), perylene (XXIV), coronene (XXV), ovalene (XXVI). (According to IUPAC rules, when a benzene ring is 'ortho-fused' to another ring the prefix 'benzo' is attached to the name of the parent ring. This is contracted to 'benz' when preceding a vowel, as in benzanthracene.)

polydentate ligand. *See* LIGAND.

polydisperse. *See* MONODISPERSE.

polyelectrolyte. Soluble macromolecule containing many ionizable groups. It may be acidic (e.g., polystyrenesulphonic acid), basic (e.g., polyethenylpyridinium bromide) or amphoteric (*see* PROTEINS).

polyene antibiotics. *See* ANTIBIOTICS.

polyenes. Hydrocarbons containing multiple double bonds (e.g., polyene ANTIBIOTICS, CAROTENOIDS). *See also* ALLENES.

polyhalide ions. Almost linear anions formed by the reaction of halide ions (in solution or as solid) with halogens (e.g., I_3^-, IBr_2^-, $BrCl_2^-$). Such ions are unstable, dissociating into monohalide and halogen or interhalogen compound. Polyhalogen cations exist in such compounds as $[BrF_2]^+$ $[SbF_6]^-$ and $[IF_4]^+$ $[SbF_6]^-$. Acids corresponding to the polyhalide ions cannot be prepared, but $HICl_4.4H_2O$ can be crystallized from a solution of iodine trichloride in aqueous hydrochloric acid.

polyhydric alcohols. ALCOHOLS with many hydroxyl groups.

polymerization. Linking of small molecules (MONOMERS) to make larger macromolecules (POLYMERS).
addition polymerization. Polymerization by successive addition of unsaturated compounds by a free radical mechanism. The molecular masses tend to be high

$(10^4–10^6)$ and are determined by the relative rates of branching and termination. Polymerization may be induced thermally, photochemically, electrochemically or by a catalyst. Initiators may be free radical generators (peroxides, azo compounds), Lewis acids (BF_3, $AlCl_3$, *see* FRIEDEL–CRAFTS REACTION), Lewis bases (Na, metal alkyls) or heterogeneous catalysts (Ni on C, SiO_2 on Al_2O_3, CrO_3, ZIEGLER–NATTA CATALYSTS). The use of catalysts allows control of the stereochemistry of the polymerization (*see* TACTICITY). In bulk (or mass) polymerization, the entire monomer is polymerized. If a latex or disperison polymer is required, the monomer is dispersed as an EMULSION in water containing the initiator. Alternatively in suspension (dispersed mass, pearl or granular) polymerization the monomer containing dissolved initiator is dispersed in a supporting medium.
condensation polymerization. Occurs with the elimination of a small molecule such as water or methanol. The formation of the polyamide polymer NYLON is an example. Linear chains are produced if there are only two reactive groups per molecule. The use of a tricarboxylic acid in the Nylon synthesis leads to cross-linking. Methanal resins (melamines) are tetrafunctional (*see* POLYMERS) and form three-dimensional structures. The formation of Nylon is also an example of COPOLYMERIZATION. Condensation polymerization occurs throughout the mixture of monomer and thus the molecular masses are low for small degrees of conversion. Average molecular masses (*see* NUMBER AVERAGE MOLECULAR MASS) are in the region 5000–30000.

polymers. Macromolecules consisting of repeated structural units (MONOMERS). Naturally occurring polymers include PROTEINS, NUCLEIC ACIDS and POLYSACCHARIDES. Manufactured polymers may be divided into thermoplastics, thermosetting resins and ELASTOMERS (synthetic RUBBERS) (*see* Table). Thermoplastics melt on heating and so may be processed by moulding and extrusion techniques. Thermosetting resins are formed irreversibly. Straight chains or sheets of molecules are held together by weak van der Waals forces in one- and two-dimensional polymers. The formation of covalent bonds between

polymers

Polymer (abbreviation)	Type[a]	Structural unit	Preparation	Properties, uses
Polyolefins				
Polyethene (PE)	TP	$-(CH_2)-$	Addition polymerization at high pressure $+ O_2$ or low pressure with Ziegler–Natta catalyst	Tough, moderate strength. Extruded and moulded for variety of goods and textiles
Polypropene (PP)	TP	$-(CH_2CHMe)-$	Addition at 50–120°C with Ziegler–Natta catalyst	Harder and stronger than PE. Isotactic. Hard-wearing carpets, etc.
Polymethylpentane	TP	$-[CH_2CH(CHMe_2)]-$	Addition over Ziegler–Natta catalyst	Hard, transparent and chemically resistant. Pipes films and laboratory ware
Polystyrene (PS)	TP	$-(CH_2CHPh)-$	Addition with peroxide initiator	Hard, transparent, low softening, high electrical resistance. Coatings, beads, construction, packaging
Natural rubber (NR)	E	$cis\text{-}[-(CH_2CMe=CHCH_2)]-$	Naturally occurring	Flexible and elastic. Wide uses
Synthetic polyisoprene (IR)	E	$cis\text{-}[-(CH_2CMe=CHCH_2)]- + trans$	Addition with Ziegler–Natta catalysts	Greater the proportion of *trans*-isomer, the less flexible the polymer. Uses as NR
Polybutadiene (BR)	E	$cis\text{-}[-(CH_2CH=CHCH_2)]- + trans$	Addition with Ziegler–Natta catalysts	As IR

Polymer (abbreviation)	Type[a]	Structural unit	Preparation	Properties, uses
Polychloroprene (NEOPRENE) (CR)	E	cis-[−(CH$_2$CHClCH=CH)]− + trans	Addition with Ziegler−Natta catalyst	As IR
Butyl rubber (BUTYL)	E	−(CH$_2$CMe=CHCH$_2$CMe$_2$CH$_2$)−	Copolymer addition of 2-methylpropene and isoprene	Less permeable than NR. Tyres and roofing
Styrene−butadiene rubber (SBR)	E	(−CH$_2$CHPhCH$_2$CH=CHCH$_2$)−	Copolymerization	High-volume synthetic rubber. Tyres, etc.
Acrylonitrile−butadiene rubbers (NBR)	E	−(CH$_2$CH=CHCH$_2$CH$_2$CHCN)−	Copolymerization	Buna-N rubbers
Polyvinyl resins				
Polyvinyl chloride (PVC)	TP	−(CH$_2$CHCl)−	Addition with peroxide initiator	Soft PVC−floor coverings, kitchenware, films, wire coatings Hard PVC−records, linings, window sashes
Polyvinyl ethanoate (polyvinyl acetate) (PVAC)	TP	−[CH$_2$CH(OCOMe)]−	Addition with peroxide initiator	Adhesive, binder for emulsion paints
Polyvinyl alcohol (PVA)	TP	−(CH$_2$CHOH)−	Hydrolysis of PVAC	Size for textiles, packaging, electrical coatings

continued overleaf

Polymer (abbreviation)	Type[a]	Structural unit	Preparation	Properties, uses
Polymethyl 2-methyl-propenoate (methacrylate) (PMMA)	TP	$-[CH_2CMe(OCOMe)]-$	Addition	Clear hard plastic. Perspex; many types of acrylate resins
Polyacrylonitrile	TP	$-(CH_2CHCN)-$	Addition with peroxide initiator. Often copolymerized	Textiles. Commercial acrylic fibre must contain 85% polyacrylonitrile
Polyacrylamide	TP	$-[CH_2CH(CONH_2)]-$	Addition with peroxide initiator	Water-soluble gels and gums. Precursor for polyelectrolytes
Polyester resins Alkyd resins (glyptal resins)	TS	3-D cross-linked structure	Condensation between phthalic anhydride and 1,2,3-trihydroxypropane	Oil-soluble paints
Unsaturated polyester	TS	$-(OC(O)CH=CHCO(O)CH_2CH_2)-$	Condensation between butenedioic anhydride and a diol	Glass fibre. Construction industry
Aromatic polyester	TP	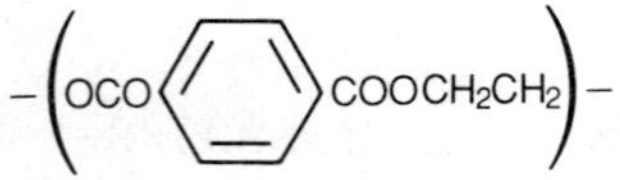	Condensation between 1,4-benzenedicarboxylic acid and an alcohol or diol	Textiles (Terylene, polyester), moulded resins

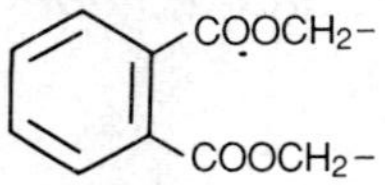

Polymer (abbreviation)	Type[a]	Structural unit	Preparation	Properties, uses
Polycarbonate	TP	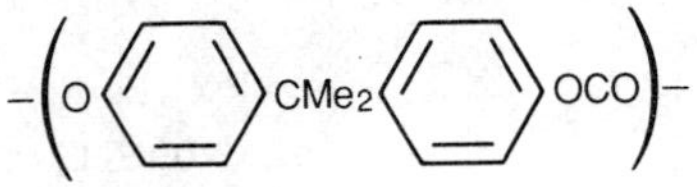	Condensation between 2,2-bis-(4-hydroxyphenyl)propane and phosgene	Transparent, tough plastic. Packaging, safety glasses, machine parts
Nitrogen polymers Urea–methanal (UF)	TS		Condensation between urea and methanal	Adhesive, foams, moulded goods
Urea–melamine (MF)	TS	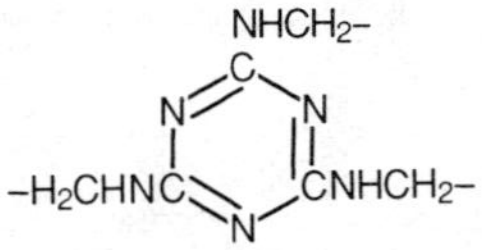	Condensation between urea and melamine	Coatings, paper and textile finishing
Polyurethane (PUR)	TS, E	$-(OCONH)-$	Condensation between a diisocyanate and a diol or diamine	Coatings, foams, elastomers
Polyamine	TP	$-(CH_2NHCH_2NHCH_2CH_2)-$	Condensation between amines and 1,1-dihaloalkanes	Flocculants

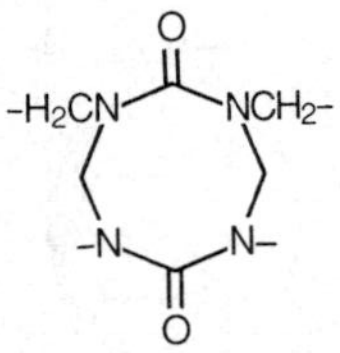

continued overleaf

Polymer (abbreviation)	Type[a]	Structural unit	Preparation	Properties, uses
Polyamide (PA)	TS	$-[NH(CH_2)_xNHCO(CH_2)_yCO]-$	Condensation between α,ω-alkanedioic acid and α,ω-alkanediamine	NYLON-x,y + 2. Textiles, coatings, engineering mountings, etc.
Polyethers				
Epoxy resins (EP)	TS	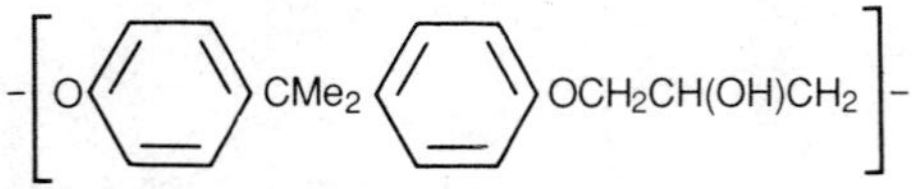	Condensation between chloropropeneoxide and 2,2-bis-(4-hydroxyphenyl)propane	Adhesives, coatings, composites
Polyethene oxide	TS	$-(CH_2CH_2O)-$	Catalytic addition of ethene oxide	Thickeners, plasticizers, coatings, cosmetics
Polyphenylene oxide (PPO)	TS	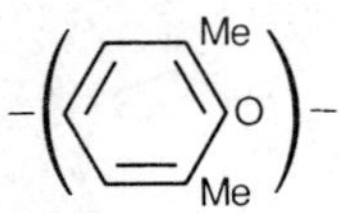	Catalytic oxidation of 2,6-dimethylphenol	Very resistant to water. Medical instruments
Polyacetal (polyoxymethylene) (POM)	TS	$-(CH_2O)-$	Addition of methanal	Flame retardant. Hard engineering plastic
Polyaldehyde	TS	$-(CHRO)-$	Addition of aldehyde	Lightweight, hard, chemically resistant. Machine parts

Polymer (abbreviation)	Type[a]	Structural unit	Preparation	Properties, uses
Other polymers				
Polytetrafluoroethene (PTFE)	TP	$-(CF_2)-$	Pyrolysis of F_2CHCl	Chemically inert, hydrophobic. Mouldings, bearings, sealant, laboratory ware
Phenolic resin	TS	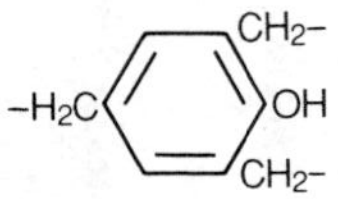	Condensation of phenol and methanal	Coatings and bonding
Polysulphone	TS	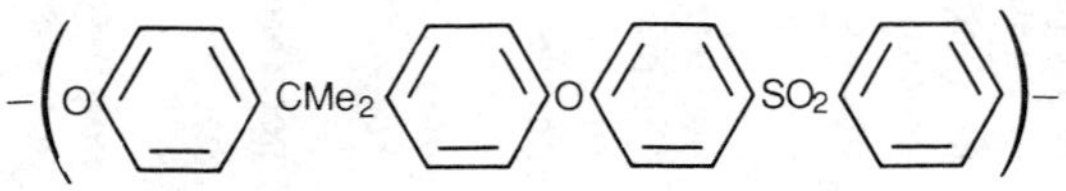	Condensation between 2,2-bis-(4-hydroxyphenyl)propane and bis-(1,4-dichlorodiphenyl) sulphone	Heavy-duty plastics, electronic and automobile parts
Polysulphide	E	$-(S-SCH_2CH_2)-$ with $S=S$	Addition of 1,1-dichloroethane and Na_2S_4	Synthetic rubber
Silicone	TS, E	$-(SiO)-$	*See* SILICON ORGANIC DERIVATIVES	Greases, oils, rubbers

[a] TP, thermoplastic; E, elastomer; TS, thermosetting resin.

chains is known as cross-linking. The properties of polymers depend on the nature of the monomer used, the length of chain and the degree of cross-linking. If the chains fit closely together or have polar groups the polymer may crystallize. Useful properties such as strength and chemical inertness increase with increasing molecular mass, but the fabrication becomes difficult and so an optimum range of molecular masses exists for most polymers. Under stress brittle plastics and fibres do not increase in length before failure, flexible plastics draw out at a yield stress and elastomers reversibly extend under a relatively low stress. Polymers are mostly electrical insulators, but a few conducting polymers (e.g., polyethyne, polythiophene) are known. Hydrocarbon polymers consist only of carbon and hydrogen (e.g., polyethene, polypropene, polystyrene). Heterochain polymers contain other atoms (e.g., polyacetal). Co-POLYMERS have more than one monomer unit, whereas homopolymers consist of a single type of 'mer'. *See also* NUMBER AVERAGE MOLECULAR MASS, POLYMERIZATION, TACTICITY.

polymolybdates. *See* MOLYBDENUM COMPOUNDS.

polymolybdenic acid. *See* MOLYBDENUM COMPOUNDS.

polymorphism. Existence of a substance in two or more crystalline forms. Physical factors (e.g., temperature, pressure, concentration, presence of foreign substances) help to modify the CRYSTAL HABIT. Not all transitions are reversible: for example, SiO_2

$$\text{quartz} \xrightarrow{870°C} \text{tridymite} \underset{}{\overset{1470°C}{\rightleftharpoons}}$$

cristobalite

Polymorphism in elements is called ALLOTROPY.

polymyxins. *See* ANTIBIOTICS.

polyols. *See* POLYHYDRIC ALCOHOLS.

polypeptide antibiotics. *See* ANTIBIOTICS.

polypeptides. *See* PEPTIDES, PROTEINS.

polyphosphates. *See* PHOSPHORUS OXOACIDS.

polyphosphoric acid. *See* PHOSPHORUS OXOACIDS.

polypropene. *See* POLYMERS.

polysaccharides. CARBOHYDRATES that are more complex than the SUGARS; they have higher relative molecular masses. Most are not crystalline substances; they are not sweet and are insoluble or less soluble in water than are the sugars. On hydrolysis they yield a large number of monosaccharides. Most widely spread polysaccharides have the general formula $(C_6H_{10}O_5)_n$. *See also* CELLULOSE, DEXTRINS, GLYCOGEN, STARCH.

polysulphides. Inorganic compounds (e.g., BaS_3, BaS_4, Ca_2S_6) containing the polymeric $[S_n]^{2-}$ ion formed by boiling solutions of sulphides with sulphur. The ions behave as dinegative chelating ligands, and are used in the production of polysulphide polymers containing the $-(SRSSRS)-$ grouping.

polytetrafluoroethene (Teflon, $-(CF_2)_n-$). Chemically resistant, high-melting POLYMER with good electrical insulating properties. *See also* FLUOROCARBONS.

polythene (polyethylene, polyethene). *See* POLYMERS.

polythionic acids. *See* SULPHUR OXOACIDS.

polytungstates. *See* TUNGSTEN COMPOUNDS.

polyunsaturates (polyunsaturated acids). FATTY ACIDS with two or more double bonds per molecule, such as linoleic (*see cis,cis-9,12*-OCTADECADIENOIC ACID) or linolenic acid (*see cis,cis,cis-9,12,15*-OCTADECATRIENOIC ACID).

polyvanadates. *See* VANADIUM COMPOUNDS.

polyvinyl acetate (polyvinyl ethanoate). *See* POLYMERS.

polyvinyl chloride. *See* POLYMERS.

Pomeranz–Fritsch synthesis. *See* ISOQUINOLINE.

***p*-orbital.** *See* ATOMIC ORBITAL.

pore size. *See* POROSITY.

porosity. Pores in solids are classed as micropores (less than 2 nm), mesopores (2–50 nm) and macropores (more than 50 nm). Micropores and mesopores may be investigated by gas adsorption when the ADSORPTION ISOTHERM for physical adsorption of, for example, nitrogen shows HYSTERESIS. Macropores are determined by forcing mercury into them and the use of LAPLACE EQUATION.

porphin nucleus. Flat ring structure (*see* Figure) consisting of four pyrrole rings

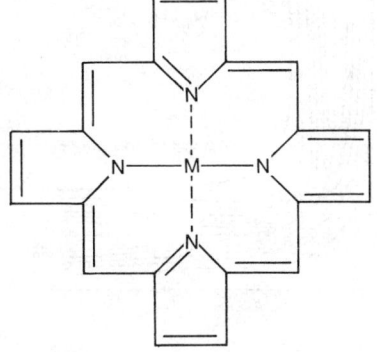

porphin

metalloporphyrin
(M = Fe, Cu, Mg, Zn, Cr and other metals)

linked by methene bridges, that occurs in biologically important pigments such as those of bile and blood and the green colour matter of plants.

porphyrins. Group of highly coloured, naturally occurring pigments containing a PORPHIN NUCLEUS with substituents in the eight β-positions of the pyrrole nuclei. The natural pigments themselves are metal chelate complexes of the porphyrins. Protoporphyrin ($C_{34}H_{34}N_4O_4$), present in HAEM, with iron(III) chloride in alkaline solution gives haemin. The CHLOROPHYLLS are magnesium complexes of porphyrins esterified with phytol (*see* TERPENES).

positron (β^+). Positive counterpart (ANTIPARTICLE) of the ELECTRON. *See also* BETA-DECAY.

positron annihilation. When a POSITRON (e.g., from sodium-22) interacts with a condensed medium (usually a solid) it annihilates on meeting an electron with the emission of an energetic photon. The life-time of the positron, and the angular distribution and momentum of the photon give information on the electronic structure of the material. Early work concentrated on metals, but ionic crystals, polymers and liquids may also be studied.

postactinide elements. *See* TRANSACTINIDE ELEMENTS.

potassium (K). Group Ia ALKALI METAL. The seventh most abundant element in the earth's crust, occurring as the minerals sylvite (KCl) or carnallite ($MgCl_2.KCl$); it is also found extensively in seawater. Potassium metal is produced by electrolysis or in a thermal process in which molten potassium chloride is reacted with sodium vapour in a column with sodium chloride being removed at the bottom, and potassium or a potassium/sodium alloy as an overhead product. Potassium is a soft, silvery-grey metal which is more reactive than sodium, igniting in air and reacting explosively with water. The metal has few uses, but compounds are extensively used in fertilizers, electrolytes in batteries, and in glass and ceramics production. It is an important element in living systems. Potassium may

be determined gravimetrically as insoluble potassium triphenylboron.
$A_r = 39.0983$, $Z = 19$, ES [Ar] $4s^1$, mp 63.7°C, bp 774°C, $\rho = 0.86 \times 10^3$ kgm^{-3}. Isotopes: ^{39}K, 93.1 percent, ^{40}K, 0.012 percent (*see* POTASSIUM–ARGON DATING), ^{41}K, 6.9 percent.

potassium–argon dating. Dating technique for rocks (micas and feldspars) containing the radioactive isotope ^{40}K, which decays (β^-) to ^{40}Ar (half-life: 1.3×10^9 yr). Measurement of ^{40}K/^{40}Ar in a sample allows the calculation of the rock's age.

potassium compounds. Potassium is a highly ELECTROPOSITIVE metal ($E^{\ominus}_{K^+,K} = -2.92$ V) and readily forms K$^+$. All salts of the common anions are soluble in water (KClO$_3$ is sparingly soluble) and are ionic. The orange superoxide (KO$_2$) is the product of the reaction of the metal with air. KO$_2$ decomposes on heating to K$_2$O$_2$ and K$_2$O. Industrially potassium chloride and potassium hydroxide are the most important compounds. Potassium iodide, iodate, permanganate and thiocyanide are used in volumetric analysis, and potassium nitrate in the manufacture of explosives. Potassium dissolves in liquid ammonia to give a blue, electrically conducting solution, which transforms to the amide (KNH$_2$). There is only a limited range of organic derivatives which, like those of sodium, are largely ionic (e.g., Ph$_3$C$^-$K$^+$). Potassium dissolves in ethanol to the ethoxide with evolution of hydrogen (compare the reaction with water).

potassium permanganate. *See* MANGANESE COMPOUNDS.

potential (*V*). Measure of the work necessary to bring a unit positive charge from an infinite distance. The potential at a point due to a charge q at a distance r in a medium of relative permittivity ε_r is

$$V = q/4\pi \, \varepsilon_0\varepsilon_r r$$

See also INTERFACIAL POTENTIAL.

potential energy. ENERGY stored in a body or system as a consequence of its position.

potential energy surface. Three-dimensional representation of the energy of a system as a function of bond lengths or bond angles, which can show the course of a reaction (*see* REACTION COORDINATE), and the position of the transition state (*see* ACTIVATED COMPLEX). Two dimensions (the third is energy) may not be enough to describe the positions of all atoms in a system, however by a suitable choice the most probable arrangement of atoms may be specified. For example, the reaction between a hydrogen atom and a molecule of hydrogen (H$_a$ + H$_b$H$_c$ → H$_a$H$_b$ + H$_c$) may be described by the distances H$_a$–H$_b$ (r_{ab}) and H$_b$–H$_c$ (r_{bc}) if it is assumed that H$_a$ approaches along the line of centres of H$_b$–H$_c$. The potential energy surface for this system is shown in perspective and as a contour map (*see* Figure). The reaction

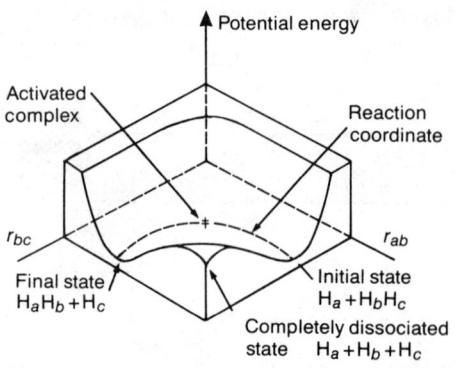

(a)

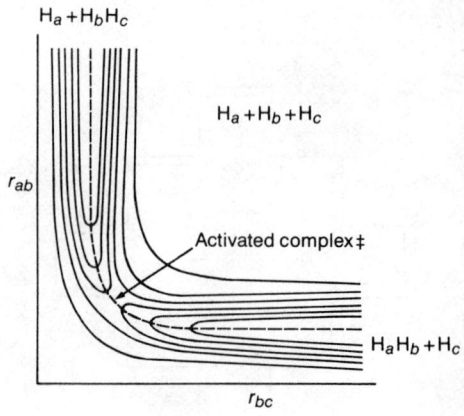

(b)

potential energy surface
(a) Perspective, (b) contour map.

coordinate is seen as the pathway of minimum energy. The transition state (‡) at which the activated complex is found occurs at a coll on the surface (i.e. at a maximum along the reaction coordinate, but a minimum on a line at right angles to the reaction coordinate).

potential of zero charge (ϕ_{pzc}). Potential at which the charge on a polarizable electrode is zero. At ϕ_{pzc}, the absolute potential difference between the electrode and electrolyte is equal to the surface potential difference (*see* INTERFACIAL POTENTIAL). *See also* ISOPOTENTIAL POINT.

potentiometric titration (electrometric titration). *See* TITRATIONS.

Pourbaix diagram. Graph of the limits of stability of different solid phases and oxidation states of a metal as a function of potential and pH (*see* Figure).

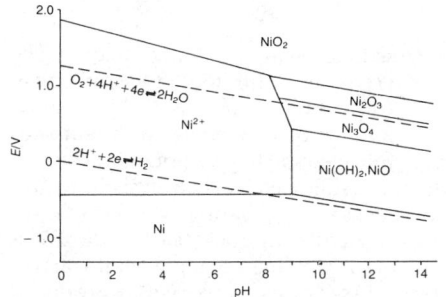

Pourbaix diagram

powder method. *See* X-RAY DIFFRACTION.

power (P; dimensions: $m\,l^2\,t^{-3}$; units: W = Js^{-1}). Rate at which WORK or ENERGY is transferred.

Pr. Praseodymium (*see* LANTHANIDE ELEMENTS), PROPYL GROUP.

praseodymium (Pr). *See* LANTHANIDE ELEMENTS.

precipitation titration. *See* INDICATORS, TITRATIONS.

pre-equilibrium. Reaction in which an intermediate attains a pseudoequilibrium (i.e. practically, if not strictly thermodynamically) with the reactants and the subsequent reaction is slow. (For examples, *see* ENZYME KINETICS, LINDEMANN THEORY.)

pre-exponential factor. *See* ARRHENIUS EQUATION.

pressure (P; dimensions: $m\,l^{-1}\,t^{-2}$; units: Pa = N m^{-2}; practical units: mmHg, atm). Force exerted by a gas per unit area.

$$P = f(V,T)$$
$$= -(\partial A/\partial V)_{T,N}$$
$$= kT\,(\partial \ln Q/\partial V)_{T,N}$$

1 atm = 760 mmHg = 101325 Pa. *See also* GAS LAWS.

pressure–composition diagram. *See* COMPLETELY MISCIBLE LIQUID SYSTEMS.

Prileschaiev's reaction. Reaction of an ALKENE with a PERACID to give an ALKENE OXIDE.

$$RCH{=}CHR' + PhCOO_2Na \rightarrow$$
$$RCH(O)CHR' + PhCO_2Na.$$

primary. Term, together with secondary and tertiary, used to specify the molecular structure of compounds that are closely similar or isomeric. Depending on the number of hydrocarbon groups attached to a carbon atom carrying a FUNCTIONAL GROUP (X) or a nitrogen atom, a compound may be classed as primary RCH_2X (RNH_2); secondary (*sec-*) $RR'CHX$ ($RR'NH$) or tertiary (*tert-*) $RR'R''CX$ ($RR'R''N$) where R,R' and R'' may be the same or different.

primary cell. *See* CELLS.

primary isotope effect. *See* ISOTOPE EFFECTS.

primary salt effect. *See* SALT EFFECTS.

primary structure of proteins. *See* PROTEINS.

primitive lattice. Lattice in which there is only one equivalent point in the unit cell. *See also* CRYSTAL LATTICE.

principal axis. *See* SYMMETRY ELEMENTS.

principal quantum number. *See* QUANTUM NUMBER.

Pro (proline). *See* AMINO ACIDS.

procaine (2-diethylaminoethyl 4-aminobenzoate, $H_2NC_6H_4COOCH_2CH_2N(C_2H_5)_2$). White, gyroscopic solid (mp 61°C); prepared, by the reduction of 2-diethylaminoethyl 4-nitrobenzoate $(C_2H_5)_2NCH_2CH_2$-$OH + p$-$O_2NC_6H_4COCl$). The hydrochloride is used as a local anaesthetic. Procaine penicillin is used as an ANTIBIOTIC and animal feed additive.

prochiral. Carbon atom linked to two identical ligands and two other different ligands (C*aabd*). If one of the identical ligands is replaced by a different ligand *e* then a chiral centre is produced (C*abde*). *See also* CHIRALITY.

producer gas. Mixture of gases formed by passing air and steam over coke or coal. Exothermic oxidation of the carbon (to carbon monoxide + some carbon dioxide) is balanced by the endothermic WATER GAS reaction. The gas is used industrially as a fuel with a CALORIFIC VALUE of about 5000 $kJ m^{-3}$.

proflavine (3,6-diaminoacridine, $C_{13}H_{11}N_3$). Prepared by the condensation of 1,3-diaminobenzene with methanoic acid. It is soluble in water giving a brown fluorescent solution. Proflavine is used as a topical antiseptic.

progesterone. *See* HORMONES.

progestin. *See* STEROIDS.

projection formula. *See* FISCHER PROJECTION FORMULA, NEWMAN PROJECTION FORMULA.

prolactin. *See* HORMONES.

prolamines. *See* PROTEINS.

proline (Pro). *See* AMINO ACIDS.

promazine. *See* CHLORPROMAZINE.

promethium (Pm). *See* LANTHANIDE ELEMENTS.

propadiene (allene). *See* ALLENES.

propagation. *See* CHAIN REACTION.

propanal (propionaldehyde, C_2H_5CHO). ALDEHYDE; colourless liquid (bp 48°C), prepared by dehydrogenation of 1-propanol, the FISCHER–TROPSCH PROCESS or by the OXO PROCESS. It is used for the preparation of 2,2-dimethylpropane, propanol and propanoic acid.

propane (C_3H_8). ALKANE (mp −190°C, bp −44.5°C), constituent of NATURAL GAS (minor), PETROLEUM GAS and liquefied petroleum gas.

propanedioic acid (malonic acid, CH_2-$(COOH)_2$). Saturated DICARBOXYLIC ACID (mp 136°C), prepared by heating potassium chloroethanoate with aqueous potassium cyanide to give potassium cyanoethanoate followed by hydrolysis in hydrochloric acid. On heating to 140–50°C or refluxing in sulphuric acid, carbon dioxide is eliminated, yielding ethanoic acid. Carbon suboxide (*see* CARBON OXIDES) is produced on heating with phosphorus pentoxide. Treatment with nitrous acid followed by hydrolysis yields 2-oxopropanedioic acid $(CO(COOH)_2)$. The diethyl ester (malonic ester) (bp 199°C) contains an active METHENE GROUP and is important for its use in synthesis. Reaction with sodium or sodium ethoxide displaces a methene hydrogen by a sodium atom which, in turn, may be displaced by an alkyl group to give RCH-$(COOEt)_2$. Further reaction gives the dialkyl-substituted ester $RR'C(COOEt)_2$, which can be hydrolyzed to the acid and decarboxylated to substituted ethanoic acids. Similar syntheses of ketones, β-ketoesters, unsaturated acids (*see* KNOEVENAGEL REACTION) and polycarboxylic acids are also possible. Reaction with urea in the presence of sodium ethoxide gives barbituric acid (*see* BARBITURATES).

propanoic acid (propionic acid, C_2H_5-COOH). CARBOXYLIC ACID (mp $-22°C$, bp $141°C$, $pK_a = 4.87$); colourless liquid with acrid odour. It is manufactured by the oxidation of 1-propanol. Propanoic acid is used in the manufacture of esters and polymers.

propanols

1-propanol (n-propyl alcohol, CH_3CH_2-CH_2OH). Liquid (bp $97.2°C$), obtained industrially by the oxidation of propane–butane mixtures or by the reduction of propenal (direct oxidation of propene). It shows typical properties of primary MONO-HYDRIC ALCOHOLS.

2-propanol. (isopropyl alcohol, $(CH_3)_2$-CHOH). Liquid (bp $82.5°C$), prepared by the hydration of propene with 80 percent sulphuric acid at temperatures below $40°C$. It is a typical secondary alcohol. It is used in the preparation of propanone (by oxidation), esters, amines and glycerol.

propanone (acetone, CH_3COCH_3). KETONE; colourless, volatile liquid (bp $56°C$) with a distinctive ethereal smell. It is manufactured by (1) the dehydrogenation (Cu or ZnO) or oxidation (silver) of 2-propanol; (2) passing propene and oxygen into an aqueous solution of palladium(II) and copper(II) chlorides; (3) oxidation of NATURAL GAS and as a byproduct in the oxidation of 2-phenylpropene to phenol. Propanone is used as a solvent for many POLYMERS (e.g., poly(methyl 2-methylpropenoate)), lacquers, etc. and for the preparation of KETEN. Reactions are typical of ketones; it undergoes CONDENSATION REACTIONS to give 4-methyl-3-penten-2-one, 2,6-dimethyl-2,5-heptadien-4-one and 4-HYDROXY-4-METHYL-2-PENTANONE. When distilled with concentrated sulphuric acid it forms 1,3,5-trimethylbenzene (*see* MESITYLENE).

propargyl alcohol. *See* 2-PROPYN-1-OL.

propellant. Chemical fuel used to provide thrust to drive rockets, etc. Liquid fuels and oxidizers are used in large rockets requiring high thrust (e.g., liquid hydrogen and oxygen, hydrazine and dinitrogen tetroxide, boranes and oxygen). Solid propellants fall into two classes: double base fuels (mixtures of nitroglycerin and cellulose nitrate) and composites of fuel and oxidant (e.g., thermosetting plastics, resins or elastomers and ammonium perchlorate or sodium nitrate). Powdered metals (e.g., boron, aluminium or beryllium) may be added to improve the burning properties.

propenal (acrolein, acryldehyde, vinyl aldehyde, $CH_2=CHCHO$). Simplest unsaturated ALDEHYDE; colourless, pungent liquid (bp $52°C$). Manufacture is by the reaction between ETHANAL and METHANAL over a lithium phosphate catalyst or by the direct oxidation of propene over copper(I) oxide supported on alumina. In the laboratory 1,2,3-trihydroxypropane is dehydrated with potassium hydrogen sulphate. Propenal readily polymerizes and is used in the manufacture of the acrolein POLYMERS and methionine (*see* AMINO ACIDS). It undergoes reactions of aldehydes and ALKENES. The latter function shows an anti-Markownikoff addition of hydrogen chloride to $ClCH_2CH_2CHO$ (*see* MARKOWNIKOFF'S RULE). The ALDOL CONDENSATION does not take place, but 2-propenyl propenate is the product of the TISCHENKO REACTION.

propene (propylene, $CH_3CH=CH_2$). ALKENE (mp $-185°C$, bp $-48°C$); manufactured by CRACKING petroleum. Its principal use is in the manufacture of POLYMERS (*see* PETROCHEMICALS, Figure 3).

propenenitrile (acrylonitrile, cyanoethene, vinyl cyanide, $CH_2=CHCN$). Volatile, toxic liquid (bp $78°C$); manufactured by the addition of hydrogen cyanide to ethyne over a copper(I) chloride catalyst, or by the reaction of propene, ammonia and air over a molybdenum catalyst. Propenenitrile is an important monomer for many vinyl POLYMERS and copolymers, and is also used as an INSECTICIDE.

propene-1,2,3-tricarboxylic acid (aconitic acid, $HC(COOH)=C(COOH)CH_2$-COOH). White crystalline solid (mp $195°C$ (decomp.)), obtained by the dehydration of citric acid (*see* 2-HYDROXYPROPANE-1,2,3-TRICARBOXYLIC ACID) with sulphuric acid. It is one of the acids in the Kreb's tricarboxylic acid cycle. It is used in the preparation of plasticizers, as an antioxidant and as a flavouring agent.

propenoic acid (acrylic acid, $CH_2=CH-COOH$). Unsaturated CARBOXYLIC ACID; colourless pungent liquid (bp 141°C), prepared by the oxidation of PROPENAL or 2-propen-1-ol with silver oxide, or by heating 3-hydroxypropanoic (hydracrylic) acid with sulphuric acid. On standing a solid polymer is formed. Acrylate polymers are formed from the methyl ester, and acrylic resins from the acid (*see* POLYMERS, Table). *See also* 2-METHYLPROPENOIC ACID.

2-propen-1-ol (allyl alcohol, $CH_2=CHCH_2OH$). Colourless liquid with pungent odour (bp 97°C); prepared by heating glycerol with ethanedioic acid or by reduction of propenal. It is manufactured by the chlorination of propene at 500°C followed by hydrolysis of the dichloride, or by the reaction of propenal and 2-propanol in the presence of a catalyst at 400°C. It has the properties of primary MONOHYDRIC ALCOHOLS, except that there is no tendency to dehydration, and the addition properties of an alkene. Oxidation ($KMnO_4$) gives glycerol, oxidation (Ag_2O) yields propenal and propenoic acid. It forms insoluble compounds with mercury salts.

5-(2-propenyl)-1,3-benzodioxole (safrole). *See* PHENOLS.

propenyl group (allyl group). The group $CH_2=CHCH_2-$.

propiolic alcohol. *See* 2-PROPYN-1-OL.

propionaldehyde. *See* PROPANAL.

propyl alcohols. *See* PROPANOLS.

propylcarbinol. *See* BUTANOLS.

propylene. *See* PROPENE.

propylene glycol. *See* 1,2-DIHYDROXYPROPANE.

propyl groups (Pr). Alkyl groups containing three carbon atoms.

   n-propyl: $CH_3CH_2CH_2-$
   iso-propyl (prop-2-yl): $(CH_3)_2CH-$

2-propyn-1-ol (propargyl alcohol, propiolic alcohol, $HC\equiv CCH_2OH$). Liquid prepared from glycerol tribromohydrin by treatment first with ethanolic potassium hydroxide, then water and finally ethanolic potassium hydroxide. It shows the group reactions of typical primary MONOHYDRIC ALCOHOLS and of a terminal alkyne linkage; with hydrogen bromide bromopropenol is formed and with phosphorus tribromide propargyl bromide (3-bromo-1-propyne). It is easily polymerized by heat or alkali, and is used as a chemical intermediate, stabilizer and corrosion inhibitor.

prostaglandins (PG). Group of chemically related hormone-like substances that are formed from essential fatty acids and occur in most mammalian tissues and body fluids. They are all oxygenated, unsaturated derivatives of prostanoic acid (a C_{20} fatty

prostanoic acid

prostaglandin E_1

prostaglandin $F_{1\alpha}$

acid in which there is a cyclopentane ring formed by connecting the C-8 and C-12 positions). The two main types differ in their oxygen function at C-9: in series E (PGE) it is carboxyl, whereas in series F (PGF) it is hydroxyl (*see* Figure). They exhibit a wide range of pharmacological activities, but are best known for their potent effect on smooth muscle, particularly in stimulating contraction and relaxation of the human uterus. Many syntheses have been developed, and these have provided a number of interesting analogues.

prostanoic acid. *See* PROSTAGLANDINS.

prosthetic group. Non-protein group of a conjugated PROTEIN. Such a group is required by some ENZYMES for their specific activity (e.g., haemin group of catalase). *See also* HAEM.

protactinium (Pa). *See* ACTINIDE ELEMENTS.

protamines. *See* PROTEINS.

protecting group. Group that is added to one functional group in a molecule to permit a chemical reaction at a different functional group. It may prevent an unwanted reaction at the site it is protecting or it may serve to direct a reagent to another site. ACETALS and KETALS are used to protect alcohol, aldehyde and ketone functions in a variety of reactions; the original function can be regenerated by acid hydrolysis after the need for protection has passed. *See also* PEPTIDE SYNTHESIS.

protective colloid. Stability of a LYOPHOBIC COLLOID may be enhanced by adsorption of a soluble lyophilic compound. The stability may be as a result of (1) the weakening of the interparticle van der Waals interaction, (2) interparticle electrostatic repulsion or (3) steric stabilization. The latter stabilizer is usually a block COPOLYMER (e.g., poly(ethene oxide)).

proteins. Chief nitrogeneous constituent of living organisms, containing about 50 percent carbon, 25 percent oxygen, 15 percent nitrogen, 7 percent hydrogen and some sulphur. Protein molecules consist of one or several long-chains (polypeptides) of AMINO ACIDS linked in a characteristic sequence. In spite of their high relative molecular mass (up to many million), many proteins have been crystallized or purified until they behave as a homogeneous substance. They are precipitated from solution by ethanol, propanone and concentrated salt solutions, and undergo denaturation (i.e. alteration of the tertiary structure of the molecule) by changes in pH, by UV radiation, by heat and by some organic solvents. Proteins are dipolar ions which migrate in an electric field and have characteristic isoelectric points. Proteins may be broadly classed as: (1) globular proteins which are compact rounded, water-soluble molecules (e.g., enzymes, antibodies, carrier proteins—haemoglobin—and storage proteins—casein); (2) fibrous or structural proteins which are insoluble in water (e.g., keratins, collagens). In classifying of proteins, a single protein may fall into more than one class (*see* Table).

structure of proteins. The grosser structure of proteins has been determined by X-ray diffraction patterns, and their shapes and sizes by sedimentation velocity, diffusion and light-scattering techniques. The three-dimensional arrangement of proteins is important in determining the properties of the protein (e.g., the specific enzyme activity).

1) primary structure. The specific covalent sequence of amino acid residues in the polypeptide chains, obtained by a quantitative analysis of acid hydrolyzates to give the relative amounts of the amino acids and sequencing of the amino acids in peptides obtained by partial hydrolysis with specific enzymes (*see* EDMAN DEGRADATION).

2) secondary structure. The manner in which the polypeptide chains are coiled according to the rules: (a) the atoms in the peptide link lie in the same plane; (b) the nitrogen, hydrogen and oxygen atoms of a hydrogen bond are in a straight line; (c) every NH and every CO group is engaged in bonding. There are two possible conformations. (i) The α-helix, in which hydrogen bonds form between a carbonyl oxygen and the NH of the fourth residue along the same chain (*see* Figure 1). This produces a stable structure with 3.5 residues per turn, each with a pitch of 0.544 nm. The majority of polypeptides have a right-handed helix on account of the preponderance of the L-configuration of naturally occurring amino acids. (ii) The β-pleated sheet in which the hydrogen bonds link different polypeptide chains, as in fibroin, a constituent of silk (*see* Figure 2).

3) tertiary structure. The three-dimensional structure established by the folding of the helical polypeptide chains due to bonding which is strong enough to overcome the hydrogen bonding responsible for the secondary structure. Such bonding includes the disulphide linkages, ionic interactions

proteins

Type	Occurrence, examples	Properties
Simple proteins		
Albumins	Present in all living tissues; ovalbumin (egg), lactalbumin (milk)	Sol. in water, dil. salt solutions
Globulins	Myosin (muscle), fibrinogen (blood), edestin (hemp seed)	Weakly acidic; insol. in water, sol. in dil. salt solutions
Protamines	Present in fish sperm (e.g., salamine) associated with nucleic acids	Simplest proteins; strongly basic, sol. in water, not coagulated on heating. Low M_r ($\sim$ 3000), contain high levels of arginine, no S-containing amino acids
Histones	Associated with nucleic acids in thymus and pancreas (not found in plants)	Similar to protamines, but wider range of amino acids; basic, sol. in water, coagulate on heating
Prolamines (gliadins)	Present in seeds of cereals; gliadin (wheat), zein (corn)	Insol. in water, sol. in 70% aq. EtOH, dil. acids and alkalis. Hydrolysis gives much Pro, Gln, but no Lys or Trp
Glutelins	Present in cereals; glutenin (wheat)	Insol. in water, sol. in dil. acids and alkalis
Scleroproteins	Present in skeletal and connective tissues of animals: (1) collagen (skin, tendons, cartilage, ligaments); (2) keratins (horn, wool); (3) elastin (elastic tissue, ligaments, arterial walls)	Insol. in water; resistant to proteolytic enzymes. (1) Most abundant protein in body, mainly Pro, Gly, Hyp. (2) Unattacked by pepsin or trypsin. (3) Structurally related to collagen, but contains about 30% Gly, 30% Leu, 15% Pro
Conjugated proteins (containing a prosthetic group)		
Phosphoproteins	Casein (milk), phosphovitin (egg yolk)	Contain phosphates, not coagulated by heat
Nucleoproteins	Nuclei of all cells, chromosomes; plant viruses and bacteriophages are pure nucleoprotein	Nucleic acid linked via salt bridges to protamines or histones; sol. in bases, insol. in acids
Lipoproteins	Present in blood and lymph, functioning in the transport of lipids from small intestine to the liver	Prosthetic group is lipid moiety of variable comp. α-lipoprotein of serum contains glyceride, phosphatide and cholesterol (35% total complex); β-lipoprotein contains some glyceride, phosphatide and cholesterol (75% of total complex)
Glycoproteins (mucoproteins)	Present in mucus (saliva, gastric juice), and as hormones, antigens, certain enzymes	Carbohydrate prosthetic group (about 45% of complex) may contain D-galactose, D-mannose, N-acetyl-D-glucosamine and sialic acid. Aqueous solutions are very viscous

(pH-dependent) and stronger hydrogen bonds. Folding is characteristic for each protein and occurs in such a way as to expose the maximum number of hydrophilic (polar) groups and enclose a maximum number of hydrophobic (non-polar) groups within its interior.

4) quaternary structure. This describes the way in which multiple subunits (not always the same) can aggregate to form a large complex. The polypeptide chains are held together by dispersion and electrostatic forces (e.g., in haemoglobin and tobacco

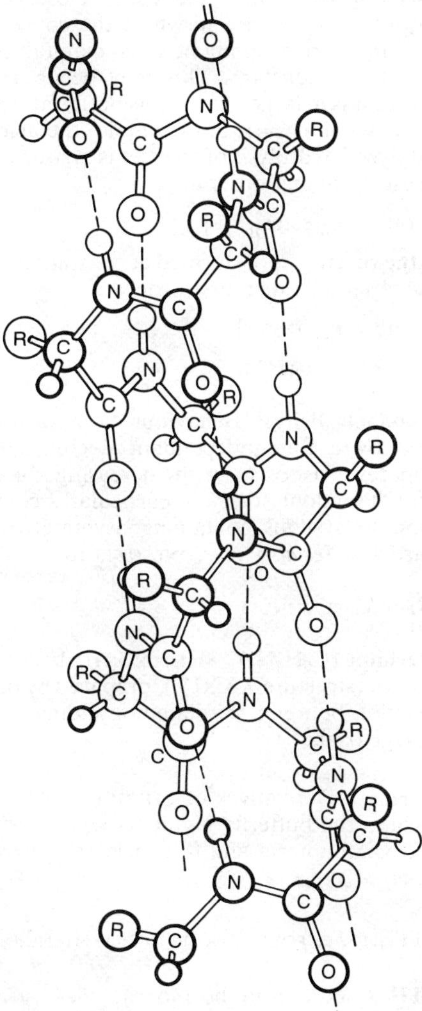

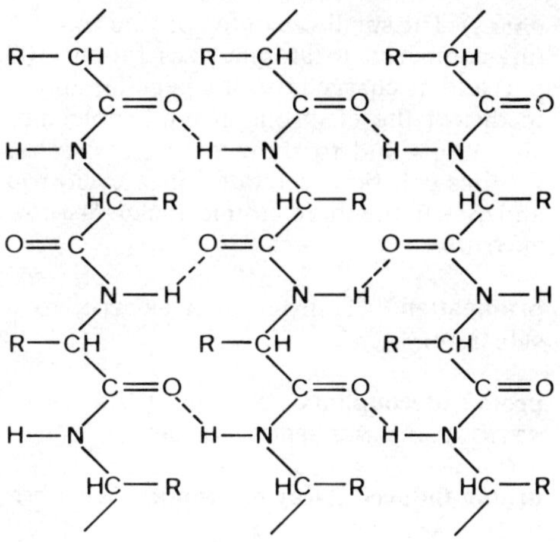

proteins
Figure 2. β-Pleated sheet.

mosaic virus). In viruses, the protein complex usually forms a protective sheath around the core nucleic acid.

protiation. Replacement of deuterium or tritium in a molecule by PROTIUM.

protide ($^1H^-$). Anion of PROTIUM.

protium. Lightest and most abundant (99.98 percent) ISOTOPE of HYDROGEN. The term is used specifically to distinguish ^1H from the other isotopes (*see* DEUTERIUM, TRITIUM).

protogenic solvent. *See* PROTOPHILIC SOLVENT.

proton (H, ^1_1p; mass: 1.672 648 5 × 10^{-27} kg). Fundamental particle with a positive charge and mass almost equal to that of a NEUTRON. Protons are constituents of all atomic nuclei, their number being the ATOMIC NUMBER of the element.

Free protons are rarely found, because of the high value of the ionization potential of H (13.6 eV, 1310 kJ mol^{-1}) they are always solvated (*see* HYDROXONIUM ION). The enthalpy of hydration of H$^+$ is –1121 kJ mol^{-1}; this provides the driving energy for the release of protons. Compounds which can donate protons are acids and those which combine with protons are bases (*see* BRØNSTED–LOWRY ACIDS AND

proteins
Figure 1. α-Helix (right-handed) of a protein chain.

BASES). The small size of the proton ($r = 1.5$ fm) compared to atomic sizes (about 100 pm) and its charge gives it a peculiar ability to distort the electrons around molecules and atoms and to show TUNNEL EFFECTS. Protons can be accelerated in a cyclotron and used to bombard atomic nuclei. *See also* PROTIUM.

protonation.   Transfer of a PROTON to a substrate.

proton decoupling.   *See* NUCLEAR MAGNETIC RESONANCE SPECTROSCOPY.

proton-induced X-ray emission.   *See* ACTIVATION ANALYSIS.

proton jump mechanism.   *See* GROTTHUS THEORY.

proton number (Z).   *See* ATOMIC NUMBER.

protophilic solvent.   Basic solvent capable of accepting a PROTON (e.g., liquid ammonia). Conversely a protogenic solvent is an acidic solvent that can donate a proton (e.g., ethanoic acid). Amphiprotic solvents can do both (e.g., water, ethanol).

protoporphyrin.   *See* PORPHYRINS.

Prussian blue.   *See* CYANOFERRATES.

prussic acid.   *See* HYDROGEN CYANIDE.

pseudoaromatics.   Term used to classify POLYCYCLIC HYDROCARBONS; those with no aromatic stability are pseudoaromatic (e.g., heptalene (structure XXXII), a bicyclic compound of $4n$ type is unstable). *See also* HÜCKEL'S RULES.

pseudohalides.   Univalent ions similar to halide ions in ionic and covalent compounds, namely CN^- from the pseudohalogen CYANOGEN $((CN)_2)$, SCN^- from thiocyanogen $((SCN)_2)$ (*see* THIOCYANATES), $SeCN^-$ from selenocyanogen $((SeCN)_2)$, $SCSN_3^-$ from AZIDOCARBONDISULPHIDE $((SCSN_3)_2)$ and OCN^-, $TeCN^-$, N_3^- and ONC^-. With hydrogen the uninegative groups form acids (weak compared with halogen acids); the silver(I) and mercury(I) salts, like the halides, are insoluble; the pseudohalogens are volatile

and react with alkali in a similar manner to halogens; and form ADDITION COMPOUNDS with alkenes. Interpseudohalogen and pseudohalogen–halogen compounds (e.g., $CNCl$, ClN_3, $SCNCl$, $CN.N_3$) and ions analogous to POLYHALIDE IONS (e.g., $NH_4(SCN)_3$) are known.

pseudohalogens.   *See*   AZIDOCARBONDISULPHIDE,   CYANOGEN,   PSEUDOHALIDES, THIOCYANATES.

pseudoionone.   *See* IONONES.

pseudo-order   of   reaction.   Observed ORDER OF REACTION in which the concentration of one component is maintained effectively constant. This may occur if a reactant is in large excess or is buffered (*see* BUFFER SOLUTION). For example, the acid-catalyzed hydrolysis of an ester is first-order in ester and first-order in H^+

$$rate = k_2[\text{ester}][H^+]$$

If the reaction is performed at constant H^+, the observed rate is first-order

$$rate = k_{obs}[\text{ester}]$$

with $k_{obs} = k_2[H^+]$.

pseudoplasticity.   Behaviour characterized by a time-independent decrease in apparent VISCOSITY with increasing shear rate. Pseudoplasticity is particularly common to systems containing asymmetrical particles. *See also* NEWTONIAN FLUID.

Pt.   *See* PLATINUM.

pteridine ($C_6H_4N_4$).   HETEROCYCLIC COMPOUND (structure XXXIV), prepared by the reaction between 4,5-diaminopyridine and polyglyoxal.

pterins.   Derivatives of PTERIDINE, which include the butterfly pigments leucopterin and xanthopterin and folic acid (*see* VITAMINS).

PTFE.   *See* POLYTETRAFLUOROETHENE.

PTH (parathyroid hormone).   *See* HORMONES.

Pu.   *See* PLUTONIUM.

purines
Occurrence and properties of purines.

Compound	Structural formula	Occurrence, isolation, preparation, properties
Purine $C_5H_4N_4$ $M_r = 120.11$		Parent compound of the group. Does not occur naturally. Prepared from uric acid Needles; mp 216−17°C, sol. in H_2O, hot EtOH; gives neutral solution
Adenine (6-aminopurine) $C_5H_5N_5$ $M_r = 135.14$		Isolated from bovine pancreas. Sublimes at 220°C; mp 360−65°C (decomp.). Loses 3 mol of H_2O at 110°C. Insol. in cold H_2O; sparingly sol. in hot H_2O, EtOH. Aq. soln neutral; combines with acids and bases
Guanine (6-oxy-2-aminopurine) $C_5H_5N_5O$ $M_r = 151.13$		Widespread in animal and vegetable tissues as constituent of nucleic acid. First isolated from guano; obtained by hydrolysis. Decomp. > 360°C. Insol. in H_2O, EtOH; sol. in acids, alkalis; forms compounds with many metals. pK 3.2; 9.92; $\varepsilon(246\ nm) = 10\,700$; $\varepsilon(275\ nm) = 8100$ at pH 6.2
Hypoxanthine (6-oxypurine) $C_5H_4N_4O$ $M_r = 136.11$		Breakdown product of nucleic acid metabolism. Synthesized by: (1) oxidation of xanthine; (2) reduction of uric acid. Decomp. 150°C. Sparingly sol. in H_2O; combines with 1 equiv. of acid or 2 equiv. of base. p$K = 8.7$
Uric acid (2,6,8-trioxypurine) $C_5H_4N_4O_3$ $M_r = 168.11$		Occurs in small quantities in urine of carnivorous animals; end-product of nitrogenous metabolism of birds. Isolated from guano with alkali and precipitated with acid. Synthesized by fusion of glycine with urea. Decomp. 250°C. Sparingly sol. in H_2O; insol. in EtOH, Et_2O; sol. in alkali. Dibasic, forms two series of salts
Xanthine (2,6-dioxypurine) $C_5H_4N_4O_2$ $M_r = 151.11$		Occurs in animal organs, yeast, potatoes, coffee beans, tea. Formed from guanine by: (1) treatment with guanase; (2) treatment with $NaNO_2$ and acid; formed from hypoxanthine by oxidation. Oxidized in body to uric acid. Decomp. without melting. Sparingly sol. in water; insol. in organic solvents; sol. in mineral acids, alkalis. p$K = 9.9, 13.2$ (40°C). 1 mol of H_2O lost at 130°C
Caffeine (1,3,7-trimethylxanthine) $C_8H_{10}N_4O_2$ $M_r = 194.19$		Alkaloid occurring in tea and coffee. Isolated from these. Prepared by: (1) methylation of theobromine; (2) condensation of cyanoethanoic acid with urea. mp 235°C. V. sol. in H_2O, EtOH, Et_2O, $CHCl_3$. Bitter taste. V. weak acid. Used medicinally as central stimulant and diuretic

continued overleaf

Compound	Structural formula	Occurrence, isolation, preparation, properties
Theobromine (3,7-dimethylxanthine) $C_7H_8N_4O_2$ $M_r = 180.17$		Obtained from cacao seeds. mp 337°C. Sparingly sol. in H_2O, EtOH, Et_2O; insol. in PhH, $CHCl_3$. Very weak base ($pK = 10.0$, 13.9). Forms compounds with alkalis. Physiologically resembles caffeine; diuretic, cardiac stimulant, vasodilator
Theophylline (1,3-dimethylxanthine) $C_7H_8N_4O_2$ $M_r = 180.17$		Occurs to a small extent in tea; chiefly prepared synthetically. Isomeric with theobromine. Sol. in hot H_2O, forms H_2O-soluble compounds with alkalis. Similar pharmacological mechanism to caffeine; used in combination with 1,2-diaminoethane as diuretic and bronchodilator

purines. Organic bases consisting of two fused heterocyclic rings (a five-and a six-membered ring). Most purines do not occur in the free state, but are combined with ribose in NUCLEOSIDES and ribose and phosphates in NUCLEOTIDES and occur in NUCLEIC ACIDS. The important compounds, based on the purine ring system are listed (*see* Table).

putrescine. *See* POLYAMINES.

PVA (poly(vinyl acetate), poly(vinyl ethanoate)). *See* POLYMERS.

PVC (poly(vinyl chloride)). *See* POLYMERS.

PV–P **curve.** For an IDEAL GAS, the product of pressure and volume (PV) is independent of the pressure. For REAL GASES, there are deviations from ideal behaviour. *See also* GAS LAW.

pyran. *See* HETEROCYCLIC COMPOUNDS (structures XX, XXI).

pyranone. *See* PYRONES.

pyranose. Stable six-membered ring form of a sugar. Pyranose derivatives are stable and crystalline (*compare* FURANOSE). The structure, applicable to nearly all sugars, is supported by X-ray analysis. Structurally the hexose GLUCOSE is a 1,5-glucopyranoside with a –CH₂OH side chain.

pyrargyrite. *See* SILVER.

pyrazine (1,4-diazine, $C_4H_4N_2$). Six-membered HETEROCYCLIC COMPOUND (structure XVIII) with two nitrogen atoms; a steam volatile solid (mp 53°C). It is prepared by the oxidation of aminoethanal with mercury(II) chloride and a base.

pyrazole (1,2-diazole, $C_3H_4N_2$). Five-membered HETEROCYCLIC COMPOUND (structure VII) containing two nitrogen atoms; a crystalline solid (mp 70°C). It is obtained by passing ethyne into a cold ethereal solution of DIAZOMETHANE. Substituted pyrazoles are prepared by the reaction between a 1,3-diketone and a phenylhydrazine. It is a weak base ($pK_a = 11.5$). It has aromatic properties undergoing electrophilic substitution in the C-4 position; nitration gives 4-nitropyrazole. The hydrogenation products pyrazoline (4,5-dihydropyrazole) and pyrazolidine (tetrahydropyrazole) are no longer aromatic, but show basic properties.

pyrazolidine (tetrahydropyrazole). *See* PYRAZOLE.

2-pyrazoline (4,5-dihydropyrazole, $C_3H_6N_2$). Five-membered HETEROCYCLIC COMPOUND. It is prepared by the partial reduction of PYRAZOLE or by the reaction between ethene and diazomethane.

pyrazolone (5-ketopyrazoline, $C_3H_4N_2O$). Amphoteric solid, soluble in acids and bases. It is prepared from sodium formylethanoic ester, hydrazine sulphate and sodium hydroxide. 1,3-Substituted pyrazolones are used extensively in dyes and pigments, colour photography and pharmaceuticals.

Pyrene. *See* TETRAHALOMETHANES.

pyrene (benzo[*def*]phenanthrene). *See* POLYCYCLIC HYDROCARBONS (structure XXIII).

pyridazine (1,2-diazine, $C_4H_4N_2$). Six-membered HETEROCYCLIC COMPOUND (structure XVI); liquid, containing two nitrogen atoms. It is prepared from 1,2-dihydro-3,6-pyridazinedione. It has aromatic character and is miscible with water, benzene and N,N-dimethylmethanamide.

pyridine (azine, C_5H_5N). Six-membered nitrogen HETEROCYCLIC COMPOUND (structure XIV); a water-miscible, poisonous liquid (bp 115.5°C) with a characteristic unpleasant odour. It occurs with methyl-pyridines in coal tar and is manufactured from ethyne and ammonia. It has basic properties ($pK_a = 5.18$), forming stable pyridinium salts with mineral acids and quaternary compounds with alkyl halides. Pyridine is reduced to PIPERIDINE by sodium and ethanol and oxidized to pyridine-N-oxide by peracids. The entire ring is deactivated towards electrophilic reagents, but the C-3 and C-5 positions are less deactivated than the C-2, C-4 and C-6 positions, which bear a partial positive charge. Electrophilic substitution takes place at the C-3 position but only under vigorous conditions (e.g., Br_2 and H_2SO_4, 300°C gives 3-bromo- and 3,5-dibromo-pyridines). Nucleophilic substitution with sodium amide or phenyllithium gives 2-amino- or 2-phenylpyridine, respectively. It is used as a solvent in the plastics industry and in the manufacture of pharmaceuticals.

pyridinium. Ion formed by N-coordination of PYRIDINE to a proton.

pyridoxal phosphate. *See* COENZYMES.

pyridoxine (vitamin B complex). *See* VITAMINS.

pyridyl group. The group C_5H_5N-.

pyrimidines (1,3-diazines). Group of related six-membered HETEROCYCLIC COMPOUNDS (structure XVII) containing two nitrogen atoms (*see* Table).

pyrocatechol (1,2-dihydroxybenzene). *See* DIHYDROXYBENZENES.

pyrogallic acid. *See* TRIHYDROXYBENZENES.

pyrogallol. *See* TRIHYDROXYBENZENES.

pyrolusite. *See* MANGANESE.

pyrolysis. Thermal decomposition of organic compounds. When applied to alkanes it is known as CRACKING.

pyrones. Keto derivatives of the pyrans. γ-Pyrone (*see* Figure), prepared by the decarboxylation of 4-pyrone-2,6-dicarboxylic acid (propanone + ethyl ethanedioate), is a basic compound showing some aromatic properties. It does not form an oxime or phenylhydrazone. Condensed pyrone systems occur in nature (e.g., ANTHOCYANINS, COUMARIN).

pyrophoric. (1) Spontaneously inflammable in air. (2) Describes alloys that emit sparks when struck (e.g., MISCH METAL).

pyrophosphate. *See* PHOSPHORUS OXO-ACIDS.

pyrophosphite. *See* PHOSPHORUS OXO-ACIDS.

pyrosilicates. *See* SILICATES.

pyrosulphuric acid. *See* SULPHUR OXO-ACIDS.

pyrosulphurous acid. *See* SULPHUR OXO-ACIDS.

pyrotechnics. Mixtures containing an oxidant (e.g., nitrate, chlorate) and a combust-

pyrimidines
Occurrence and properties of pyrimidines.

Compound	Structural formula	Occurrence, isolation, preparation, properties
Pyrimidine $C_4H_4N_2$ $M_r = 80.09$		Parent compound of the group. Prepared by action of Zn dust on 2,4,6-trichloropyrimidine (barbituric acid + $POCl_3$). Crystalline compound, penetrating smell; mp 21°C; bp 120°C. Sol. in H_2O, EtOH, Et_2O
Cytosine (2-oxy-4-amino-pyrimidine) $C_4H_5N_3O$ $M_r = 110.10$		Hydrolysis product of ribonucleic acid. Isolated from thymus nucleic acid. Sol. in H_2O, EtOH; insol. in Et_2O. Forms salts with acids; with NaOCl and NH_4OH gives red coloration
Uracil (2,6-dioxytetra-hydropyrimidine) $C_4H_4N_2O_2$ $M_r = 112.09$		Hydrolysis product of ribonucleic acid. Colourless crystalline powder turning brown on heating, mp 338°C (decomp.). Sol. in hot H_2O, alkalis ($pK = 9.45$); insol. in EtOH, Et_2O. Used as a diuretic; derivatives have pharmaceutical importance (e.g., 5-fluorouracil used in cancer treatment)
Thymine (5-methyluracil) $C_5H_6N_2O_2$ $M_r = 126.12$		Hydrolysis product of deoxyribonucleic acid. White solid; mp 321−5°C. Sol. in hot H_2O; slightly sol. in EtOH; insol. in Et_2O. Weak acid ($pK = 9.94$). Forms salts with alkalis. Oxidation gives urea, ethanal, 2-oxopropanoic acid and methanoic acid. With hydrazine forms urea and 4-methylpyrazolone

ible substance (e.g., charcoal, sulphur, antimony sulphide) which on ignition give rise to highly exothermic reactions. Illuminating effects are enhanced by the addition of metallic (e.g., magnesium) powders, coloured effects by metallic salts and smoke production by the inclusion of phosphorus. They are used as warning flares, incendaries and for smoke production.

pyrrholite. *See* NICKEL.

pyrrole (azole, C_4H_5N). Five-membered nitrogen HETEROCYCLIC COMPOUND (structure I); a colourless oil (bp 130°C). It is prepared by heating ammonium 4,5-tetra-hydroxyhexanedioate (mucate) (oxidation of galactose) and industrially by passing a mixture of FURAN, ammonia and steam over hot alumina. Substituted pyrroles can be prepared by the reaction of a 1,4-diketone with ammonia or an amine (*see* HANTZSCH SYNTHESIS). It is reduced to PYRROLIDINE with hydrogen and a catalyst. Pyrrole is aromatic in character; on protonation it forms a non-aromatic reactive cation that polymerizes easily. It has amphoteric properties; the imino hydrogen can be replaced by potassium. With methyl iodide it forms 1-methylpyrrole. It is extremely reactive towards electrophilic substitution (*compare* PHENOL, ANILINE) (e.g., it reacts readily with iodine to give tetraiodopyrrole, couples with benzene-diazonium chloride and undergoes Friedel–Crafts acylation (*see* FRIEDEL–CRAFTS REACTION) with ethanoic anhydride in the absence of a catalyst). The 2-positions are most readily attacked. Many derivatives occur naturally (e.g., proline, indican, haem and chlorophyll).

pyrrolidine (tetrahydropyrrole, C_4H_9N). Five-membered saturated HETEROCYCLIC COMPOUND (structure IV); colourless liquid (bp 89°C), with strong basic properties, that fumes in air. It occurs naturally in tobacco leaves, but is made industrially by the catalytic hydrogenation of pyrrole.

pyruvic acid. *See* 2-OXOPROPANOIC ACID.

Q

Q. *See* PARTITION FUNCTION.

q. Quantity of HEAT.

Q-branch. *See* VIBRATION–ROTATION SPECTRA.

quadrupole moment. *See* NUCLEAR SPIN.

qualitative analysis. Identification of constituents. Various chemical 'wet' tests exist for different elements, radicals and types of compound, and systematic methods are available for the analysis of mixtures. Modern instrumental methods (e.g., chromatographic and spectroscopic techniques) are available for both the identification and the quantitative determination of constituents.

quality factor. *See* DOSE, RADIATION UNITS.

quantitative analysis. Determination of the composition of pure compounds, mixtures and solutions. Typical methods include volumetric (*see* TITRATIONS) and GRAVIMETRIC ANALYSIS, spectroscopic techniques, MASS SPECTROMETRY, ACTIVATION ANALYSIS, electrical methods, POLAROGRAPHY.

quantum. Basic discrete unit of energy. For radiation of frequency v the quantum of energy is hv, where h is the PLANCK CONSTANT.

quantum efficiency (Φ). Ratio of the number of molecules that react to the number of photons absorbed. *See also* PHOTOCHEMISTRY.

quantum mechanics. Branch of science dealing with the movements and energies of very small particles that do not obey the laws of classical mechanics; in chemistry this is largely of ELECTRONS. The equations of motion may be expressed as a wave equation (*see* SCHRÖDINGER EQUATION) from which comes the quantization of energy (*see* QUANTUM). *See also* QUANTUM THEORY, WAVEFUNCTION.

quantum number. By QUANTUM THEORY energy may only change by a discrete QUANTUM. The particular ENERGY LEVEL of an atom or molecule may be described by a number, or series of numbers, which arises out of the solution of the wave equation (*see* SCHRÖDINGER EQUATION) for that system. The solution for a hydrogen atom gives a WAVEFUNCTION with three quantum numbers $\phi_{n,l,ml}$ (*see* ATOMIC ORBITAL). The principal quantum number (n) determines the energy of the system (*see* BALMER SERIES). The energy levels corresponding to $n = 1, 2, 3$ are also referred to as the K-shell, L-shell and M-shell. This nomenclature is used in X-RAY SPECTROMETRY and ELECTRON SPECTROSCOPY. The orbital angular quantum number l takes values $0, 1, \ldots, (n-1)$ and the azimuthal quantum number m_l which determines the momentum about the z-axis takes values $0, \pm 1, \pm 2, \ldots, \pm l$. An electron in an orbital has spin angular momentum $sh/2\pi$, where s is the spin quantum number $= 1/2$. As with orbital angular momentum, spin angular momentum has a component about the z-axis $m_s h/2\pi$ where $m_s = \pm 1/2$. The coupling between s and l leads to a total angular momentum quantum number j which has values $l+s, l+s-1, \ldots, |l-s|$. The description of a many-electron atom leads to quantum numbers L, S and J which are vector sums of the single electron numbers. *See also* TERM SYMBOLS.

quantum theory. Theory of matter which, at the microscopic level of atoms and electrons, predicts that changes in the state of a system may only occur in discrete energy

402

steps known as quanta (*see* QUANTUM). *See also* SCHRÖDINGER EQUATION, WAVE-FUNCTION.

quark. *See* ELEMENTARY PARTICLES.

quartz. *See* SILICA.

quaternary ammonium compounds. Compounds containing the ion R_4N^+ (where R = alkyl, aryl, H). They are prepared as salts of AMINES by reaction with an acid. For example, methylamine chloride is the salt of methylamine and hydrochloric acid. If an amine is treated with an alkyl halide, the quaternary salts formed have more alkyl groups than the starting amine. Conversely on pyrolysis alkyl groups are lost and the ultimate product is the ammonium salt. Heating *in vacuo* gives good yields of the tertiary amine and alkenes if elimination can occur (*see* HOFMANN ELIMINATION). They are used as fabric softeners, in cationic emulsions, non-aqueous electrolytes, germicides and antistatic agents.

quaternary structure of proteins. *See* PROTEINS.

quenching. Decrease in FLUORESCENCE by the addition of a chemical that can transfer the energy of an excited state before it may be radiated. For example, molecular oxygen is an efficient quenching agent for the fluorescence of aromatic molecules.

quicklime. *See* CALCIUM COMPOUNDS.

quinhydrone ($C_{12}H_{10}O_4$). Green–black 1:1 complex (mp 171°C) crystallized from a mixture of alcoholic solutions of quinone and hydroquinone. It is a charge-transfer complex, the diol acting as the electron donor and the dione as the acceptor. It is very sparingly soluble in cold water, decomposing into its constituents on boiling. Prior to the glass electrode it was used as a hydrogen-indicating electrode based on the redox reaction

quinone $+ 2H^+ + 2e \rightleftharpoons$ hydroquinone.

quinidine. *See* ALKALOIDS.

quinine. *See* ALKALOIDS.

quinol (1,4-dihydroxybenzene). *See* DIHYDROXYBENZENES.

quinoline (1-benzazine, C_9H_7N). Nitrogen-containing HETEROCYCLIC COMPOUND (structure XXXII) occurring in the high-boiling fraction of coal tar. It is a colourless, refractive liquid (bp 238°C) with a disagreeable odour, prepared by the SKRAUP REACTION. It is very hygroscopic and weakly basic ($pK_a = 4.9$), giving stable salts with mineral acids and quaternary compounds with alkyl halides (quinolinium salts). It is reduced (Sn + HCl) to di- and finally tetrahydroquinoline and oxidized (alkaline $KMnO_4$) to quinolinic acid (pyridine-2,3-dicarboxylic acid). In electrophilic substitution reactions, the C-8 is most reactive; for example, sulphonation at 220°C yields quinoline-8-sulphonic acid. With the strong nucleophile sodamide 2-aminoquinoline is formed. The most important quinoline alkaloid is quinine (*see* ALKALOIDS). Quinoline is used as an acid scavenger and as a high-boiling solvent, and in the manufacture of dyestuffs and antimalarials (e.g., chloroquine and primaquine).

quinones. Cyclic diketones (*see* Figure 1). formed by the replacement of two hydrogen atoms in a benzene ring by two oxygen atoms. A variety of quinone-like structures have been prepared, the most common are the 1,2- and 1,4-quinones (no 1,3-quinones are known), usually the 1,2-quinones are more difficult to prepare and are more reactive than the 1,4-quinones. A number of examples are known where the quinone arrangement extends over more than one ring. Many quinones (e.g., coenzyme Q_{10}) play an important role in oxidation–reduction processes in living organisms. They occur widely as natural pigments found in plants, fungi, lichens, marine organisms and insects. Quinones are reduced (electrochemically, metal + acid, catalytic hydrogenation) to the corresponding dihydroxy compound. As α,β-unsaturated ketones, quinones have the potential of forming 1,4-addition compounds, these are unstable and undergo enolization giving substituted 1,4-benzenediols. They undergo DIELS–ALDER REACTIONS provided there is one double bond that is not part of an aromatic ring.

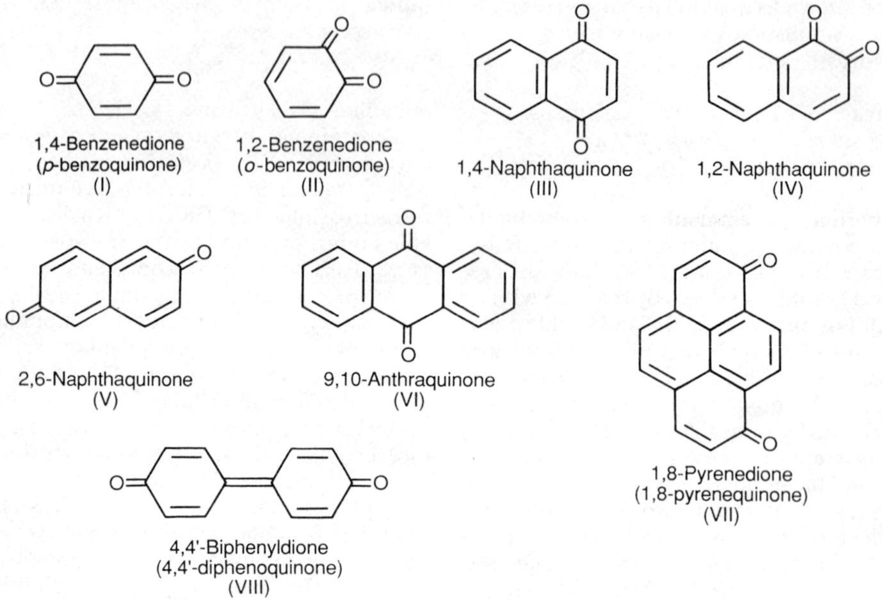

1,4-Benzenedione
(*p*-benzoquinone)
(I)

1,2-Benzenedione
(*o*-benzoquinone)
(II)

1,4-Naphthaquinone
(III)

1,2-Naphthaquinone
(IV)

2,6-Naphthaquinone
(V)

9,10-Anthraquinone
(VI)

1,8-Pyrenedione
(1,8-pyrenequinone)
(VII)

4,4'-Biphenyldione
(4,4'-diphenoquinone)
(VIII)

quinones
Figure 1.

1,4-benzoquinone (*p*-benzoquinone, I). Yellow solid (mp 116°C) with sharp smell, slightly soluble in water, steam volatile, prepared by the oxidation of 1,4-dihydroxybenzene ($FeCl_3$, MnO_2 + H_2SO_4 or lead tetraethanoate) or aniline (CrO_3 + H_2SO_4). It is reduced electrolytically to 1,4-dihydroxybenzene and with hydrogen sulphide to hydroquinone (*see* DIHYDROXYBENZENES). It dehydrogenates many hydroaromatic compounds. I is a useful oxidizing agent in reactions where inorganic oxidizing agents must be avoided; it liberates iodine from acidified potassium iodide. It forms mono- and dioximes and acts as a dienophile in Diels–Alder reactions (*see* Figure 2).

1,2-benzoquinone (*o*-benzoquinone, II). Exists in two forms, one unstable green needles and the other stable red crystalline plates, not steam volatile. II is prepared by the oxidation of 1,2-dihydroxybenzene in dry ethereal solution in the presence of anhydrous sodium sulphate. It is a strong oxidizing agent, reduced to 1,2-dihydroxybenzene with sulphur dioxide.

1,4-naphthaquinone (III). Volatile yellow solid (mp 125°C) resembling 1,4-benzoquinone in many properties. It is prepared by the oxidation of 1,4-diamino-, dihydroxy- or aminohydroxynaphthalene. III is reduced (Sn + HCl) to 1,4-dihyroxynaphthalene and oxidized (HNO_3) to BENZENE-1,2-DICARBOXYLIC ACID. It forms a mono-

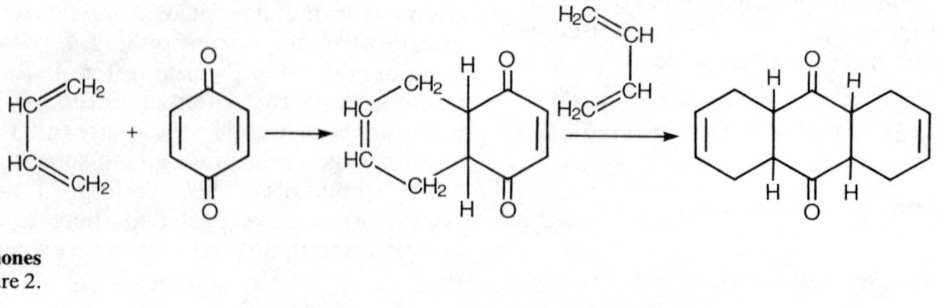

quinones
Figure 2.

oxime with hydroxylamine. Vitamin K is a derivative of 1,4-naphthaquinone.

1,2-naphthaquinone (IV). Non-volatile red solid (decomp. 125°C) prepared by oxidation ($FeCl_3$ + HCl) of 1-amino-2-hydroxy-naphthalene hydrochloride.

2,6-naphthaquinone (V). Non-volatile orange solid (mp 135°C), obtained by the oxidation of a benzene solution of 2,6-dihydroxynaphthalene with lead(IV) oxide.

9,10-anthraquinone (VI). Colourless solid (mp 285°C), prepared by: (1) condensing benzene with phthalic anhydride ($AlCl_3$) to give *o*-benzylbenzoic acid followed by heating with sulphuric acid at 120°C; (2) condensation of 1,4-naphthaquinone (III) with butadiene. Reduction (HI) gives anthracene and its 9,10-dihydro compound, and reduction (Sn + HCl) gives anthrone. Nitration, with mixed acid, gives 1-mono- and the 1,5- and 1,8-dinitroanthraquinones; sulphonation with oleum at 160°C gives the 2- and 2,6- and 2,7- derivatives. It does not undergo the FRIEDEL–CRAFTS REACTION and is halogenated with great difficulty.

R

R. *See* ROENTGEN.

R. *See* GAS CONSTANT.

R- (configurational symbol). *See* OPTICAL ACTIVITY.

ρ. *See* DENSITY, RESISTIVITY.

Ra. *See* RADIUM.

racemic acid. *See* 2,3-DIHYDROXYBUTANEDIOIC ACID.

racemic mixture. Mixture of a pair of ENANTIOMERS in equimolar amounts. This is optically inactive by external compensation and is represented by DL- or (±)-.

rad. Absorbed DOSE of ionizing radiation when the energy per unit mass imparted to matter by the radiation is 10^{-2} J kg^{-1}. *See also* RADIATION UNITS.

radiation. (1) Energy in the form of electromagnetic waves, radiant energy (*see* ELECTROMAGNETIC RADIATION). (2) Energy liberated from a radioactive source or nuclear reactor in the form of particles which possess mass and which may or may not be electrically charged (alpha-, beta-particles and neutrons). Radiation is used in medicine in the form of X-rays, and in industry as a sterilizing agent, and as a vitamin activator and polymerization initiator. Radiation forms the basis of all types of spectroscopic analysis.

radiationless transition. *See* FLUORESCENCE.

radiation units. Measure of activity or dose of radiation (*see* Table).

radical. Chemical entity having unpaired electrons or free valency. In a compound the term refers to a group. When generated free by HOMOLYTIC FISSION of a bond FREE RADICALS are formed.

radical ion. *See* MASS SPECTROMETRY.

radioactivation analysis. Nondestructive technique which depends on the proportionality between the induced activity (A) and the mass of the isotope (w)

$$A = kw$$

A known mass of the sample containing an element M is irradiated and counted under identical conditions to a known mass of pure M. The mass of M in the sample is given by

$$A_{unknown} \times w_{standard}/A_{standard}$$

hence the percentage composition of M in the sample can be calculated. The limits of detection are of the order 10^{-10}–10^{-8}g.

radioactive age. Age of a geological or archaeological specimen determined by the decay of a particular RADIONUCLIDE. *See also* CARBON-14 DATING, POTASSIUM–ARGON DATING.

radioactive decay series. Series of RADIONUCLIDES in which each member of the series is formed by the decay of the previous nuclide (*see* Figure). A series ends with a stable nuclide. The three naturally occurring series are the actinium, thorium and uranium decay series.

radioactivity. Spontaneous nuclear transformation in which energy is emitted. Natural radioactivity arises from the disintegration of naturally occurring radioisotopes (*see* RADIOACTIVE DECAY SERIES). The rate of disintegration cannot be altered by chemical changes or changes in the environment. Radioactivity can be induced in many nuclides by bombardment with

406

radiation units

	SI unit	Pre-SI unit	Conversion factor
Activity, A	Bq	Ci	1 Bq = 27 pCi
			1 Ci = 3.7×10^{10} Bq
Specific activity, a			
Solids	Bq mol^{-1}	Ci mol^{-1}	
	Bq g^{-1}	Ci g^{-1}	
Solutions	Bq dm^{-3}	Ci dm^{-3}	
Dose	Gy	rad	1 Gy = 10^2 rad
			1 rad = 10^{-2} Gy
Dose equivalent	Sv	rem	1 Sv = 10^2 rem
(dose $\times$ QF^a)	(Gy $\times$ QF)	(rad $\times$ QF)	1 rem = 10^{-2} Sv
Exposure dose	*	R^b	R = 2.58×10^{-4} C kg^{-1}

^{a} QF, quality factor, a dimensionless factor used in an attempt to quantify the biological effect of different radiation.
^{b} R, roentgen.
* Not named at present, but is radiation unit that produces a charge of 1 C kg^{-1}.

radioactivity
Types of radioactivity.

Type	Symbol	Particles emitted	Change in atomic number/ΔZ	Change in atomic mass number/ΔA	Example
Alpha	α	Helium nucleus	-2	-4	$^{226}_{88}\text{Ra} \rightarrow {}^{222}_{86}\text{Rn} + {}^4_2\text{He}$
Beta-negatron	β^-	Negative electron	$+1$	0	$^{24}_{11}\text{Na} \rightarrow {}^{24}_{12}\text{Mg} + {}^0_{-1}e$
Beta-positron	β^+	Positive electron	-1	0	$^{22}_{11}\text{Na} \rightarrow {}^{22}_{10}\text{Ne} + {}^0_1e$
Electron capture	EC	Neutrino	-1	0	$^7_4\text{Be} \rightarrow {}^7_3\text{Li} + \nu$
Isomeric transition	IT	γ-rays	0	0	$^{137}_{56}\text{Ba*} \rightarrow {}^{137}_{56}\text{Ba}$
Proton	p	Proton	-1	-1	$^{53}_{27}\text{Co*} \rightarrow {}^{52}_{26}\text{Fe} + {}^1_1p$
Spontaneous fission	f	Heavy fragments	Various	Various	$^{238}_{92}\text{U} \rightarrow {}^{133}_{50}\text{Sn} + {}^{105}_{42}\text{Mo}$

* excited state.

neutrons or other high-energy particles, for example

$$^{27}_{13}\text{Al} + {}^4_2\text{He} \rightarrow {}^{30}_{15}\text{P} + {}^1_0n$$

the phosphorus isotope disintegrating to a silicon nucleus

$$^{30}_{15}\text{P} \rightarrow {}^{30}_{14}\text{Si} + {}^0_1e$$

thus producing the observed artificial radioactivity.

radiocarbon dating. *See* CARBON-14 DATING.

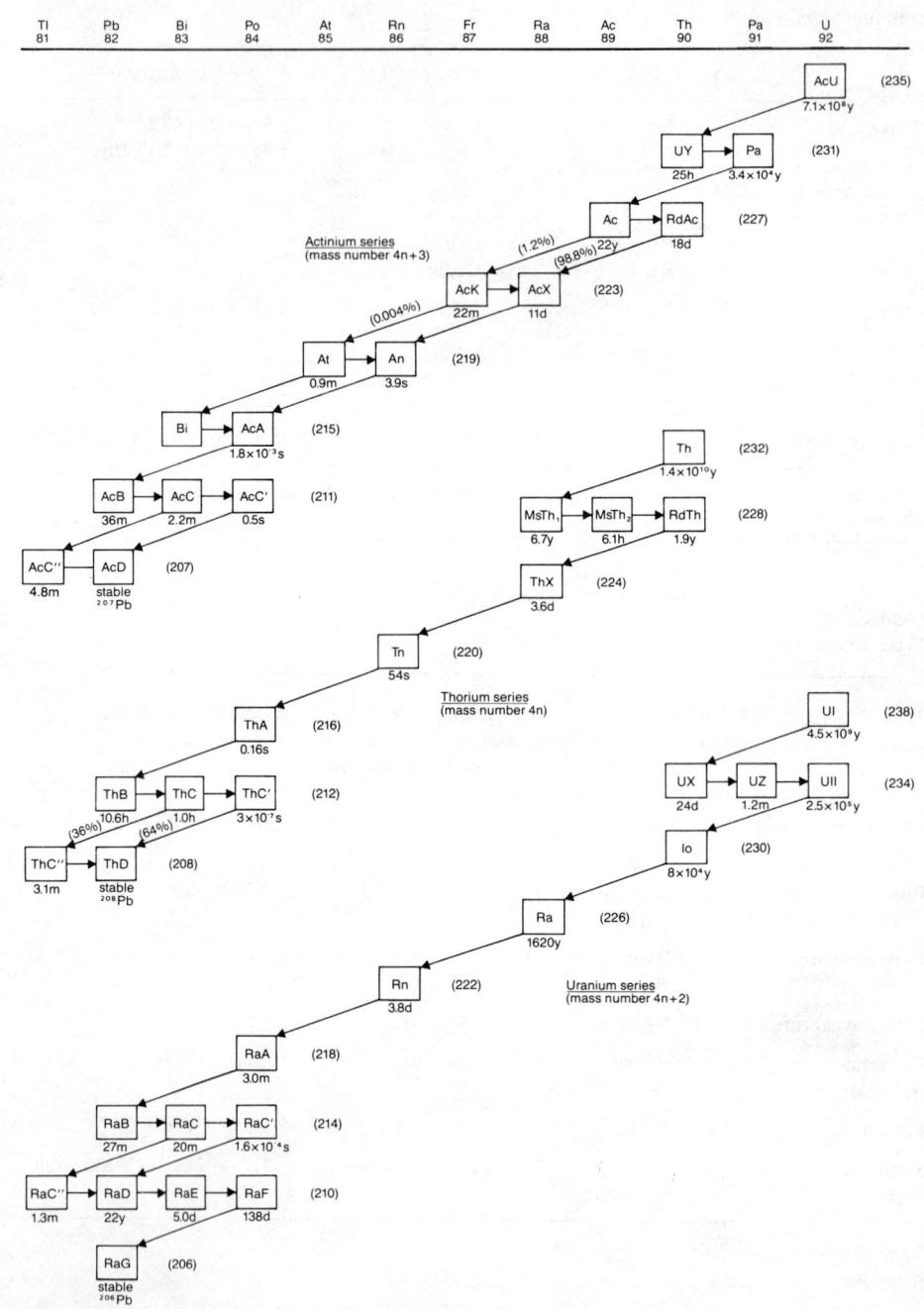

Tl	Pb	Bi	Po	At	Rn	Fr	Ra	Ac	Th	Pa	U
81	82	83	84	85	86	87	88	89	90	91	92

radioactive decay series

Main lines of decay for the actinium, thorium and uranium series.

Diagonal arrows denote α-decay, the atomic number decreasing by two and the mass number by four. Horizontal arrows denote a negatron decay, the atomic number increasing by one, but mass number unchanged. Life times are quoted: y, year; d, day; m, minute; s, second. Mass numbers (in parentheses) are for the horizontal row of isotopes. Names in boxes are isotopes of the elements listed at the top of the table; thus $MsTh_1$ is $^{228}_{88}Ra$.

radiofrequency. *See* ELECTROMAGNETIC SPECTRUM.

radioisotope. ISOTOPE of an element that is radioactive.

radionuclide. NUCLIDE that is radioactive.

radium (Ra). Radioactive ALKALINE EARTH METAL found in small amounts in uranium minerals. The metal is extracted as an AMALGAM by electrolysis of a radium chloride ($RaCl_2$) solution at a mercury cathode. The silvery–white metal reacts readily with water, oxygen, hydrogen and other non-metals. Its uses are as a radioactive source (different isotopes give alpha-, beta- and gamma-radiation) for the treatment of cancer, in luminous paints and as a gamma-source. The chemistry of radium is typical of alkaline earth metals. It is the most ELECTROPOSITIVE alkaline earth metal ($E^{\ominus}_{Ra^{2+},Ra} = -2.92$ V) and forms salts containing Ra^{2+}, which are isomorphous with those of Ba^{2+}.
$A_r = 226.025$, $Z = 88$, ES $[Rn]7s^2$, mp 700°C, bp 1140°C, $\rho = 5.0 \times 10^3$ $kg\,m^{-3}$.

radius–ratio rule. For substances of general formula MX that form ionic crystals, not only do the ions have different radii but they have opposite charges. This makes packing of the two lattices difficult. The rule states eight-coordination (caesium chloride-type) should be expected when the ratio of cation and anion radii (r_+/r_-) exceeds 0.73, six-coordination (sodium chloride-type) for ratios between 0.73 and 0.41, and for ratios less than 0.41 the most efficient packing leads to four-coordination (wurtzite-type) (*see* CRYSTAL FORMS). Although founded on geometric considerations, the rule is well supported experimentally. Deviations of structure from the predicted form indicates a shift from ionic towards covalent bonding.

radon (Rn). Radioactive gas belonging to group 0; the product of radioactive decay of the heavy elements (*see* RADIOACTIVE DECAY SERIES). Of the 20 isotopes known, the important ones—^{222}Rn (radon) from radium (half-life: 3.8 days), ^{220}Rn (thoron, Th) from thorium (half-life: 54.5 s) and ^{219}Rn (actinon, An) from actinium (half-

life: 3.92 s)—are all alpha-emitters. A radon fluoride has been prepared; oxides and chlorides are probably capable of synthesis, but the alpha-activity may render the compounds too short-lived to be isolated. Radon is used as a radiation source and as a gaseous tracer. It creates a hazard in uranium mines.
$A_r \sim 222.0$, $Z = 86$, ES $[Xe]\,4f^{14}\,5d^{10}\,6s^2$ $6p^6$, mp −71°C, bp −61.8°C, $\rho = 9.73$ $kg\,m^{-3}$.

raffinate. *See* LIQUID–LIQUID EXTRACTION.

raffinose ($C_{18}H_{32}O_{16}$). Best known trisaccharide, a glycoside of GLUCOSE, GALACTOSE and FRUCTOSE. It is obtained as a crystalline powder (mp 118–9°C) from sugar beet or cotton seed meal. It has no reducing properties.

Raman spectroscopy. Raman effect is the scattering of light with a change in energy equal to the excitation energy of a rotational or vibrational mode which occurs with a change in polarization of the molecule. A spectrum is produced in a way similar to FLUORESCENCE spectroscopy. The sample is irradiated with an intense monochromatic beam (e.g., helium–neon LASER 632.8 nm, argon laser 488 nm) and the weak scattered light monitored by photon counting around ±50 nm of the exciting wavelength. Lines of energy lower than the exciting radiation are called Stokes lines and those of higher energy, but usually lower intensity, anti-Stokes lines. Rotational Raman transitions occur with the SELECTION RULE of $\Delta J = \pm 2$, and are also seen in the fine structure of vibrational spectra (*see* VIBRATION–ROTATION SPECTRA). Vibrational excitation has the selection rule that the vibrational quantum number changes by one with a change in POLARIZABILITY of the molecule. Thus the stretch of a diatomic molecule is Raman active. Molecular structure and shape may thus be determined from a comparison of Raman and IR spectra.

The resonance Raman effect occurs when Raman emission is enhanced by orders of magnitude by the absorption of the exciting radiation. The hyper-Raman effect is caused by the simultaneous arrival of two photons at the molecule. Selection rules are different from the normal Raman rules. Stimulated Raman effect is an enhanced Raman effect by coherent pumping of vibrational levels of the molecule by the intense electric field of the laser beam.

In coherent anti-Stokes Raman scattering (CARS), two lasers are used, one at fixed frequency and one tunable allowing the sample to be scanned in stimulated emission. Surface-enhanced Raman spectroscopy (SERS) is based on a phenomenon of enhanced emission arising from certain species strongly adsorbed at a surface.

Ramsay–Shields equation. Equation that relates the SURFACE TENSION (γ) of a liquid to the temperature T

$$\gamma = k(T_c - T - 6)(M/\rho)^{-2/3}$$

where M is the molecular mass (kg), ρ the density (kg m^{-3}) and T_c the critical temperature (see CRITICAL CONSTANTS). k is the Ramsay–Shields constant and for undissociated liquids has a value about 2.12×10^{-7}.

random coil. Various segments of a flexible linear POLYMER are subjected to independent thermal agitation, causing the molecule to adopt a random configuration in a 'self-avoiding walk'. The distance between the ends of the chain made up of n segments each of length l is $l(n)^{1/2}$. If the angle between the segments is $109°28'$ (i.e. tetrahedral) the distance is $l(2n)^{1/2}$. The randomness of the coil may be determined by measurement of viscosity.

Raney nickel. High-surface-area form of nickel. A nickel–aluminium alloy is treated with sodium hydroxide to leach out the aluminium. The resulting spongy metal shows great catalytic activity for HYDROGENATION.

Raoult's law. For IDEAL SOLUTIONS, the partial vapour pressure of one component in a mixture is equal to the vapour pressure of the pure component multiplied by its MOLE FRACTION in the liquid.

$$p_A = p_A^{\ominus} x_A \quad \text{and} \quad p_B = p_B^{\ominus} x_B$$

If B is non-volatile, $p_B^{\ominus} = 0$ and hence

$$x_B = (p_A^{\ominus} - p)/p_A^{\ominus}$$

Total vapour pressure above an ideal mixture given by

$$p = p_A^{\ominus} x_A + p_B^{\ominus} x_B$$

is linear between the vapour pressures of the two components. The composition of the vapour in equilibrium with the liquid mixture is given by

$$y_A = p_A^{\ominus} x_A/(p_A^{\ominus} x_A + p_B^{\ominus} x_B)$$

If $p_A^{\ominus} > p_B^{\ominus}$, then $y_A > x_A$ and the vapour is richer in the more volatile component. Deviations from Raoult's law for liquid mixtures cause the formation of an azeotrope (see AZEOTROPIC MIXTURE). See also REAL SOLUTION.

Rappe reaction. See ETHYNE.

rare earths. See LANTHANIDE ELEMENTS.

rare gases. See GROUP O ELEMENTS.

Raschig synthesis. See NITROGEN HYDRIDES.

rasorite. See BORON.

Rast's method. Rapid micro method of determining the relative molecular mass (accurate to 10 percent) by the depression of the FREEZING POINT of camphor, which has a large value of the cryoscopic constant ($k_c \sim 40$ K kg mol^{-1}).

rate constant (rate coefficient, k). Constant of proportionality in the RATE EQUATION. For reactions having an integral order (see ORDER OF REACTION), the units of an nth order rate constant (k_n) are (concentration units)$^{1-n}$ s^{-1}. See also HALF-LIFE.

rate-determining step (rate-limiting step). Individual reaction in a sequence of steps having the smallest rate, and thus the one that determines the overall progress of the reaction. See also RATE OF REACTION, RATE EQUATION.

rate equation. Relationship between the RATE OF REACTION and the concentrations of species in the system. In general, for reactants A, B, C, . . .

$$d\xi/dt = k[A]^a[B]^b[C]^c \ldots$$

a, b, etc. are ORDERS OF REACTION with

respect to A, B, etc. *See also* INTEGRATED RATE EQUATION.

rate law. *See* RATE EQUATION.

rate-limiting step. *See* RATE-DETERMINING STEP.

rate of reaction ($d\xi/dt$). Rate of advancement of a chemical reaction. In general for a reaction of STOICHIOMETRY

$$aA + bB + \ldots \rightarrow rR + sS + \ldots$$

$$\text{rate} = -1/a \, d[A]/dt = -1/b \, d[B]/dt = \ldots$$

$$= +1/r \, d[R]/dt = +1/s \, d[S]/dt = \ldots$$

See also RATE EQUATION, REACTION CO-ORDINATE.

ratio of heat capacities (γ). Ratio of the HEAT CAPACITY of a gas at constant pressure to that at constant volume ($\gamma = C_P/C_V$). The ratio is characteristic of the atomicity of the gas molecule: monatomic, 1.67; diatomic, 1.40; triatomic, 1.33.

Rayleigh scattering. Elastic (i.e. with no change in wavelength) scattering of radiation in a transparent medium. About 0.1 percent of the light is scattered in this way. A smaller part is scattered inelastically due to the Raman effect (*see* RAMAN SPECTRO-SCOPY).

rayon. *See* FIBRES.

Rb. *See* RUBIDIUM.

R-branch. *See* VIBRATION–ROTATION SPECTRA.

RDE. *See* ROTATING DISC ELECTRODE.

RDX (1,3,5-trinitrohexahydro-1,3,5-triazine). *See* EXPLOSIVES.

Re. *See* REYNOLD'S NUMBER, RHENIUM.

reaction coordinate (ξ). Pathway of minimum energy for a reaction. The energy of the system passes through a maximum at the transition state (*see* ACTIVATED COM-PLEX). *See also* POTENTIAL ENERGY SURFACE.

reaction cross-section. *See* COLLISION THEORY.

reaction order. *See* ORDER OF REACTION.

reactive dyes. DYES that form covalent bonds with the substrate. Such dyes are synthesized with a reactive group that can combine with amino or hydroxyl groups on cellulose, wool, silk and other fabrics. For example, activated halide atoms or ethenyl groups react with cellulose to form ester or ether links. Acid acceptors and heat promote the reaction.

realgar. *See* ARSENIC.

real gas. Gas that does not have the properties defined for an IDEAL GAS. It does not obey the ideal gas equation because the molecules are of finite size and there are forces between them. At high temperatures, the pressure–volume curve (*see* Figure) is nearly that for an ideal gas. At

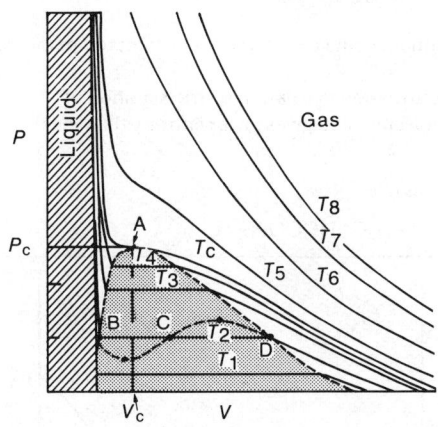

real gas
Typical isotherms for real gas (Andrew's isothermals)
$T_8 > T_7 > T_6 > T_5 > T_c > T_4 > T_3 > T_2 > T_1 >$.

lower temperatures, the isotherms have more complicated shapes, and at the critical temperature (T_c) there is an inflexion point A where the pressure and volume have the critical values P_c and V_c, respectively (*see* CRITICAL CONSTANTS). This is the critical state at which the gas and liquid phases have the same density. Below T_c, there is a two-phase region where gas (generally referred to as vapour) and liquid coexist in equilibrium. In the liquid region, to the left of the two-phase region, the isotherms are much steeper because the liquid is less compressible than the gas. The line BCD, a plot of the VAN DER WAALS EQUATION (a cubic in V), appears to show a continuous transition from the gaseous to the liquid state. In reality, this transition is abrupt and discontinuous, both phases existing along the horizontal line. The pressure at which liquid and vapour are in equilibrium is the saturated vapour pressure of the liquid.

real solution.   Solutions that shows deviations from RAOULT'S LAW and HENRY'S LAW; some showing negative and some showing positive deviations (*see* Figure and Table). Real solutions can be discussed in terms of excess functions, G^E, S^E, etc. defined as the difference between the observed value of the function of mixing and the calculated function for mixing of ideal solutions. Thus

$$S^E = \Delta S_{obs} - nR(x_A \ln x_A + x_B \ln x_B)$$

real solution

Negative deviations	Positive deviations
Strong forces of attraction between unlike molecules, $A-B > A-A, B-B$	Cohesive forces greater than forces between unlike molecules, $A-A, B-B > A-B$
$\Delta V_{mixing} < 0$	$\Delta V_{mixing} > 0$
$\Delta H_{mixing} < 0$	$\Delta H_{mixing} > 0$
Escaping tendency of each reduced	Escaping tendency of each increased
Minimum in vp−c curve at constant temperature	Maximum in vp−c curve at constant temperature
Association in solution restricts motion of molecules and gives lower entropy than ideal	Mixing breaks up association, heat is absorbed to accomplish this, entropy of mixing is less than ideal
Example:	Example:
$HCl-H_2O$, $HCOOH-H_2O$	$EtOH-H_2O$, dioxan$-H_2O$

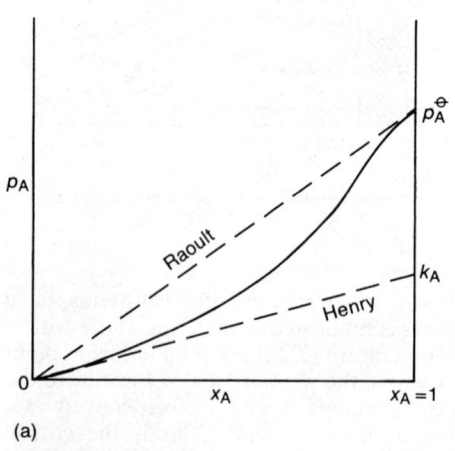

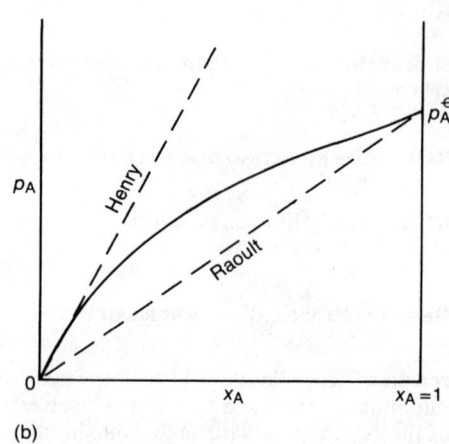

(a)   (b)

real solution
(a) negative deviation; (b) positive deviation.

Deviations of these excess functions from zero indicates the extent to which the solution is nonideal. A regular solution, a model in which $H^E \neq 0$, but S^E, can be visualized as a mixture in which the molecules are randomly distributed even though the interactions between A–A, B–B and A–B differ. *See also* COMPLETELY MISCIBLE LIQUID SYSTEMS, IDEAL SOLUTION.

Rebaudioside A. Sweet glycoside used as a SWEETENING AGENT.

reciprocal ohm. *See* MHO.

reciprocal proportions, law of. *See* EQUIVALENT PROPORTIONS, LAW OF.

rectified spirit. Aqueous ethanol containing 90 percent by volume ethanol.

redox electrode system. Inert platinum electrode in contact with a solution containing the oxidized and reduced forms of a redox couple (e.g., Fe^{3+}, $Fe^{2+}|Pt$). The ELECTRODE POTENTIAL of the half-cell is related to the standard electrode potential and the activities of oxidized and reduced forms by the NERNST EQUATION. The formal electrode potential (E') is defined in terms of concentrations; for the example given above

$$E = E' + (RT/F) \ln c_{Fe^{3+}}/c_{Fe^{2+}}$$

thus $E' = E^{\ominus} + (RT/F) \ln \gamma_{Fe^{3+}}/\gamma_{Fe^{2+}}$, where γ is an ACTIVITY COEFFICIENT, and thus E' depends on the IONIC STRENGTH of the solution. Standard redox potentials may be measured from the emf of suitably constructed cells in which the activity coefficients are estimated from the DEBYE–HÜCKEL ACTIVITY COEFFICIENT EQUATION. Values of E' may also be obtained from electrometric TITRATIONS.

redox indicator. *See* INDICATORS.

redox potential. *See* REDOX ELECTRODE SYSTEM.

redox titration. *See* TITRATIONS.

reduced equation of state. Defining a set of reduced variables as the actual variable divided by the corresponding critical constant (i.e. $P_r = P/P_c$; $V_r = V/V_c$; $T_r = T/T_c$), the reduced van der Waals equation of state can be written as

$$(P_r + 3/V_r^2)(V_r - 1/3) = 8T_r/3$$

If the van der Waals equation was exact, all gases would follow the same equation of state, since all the constants associated with the individual nature of the gas are removed. Reduced equations of state lead to the idea that if different substances are compared at equal fractions of their critical constants, they will appear to behave similarly; this is known as the principle of corresponding states. It is approximately valid for gases composed of spherical non-polar molecules, but breaks down when the molecules are non-spherical or polar.

reduced mass (μ; units: kg). Inertial mass of rotating or vibrating atoms. For two atoms mass m_1 and m_2, $\mu = m_1 m_2/(m_1 + m_2)$. *See also* HARMONIC OSCILLATOR, ROTATIONAL SPECTRA.

reducing agent. Substance that brings about REDUCTION of another substance; in the process it is oxidized. Reducing agents contain atoms with a low OXIDATION STATE, losing electrons during the process. Reducing agents have low standard ELECTRODE POTENTIALS.

reducing sugars. *See* SUGARS.

reduction. Reverse process to OXIDATION in which there is a gain of electrons by the reducing agent and a decrease in the oxidation state. In organic chemistry, the most obvious effect of reduction is the increase in the proportion of hydrogen and the decrease in the number of multiple bonds in the molecule. Specific reducing agents include hydrogen and lithium tetrahydridoaluminate.

reduction potential. *See* ELECTRODE POTENTIAL.

reference electrode. Reversible, nonpolarizable HALF-CELL of which the potential remains in equilibrium, against which the potential of a test half-cell may be measured or controlled. Common refe-

reference electrode
Potentials of some common reference electrodes in aqueous solution at 25°C.

Electrode	Half cell reaction	$E^{\ominus}$ $(25°C)/V^a$
Hydrogen	$H^+|H_2,Pt$	0.0
Saturated calomel	$Cl^-|Hg_2Cl_2,Hg$	+ 0.242
Normal calomel	$Cl^-|Hg_2Cl_2,Hg$ (1 mol dm^{-3})	+ 0.281
0.1 Normal calomel	$Cl^-|Hg_2Cl_2,Hg$ (0.1 mol dm^{-3})	+ 0.334
Silver–silver chloride	$Cl^-|AgCl,Ag$	+ 0.222
Silver–silver bromide	$Br^-|AgBr,Ag$	+ 0.071
Silver–silver iodide	$I^-|AgI,Ag$	– 0.152
Silver–silver oxide	$OH^-|Ag_2O,Ag$	+ 1.170
Mercury–mercury oxide	$OH^-|HgO,Hg$	+ 0.926
Antimony–antimony oxide	$OH^-|Sb_2O_3,Sb$	+ 0.255
Lead–lead sulphate	$SO_4^{2-}|PbSO_4,Pb(Hg)$	+ 0.965
Quinhydrone	$H^+, QH_2,Q|Pt$	+ 0.699
	(where $Q = C_6H_4O_2$)	

[a] All electrode potentials are given versus a standard hydrogen electrode except the metal–metal oxide electrodes which are versus a hydrogen electrode in the same electrolyte.

rence electrodes are tabulated (*see* Table). *See also* ELECTRODE POTENTIAL, TITRATIONS.

refining. (1) Production of pure metal from ORES or ALLOYS. (2) Conversion of crude oil into useful products. *See also* CRACKING, DISTILLATION, PETROLEUM, REFORMING.

refluxing. Process in which a liquid is boiled in a vessel attached to a (reflux) condenser so that the vapour is condensed and the condensate continuously returned to the vessel. It is a useful technique for carrying out reactions over extended periods of time.

Reformatsky reaction. CONDENSATION REACTION between ALDEHYDES or KETONES and α-haloesters in the presence of zinc. Subsequent working up with dilute acid gives the β-hydroxyester, for example

$R''CHBrCOOEt + Zn \rightarrow$

$BrZnCHR''COOEt \xrightarrow{RR'CO}$

$RR'C(OZnBr)CHR''COOEt \xrightarrow{H^+}$

$RR'C(OH)CHR''COOEt$

reforming. Thermal or catalytic process by which GASOLINE distilled from PETROLEUM is converted to higher-octane derivatives (*see* OCTANE NUMBER). Branched isomers are produced from straight-chain hydrocarbons, and cycloalkanes are dehydrogenated to aromatic compounds. Thermal reforming takes place at higher temperatures (600°C) and pressures (70 atm) than thermal CRACKING. Reforming over platinized aluminas and similar catalysts at lower temperatures and pressures is now the more important process.

refractory material. Substance that is stable at temperatures in excess of 1500°C in a clean atmosphere. It may be acidic (e.g., kaolin, silica, titania), basic (e.g., lime, magnesia, alumina) or neutral (e.g., graphite, carbide, chromite).

refrigeration. Process of cooling a substance and maintaining it at a temperature below that of its surroundings. Most refrigerators use a vapour compression cycle, equivalent to that of a heat pump. The refrigerant vapour is compressed and condensed, and the high-pressure liquid passed through a reducing valve. Reduction of pressure causes partial vaporization of the liquid causing a lowering of the temperature. The cold liquid passes to an evaporator where it is vaporized by heat exchange with the material to be cooled; the vapour passes back to the compressor for a further cycle. Chlorofluorocarbons (*see* FREONS) are widely used as refrigerants; ammonia and sulphur dioxide are not used extensively owing to their toxicity.

regular solution.   *See* REAL SOLUTION.

regular system.   *See* CRYSTAL SYSTEMS.

Reimer–Tiemann reaction.   Formation of phenolic aldehydes by refluxing an alkaline solution of the phenol with trichloromethane; with PHENOL a mixture of the 2- and 4-hydroxybenzaldehydes is formed (*see* Figure). If tetrachloromethane is used instead of trichloromethane, a phenolic acid (e.g., 2-hydroxybenzoic acid) is formed.

relative atomic mass (relative atomic weight, A_r).   Ratio of the average mass per atom of the naturally occurring element to 1/12 of the mass of a carbon-12 atom.

relative density (specific gravity).   Dimensionless quantity; ratio of the DENSITY of a substance to that of a reference substance under specified conditions. For solids and liquids, the reference is water and the density of the solid or liquid is taken at a specified temperature (usually 20°C) and the density of water at 4°C.

relative molecular mass (relative molecular weight, M_r).   Ratio of the average mass per molecule of the naturally occuring element or compound to 1/12 of the mass of a carbon-12 atom. It is equal to the sum of the RELATIVE ATOMIC MASSES of the constituent atoms.

relative permittivity.   *See* PERMITTIVITY.

relaxation.   Movement of a system from an EXCITED STATE to the ground state. In kinetics, fast reactions are studied by suddenly exciting an equilibrium by light

(*see* FLASH PHOTOLYSIS) or heat (*see* TEMPERATURE JUMP) and then by following the subsequent re-establishment of equilibrium. (For kinetic equation of a system near equilibrium—*see* INTEGRATED RATE EQUATION, Table.)

rennin.   *See* ENZYMES.

representation.   *See* GROUP THEORY.

repulsive forces.   *See* INTERMOLECULAR FORCES, LENNARD–JONES (12,6) POTENTIAL.

reserpine.   *See* ALKALOIDS.

residual entropy.   *See* SPECTROSCOPIC ENTROPY.

resin.   Any of a class of solid or semi-solid organic materials of natural (vegetable products, e.g., rosin) or synthetic (*see* POLYMERS) origin with no definite melting point. They have a high and variable molecular mass.

resistance (Ω; units $V A^{-1}$).   A conductor has a resistance of 1 ohm when a current of 1 ampere flows through a potential difference of 1 volt. It is the reciprocal of CONDUCTANCE.

resistance thermometer.   *See* THERMOMETER.

resistivity (ρ; units: Ωm).   The resistivity of a conductor length l, cross-sectional area A of resistance R is $\rho = RA/l$. It is the reciprocal of CONDUCTIVITY.

resolution.   (1) In optical activity, separation of a RACEMIC MIXTURE into its two enantiomorphic forms; for example by (a) mechanical separation of the crystals of the two forms, (b) biochemical separation using bacteria or moulds which preferentially destroy one enantiomer, (c) salt formation with a chiral reagent (*see* Figure).

Bases used for resolution of acids include quinine, brucine, strychnine and morphine (*see* ALKALOIDS) and acids for resolution of bases include 2,3-dihydroxybutanedioic (tartaric) acid and 7,7-dimethyl-2-oxobicyclo[2.2.1]heptane-1-methanesulphonic

$$(+)\text{-B} + (-)\text{-B} + 2(+)\text{-A} \underset{\searrow}{\overset{\nearrow}{}} \begin{matrix} (+)\text{-B} - (+)\text{-A} & \xrightarrow{\hspace{2cm}} & (+)\text{-B} \\ & \text{decomposition} & \\ (-)\text{-B} - (+)\text{-A} & \xrightarrow{\hspace{2cm}} & (-)\text{-B} \end{matrix}$$

racemic mixture resolving agent mixture of diastereomers separated by fractional crystallization chromatography, etc.

resolution
Salt formation with a chiral agent.

(camphorsulphonic) acid. (2) In chromatography, ability of chromatographic technique to separate two peaks on a CHROMATOGRAM.

resolving power. Measure of the ability of an optical instrument to produce detectably separate images of close objects or to separate spectral lines whose wavelengths are close together. For a spectrometer, it is $\lambda/\delta\lambda$, where $\delta\lambda$ is the difference in wavelength of two equally strong spectral lines that can barely be separated and λ is the average wavelength of the two lines.

resonance. Method of representing the structure of a molecule by two or more conventional formulae. If the valence electrons are capable of several different arrangements which differ by only a small amount of energy, then the actual arrangement will be a hybrid of these various alternatives, canonical forms (e.g., Kekulé structures of BENZENE, and for methanal: $H_2C{=}O \longleftrightarrow H_2C^+O^-$).

resonance energy. Difference between the observed and the calculated (from BOND DISSOCIATION ENTHALPY values) enthalpy of atomization. For benzene the resonance energy is 169 J mol^{-1}, thus more heat is evolved in the formation of BENZENE from gaseous atoms than would be expected on the basis of the Kekulé structure.

resorcinol. *See* DIHYDROXYBENZENES.

retention of configuration. *See* CONFIGURATION.

retention time. Time interval from the time of injection to the appearance of the peak maximum in gas–liquid chromatography and high-pressure liquid chromatography. *See also* CHROMATOGRAM.

retention volume. Volume of MOBILE PHASE required to elute a substance from a chromatograph column (i.e. the flow rate multiplied by the retention time). *See also* CHROMATOGRAM.

retinol (vitamin A). *See* VITAMINS.

retro organic synthesis. Logical analysis of possible synthetic pathways to a target molecule which involves considering the different reasonable routes by which its structure may be broken down into a number of simple reagents and small molecules.

reversed-phase chromatography. Technique in which the MOBILE PHASE is hydrophilic and the STATIONARY PHASE hydrophobic (*compare* PAPER CHROMATOGRAPHY). The mobile phase (commonly a mixture of water with methanol, ethanol or ethanenitrile) is the more polar phase, and polar solutes will move with this phase rather than remain with the less polar stationary phase. It is used in the separation of hydrocarbons and related compounds and in high-pressure liquid chromatography. In paper chromatography, the paper is soaked in the appropriate solvent before using the aqueous solvent as mobile phase, and in thin-layer chromatography the silica gel is bonded with octadecylsilane to give the hydrophobic stationary phase.

reversed-phase liquid chromatography (RPLC). Powerful method of separation of peptides and proteins, using a stationary hydrophobic phase and a polar mobile phase (*see* REVERSED-PHASE CHROMATOGRAPHY). The extent of retention of a molecule depends on the number, size and stereochemistry of its hydrophobic and hydrophilic groups, and is manifested by the combined interactions of each group with the stationary and mobile phases and even with each other. Reversed phases such as octadecylsilyl (ODS), octylsilyl, butylsilyl and propylsilyl functions bonded to silica supports, and mobile phases consisting of aqueous solutions of various acids with ethanenitrile, methanol or propanol have been used with success in the separation of a range of proteins.

reverse osmosis. Separation of a solute from a solution by causing the solvent to flow through a semipermeable membrane under an applied pressure in excess of the normal OSMOTIC PRESSURE (> 25 atm). The method is used for the desalination of seawater, the treatment of water for recycling in chemical plants and the concentration of food products (e.g., milk and fruit juices).

reversible galvanic cell.   *See* CELLS.

reversible process. Process that can be reversed by an infinitesimal change in a variable (e.g., temperature or pressure); in the limit the process proceeds infinitely slowly through a sequence of equilibrium states. For the process to occur, the driving (external) force must be slightly greater than the opposing (internal) force (e.g., the balancing of the potential of a cell in a potentiometer circuit). Reversible processes do not generate ENTROPY, but may transfer it from one part of the universe to another. The maximum amount of WORK is obtained when a process is carried out reversibly. *See also* ADIABATIC PROCESS, IRREVERSIBLE PROCESS, ISOBARIC PROCESS, ISOCHORIC PROCESS, ISOTHERMAL PROCESS.

Reynold's number (Re). Dimensionless number, quantity used in fluid dynamics to determine the type of flow through a pipe. It is defined by

$$Re = v \rho l / \eta$$

where v is the flow velocity and l a characteristic linear dimension (e.g., the diameter of the pipe). In a pipe, laminar flow usually occurs if Re < 2000 and turbulent flow if Re > 3000.

R_f value. Measure of the movement of a solute relative to the solvent front in paper and thin-layer chromatography. It is a constant for a particular compound under specified conditions of stationary and mobile phases and temperature.

Rh.   *See* RHODIUM.

rH. Term used in redox reactions

$$rH = -\lg p_{H_2}$$

where p_{H_2} is the partial pressure of hydrogen in equilibrium with the system.

rhamnose ($CH_3(CHOH)_4CHO$). 6-Deoxyhexose. β-L-(–)-Rhamnose is a constituent of many glycosides particularly in combination with 3-hydroxyflavone derivatives. It exhibits the usual properties of HEXOSES most closely resembling MANNOSE. It undergoes mutarotation.

rhenium (Re). Group VIIb, third row TRANSITION ELEMENT. It is found in small quantities associated with many minerals (e.g., in molybdenum glance, MoS_2). The metal is obtained by reduction of rhenium compounds by hydrogen. Rhenium is used in alloys with tungsten and molybdenum, in thermocouples and as a catalyst.
$A_r = 186.207$, $Z = 75$, ES [Xe]$5d^5 6s^2$, mp 3180°C, bp 5627°C, $\rho = 20.53 \times 10^3 \, kg\,m^{-3}$.

rhenium compounds. Oxidation states –1 to +7 are found; +4 and +5 are common. The higher states are less oxidizing than the corresponding MANGANESE COMPOUNDS. Known halides are ReF_n ($n = 4, 5, 6, 7$), $ReCl_n$ ($n = 3, 4, 5, 6$), $ReBr_n$ ($n = 3, 4, 5$), ReI_n ($n = 1, 2, 3, 4$). The higher halides are hydrolyzed in water, and the lower halides have rhenium–rhenium bonds (e.g., $ReCl_3$ has a triangle of rhenium atoms). High-temperature oxidation of the metal gives volatile Re_2O_7, which may be reduced to ReO_2. ReO_3, Re_2O_3 and Rh_2O are also known. Perrhenic acid ($HReO_4$, compare $HMnO_4$) is formed when Re_2O_7 dissolves in water.

rheology. Study of the deformation and flow of matter, that is flow–stress relationships. It is a subject of increasing technological importance in pharmaceutical, rubber, plastic, food, paint and textile manufacture where the suitability of the products is, to a large extent, judged in terms of their mechanical properties.

rheopexy. Phenomenon whereby the gelation of some thixotropic sols (*see* THIXOTROPY) is accelerated by gentle agitation: for example, bentonite clay suspensions often set slowly on standing, but quite rapidly when gently disturbed. *See also* NEWTONIAN FLUID.

rhodium (Rh). PLATINUM GROUP METAL, second row TRANSITION ELEMENT. It occurs naturally alloyed with other noble metals. Insoluble in AQUA REGIA, rhodium is smelted out with lead and extracted by fusing with potassium hydrogen sulphate. The chloride (precipitated rhodium hydroxide treated with hydrochloric acid) is purified using an ION EXCHANGER and reduced to the metal by hydrogen. Rhodium is a hard white metal, the principal use of which is to alloy with platinum to increase its hardness and melting point. Rhodium is an excellent HYDROGENATION catalyst and is used in thermocouples and other high-temperature devices. It is inert to most acids and halogens at moderate temperatures, but is attacked by hot sulphuric acid, hot hydrobromic acid and hot sodium chlorate(I). Rhodium dissolves in molten lead and bismuth, but is insoluble in mercury, molten gold, silver, sodium or potassium.
$A_r = 102.9055$, $Z = 45$, ES $[Kr]4d^8 5s^1$, mp 1966°C, bp 3727°C, $\rho = 12.4 \times 10^3$ kg m^{-3}. Isotopes: ^{103}Rh, 100 percent.

rhodium compounds. Oxidation states -1 to $+6$ are found, $+3$ is most common; higher oxidation states are strongly oxidizing and lower states are unstable. Octahedral complexes of rhodium(III) are numerous in aqueous solution, ligands successively replacing water in the hexa-aquo rhodium(III) ion. A number of halides are known (e.g., RhF_6 ($Rh+F_2$), $(RhF_5)_4$, RhF_4, RhF_3). Anhydrous $RhCl_3$ gives $[RhCl_6]^{3-}$ in aqueous solution. Rh_2O_3 is the most common oxide. Solid rhodates(IV) and rhodates(III) form with alkali metals and alkaline earths (e.g., Na_2RhO_3 and $BaRh_2O_4$). Square planar Rh(I) complexes are active catalysts (*see* WILKINSON'S CATALYST) and readily react to rhodium-(III). Several carbonyls are known (e.g., $Rh_2(CO)_8$, $Rh_6(CO)_{16}$, $[Rh(CO)_4]^-$).

rhombic system. *See* CRYSTAL SYSTEMS.

rhombohedral system. *See* CRYSTAL SYSTEMS.

riboflavin (vitamin B complex). *See* VITAMINS.

ribonuclease. *See* ENZYMES.

ribonucleic acid. *See* NUCLEIC ACIDS.

ribose. *See* PENTOSES.

ribosome. Particle consisting of roughly equal amounts of RNA (*see* NUCLEIC ACIDS) and protein (i.e. a ribonucleoprotein) which occurs in virtually all living cells and is the site of protein synthesis. Two types are known: (1) those in chloroplasts and mitochondria of eukaryotes, type 70S with 50S and 30S subunits (18 nm diameter); (2) those in cytoplasm, type 80S with 60S and 40S subunits (22 nm diameter).

ribulose. *See* PENTOSES.

Rice–Ramsperger–Kassel–Marcus theory. *See* RRKM THEORY.

ricinine. *See* ALKALOIDS.

ricinoleic acid. *See* CASTOR OIL.

Rideal–Eley mechanism. Surface reaction in which the total number of bonds to the surface remains constant. This usually involves both covalent bonds and weak van der Waals interactions (*see* INTER-MOLECULAR FORCES). The theory has been applied to the catalyzed *ortho–para*-hydrogen conversion, the hydrogenation of ethene on nickel and the exchange between hydrogen and deuterium on tungsten. *See also* HETEROGENEOUS CATALYSIS.

ring cleavage. Breaking of a carbocyclic or heterocyclic ring to give aliphatic products. Examples include: (1) catalytic reduction or reaction with hydrogen iodide (e.g., piperidine to n-pentane); (2) oxidation of cyclic alcohol, ketone or alkene; (3) HOFMANN ELIMINATION (e.g., treatment of piperidine with chloromethane, and silver oxide followed by elimination of water to dimethyl-4-pentenylamine; repeat operation yields 1,3-pentadiene).

ring contraction/enlargement
Contraction/enlargement of cyclic amines.

ring closure. Ease of ring closure depends on the distance between the two groups involved and the strain of the ring. (For selection of typical methods—*see* Table.)

ring contraction/enlargement. Any reaction that involves rearrangement of the carbon skeleton of a molecule can in principle be used for ring contraction or enlargement. Thus

will give a contraction if C′ and C‴ are connected in a ring and expansion if C′ and C″ are connected in a ring. Certain cyclic amines with nitrous acid (HNO_2) can undergo ring contraction or expansion through an intermediate carbonium ion (*see* Figure). Alteration of ring size may occur during a PINACOL–PINACOLONE REARRANGEMENT. Some cyclic ketones suffer ring enlargement when they react with diazomethane, thus cyclohexanone gives cycloheptanone and higher cycloalkanones. The BECKMANN REARRANGEMENT of cyclic ketoximes results in ring enlargement to lactans.

ring current. Effect found in NUCLEAR MAGNETIC RESONANCE SPECTROSCOPY in which electrons delocalized in unsaturated and aromatic rings deshield (*see* SHIELDING) the protons in the ring. Absorption thus occurs at a lower field.

Ringer's solution (physiological saline). Aqueous solution of calcium, potassium and sodium chlorides used as medium for the *in vitro* study of living cells and tissue. The solution must be ISOTONIC with the cells and at the correct pH. This is achieved by buffering with bicarbonate or phosphate.

ring enlargement. *See* RING CONTRACTION/ENLARGEMENT.

ring strain. Strain imposed in ring compounds when the $C\hat{C}C$ angle is less than the tetrahedral angle (109°28′). Assuming planar molecules, a measure of the total strain may be obtained from the difference between the enthalpy of combustion of the cycloalkanes and n-alkanes. The values for three-, four-, five-, six- and seven-membered ring compounds are 120, 112, 35, 12 and 35 kJ, respectively, indicating an increase in stability up to the six-membered ring. From seven to 11 carbons, the stability decreases slightly; larger ring compounds have a stability comparable to that of a six-membered ring compound. In cyclopropane, the carbon hybridized orbitals are not pointing towards one another (the angle of 60° is impossible since the carbon valence angle can never be less than 90°, i.e. pure *p*-orbitals) in the same straight line. Hence there is loss of overlap and instability, and the molecule is in a state of strain through 'bent' bonds.

Ritz combination principle. Position of any lines in the spectrum of atomic hydrogen may be expressed as the difference between two terms of the form R/n^2, where n is an integer. *See also* ATOMIC SPECTROSCOPY, BALMER SERIES.

RMS (root mean square velocity). *See* KINETIC THEORY OF GASES.

Rn. *See* RADON.

RNA (ribonucleic acid). *See* NUCLEIC ACIDS.

Rochelle salt (potassium sodium tartrate). *See* 2,3-DIHYDROXYBUTANEDIOIC ACID.

ring closure
Some typical methods of ring closure

---

1) Treatment of the α,ω-dihalide with sodium or zinc (suitable up to 1,6-dihalide)

2) Condensation reaction between α,ω-dihalide and sodium diethylpropanedioate

$$Br(CH_2)_nBr + CH_2(COOEt)_2 + NaOEt \longrightarrow (CH_2)_nC(COOEt)_2$$

3) Intramolecular FRIEDEL–CRAFTS REACTION

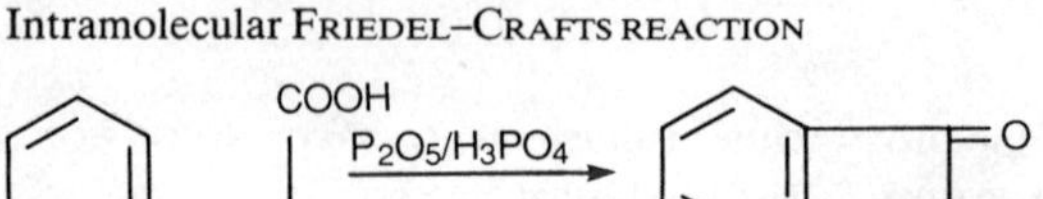

4) Dieckmann synthesis. Intramolecular cyclization of a dicarboxylic ester in the presence of a strong base

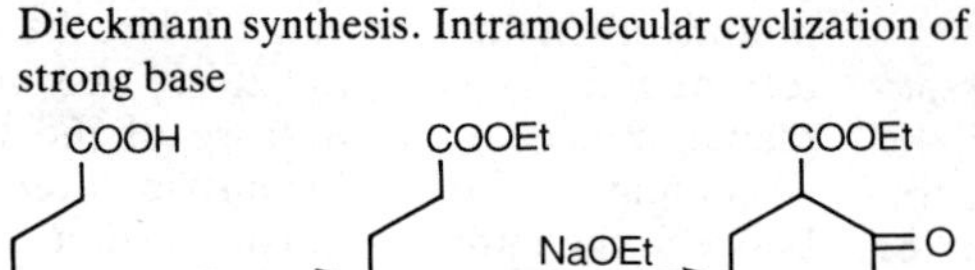

5) Intramolecular aldol-type reaction. 1,4-, 1,5- and 1,6-dicarbonyl compounds are capable of base-catalyzed cyclizations

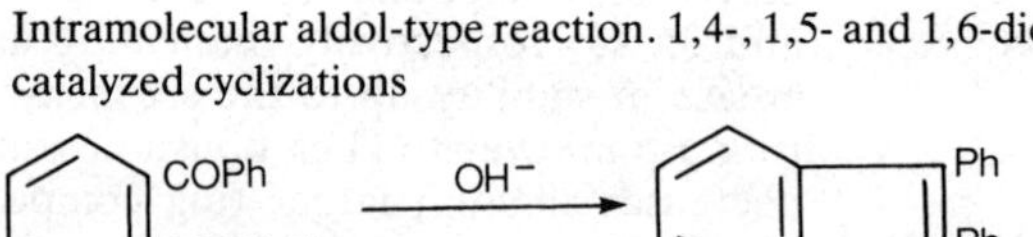

6) DIELS–ALDER REACTION

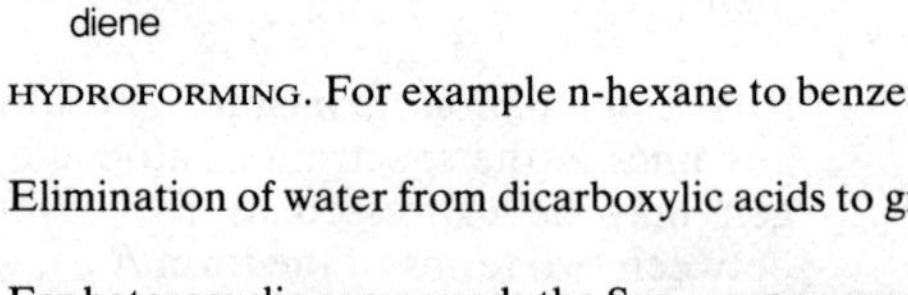

open chain   dienophile
diene

7) HYDROFORMING. For example n-hexane to benzene

8) Elimination of water from dicarboxylic acids to give anhydrides

9) For heterocyclic compounds the SKRAUP REACTION, the FISCHER INDOLE SYNTHESIS, and many others for specific compounds

10) Dimerization of butadiene in the presence of organometallic catalysts to 1,5-cyclooctadiene and 1,5,9-cyclododecatriene

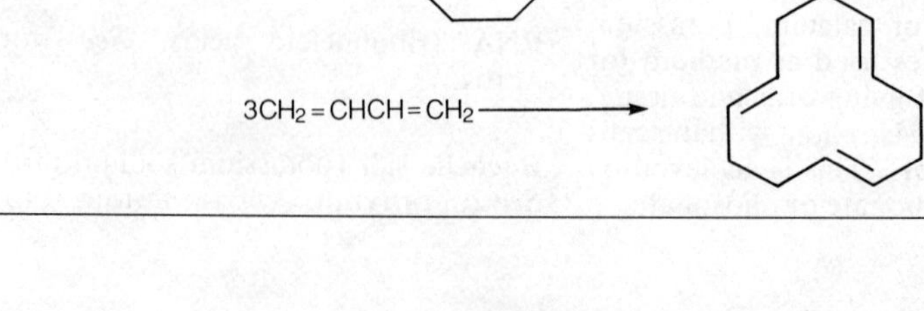

$2CH_2 = CHCH = CH_2 \longrightarrow$

$3CH_2 = CHCH = CH_2 \longrightarrow$

rocking vibration. *See* NORMAL MODE.

rock salt lattice. *See* CRYSTAL FORMS.

rodenticide. Chemical used to kill rats and other rodents (e.g., WARFARIN).

rodinol. *See* AMINOPHENOLS.

roentgen (R). Unit of exposure dosage that will produce ions carrying a charge of 2.58×10^{-4} C in 1 kg of dry air.

root mean square velocity. *See* KINETIC THEORY OF GASES.

Rosenmund reaction. *See* BENZALDEHYDE.

rotary reflection. *See* SYMMETRY ELEMENTS.

rotating disc electrode (RDE). Electrode that is rotated about its axis at a known speed in order to control the transport of material to it. The LIMITING CURRENT DENSITY is given by the Levich equation

$$i_L = 1.554\omega^{1/2}\nu^{-1/6}D^{2/3}c_{bulk}$$

where ω is the rotation speed, ν the KINEMATIC VISCOSITY, D the diffusion coefficient (*see* FICK'S LAW OF DIFFUSION) and c_{bulk} the concentration of the reactant in the bulk solution.

rotational axis. *See* SYMMETRY ELEMENTS.

rotational partition function. *See* PARTITION FUNCTION.

rotational selection rules. *See* SELECTION RULES.

rotational spectra (microwave spectra). Absorption or emission in the microwave part of the ELECTROMAGNETIC SPECTRUM due to rotational transitions of molecules. For a heteronuclear diatomic molecule, a transition from rotational state J to $J+1$ requires energy ΔE_{rot} equal to $(J+1)h^2/4\pi^2 I$. I is the MOMENT OF INERTIA, $I = \mu r^2$, where μ is the REDUCED MASS and r the bond length. The frequency of a rotational transition ($\Delta E_{rot}/h$) is $2B(J+1)$. The rotational constant (B, Hz) is equal to

$h/8\pi^2 I$. The SELECTION RULES for rotational transitions require the molecule to have a permanent DIPOLE MOMENT and $\Delta J = \pm 1$. Rotating molecules are classed in terms of their moments of inertia about three orthogonal axes (I_A, I_B, I_C, such that $I_A \leqslant I_B \leqslant I_C$). A linear molecule has $I_A = 0$, $I_B = I_C$, symmetric tops have $I_A = I_B < I_C$ (oblate) and $I_A < I_B = I_C$ (prolate), spherical tops have $I_A = I_B = I_C$ and asymmetric tops $I_A < I_B < I_C$. Rotational spectra of low-pressure gases (1–10 Pa) are obtained in a microwave spectrometer (*see* SPECTROPHOTOMETER). The output signal is modulated at about 50 000 Hz or by an electrostatic field (*see* STARK EFFECT). The measurement of the rotational constant (B) from the spacing of rotational lines allows calculation of the bond lengths in molecules. Allowance for ANHARMONICITY, centrifugal distortion and coriolis forces gives added information. Rotational energy levels may be split by coupling with NUCLEAR SPIN, or nuclear quadrupole moments, by an electric field or a magnetic field (*see* ZEEMAN EFFECT).

rotatory power. *See* SPECIFIC OPTICAL ROTATORY POWER, SPECIFIC ROTATION.

rotenone ($C_{23}H_{22}O_6$). Principal constituent of derris root. Insoluble in water, soluble in most organic solvents. Very toxic, powerful inhibitor of mitochondrial electron transport. It is used in flea powders and fly sprays.

rotenone

RPLC. *See* REVERSED-PHASE LIQUID CHROMATOGRAPHY.

RRKM theory (Rice–Ramsperger–Kassel–Marcus theory). Statistical mechanical treatment of the kinetics of unimolecular reactions (*see* LINDEMANN THEORY) which takes into account the vibrations, NORMAL MODES and ZERO-POINT ENERGY of the

ACTIVATED COMPLEX. It combines an earlier theory, the RRK theory, with ABSOLUTE RATE THEORY.

R, S-system.   See OPTICAL ACTIVITY.

Ru.   See RUTHENIUM.

rubber.   POLYMER that is ELASTIC at room temperature. Rubbers are based on polymers of butadiene and its derivatives. Before use many additives are compounded with the raw rubber, such as vulcanizing agents (see VULCANIZATION), accelerators, fillers, EXTENDERS, blowing agents. (For the composition of natural and synthetic rubbers—see POLYMERS, Table.) See also ELASTOMER.

rubidium (Rb).   Group Ia ALKALI METAL. It occurs naturally in reasonable abundance in complex minerals (e.g., lepidolite, 3 percent Rb_2O) and in seawater. Rubidium metal is not prepared commercially, but may be extracted from the chloride with calcium or by electrolysis. Rubidium compounds are purified from lithium by complex formation with, for example, $[SnCl_3]^-$ or $[Fe(CN)_6]^{4-}$. The metal is silvery–white and is very reactive with air and water. It is used as a getter and in photocells. The naturally occurring isotope ^{87}Rb, which decays to ^{87}Sr by β^- (half-life: 4.7×10^{11} yr), is used in dating rocks (compare POTASSIUM–ARGON DATING).
$A_r = 85.4678$, $Z = 37$, ES $[Kr]5s^1$, mp $38.9°C$, bp $688°C$, $\rho = 1.53 \times 10^3$ kg m^{-3}. Isotopes: ^{85}Rb 72.2 percent, ^{87}Rb 27.8 percent.

rubidium compounds.   Rubidium is a very ELECTROPOSITIVE metal ($E^\ominus_{Rb^+,Rb} = -2.92$ V) forming Rb^+ in all of its ionic salts. Reactions are typical of the ALKALI METALS. Some organic compounds (e.g., with CROWN ETHERS) are known. These are ionic and similar to those of sodium and potassium.

ruby.   See CORUNDUM.

Russell–Saunders coupling.   See TERM SYMBOLS.

rust.   Hydrated iron(III) oxide formed by the oxidation of iron in the atmosphere; the presence of water is critical. The process is accelerated by atmospheric pollutants (e.g., sulphur dioxide, salt spray). See also CORROSION.

ruthenium (Ru).   Second row TRANSITION ELEMENT. One of the noble or PLATINUM GROUP METALS which occurs naturally alloyed with other metals of the group (see OSMIRIDIUM). Pure ruthenium is separated from osmium (from which platinum, palladium and rhenium have been extracted previously) by fusion with sodium oxide to yield soluble species, which are converted to the chloride (Cl_2, HCl). The metal is obtained from the reduction of compounds by hydrogen. Ruthenium, a hard silvery-white metal, is resistant to acids up to 100°C and is moderately resistant to halogens at room temperature. The metal is mainly used in CATALYSIS.
$A_r = 101.07$, $Z = 44$, ES $[Kr]4d^7 5s^1$, mp $2310°C$, bp $4080°C$, $\rho = 12.45 \times 10^3$ kg m^{-3}. Isotopes: ^{96}Ru, 5.46 percent; ^{98}Ru, 1.87 percent; ^{99}Ru, 12.63 percent; ^{100}Ru, 12.53 percent; ^{101}Ru, 17.02 percent; ^{102}Ru, 31.60 percent; ^{104}Ru, 18.87 percent.

ruthenium compounds.   Oxidation states -2 to $+8$ are known. In aqueous solution $+3$ and $+2$ are common as the hexaaquo complex. Nitrogen complexes readily, dinitrogen and nitrosyl complexes of ruthenium(II) and ruthenium(III) are known. Of the halides, RuF_n ($n = 3, 4, 5, 6$) have been synthesized. Complex halides (e.g., K_2RuCl_6) are formed from RuO_3 and a mixture of the acid halide and alkali metal salt. $Ru(CO)_5$ is formed by the action of carbon monoxide on the metal. Ruthenates(VII) (RuO_4^-) decompose in water to orange ruthenates(VI) (RuO_4^{2-}) and oxygen. RuO_2 is conducting and is the major constituent of dimensionally stable anodes used in the production of chlorine.

rutherfordium.   See TRANSACTINIDE ELEMENTS.

rutile.   See TITANIUM.

rutile lattice.   See CRYSTAL FORMS.

Rydberg constant.   See BALMER SERIES.

Rydberg series.   See ELECTRON SPECTROSCOPY.

S

S. *See* SIEMENS, SULPHUR, SVEDBERG UNIT.

S. *See* ENTROPY, SPECIFIC ACTIVITY.

S-. Configurational symbol (*see* OPTICAL ACTIVITY.

s. *See* SECOND.

s. Spin quantum number (*see* QUANTUM NUMBER).

σ. *See* COLLISION CROSS-SECTION, SYMMETRY NUMBER.

σ-bonding. *See* MOLECULAR ORBITAL.

σ-donor ligand. *See* CARBON-TO-METAL BOND.

σ-orbital. *See* MOLECULAR ORBITAL.

Sabatier–Senderens reduction. Catalytic reduction, with hydrogen, of organic compounds in the vapour phase over hot, finely divided nickel.

sabinene. *See* TERPENES.

saccharin. *See* SWEETENING AGENTS.

Sackur–Tetrode equation. Equation permitting the calculation of the total molar translational ENTROPY of a gas (for a monatomic gas the total entropy) from a knowledge of relative molecular mass, temperature and pressure. Substituting the value of Q_t and the values of the fundamental constants in

$$S = N_A k \ln Q + N_A k T \, (\partial \ln Q / \partial T)_V$$

(*see* PARTITION FUNCTION).

S_t (J K^{-1} mol^{-1}) =

   28.72 lg $(M_r/$kg mol$^{-1})$ + 47.87 lg T

   − 19.14 lg $(P/$N m$^{-2})$ + 172.79

S_t of a gas at normal temperatures provides the greatest contribution to the total entropy. *See also* SPECTROSCOPIC ENTROPY.

sacrificial protection. *See* CORROSION.

saddle point. *See* ABSOLUTE RATE THEORY, ACTIVATED COMPLEX.

safrole. *See* PHENOLS.

salicin. *See* GLYCOSIDES.

salicylic acid. *See* 2-HYDROXYBENZOIC ACID.

salt bridge. Method of connecting a REFERENCE ELECTRODE with the test solution or two solutions such that the LIQUID JUNCTION POTENTIAL is reduced as far as possible. It consists of an inverted U-tube plugged at the ends containing a concentrated solution of an electrolyte of which the ions have similar TRANSPORT NUMBERS. The bridge solution may be made with agar (*see* SEAWEED COLLOIDS) to reduce diffusion of the bridge electrolyte into the cell solutions. Common bridge electrolytes are potassium chloride, potassium nitrate or ammonium nitrate.

salt effect. *See* KINETIC SALT EFFECT.

salting out. (1) COAGULATION of hydrophilic colloids (*see* LYOPHILIC COLLOID) by the addition of large amounts of electrolyte. The electrolyte competes for water of hydration so causing precipitation (*see* LYOTROPIC SERIES). Non-electrolytes may also salt out lyophilic sols by desolvating the colloid. For example, propanone and ethanol will coagulate aqueous gelatin, and petroleum ether will coagulate a benzene solution of rubber. (2) Precipitation of a non-electrolyte by the addition of a strong electrolyte.

salvarsan. *See* ARSENIC ORGANIC DERIVATIVES.

samarium (Sm). *See* LANTHANIDE ELEMENTS.

sand. Fine particles of SILICA found naturally after the weathering of sandstone. It is used in building and the manufacture of GLASS.

Sandmeyer reaction. Replacement of diazonium groups in aromatic compounds by halogen or pseudohalogen groups under catalytic influence of the corresponding copper(I) salt: for example,

$$PhN_2^+ Cl^- + KCN \longrightarrow PhCN + N_2$$

It permits the conversion of an amine to the corresponding halogen compound. *See also* GATTERMANN REACTION.

sandwich compound. Transition metal complex in which the metal atom is located between two cyclic π-donor organic ligands (*see* Figure), for example, cyclopentadienyl (*see* METALLOCENE), BENZENE, cyclohexadienyl, cyclobutadienyl, tropylium, cyclooctatetraene (*see* CYCLOALKENES). A compound having only one ring is known as a 'half sandwich' compound. Mixed sandwich compounds are known, for example the isoelectronic series based on ferrocene has been synthesized. The bonding is *hapto* in each case.

Sanger's reagent. *See* 1-FLUORO-2,4-DINITROBENZENE.

sapogenins. AGLYCONES obtained on the hydrolysis of plant GLYCOSIDES or SAPONINS. *See also* DIGITALIS.

saponification. Alkaline hydrolysis of an ESTER to yield a metal salt of the acid and an alcohol.

saponins. GLYCOSIDES, occurring in a wide variety of plants; characterized by the formation of colloidal soapy solutions in water. Even at high dilutions they lyze red blood cells. On hydrolysis, the saponins give a well-characterized sapogenin and a variety of sugars (glucose, galactose and arabinose are the most common). *See also* DIGITALIS.

sapphire. *See* CORUNDUM.

saturated compound. Organic compound having no multiple bonds in the molecule, or no multiple bonds in an alkyl chain. For example, 'saturated' carboxylic acids refers to the alkyl chain not to the CARBOXYL GROUP. Unsaturated compounds, therefore, have reactive multiple bonds.

saturated solution. Solution in which dissolved and undissolved solute are in EQUILIBRIUM with one another. The concentration of a saturated solution, termed the solubility, depends on the temperature. Breaks in the solubility/temperature curve are due to changes in the nature of the solid phase, dehydration, crystal structure, etc. (e.g., $Na_2SO_4.10H_2O$ in water).

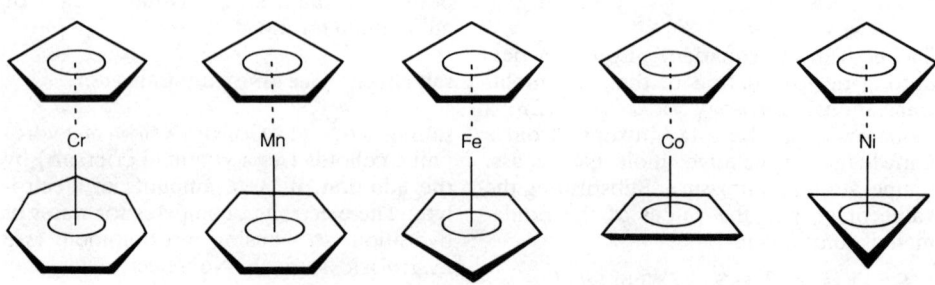

sandwich compounds

sawhorse projection formula. *See* NEW MAN PROJECTION FORMULA.

Sb. *See* ANTIMONY.

s-band. *See* BAND THEORY OF SOLIDS.

S-branch. *See* VIBRATION–ROTATION SPECTRA.

Sc. *See* SCANDIUM.

scandium (Sc). Group IIIb, first row TRANSITION ELEMENT. It occurs naturally as the ore thortveitite (34 percent Sc_2O_3), which may be reduced to the metal by calcium in an inert atmosphere, further refinement being by ZONE REFINING. The metal may also be produced by ELECTROLYSIS of the chloride. $A_r = 44.95591$, $Z = 21$, ES [Ar] $3d^1 4s^2$, mp 1811°C, bp 2870°C, $\rho = 3.02 \times 10^3$ kg m^{-3}. Isotope: ^{45}Sc, 100 percent.

scandium compounds. Scandium in oxidation state +3 forms complexes in aqueous solution: for example, $[ScF]^{2+}$, $[ScSO_4]^+$, $[Sc(S_2O_3)_3]^{3-}$, $[Sc(C_2O_4)]^+$, $[Sc(C_2O_4)_2]^-$, $[Sc(CH_3COO)_n]^{(3-n)+}$ ($n = 1, 2, 3, 4$). All halides are known together with the nitrate, monophosphide (ScP), silicides (Sc_nSi_m, $n,m = 1,2; 3,5; 1,1; 5,3$) and numerous organic complexes.

scattering. Radiation or particles that impinge on a medium and recoil in all directions. The scattering is dependent on the energy of the radiation or particle and the nature of the medium. *See also* RAYLEIGH SCATTERING, TYNDALL EFFECT.

scavenger. Species that reacts with, and thus removes, a particular molecule; for example, short-lived RADICALS may be isolated by reacting with a spin trap (*see* SPIN TRAPPING). Unwanted species may be scavenged (e.g., dissolved oxygen by hydrazine). In combustion engines BROMOETHANE and CHLOROETHANE are added to scavenge lead from ANTIKNOCK ADDITIVES, volatile lead halides being removed in the exhausts.

SCF. *See* SELF-CONSISTENT FIELD.

scheelite. *See* TUNGSTEN.

Schiemann reaction. Method for the preparation of aromatic fluoro compounds by treating diazonium compounds with borofluoric acid (HBF_4) to give the insoluble diazonium borofluoride. On gentle heating this decomposes to the aromatic fluoro compound. Diazonium hexafluorophosphate (diazonium salt + HPF_6) is an alternative intermediate.

Schiff's bases (anils, *N*-arylimines, ArN=CR$_2$). Prepared by the reaction between aromatic amines (*see* ANILINE) and aldehydes and ketones. The reaction is reversible in acid and thus the formation of a Schiff's base is used to protect an amino group

$$RCHO + ArNH_2 \rightleftharpoons RCH{=}NAr + H_2O$$

They are crystalline solids with characteristic melting points.

Schiff's reagent. Solution of rosaniline hydrochloride decolorized by sulphur dioxide. Aliphatic aldehydes and aldose sugars rapidly restore the magneta colour; aliphatic ketones do so more slowly.

Schotten–Baumann reaction. Simple method of ACYLATION in which an amine can be quantitatively converted into an amide. A mixture of the amine, cold aqueous base and excess of acyl halide (CH_3COCl, $PhCOCl$, etc.) is shaken together, preferably in the presence of a solvent. The method can also be applied to alcohols, the 4-nitrobenzoyl and 3,5-dinitrobenzoyl derivatives of the alcohols are well-defined crystalline substances used to characterize the alcohol.

Schottky defect. *See* DEFECT STRUCTURE.

Schrödinger equation

$$\mathcal{H}\psi = E\psi,$$

where $\mathcal{H}$ is the Hamiltonian operator, E the energy (the EIGENVALUE of the equation) and ψ the WAVEFUNCTION. The Hamiltonian is made up of two parts: a kinetic energy operator and a potential energy operator. The kinetic energy operator is $-h^2/8\pi^2 m \nabla^2$, ∇^2 for different coordinate systems depends on the system. The potential energy operator depends on the system being studied. The solution of the

Schrödinger equation is fundamental to all descriptions of the electronic structures of atoms and molecules. *See also* ATOMIC ORBITAL, MOLECULAR ORBITAL.

Schultze–Hardy rule. Critical coagulation concentration of an indifferent electrolyte for a LYOPHOBIC COLLOID is extremely dependent on the charge of the counterions, but practically independent of the particular nature of the ions, the charge of the co-ions and the concentration of the sol. It is only moderately dependent on the nature of the sol.

scintillation. Light emitted when ionizing radiation interacts with a suitable material (e.g., zinc sulphide for alpha-particles, lithium-doped sodium iodide for gamma-rays). Measurement of the light by a sensitive photomultiplier in a scintillation counter gives the amount of radiation. Scintillation counters are used to follow the incorporation of radioactive TRACERS in molecules of, for example, a biological origin.

scintillation counter. Equipment used for the detection and estimation of radioactivity by flashes of light emitted by excited atoms when they fall back to their ground state. The excited atoms arise from the interaction of the ionizing radiation with the scintillation medium. The scintillation medium may be a screen of material such as zinc sulphide for alpha-particles, thallium-doped sodium iodide crystals, NaI(Tl), for gamma-rays or liquid scintillators such as *p*-terphenyl (PPP), 1,4-diphenyloxazole (PPO) and 1,4-di-2-(5-phenyloxazole) (POPOP) for beta-particles.

(–)-scopolamine. · *See* ALKALOIDS.

screw axis. SYMMETRY OPERATION in which a rotation is followed by a lattice TRANSLATION parallel to the axis. *See also* SPACE GROUP.

Se. *See* SELENIUM.

seawater. Aqueous solution containing (in ppm) Cl^- (19 000), Na^+ (11 000), SO_4^{2-} (2600), Mg^{2+} (1300), Ca^{2+} (400), K^+ (400), HCO_3^- (140), Br^- (70), H_3BO_3 (30) and Sr^{2+} (10); buffered to pH 8.0–8.4 by the presence of Ca^{2+} and HCO_3^-. Magnesium and bromine are extracted from seawater.

seawater battery. *See* BATTERY.

seaweed colloids. Complex POLYSACCHARIDES obtained by the extraction of algae, particularly seaweed.
agar. Mixture of two polysaccharides: agarose a neutral polysaccharide with 3,6-anhydro-L-galactose and D-galactose as repeating units and agaropectin which carries carboxyl and sulphate groups. It is extracted from the red marine algae *Gelidium amansii* using hot water and is marketed as a powder. It is soluble in hot water, forming a sol which cannot be coagulated by salts, but which sets to a jelly on cooling. It is used extensively as a solid medium for the cultivation of bacteria and as a thickener, stabilizer and gelling agent in the food industry (E406).
algin (alginic acid). Polymer of D-mannuronic acid (*see* URONIC ACIDS); extracted from brown seaweeds, mainly *Laminaria*, growing off the west coasts of Scotland and Ireland. A solution of the sodium salt is very viscous. The salts and esters are used as emulsifiers, gelling agents and thickeners for food products (E400–E405) and in pharmaceuticals.
carrageenan (Irish moss). Natural extract of several seaweeds, notably *Chondrus crispus*. It is used extensively in the food industry as an emulsifying, thickening, and suspending and gelling agent (E407).

sec- (secondary). *See* PRIMARY.

secobarbital. *See* BARBITURATES.

seconal. *See* BARBITURATES.

second (s). SI unit of time; the duration of 9 192 631 770 periods of the radiation corresponding to the transition between the two hyperfine levels of the ground state of the caesium-133 atom.

secondary (*sec-*). *See* PRIMARY.

secondary cell. *See* CELLS.

secondary electron emission. Electron emission from a solid under the stimulus of a primary electron beam with energy of between 10 and 500 eV. True secondary electrons are generated by a cascade process and have energies characteristic of the surface in the range 0–20 eV. *See also* AUGER EFFECT, ELECTRON SPECTROSCOPY.

secondary ion mass spectrometry. *See* MASS SPECTROMETRY.

secondary isotope effect. *See* ISOTOPE EFFECTS.

secondary salt effect. *See* KINETIC SALT EFFECT.

secondary structure of proteins. *See* PROTEINS.

second law of thermodynamics. There are many statements of this law: (1) the ENTROPY of an isolated system increases during any natural process (Clausius); (2) heat will not pass spontaneously from a colder to a hotter body; (3) spontaneous processes are those that when carried out under proper conditions can be made to do WORK; (4) the energy of the universe is constant and the entropy (disorder) of the universe is tending to a maximum. It is a law of nature which distinguishes changes that occur of their own accord from those that do not occur under given circumstances and introduces the concept of entropy.

second-order reaction. *See* ORDER OF REACTION.

second-order transition. *See* PHASE TRANSITION.

sedimentation. Settling of solid particles in a liquid under gravity or a centrifugal force. Measurement of the sedimentation velocity (*see* SEDIMENTATION COEFFICIENT) under gravity may be used to determine particle size with a lower limit of about 1 μm. For smaller particles (e.g., nucleic acids, proteins, etc.) an ULTRACENTRIFUGE is used.

sedimentation coefficient. For a dispersion of a macromolecule subjected to a force, the movement of the boundary of the solid material reaches a constant velocity dx/dt when the driving force on the particles

(mg) is balanced by the resistance of the liquid.

$$(1-\nu\rho)mg = f\,dx/dt$$

$(1-\nu\rho)$ is a buoyancy correction and f is the frictional coefficient $= 6\pi\eta r$ for a spherical particle of radius r, in medium of viscosity η (*see* STOKES' LAW). For a centrifugal force the molecular mass of the particles is given by

$$M = kTs/D(1-\bar{v}\rho)$$

where D is the diffusion coefficient and s the sedimentation coefficient $= (dx/dt)/\omega^2 x$, ω is the rotation velocity. *See also* COLLOID, SVEDBERG UNIT, ULTRACENTRIFUGE.

sedimentation potential. SEDIMENTATION of a charged colloid leads to separation of the lighter COUNTER IONS from the heavier colloidal particles which gives rise to a potential gradient. This acts counter to the process of sedimentation.

sedimentation velocity. *See* SEDIMENTATION COEFFICIENT.

selection rules. Conditions that must be fulfilled for a transition between energy levels of an atom or molecule. 'Allowed' transitions occur; 'forbidden' ones do not. Gross selection rules are the general symmetry requirements of the transition. Specific selection rules are statements of the changes in QUANTUM NUMBERS (*see* Table).

selectivity coefficient $(k_{i,j})$. Ratio of the response of an ION-SELECTIVE ELECTRODE (for ion i) to an interfering ion j relative to that of the ion i.

selenates(IV), (VI). *See* SELENIUM OXOACIDS.

selenic acids. *See* SELENIUM OXOACIDS.

selenides. Binary compounds of selenium with other elements, generally prepared by direct reaction of the elements or by the action of hydrogen selenide (H_2Se) on the element or a derivative. Selenides are less stable than the SULPHIDES, being more easily oxidized and more readily hydrolyzed to H_2Se.

β-selenine. *See* TERPENES.

selection rules
Selection rules for energy transitions.

Transition	Spectroscopy	Gross selection rule	Specific selection rule
Electronic			
Hydrogen atomic levels	ATOMIC (*see also* BALMER SERIES)	Change in dipole	$\Delta l = \pm 1$ $\Delta n =$ any integer
Atomic levels when Russell–Saunders coupling occurs (*see* TERM SYMBOLS)	Atomic	Change in dipole	$\Delta S = 0$ $\Delta L = \pm 1, 0$ $\Delta J = \pm 1, 0$ (but $J = 0$ to $J = 0$ is forbidden)
Molecular levels	ELECTRONIC	Change in dipole	$\Delta \Sigma = 0$ $\Delta \Lambda = \pm 1, 0$
Complexes with centre of symmetry	Electronic	Change in parity (Laporte selection rules)	
Vibrational			
	INFRARED	Change in dipole	$\Delta v = \pm 1$ HARMONIC OSCILLATOR $\Delta v = \pm 2$ ANHARMONICITY
	RAMAN	Change in polarizability	$\Delta v = \pm 1$
Rotation			
Pure rotation	ROTATION	Permanent dipole moment	$\Delta J = \pm 1$
	Rotational Raman	Anisotropic polarizability	$\Delta J = \pm 2$
Vibration–rotation	Infrared	See above	$\Delta v = \pm 1$ $\Delta J = -1$ P-branch $\Delta J = +1$ R-branch $\Delta J = 0$ Q-branch
	Raman	See above	$\Delta v = \pm 1$ $\Delta J = -2$ O-branch $\Delta J = +2$ S-branch $\Delta J = 0$ Q-branch

selenium (Se).   Grey, non-metallic GROUP VI ELEMENT. The element (0.09 ppm of the earth's crust) occurs as selenide impurities in sulphide ores and flue dust. It is extracted as a solution in potassium cyanide solution to give KCNSe from which selenium is precipitated by acid. It also occurs in the anode sludge of electrolytic refining plants. Selenium is a *p*-type SEMICONDUCTOR; the grey allotrope, containing infinite spiral chains of atoms, is light-sensitive. The red allotrope, precipitated from solution by sulphur dioxide, contains Se_8 units and is soluble in carbon disulphide. Oxidation states +6, +4, +2, 0, –2 exist. Selenium reacts with many elements to give selenides,

halides, oxides and sulphides, and is oxidized to selenous acid by nitric acid and to selenic acid by sulphuric acid (*see* SELENIUM OXOACIDS). Selenium(IV) compounds act as acceptors and selenium(II) compounds as donors. The organic derivatives of selenium, mainly selenium(II), are similar to the simpler sulphur derivatives. Elemental selenium is non-toxic, but hydrogen selenide and other compounds are extremely toxic. It is used in photocells, xerography (a reprographic process) and as a dehydrogenating agent in organic syntheses, as a decolorizing agent in the glass industry and as an additive to stainless steel. It is a nutritional factor (interrelationship with vitamin E), preventing muscle degenerative diseases.
A_r = 78.96, Z = 34, ES [Ar] $3d^{10} 4s^2 4p^4$, mp 217°C, bp 684.9°C, ρ = 4.79 × 10^3 kg m^{-3}. Isotopes: ^{74}Se 0.89 percent, ^{76}Se 9.02 percent, ^{77}Se 7.58 percent, ^{78}Se 23.52 percent, ^{80}Se 49.82 percent, ^{82}Se 9.19 percent.

selenium halides.

selenium hexafluoride (SeF$_6$). Chemically fairly inert, colourless gas (direct combination of selenium and fluorine).

selenium tetrafluoride (SeF$_4$). Colourless liquid (F$_2$ on Se surface, 0°C). Forms complex fluorides (MSeF$_5$) with alkali metal fluorides, and 1:1 addition compounds with boron trifluoride and arsenic pentafluoride.

selenium tetrachloride (SeCl$_4$). Colourless crystalline solid (Cl$_2$ + Se$_2$Cl$_2$). In concentrated hydrochloric acid it forms hexachloroselenates ([SeCl$_6$]$^{2-}$). It is thermally more stable than SCl$_4$, but is hydrolyzed by water.

selenium monochloride (Se$_2$Cl$_2$). Reddish liquid (Cl$_2$+Se). It is a good chlorinating agent.

selenium tetrabromide (SeBr$_4$). Yellow solid (Se + excess Br$_2$) which decomposes to selenium monobromide and bromine at room temperature and is hydrolyzed by water to give selenous acid and hydrobromic acid.

selenium monobromide (Se$_2$Br$_2$). Dark red liquid (Se + equivalent amount of Br$_2$) which decomposes on heating and is hydrolyzed by water.

other selenium halides. No selenium iodides are known. The oxohalides (SeOF$_2$, SeOCl$_2$ (heating SeO$_2$ + SeCl$_4$), SeOBr$_2$ (distilling SeOCl$_2$ with NaBr)) are known.

selenium oxides.

selenium dioxide (SeO$_2$). White solid (burn Se in air) with polymeric structure, forms selenous acid with water. It used as an oxidizing agent in organic chemistry. The mixed oxide (SSeO$_3$) is formed from the reaction of selenium with molten sulphur trioxide.

selenium trioxide (SeO$_3$). Strong oxidant, formed with much selenium dioxide, when selenium vapour in oxygen is subjected to an electric discharge. The anhydride of selenic acid, it dissolves in liquid hydrogen fluoride to give fluoroselenic acid (FSeO$_3$H).

selenium oxoacids.

Very simple system by comparison with the SULPHUR OXOACIDS. No peroxoacids are known.

selenic(VI) acid (H$_2$SeO$_4$). Strong acid, similar to sulphuric acid obtained by the vigorous oxidation of selenites (selenates-(IV)) or by the fusion of selenium with potassium nitrate. The acid forms two series of salts which are isomorphous with the sulphates and bisulphates.

selenous acid (selenic(IV) acid, H$_2$SeO$_3$). Solution of selenium dioxide in water gives a solution containing the acid OSe(OH)$_2$. The acid reacts with oxides, carbonates, etc. to give salts containing the ions HSeO$_3^-$ and SeO$_3^{2-}$ which can be isolated.

selenosulphuric acid (H$_2$SeSO$_3$). Analogue of thiosulphuric acid. The free acid does not exist. Salts are prepared by the action of selenium on a sulphite; they readily decompose.

selenosulphuric acid.   *See* SELENIUM OXOACIDS.

self-consistent field (SCF).

Model of a multi-electron system in which each electron is considered to be affected by the average field of all the other electrons. Self-consistency is achieved when the orbitals used to calculate the field are generated as solutions to the SCHRÖDINGER EQUATION. The method is also known as the Hartree–Fock method.

self-ionization.   *See* AUTOPROTOLYSIS.

semicarbazide (aminourea, $NH_2CONH-NH_2$). White crystals (mp 96°C), prepared by heating hydrazine sulphate with potassium cyanate or by the electrochemical reduction of nitrourea (20 percent H_2SO_4, 10°C, Pb cathode). It is used to characterize oxo compounds with which it forms SEMICARBAZONES.

semicarbazones. Organic compounds containing the group $>C=NNHCONH_2$. They are formed by the condensation of oxo compounds with SEMICARBAZIDE in ethanol or pyridine

$$NH_2CONHNH_2 + O=CR_2 \rightarrow$$

$$R_2C=NNHCONH_2$$

This reaction is used to characterize oxo compounds. When heated with sodium ethoxide at 180°C, nitrogen is eliminated and a hydrocarbon (R_2CH_2) is formed.

semiconductor. Material having a conductivity between that of a METAL and an INSULATOR and a positive temperature coefficient of conductivity. Semiconductors may be classed as intrinsic or doped. In terms of the BAND THEORY OF SOLIDS, an intrinsic semiconductor is a material with a band gap which allows a significant population of electrons in the conduction band by thermal excitation. An impurity semiconductor contains a small concentration of foreign atoms which may either accept electrons from the valence band or donate them to the conduction band. In the former case the conductor is the positive hole remaining in the valence band and the material is said to be a p-type semiconductor, and in the latter case the impurity electrons in the conduction band conduct to give a n-type semiconductor. For example, silicon is an intrinsic semiconductor which may be doped with relatively electron-deficient boron (p-type) or electron-rich phosphorus (n-type) (*see* Figure).

semiempirical molecular orbital theory. MOLECULAR ORBITAL CALCULATIONS carried out within a framework in which the forms and energies of atomic orbitals (ao) are given by a set of predetermined parameters. A SELF-CONSISTENT FIELD calculation is then used to determine the minimum energy. The different methods are classed according to the form and number of integrals used to describe the ao; the simpler methods have little directionality. (1) CNDO (complete neglect of differential overlap) is the most simple method in which the ao are all spherically symmetrical. (2) INDO (intermediate neglect of differential overlap) includes one-centre repulsion integrals between ao on the same atom. (3) NDDO (neglect of diatomic differential overlap) has three- and four-centre integrals between ao on the same atom. (4) MNDO (modified NDO) was developed independently of INDO. (5) MNDO/C, a variant of MNDO, includes a correlation correction to fit experimental data. It gives improved results for excited states. (6) MINDO/3 is a modified INDO method which runs much more quickly because of parameterized one-centre integrals.

semipermeable membrane (spm). Membrane that is permeable to molecules of solvent, but not solute molecules or ions. Examples include colloidon, swollen cellophane, pig's bladder and copper(II) cyanoferrate(II) supported on wire gauze or porous pot. *See also* OSMOTIC PRESSURE.

Sephadex. Trade name for dextran gels, consisting of fractions of dextran crosslinked with 1-chloro-2,3-epoxypropane. They are used in GEL FILTRATION CHROMATOGRAPHY. The gels can be linked to acidic or basic groups for ion exchange.

sequence rules. Designation of priority of groups or atoms attached to a chiral centre

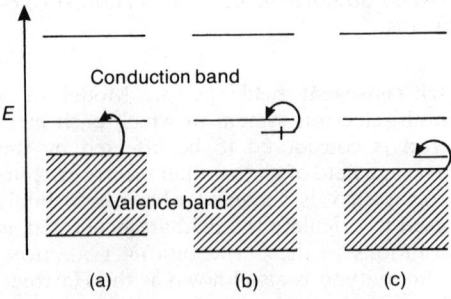

semiconductor
(a) Intrinsic semiconductor, (b) n-type doping, (c) p-type doping.

shapes of molecules

General classification of the shapes of molecules (bond angles normally apply to a molecule in the gaseous phase or to an ion in a crystal lattice)

Shape	Coordination number	Examples
Linear  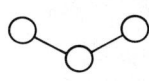	2	CO_2, HCN, HCCH, NCCN, NNO, XeF_2, I_3^-
Bent 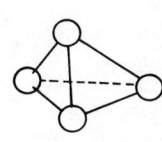	2	$ClO_2(117°)$, $SO_2(120°)$, $H_2O(105°)$, $NO_2(134°)$, $O_3(117°)$, $NO_2^-(115°)$, $CH_3OCH_3(111°)$
Trigonal planar (120°) 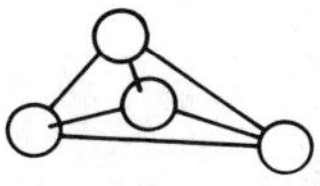 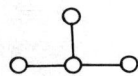	3	BX_3 (X = Br, Cl, F), NO_3^-, SO_3, CO_3^{2-}
Trigonal pyramidal	3	$AsBr_3(100°)$, SO_3^{2-}, $NH_3(107°)$, $PCl_3(100°)$, $ClO_3^-(110°)$, $PH_3(94°)$, XeO_3
T-shape	3	$BrF_3(86°)$, $ClF_3(89°)$

continued overleaf

Shape	Coordination number	Examples
Tetrahedral 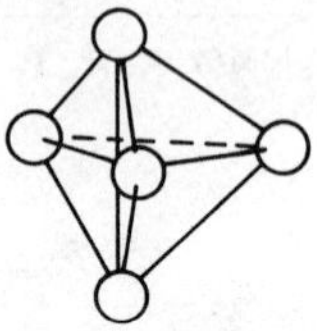	4	CX_4, SiX_4 (X = Br, Cl, F, H, I), NH_4^+, BF_4^-, MnO_4^-, PO_4^{3-}, SO_4^{2-}, $Ni(CO)_4$
Square planar (90°) 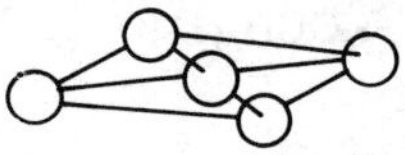	4	$AuCl_4^-$, $Pt(NH_3)_4^{2+}$, ICl_4^-, XeF_4
Trigonal bipyramidal (90° and 120°) 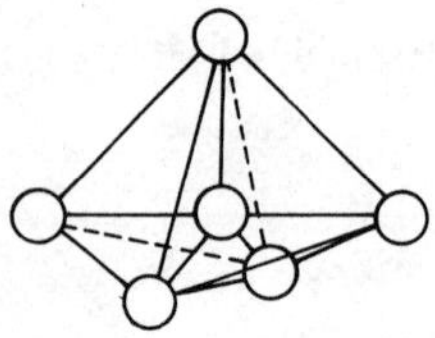	5	AsF_5, PCl_5, $MoCl_5$, PF_5
Octahedral (90°) 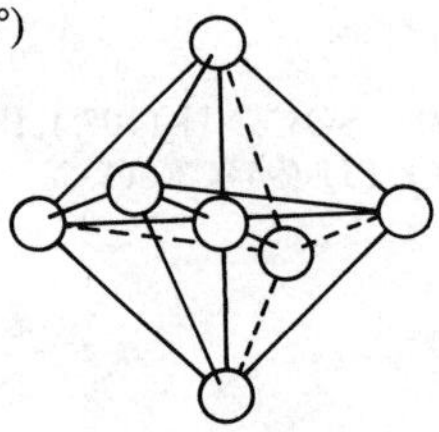	6	SF_6, UF_6, XeF_6, PCl_6^- (six-coordinated complexes of many metals)
Trigonal prismatic 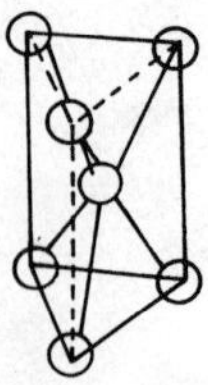	6	MoS_2, WS_2

Shape	Coordination number	Examples
Dodecahedral 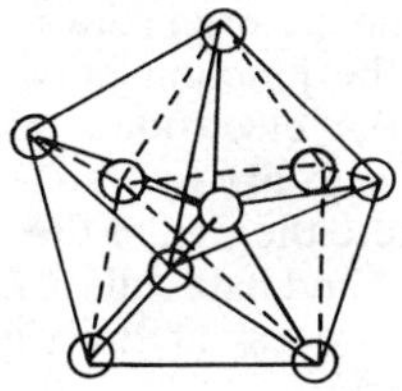	8	$Mo(CN)_8^{4-}$

| Icosahedral 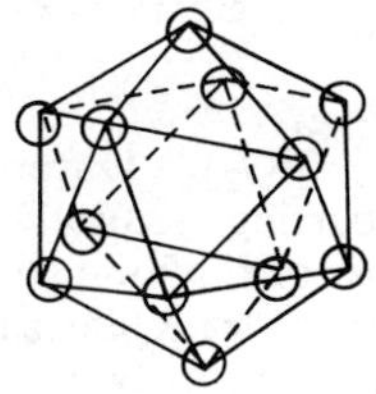 | 12 | All forms of elemental boron, $B_{12}H_{12}^{2-}$, carboranes of $B_{10}C_2H_{12}$-type |

to unambiguously describe the three-dimensional structure of a molecule (*see* CHIRALITY). (1) Groups are arranged in decreasing atomic number of the atoms by which they are bound to the asymmetric carbon. (2) If two or more of these atoms have the same atomic number, then the relative priority is determined by a similar comparison of the atomic numbers of the next atoms in the groups. If this fails the next atoms in the group are considered. (3) Isotopes are arranged in order of decreasing mass number. (4) A lone pair of electrons has lower priority than ^1H. The order of priority sequence for some common substituents is: I, Br, Cl, SO_3H, SH, F, OCOR, OR, OH, NO_2, NHR, NH_2, COOR, COOH, COR, CHO, CH_2OH, CN, Ph, CR_3, CHR_2, CH_2R, CH_3, D, H.

sequestering agent. Compound that can form complexes with metal ions. A sequestering agent may be used to make metal ions (e.g., iron and copper) available to plants in horticulture. *See also* CHELATE LIGAND, COMPLEX COMPOUND, COORDINATION COMPOUND.

Ser (serine). *See* AMINO ACIDS.

serine (Ser). *See* AMINO ACIDS.

SERS (surface-enhanced Raman spectroscopy). *See* RAMAN SPECTROSCOPY.

sesquiterpenes. *See* TERPENES.

shapes of molecules. Molecular structures may be deduced from spectroscopic data (IR, Raman), electron and X-ray diffraction patterns, etc. (For classification of structures—*see* Table.)

SHE (standard hydrogen electrode). *See* HYDROGEN ELECTRODE.

shear rate. *See* NEWTONIAN FLUID.

shear stress. *See* NEWTONIAN FLUID.

shell. *See* ELECTRONIC CONFIGURATION.

shielding. Effect found in NUCLEAR MAGNETIC RESONANCE SPECTROSCOPY in which adjacent electropositive atoms and groups create high electron density at a given nucleus and thus cause absorption at higher fields. The effective magnetic field at a nucleus is $H_0-\sigma H_0$, where σH_0 is the extra field induced by the electron density.

Conversely electronegative groups tend to deshield the nucleus.

shift reagent. *See* NUCLEAR MAGNETIC RESONANCE SPECTROSCOPY.

short-range forces. *See* INTERMOLECULAR FORCES.

Si. *See* SILICON.

sialic acids. *N*- and *O*-acyl derivatives of the AMINO SUGAR neuraminic acid. They are widely distributed and have been isolated from milk, urine, red blood cells (responsible for 67 percent of the surface charge) and bacterial cell wall structures.

side chain. Group of similar atoms (two or more, generally carbons, e.g., C_2H_5-) which branches off from a straight-chain or cyclic molecule.

siderite. *See* IRON.

siemens (S; units: Ω^{-1}, A V^{-1}). SI unit of CONDUCTANCE, equal to one reciprocal ohm.

Siemens–Martin process. Manufacture of STEEL by the addition of iron and other ores, and ferromanganese (*see* FERROALLOY) to pig iron.

sievert (Sv). SI unit of DOSE equivalent (i.e. the dose equivalent) when the product of the absorbed dose of radiation and a dimensionless factor (quality factor) is 1 J kg^{-1}).

sigma-bonding. σ-Bonding (*see* MOLECULAR ORBITAL).

sigma-donor ligand. σ-Donor ligand (*see* CARBON-TO-METAL BOND).

sigma-orbital. σ-Orbital (*see* MOLECULAR ORBITAL).

sigmatropic reaction. *See* WOODWARD–HOFFMANN RULE.

sign convention of electrochemical cells. IUPAC or Stockholm convention of emf values and electrode potentials of electrochemical cells is: (1) electrode (half-cell) potentials are tabulated for the reduction reaction ox + $ne \rightarrow$ red; (2) the STANDARD HYDROGEN ELECTRODE is assigned a zero potential for all temperatures; (3) a half-cell is written ion,ion,. . .|gaseous reactant, solid reactant,electrode material (the vertical line | represents the junction between the electrolyte and the electrode, gases, etc.); (4) reactions of cells are written (half cell)$_L$ ⋮ (half cell)$_R$ where the sign of the cell is the sign of the right-hand half-cell. If E_{cell} is positive, then the process at the right-hand electrode is reduction.

silanes. *See* SILICON INORGANIC COMPOUNDS.

silica (silicon dioxide, SiO_2). Different forms are known, occurring naturally or manufactured. All consist of infinite structures of linked SiO_4 tetrahedra. Quartz is a low-temperature form found in two main crystal habits: α-quartz tetrahedral hexagonal < 573°C and β-quartz hemihedral hexagonal 573–870°C. Other forms are an α- and β-tridymite and α- and β-cristobalite. Quartz is used extensively as an optical material, especially for the transmission of light in the ultraviolet part of the ELECTROMAGNETIC SPECTRUM. Naturally occurring amorphous silicas are obsidian and flint. High-pressure forms are coesite, heatite and stishovite, the last containing SiO_6 octahedra. Vitreous silica or quartz glass is formed on the slow cooling of liquid silica. Silica is also used in optical apparatus and also in high-vacuum and high-temperature apparatus where its low coefficient of thermal expansion confers resistance to sudden thermal shock.

silica gel. Amorphous hydrated silicon dioxide (*see* SILICA), prepared by the coagulation of silica sol (sodium silicate solution + HCl) or decomposition of some silicates. When dehydrated the material is HYGROSCOPIC and is used as a drying agent. The extent of its moisture content is demonstrated by the incorporation of cobalt(II) salts which are blue when anhydrous and turn pink when hydrated. The dehydrated form is easily regenerated by heating. Dehydrated silica gel is used as an adsorbent in the recovery of solvents, for drying gases, refining mineral oils, filtration and as a catalyst support. It is extensively employed

silicates
Naturally occurring silicates.

Mineral	Formula	Structure, type
Phenacite	Be_2SiO_4	Simple discrete $[SiO_4]^{2-}$
Willemite	Zn_2SiO_4	Cations surrounded by four oxygen atoms
Thortveitite	$Sc_2Si_2O_7$	Two tetrahedra sharing oxygen in pyrosilicate ion
Hemimorphite	$Zn_4(OH)_2Si_2O_7$	Two tetrahedra sharing oxygen in pyrosilicate ion
Benitoite	$BaTiSi_3O_9$	Cyclic—three tetrahedra
Beryl	$Be_3Al_2Si_6O_{18}$	Cyclic—six tetrahedra
Enstatite	$MgSiO_3$	Chains of $(SiO_3^{2-})_n$ (pyroxene)
Diopside	$CaMg(SiO_3)_2$	Chains of $(SiO_3^{2-})_n$ (pyroxene)
Spodumene	$LiAl(SiO_3)_2$	Chains of $(SiO_3^{2-})_n$ (pyroxene)
Tremolite	$Ca_2Mg_5(Si_4O_{11})_2(OH)_2$	Cross-linked chains (amphibole) (asbestos minerals)
Muscovite	$K_2Al_4(Si_6Al_2O_{20})(OH,F)_4$	Sheets of linked tetrahedra (mica)
Orthoclase	$KAlSi_3O_8$	Three-dimensional linked (feldspar)
Anorthite	$CaAl_2Si_2O_8$	Three-dimensional linked (feldspar)
Chabazite	$(Ca,Na_2)OAl_2O_3.4SiO_2.6H_2O$	Three-dimensional linked (zeolite)
Erionite	$(Ca,Mg,Na_2,K_2)O.Al_2O_3.6SiO_2.6H_2O$	Three-dimensional linked (zeolite)

as an adsorbant in liquid thin-layer, gas–liquid and high-performance chromatography.

silicates. Derivatives of silicon dioxide (*see* SILICA) acting as an acid, formed by fusing alkali metal carbonates with silica at 1300°C. They also occur naturally in rocks, (e.g., zircon, micas, feldspars, zeolites and aluminosilicates). Silicates consist of single SiO_4 tetrahedra or extensive polymeric structures. The former are soluble in water (e.g., sodium silicate, 'water glass'). In complex structures the silica tetrahedra may be in chains, rings, sheets, etc (*see* Table).

silicides. Compounds containing Si^{4-}; similar to CARBIDES. They are formed with very ELECTROPOSITIVE metals (e.g., Mg_2Si). Silicides dissolve in acids to give silanes (*see* SILICON INORGANIC COMPOUNDS) and hydrogen.

silicon (Si). GROUP IV ELEMENT; the second most abundant element, occurring naturally as SILICA (SiO_2) and SILICATES. The element may be extracted by reducing potassium fluorosilicate with potassium at high temperatures followed by removal of potassium fluoride with water. Industrially SiO_2 is reduced by carbon in an electric furnace to give 98 percent pure silicon. Very pure material for the semiconductor industry (impurities < 1 ppb) is obtained by ZONE REFINING silicon from silicon tetrachloride. Silicon is a semiconducting METALLOID with a diamond structure (*see* CRYSTAL FORMS). Doping with group III elements (e.g., boron) leads to a p-type SEMICONDUCTOR and with group V elements (e.g., phosphorus) gives a n-type semiconductor. Its chemical behaviour is that of a METAL. It forms salts of Si^{4+} and covalent compounds in which silicon is the more ELECTROPOSITIVE element. It does occur as a negative ion in SILICIDES. It is a relatively unreactive element, forming a protective film of silica in air, being unattacked by acids except hydrofluoric acid and slowly dissolving in concentrated alkalis. It reacts with chlorine and fluorine. Silicon is used in metallurgy to improve the strength of metals (e.g., aluminium), in the semiconductor industry as the basic component of microchips, in photovoltaic devices, as quartz in glassware, electronic watches, piezoelectric devices and combined in silicon greases and oils.

$A_r = 28.0855$, $Z = 14$, ES [Ne] $3s^2 3p^2$, mp 1410°C, bp 2360°C, $\rho = 2.33 \times 10^3$ kg m^{-3}. Isotopes: ^{28}Si, 92.2 percent; ^{29}Si, 4.7 percent; ^{30}Si, 3.1 percent.

silicon dioxide.   *See* SILICA.

silicone polymers.   *See* POLYMERS.

silicones.   *See* SILICON ORGANIC DERIVATIVES.

silicon inorganic compounds.   Silicon forms compounds of silicon(II) and silicon-(IV), but unlike carbon does not form $p\pi$–$p\pi$ bonds, although with oxygen and nitrogen there is evidence for $p\pi$–$d\pi$ bonding. Silicon monoxide is the product of the reduction of silica by carbon. The dioxide (SiO$_2$) is the naturally occurring compound (*see* SILICA). It is unreactive, being attacked only by fluorine, hydrofluoric acid 'and alkalis. Anions of SiO$_2$, the SILICATES, are also extensively found in nature. Silicon forms hydrides (silanes), but due to the weakness of silicon–silicon bonds (approximately half that of carbon–carbon), a maximum of Si$_8$H$_{18}$ is found. Silanes are prepared by the acid hydrolysis of magnesium silicide. Monosilane (SiH$_4$), prepared from silicon tetrachloride and lithium tetrahydridoaluminate, reacts with alkenes to give alkylsilanes (*see* SILICON ORGANIC DERIVATIVES) and is oxidized explosively by halogens. SiHCl$_3$ is formed by the reactions of silicon with hydrochloric acid. Halides up to Si$_2$I$_6$, Si$_2$Br$_6$, Si$_6$Cl$_{14}$ and Si$_{14}$F$_{30}$ are known, plus several mixed halides. All halides, except SiF$_4$, are hydrolyzed to hydrated silicon dioxide (silicic acid). SiF$_4$ forms complexes (e.g., [SiF$_6$]$^{2-}$ and [SiF$_5$]$^-$), oxidizes in air and with silicon at 1150°C forms gaseous SiF$_2$ which condenses to a polymer. Oxohalides from the action of halogen and oxygen on silicon, or from the controlled hydrolysis of silicon halides have the formula X$_3$SiO(SiOX$_2$)$_n$-SiCl$_3$. These contain the siloxane linkage (*see* SILICON ORGANIC DERIVATIVES). Silicon nitride (Si$_3$N$_4$) is formed from the elements. It reacts with alkalis to give ammonia and silicates. Esters of silicic acid (e.g., methyl silicate, Si + CH$_3$OH over Cu) are used as sources of pure silicon.

Ethyl silicate (silicon ester, Si(OEt)$_4$) is used to waterproof stonework and in adhesives. Silicon carbide (carborundum, SiC) is a hard refractory compound with covalent C–Si bonds. It is prepared by heating carbon and silicon dioxide at 2000°C and is used as an abrasive.

silicon organic derivatives.   Silicon forms an extensive series of compounds containing carbon–silicon σ-bonds, but no π-bonds are found. Alkylsilanes are prepared by the reaction of silicon tetrafluoride with stoichiometric amounts of GRIGNARD REAGENTS. Hydrogen is introduced by lithium tetrahydridoaluminate (e.g., SiCl$_4$ + 2CH$_3$MgI → (CH$_3$)$_2$SiCl$_2$ + LiAlH$_4$ → (CH$_3$)$_3$SiH$_2$). They are colourless oils and are mostly stable in air. Compounds containing silicon–hydrogen bonds are oxidized to silanols (R$_3$Si–OH) by benzoyl peroxide and oxygen. Tetraalkylsilanes are oxidized at the alkyl group. Dimethylsilene ((CH$_3$)$_2$Si) has been generated in the gas-phase reaction between (CH$_3$)$_2$SiCl$_2$ and potassium vapour. Alkylchlorosilanes are used in synthesis. Methylchlorosilanes are manufactured by passing chloromethane over a mixture of silicon and copper at 350°C. Chlorine is readily displaced by nucleophiles (e.g., water) to give silanols and alcohols to give alkoxysilanes. Siloxanes (R$_3$Si–O–SiR$_3$), characterized by the Si–O–Si linkage, are formed when silanols are heated alone or with the acid. When diols are heated (or formed by the hydrolysis of dialkyldichlorosilanes), they polymerize to silicones ((R$_2$SiO)$_n$). These are oils containing linear and cyclic units. Cross-linked polymers are obtained by using mixtures of the dialkyldichlorosilanes and monoalkyltrichlorosilanes. Silicone rubbers, oils, greases, resins, etc. may be formed by controlling the degree of cross-linking. They are exceptionally inert and are used extensively as substitutes for the corresponding carbon-based polymers (*see also* POLYMERS). Nitrogen analogues of silanols and siloxanes, silanamines and silazines are known. For example, Et$_3$SiCl + Me$_2$NH → Et$_3$SiNMe$_2$ and Me$_3$SiNHEt + Me$_3$SiCl → Me$_3$Si–N(Et)–SiMe$_3$.

siloxanes.   *See* SILICON ORGANIC DERIVATIVES.

silver (Ag).   Group Ib, second row TRANSITION ELEMENT. It is found naturally as the metal, as sulphide ores argentite (silver glance, Ag_2S), pyrargyrite (ruby silver, Ag_3SbS_3), silver–copper glance $((Cu,Ag)_2S)$ and as the chloride cerargyrite (horn silver, AgCl). Silver is also recovered from copper and lead ores, and from scrap metals when it is complexed with cyanide and reduced by zinc. Silver is purified by ELECTROLYSIS of cyanide solutions. The metal is soft, white, reflective, ductile and has high thermal and electrical conductivity. It is used in jewellery, coinage, electrical devices and photography. Sterling silver is alloyed with 7.5 percent copper. Silver is the most reactive of the noble metals. Although silver does not oxidize in air, it dissolves in oxidizing acids giving compounds of Ag(I). Ag(I) compounds are readily reduced (e.g., by methanoic acid), giving a silver mirror.
$A_r = 107.8682$, $Z = 47$, ES [Kr] $4d^{10} 5s^1$, mp 962°C, bp 2212°C, $\rho = 10.5 \times 10^3$ kg m^{-3}. Isotopes, ^{107}Ag, 51.35 percent; ^{109}Ag, 48.65 percent.

silver compounds.   Silver is almost always monovalent in its compounds (silver(I), argentous), but AgO, AgF_2 and AgS, strongly oxidizing compounds of silver(II) (argentic), are known ($E^{\ominus}_{Ag^{2+},Ag} = 1.98$ V; $E^{\ominus}_{Ag^+,Ag} = 0.799$ V). Some complexes of silver(III) (e.g., $[AgF_4]^-$ and $[AgF_6]^{3-}$) and silver(IV) (e.g., Cs_2AgF_6) exist. Many silver compounds are sensitive to light and are used in photography (e.g., AgCl, AgBr, AgI, Ag_3PO_4). Silver nitrate (lunar caustic) is a useful soluble silver salt used in forming silver mirrors and as a starting point for the preparation of other silver compounds. The silver–silver chloride electrode is one of the common REFERENCE ELECTRODES ($E^{\ominus}_{AgCl,Ag,Cl^-} = 0.222$ V). Silver oxide is a rechargeable battery anode of high-power density. Silver complexes are frequently two-coordinate (e.g., $[Ag(NH_3)_2]^+$, $[Ag(CN)_2]^-$), but four-coordinate complexes are known (e.g., $[Ag(CN)_4]^{3-}$).

silver–copper glance.   See SILVER.

silver electrode.   See HALF-CELL.

silver glance.   See SILVER.

silver–silver halide electrode.   See HALF-CELL, SILVER COMPOUNDS.

silyl group.   The group R_3Si-, where R = organic group or hydrogen. See also SILICON ORGANIC DERIVATIVES.

simazine (2-chloro-4,6-bis-(ethylamino)-s-triazine, $ClC_3N_3(NHC_2H_5)_2$). Pre-emergence HERBICIDE, toxic by inhalation and ingestion.

SIMS (secondary ion mass spectrometry). See MASS SPECTROMETRY.

single bond.   See COVALENT BOND.

sintered glass.   Porous GLASS formed by SINTERING glass fibre, usually in the form of discs. The degree of sintering determines the porosity, and the resulting glass is used as a separator for electrochemical cells, for gas effusion and for filtering.

sintering.   Aggregation of solids by heating, possibly under pressure, below the melting point. The temperature at which atomic and molecular solid state diffusion becomes important is known as the Tamman temperature, which is equal to half the melting point. Sintering is used to form high-melting-point POLYMERS (e.g., tetrafluoroethene) and porous glasses (see SINTERED GLASS). It may also occur to the detriment of catalytic processes when the attendant loss of surface area causes a decrease in performance.

sitosterol.   See STEROLS.

SI units (Système International d'Unités). Coherent system based on the seven base units (METRE, KILOGRAM, SECOND, AMPERE, KELVIN, MOLE and CANDELA), together with two supplementary dimensionless units (radian and steradian). The SI units for derived physical quantities are constructed by multiplication and division of the basic units without the introduction of any numerical factor, thus the SI unit of energy is kg m^2 s^{-2}, which has the special name, joule. A few derived SI units have special names (see Appendix, Table 2). Conversion factors for cgs and other obsolete units to SI units are given in Appendix, Table 3.

sizing. *See* PAPER, STARCH.

skatole (3-methylindole, C_9H_8N). White crystalline solid (mp 205°C), chief volatile component of faeces; formed by bacterial degradation of tryptophan in the intestine. It is prepared by the FISCHER INDOLE SYNTHESIS from propanal phenylhydrazine.

Skraup reaction. Reaction for the synthesis of QUINOLINE in which a mixture of aniline, glycerol and sulphuric acid is heated in the presence of a mild oxidizing agent (e.g., nitrobenzene). Substituted anilines give substituted quinolines.

slag. Melt of OXIDES formed during the SMELTING and REFINING of metals which removes the GANGUE and other impurities leaving the pure liquid metal. Fluxes (e.g., alumina, lime, silica, iron(III) oxide) are added to lower the melting point of the slag. Solidified slag may be used as aggregate for road building, to manufacture cement or, if the slag has a high content of phosphorus-(v) oxide, as a fertilizer.

slaked lime. *See* CALCIUM COMPOUNDS.

Slater determinant. Way of writing a many-electron WAVEFUNCTION which ensures that it is antisymmetric with respect to exchange of electrons. A matrix of spin orbitals (*see* ORBITAL) is constructed with the element of the ith row and jth column being the spin orbital j containing electron i. For example the ground state wavefunction for helium (two electrons in a $1s$ ATOMIC ORBITAL) is

$$\psi = N \begin{vmatrix} \phi_{1s}\alpha(1) & \phi_{1s}\beta(1) \\ \phi_{1s}\alpha(2) & \phi_{1s}\beta(2) \end{vmatrix}$$

$$= N\phi_{1s}\alpha(1)\,\phi_{1s}\beta(2) -$$
$$N\phi_{1s}\alpha(2)\,\phi_{1s}\beta(1)$$

where α and β are the two spin states of the electron.

Slater-type orbital (STO). ATOMIC ORBITALS for many electron atoms, the radial part being of the form

$$R(r) = Nr^{n'} \exp(-\zeta r)$$

where n' and ζ are constants which may be determined by the variational method (*see* VARIATION PRINCIPLE). For ease of calculation of molecular integrals, the STOs are written as the sum of a number of gaussian functions having exponents in r^2. These are referred to as STO-nG where n (minimally $n=3$) is the number of gaussians. Further refinements are possible by splitting the valence set and adding extra polarization functions.

slime. *See* CAPSULE.

slimicide. Chemical (e.g., CHLORINE, PHENOLS) toxic to bacteria and/or fungi which form aqueous slimes. Slimicides are used largely in paper mills and in textile and leather industries.

slip planes. *See* CRYSTAL STRUCTURE.

Sm. Samarium (*see* LANTHANIDE ELEMENTS).

smaltite. *See* COBALT.

smelting. Melting of metallic ores with a reducing agent (e.g., carbon, hydrogen, electropositive metals) as part of metal REFINING.

smectic liquid crystals. *See* LIQUID CRYSTALS.

smithsonite. *See* ZINC.

smoke. *See* AEROSOL.

Sn. *See* TIN.

S_N1 mechanism (substitution nucleophilic unimolecular mechanism). Reaction in which the rate-determining step is the cleavage of a carbon–halogen bond and the formation of a CARBONIUM ION, for example

$$CR_3Cl \rightarrow CR_3^+ + Cl^-$$

which then reacts with the NUCLEOPHILE

$$CR_3^+ + OH^- \rightarrow CR_3OH$$

The carbonium ion is planar and the nucleophile can attack from either side; thus if the original molecule is optically active (three different R groups) a racemic mixture of products is formed. The rate of a S_N1 reaction is proportional to the concentration of the substrate and independent of the concentration of the nucleophile. The order of reactivity of halides in a S_N1 reaction is $I > Br > Cl > F$ and for the group (for a given halide) the order is tertiary, allylic, benzylic $\geqslant$ secondary $>$ primary.

S_N2 mechanism (substitution nucleophilic bimolecular mechanism). Concerted reaction in which the NUCLEOPHILE (Z^-) approaches from the side of the substrate opposite to that of the leaving group (X^-)

$$Z^- + R-X \rightarrow Z-R + X^-$$

The reaction is bimolecular and shows second-order kinetics. A S_N2 reaction is always accompanied by an inversion of configuration (*see* WALDEN INVERSION) at the carbon atom originally bonded to X. This phenomenon can be observed experimentally if this carbon is asymmetric and the reaction is carried out on a single enantiomer.

2-(*S*)-chlorobutane   2-(*R*)-butanol

S_N2 reactions are sensitive to steric hindrance and insensitive to the polarity of the solvent. The order of reactivity of halides is methyl $>$ primary $>$ secondary $\geqslant$ neopentyl, tertiary.

soap. Sodium or potassium salt of n-OCTADECANOIC ACID, n-HEXADECANOIC ACID *cis*-9-OCTADECENOIC ACID or other fatty acid. Soaps are manufactured from animal and vegetable fats by heating with sodium hydroxide and salted out by sodium chloride (*see* SALTING OUT). Metallic soaps (salts of alkaline earth metals, lithium, copper, zinc, magnesium) are used in greases and lubricating oils, as pharmaceuticals and fungicides. The incendiary Napalm is an aluminium soap of octadecanoic acid and carboxylic acids derived from petroleum.

soda. *See* SODIUM.

soda lime. Dried product of the reaction between calcium oxide and sodium hydroxide. It is used as an adsorbent for carbon dioxide and as an alkali.

sodium (Na). Group Ia ALKALI METAL. It is the sixth most abundant element, being found dissolved in seawater as the chloride, also as a solid (rock salt), as the carbonate (soda, trona), borate (borax), nitrate (Chile saltpetre) and sulphate. Sodium metal is extracted by electrolysis of fused sodium chloride–calcium chloride in a Down's cell. Sodium is a soft, silvery metal which reacts rapidly with air, water and halogens. It dissolves in liquid ammonia to give a blue, conducting solution ($Na^+ + e$) and then the amide ($NaNH_2$), and in mercury to give an amalgam. Some 60 percent of the sodium metal produced is used in the production of tetraethyllead (*see* LEAD ORGANIC DERIVATIVES). It is also used as a heat exchanger in nuclear power stations ($C_P = 1.23$ kJ kg^{-1} K^{-1}) and as a reducing agent. $A_r = 22.989\ 77$, $Z = 11$, ES [Ne] $3s^1$, mp 97.8°C, bp 890°C, $\rho = 0.97 \times 10^3$ kg m^{-3}. Isotope: ^{23}Na 100 percent.

sodium alginate. *See* SEAWEED COLLOIDS.

sodium amalgam. *See* AMALGAMS.

sodium chloride lattice (rock salt lattice). *See* CRYSTAL FORMS.

sodium compounds. Sodium is a highly ELECTROPOSITIVE metal ($E^{\ominus}_{Na^+,Na} = -2.71$ V). The only oxidation state found is +1, with compounds being almost entirely ionic (Na_2 vapour and some chelated compounds are covalent). Sodium readily forms salts with most anions, all of which are soluble in water. They are characterized by high melting points and high conductivities of the melts. Sodium hydride (NaH) is known, but of greater importance in organic synthesis is sodium tetrahydridoborate (*see* BORON INORGANIC COMPOUNDS). Industrially sodium chloride is the most important compound, being the starting point for the production of chlorine and chlorates and sodium hydroxide (caustic soda). Sodium carbonate is produced by the SOLVAY PROCESS. Sodium does not have such an extensive organic chemistry as lithium. It does not react with alkanes, but forms addition compounds with highly conjugated and aromatic molecules which are ionic in character (e.g., $C_6H_6^- Na^+$). The addition to dienes is the basis of a synthetic rubber process. ACETYLIDES are formed by direct reaction. Sodium reacts with alcohols to form ALKOXIDES and hydrogen (compare the reaction with water).

sodium cyclamate. *See* SWEETENING AGENTS.

sodium dichromate. *See* CHROMIUM COMPOUNDS.

sodium pump. Active transport system in cells that moves sodium ions out of the cell against the concentration gradient; ATP supplies the necessary energy. The pump maintains the separation of ions required to produce a resting potential across a cell membrane.

sodium silicate. *See* SILICATES.

sodium–sulphur battery. High-temperature (300–400°C) battery in which the overall reaction is the reaction of sodium and sulphur to give sodium polysulphide ($E = 2.08$ V). The SOLID ELECTROLYTE is sodium β-alumina.

soft acids and bases. *See* HARD AND SOFT ACIDS AND BASES.

soft water. *See* HARDNESS OF WATER.

sol. COLLOID SOLUTION, commonly of an inorganic dispersion (e.g., a gold sol is prepared by reducing $HAuCl_4$). Sols are termed monodisperse if the particle size is constant, or polydisperse if a range of sizes are observed (*see* NUMBER AVERAGE MOLECULAR MASS).

solder. Low-melting ALLOY used for bonding metals. Hard solder and silver solder are BRASSES melting at about 800°C. Soft solders are eutectics of lead and tin.

solid electrolyte (superionic conductor, fast ion conductor). Compound that exhibits high ionic conductivity (>10 S m^{-1}) in the solid state with a low activation energy for conduction. Conduction may be by the movement of defects (e.g., AgCl doped with Cd^{2+}), by the internal melting of one

ion (e.g., AgI, $RbAg_4I_5$) or by an ion exchange mechanism (sodium β-alumina, NASICON, $Na_3Zr_2PSi_2O_{12}$). They are used for solid-state batteries and fuel cells (for example, *see* SODIUM–SULPHUR BATTERY).

solid foam. *See* FOAM.

solid–gas equilibrium. *See* ONE-COMPONENT SYSTEM.

solid–liquid equilibrium. *See* ONE-COMPONENT SYSTEM.

solid solution. Product when two or more atoms or compounds share a common lattice. Solid solutions may be substitutional (e.g., Al_2O_3/Cr_2O_3) in which either atom may be found at a particular lattice site or interstitial in which a small atom is accommodated in a host lattice (e.g., hydrogen in palladium). The lattice parameters (*see* X-RAY DIFFRACTION) may vary with composition. If this is linear, Vegard's law is said to hold. PHASE DIAGRAMS are used to show the variation in the composition of solid solutions with temperature. *See also* ALLOY, INTERCALATION COMPOUND, TWO-COMPONENT CONDENSED SYSTEM.

solid-state reaction. Reaction between solids only takes place at an appreciable rate at high temperatures (greater than 1000°C) when sufficient energy is available to cause the diffusion of ions and atoms. The reaction rate often shows parabolic kinetics if the reacting atoms have to diffuse through the growing layer of product (i.e. $d[\text{product}]/dt = k/[\text{product}]$ or $[\text{product}] = (kt)^{1/2}$. The rate of a solid-state reaction may also depend on the formation of nuclei of product around which the reaction occurs. Solid-state reactions are used to prepare mixed oxides for catalysis, ceramics and glasses.

solidus curve. *See* TWO-COMPONENT CONDENSED SYSTEM.

solubility. *See* SATURATED SOLUTION.

solubility curve. Graphical representation (PHASE DIAGRAM) of the variation of the solubility of a compound with temperature. Breaks or discontinuities in the curve are due to changes of state (e.g., dehydration of hydrates). *See also* TWO-COMPONENT CONDENSED SYSTEM.

solubility product (K_s; units: $mol^v \, dm^{-3v}$). For a sparingly soluble 1:1 salt in which the solid is in equilibrium with its solution

$$MA_{solid} \rightleftharpoons MA_{dissolved} \rightleftharpoons M^+ + A^-$$

$$K_s = a_{M^+} \, a_{A^-}$$

$$= c_{M^+} \, c_{A^-} \, \gamma_{M^+} \, \gamma_{A^-}$$

$$= a_{\pm}^2 = c_{\pm}^2 \, \gamma_{\pm}^2 \, (a_{MA,solid} = 1)$$

If an electrolyte with a common ion is added to such a solution, the increase in ionic strength causes $\gamma_{\pm}$ to decrease, but this is outweighed by the increase in c_{A^-}, hence the solubility decreases to keep K_s constant. If the added electrolyte contains no common ion, $\gamma_{\pm}$ decreases, and hence the solubility increases to keep K_s constant. For a solute that dissociates into v^+ (v^-) positive (negative) ions of charge z^+ (z^-) respectively

$$K_s = a_{z+}^{v^+} \, a_{z-}^{v^-}$$

soluble. Denoting that a substance dissolves in a specified solvent.

solute. *See* SOLUTION.

solution. Homogeneous dispersion of two or more substances in each other (e.g., gases in liquids, liquids in liquids, solids in liquids, solids in solids). Although there is no fundamental difference between the COMPONENTS, the component present in excess is known as the solvent and the other the solute. *See also* BINARY LIQUID MIXTURE, COLLOID, IDEAL SOLUTION, REAL SOLUTION, SOLID SOLUTION.

solvated electron. Free electron which is associated with one or more solvent molecules formed by the irradiation of the solvent by X- or gamma-rays, electrochemical injection or, in the case of liquid ammonia, by the dissolution of an alkali metal.

solvation. Association of a solvent with a solute (*see* SOLUTION). In aqueous solutions, this is referred to as HYDRATION. In lyophilic sols (*see* LYOPHILIC COLLOID),

solvation is responsible for the stability of the sol.

Solvay process (ammonia soda process). Major industrial preparation of sodium carbonate. Sodium chloride is converted to sodium bicarbonate by reaction with ammonia and carbon dioxide

$$NaCl + NH_3 + CO_2 + H_2O \rightarrow$$

$$NaHCO_3 + NH_4Cl$$

Heating the bicarbonate gives sodium carbonate and regenerates some carbon dioxide

$$2NaHCO_3 \rightarrow Na_2CO_3 + CO_2 + H_2O$$

Ammonia is regenerated by treating ammonium chloride with calcium hydroxide slurry.

solvent. *See* SOLUTION.

solvent extraction. *See* EXTRACTION, LIQUID–LIQUID EXTRACTION.

solvolysis. Reaction between a molecule and solvent (*see* SOLUTION). In water this is known as HYDROLYSIS.

Sommelet reaction. *See* BENZALDEHYDE.

sonochemistry. *See* CHEMICAL ULTRASONICS.

sorbent. *See* ADSORBENT.

s-orbital. *See* ATOMIC ORBITAL.

sorbitol. *See* HEXAHYDRIC ALCOHOLS, SWEETENING AGENTS.

sorbose. *See* HEXOSES.

sorption. *See* ADSORPTION.

space group. Whole array of SYMMETRY ELEMENTS in a CRYSTAL LATTICE. The introduction of TRANSLATION into crystals leads to more SYMMETRY OPERATIONS and an increase in the number of possible space groups to 230. Two new kinds of symmetry operation are required, the GLIDE PLANE and the SCREW AXIS. The atoms in any crystal must be arranged so that they fit into one of these groups.

space lattice. *See* CRYSTAL SYSTEMS.

specific activity (S, units: disintegrations s^{-1} kg^{-1}). Induced specific activity of a sample can be calculated from

$$S = N_A \cdot \sigma \cdot \phi \cdot f \cdot (0.5)^{T/t,1/2} \cdot (1 - 0.5^{t/t,1/2})/A$$

where σ is the capture CROSS-SECTION, ϕ the NEUTRON FLUX, f the fractional abundance of the particular isotope, t the irradiation time, T the delay between the end of irradiation and start of counting, $t_{1/2}$ the half-life and A/kg mol^{-1} the atomic mass of the isotope.

specific conductance. *See* CONDUCTIVITY.

specific gravity. *See* RELATIVE DENSITY.

specific heat capacity. *See* HEAT CAPACITY.

specific optical rotatory power (α_m; units: rad m^2 kg^{-1}). For a pure substance, particularly in the liquid state at a specified temperature, usually for the sodium-D line

$$\alpha_m = \alpha/\rho l$$

where α is the angle of rotation, measured in radians, of the plane of polarization produced by a column of liquid of length l m and density ρ kg m^{-3}. For a solution containing m kg of substance in V m^3

$$\alpha_m = \alpha V/ml = 1.745 \times 10^{-4} [\alpha]_D^t$$

The molar optical rotatory power (α_n; units: rad m^2 mol^{-1}) is defined by

$$\alpha_n = \alpha V/nl = \alpha/cl = 1.745 \times 10^{-2} [M]_D^t$$

where c mol m^{-3} is the concentration. *See also* SPECIFIC ROTATION.

specific resistance. *See* RESISTIVITY.

specific rotation ($[\alpha]_D^t$; units: deg dm^{-1} cm^3 g^{-1}). Pre-SI unit, still widely used, to express the optical activity of sugars and amino acids (usually quoted as degrees) and defined as

$$[\alpha]_D^t = 100\alpha/lc$$

where α, the angle of rotation, is now in degrees per decimetre, l the path length in decimetres and c the concentration of the optically active substance in gram per 100 cubic centimetres. The molar rotation ($[M]_D^t$; units: deg dm^{-1} l mol^{-1}) is defined as

$$[M]_D^t = \alpha/cl$$

where c is the concentration in mol dm^{-3}. *See also* SPECIFIC OPTICAL ROTATORY POWER.

specific surface area (ssa; units: m^2 g^{-1}). Surface area of 1 g of a solid. This may be measured by adsorption of nitrogen (cross-sectional area on a surface 0.162 nm^2), and the use of the BET isotherm (*see* ADSORPTION ISOTHERM). The average diameter of spherical particles (d μm) of density ρ g cm^{-3} is related to the specific surface area by ssa = $6/\rho d$.

spectral series. *See* BALMER SERIES.

spectrochemical series. LIGANDS arranged in increasing order of the value of Δ found in complexes with transition elements: $I^- < Br^- < Cl^- < F^- < OH^- < H_2O < NCS^- <$ pyridine $< NH_3 <$ diaminoethene $<$ 1,10-phenanthroline $< NO_2^- < CN^-$. *See* HIGH-SPIN COMPLEX, LIGAND FIELD THEORY.

spectrograph. Device for recording a spectrum on a photographic film.

spectrometer. *See* SPECTROPHOTOMETER.

spectrophotometer (spectrometer). Device for recording the frequency and intensity of absorption or emission of ELECTROMAGNETIC RADIATION. The exact nature of the instrument depends on the range of wavelengths and the form of the material to be studied (*see* Table). The output of a spectrophotometer is usually in the form of a graph of intensity against wavelength or frequency. In the case of ELECTRON SPIN RESONANCE SPECTROSCOPY, it is a derivative spectrum. Different approaches to the measurement of spectra include use of FOURIER TRANSFORM ANALYSIS (FT analysis) of an interference experiment, and the direct measurement of absorption in PHOTOACOUSTIC SPECTROSCOPY. In FT spectroscopy, the entire frequency range is passed through an interferometer. The two beams produced fall on the sample and are subsequently detected. Computer analysis of the interferogram yields the normal

spectrophotometer

Technique spectroscopy	Radiation	Source	Intensity control	Frequency control	Sample holder	Detector
Mössbauer	Gamma	Radioactive nuclei	Amount of source	Change source	Supported solid	Geiger counter, scintillation counter
X-ray	X	e-bombarded metal	Flux of electrons	Target metal + crystal grating	Supported solid	Geiger counter, scintillation counter
Atomic Fluorescence Electronic	Vacuum UV / UV / Visible	Discharge lamp, D$_2$-discharge lamp, Tungsten lamp, laser			Quartz cuvette / Glass cuvette	Photomultiplier, photographic plate, photocell
Raman						
Vibration	IR	Nernst glower, glow bar	Energy into source + slits, diaphragm	Filters + grating monochromator	KBr disc, mull	Thermocouple, bolometer
Vibration–rotation	Far-IR	IR laser, mercury discharge			KBr disc, mull	Thermocouple, bolometer
Rotation Electron spin resonance	Microwave	Klystron magnetron, Backward wave oscillator	Energy into source	Sweep magnetic or electric field	Low P gas / Quartz tube	Silicon-tungsten crystal
Nuclear magnetic resonance	Radio	Radio transmitter			Quartz tube	Radio receiver

absorption spectrum. FT methods are used in INFRARED SPECTROSCOPY and NUCLEAR MAGNETIC RESONANCE SPECTROSCOPY.

spectrophotometric titration. *See* TITRATIONS.

spectroscope. Device by which a spectrum is made visible to the eye.

spectroscopic entropy (S_{spec}). Total ENTROPY of a perfect gas

$$S_{spec} = S_t + S_r + S_v + S_e$$

is calculated from the translational, rotational, vibrational and electronic PARTITION FUNCTIONS using spectroscopic data. For each type of energy

$$S = U/T + Nk \ln Q$$

The dominant contribution is associated with translational motion; vibrational and electronic contributions are negligible at 298 K. Although the difference between S_{therm} and S_{spec} for a large number of molecules is negligible, there are many molecules (e.g., CO, NO, N_2O, CH_3D) for which the difference is greater than can be accounted for by experimental error. These discrepancies are due to the presence of some disorder in the solid at 0 K and the departure of the probability (W) from unity. Thus the entropy of a solid at 0 K is greater than zero. This difference is the residual entropy at 0 K; that is

$$S_{spec} = S_{therm} + k \ln W$$

In a crystal composed of molecules AB, where A and B are of similar size, there may be little energy difference between the energies of the arrays AB AB AB AB and AB BA AB BA, so that the molecules can adopt either orientation in the solid, and there is insufficient energy at the low temperatures for the molecules to be reorientated. The residual entropy of such a crystal, on probability grounds, might be expected to be $k \ln (2)^{N_A} = 5.74$ J K^{-1} mol^{-1}, a value close to that observed for carbon monoxide, nitrogen monoxide and dinitrogen oxide. for CH_3D, which can adopt four different orientations in the lattice, the calculated residual entropy ($k \ln (4)^{N_A}$) is 11.59, which is in close agreement with the experimental value of 11.52 J K^{-1} mol^{-1}. The residual entropy of

water is due to the hydrogen-bonded nature of the ice crystal and the tetrahedral arrangement of the hydrogen atoms around the oxygen atoms. *See also* BOLTZMANN EQUATION, STATISTICAL MECHANICS, THIRD LAW OF THERMODYNAMICS.

spectroscopic splitting factor. *See* LANDÉ g-FACTOR.

spectroscopic term symbol. *See* TERM SYMBOLS.

spectrum. *See* ELECTROMAGNETIC SPECTRUM.

speed (dimensions: $l\ t^{-1}$; units: m s^{-1}). Scalar quantity; defined as the ratio of the distance travelled by a body to the time taken, but no direction is specified. *See also* VELOCITY.

spermidine. *See* POLYAMINES.

spermine. *See* POLYAMINES.

sphalerite. *See* ZINC.

spherocolloid. *See* MICELLE.

sphingomyelins. *See* PHOSPHOLIPIDS.

sphingosine ($CH_3(CH_2)_{12}CH=CHCH-(OH)CH(NH_2)CH_2OH$). Base obtained on the hydrolysis of sphingomyelins (*see* PHOSPHOLIPIDS) and CEREBROSIDES.

***sp*-hybrid.** *See* HYBRIDIZATION.

spinel. *See* SPINELS.

spinels. Group of oxide minerals, occurring in high-temperature igneous rocks, of general formula $M^{2+}M_2^{3+}O_4$, where M^{2+} = Mg, Fe, Co, Mn or Ni and M^{3+} = Al, Fe or Cr. They crystallize in the cubic system with close-packed oxygen atoms. In a normal spinel, M^{2+} is in tetrahedral coordination, and in an inverse spinel half the M^{3+} are in tetrahedral and half in octahedral coordination. There are three series: spinel, $MgAl_2O_4$; magnetite, Fe_3O_4; chromite, $FeCr_2O_4$.

spin labelling. Attachment of a paramagnetic (*see* PARAMAGNETISM) atom or radical to a macromolecule. ELECTRON SPIN RESONANCE SPECTROSCOPY then gives information about the structure or function of the environment of the spin label. Typical spin labels are Cu^{2+}, Mn^{2+}, RNO.

spin moment (μ_s). MAGNETIC MOMENT of an electron. In an atom

$$\mu_s = g[S(S+1)]^{1/2}$$

where g is the gyromagnetic ratio and S the total spin QUANTUM NUMBER.

spin multiplicity. *See* MULTIPLICITY.

spin orbital. *See* ORBITAL.

spin quantum number. *See* QUANTUM NUMBER.

spin–spin coupling. *See* NUCLEAR MAGNETIC RESONANCE SPECTROSCOPY.

spin trapping. Method of detecting and identifying FREE RADICALS by reaction with a molecule to produce a more stable radical that may be studied by ELECTRON SPIN RESONANCE SPECTROSCOPY. Nitroxides and nitrones are often used as traps. The ESR spectrum of a radical formed from a nitroxide is split into three by the nitrogen-14 nucleus and then further by protons of the trapped radical.

$$t\text{-RCNO} + (R')^\cdot \rightarrow t\text{-RC}(R')\text{CNO}^\cdot$$

Nitrones trap the radical one atom further removed from the site of the unpaired electron (oxygen atom) which makes the interpretation of the resulting spectrum more difficult

$$R_2C = \overset{+}{N}(R)\overline{O} + (R')^\cdot \rightarrow R_2R'CN(R)O^\cdot$$

spirane systems (spiro-compounds). Bicyclic compounds with one atom common to both rings; the common atom may be carbon, nitrogen, boron, etc. The numbering starts next to the junction in the smaller ring.
spiropentane. Liquid (mp −107°C, bp 39°C) with a strain energy of 234 kJ mol⁻¹ (compare with cyclopentane, 27 kJ mol⁻¹); formed when pentaerythritol bromide ($C(CH_2Br)_4$) is heated with zinc dust in the presence of EDTA.

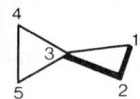 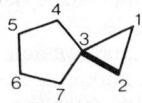

spiro(2.2)pentane   spiro(4.2)heptane

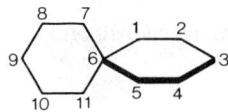

spiro(5.5)undecane

spirane systems

spiro-. Prefix indicating a polycyclic compound containing an atom which is a member of two rings.

spiropentane. *See* SPIRANE SYSTEMS.

spodumene. *See* LITHIUM.

spontaneous fission. *See* RADIOACTIVITY.

spontaneous process. IRREVERSIBLE PROCESS that is capable of doing WORK (e.g., nuclear fission, the flow of heat from a hotter to a colder body, the expansion of a gas). Such processes are always accompanied by a decrease in the GIBBS FUNCTION and increase in the ENTROPY or disorder of the universe.

sputtering. Process in which atoms are ejected from a solid surface by the impact of energetic ions (usually Ar^+). Sputtered thin films are formed by creating a radiofrequency discharge in an argon atmosphere with a metal electrode (the target) and a cooled surface to condense the film. In ultrahigh-vacuum experiments, argon ion bombardment is used to clean the surface of metals. If the surface is being analyzed by Auger electron spectroscopy (*see* AUGER EFFECT) or electron spectroscopy for chemical analysis (*see* ELECTRONIC SPECTROSCOPY) then the composition of the material as successive layers are uncovered may be determined. This technique is known as depth profiling. Sputter ion pumps (GETTER pumps) sputter a film of titanium which efficiently adsorbs gas molecules.

squalene.   *See* TERPENES.

square planar.   Describing a coordination compound in which four ligands are positioned at the corners of a square coordinated to a metal ion at the centre (e.g., $Pt(NH_3)_4^{2+}$).

square planar coordination.   *See* SHAPES OF MOLECULES.

Sr.   *See* STRONTIUM.

ssa.   *See* SPECIFIC SURFACE AREA.

St.   *See* STOKES.

stability constant.   Reciprocal of the FORMATION CONSTANT.

stability of sols.   *See* DLVO THEORY.

stabilization energy.   Amount by which the energy of a delocalized structure is less than the calculated theoretical energy of a structure with localized bonds.

stabilizers.   Additives that prevent deterioration or decomposition. For example, 1,4-dihydroxybenzenes are added to methyl 2-methylpropenoate monomer to prevent polymerization.

staggered conformation.   *See* CONFORMERS, NEWMAN PROJECTION FORMULA.

stainless steel.   *See* STEEL.

standard cell.   CELL used for purposes of comparison and in calibration of measuring circuits. A standard cell should provide a constant and accurately reproducible emf and should not deteriorate on standing. The Weston cell (*see* Figure) has a low temperature coefficient; at T K the emf is given by

$$E_T = 1.01830 - 4.06 \times 10^{-5}\,(T - 293)$$
$$-9.5 \times 10^{-7}\,(T - 293)^2$$

The Clark cell (*see* Figure) has a higher temperature coefficient; at T K the emf is given by

$$E_T = 1.4330[1 - 8.4 \times 10^{-3}(T - 288)]$$

standard electrode potential.   *See* ELECTRODE POTENTIAL.

standard emf.   *See* ELECTROMOTIVE FORCE.

standard hydrogen electrode.   *See* HYDROGEN ELECTRODE.

standard molar enthalpy.   *See* ENTHALPY.

standard molar entropy.   *See* ENTROPY.

standard molar Gibbs function.   *See* GIBBS FUNCTION.

standard solution.   Solution of known concentration for use in an analytical process.

standard state.   Stable form of the substance at a pressure of 1 atm and the specified temperature (usually, but not necessarily 298.15 K). For a real gas, the standard state is some hypothetical state in which the gas is behaving ideally; little error is involved if the standard pressure is taken as 1 atm. For a BINARY LIQUID MIXTURE, the standard state of the solvent, using RAOULT'S LAW, is as $x_A \rightarrow 1$ (i.e. pure solvent) and for the solute, using HENRY'S LAW, as $x_B \rightarrow 0$ (i.e. infinite dilution). For an ionic solution, the standard state is a hypothetical state when $a = 1$. At low molalities, $\gamma_i \rightarrow 1$ as $m_i \rightarrow 0$. Thermodynamic functions are designated 'standard' when they refer to changes in which

$$\ominus Cd(Hg) \mid CdSO_4.\tfrac{2}{3}H_2O(s) \mid \begin{array}{c} \text{saturated solution} \\ CdSO_4 \end{array} \mid \begin{array}{c} Hg_2SO_4 \\ \text{paste} \end{array} \mid Hg \oplus$$

(a)

$$\ominus Zn(Hg) \mid ZnSO_4.7H_2O(s) \mid \begin{array}{c} \text{saturated solution} \\ ZnSO_4 \end{array} \mid \begin{array}{c} Hg_2SO_4 \\ \text{paste} \end{array} \mid Hg \oplus$$

(b)

standard cell
(a) Weston cell, (b) Clark cell.

reactants and products are in their standard states. *See also* CHEMICAL POTENTIAL, FUGACITY.

standard temperature and pressure (stp). Conditions used for the comparison of the properties of gases: 0°C (298.15 K) and 1 atmosphere (760 mmHg, 101 325 Pa) pressure.

stannanes. *See* TIN INORGANIC COMPOUNDS.

stannates. *See* TIN INORGANIC COMPOUNDS.

starch. POLYSACCHARIDE composed of repeating glucopyranose units (*see* GLUCOSE), attached by glycosidic linkages. All plants use starch as their principal food reserve; starches from different plants differ in their composition. Hydrolysis with amylase yields maltose, whereas complete acid hydrolysis yields glucose. It is insoluble in cold water, but in hot water it gives an opalescent dispersion. It is prepared from wheat, maize, rice, etc. by physical processes including steeping, milling and sedimentation. It is used as an adhesive, for sizing paper and as a diluent in foods and drugs. Starch consists of two fractions (*see* Figure). (1) Amylose (10–20 percent of starch) is composed of 250–300 glucose units attached via α-1,4-glycosidic linkages. It is soluble in hot water and gives a blue

colour with iodine solution. (2) Amylopectin (80–90 percent of starch) is composed of about 1000 glucose units in a structure that is branched about every 25 units. The branches are attached through α-1,6-glycosidic linkages; these lead to a small amount of isomaltose on hydrolysis. Amylopectin, which gives a violet colour with iodine, is insoluble in hot water; the differing solubilities provide a method of separating the two components. Amylose is used in the preparation of edible films and amylopectin as a thickener in foodstuffs, and for sizing paper and fabrics.

Stark effect. Splitting of rotational lines into $(2J+1)$ components by an electric field (*see* ROTATIONAL SPECTRA). The first-order effect is linearly proportional to the electric field, typical splittings being 30 MHz at a field of 100 V cm^{-1}. The second-order effect arises from the interaction of a field-induced dipole with the field. These splittings are about ten times smaller than the first-order effects.

state function. Property of the present state of the system which is independent of the way the state was prepared. The change of any state function depends only on the value of the function in the initial and final states:

$$\Delta X \quad = \quad X_2 \quad - \quad X_1$$

$$\begin{array}{ccc} \text{increase in} & = \text{final value} & - \text{ initial value} \\ \text{value of } X & \text{of } X & \text{of } X \end{array}$$

amylose
n = 250-300

amylopectin

branch point

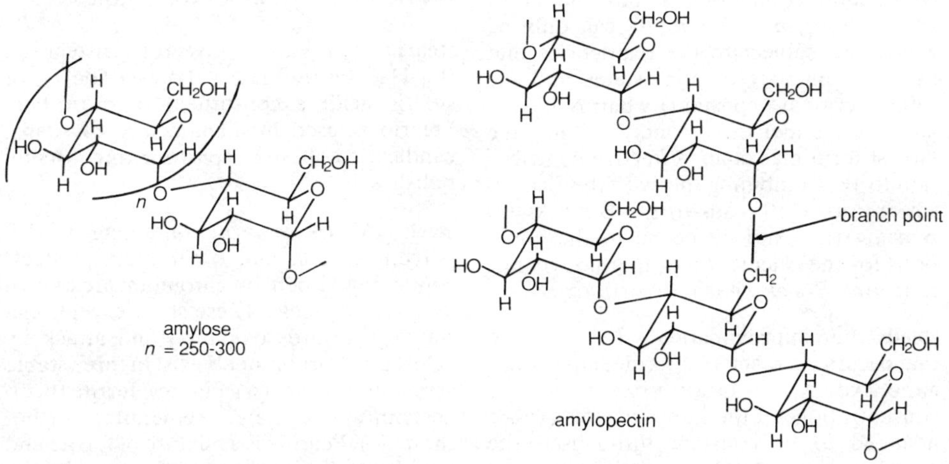

starch

State functions (e.g., P, T, V, S, U, H, G and A) are not necessarily independent of one another; once the state of the system is specified by the values of a few state functions the values of all the others are fixed. HEAT and WORK are not state functions.

states of matter. Three physical states in which matter can exist: solid, liquid, gas; plasma is often regarded as a fourth state of matter.

stationary phase. Immobile phase in all forms of CHROMATOGRAPHY. The liquids selected for GAS–LIQUID CHROMATO-GRAPHY must be non-volatile and thermally stable at operating temperature and inert with respect to the mobile phase, the solute and support, and good solvents for a wide range of solutes. The most common substances used include apeizons, silicones, squalane, polyethylene glycols, diethylene-glycol succinate. *See also* MOBILE PHASE.

statistical mechanics. Bridge between the dynamical properties of microstates (particles, molecules) and the thermodynamic properties of the entire assembly. For an assembly of a large number of microstates, statistical methods can be used to calculate the fraction of the total number of micro-states that have an assigned energy. For an assembly of N particles, if n_1 are in energy level ε_1, n_2 in level ε_2, etc. (n_1, n_2, . . ., n_i are known as distribution numbers) then

$$W = \sum \frac{N!}{\prod_i n_i!}$$

where summation is over all possible sets of values of n_1, n_2, . . ., n_i. W can only be evaluated (subject to the restrictions that the total number of microstates and the total energy are constant) when N is very large, and under these conditions only the largest term makes any effective contribution to W. Combining this with the BOLTZ-MANN EQUATION leads to the BOLTZMANN DISTRIBUTION and hence on to the equations for the calculation of thermodynamic functions. *See also* PARTITION FUNCTION.

steady-state approximation. In a reaction that occurs via reactive intermediates (e.g., FREE RADICALS in CHAIN REACTIONS), the concentrations of the intermediates (I) are assumed to be constant throughout the reaction (i.e. d[I]/dt = 0). Using this

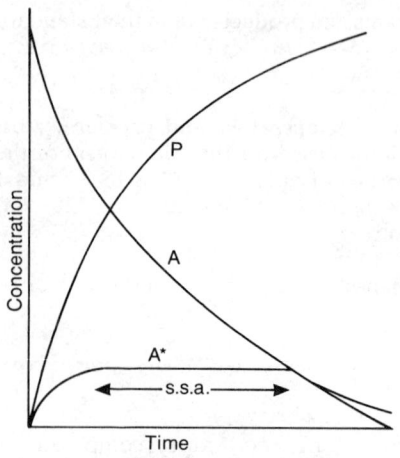

steady-state approximation
Concentration of species during the reaction A→A*→P showing the region of validity of the steady-state approximation where d[A*]/dt = 0.

assumption an expression for the steady-state concentrations of the intermediates may be obtained. *See also* LINDEMANN THEORY.

steam distillation. *See* IMMISCIBLE LIQUIDS.

steam point. Temperature at which the maximum VAPOUR PRESSURE of water is 1 atm (101 325 N m^{-2}) (i.e. 100°C).

steam reforming. *See* GASIFICATION.

stearic acid. *See* n-OCTADECANOIC ACID.

stearin (tristearin, glycerol tristearate, $(C_{17}H_{35}COO)_3C_3H_5$). Triglyceride of stearic acid; a constituent of most fats. Stearin is used in manufacture of soaps, candles, adhesive pastes and metal polishes.

steel. ALLOY of IRON containing 0.5–1.7 percent carbon and other metals. Steels containing 12 percent chromium are known as stainless steels. These show exceptional stability towards oxidation and attack by acids, etc. Carbon steels exist in three stable crystalline phases (*see* Figure): ferrite (bcc), austenite (fcc) and cementite (ortho-rhombic). Pearlite is a mixture of ferrite and cementite. Steels are manufactured by the

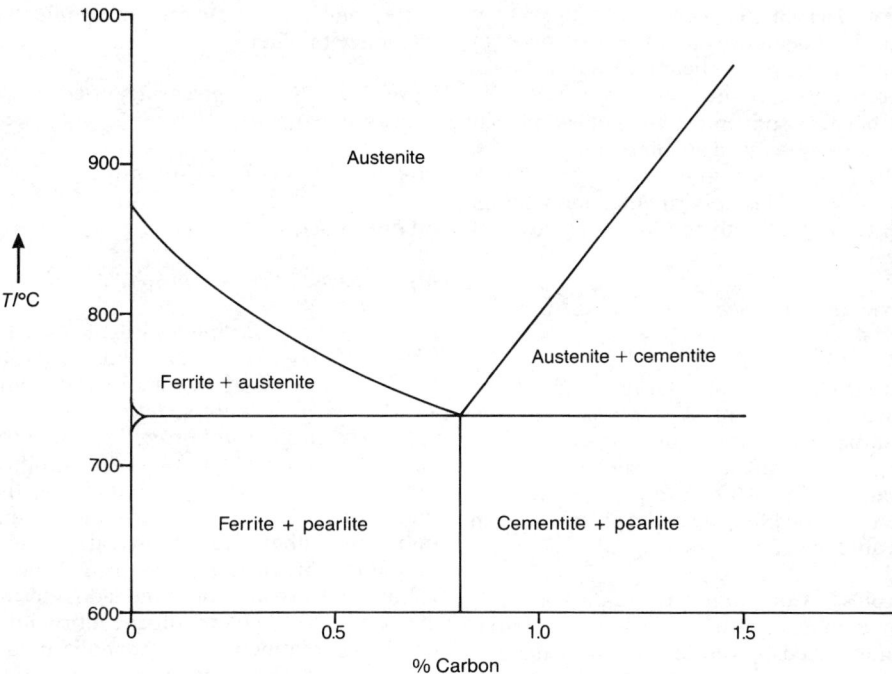

steel
Phase diagram of steel

BASIC OXYGEN PROCESS in electric furnaces and by the SIEMENS–MARTIN PROCESS. *See also* CAST IRON.

Stefan's law. *See* BLACK BODY RADIATION.

Stephan reaction. *See* BENZALDEHYDE, KETONES.

stereochemistry. Study of the spatial arrangements of atoms in molecules and complexes. *See also* ISOMERISM, OPTICAL ACTIVITY.

stereoisomers. Compounds whose molecules have the same number and kinds of atoms with the same atomic arrangement, but which differ in their spatial relationship.

stereoselective reaction. Reaction in which a starting material gives one DIASTEREOISOMER at a higher rate (preferentially) than another diastereoisomer. Reactions may be completely or highly stereoselective or show low stereoselectivity (e.g.,

the addition of halogens to alkenes or alkynes is predominantly *trans*).

stereospecific reaction. Reaction in which two stereoisomeric starting materials give different DIASTEREOISOMERS (i.e. one product is formed from each geometric isomer). An example is the *cis*-hydroxylation of *trans*-2-butene in the presence of osmium tetroxide which gives the DL-isomer, whereas *cis*-2-butene gives the *meso*-isomer.

steric factor. *See* ABSOLUTE RATE THEORY, ACTIVATED COMPLEX.

steric hindrance. Prevention or hindrance of a chemical reaction because of neighbouring groups on the same molecule; for example, the esterification of alcohols and substitution in the *para*-position in preference to the *ortho*-position if bulky groups are involved, nucleophilic displacements in S_N2 reactions. Groups in the 2,2'-positions of BIPHENYL compounds prevent free rotation of the rings.

Stern–Gerlach experiment. Experiment that first demonstrated the existence of ELECTRON SPIN. A beam of silver atoms passed between the poles of a powerful magnet was split into two, indicating that the atoms possessed an angular momentum with QUANTUM NUMBER 1/2 ($S = 2s + 1 = 2$, if $s = 1/2$). This spin angular momentum was associated with the lone s-electron of the silver atom.

Stern model. *See* ELECTRICAL DOUBLE LAYER.

Stern–Volmer plot. Graph of ϕ_0/ϕ_q against [Q] where ϕ_0 and ϕ_q are the quantum yields for FLUORESCENCE in the absence and presence, respectively, of a quenching agent Q. $\phi_0/\phi_q = 1 + \tau k_q[Q]$, where τ is the life-time of the excited state in the absence of quenching agent.

steroids. Compounds containing a cyclopentenophenanthrene carbon skeleton, characterized by the formation of methylcyclopentenophenanthrene (Diles hydrocarbon) on dehydrogenation with selenium. To this large group of compounds belong the STEROLS, BILE ACIDS, cardiac aglycones (*see* DIGITALIS), sex HORMONES, CORTICOSTEROIDS, toad poisons and steroid SAPOGENINS. Many synthetic steroids are known; some are valuable medicinals. The accepted numbering system for the steroids and attached side chains is illustrated for cholesterol (*see* STEROLS, Table). The methyl groups at the junction of rings A and B (C-10) and rings C and D (C-13) are called angular methyls. Methyl groups and hydrogen atoms at ring junctions are explicitly written as CH_3 or H and the stereochemistry is specified by a solid line if the atom or group is above the plane of the ring (β) and a dashed line if below the plane of the ring (α).

sterols. Alcohols containing the C-19 ring structure (as shown for cholesterol) that vary considerably in their peripheral structural features, stereochemistry and degree of ring saturation (*see* Table). They are present in all animal and plant cells (also in some bacteria), often partially esterified with higher fatty acids. They are isolated by hydrolyzing the fat fraction (ethanolic KOH) and extracting the unsaponifiable fraction with ether.

Stevioside. Sweet glycoside used as a SWEETENING AGENT.

stibine. *See* ANTIMONY HYDRIDE.

stibnite. *See* ANTIMONY.

stigmasterol. *See* STEROLS.

stilbene (*trans*-1,2-diphenylethene, C_6H_5-$CH=CHC_6H_5$). White crystalline solid (mp 124°C), prepared by heating benzal chloride ($C_6H_5CHCl_2$) with sodium or by treating benzyl magnesium bromide with benzaldehyde and dehydrating the product (benzylphenylmethanol). Stilbene (the stable *trans*-isomer) is converted to the unstable iso-stilbene (a yellow oil) by UV irradiation. Reduction (Na + EtOH) gives 1,2-diphenylethane; bromine adds across the double bond giving stilbene dibromide, which on treatment with ethanolic potassium hydroxide yields DIPHENYLETHYNE. *See also* ACYLOINS.

stilboestrol (4,4'-dihydroxy-α,β-diethylstilbene, $HOC_6H_4C(C_2H_5)=C(C_2H_5)C_6H_4$-OH). Synthetic oestrogenic compound (mp 168°C), structurally unrelated to steroidal oestrogens, that is highly active when administered orally. It is used for menopausal symptoms and the suppression of lactation. It has been implicated in uterine cancer. It is used as an additive in cattle and chicken feed to stimulate weight gain. An even more important oestrogenic substitute is dihydrostilboestrol (anethole hydrobromide + PhMgBr in presence of $CoCl_2$, followed by demethylation).

stishovite. *See* SILICA.

STO. *See* SLATER-TYPE ORBITAL.

stoichiometric. Describes chemical reactions in which the reactants combine in simple, whole-number ratios. A stoichiometric compound is one in which the atoms are combined in exact whole number ratios. In NON-STOICHIOMETRIC COMPOUNDS, there are defects in the crystal lattice or part replacement of one element by another (e.g., rutile has the formula $TiO_{1.8}$).

sterols

Compound	Structural formula	Occurrence/properties/comments
Cholesterol (5-cholesten-3β-ol, $C_{27}H_{46}O$)	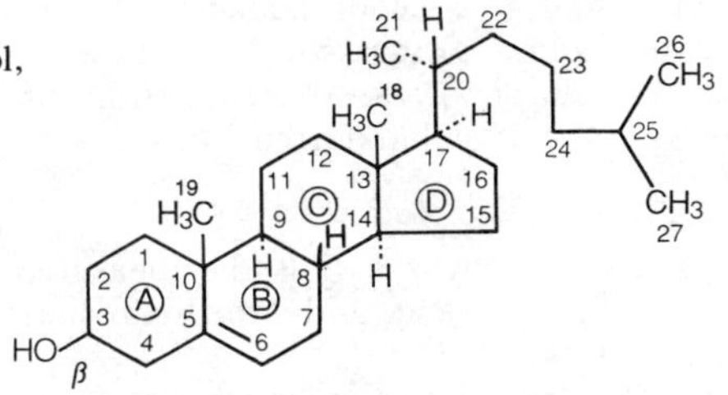	Present in all parts of animal body; concentrated in spinal cord, brain, skin secretions and gallstones. Unsaturated, unsaponifiable alcohol (mp 149°C). Synthesized in body from ethanoate units; its metabolism is regulated by a specific set of enzymes. Reduced (H_2/Pt) to 5α-cholestan-3β-ol. Parent compound of many other steroids; its presence in excess blood is suspected as being contributory factor in cardiovascular disease
Ergosterol (24-methyl-5,7,22-cholestatrien-3β-ol, $C_{28}H_{44}O$)	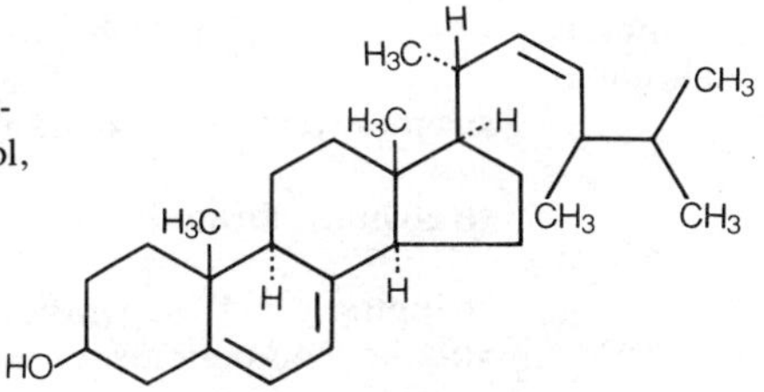	Plant sterol (mp 168°C) isolated from the ergots of rye and many fungi and yeasts. UV radiation gives vitamin D_2.
Stigmasterol (24-ethyl-5,22-cholestadien-3β-ol, $C_{29}H_{48}O$)	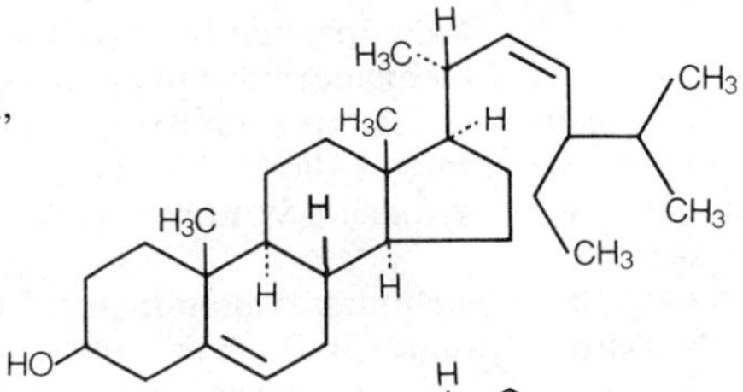	Plant sterol (mp 170°C); isolated from soya bean oil
Sitosterol (24-ethylcholesterol, $C_{29}H_{50}O$)	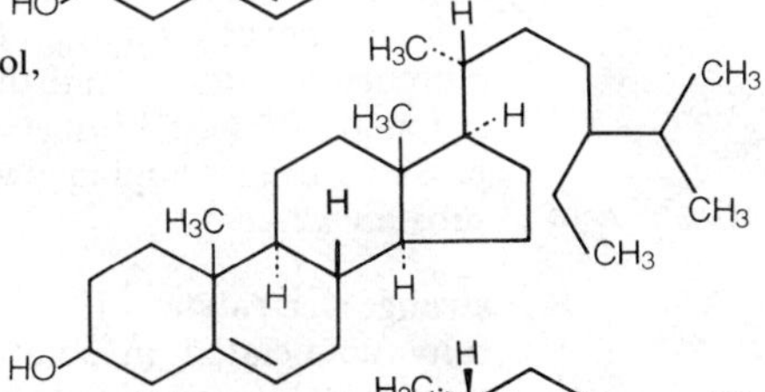	One of at least 6 sterols present in wheat germ oil (mp 137°C)
Lanosterol (isocholesterol, $C_{30}H_{50}O$)	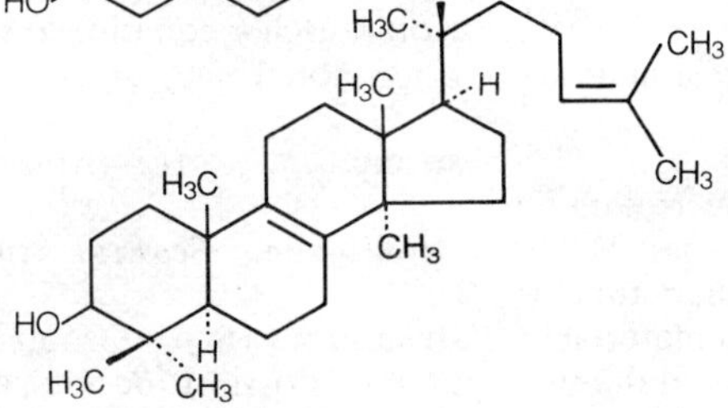	Trimethyl sterol or triterpenoid (mp 138°C) found in non-saponifiable fraction of wool wax. Formed by oxidative cyclization of squalene, followed by rearrangement of the methyl group. The precursor of sterols such as cholesterol in animals and fungi

Stoichiometric mixture is a mixture of reactants which give products with no excess reactant.

stoichiometry.   Relative numbers of atoms that are combined together in a compound, or the relative numbers of reactants and products in a chemical reaction. *See also* NON-STOICHIOMETRIC COMPOUND.

stokes (St).   CGS UNIT of KINEMATIC VISCOSITY; $St = 10^{-4} m^2 s^{-1}$.

Stokes–Einstein equation.   Relationship between the diffusion coefficient (D) (*see* FICK'S LAW OF DIFFUSION) and the VISCOSITY (η) of the medium through which a molecule of radius r moves

$$D = kT/6\pi\eta r$$

Stokes' law.   Law predicting the frictional force on a spherical particle, of radius r, falling through a viscous medium under gravity with a velocity v.

$$f = 6\pi\eta rv$$

See also VISCOSITY.

Stokes' lines.   *See* RAMAN SPECTROSCOPY.

stolzite.   *See* TUNGSTEN.

stopped flow method.   Method of measuring fast reaction rates in solution by injecting separate solutions containing each reactant into a chamber giving fast mixing. The reaction is then followed by spectrophotometry or measurement of conductance or pH changes, etc.

storage battery.   *See* BATTERY.

stp.   *See* STANDARD TEMPERATURE AND PRESSURE.

strain energy.   *See* RING STRAIN.

streaming birefringence.   *See* DOUBLE REFRACTION.

streaming potential.   Electrokinetic phenomenon (*see* ELECTROKINESIS). When a liquid is forced through a capillary tube or through a plug of finely divided material, a potential difference arises between the two ends of the material. The hydrostatic pressure is measured by the difference in level of the liquid in the two reservoirs. The potential difference is measured between the two platinum electrodes which enclose the material under investigation. The effect is a consequence of the ELECTRICAL DOUBLE LAYER which exists at an interface. The ZETA-POTENTIAL at the surface may be calculated from

$$E = \zeta\varepsilon P/4\pi\eta\kappa$$

where E is the measured potential difference and P the hydrostatic pressure.

Strecker synthesis.   Synthesis of an α-AMINO ACID by the simultaneous reaction of an aldehyde with ammonia and hydrogen cyanide followed by hydrolysis of the resulting amino nitrile

$$RCHO + HCN + NH_3 \longrightarrow$$
$$RCH(NH_2)CN \xrightarrow{H_2O} RCH(NH_2)COOH$$

streptomycin.   *See* ANTIBIOTICS.

stretching vibration.   *See* NORMAL MODE.

stripping.   (1) Separation of the more volatile components of a mixture which leaves a pure residuum. This may be accomplished by VACUUM DISTILLATION or by evaporation in an inert atmosphere. (2) In electrochemistry, oxidation of an adsorbed layer on an electrode. *See also* STRIPPING VOLTAMMETRY.

stripping voltammetry.   Analytical technique in which the substance to be analyzed is first concentrated at an electrode or in a mercury AMALGAM and then is anodically stripped off (*see* STRIPPING). The charge passed during stripping gives the amount of substance.

strong electrolyte.   ELECTROLYTE that is fully dissociated in solution and follows Kohlrausch's equation (*see* CONDUCTANCE EQUATIONS).

strontia.   *See* STRONTIUM COMPOUNDS.

strontianite.   *See* STRONTIUM.

strontium (Sr).   Group IIa ALKALINE EARTH METAL. It occurs naturally as strontianite ($SrCO_3$) and celestine ($SrSO_4$). It is

extracted by electrolysis of strontium chloride/potassium chloride melt or the vacuum reduction of strontium monoxide by aluminium. It is a silvery-white metal reacting with water, oxygen, hydrogen and other non-metals. It is used in magnet alloys (strontium ferrates) and GETTERS. Its compounds are used in pyrotechnics ($Sr(NO_3)_2$) and in soaps and greases ($Sr(OH)_2$). The radioactive isotope (strontium-90; half-life: 28 yr) is present in fall-out and can substitute calcium in bones.
$A_r = 87.62$, $Z = 38$, ES [Kr] $5s^2$, mp 786°C, bp 1380°C, $\rho = 2.62 \times 10^3$ kg m^{-3}, Isotopes: ^{84}Sr, 0.6 percent; ^{86}Sr, 9.9 percent; ^{87}Sr, 7.0 percent; ^{88}Sr, 82.6 percent.

strontium compounds. Strontium is an ELECTROPOSITIVE metal ($E^{\ominus}_{Sr^{2+},Sr} = -2.89$ V) that forms Sr^{2+} in its predominantly ionic compounds. Most salts are soluble with the exception of strontium fluoride (SrF_2), strontium sulphate ($SrSO_4$) and strontium hydroxide ($Sr(OH)_2$). The chlorate(VII) is soluble in organic solvents. The nitride and carbide react with water to give ammonia and ethyne, respectively. Two oxides (SrO_2 and SrO) are formed; the higher SrO_2 is prepared by the reaction between SrO and oxygen at 800°C. The boride (SrB_6) has metallic conductivity. Some unstable and very reactive strontium alkyls are known (e.g., $(C_2H_5)_2Sr$).

structural formula. *See* FORMULA.

structural isomerism. *See* ISOMERISM.

structure-sensitive reaction. *See* SUPPORTED CATALYST.

strychnine. *See* ALKALOIDS.

styrene (ethenylbenzene, vinylbenzene, $C_6H_5CH=CH_2$). Only important aromatic hydrocarbon with an unsaturated side chain; it occurs in storax (a balsam); colourless liquid (bp 145°C), prepared industrially by the DEHYDROGENATION of ETHYLBENZENE at 600°C in the presence of steam and zinc oxide as catalyst. It is formed when 3-phenylpropenoic acid is heated in the presence of hydroquinone. It polymerizes slowly on standing and rapidly in sunlight to metastyrene ($(C_8H_8)_n$) which is depolymerized by heat. Reduction (H_2 + Ni at 25°C) gives ethylbenzene; hypochlorous acid adds across the double bond giving $C_6H_5CH(OH)CH_2Cl$. Styrene is used extensively in the manufacture of POLYMERS (e.g., polystyrene, sulphonation produces ion-exchange resins, styrene–butadiene rubber).

styrene-butadiene rubbers. *See* POLYMERS, RUBBER.

styrene polymers. *See* POLYMERS.

sublimation. Direct change of state from solid to vapour without passing through the liquid state.

substituent. Atom or group of atoms that replaces another in a substitution reaction.

substitution nucleophilic bimolecular mechanism. *See* S_N2 MECHANISM.

substitution nucleophilic unimolecular mechanism. *See* S_N1 MECHANISM.

substitution reaction. Reaction in which an atom or group in a compound is replaced by another atom or group (e.g., the chlorination of benzene). *See also* ELECTROPHILIC SUBSTITUTION, NUCLEOPHILIC SUBSTITUTION, S_N1 MECHANISM, S_N2 MECHANISM.

substrate. General term for the molecule undergoing the reaction in question; specifically in biochemical reactions it is the molecule on which an enzyme acts.

succinic acid. *See* BUTANEDIOIC ACID.

succinic anhydride. *See* BUTANEDIOIC ANHYDRIDE.

succinimide. *See* BUTANIMIDE.

sucrose (cane sugar). DISACCHARIDE (structure I) obtained commercially from the sugar cane and sugar beet; in addition, a syrup, known as molasses, is always obtained. This commercial product, which will not crystallize, is used in the fermentation industry. It caramelizes above its melting point (180°C) and with concentrated sulphuric acid yields almost pure carbon. In solution $[\alpha]_D^{25} = +66.5°$, but on hydrolysis with acid or the enzyme invertase it gives an equimolar mixture of D-(+)-GLUCOSE and D-(−)-FRUCTOSE (invert sugar) with a negative optical rotatory power. Sucrose does not undergo mutarotation, is not reducing and does not form an OSAZONE, indicating that the aldehyde and ketone groups of glucose and fructose, respectively, are not free. It is used in the food industry. *See also* SWEETENING AGENTS.

sugars. Crystalline carbohydrates with a sweet taste and very soluble in water. Excluding hydroxyethanal (glycolaldehyde) from the group; sugars may be defined as optically active polyhydroxyaldehydes or polyhydroxyketones. Two main classes are recognized.
monosaccharides. Sugars that cannot be hydrolyzed into smaller molecules, of general formula $C_nH_{2n}O_n$, where n = 2–10. The most important are the HEXOSES and PENTOSES.
oligosaccharides. On hydrolysis, these sugars yield two to nine monosaccharide units (e.g., DISACCHARIDES yield two monosaccharide units which may, or may not, be the same; trisaccharides yield three monosaccharide units).

sulpha drugs. *See* SULPHONAMIDES.

sulphamic acid (HSO_3NH_2). White crystalline solid (mp 205°C), prepared by the action of fuming sulphuric acid on urea. It is a strong acid, soluble in water forming a series of salts, the sulphamates, and is a primary volumetric standard. Sulphamic acid is used to clean metals and ceramics, as a plasticizer, fire retardant, and textile and paper bleach; ammonium sulphamate is a HERBICIDE.

sulphanes. Compounds of hydrogen and sulphur containing sulphur chains. Compounds H_2S_2 to H_2S_6 have been isolated in the pure state (Na_2S_n with HCl, S_2Cl_2 with H_2S or H_2S_2). They are yellow oils, which are stable for a time, eventually decomposing to hydrogen sulphide and sulphur.

sulphanilamides. *See* SULPHONAMIDES.

sulphates. Formally compounds (salts or esters) containing an anionic oxo sulphur species; practically they are limited to the

derivatives of sulphuric and sulphurous acids (*see* SULPHUR OXOACIDS).
sulphates(VI). Contain the SO_4^{2-}, which may be ionic or coordinated. There are two series of salts ($RHSO_4$ and R_2SO_4). Many sulphates occur as minerals (e.g., gypsum, anhydrite, barytes). Most sulphates are crystalline and soluble in water. Sulphates of lead, calcium and strontium are sparingly soluble, whereas that of barium is insoluble in water.
organic sulphates. Esters obtained by sulphonation of primary alcohols. The sodium salts of long-chain alkyl sulphates (e.g., sodium lauryl sulphate, $CH_3(CH_2)_{10}$-OSO_3^- Na^+) are excellent anionic DETERGENTS.
sulphates(IV) (sulphites). Contain the SO_3^{2-} ion. Two series of salts are prepared by passing sulphur dioxide through a solution of the corresponding hydroxide. They react with acids to give sulphur dioxide.

sulphenes ($RCH=SO_2$). Reactive intermediates formed when a SULPHONYL CHLORIDE, with α-hydrogens, reacts with a base (e.g., CH_3SO_2Cl + Et_3N → $CH_2=$ SO_2). In the presence of alcohols and amines addition occurs, producing sulphonates and SULPHONAMIDES.

sulphides. (1) Inorganic compounds of sulphur with more electropositive elements prepared by direct interaction of the elements or by precipitation from a salt solution with hydrogen sulphide. Metal sulphides, which often occur as minerals, are ionic, reacting with acids to give hydrogen sulphide. The very low solubility of heavy metal sulphides is used in qualitative analysis for the separation of the metals. Covalent derivatives are formed with non-metals (e.g., CARBON DISULPHIDE, SULPHUR HALIDES). (2) Organic sulphides (*see* THIOETHERS) contain the –S– group linked to two hydrocarbon groups. *See also* POLYSULPHIDES.

sulphites. *See* SULPHATES, SULPHUR OXOACIDS.

sulpholane. *See* TETRAHYDROTHIOPHEN-1,1-DIOXIDE.

sulphonamides. Aromatic compounds containing the group –SO_2NR_2, prepared by the action of ammonia or appropriate amine on 4-acetamidobenzenesulphonyl chloride followed by hydrolysis of the acetyl derivative to give the base. They are colourless crystalline substances of varying solubility in water. Sulphonamides, the so-called 'sulpha drugs', have been used as bacteriostatic agents against bacteria since the 1930s, when the dye Prontosil was first shown to be an effective antibacterial agent. They act by inhibiting the uptake of 4-aminobenzoic acid by bacterial cells which require this as a precursor of folic acid (*see* VITAMINS). Their value as antibacterial agents is threatened by the emergence of resistant strains when recourse is made to other antibacterial agents and ANTIBIOTICS. Many substituted sulphonamides (usually in the form of the sodium salts) have been synthesized; they differ in their specificity, rates of absorption, etc. Typical sulphonamides include:

sulphanilamide ($4\text{-}H_2NC_6H_4SO_2NH_2$). The original sulphonamide, active against streptococci and gonococci (causes unpleasant side reactions).

sulphaguanidine ($4\text{-}H_2NC_6H_4SO_2NHC\text{-}(NH)NH_2$). Active against intestinal infections.

sulphadiazine ($4\text{-}H_2NC_6H_4SO_2NH\text{-}(C_4H_3N_2)$). Least toxic of the most potent sulphonamides.

sulphathiazole ($4\text{-}H_2NC_6H_4SO_2NH\text{-}(C_3H_2NS)$). Active against streptococci, pneumococci, staphylococci and gonococci.

sulphapyridine ($4\text{-}H_2NC_6H_4SO_2NH\text{-}(C_5H_4N)$).

sulphonation. Introduction of a sulphonic group ($-SO_3H$) into an organic compound. Common sulphonating agents include concentrated and fuming sulphuric acids, sulphur trioxide, chlorosulphonic acid.

sulphones. Organic compounds containing a $>SO_2$ group attached directly to two carbon atoms (e.g., TETRAHYDROTHIOPHEN-1,1-DIOXIDE), prepared by the oxidation of organic SULPHIDES or SULPHOXIDES. Sulphones are unreactive to attack on the sulphur or oxygen, but are reduced to sulphides by lithium tetrahydridoaluminate.

sulphonic acids. Organic compounds (e.g., BENZENESULPHONIC ACID) containing the $-SO_3H$ group. Alkylsulphonic acids are prepared by the oxidation of thiols or treatment of an alkyl halide with an alkali sulphite. Arylsulphonic acids are obtained by heating an aromatic hydrocarbon with concentrated or fuming sulphuric acid. They are strong acids, very soluble in water readily forming water-soluble salts. Fusion of an aryl acid with sodium hydroxide gives a phenol and with potassium cyanide a nitrile. The esters of 4-toluenesulphonic acid are frequently called tosylates; many tosylates are crystalline compounds which can be used for identification. The sulphonate group is a good leaving group from a carbon atom in nucleophilic substitution reactions. Sulphonation of long-chain alkylbenzenes yields alkyl arylsulphonates (e.g., $4\text{-}C_{12}H_{25}C_6H_4SO_3Na$), an important group of detergents; straight-chain alkyl detergents are BIODEGRADABLE

sulphonium salts. Organic compounds containing R_3S^+ ion formed by the reaction of an organic SULPHIDE with alkyl halide. Pyrolysis gives sulphide and alkyl halide.

sulphonyl chlorides. Organic compounds (RSO_2Cl, where R is an alkyl or aryl group), prepared from corresponding sulphonic acid with phosphorus pentachloride. They have the properties of an acid chloride, reacting with water, bases, alcohols, ammonia and amines to give RSO_3H, RSO_3Na, RSO_3R', RSO_2NH_2, $RSO_2\text{-}NHR'$, $RSO_2NR'R''$. 4-Toluenesulphonyl chloride (toluene + chlorosulphonic acid) is a white crystalline powder which forms characteristic crystalline derivatives with many alcohols, phenols, sugars and amines.

sulphoxides. Organic compounds containing the $>SO$ group linked to two carbon atoms (e.g., DIMETHYLSULPHOXIDE). Sulphoxides are prepared by the oxidation of organic SULPHIDES with one equivalent of hydrogen peroxide. They are hygroscopic liquids with large dipole moments; they act as Lewis acids primarily at the oxygen atom. On oxidation they yield SULPHONES.

sulphoxylic acid. *See* SULPHUR OXOACIDS.

sulphur (S). Yellow, non-metallic GROUP VI ELEMENT. It occurs (0.05 percent earth's crust) in the free state (FRASCH PROCESS), as sulphates (gypsum) and sulphides (galena, iron pyrites and copper pyrites) and as hydrogen sulphide in natural gas and oil deposits. It exists in various allotropic forms: rhombic is stable at low temperatures transforming into monoclinic at 95.5°C (*see* ENANTIOTROPY); these crystalline forms contain cyclic S_8 molecules. Other allotropic forms include ρ-S (S_6 chair rings), λ-S and μ-S (spiral chains). In addition to the mainly S_2 (paramagnetic) species, the vapour also contains S_4, S_6 and S_8 species. Sulphur is very reactive, combining with most other elements (*see* Figure). It is insoluble in water, but soluble in carbon disulphide and other non-polar solvents. Oxidation states +6, +4, +2, 0 and −2 exist. Saturated hydrocarbons are dehydrogenated. The reaction with AL-KENES and other unsaturated hydrocarbons is important in the VULCANIZATION (formation of sulphur bridges) of natural and synthetic rubbers. There are a wide range of sulphur-containing organic compounds (e.g., PROTEINS, SULPHONIC ACIDS, SULPHONES, SULPHOXIDES, THIOLS, THIOPHENE). The principal use of sulphur is in the manu-facture of sulphuric acid (*see* CONTACT PROCESS), but it is also used in the wood pulp, rubber and dyestuff industries and as an insecticide and fungicide.

A_r = 32.066, Z = 16, ES [Ne] $3s^2\ 3p^4$, mp 112.8°C, bp 444.67°C, ρ = 2.07 × 10^3 kg m^{-3}. Isotopes: ^{32}S, 95 percent; ^{33}S, 0.76 percent; ^{34}S, 4.2 percent; ^{36}S, 0.01 percent.

sulphur halides.
sulphur hexafluoride (SF_6). Very inert gas (S + F_2), with a high dielectric strength; used as a gaseous insulator in high-voltage generators.
sulphur tetrafluoride (SF_4). Extremely reactive gas (S_2Cl_2 + NaF), instantly hydrolyzed to sulphur dioxide and hydrogen fluoride; powerful, but selective fluorinating agent for organic (>CO and –P=O to >CF$_2$, and –PF$_2$ and –COOH to –CF$_3$ without attack on other functional groups) and inorganic compounds (NaF + TiO_2 + $SF_4 \rightarrow Na_2TiF_6$).
sulphur monofluoride (S_2F_2). Silver fluoride and sulphur in vacuum produces two isomers, the more stable and abundant is

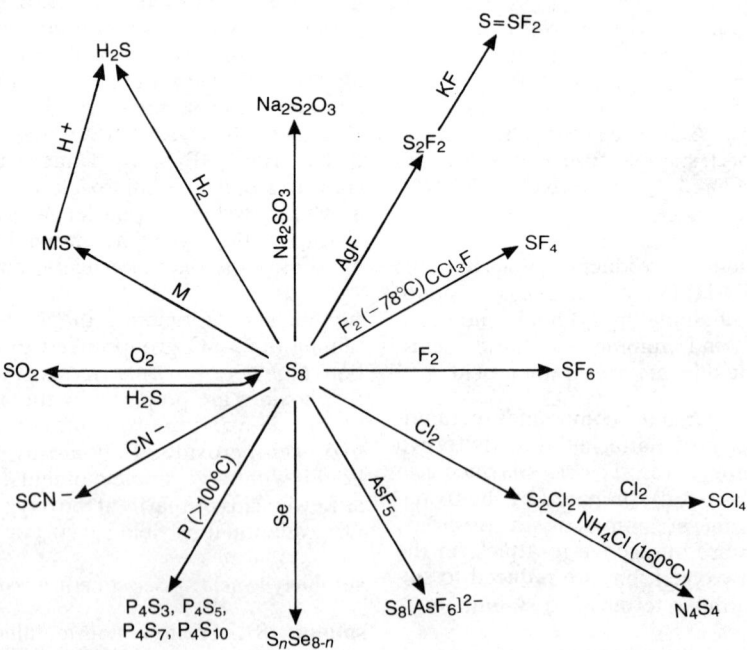

sulphur
Some reactions of sulphur.

$F_2S=S$, the other is F–S–S–F. Very reactive compound.

disulphur decafluoride (S_2F_{10}). Inert gas (radiation of mixture of $SF_5Cl + H_2$), not hydrolyzed by water or alkalis.

sulphur tetrachloride (SCl_4). Unstable compound (sulphur chlorides + Cl_2 at –80°C), dissociating above –31°C.

sulphur dichloride (SCl_2). Red, oily liquid ($S_2Cl_2 + Cl_2$); readily hydrolyzed by water.

sulphur monochloride (S_2Cl_2). Chlorination of molten sulphur yields a yellow liquid, readily hydrolyzed by water. It is used as a solvent for sulphur in vulcanization of rubber.

other sulphur halides. Sulphur monobromide is the only known bromide, no iodides are known. The mixed halide (SF_5Cl) (SF_4 + ClF in presence of CsF) is a colourless gas which is more reactive than SF_6. It is hydrolyzed by alkalis, but not by acids.

sulphur halooxoacids.

fluorosulphurous acid (HSO_2F). Free acid does not exist. The salts (e.g., KF + SO_2) are mild fluorinating agents (e.g., –COCl → –COF).

fluorosulphuric acid (HSO_3F). Stable acid obtained by the distillation of KHF_2 with oleum. Salts are similar to the chlorates.

chlorosulphuric acid (chlorosulphonic acid, HSO_3Cl). Colourless fuming liquid (SO_3 + dry HCl). It undergoes explosive reaction with water, but forms no salts. The acid is used extensively as a chlorinating and sulphonating agent.

bromosulphuric acid. (HBr + SO_3 in liquid SO_2). Decomposes at the melting point.

sulphuric(VI) acid. *See* SULPHUR OXO-ACIDS.

sulphur–nitrogen derivatives.
There is an extensive chemistry of sulphur–nitrogen compounds (many of them being cyclic derivatives), comprising species of the type SN^+, S_2N_2, $S_3N_3^+$, S_4N_4, $S_5N_5^+$, . . .; the cationic species are known as thiazyl ions. The most important compound, tetrasulphur tetranitride (S_4N_4) (S_2Cl_2 + NH_4Cl, or S + NH_3 in CCl_4), exists as yellow crystals which are detonated by shock. It is diamagnetic and has a cage structure (D_{2d} symmetry). It undergoes two types of reaction: (1) those in which the sulphur–nitrogen ring is preserved (e.g., addition reaction with boron trifluoride or reduction to $S_4N_4H_2$); (2) those in which ring cleavage occurs (e.g., with thionyl chloride to give the planar thiotrithiazyl ion, $S_4N_3^+$, and with sulphuric acid to give $S_2N_2^+$). It is reduced by tin(II) chloride to tetrasulphur tetraimide ($S_4(NH)_4$), and with phosphorus trichloride gives the compound $[Cl_3PNP(Cl)_2NPCl_3]^+PCl_6^-$.

sulphurous acid. *See* SULPHUR OXOACIDS.

sulphur oxides.

disulphur monoxide (S_2O, SSO). Unstable, with many of the properties of a free radical. S_2O is prepared by subjecting a mixture of sulphur and sulphur(IV) oxide to an electric discharge.

sulphur(IV) oxide (sulphur dioxide, SO_2). Colourless gas with a choking smell, prepared by burning sulphur, metal sulphides or hydrogen sulphide in air or by the action of an acid on a sulphite (sulphate(IV)). It is a powerful reducing agent particularly in water, where it behaves also as an acid (sulphuric(IV) acid). With lone pairs of electrons, it can act as a Lewis base or acid. It reacts with phosphorus pentachloride to give thionyl chloride and with chlorine to give sulphuryl chloride (*see* SULPHUR OXOHALIDES), and it forms stable 1:1 charge-transfer complexes with amines (e.g., $Me_3N.SO_2$). Liquid sulphur dioxide is a useful non-aqueous solvent ($\varepsilon_r = 15.1$) dissolving many organic and inorganic compounds. SO_2 is used in bleaching and as a fumigant and food preservative (E220). Large quantities are used for the manufacture of sulphuric acid.

sulphur(VI) oxide (sulphur trioxide, SO_3). White, fuming solid prepared by the reaction of sulphur dioxide with oxygen in the presence of a catalyst (*see* CONTACT PROCESS). It reacts vigorously with water to give sulphuric acid and is absorbed by sulphuric acid to give oleum (($SO_3)_n.H_2O$). SO_3 functions as a fairly strong Lewis acid towards bases that it does not oxidize; it is a powerful, indiscriminate oxidizing agent. It is used in sulphonation reactions and for the manufacture of sulphonated oils and detergents (*see* SURFACE-ACTIVE AGENT).

sulphur oxoacids. Sulphur forms an extensive range of oxoacids, some of which only exist in the form of their anions and salts.

sulphoxylic acid (H_2SO_2, $HS(O)OH$). The acid is unknown; the cobalt salt is made by the action of sodium dithionate on cobalt(II) ethanoate, followed by treatment with ammonia.

sulphuric(IV) acid (sulphurous acid, H_2SO_3, $OS(OH)_2$). Sulphur dioxide (*see* SULPHUR OXIDES) is soluble in water giving a gas hydrate ($SO_2.7H_2O$); the free acid does not exist. The solution reacts with bases to give two series of salts, the sulphates(IV) containing the SO_3^{2-} and the HSO_3^- ions.

sulphuric(VI) acid (sulphuric acid, tetraoxosulphuric(VI) acid, H_2SO_4, $O_2S(OH)_2$). Strong dibasic acid, manufactured from sulphur dioxide and air by the CONTACT PROCESS; forms two series of salts (SULPHATES). The pure acid, obtained from the commercial 98 percent acid by the addition of sulphur trioxide (*see* SULPHUR OXIDES), is rarely used. Strongly protogenic solvent, most compounds of oxygen and nitrogen (e.g., ethers, amines) accept protons from it, carboxylic acids dissociate as bases. Nitric acid gives the NITRONIUM ION (NO_2^+), the active agent in aromatic nitrations. The acid is miscible with water in all proportions, forming an azeotrope (*see* AZEOTROPIC MIXTURE) with water (98.3 percent acid). It is a very powerful dehydrating agent (reaction with water is violent and exothermic) for both organic and inorganic compounds (causes charring of carbohydrates), and is a strong oxidizing agent. It has a low volatility and hence displaces most other acids from their salts. It is an important industrial acid, used in organic syntheses, the production of fertilizers, explosives, pigments and petrochemicals.

thiosulphuric acid ($H_2S_2O_3$, $S(O)S(OH)_2$). Free acid is only known ($SO_3 + H_2S$) at $-78°C$ as an etherate. Thiosulphates, containing the $S_2O_3^{2-}$ ion, are formed by boiling a solution of a sulphite with sulphide. They decompose in acid solution to sulphur dioxide and sulphur, and react quantitatively with iodine, a reaction used in volumetric analysis. Alkali thiosulphates are used extensively in photography to remove unreacted silver bromide from the emulsion as complexes.

dithionous acid (hyposulphurous acid, hydrosulphurous acid, $H_2S_2O_4$, $O(HO)SS(OH)O$). The acid is unknown. Dithionates (reduction of sulphites with Zn or SO_2) are unstable in water giving $S_2O_3^{2-}$ and HSO_3^-. They are strong reducing agents.

pyrosulphurous acid (disulphurous acid, $H_2S_2O_5$, $O_2(HO)SS(OH)O$). The acid is unknown. The salts, obtained on heating the hydrogen sulphites ($MHSO_3$) are strong reducing agents (e.g., sodium pyrosulphite, so-called sodium metabisulphite).

dithionic acid ($H_2S_2O_6$, $O_2(OH)SS(OH)O_2$). The acid (oxidation of a solution of sulphur dioxide with manganese(IV) oxide) is a moderately stable strong acid that decomposes on warming. Dithionates are soluble in water and are not decomposed by heating or by sulphites and sulphides.

pyrosulphuric acid (disulphuric acid, $H_2S_2O_7$, $O_2(OH)S–O–S(OH)O_2$). Present in fuming sulphuric(VI) acid (oleum) (i.e. $SO_3 + H_2SO_4$).

polythionic acids ($H_2S_nO_6$, $O_2(OH)SS_nS(OH)O_2$, where $n = 1–4$). These acids are named according to the total number of sulphur atoms (e.g., trithionate, $S_3O_6^{2-}$, tetrathionate, $S_4O_6^{2-}$). They are a series of unstable dibasic acids; no acid salts are known. Mixtures of polythionates are obtained by reduction of thiosulphate solutions with sulphur dioxide, or by the reaction of hydrogen sulphide with sulphur dioxide in aqueous solution. Tetrathionates (reduction of thiosulphates with iodine) are thermally unstable in the solid state.

peroxomonosulphuric acid (Caro's acid, H_2SO_5, $O_2S(OH)(OOH)$). Obtained by the hydrolysis of peroxodisulphuric acid or by the action of concentrated hydrogen peroxide on sulphuric acid or chlorosulphuric acid. Salts (e.g., $KHSO_5$) cannot be obtained pure, decomposing in aqueous solution.

peroxodisulphuric acid ($H_2S_2O_8$, $O_2(HO)SOOS(OH)O_2$). Sodium and ammonium salts are prepared by the electrolysis of the sulphate at low temperatures and high current densities; the acid can be obtained from the salts. The peroxodisulphate ion is a powerful and strong oxidizing agent ($E^⊖ = +2.01$ V); silver is required as a catalyst.

sulphur oxohalides. Three main types exist.

thionyl halides (SOX$_2$). All halides are known; with the exception of the fluoride they all react violently with water giving sulphur dioxide and hydrogen halide. Thionyl chloride (PCl$_5$ + SO$_2$) decomposes on heating and reacts with hydrofluoric or hydrobromic acid to give the corresponding thionyl halide. It is used to prepare anhydrous metal chlorides and as a chlorinating agent in organic syntheses.

sulphuryl halides (SO$_2$X$_2$). The chloride (SO$_2$ + Cl$_2$ in presence of a catalyst) and the fluoride (F$_2$ + SO$_2$Cl$_2$) are the most important. SO$_2$F$_2$ is an inert gas unaffected by water, but slowly hydrolyzed with strong aqueous alkalis. SO$_2$Cl$_2$, a fuming liquid, decomposes below 300°C and reacts readily with water. It is used as a chlorinating agent. Mixed halides (e.g., SO$_2$FCl, SO$_2$FBr) are known.

complex oxochlorides and oxofluorides. Compounds such as S$_2$O$_5$F$_2$, S$_2$O$_6$F$_2$ (SO$_3$ + F$_2$), SOF$_4$ and SO$_3$F$_2$ are known.

sulphuryl. Containing the >SO$_2$ group.

sulphuryl halides. *See* SULPHUR OXO-HALIDES.

Sunnett. Trade name for a SWEETENING AGENT.

superacid. Acid (*see* ACIDS AND BASES) with an ability to donate a proton greater than that of anhydrous sulphuric acid. The HAMMETT ACIDITY FUNCTION (H_0) is used as a measure of the acidity, some values being H$_2$SO$_4$ –11.9, HSO$_3$Cl –13.8, H$_2$S$_2$O$_7$ –14.4, HF –15.1, 1 percent SbF$_5$/HF –21. Many compounds are protonated in superacids (e.g., CH$_3$COOH$_2^+$).

superconductivity. At low temperatures (less than 10 K), some metals, alloys and molecular conductors have zero resistance. It is possible therefore to keep large currents flowing in a loop and thus generate magnetic fields. It is used in superconducting magnets for NUCLEAR MAGNETIC RESONANCE SPECTROSCOPY. A range of ceramic materials based on copper oxide doped with rare earth elements have shown superconductivity at above 77 K.

supercooling. Cooling of a system to a temperature below that at which a phase transition should occur without any change taking place, for example, cooling of: (1) a liquid below its freezing point without freezing; (2) a solution to a temperature below which the concentration is greater than that of a SATURATED SOLUTION without crystallization (i.e. SUPERSATURATED); (3) a vapour below its normal DEW POINT without condensation. All such states are unstable and, if disturbed, will revert to the form stable at that temperature.

superfluid. Liquid (e.g., HELIUM) which flows without friction and exhibits very high thermal conductivity at very low temperatures. Superfluids are sometimes called superconductors.

superheating. Heating a liquid to a temperature above its normal BOILING POINT without boiling occurring. It gives rise to the phenomenon of 'bumping' when there are no nuclei or rough surfaces on which bubbles can form and grow.

superheavy elements (supertransuranics). *See* TRANSACTINIDE ELEMENTS.

superionic conductor. *See* SOLID ELECTROLYTE.

superoxides. Group of inorganic compounds containing the paramagnetic O$_2^-$ ion with an unpaired electron. The action of oxygen on potassium, rubidium or caesium gives yellow to orange crystalline compounds of formula MO$_2$. They are powerful oxidizing agents reacting vigorously with water to give oxygen.

superphosphate. Most important phosphorus FERTILIZER, made by the action of concentrated sulphuric acid on insoluble phosphate rock (Ca$_3$(PO$_4$)$_2$) to form a mixture of calcium sulphate and soluble monobasic calcium phosphate (Ca(H$_2$PO$_4$)$_2$).

supersaturated. Denotes (1) a solution of higher concentration than that of a SATURATED SOLUTION at the same temperature, (2) a vapour at a higher vapour pressure than a saturated vapour at the same temperature. Both systems are thermodynamically unstable. Supersaturation is usually a consequence of SUPER-COOLING.

supertransuranics. See TRANSACTINIDE ELEMENTS.

supported catalyst. Heterogeneous catalyst (*see* HETEROGENEOUS CATALYSIS) often a precious metal, prepared in a highly dispersed form on a support (e.g., SILICA GEL, alumina (*see* ALUMINIUM INORGANIC COMPOUNDS) or CARBON). The catalyst is cheaper, more easily handled, has a high surface-to-volume ratio and is less prone to SINTERING. The particle size may be measured by transmission electron microscopy, the broadening of X-RAY DIFFRACTION lines and adsorption measurements (*see* ADSORPTION ISOTHERM). Some reactions (e.g., HYDROGENOLYSIS of hydrocarbons) show a specific activity that increases with decreasing particle size of the catalyst. These are known as demanding or structure-sensitive reactions. Those reactions that have constant specific activity are called facile or structure-insensitive.

support medium. Granular material on which the STATIONARY PHASE is deposited. The necessary properties are chemical inertness, high surface area, porosity and thermal stability. Typical supports include celite, firebrick, kieselguhr, carbon, polymers, glass beads and silica gel. For gas–liquid chromatography columns, particle sizes are in the range 125–250 μm and for high-performance liquid chromatography 20–150 μm.

suprafacial. On the same face of a molecule (i.e. *cis-*) or reaction site of a catalyst; antarafacial is the opposite (*see* Figure). It is used to classify the symmetry of cyclo-addition reactions when applying the WOODWARD–HOFFMANN RULE.

surface. See INTERFACE.

surface-active agent (surfactant). Solute (*see* SOLUTION) that accumulates at an interface and in consequence (by the Gibbs ADSORPTION ISOTHERM) strongly lowers the SURFACE TENSION. An aqueous surfactant is usually a large molecule with hydrophobic and hydrophilic parts, the hydrophobic part being oriented out of the liquid (*see* LANGMUIR–BLODGETT FILM). Anionic surfactants (e.g., carboxylic or sulphonic acids and their salts such as sodium dodecyl sulphate) are the constituents of many common SOAPS and detergents. Cationic surfactants contain salts of amines (e.g., hexadecanyl trimethyl ammonium bromide (cetramide, CTAB)) and non-ionic surfactants rely on groups of different polarity. At high concentrations surfactants may form MICELLES.

surface charge (σ; units: C m^{-2}). Charge at the surface of a material (solid, colloidal particle, biological cell) in contact with a polar medium may be due to the adsorption of ions at the surface—a non-ionogenic surface (e.g., Cl$^-$ on an AgCl sol)—or to ionization of groups at the surface—an ionogenic surface (e.g., –SO$_3^-$ at a sulphonated polystyrene latex particle). The

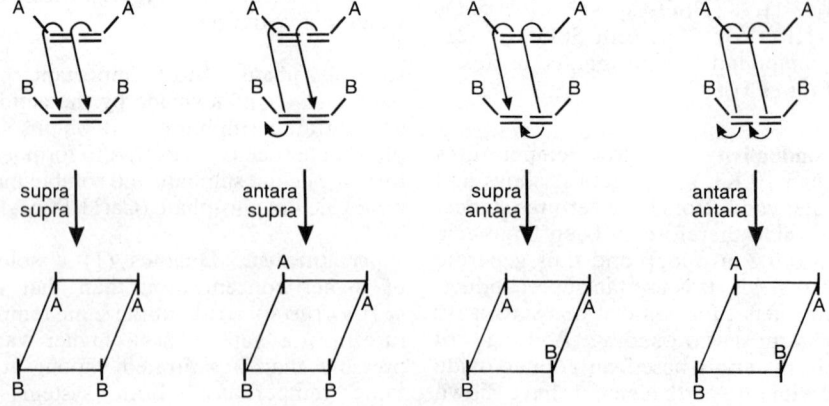

suprafacial

charge of a non-ionogenic surface is usually negative because anions are overall smaller, less solvated and more polarizing than cations. The ZETA-POTENTIAL becomes more negative with increasing pH due to the adsorption of OH^- and the desorption of H^+. Proteins are positively charged below the ISOELECTRIC POINT, and negatively charged at pH values above it. *See also* COLLOID.

surface energy.   *See* SURFACE TENSION.

surface-enhanced Raman spectroscopy. *See* RAMAN SPECTROSCOPY.

surface excess.   *See* ADSORPTION, ADSORPTION ISOTHERM.

surface film.   Thin layer accumulated at an INTERFACE. Monomolecular films are used to coat surfaces to reduce friction, wear, corrosion or to stabilize emulsions, foams, etc., or to reduce evaporation from liquids. Surface films containing PROTEINS mimic lipid membranes in biological systems. The nature of films is investigated by the measurement of SURFACE PRESSURE versus area curves in a Langmuir trough or using a film balance. From this the orientation, cross-sectional area and the nature of the intermolecular forces in the film (*see* SURFACE VISCOSITY) are determined. An oriented monolayer may be picked up on to a solid surface (*see* LANGMUIR–BLODGETT FILM). *See also* SURFACE-ACTIVE AGENT.

surface potential (dipole potential, chi-potential, χ). Potential associated with work done on crossing a surface dipole. An absorbed layer on the electrode creates a solution surface potential (χ_S). The surface potential of a metal (χ_M) arises from the finite chance of finding an electron outside the metal surface. Thus across an electrode/electrolyte interface a surface potential is defined as

$$\Delta\chi = \chi_M - \chi_S$$

surface pressure (Π; units: N m^{-1}). Two-dimensional pressure exerted between regions of different SURFACE TENSION.

$$\Pi = \gamma_1 - \gamma_2$$

The surface pressure is related to the area available per molecule in an analogous way to the three-dimensional pressure–volume characteristics of gases and vapours (*see* GAS LAW). An 'ideal' surface film obeys an equation of the form $\Pi A = nkT$. The analogue of the VAN DER WAALS EQUATION is the Hill–de Boer equation

$$(\Pi + \alpha/A^2)(A - A_0) = kT$$

where A_0 and α are constants. *See also* SURFACE FILM.

surface tension (γ; units: N m^{-1}). Force experienced by a molecule at an interface in the direction of the bulk. The work (dw) required to create surface area $d\sigma$ is given by $dw = \gamma d\sigma$. The GIBBS–DUHEM EQUATION is modified when the surface area is changing

$$d\mu = -SdT + VdP + \gamma d\sigma$$

See also RAMSEY–SHIELDS EQUATION.

surface viscosity.   Describes the state of a SURFACE FILM which may be classed as gaseous, liquid-expanded, liquid or condensed. *See also* SURFACE PRESSURE.

surfactant.   *See* SURFACE-ACTIVE AGENT.

surroundings.   *See* SYSTEM.

suspension.   Dispersion of small particles of solid in a liquid (*see* COLLOID) or gas (called a smoke).

suspensoid.   *See* LYOPHILIC COLLOID.

Sv.   *See* SIEVERT.

Svedberg unit (Svedberg or S unit). SEDIMENTATION COEFFICIENT equal to 1×10^{-13} s. The sedimentation coefficient (s) is the normal means of expressing the rate of sedimentation in an ultracentrifuge. S values are quoted as indicators of the size or molecular mass in the comparison of cellular particles and macromolecules (e.g., 70S RIBOSOMES have 30S and 50S subunits).

sweetening agents.   Classical calorific (nutritive) sweetening agents (sucrose, lactose, maltose and fructose) are usually extracted from plants or other natural sources (e.g., honey). There is currently no

sweetening agents

Compound	Sweetening potency (sucrose=1)	Taste characteristics	Preparation/other comments
Sorbitol	0.5	Sweet taste similar to sucrose	Relatively safe for diabetics. Calorific. Often used in combination with other sweeteners to provide bulk. Metabolism does not involve insulin
Xylitol	1.0	Sweet taste similar to sucrose	Calorific. Metabolism does not involve insulin
Sodium cyclamate $C_6H_{11}NHSO_3^- Na^+$	40	Metallic and bitter after taste	Sulphonation of cyclohexylamine. Non-calorific. Converted by gut flora to highly toxic cyclohexylamine. Banned in UK, USA and Canada, use permitted in Australia and Asian countries
DL-Aminomalonyl-D-alanine isopropyl ester (RTl-001)	58	No side or after taste	Racemic mixture; likely that only one isomer is sweetly active. More stable than aspartame at pH 3.5. Non-carcinogenic in mice. At present under development
Aspartame (Nutrasweet, Canderel) $PhCH_2CHNHCOCHNH_2$ | | $COOCH_3$ CH_2COOH	200	Similar to sugar, slight bitterness develops later. Enhances fruit flavour	Same calorific value as sugar, but insignificant in view of intense sweetness. Instability in solution depends on time, temperature and pH. Cannot be used in baked or fried products. Often used in combination with saccharin
Acesulphame K (Sunnett)	200	Bitter and metallic after taste (better than saccharin)	Not metabolized. Does not decompose until about 225°C. Above pH 3 there are no shelf life problems at room temperatures. Used in combination with other sugar substitutes, e.g. sorbitol for diabetic products

Sorbitol structure:
CH_2OH — HO—C—H — H—C—OH — HO—C—H — HO—C—H — CH_2OH

Xylitol structure:
CH_2OH — H—C—OH — HO—C—H — H—C—OH — CH_2OH

Compound	Sweetening potency (sucrose=1)	Taste characteristics	Preparation/other comments
Sweet glycosides 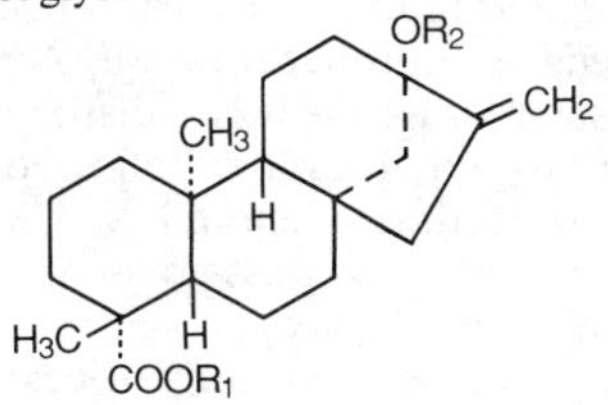			Natural sweet diterpenoid glycosides extracted from a wild South American shrub, *Stevia rebaudiana*. Non-calorific, non-fermentable. Marketed in China and Japan
Stevioside R_1 = glucosyl R_2 = glucosyl−glucosyl	250−300	Slow latent sweetness, bitter and metallic after taste	
Rebaudioside A R_1 = glucosyl R_2 = glucosyl−glucosyl−glucosyl	350−400	After taste better than Stevioside	
Sodium saccharin 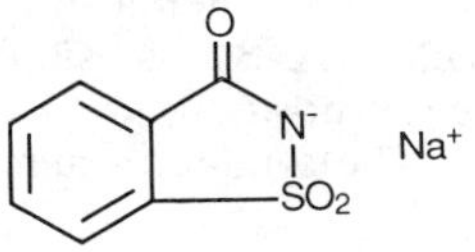	450	Metallic and bitter after taste	Oxidation of 2-methyl benzenesulphonamide to benzenesulphonamide-2-carboxylic acid followed by cyclic dehydration. Non-calorific. Most widely used sugar substitute
1',4,6'-Trichlorogalactosucrose 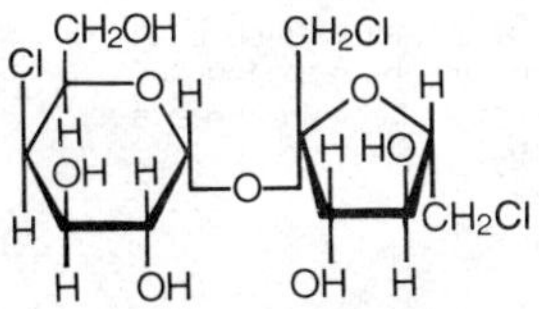	2000	Sweet taste close to sucrose, no unpleasant after taste	Non-calorific. Can be used at high temperatures required for cooking. Stable under acidic conditions. At present under test and development. 1',4,6,6'-Tetrachlorogalacto-sucrose has similar sweet properties

sugar substitute that completely replicates sucrose for its taste, stability, bulking and general preservative properties. Synthetic non-nutritive sweeteners must: (1) have a taste profile similar to that of sucrose; (2) be non-calorific or of low energy value; (3) be non-toxic and devoid of side effects; (4) be soluble in water and preferably stable to heat and acidic conditions. (For main sugar substitutes—*see* Table.)

swelling of colloids. Increase in the volume of a dry hydrophilic colloid (*see* LYOPHILIC COLLOID) on exposure to water or moist air. This effect is seen with many natural colloids (e.g., wood and carbo-hydrates). The effect is at a maximum at the ISOELECTRIC POINT and is influenced by the presence of electrolytes.

sylvite. *See* CHLORINE, POTASSIUM.

symmetry elements. Points, lines and planes with respect to which a SYMMETRY OPERATION is carried out. Objects are classified into groups according to their symmetry by listing all their symmetry elements; thus a sphere is in a different group to a cube and the ammonia molecule in a different group to the water molecule. There are five categories of symmetry operation and element.

1) Rotation (the operation) about an axis of symmetry (the corresponding element). A rotation through an angle of $360°/n$ $(2\pi/n)$ which leaves the body (or molecule) in an indistinguishable condition has a n-fold axis of symmetry—designated C_n (see Figure). Thus water has a two-fold axis of symmetry, C_2, and ammonia a C_3 axis. The cube has three C_4 axes, four C_3 axes and six C_2 axes. For a molecule with several rotation axes, that with the greatest value of n is the principal axis.

2) Reflection in a plane of symmetry, the reflection is indistinguishable from its initial form—designated σ. Mirror planes may take various orientations with respect to the axes of symmetry present. When a mirror plane contains the principal axis of symmetry it is denoted σ_v, the planes in water (two) and ammonia (three) are all σ_v. A mirror plane perpendicular to the principal axis is denoted σ_h; the benzene molecule with a C_6 axis has a σ_h plane.

3) Inversion through a centre of symmetry. A centre of symmetry (inversion) is a point through which there is reflection to an identical point in the pattern. Neither water nor ammonia possesses a centre of symmetry, whereas the sphere, cube, benzene molecule and regular octahedron do.

4) Improper rotation or rotary reflection about an axis of improper rotation. A body possesses such an axis if it is indistinguishable after n-fold rotation followed by horizontal reflection—denoted S_n. Water and ammonia do not possess a S_n axis; methane has three S_4 axes.

5) Identity. Present in all lattices and molecules. *See also* POINT GROUP, SCREW AXIS, SPACE GROUP, TRANSLATION.

symmetry number (σ). Number of indistinguishable positions into which a molecule can be turned by simple rigid rotation through 360°. For a HOMONUCLEAR molecule (A–A), the position after rotation through 180° is identical to that initially (i.e. $\sigma = 2$). In contrast, when a HETERONUCLEAR molecule (A–B) is similarly rotated, the same two positions are distinguishable ($\sigma = 1$). The symmetry number,

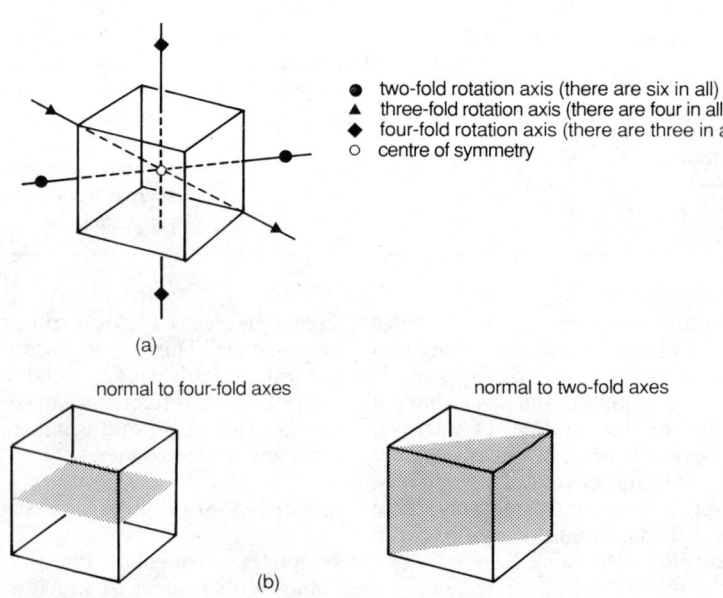

● two-fold rotation axis (there are six in all)
▲ three-fold rotation axis (there are four in all)
◆ four-fold rotation axis (there are three in all)
○ centre of symmetry

(a)

normal to four-fold axes   normal to two-fold axes

(b)

symmetry elements
(a) Symmetry axes in a cube, (b) symmetry planes in a cube.

introduced into the equations for the calculation of the rotational PARTITION FUNCTION (*see* Table) makes the value of Q_r for a homonuclear molecule half the value for an otherwise identical heteronuclear molecule.

For diatomic molecules

NO, CO: $\sigma = 1$; N_2, O_2: $\sigma = 2$

For polyatomic molecules (linear):

HCN, N_2O: $\sigma = 1$; C_2H_2, CO_2: $\sigma = 2$

spherical rotators ($I_A = I_B = I_C$):

CH$_4$, CCl$_4$: $\sigma = 12$

spherical tops ($I_A = I_B \neq I_C$):

NH$_3$, CHCl$_3$: $\sigma = 3$

asymmetrical tops ($I_A \neq I_B \neq I_C$):

H$_2$O: $\sigma = 2$;   C$_6$H$_6$: $\sigma = 12$

See also MOMENT OF INERTIA.

symmetry operation. Operation that leaves an object (body or molecule) looking the same.

syn-addition. In >C=C<, –C≡C– and –C=NR compounds, two stereochemically different modes of addition are structurally discernable if the atoms are unsymmetrically substituted. *syn*-Addition occurs when the groups are added on the same face of the multiple bond and *anti*-addition if groups are at opposite faces of the bond. The degree to which one stereochemistry predominates is described as the stereoselectivity of the reaction.

syndiotactic polymer. *See* TACTICITY.

syneresis. Spontaneous separation of a liquid from a gel or colloidal suspension due to contraction of the gel.

synergism. Cooperative effect in which the activity of two substances in a system is greater than the sum of each acting alone. The term is also applied to the action of some supported catalysts in which an intermediate can spill over onto the support and thus apparently enhance the reactivity.

syn-isomer. *See* ISOMERISM.

syntactic polymer. *See* TACTICITY.

synthesis. Formation of more complex molecules from simpler ones. *See also* RETRO ORGANIC SYNTHESIS.

synthesis gas. Mixture of carbon monoxide and hydrogen (1:3) produced by steam REFORMING of NATURAL GAS at 900°C. It is used as a precursor for industrial processes (*see* FISCHER–TROPSCH PROCESS, HABER PROCESS, OXO PROCESS).

synthetic fibres. *See* FIBRES.

system. The physical universe is, for convenience, divided into the system, that is the part under study and the surroundings. A system may be simple or complex, homogeneous or heterogeneous, and the boundaries may be real (walls of a vessel) or imaginary. An adiabatic system is one in which heat cannot pass across the boundaries. A closed system is one in which matter cannot pass across boundaries (e.g., liquid and vapour in a sealed container). An open system is one in which matter can pass across boundaries (e.g., liquid in a vessel from which vapour can escape). An isolated system is one in which there is no interaction between the system and its surroundings.

T

2,4,5-T (2,4,5-trichlorophenoxyacetic acid). *See* AUXINS.

T. *See* TRITIUM.

T. *See* TEMPERATURE.

t. *See* TRANSPORT NUMBER.

$t_{1/2}$. *See* HALF-LIFE.

$\tau_{1/2}$. *See* HALF-LIFE.

τ-scale. Tau-scale (*see* NUCLEAR MAGNETIC RESONANCE SPECTROSCOPY).

Ta. *See* TANTALUM.

tacticity. Describes the stereochemical arrangement of pendant groups about the backbone of a POLYMER. Atactic polymers have a random arrangement, in syndiotactic (syntactic) polymers the groups alternate from side to side and in isotactic polymers they are all on the same side. The tacticity of a polymer is an important property which affects the way chains fit together. Isotactic polypropene is the form produced commercially; syntactic polypropene has no value at all. Poly(methyl 2-methylpropenoate) (PMMA) is largely syntactic.

tactosol. COLLOID containing particles that can be oriented in a particular direction (e.g., by a magnetic or an electric field).

Tafel equation. *See* BUTLER–VOLMER EQUATION.

talose. *See* HEXOSES.

Tamman temperature. *See* SINTERING.

tannic acid (gallotannic acid, tannin, $C_{76}H_{52}O_{46}$). Yellow powder (mp 210–15°C (decomp.)) with an astringent taste; obtained by the fermentation of oak galls and extraction with ether. It forms precipitates with metallic salts, proteins and alkaloids. Hydrolysis gives glucose and gallic acid (3,4,5-trihydroxybenzoic acid).

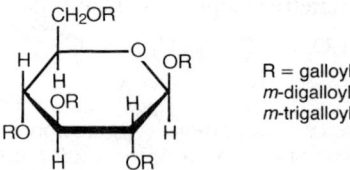

R = galloyl,
m-digalloyl,
m-trigalloyl

tannic acid

tanning. Process of converting animal hides into useful leather. After cleaning and neutralizing, the skins are finally treated either with tannin which combines with the collagen and other proteins present or with basic chromium(III) sulphate which forms complexes with the collagen.

tannins. Heterogeneous group of plant phenolic derivatives (catechol, *see* DIHYDROXYBENZENES; pyrogallol, *see* TRIHYDROXYBENZENES) that are widely distributed in higher plants, leaves, bark, unripe fruits, galls and seed coats. They have an astringent taste. They give a green coloration with iron salts and form precipitates with proteins and alkaloids. Their structures are similar to that of TANNIC ACID. They all combine with animal hides to form leather and are used in tanning, as mordants in the dye industry and in the preparation of ink.

tantalite. *See* TANTALUM.

tantalum (Ta). Group Vb, third row TRANSITION ELEMENT. It occurs naturally with NIOBIUM in tantalite and other ores. It is separated from solutions (ores are fused with sodium hydroxide and washed with acid) by solvent extraction with 2-methyl-2-pentanone. The organic layer is back-extracted with water, from which tantalum oxides are precipitated with boric acid and ammonia. Metallic tantalum is prepared by the ELECTROLYSIS of fused K_2TaF_7 or by the reduction of Ta_2O_5 by carbon or alkali metals. Tantalum is mainly used in ALLOYS (e.g., ferrotantalum) to which it imparts chemical resistivity. The metal is inert to attack by acids, except hydrofluoric, and alkalis at room temperature. Most non-metals react at high temperatures.
$A_r = 180.9479$, $Z = 73$, ES [Xe] $5d^36s^2$, mp 2996°C, bp 5425°C, $\rho = 16.6 \times 10^3$ kg m^{-3}. Isotopes: ^{180}Ta, 0.01 percent; ^{181}Ta, 99.99 percent.

tantalum compounds. The chemistry of tantalum is similar to that of NIOBIUM, largely that of oxidation state +5. Although lower oxidation states have not been isolated, oxidation of tantalum metal gives NON-STOICHIOMETRIC COMPOUNDS (TaO$_x$, $x < 2.5$). Ta_2O_5 forms tantalates, compounds of TaO$_4^{3-}$ by fusion with metal oxides, and polytantalates $Ta_6O_{19}^{8-}$ by ethanol extraction of the product of alkali metal hydroxide fusion. A series of tantalum halides TaX$_5$ and TaX$_4$ (but not TaF$_4$) is formed from the metal and halogen. Complex halide ions are found in acid solution (TaF$_6^-$, TaF$_7^{2-}$, TaF$_8^{3-}$). Reduction of the pentahalides yields metal cluster compounds (e.g., $[Ta_6Cl_{12}]^{2+}$).

tar. Condensate from the destructive distillation of organic materials (e.g., wood, coke, coal and oil). See also PETROLEUM, PITCH.

tartar emetic. See ANTIMONY OXIDES AND OXOACIDS.

tartaric acid. See 2,3-DIHYDROXYBUT-ANEDIOIC ACID.

tau-particle. See ELEMENTARY PARTICLES.

taurine ($H_2NCH_2CH_2SO_3H$). Sulphonic acid-containing AMINO ACID, formed in the liver from cysteine. It is a component of the BILE ACID, taurocholic acid.

taurocholic acid. See BILE ACIDS.

tau-scale. See NUCLEAR MAGNETIC RESONANCE SPECTROSCOPY.

tautomerism. See ISOMERISM.

Tb. Terbium (see LANTHANIDE ELEMENTS).

Tc. See TECHNETIUM.

TCP. See TRICRESYL PHOSPHATE.

Te. See TELLURIUM.

tear gas (lachrymator). Compound which when dispersed in the atmosphere blinds the eyes with tears. Common tear gases include bromopropanone and benzyl bromide. See also CS GAS.

technetium (Tc). Group VII, second row TRANSITION ELEMENT. There are no stable isotopes of technetium (from greek 'artificial'). Technetium-99 (half-life: 2×10^5 yr) is produced by the action of neutrons on molybdenum-98

$$^{98}\text{Mo}(n,\gamma) \rightarrow {}^{99}\text{Mo} \rightarrow {}^{99}\text{Tc}$$

The metal is prepared by reduction of the sulphate by hydrogen at 1000°C. It is silvery-grey and is isomorphous with rhenium, osmium and rhodium (hcp). The chemistry of technetium is similar to that of rhenium (see RHENIUM COMPOUNDS). The known halides are TcF$_6$, TcF$_5$ and TcCl$_4$. The metal reacts with oxygen to form Tc_2O_7 (compare with Re_2O_7). Other oxides are TcO_2 and TcO_3. Several complexes are known which are analogues of those of rhenium.
$A_r = 98.9062$, $Z = 43$, ES [Kr] $4d^6 5s^1$, mp 2172°C, bp 4877°C, $\rho = 11.5 \times 10^3$ kg m^{-3}.

Teflon. See POLYTETRAFLUOROETHENE.

teichoic acids. Negatively charged matrix structural polymers (see Figure) present in the cell walls of gram-positive bacteria. Although the essential details of their composition for a range of organisms is known, there is a lack of information on their fine structure. In ribitol teichoic acids,

R = α or β-*N*-acetylglucosaminyl

Staphylococcus aureus

R = H or D-Ala

Lactobacillus casei

teichoic acids
Structural formulae of different types of teichoic acids from walls of representative microorganisms. Ala indicates a D-alanine residue.

a phosphodiester linkage joins C-1 and C-5 of adjacent residues, labile D-alanine residues may be present on C-2, whereas various sugar substituents may be present on C-4 and C-3. The more widespread glycerol teichoic acids are also found in bacterial membranes. The adjacent glycerol residues are joined by a phosphodiester linkage and the hydroxyl group at C-2 may be substituted with a D-alanyl ester residue or a glycosyl residue (e.g., α-D-glucose, *N*-acetyl-D-glucosamine). Teichoic acids maintain a high concentration of divalent ions (e.g., Mg^{2+}) in the region of the membrane and may be involved in the action of autolytic enzymes. They participate in the binding of phages by several species of organisms.

teichuronic acids. Structural polymers which appear when bacteria, which normally contain glycerol teichoic acid as part of their cell wall matrix, are grown in phosphate-deficient medium. These phosphate-free, anionic polymers contain glucuronic acid residues and like the TEICHOIC ACIDS have the ability to bind divalent ions.

tellurium (Te). Greyish-white crystalline GROUP VI ELEMENT. The element (0.002 ppm of earth's crust) occurs as tellurides in sulphide ores and is recovered from flue dust after the combustion of the ores (taken into solution with sulphuric acid and reduced to tellurium with zinc). The grey 'metallic' form has a chain structure with a low electrical conductivity. Tellurium combines readily with oxygen, halogens and metals (not with sulphur or selenium) and dissolves in oxidizing agents. Grey tellurium and selenium form a range of solid solutions containing chains in which selenium and tellurium alternate almost randomly. Oxidation states +6, +4, +2, 0, −2 exist. Tellurium is more electropositive than sulphur giving Te_4^{2+} compounds and almost cationic character in tellurium dioxide. Organic derivatives of tellurium in +2 state are similar to the divalent sulphur derivatives, but they are less stable. Compounds of tellurium are very poisonous. Tellurium is used in alloys with lead, copper and steel, and as a colouring agent in chinaware, porcelain, etc.
A_r = 127.60, Z = 52, ES [Kr] $4d^{10} 5s^2 5p^4$, mp 449.5°C, bp 989.9°C, ρ = 6.24 × 10^3 kg m^{-3}. Isotopes: ^{120}Te 0.09 percent, ^{122}Te 2.46 percent, ^{123}Te 0.87 percent, ^{124}Te 4.61 percent, ^{125}Te 6.99 percent, ^{126}Te 18.71 percent, ^{128}Te 31.79 percent, ^{130}Te 34.48 percent (artificial radioisotopes: 114–119, 121, 127, 129, 131–134).

tellurium halides. Tellurium forms a complete range of halides including an iodide (*compare* SULPHUR and SELENIUM).

tellurium hexafluoride (TeF_6). Colourless gas, formed by direct reaction of tellurium with fluorine, that is more reactive than sulphur hexafluoride and readily hydrolyzed by water.

tellurium tetrafluoride (TeF_4). Colourless, deliquescent solid (SeF_4 + TeO_2), readily hydrolyzed by water.

tellurium tetrachloride ($TeCl_4$). White, hygroscopic solid obtained by direct chlorination of tellurium. Fused $TeCl_4$ is a good conductor of electricity, possibly existing as $TeCl_3^+$ and $TeCl_5^-$. With hydrochloric acid it gives H_2TeCl_6, the salts are isomorphous with SiF_6^{2-} and $SnCl_6^{2-}$.

tellurium dichloride ($TeCl_2$). Black, unstable solid made by the action of dichlorodifluoromethane on tellurium at 500°C.

tellurium tetrabromide ($TeBr_4$). Red solid prepared by the reaction of tellurium with excess bromine. The vapour dissociates into tellurium dibromide and bromine. It forms many complexes including the $TeBr_6^{2-}$ ion.

tellurium dibromide ($TeBr_2$). Unstable compound (Te + $TeBr_4$ in ether).

tellurium tetraiodide (TeI_4). Iron-grey crystals made by direct union or by the reaction of tellurium dioxide with hydrogen iodide in aqueous solution. It is hydrolyzed slowly by cold, but rapidly by hot, water.

tellurium oxides.

tellurium monoxide (TeO). Black solid obtained by heating $TeSO_3$ (Te + SO_3).

tellurium dioxide (TeO_2). White crystalline substance with rutile structure (*see* CRYSTAL FORMS) obtained by burning tellurium in air or by heating tellurates(IV). An amphoteric oxide almost insoluble in water, but soluble in bases forming tellurates(IV) (TeO_3^{2-}) and in acids.

tellurium trioxide (TeO_3). Orange compound (dehydration of telluric acid, $Te(OH)_6$), which reacts only slowly with water but with hot concentrated alkalis gives tellurates(VI). It is a strong oxidizing agent.

tellurium oxoacids.

tellurous acid (H_2TeO_3). The free acid does not exist (tellurium dioxide (TeO_2) is insoluble in water). Tellurites, or tellurates(IV), may be crystallized from solutions of TeO_2 in alkali metal hydroxides.

telluric acid ($Te(OH)_6$). Colourless crystals obtained by dissolving tellurium in

AQUA REGIA in the presence of a chlorate-(V). It is a weak dibasic acid which gives salts (e.g., $NaTeO(OH)_5$, $Na_2TeO_2(OH)_4$).

other tellurium oxoacids. No peroxoacids are known.

Temkin adsorption isotherm. *See* ADSORPTION ISOTHERM.

temperature (T; units: K, °C). Condition of a body which determines the transfer of HEAT to or from other bodies. If there is no heat flow the bodies are in thermal equilibrium, if there is a heat flow the direction of flow is from the body of higher temperature. Two methods of quantifying this property. (1) The empirical method which relies on a reproducible temperature-dependent event and assigns fixed points on a scale of values to those events (e.g., the Celsius scale assigns the freezing point and boiling point of water as 0 and 100, respectively, and divides the scale between them into 100 degrees). The perfect gas scale, in which the pressure is monitored at constant volume, is almost independent of the monitoring substance (compare liquid in glass thermometers). (2) Thermodynamic temperature scale (using the concepts of the CARNOT THEOREM) is defined by the equation

$$T(X) = 273.16(K)\ X/X'$$

where X is the value of a thermodynamic property (P, V, . . .) at T and X' the corresponding value at the TRIPLE POINT of water (273.16 K). The perfect gas temperature (numerically equal to the kelvin scale) is thus defined

$$T = 273.16 \lim_{P' \to 0} P/P' \text{ at constant volume}$$

See also ZEROTH LAW OF THERMODYNAMICS.

temperature–composition diagram. *See* COMPLETELY MISCIBLE LIQUID SYSTEMS.

temperature jump. RELAXATION method for following the kinetics of fast reactions by causing a sudden increase in temperature on the discharge of a capacitor in a system at equilibrium. The relaxation back to equilibrium is followed spectrophotometrically or by conductance measurements.

temperature programming. Process in which the column temperature in GAS–LIQUID CHROMATOGRAPHY is increased in a programmed manner during the running of the chromatogram. It reduces the running time and reduces tailing of the peaks.

temperature scales. The fundamental scale is the absolute thermodynamic or kelvin scale; the zero is –273.15°C (*see* Table 1). The temperature scale adopted by

temperature scales
Table 1. Conversions.

Scale	fp_{water}	bp_{water}	Conversion
kelvin	273.15	373.15	
Celsius	0	100	
Fahrenheit	32	212	
			$°C = 5(°F - 32)/9$
			$°F = (9/5)°C + 32$

temperature scales
Table 2. International practical temperature scale.

Fixed point	T/K
Triple point of hydrogen	13.81
Temperature of hydrogen with vapour pressure of 25/76 atm	17.042
Boiling point of hydrogen	20.28
Boiling point of neon	27.102
Triple point of oxygen	54.361
Boiling point of oxygen	90.188
Triple point of water	273.15
Boiling point of water	373.15
Melting point of zinc	692.73
Melting point of silver	1235.08
Melting point of gold	1337.58

the Bureau International des Poids et Mesures is the constant volume hydrogen gas thermometer. The magnitude of the degree on both scales is 1/100 of the difference between the boiling point (bp) and freezing point (fp) of water at 1 atm (*see* Table 2). The Rankine scale is the absolute Fahrenheit scale (0°R = –459.69°F).

temporary hardness. *See* HARDNESS OF WATER.

terbium (Tb). *See* LANTHANIDE ELEMENTS.

terdentate ligand. *See* CHELATE LIGAND, LIGAND.

terephthalic acid. *See* BENZENEDICARBOXYLIC ACIDS.

termination reaction. *See* CHAIN REACTION.

term symbols. Nomenclature for electronic configurations of atoms. A term symbol is of the form $^{2s+1}L_J$. L is a letter corresponding to the total orbital angular momentum (i.e. the vector sum of the l values of the individual ATOMIC ORBITALS) (for two orbitals = l_1+l_2, l_1+l_2-1, . . ., $|l_1-l_2|$). $L = S, P, D, F, G$ for sum = 0, 1, 2, 3, 4, 5. The total spin MULTIPLICITY $(2S+1)$ is calculated in a similar manner from the spin quantum numbers (s) of each orbital. The right subscript J is the total angular momentum given by the coupling (Russell–Saunders or R–S coupling) between L and S, taking the values $L+S$, $L+S-1$, . . ., $|L-S|$. R–S coupling assumes the independence of L and S and is valid for light atoms (to the end of the first ROW TRANSITION ELEMENTS). Each term is $(2J+1)$ $(M_J = J, J-1, . . ., 0, . . ., -J+1, -J)$ degenerate which is removed in a magnetic field by the ZEEMAN EFFECT. *See also* J–J COUPLING.

ternary system. *See* THREE-COMPONENT SYSTEM.

terpenes. Terpenes and related terpenoid compounds, among the most widely distributed of naturally occurring compounds, are often referred to as 'isoprenoid' compounds because of the occurrence in their structure of the five-carbon skeleton of isoprene (C_5H_8) (*see* 2-METHYL-1,3-BUTADIENE). In the strict sense terpenes are volatile aromatic hydrocarbons of empirical formula $C_{10}H_{16}$; in the wider sense they include sesquiterpenes $(C_{15}H_{24})$, diterpenes $(C_{20}H_{32})$ and higher polymers as well as various oxygen-containing compounds (alcohols, ketones and camphors) derived from the terpene hydrocarbons. Many are of pharmacological importance. The terpene hydrocarbons, considered as polymers

terpenes

Table 1. Some isoprenoid hydrocarbons. In the conventional representation used in formulae of this type, terminal single lines and terminal double lines refer, respectively, to methyl and methene groups. Functional groups (carboxyl, hydroxyl, aldehyde, acid, etc.) are written in the normal way.

Compound	Structural formula	Occurrence/properties/uses
Terpenes (C$_{10}$H$_{16}$)		
Myrcene		Acyclic monoterpene. Present in bayberry wax, oils of bay, verbena, hop. Oil with pleasant odour; bp 167°C. Used in perfumery and as a flavouring
Ocimene (3,7-dimethyl-1,3,7-octatriene)		Acyclic monoterpene. Present in leaves of *Ocimum basilicum*; bp 81°C/30 mmHg). Isomerizes to 2,6-dimethyl-2,4,6-octatriene on heating
Limonene (citrene, carvene)		Optically active cyclic terpene. Present in oils of lemon, orange, carraway, dill; bp 176°C. Racemic form known as depentene. H$_2$SO$_4$ converts it to α-terpineol and cineole. Skin irritant. Used as solvent, wetting and dispersing agent
Terpinolene		Monocyclic terpene obtained by the action of acids on α-pinene; converted to its isomer terpinene in presence of mineral acids; bp 185°C
α-Terpinene (*p*-menthadiene)		Present in oils of cardamom, marjoram, coriander; also present in mixture of terpenes on treatment of geraniol with conc. ethanoic acid or α-pinene with H$_2$SO$_4$. Pleasant smell of lemons; bp 173°C.
β-Terpinene		Not obtained from natural sources, synthesized from sabinene
γ-Terpinene		Present in oils of coriander, lemon, ajowan, samphire; bp 182°C
(+)-Sabinene (4(10)-thujene)		Dicyclic monoterpene. Present in oil of savin from *Juniperus sabina*; bp 163°C/60 mmHg. With acids undergoes addition to double bond and opening of cyclopropane ring
α-Pinene		Dicyclic monoterpene. Most essential oil derived from the Coniferae; main constituent of oil of turpentine; mp –55°C, bp 156°C. Contains two asymmetric carbon atoms. (+)-Form is obtained by fractional distillation of Greek turpentine. Heating under pressure gives dipentene, reduced (H$_2$/Ni) to pinane, oxidized (KMnO$_4$) to pinonic acid (C$_{10}$H$_{16}$O$_3$). Reacts with conc. H$_2$SO$_4$ giving mixture of limonene, dipentene, terpinolene, terpinene, camphene and *p*-cymene. Bornyl chloride formed on treatment with HCl. Skin irritant. Used in manufacture of camphor, insecticides, perfume bases, plasticizers

continued overleaf

Compound	Structural formula	Occurrence/properties/uses
Camphene		Chiral compound. Present in oils of ginger, citronella, rosemary, turpentine, cypress, camphor. Prepared synthetically by elimination of HCl from bornyl chloride or dehydrogenation of borneol; mp 51°C. Used in manufacture of synthetic camphor
Sesquiterpenes ($C_{15}H_{24}$)		
α-Farnesene		Acyclic sesquiterpene. Present in oil of citronella, natural coating of apples. Synthesized by heating neridol with ethanoic anhydride. Thin oil; bp 129°C/12 mmHg
Zingiberene		Present in oil of ginger; always accompanied by small amounts of bisaboline. Oil; bp 134°C/14 mmHg
β-Selinine		Present in oil of celery; bp 121°C/6 mmHg
β-Carophyllene		Present in oil of cloves, cinnamon, pimento; bp 118°C/10 mmHg
Triterpenes ($C_{30}H_{50}$)		
Squalene		Acyclic triterpene. Present in shark liver oil, human sebum. Precursor of cholesterol and other sterols, and triterpenoids of plants and animals. All *trans*-isomer; bp 261°C/9 mmHg. Squalene and saturated squalene (perhydrosqualene, $C_{30}H_{62}$) used as non-toxic vehicles for cosmetics and for promoting absorption of drugs applied to skin
Tetraterpenes ($C_{40}H_{50}$)		
Lycopene (E160(d))		Red carotenoid pigment of tomatoes, rose hips, paprika; mp 175°C. Catalytic hydrogenation adds 13 H_2; in dry O_2 absorbs 11 O atoms and becomes colourless. Gives deep blue colours with conc. H_2SO_4 and $SbCl_3$

terpenes

Table 2. Some oxygen-containing isoprenoid compounds.

Compound	Structural formula	Occurrence/properties/uses
Geraniol (3,7-dimethyl-2,6-octadien-1-ol)	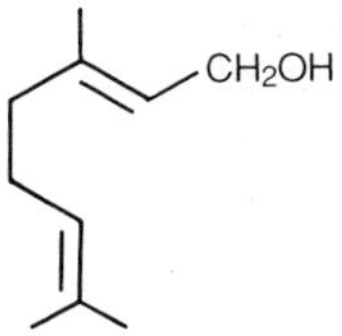 CH_2OH	Present in many essential oils. Manufactured from turpentine (bp 230°C). Unstable in air, oxidized to geranial; HCl (g) gives limonene (*see* above table). Odour of roses; used in perfumery
Nerol (*cis*-3,7-dimethyl-2,6-octadien-1-ol)	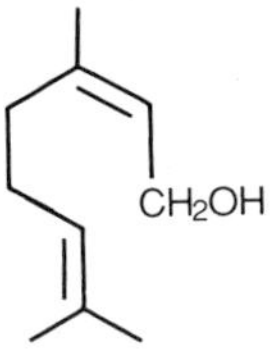 CH_2OH	Present in oils of neroli, petit-grain, bergamol and other essential oils. Isomer of geraniol; bp 225°C. Blander smell than geraniol; used in perfumery
Linalool (3,7-dimethyl-1,6-octadien-3-ol)	 HO	Present in oil in linaloe; bp 198–200°C. Isomerizes to geraniol in presence of acids; oxidized (CrO_3) to citral; distillation with I_2 gives myrcene (*see* above table)
Geranial (3,7-dimethyl-2,6-octadienal, citral a)	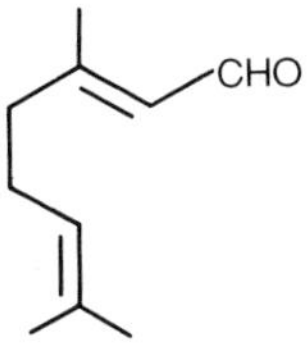 CHO	Present in lemon grass oil from *Cymbopogon flexuosus*; bp 120°C/20 mmHg. *p*-Cymene formed with acids; propanone gives pseudoionine ($C_{13}H_{20}O$) which is readily converted into α- and β-IONONE
Neral (citral b)	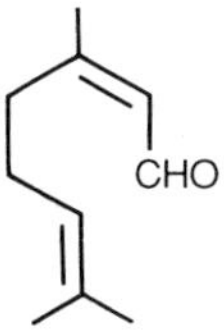 CHO	Isomer of geranial. Lemon odour less intense than geranial
Menthone (2-isopropyl-5-methyl-cyclohexanone)	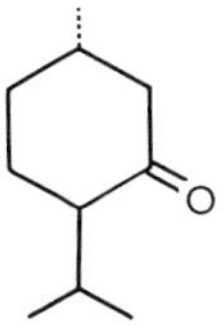 O	Optically active ketone obtained by oxidation of menthol. (–)-Form occurs in some oils; bp 204°C/750 mmHg. Reduced to 4 optically active alcohols

continued overleaf

Compound	Structural formula	Occurrence/properties/uses
Menthol (2-isopropyl-5-methyl-cyclohexanol)	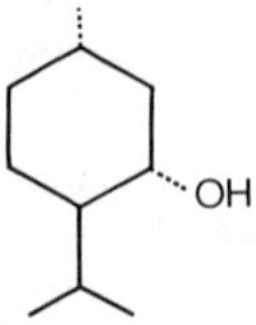	Present in peppermint and other oils. Optically active primary alcohol. Obtained from other terpene derivatives (e.g., reduction of menthone); mp 43°C. Used as analgesic in rheumatism and, by inhalation, for relief of nasal congestion
1,8-Cineole (eucalyptol)	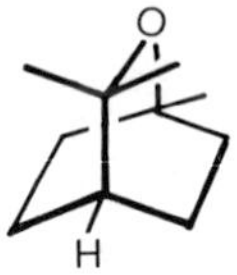	Present in oils of eucalyptus, wormseed, cajuput. Colourless viscous oil with camphor-like odour; mp 1°C, bp 173°C
Borneol		Prepared by the reduction of camphor; inactive form obtained by acid hydration of pinene or camphene. Oxidized to camphor; dehydrated to camphene; mp 208–10°C
Isoborneol	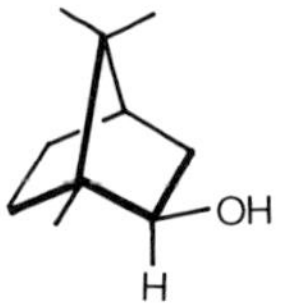	Isomer of borneol
Camphor		Present in wood of camphor tree *Cinnamonum camphora*. Manufactured by oxidation of camphene, from pinene:

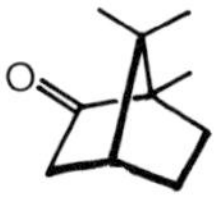

$$\text{pinene} \xrightarrow[\text{HCl}]{\text{dry}} \text{bornyl chloride} \xrightarrow{-\text{HCl}} \text{camphene} \xrightarrow{[\text{O}]} \text{camphor}$$

Exhibits the normal properties of a ketone, reduced to secondary alcohol (borneol and isoborneol); dehydrated (P$_2$O$_5$) to *p*-cymene; reduced with I$_2$ to 2-hydroxycymene. Halogen, nitro, hydroxy and sulphonic acid derivatives are known; mp 179°C. Used for manufacture of celluloid and explosives, and medicinally as stimulant and popular remedy for colds (camphorated oil, a 20% solution in olive oil)

Thujone	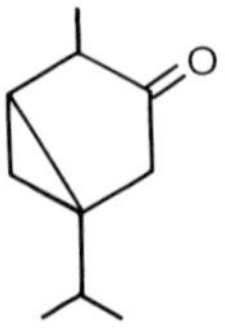	Present in oils of thuja, tansy, wormwood. Oil with smell resembling that of menthol. Reduced (Na + EtOH) to secondary dicyclic alcohol (thujyl alcohol) which is also found free and combined in wormwood oil; bp 75°C/10 mmHg

Compound	Structural formula	Occurrence/properties/uses
Verbenone	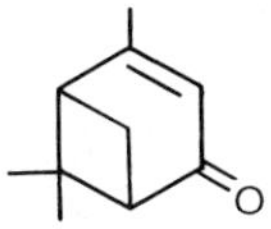	Present in Spanish verbena oil associated with geranial and neral. (+)-Isomer (bp 228°C) and (–)-isomer (bp 253°C) produced by the auto-oxidation of (+)- and (–)-pinene, respectively. Odour of both camphor and peppermint
Fenchone	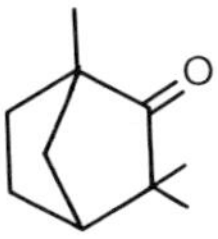	Present in fennel and lavender oils; (–)-Isomer in thuja oil. Reduced (Na + EtOH) to stereoisomeric secondary α and β-fenchyl alcohols; bp 193°C. Camphor-like smell
Farnesol (3,7,11-trimethyldodeca-*trans*-2-*trans*-6,10-trienol)	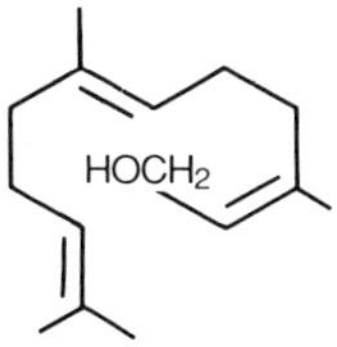	Present in ambrette seed and other essential oils. Sesquiterpene primary alcohol. Oxidized to aldehyde farnesal; dehydrated to α-farnesene. Odour of lily of the valley; bp 120°C/0.3 mmHg. Important precursor of squalene, key intermediate in steroid and triterpenoid biogenesis. Used in perfumery
Phytol (3,7,11,15-tetramethyl-2-hexadecan-1-ol)	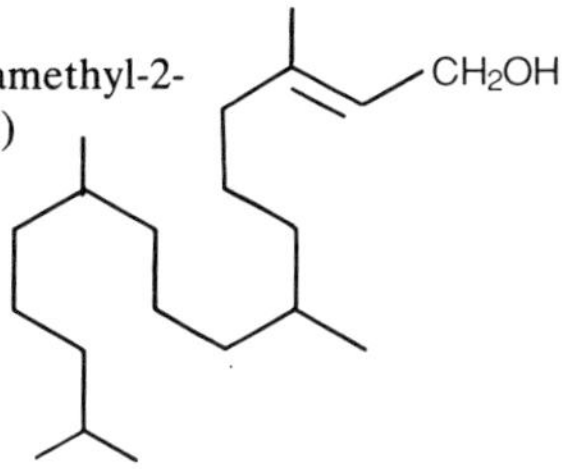	Diterpenic alcohol; occurs as ester of the propanoic acid side chain of CHLOROPHYLL. Obtained as colourless oil (bp 202°C/10 mmHg) by alkaline treatment of chlorophyll. Oxidation yields ketone. Phytyl group is also a side chain of vitamin K
Vitamin A		*See* VITAMINS
Abietic acid	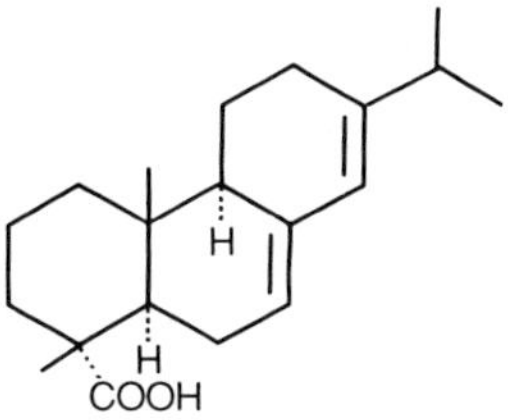	Diterpene acid; major constituent of rosin (non-volatile residue in the manufacture of turpentine by steam distillation of pure oleoresin). Used extensively in varnishes and as its sodium salt in laundry soaps
Lanosterol		Triterpenoid alcohol (*see* STEROLS)

terpenes
Some reactions of terpenes.

of isoprene, may be acyclic, monocyclic, dicyclic, etc. (*see* Table 1). They have a characteristic, usually pleasant odour and are obtained from vegetable materials, oils, etc. by steam distillation or by chromatographic techniques. They show olefinic properties, being easily oxidized, ozonized and polymerized (*see* Figure). They form addition products with hydrogen halides, bromine and nitrosyl chloride, the highly crystalline derivatives of the last compound are used for separating and identifying the terpenes. Those capable of optical activity occur in both forms.

A great number of oxygen-containing isoprenoid compounds are known (*see* Table 2); in addition to their olefinic properties they have the characteristics of the functional group, primary or secondary alcohol, aldehyde or ketone. The alcohols undergo oxidation to the aldehyde or ketone and the carbonyl groups form oximes, etc.

terpinenes. *See* TERPENES.

terpinolene. *See* TERPENES.

terramycin. *See* ANTIBIOTICS.

tert- (tertiary). *See* PRIMARY.

tertiary (*tert-*). *See* PRIMARY.

tertiary structure of proteins. *See* PRO-TEINS.

Terylene. *See* FIBRES, POLYMERS.

testosterone. *See* HORMONES.

1,3,5,7-tetraazaadamantane. *See* HEX-AMINE.

tetrabromomethane. *See* TETRAHALO-METHANES.

1,1,2,2-tetrachloroethane (acetylene tetrachloride, $CHCl_2CHCl_2$). Liquid (bp 146°C), manufactured by the chlorination of ethyne at 80°C over iron(III) chloride, or in a solution of antimony(V) chloride in tetrachloroethane. It is inflammable and used as an industrial solvent for paints, oils, fats and rubbers, and is the starting material for the preparation of trichloroethene and tetrachloroethene.

tetrachloroethene ($CCl_2{=}CCl_2$). Liquid (bp 121°C), prepared from other chlorinated ethenes and ethanes; for example from 1,1,2,2-tetrachloroethane the sequence of reactions is

$$CHCl_2{-}CHCl_2 \xrightarrow[\text{catalysis}]{400°C} CCl_2{=}CHCl \xrightarrow[\text{light/Ca(OH)}_2]{Cl_2}$$

$$CCl_3CHCl_2 \xrightarrow[\text{catalysis}]{300°C} CCl_2{=}CCl_2$$

It is used as a dry-cleaning solvent.

tetrachloromethane. *See* TETRAHALOMETHANES.

tetracosanoic acid (lignoceric acid, $C_{23}H_{47}$-COOH). Fatty acid present free and combined in many oils, fats and waxes. *See also* PHOSPHOLIPIDS.

tetracyclines. *See* ANTIBIOTICS.

tetrad axis. *See* CRYSTAL SYMMETRY.

tetraethyllead. *See* LEAD ORGANIC DERIVATIVES.

tetrafluorohydrazine. *See* NITROGEN HALIDES.

tetrafluoromethane. *See* TETRAHALOMETHANES.

tetragonal system. *See* CRYSTAL SYSTEMS.

tetrahalomethanes (carbon tetrahalides; CX_4, where $X = F, Cl, Br, I$).
tetrabromomethane (carbon tetrabromide). White solid (mp 190°C).
tetrachloromethane (carbon tetrachloride). Volatile liquid (bp 76.5°C), manufactured by: (1) action of chlorine on carbon disulphide over aluminium trichloride

$$CS_2 + 3Cl_2 \rightarrow CCl_4 + S_2Cl_2$$

(2) chlorination of methane; (3) chlorinolysis of hydrocarbons with cleavage of carbon–carbon bonds. It is used as a solvent, fumigant and fire extinguisher (under the trade name Pyrene). Hot vapour in contact with water vapour gives CARBONYL CHLORIDE. It is reduced by iron filings to trichloromethane.
tetrafluoromethane (carbon tetrafluoride). Gas (bp −128°C).
tetraiodomethane (carbon tetraiodide). Red solid (mp 171°C), decomposes on heating.

tetrahedral coordination. *See* SHAPES OF MOLECULES.

tetrahydrocannabinols. Active components of hashish, the resin of hemp (*Cannabis sativa*), and marijuana, the dried flowering tops of hemp. They are pharmacologically active and addictive narcotics. *See also* PHENOLS.

tetrahydrofuran (diethylene oxide, oxolane, THF, C_4H_8O). Fully saturated five-membered oxygen-containing HETEROCYCLIC COMPOUND (structure V). Colourless liquid (bp 67°C), prepared by the catalytic dehydration of 1,4-butanediol or by the cyclization of halohydrins (2-halo-alcohols) in the presence of sodium hydroxide. It undergoes ring-opening and polymerization with carbonium, diazonium and trialkoxonium ions and with Lewis acid halides giving low-molecular-mass diol polymers. With acid catalysts at high temperatures, butadiene is formed. It is widely used as an ethereal-type solvent for chemical (GRIGNARD REAGENT and metal hydride) reactions and industrially as a solvent for resins, plastics and elastomers.

1,2,3,4-tetrahydronaphthalene (tetralin, $C_{10}H_{12}$). Colourless liquid (bp 207°C), obtained by the catalytic hydrogenation of NAPHTHALENE; more vigorous treatment gives BICYCLO[4.4.0]DECANE. Oxidation gives 1,2-benzenedicarboxylic acid; bromine substitutes in the alicyclic ring. The aromatic ring can be nitrated and sulphonated. It is non-toxic and is widely used as a solvent.

tetrahydropyran. *See* HETEROCYCLIC COMPOUNDS (structure XXII).

tetrahydrothiophen-1,1-dioxide (tetramethylenesulphone, sulpholane, $C_4H_8O_2S$). Polar ($\varepsilon_r = 43.3$), non-hydrogen-bonded (dipolar aprotic) solvent (bp 285°C) for polar and non-polar compounds; miscible with water. A selective solvent for liquid–vapour extractions (*see* GAS–LIQUID CHROMATOGRAPHY).

tetrahydrothiophen-1,1-dioxide

tetrahydrothiophene (C_4H_8S). Fully saturated five-membered sulphur HETEROCYCLIC COMPOUND (structure VI), readily prepared from $(CH_2)_4Br_2$ and sodium monosulphide. It has the properties of an ordinary sulphide giving TETRAHYDRO-

THIOPHEN-1,1-DIOXIDE (sulpholane) on oxidation and addition products with methyl iodide, mercury(II) chloride, etc.

2,3,4,5-tetrahydroxyhexanedioic acid (galactaric acid, mucic acid). Optically inactive compound (mp 206°C); prepared by the oxidation (HNO_3) of either galactose or the galactans obtained from wood. On heating it decomposes to furoic acid. Distillation of the ammonium salt gives PYRROLE.

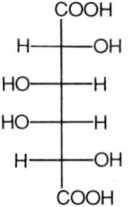

2,3,4,5-tetrahydroxyhexanedoic acid

tetrahydroxymethylmethane. *See* PENTAERYTHRITOL.

tetraiodomethane. *See* TETRAHALOMETHANES.

3,5,3',5'-tetraiodothyronine (L-thyroxine). *See* HORMONES.

tetralin. *See* 1,2,3,4-TETRAHYDRONAPHTHALENE.

tetramethylene diamine (putrescine). *See* POLYAMINES.

tetramethylenesulphone. *See* TETRAHYDROTHIOPHEN-1,1-DIOXIDE.

tetramethylsilane (TMS, $(CH_3)_4Si$). Used as an internal reference standard for proton NUCLEAR MAGNETIC RESONANCE SPECTROSCOPY. All 12 hydrogen atoms are equivalent.

tetrazolium blue. Lemon-yellow crystals (*see* Figure), slightly soluble in water ($E^{\ominus'} = -0.08$ V). It forms dark blue diformazan pigment with reducing agents, and is used as a bacterial stain and for demonstration of redox systems in tissues.

tetrazolium blue

tetroses ($C_4H_8O_4$). CARBOHYDRATES with four carbon atoms (*see* Figure). There are four aldotetroses, the stereoisomers of erythrose (a liquid, very soluble in water and ethanol) and threose (a hygroscopic,

D-erythrose   D-threose   D-erythrulose

crystalline solid, soluble in water and ethanol) and two stereoisomers of the ketotetrose, erythrulose.

TGA. *See* THERMOGRAVIMETRIC ANALYSIS.

Th. Thorium (*see* ACTINIDE ELEMENTS).

thallium (Tl). Rare group IIIb element that occurs naturally in some sulphur and selenide ores of copper and iron. The metal is recovered electrolytically from solutions of the sulphate. Thallium is the least ELECTROPOSITIVE metal in group III ($E^{\ominus}_{Tl^+,Tl} = 0.366$ V). In aqueous solution, thallium(I) is more stable than thallium(III); $E^{\ominus}_{Tl^{3+},Tl^+} = 1.25$ V (*see* INERT PAIR EFFECT). However complexes are formed more readily with thallium(III) than thallium(I). The metal is not attacked by dilute acids, but dissolves in strong oxidizing acids.

Thallium(I) halides, with the exception of the fluoride, are similar to silver halides in their sensitivity to light and their insolubility. Most thallium(I) salts are insoluble in water. Thallium(I) forms a triiodide, which readily loses iodine. TlX_3 (X=F, Cl, Br) are

known. $TlBr_3$ (Tl + Br_2) is unstable and decomposes to $TlBr_2$, which is $Tl^I[Tl^{III}Br_4]$. In solution the chloride forms complex ions (e.g., $[TlCl]^+$, $[TlCl_4]^-$). The metal reacts with oxygen to give thallic oxide (Tl_2O_3). Black thallium oxide (TlO) is formed by heating thallous hydroxide.

Organothallium compounds of general formula R_nTlX_{3-n} ($n = 1, 2, 3$) may be prepared from GRIGNARD REAGENTS. R_2TlX are ionic and stable towards oxygen and water. The trialkyls are very reactive. One of the few thallium(I) organic derivatives, $Tl_4(OR)_4$, is formed by dissolving thallium in the alcohol ROH. Thallium has an odd nuclear spin and so may be investigated by NUCLEAR MAGNETIC RESONANCE SPECTROSCOPY.

Thallium compounds are used as poisons, in glass-making and in electronic devices. $A_r = 204.383$, $Z = 81$, ES [Xe] $4f^{14} 5d^{10} 6s^2 6p^1$, mp 304°C, bp 1460°C, $\rho = 11.8 \times 10^3$ kg m^{-3}. Isotopes: ^{203}Tl, 29.5 percent; ^{205}Tl, 70.5 percent.

theobromine. *See* ALKALOIDS, PURINES.

theophylline. *See* PURINES.

theoretical plate. In a distillation or absorption column, a theoretical plate is one in which perfect liquid–vapour contacting occurs, so that the liquid and vapour streams leaving it are in equilibrium. For a PLATE COLUMN, the performance of real plates is related to that of a theoretical one by the PLATE EFFICIENCY. For a packed column, the HEIGHT EQUIVALENT OF A THEORETICAL PLATE gives a measure of the efficiency of the packing.

thermal analysis. Study of cooling curves (temperature–time plots) of melts of simple substances or mixtures. When a melt is cooled at a uniform rate, the temperature decreases regularly; when a solid separates from a liquid of different composition, there is a decrease in the rate of cooling due to heat evolved during the changes. When a pure compound or EUTECTIC MIXTURE separates from a liquid of the same composition, the temperature remains constant until the whole mass solidifies; allotropic transformations give a similar change. Phase diagrams for a TWO-COMPONENT CONDENSED SYSTEM can be constructed from such curves.

thermal conductivity detector. *See* DETECTION SYSTEMS.

thermal equilibrium. *See* TEMPERATURE.

thermistor. Semiconductor device, the resistance of which decreases with increase in temperature. It consists of a bead or rod of various metal oxides (e.g., Mn, Ni, Co, Cu, Fe). Thermistors are used as THERMOMETERS, often forming one element of a resistance bridge, of particular use in the determination of small changes of temperature (e.g., elevation of boiling point, depression of freezing point).

thermite reaction. Reduction of a metal oxide (e.g., Fe_2O_3) by aluminium, ignited by magnesium ribbon. The heat generated is used to weld metals.

thermochemistry. Branch of physical chemistry concerned with the study of heat changes during chemical reactions. *See also* ENTHALPY, FIRST LAW OF THERMODYNAMICS, HESS'S LAW.

thermocouple. Device consisting of two dissimilar metal wires welded together at their ends (e.g., copper–constantan–copper) to create two junctions ('hot' and 'cold'). A thermoelectric emf is generated when the two junctions are at different temperatures; its magnitude is related to the temperature difference. Thermocouples are used as THERMOMETERS, with limited range.

thermodynamic equations of state.

$$(\partial U/\partial V)_T = T(\partial P/\partial T)_V - P$$

for $U = f(T, V)$

$$(\partial H/\partial P)_T = V - T(\partial V/\partial T)_P$$

for $H = f(T, P)$

These equations apply equally well to gases, liquids and solids, and are useful in interpreting equations of state for gases. For an ideal gas

$$(\partial P/\partial T)_V = P/T$$

hence $(\partial U/\partial V)_T = 0$

$$(\partial V/\partial T)_P = V/T$$

hence $(\partial H/\partial P)_T = 0$

It is possible to evaluate $(\partial H/\partial P)_T$, which is related to the Joule–Thomson coefficient and $(\partial U/\partial V)_T$ from these equations for non-ideal gases.

thermodynamic feasibility. *See* GIBBS FUNCTION.

thermodynamic identities. From the FUNDAMENTAL THERMODYNAMIC EQUATIONS the following are the more important identities that can be obtained.

$$T = (\partial U/\partial S)_{V,n_j} = (\partial H/\partial S)_{P,n_j}$$

$$P = -(\partial U/\partial V)_{S,n_j} = -(\partial A/\partial V)_{T,n_j}$$

$$S = -(\partial G/\partial T)_{P,n_j} = -(\partial A/\partial T)_{V,n_j}$$

$$V = (\partial G/\partial P)_{T,n_j} = (\partial H/\partial P)_{S,n_j}$$

thermodynamics. Study of laws that govern the conversion of energy from one form to another, the direction of the flow of heat and the availability of energy to do work. The ZEROTH LAW OF THERMODYNAMICS establishes the principle of thermal equilibrium. The FIRST LAW OF THERMODYNAMICS deals with the conservation of energy and the equivalence of different forms of energy. The SECOND LAW OF THERMODYNAMICS, which introduces the concept of ENTROPY, identifies the direction in which a reaction will proceed. The THIRD LAW OF THERMODYNAMICS provides an absolute scale of values for entropy. These are all laws of nature for which there is no *a priori* proof.

thermodynamics of cells. The emf (E) of a reversible cell in which n electrons are passed per mole of reactants and the temperature coefficient of E $(\partial E/\partial T)_P$ may be related to the thermodynamic functions ΔG, ΔH and ΔS by

$$\Delta G = -nFE$$

$$\Delta S = nF(\partial E/\partial T)_P$$

$$\Delta H = -nF[E - T(\partial E/\partial T)_P]$$

The thermodynamic EQUILIBRIUM CONSTANT is obtained from $\Delta G^{\ominus}$ and thus $E^{\ominus}$

$$\ln K_{\text{therm}} = -\Delta G^{\ominus}/RT = nFE^{\ominus}/RT$$

thermodynamic temperature scale. *See* TEMPERATURE.

thermogram. *See* THERMOGRAVIMETRIC ANALYSIS.

thermogravimetric analysis (TGA). Quantitative measure of the change of weight associated with a transition as the temperature of a sample is increased with time. Dehydration and decomposition are typical reactions studied. The derivative of the weight versus temperature relationship (thermogram) is often used to increase resolution of overlapping weight losses.

thermometer. Temperature-measuring instrument. There are many different types, but all depend on the variation, with temperature, of a property of a substance; the choice depends on the degree of accuracy and the temperature range: (1) liquid-in-glass, the expansion of a liquid (mercury, –39 to 450°C; ethanol, –130 to 75°C); (2) gas, the variation of the pressure of a gas at constant volume; (3) bimetallic strip, the unequal expansion of two dissimilar metals bonded together; (4) resistance, based on the change of resistance of metals (e.g., platinium, nickel, –200 to > 1200°C); (5) pyrometers using IR-sensitive photocells for very high temperatures; (6) LIQUID CRYSTALS; (7) THERMOCOUPLES and THERMOPILES, thermoelectric emf. THERMISTORS are mainly used for measuring small temperature differences.

thermometric titration. *See* TITRATIONS.

thermoneutral potential. Potential of a HALF-CELL at which the reaction will procede without supply of external heat. This is $-\Delta H^{\ominus}/nF$ and thus differs from the standard ELECTRODE POTENTIAL by the factor $T\Delta S^{\ominus}/nF$. In the electrolysis of water, $\Delta G^{\ominus}$ is 1.23 V at 298 K and the thermoneutral potential is 1.47 V. Above 1.23 V therefore the reaction is spontaneous, but between 1.23 and 1.47 V heat is extracted from the surroundings.

thermopile. Device used to measure the intensity of radiant energy, consisting of a number of THERMOCOUPLES connected in series to increase the sensitivity. The hot junctions are blackened and exposed to the radiation, the cold junctions are shielded.

thermoplastics. *See* POLYMERS.

thermosetting resins. *See* POLYMERS.

THF. *See* TETRAHYDROFURAN.

thiamine (vitamin B_1). *See* VITAMINS.

thiazole (C_3H_3NS). Five-membered HETEROCYCLIC COMPOUND (structure X); liquid (bp 115–18°C) resembling PYRIDINE, obtained by heating the diazonium salt of 2-aminothiazole ($ClCH_2CHO$ + thiourea) with ethanol. It forms compounds with gold(III), mercury(II) and platinum(IV) chlorides. It is used as an intermediate for fungicides, dyes and rubber accelerators.

thiazyl ions. *See* SULPHUR–NITROGEN DERIVATIVES.

thienyl group (C_4H_3S-). *See* HETEROCYCLIC COMPOUNDS.

thin film. *See* SURFACE FILM.

thin-layer chromatography (TLC). Type of adsorption chromatography in which the adsorbant is spread in layers (150–250 μm thick) on glass, aluminium foil or plastic sheets; these are available commercially. Most chromatograms are run in the ascending form and when dry the various components are located by suitable means. It is essentially a microanalytical method with the virtues of speed, convenience and cheapness that has largely superseded PAPER CHROMATOGRAPHY.

thio-. Prefix indicating that sulphur has replaced oxygen in an organic compound (e.g., methanethiol, CH_3SH, compared with methanol, CH_3OH).

thioacetic acid. *See* ETHANETHIOIC ACID.

thioalcohols. *See* THIOLS.

thiocarbamide. *See* THIOUREA.

thiocarbonic acid (H_2CS_3). Formed by the acidification of thiocarbonates

$$Na_2S + CS_2 \rightarrow Na_2CS_3$$

Thiocarbonates are slowly hydrolyzed and react with carbon dioxide to give carbon disulphide. They are used as fungicides.

thiocyanates (thiocyanic acid, HSCN). Gas at room temperature ($KHSO_4$ + KSCN) readily polymerizes. Forms thiocyanic esters (RSCN) and isothiocyanic esters (mustard oils, RNCS). Salts formed by fusing a cyanide with sulphur. SCN^- is a good ligand, the complexes may be sulphur- or nitrogen-bonded. Oxidation gives thiocyanogen ($(SCN)_2$), which polymerizes above 0°C.

thiocyanogen. *See* THIOCYANATES.

thioethers. Organic sulphides, accessible through nucleophilic or radical reactions (e.g., $RS^- + CH_3Br \rightarrow RSCH_3$). Weak Lewis bases, highly nucleophilic, easily oxidized to SULPHOXIDES and SULPHONES. Alkylation yields sulphonium salts (R_3S^+ X^-). They form an insoluble complex with mercury(II) chloride ($R_2S^{+-}HgCl_2$). Disulphides (RSSR) are prepared by oxidation of thiols (RSH) with iodine.

thiols (mercaptans, thioalcohols, RSH). Sulphur analogues of alcohols, which are more acidic and volatile than the corresponding alcohol, with disagreeable odours (butanethiol, skunk; propanethiol, onions). They are prepared by action of an alkyl or aryl halide on potassium bisulphide or the reaction between sulphur and a GRIGNARD REAGENT. They react with heavy metal ions to give mercaptides (e.g., $Pb(SR)_2$). Mild oxidation yields disulphides (RSSR); stronger oxidation gives SULPHONIC ACIDS (RCH_2SO_3H). MERCAPTALS

($RR'C(SR'')_2$) are formed with aldehydes and ketones. Thiols are used in polymerization reactions to control the length of the polymer chain and in the synthesis of agricultural chemicals.

thionic acids. *See* SULPHUR OXOACIDS.

thionyl. Containing the group $> SO$.

thionyl halides. *See* SULPHUR OXOHALIDES.

thiopental sodium. *See* BARBITURATES.

thiophene (C_4H_4S). Five-membered sulphur HETEROCYCLIC COMPOUND (structure III). It is a colourless liquid (bp 84.4°C) with a smell resembling that of benzene; manufactured by high-temperature gasphase reaction of butane with sulphur (butadiene is an intermediate). Substituted thiophenes are synthesized by heating a 1,4-dicarbonyl compound with phosphorus pentasulphide. Thiophene is more stable than is FURAN, but polymerizes in the presence of a strong acid. It undergoes electrophilic aromatic substitution more readily than does benzene. It can be nitrated, sulphonated and brominated, the 1-position being most readily attacked. Simple amino and hydroxy derivatives are more stable than those of furan, but not as stable as aniline and phenol. It is used as a chemical intermediate and in the manufacture of condensation polymers.

thiosulphuric acid. *See* SULPHUR OXOACIDS.

thiourea (($NH_2)_2CS$). Colourless, crystalline compound (mp 176–8°C), manufactured by the action of hydrogen sulphide on CYANAMIDE. It decomposes on heating with water to give ammonium thiocyanate and forms addition complexes with many metal salts. It resembles UREA in many of its properties. Alkylation gives a THIOL and a polymer ($NH_2CN)_n$. It is used as a photographic sensitizer.

third law of thermodynamics. If the ENTROPY of every element in a stable crystalline state at 0 K is taken as zero, every substance has a finite positive entropy, but at 0 K the entropy may become zero, and does so for perfectly crystalline substances. Entropy is related to the thermodynamic probability by the BOLTZMANN EQUATION; thus the more randomly the molecules are arranged the greater are the values of the probability (W) and entropy. For perfect order, $W = 1$ and $S = 0$ (i.e. the lattice sites are completely fixed). Solutions and glasses are excluded from the third law. It is a law of nature and provides a method for obtaining the thermal value of the entropy of the system

$$S^{\ominus}_T = S^{\ominus}_0 + \int_0^T C^{\ominus}_P \, dT/T + \Sigma \, \Delta H_{tr}/T_{tr}$$

Some values of S_{therm} are significantly less than the SPECTROSCOPIC ENTROPY (S_{spec}).

third-order reaction. *See* ORDER OF REACTION.

thixotropy. Time-dependent increase of apparent VISCOSITY with shear rate. At constant shear rate, the apparent viscosity decreases with time until a balance between structural breakdown and structure reformation is reached. If the sheared system is allowed to stand, it eventually regains its original structure. Classic examples include weak gel systems such as flocculated sols of iron(III) oxide, alumina and many clays, which can be liquefied on shaking and solidified on standing. Thixotropy is important in the paint industry. *See also* NEWTONIAN FLUID.

thorium (Th). *See* ACTINIDE ELEMENTS.

thorium decay series. *See* RADIOACTIVE DECAY SERIES.

thoron (Tn). Obsolete name for $^{220}_{86}\text{Rn}$ (*see* RADIOACTIVE DECAY SERIES).

thortveitite. *See* SCANDIUM, SILICATES.

Thr (threonine). *See* AMINO ACIDS.

three-component system. System, which according to the PHASE RULE, has a maximum of four DEGREES OF FREEDOM; thus in one PHASE there are four independent variables: temperature, pressure and the concentration of two components. These are most easily represented by plotting composition diagrams on triangular graph

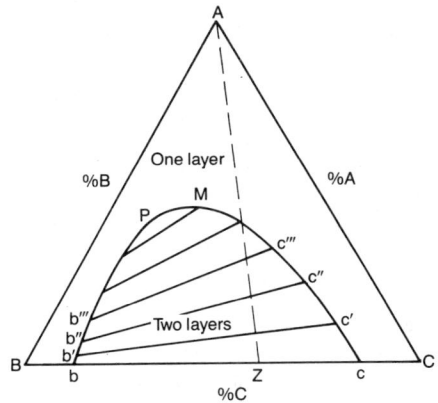

three-component system
Figure 1. System of three liquids, one pair being partially miscible.

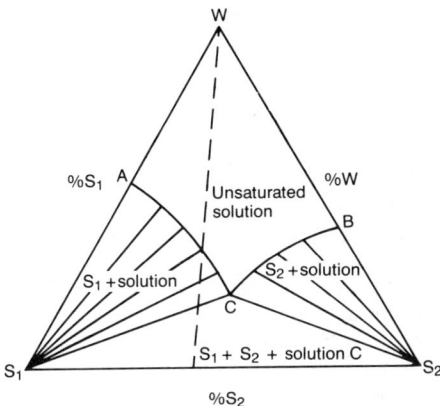

three-component system
Figure 2. Three-component system of two salts with a common ion and water.

paper at constant temperature and pressure. Several distinct systems are recognized: (1) Three liquids. (a) One pair partially miscible (*see* Figure 1). A is distributed between B and C, giving rise to ternary conjugate solutions b and c joined by a TIE LINE. With additional A the compositions of the two layers approach each other until at P, the plait point, they are equal and only one layer exists. The binodal curve has a maximum at M which, in general, does not coincide with P (e.g., ethanoic acid/water/trichloromethane). The binodal curve progressively encloses a smaller area with increasing temperature finally giving a ternary consolute point. (b) Two pairs partially miscible. Two binodal curves coalescing at lower temperatures (e.g., water/ethanol/butanol). (c) Three pairs partially miscible. Three binodal curves present; areas of one, two or three liquid layers can exist (e.g., water/diethyl ether/butanedinitrile). (2) Systems with solid phases (ternary eutectic systems). Three binary eutectics and one ternary eutectic are formed. (3) Systems of two salts with a common ion and water (*see* Figure 2). Simplest system is one in which components S_1 and S_2 separate out with no compound formation (e.g., $NaCl/NH_4Cl/H_2O$). The phase diagrams become more complicated when hydrates, compounds and double salts are formed by the components.

threo-.   *See* ERYTHRO-.

threonine (Thr).   *See* AMINO ACIDS.

threose.   *See* TETROSES.

threshold limit values (TLV).   Set of standards established for the concentration of air-borne substances; they are time-weighted average concentrations, based on conditions which it is believed that workers may be repeatedly exposed to day after day without adverse effects. They are intended to serve as a guide in the control of health hazards rather than as definitive markers between safe and dangerous concentrations: for example, pyridine 5 ppm, hydrogen sulphide 10 ppm, benzene 25 ppm, hexane 500 ppm.

4(10)thujene.   *See* TERPENES.

thujone.   *See* TERPENES.

thulium (Tm).   *See* LANTHANIDE ELEMENTS.

thymidine.   *See* NUCLEOSIDES.

thymine.   *See* PYRIMIDINES.

thymol.   *See* PHENOLS.

thyrocalcitonin.   *See* HORMONES.

thyroid-stimulating hormone.   *See* HORMONES.

thyrotrophin.   *See* HORMONES.

L-thyroxine.   *See* HORMONES.

Ti.   *See* TITANIUM.

tie line.   Line connecting the composition of two phases in equilibrium at constant temperature (pressure) in a PHASE DIAGRAM determined at constant pressure (temperature). *See also* BINARY LIQUID MIXTURE, THREE-COMPONENT SYSTEM, TWO-COMPONENT CONDENSED SYSTEM.

time-of-flight spectrometer.   Device for detection of gaseous ions by the measurement of the time taken to traverse a given distance (drift path). For ions of mass-to-charge ratio m/e accelerated to voltage V, $m/e = 2V/L^2 t^2$, where L is the length of the drift path and t the time of traverse. It is used for kinetic studies of fast reactions, in MASS SPECTROMETRY and in the atom probe FIELD ION MICROSCOPY.

tin (Sn).   GROUP IV ELEMENT. Tin occurs naturally as the dioxide ore cassiterite (tinstone), which is easily reduced to the metal by carbon. Three allotropes of the metal are known. Below 13.6°C grey or α-tin has a diamond structure, at 13.6–161°C white or β-tin and above 161°C γ-tin; both have distorted close-packed arrangements. β- and γ-tin are metals. Tin is a reactive metal which dissolves in acids, strong alkalis and combines readily with oxygen and chlorine. Its reactivity towards acids is limited by the high OVERPOTENTIAL required for hydrogen evolution. Therefore its corrosion in weakly acidic solutions is limited by the rate of supply of oxygen and not the evolution of hydrogen. Tin remains stable in air because of a thin film of oxide. The major use of tin is in low-melting alloys (SOLDER, Britannia metal), other ALLOYS (BRONZE, gun metal, pewter) and as a coating for STEEL in 'tin' cans (*see* TIN PLATE). Uses of compounds are as transparent, semiconducting coats (SnO_2), manufacture of capacitors ($Pb(SnO_3)_2$) and as fungicides and plastic stabilizers (*see* TIN ORGANIC DERIVATIVES).
$A_r = 118.710$, $Z = 50$, ES [Kr] $4d^{10} 5s^2 5p^2$,

mp 232°C, bp 2270°C, $\rho = 7.28 \times 10^3$ kg m^{-3} (β), 5.75×10^3 kg m^{-3} (α). Isotopes: ^{112}S, 1.0 percent; ^{114}Sn, 0.65 percent; ^{115}Sn, 0.35 percent; ^{116}Sn, 14.2 percent; ^{117}Sn, 7.6 percent; ^{118}Sn, 24.0 percent; ^{119}Sn, 8.6 percent; ^{120}Sn, 32.8 percent; ^{122}Sn, 4.8 percent; ^{124}Sn, 6.0 percent.

tincture. Solution, usually of a medicinal compound, in which ethanol is the solvent (e.g., tincture of iodine).

tin inorganic compounds. Tin forms compounds in the +2 and +4 oxidation states. Tin(II) compounds are reducing ($E^{\ominus}_{Sn^{4+},Sn^{2+}} = 0.15$ V, $E^{\ominus}_{Sn^{2+},Sn} = -0.14$ V) and hydrolyze in aqueous solution to give complex species such as $[Sn_3(OH)_4]^{2+}$. The dihalides are prepared by heating tin with HX. They act as acceptors in complexes (e.g., $[SnCl_3]^-$) which then functions as a donor π-bonding ligand. Tin dichloride (SnCl$_2$) is the major component in tin-plating baths. The action of alkali on SnCl$_2$ followed by controlled heating of the precipitate yields blue–black crystals of tin monoxide. Tin-(IV) compounds are largely covalent with the exception of the dioxide (SnO$_2$). This is the natural form of tin and may be prepared by the oxidation of the metal or dehydration of Sn(OH)$_4$. It is used in ceramics, as a polisher and as a refractory. Stannates(IV) contain the complex $[Sn(OH)_6]^{2-}$ in solution, and are found in mixed oxides (e.g., K$_2$SnO$_3$.3H$_2$O). They are used in alkaline electroplating baths. The tin(IV) halides are all known. They are hygroscopic and hydrolyze in water. In acids SnCl$_4$ gives $[SnCl_6]^{2-}$. The halides form six coordinate adducts with nitrogen, oxygen, phosphorus, and sulphur ligands. Tin tetrachloride (SnCl$_4$) is a Friedel–Crafts catalyst (see FRIEDEL–CRAFTS REACTION). The stannanes SnH$_4$ and Sn$_2$H$_6$ are unstable and highly reactive.

tin organic derivatives. Compounds R$_n$SnX$_{4-n}$ (R=aryl, alkyl; X=halogen, O, OH, COOH; n=1–4) are known. R$_3$SnCl is a Lewis acid and forms trigonal and bipyramidal adducts (e.g., (CH$_3$)$_3$SnCl.py). (CH$_3$)$_2$SnCl$_2$ with sodium in liquid ammonia gives catenated and cyclic (Sn(CH$_3$)$_2$)$_n$ (where n = 12–20). Dioctyltin compounds are used to stabilize PVC plastics against degradation by heat or light.

They may be prepared from GRIGNARD REAGENTS. Tin salts of organic acids (e.g., (C$_8$H$_{17}$COO)$_2$Sn) are important catalysts in the manufacture of polyurethane (see POLYMERS).

tin plate. Protective covering of tin, about 1 μm thick, on a corrodable metal. Formerly the metal was dipped into molten tin (hot tinning), but the majority of small items are coated electrochemically from acid or alkaline baths. Tin alloy coatings (with copper, lead, nickel, cobalt, cadmium) are finding increasing use for decorative and protective finishes. Tin plate is widely used in the food industry for protecting steel cans.

Tischenko reaction. CANNIZZARO REACTION of an ALDEHYDE in the presence of aluminium ethoxide which yields an ESTER. ETHYL ETHANOATE is manufactured from ETHANAL by this reaction

$$2CH_3CHO \rightarrow CH_3COOC_2H_5.$$

titanates. See TITANIUM COMPOUNDS.

titanium (Ti). Group IVb, first row TRANSITION ELEMENT. The dioxide (TiO$_2$) occurs as rutile (see CRYSTAL FORMS), anatase and brookite, and as titanates (see TITANIUM COMPOUNDS), principally ilmenite (FeTiO$_3$). Metal is produced by sodium or magnesium reduction of titanium tetrachloride, itself derived by treating ilmenite, dissolved in sulphuric acid with potassium chloride plus hydrochloric acid. Important use of the metal is in alloys (with aluminium, manganese, chromium, iron) employed in aircraft, engines, etc. The metal burns in air to give the oxide and reacts with halogens (X$_2$) to give TiX$_4$. Hydrogen is absorbed under pressure to give a compound of stoichiometry approaching TiH$_2$. $A_r = 47.88$, $Z = 22$, ES [Ar] $3d^2 4s^2$, mp 1660°C, bp 3287°C, $\rho = 4.50 \times 10^3$ kg m^{-3}. Isotopes: ^{46}Ti, 7.99 percent; ^{47}Ti, 7.32 percent; ^{48}Ti, 73.99 percent; ^{49}Ti, 5.46 percent; ^{50}Ti, 5.25 percent.

titanium compounds. Oxidation states +4 (titanic) and +3 (titanous) are most common, forming a number of simple salts; titanium(II) is also known (e.g., in TiCl$_2$). Titanium(IV) forms octahedral complexes (e.g., $[TiCl_6]^{2-}$) and complex oxides (metatitanates, $MTiO_3$, where M=Ca) (perovskite), Sr, Ba, Mg (*see* PEROVSKITE STRUCTURE). Orthotitanates have the general formula M_2TiO_4, where M = alkali metal. TiO_2 is widely used as a white pigment in paint. Titanium(III) compounds are reducing agents and titanium(II) compounds are very reducing, immediately reacting with water. Organic titanates of the formula $Ti(OR)_4$ are known. Low-temperature polymerization of ethene is accomplished by a ZIEGLER–NATTA CATALYST, a constituent of which is titanium tetrachloride.

titrand. Substance that is analyzed in a TITRATION.

titrant. Standard solution of known concentration and composition used for analytical TITRATIONS.

titrations (volumetric analysis). Methods for determining the chemical equivalence of one reagent for another in a well-characterized reaction by observing the change in some property of the solution. The equilibrium of the reaction should lie far towards completion so that the equivalence point, at which neither reagent is present in excess, is accompanied by a large and sudden charge in the property. For any titration, represented by the balanced stoichiometric equation:

$$aA + bB \rightarrow products$$

if V_A cm^3 of solution A of known concentration c_A mol dm^{-3} is equivalent to V_B cm^3 of solution B of unknown concentration c_B mol dm^{-3}, then

$$ac_BV_B = bc_AV_A$$

thus permitting the determination of c_B. In addition to the use of INDICATORS for locating the end-point, various physical methods are available (*see* Table). Some of these have certain advantages, for example, for titrations for which the equilibrium constant is not favourable, for titrations of coloured or very dilute solutions. *See also* COULOMETRIC TITRATION.

Tl. *See* THALLIUM.

TLC. *See* THIN-LAYER CHROMATO-GRAPHY.

TLV. *See* THRESHOLD LIMIT VALUES.

Tm. Thulium (*see* LANTHANIDE ELEMENTS).

TMS. *See* TETRAMETHYLSILANE.

T_N. Néel temperature (*see* ANTIFERRO-MAGNETISM).

Tn. *See* THORON.

TNB (trinitrobenzene). *See* EXPLOSIVES.

TNT (trinitrotoluene). *See* EXPLOSIVES.

α-tocopherol (vitamin E). *See* VITAMINS.

Tollen's reagent. Ammoniacal solution of silver oxide ($[Ag(NH_3)_2]^+$), used as a test for aldehydes and reducing sugars, which cause the deposition of a film of metallic silver. Ketones do not give a similar reaction.

toluene (methylbenzene, $C_6H_5CH_3$). Obtained commercially from coal tar, but mainly by catalytic HYDROFORMING of n-heptane or methylcyclohexane. It is prepared by the WURTZ–FITTIG REACTION (PhBr + MeI + Na) or the FRIEDEL–CRAFTS REACTION (PhH + MeCl). Toluene is a colourless liquid (bp. 110.8°C) and it has a characteristic smell, burns with a smoky flame and is steam volatile. Toluene resembles BENZENE in many of its properties, but since it contains a side chain it can undergo further reactions. Oxidation, in stages, gives benzaldehyde and benzoic acid; chlorine (and bromine) react with boiling toluene substituting in the methyl group giving benzyl, and benzal chlorides and benzotrichloride; these behave like aliphatic halides, but are more active because of the benzene ring in the α-position. Chlorination in the cold in the presence of iron yields a mixture of 2- and 4-chlorotoluenes; nitration, sulphonation and MERCURATION gives mixtures of the 2- and 4-derivatives. Vigorous nitration gives 2,4,6-trinitrotoluene (TNT), used as an EXPLOSIVE. Reduction (Li in methylamine)

titrations Different types of titrations.

Type	Principles	Typical titration plots	Applications/comments
Potentiometric	Measurement of variation of potential of an indicator electrode in equilibrium with its ions (1) Acid–base titrations glass $\mid$ H$^+$ $\vdots$ KCl $\mid$ Hg$_2$Cl$_2$,Hg electrode $E_{cell} = E' + 0.0591$ pH (2) Redox titrations (Fe^{2+} vs KMnO$_4$) Pt $\mid$Fe^{2+},Fe^{3+} $\vdots$ KCl $\mid$ Hg$_2$Cl$_2$,Hg $E_{cell} = E' - (RT/F) \ln c_{Fe^{2+}}/c_{Fe^{3+}}$ (3) Precipitation titrations (NaCl vs AgNO$_3$) Hg,Hg$_2$Cl$_2$ $\mid$ KCl $\vdots$ NH$_4$NO$_3$ $\vdots$ Cl$^-$ $\mid$ Ag $E_{cell} = E' + (RT/F) \ln c_{Ag^+}$ The first differential plot (– – –) is a maximum at the equivalence point	 (I) Strong acid, strong reducing agent (II) Strong base, strong oxidizing agent (III) Weak acid, poor reducing agent (IV) Weak base, poor oxidizing agent A is equivalence point for I–IV, B for I–II, C for II–III	(1) Not suitable for titration of weak acid/weak base (III–IV) (2) Not suitable for redox systems where $E_{0,R}^{\ominus}$ values are close together (III–IV) (3) Curve I–II is typical. Useful for simultaneous determination of Br$^-$, Cl$^-$, I$^-$
Conductometric	Change of conductance (G) of electrolyte (A$^+$B$^-$) on addition of C$^+$D$^-$, especially when one of the products CB is only slightly ionized or insoluble A$^+$B$^-$ + C$^+$D$^-$ $\longrightarrow$ A$^+$D$^-$ + CB A$^+$ ions replace C$^+$ ions and the conductance may increase or decrease depending on the relative mobility values of the ions	 (I) Strong acid – strong base; precipitation (II) Strong acid – weak base (III) Weak acid – weak base	The more acute the angle at which the lines intersect the greater the accuracy of location the end-point. Volume correction necessary unless concn of titrant is 100 times that of titrand. Useful for titrations in which there is hydrolysis at end-point, weak acid/weak base

continued overleaf

Type	Principles	Typical titration plots	Applications/comments
Amperometric	Diffusion current (i_d) is directly proportional to the concn of electroactive species. Removal of some of this material during a titration causes a lowering of i_d. i_d is measured at a suitable applied voltage and plotted as a function of the volume of the added titrant	 Volume of reagent Volume of reagent Volume of reagent (I) (II) (III) (I) Analyte (Pb^{2+}) electroactive, titrant (H_2SO_4) not (II) Titrant (Ba^{2+}) electroactive, titrant (SO_4^{2-}) not (III) Both analyte (Pb^{2+}) and titrant ($K_2Cr_2O_7$) electroactive	There may be a degree of curvature near the end-point, which may be accurately located by extrapolation of linear portions of the curves. Volume correction must be made
Thermometric	Depends on enthalpy change during a chemical reaction. Temperature change during titration in an insulated vessel is measured. Graph of T against volume shows marked break at end-point. In modern instruments, titrant is added continuously from motor-driven syringe		Useful for titrations of weak acids and bases, compleximetric and precipitation titrations and titrations in non-aqueous solutions
Spectrophotometric	Absorbance of analyte solution, in special cell in spectrophotometer, is measured at a fixed wavelength during addition of titrant. Applicable to reactions in which there is a change of colour (e.g., permanganate, iodine and titrations using an indicator)	 $KMnO_4$ added to (I) water, (II) iron(II) solution	Concn of titrant must be much greater than that of analyte to minimize volume changes. Advantage, shared with conductimetric, amperometric and thermometric titrations, is that measured property is directly proportional to the concentration of substance being followed

gives tetrahydrotoluene, but in the presence of ethanol dihydrotoluene is formed. Toluene is used extensively as a solvent and as a raw material for caprolactam, phenol and dyestuffs and the sulphonate for saccharin and chloramine-T.

4-toluenesulphonic anhydride $((CH_3C_6H_4-SO_2)_2O)$. More powerful sulphonating agent than sulphuryl halides, particularly for hindered alcohols and sugars.

4-toluenesulphonyl chloride (tosyl chloride). *See* SULPHONYL CHLORIDES.

toluidines. *See* AMINOTOLUENES.

toluyl group (phenylethanoyl group). The group $C_6H_5CH_2CO-$.

tolyl group. The group $CH_3C_6H_4-$.

topaz. *See* CORUNDUM.

torr. Unit of pressure used in high-vacuum technology (1 torr = 133.322 Pa).

tosylates. *See* SULPHONIC ACIDS.

tosyl chloride (4-toluenesulphonyl chloride). *See* SULPHONYL CHLORIDES.

tosyl group. The group $CH_3C_6H_4SO_2-$.

town gas. *See* COAL GAS.

tracer. Chemical entity, commonly a radioisotope (e.g., TRITIUM) or stable isotope (e.g., ^2_1H, $^{18}_8O$), added to the system (e.g., chemical reaction, diffusion process) which can be traced through the process by appropriate sensitive detection methods. The addition of small amounts of the tracer does not alter the experimental conditions. *See also* LABEL.

transactinide elements. Elements of atomic number (Z) greater than 103 (lawrencium). Elements 104 to 118 (not all are known) would be expected to form a row of the periodic table under HAFNIUM $(Z = 72)$ to RADON $(Z = 86)$.
rutherfordium (Rf; unniqualium) $(Z = 104$, known isotopes 256–261). Produced by bombardment of lighter elements with

charged heavy particles. The aqueous chemistry of $^{261}Rf^{4+}$ has been studied.
hahnium $(Z = 105$, unnipentium), isotopes 260–262). Synthesized in a similar manner to rutherfordium.
element 106 (unnihexium). Discovered following bombardment of ^{249}Cf with ^{18}O. $^{263}106$ has a half-life of 0.9 seconds.
supertransuranic elements (superheavy elements, $Z > 108$). Although not discovered or synthesized include, in theory, very stable elements. Element 110 has a predicted half-life of 10^8 years and $^{296}112$ of 100 years. Naturally occurring superheavy elements may yet be discovered.

transcription. Transfer of information, in the form of nucleotide sequence, from DNA to RNA. A single strand of DNA acts as a template for the synthesis of a complementary RNA sequence. There is no change in the coding language (i.e. the sequence of bases in the coding information). This is in contrast to TRANSLATION which involves conversion of a nucleotide code into an amino acid code.

***trans*-effect.** In the reactions of square planar COORDINATION COMPOUNDS (e.g., platinum(II))

$$L_2PtA_2 + B \rightarrow L_2PtAB$$

The proportion of *cis*- and *trans*-isomers depends on the nature of the ligand L. Ligands may be arranged in a series of increasing tendency to force incoming ligands *trans* to them: $H_2O < OH^- < NH_3 < Cl^- < Br^- < I^- = NO_2 = PR_3 \ll CO = C_2H_4 = CN^-$. Therefore substitution of $[Pt(NH_3)_4]^{2+}$ by Cl^- leads to *trans*-$[Pt(NH_3)_2Cl_2]$, whereas $[PtCl_4]^{2-} + 2NH_3$ gives *cis*-$[Pt(NH_3)_2Cl_2]$.

transesterification. *See* ESTERS.

transferases. *See* ENZYMES.

transference number (n_i). Number of moles of a given ionic species transferred in the direction of the positive current for the passage of 1 faraday of electricity. For a fully dissociated 1:1 electrolyte, the transference number is the same as the TRANSPORT NUMBER.

transient. Short-lived species, signal, etc.

trans-isomer. *See* ISOMERISM.

transition. Movement of a system from one state to another. This may occur with an energy change (*see* SELECTION RULES) in a spectroscopic experiment, a phase change (*see* PHASE TRANSITION) or of a species in the course of a chemical reaction (*see* ABSOLUTE RATE THEORY, ACTIVATED COMPLEX).

transition elements. Elements which, in at least one commonly occurring oxidation state, have partially filled *d*- or *f*-shells. The main transition group or *d*-block (*see* Table) consists of three transition series (rows) comprising elements $Z = 21$–29 (scandium to copper), $Z = 39$–47 (yttrium to silver) and $Z = 72$–79 (hafnium to gold). Copper and silver are included by virtue of the $+2$ oxidation state and gold by virtue of the $+3$ oxidation state. The 15 elements lanthanum ($Z = 57$) to lutetium ($Z = 71$) have partially filled 4*f*-orbitals and are known as LANTHANIDE ELEMENTS. Electronically lutetium ($4f^{14} 5d^{6} 6s^{2}$) and lanthanum ($5d^{1} 6s^{2}$) are third-row transition elements, but because of very similar properties to other lanthanides they arc usually included with them. A third set of transition elements, the ACTINIDE ELEMENTS, includes the 15 elements starting with actinium ($Z = 89$). All transition elements are hard, strong, dense metals with high melting and boiling points and low vapour pressures, and are good conductors of heat and electricity. They form alloys with one another and with other metallic elements. They mostly exhibit multiple valence with a maximum valence found in the middle of a series (ruthenium, osmium $+8$). Complex ions are formed easily (*see* CRYSTAL FIELD THEORY, LIGAND FIELD THEORY). These are coloured in at least one oxidation state. The presence of unpaired electrons leads to PARAMAGNETISM. Transition elements are often good catalysts (*see* CATALYSIS).

transition metal complex. *See* COMPLEX COMPOUND, COORDINATION COMPOUND.

transition state theory. *See* ABSOLUTE RATE THEORY, ACTIVATED COMPLEX.

transition temperature. *See* PHASE TRANSITION.

translation. (1) SYMMETRY OPERATION in which an object moves through some distance in a straight line. *See also* SPACE GROUP. (2) Conversion of the sequence of NUCLEOTIDES in a molecule of messenger RNA (*see* NUCLEIC ACIDS) into a corresponding sequence of amino acids in a polypeptide chain. This process of PROTEIN synthesis occurs on the ribosomes which are usually attached to a single molecule of mRNA. Each triplet of nucleotides in the mRNA determines a single amino acid (*see* GENETIC CODE).

translational energy. *See* KINETIC ENERGY.

translational partition function. *See* PARTITION FUNCTION.

transmittance (T). Fraction of radiation passing through a sample. This equals I/I_0, where I is the intensity of the radiation passed through the sample and I_0 the intensity falling on the sample. *See also* ABSORBANCE, BEER–LAMBERT LAW.

transmutation. Transformation of atoms of one element into atoms of a different element as a result of a nuclear reaction. The reaction may be one in which two nuclei interact (e.g., formation of oxygen from nitrogen and a helium nucleus) or one in which a nucleus reacts with an elementary particle (e.g., sodium atom plus a proton forming a magnesium atom). Many new elements and synthetic isotopes have been obtained by transmutation (e.g., neutron bombardment of uranium to give neptunium and plutonium). The spontaneous RADIOACTIVE DECAY SERIES are not regarded as transmutations since they are out of control of the experimenter.

transport number (t_i). Fraction of the current carried by a given ion. In terms of the IONIC MOBILITIES (u_i) and MOLAR IONIC CONDUCTIVITIES (λ_i) for a simple 1:1 electrolyte

$$t_i = u_i / (u_+ + u_-) = \lambda_i / (\lambda_+ + \lambda_-)$$

In mixed electrolytes the relative concentrations influence the quantity of electricity carried by the ions, and it is customary to calculate the TRANSFERENCE NUMBER.

transition elements

First row

Z	21 Sc	22 Ti	23 V	24 Cr	25 Mn	26 Fe	27 Co	28 Ni	29 Cu
ES [Ar]+	$3d^1 4s^2$	$3d^2 4s^2$	$3d^3 4s^2$	$3d^5 4s^1$	$3d^5 4s^2$	$3d^6 4s^2$	$3d^7 4s^2$	$3d^8 4s^2$	$3d^{10} 4s^1$
Oxidation states	3	(2) 3, 4	(−1, 1, 2) 3, 4, 5	(−1, −2, 1) (2) 3 (4, 5, 6)	(1) 2, 3 (4, 5, 6) 7	(1) 2, 3 (4, 6)	(−1, 1) 2, 3 (4, 5)	(1) 2 (3, 4)	1, 2 (3)

Second row

Z	39 Y	40 Zr	41 Nb	42 Mo	43 Tc	44 Ru	45 Rh	46 Pd	47 Ag
ES [Kr]+	$4d^1 5s^2$	$4d^2 5s^2$	$4d^4 5s^1$	$4d^5 5s^1$	$4d^6 5s^1$	$4d^7 5s^1$	$4d^8 5s^1$	$4d^{10}$	$4d^{10} 5s^1$
Oxidation states	3	(1, 3) 4	(1, 2, 4) 5	(1, 2, 3, 4, 5) 6	(1, 2, 3) 4, 5 (6) 7	(1) 2, 3, 4 (5, 6, 7, 8)	1, 2, 3 (4, 6)	2 (4)	1 (2, 3)

Third row

Z	(71) Lu[a]	72 Hf	73 Ta	74 W	75 Re	76 Os	77 Ir	78 Pt	79 Au
ES [Xe]+	$5d^1 6s^2$	$5d^2 6s^2$	$5d^3 6s^2$	$5d^4 6s^2$	$5d^5 6s^2$	$5d^6 6s^2$	$5d^9$	$5d^9 6s^1$	$5d^{10} 6s^1$
Oxidation states	3	(1, 3) 4	(1, 2, 4) 5	(1, 2, 3, 4, 5) 6	(1, 2, 3) 4, 5 (6, 7)	(1, 2) 3, 4 (5) 6 (7, 8)	1, 3, 4 (5, 6)	2, 4 (5, 6)	1, 3

[a] Usually included in the LANTHANIDE ELEMENTS.

Transport numbers are determined by: (1) the Hittorf method in which the changes in concentration around the anode and cathode are measured; (2) the moving boundary method in which the movement of the boundary between two solutions, the leading solution (containing the ion under study) and an indicator solution, is measured under conditions of constant current; (3) from the ratio of the emfs of CONCENTRATION CELLS with and without transport.

transuranic elements. *See* TRANSACTINIDE ELEMENTS.

tremolite. *See* SILICATES.

triad axis. *See* CRYSTAL SYMMETRY.

2,4,6-triamino-1,3,5-triazine (melamine). Solid (mp 354°C), prepared from urea or by heating dicyandiamide ($H_2NC(NH)$-NHCN). It is used in the manufacture of thermosetting resins by condensation with methanal (*see* POLYMERS).

2,4,6-triamino-1,3,5-triazine

triazines. Compounds containing the $C_3N_3H_3$ ring system (*see* HETEROCYCLIC COMPOUNDS, structures XXIV–XXVI). The parent compound is unknown, all derivatives are stable, high-melting solids. Derivatives of *s*-triazine (e.g., simazine, atrazine) are used as selective herbicides.

triazoles. Compounds containing the $C_2N_3H_3$ ring system (*see* HETEROCYCLIC COMPOUND, structures XI, XII).
1,2,3-triazole. Prepared by the direct action of hydrazoic acid on ethyne. Substituted triazines are prepared by the reaction between ethynedicarboxylic acid and benzyl azide. They can be reduced to the parent compound.
1,2,4-triazole. Prepared by the condensation of two molecules of aminomethanal with hydrazine followed by cyclization.

Substituted triazoles are prepared by the condensation of a substituted hydrazine with methanoic amide.

tribenzobicyclo[2.2.2]octatriene. *See* TRIPTYCENE.

trichloroacetic acid. *See* CHLOROETHANOIC ACIDS.

1,1,1-trichloro-2,2-bis(4-chlorophenyl)-ethane (4,4′-dichlorodiphenyltrichloroethane, DDT). Solid (mp 109°C), manufactured by heating chlorobenzene and trichloroethanal in concentrated sulphuric acid

$$2PhCl + Cl_3CCHO \rightarrow$$
$$Cl_3CCH(ClC_6H_4)_2$$

It is a powerful insecticide. Commercial DDT contains about 75 percent of the 4,4′-isomer and 20 percent of the 2,4′-isomer.

trichloroethanal (chloral, CCl_3CHO). ALDEHYDE; colourless, pungent oil (bp 98°C), manufactured from the chlorination of ethanol. Its uses are as a source of trichloromethane (*see* TRIHALOMETHANES), when reacted with concentrated potassium hydroxide, and as an hypnotic. Oxidation by nitric acid yields trichloroethanoic acid (*see* CHLOROETHANOIC ACIDS). It polymerizes on standing to a white solid. The hydrate ($CCl_3C(OH)_2$, mp 57°C) is produced by reaction with water, trichloroethanal being regenerated in concentrated sulphuric acid.

trichloroethanoic acid. *See* CHLOROETHANOIC ACIDS.

trichloroethene (trichloroethylene, $CHCl=CCl_2$). Colourless, toxic liquid (bp 87°C) with trichloromethane-like smell, manufactured by the vapour-phase cracking of tetrachloroethane which is derived from the chlorination of ethyne (*see* PETROCHEMICALS). Trichloroethene gives sodium hydroxyethanoate when heated with sodium hydroxide under pressure. Oxygen in a photochemical reaction gives dichloroethanoyl chloride. It is used as a solvent for degreasing metals, for extracting oils, as an insecticide, paint stripper and fire extinguisher.

1′,4,6′-trichlorogalactosucrose. *See* SWEET-
ENING AGENTS.

trichloromethane. *See* TRIHALOMETHANES.

trichloronitromethane (chloropicrin, Cl_3-
CNO_2). Colourless lachrymatory, toxic
liquid (bp 112°C), prepared by the oxidation of picric acid (*see* NITROPHENOLS)
with sodium hypochlorite.

2,4,5-trichlorophenoxyacetic acid. *See* AUX-
INS.

triclinic system. *See* CRYSTAL SYSTEMS.

tricresyl phosphate (TCP, $(CH_3C_6H_4O)_3$-
PO). Colourless liquid containing a
mixture of all isomers (bp 420°C), insoluble
in water, soluble in most common organic
solvents. TCP is used as a plasticizer, as a
plastic fire retardant and lubricant additive.

tridymite. *See* SILICA.

triethanolamine. *See* ETHANOLAMINES.

triethylamine. *See* AMINES.

triglycerides. *See* GLYCERIDES.

trigonal coordination. *See* SHAPES OF
MOLECULES.

trigonal system. *See* CRYSTAL SYSTEMS.

trihalomethanes (haloforms; HCX_3, where
X = F, Cl, Br, I). Prepared by the haloform reaction of ethanal or methyl ketones
(i.e. compounds CH_3COR) with the halogen dissolved in aqueous alkali. The
halogenating agent is thus the hypohalite

   $RCOCH_3 + 3NaOX \rightarrow$

   $RCOCX_3 + 3NaOH$

then

   $RCOCX_3 + NaOH \rightarrow RCOONa +$

   HCX_3

A particular test for the ETHANOYL GROUP
or groups which may be oxidized to it is the
iodoform test (X = I). Haloforms are oxidized to COX_2 (*see* CARBONYL CHLORIDE)
and, except trifluoromethane, undergo the
CARBYLAMINE REACTION.

trichloromethane (chloroform). Liquid
(bp 60°C), manufactured by heating ethanal
or propanone with bleaching powder (*see*
BLEACHING AGENTS) or the chlorination of
methane. It is used as a solvent, erstwhile
anaesthetic and as a precursor of FREONS.
tribromomethane (bromoform). Liquid
(bp 151°C), prepared by the action of bromine and sodium hydroxide on ethanol. It
reacts with potassium hydroxide to potassium bromide and carbon monoxide.
triiodomethane (iodoform). Yellow crystals (mp 119°C), prepared by electrolysis of
iodine in ethanol or propanone. It reacts
with potassium hydroxide to give diiodo-
methane.

trihydroxybenzenes (benzenetriols, C_6H_3-
$(OH)_3$).
1,2,3-(vic)-trihydroxybenzene (pyrogallic
acid, pyrogallol). Colourless solid
(mp 133°C), soluble in water, ethanol and
ether. It is obtained by decarboxylation of
gallic acid (3,4,5-trihydroxybenzoic acid,
prepared by boiling tannin with dilute
acids) at its melting point (253°C). Alkaline
solutions oxidize rapidly in air and become
dark brown in colour. It reduces salts of
silver, gold, platinum and mercury to their
metals, and is used as a photographic
developer and as an absorbant for oxygen in
gas analysis.
1,2,4-(as)-trihydroxybenzene (hydroxyquinol). Solid (mp 140°C), prepared by
the hydrolysis of its triethanoate (heat 4-
benzoquinone with ethanoic anhydride and
conc. H_2SO_4).
1,3,5-(s)-trihydroxybenzene (phloroglucinol). Occurs in many natural glucosides
from which it may be obtained by alkali
fusion. It is prepared by the reduction (Sn +
HCl) of 1,3,5-trinitrobenzoic acid and heating the resulting amino derivative with
hydrochloric acid. It is a colourless solid
(mp 218°C), fairly soluble in water; alkaline
solutions darken rapidly due to oxidation.
In some reactions it behaves like a triketo
compound (e.g., the formation of a tri-
oxime). Alkylation in the presence of an
alkali forms only *C*-alkyl derivatives. It is
used in photography, printing and pharmaceuticals.

3,4,5-trihydroxybenzoic acid (gallic acid,
$(HO)_3C_6H_2COOH$). Colourless powder
(mp 253°C), sparingly soluble in water;

ninhydrin   amino acid

1,2,3-triketohydrindene hydrate
Reaction with amino acids.

obtained by the hydrolysis of TANNIC ACID. It decomposes on heating to pyrogallol (*see* TRIHYDROXYBENZENES), gives blue–black colour with iron(III) salts, and is used in the manufacture of inks.

tri-(2-hydroxyethyl)amine. *See* ETHANOLAMINES.

1,2,3-trihydroxypropane. *See* GLYCEROL.

2,4,6-trihydroxy-1,3,5-triazine. *See* CYANURIC ACID.

trihydric alcohols. *See* TRIOLS.

triiodomethane. *See* TRIHALOMETHANES.

3,5,3′-triiodothyronine. *See* HORMONES.

1,2,3-triketohydrindene hydrate (ninhydrin). Pale brown crystals (mp 242°C (decomp.)); prepared in low yield by oxidation (SeO$_2$) of diketohydrindene or from diethyl 1,2-benzenedicarboxylate and dimethylsulphoxide. It is a sensitive reagent for colorimetric detection and determination of amino acids (except proline and hydroxyproline), reacting to form a SCHIFF'S BASE whose anion is deep blue (*see* Figure).

trimer. *See* OLIGOMER.

trimethylamine. *See* METHYLAMINES.

1,2,3-trimethylbenzene. *See* HEMIMELLITINE.

1,3,5-trimethylbenzene. *See* MESITYLENE.

trimethylcarbinol. *See* BUTANOLS.

4-(2,6,6-trimethyl-2-cyclohexenyl)-3-buten-2-ones. *See* IONONES.

trimethylene. *See* CYCLOALKANES.

trimethylglycine. *See* BETAINE.

1,3,7-trimethylxanthine (caffeine). *See* ALKALOIDS, PURINES.

trinitrobenzene. *See* EXPLOSIVES.

1,3,5-trinitrohexahydro-1,3,5-triazine (RDX). *See* EXPLOSIVES.

2,4,6-trinitrophenol. *See* NITROPHENOLS.

trinitrotoluene. *See* EXPLOSIVES.

triolein. *See* OLEIN.

triols (trihydric alcohols). ALCOHOLS containing three hydroxyl groups per molecule, such as GLYCEROL.

trioses (C$_3$H$_6$O$_3$). CARBOHYDRATES with three carbon atoms (e.g., D- and L-glyceraldehyde, HOCH$_2$CH(OH)CHO, chosen as the standard configuration in sugar chemistry) (*see* OPTICAL ACTIVITY). The isomeric dihydroxyacetone is not optically active and hence is not a SUGAR.

trioxygen. *See* OZONE.

2,6,8-trioxypurine (uric acid). *See* PURINES.

tripalmitin. *See* PALMITIN.

triphenylcarbinol (Ph$_3$COH). Tertiary alcohol (Ph$_3$CH + CrO$_3$ or by Grignard reaction, Ph$_2$CO + PhBr), which reacts readily with hydrochloric acid in ethanoic acid or with ethanoyl chloride giving triphenylmethyl chloride (Ph$_3$CCl) and with alcohols in the presence of acid to give ethers.

triphenylene   (9,10-benzophenanthrene, $C_{18}H_{12}$). Benzenoid-type   POLYCYCLIC HYDROCARBON (structure XI), solid (mp 199°C) which occurs in coal tar; solutions exhibit a blue fluorescence.

triphenylmethane   (tritane,   Ph₃CH). Parent compound of the triphenylmethane dyes (see DYES); solid (mp 78.2°C), prepared by the reaction of benzene and trichloromethane ($AlCl_3$ catalyst). It reacts with bromine to give triphenylmethyl bromide (Ph_3CBr) and with sodium in ethereal solution to give triphenylmethylsodium (Ph_3C^- Na$^+$) and is oxidized (CrO_3 in ethanoic acid) to triphenylcarbinol (Ph_3COH).

triphenylmethane dyes.   See DYES.

triphenylmethyl (Ph_3C).
free radical ($Ph_3C^·$).   Isolated only in the dimeric form. In solution an equilibrium is established between the colourless dimer and the yellow paramagnetic radical, the proportion of the radical increasing with dilution or increase in temperature. The equilibrium mixture is obtained by heating TRIPHENYLMETHYL   CHLORIDE   (Ph_3CCl) with sodium, silver or zinc in the complete absence of air. The free radical is very reactive, combining with oxygen to give the colourless   peroxide $Ph_3C–O–O–CPh_3$, iodine   to   give   triphenylmethyl   iodide (Ph_3CI) and sodium to give the red triphenylmethyl sodium (Ph_3C^- Na$^+$). It is a powerful reducing agent.

$$2Ph_3C^· \rightleftharpoons Ph_2C = \text{(cyclohexadiene ring)} \begin{array}{c} H \\ CPh_3 \end{array}$$

anion (Ph_3C^-).   Formed when an ethereal solution of Ph_3CCl or triphenylmethane is treated with sodium, the solution contains Ph_3C^- Na$^+$.
carbonium ion (Ph_3C^+).   Triphenylmethyl salts (Ph_3C^+ X$^-$, where X$^-$ = BF_4^-, ClO_4^-), orange–red solids (the appropriate strong acid + TRIPHENYLCARBINOL in ethanoic anhydride) are used for hydride ion

abstraction (see TROPYLIUM ION). They are readily   hydrolyzed   to   triphenylcarbinol (Ph_3COH).

The radical and ions are stable due to resonance because of the high degree of delocalization of the electrons or charge over the benzene rings.

triphenylmethyl chloride (trityl chloride, Ph_3CCl). Solid prepared by the FRIEDEL–CRAFTS REACTION between benzene and tetrachloromethane. The halogen atom is highly reactive, boiling with water gives the alcohol and with ethanol the ethyl ether, reaction with sugar derivatives gives triphenylmethyl ethers. A solution in liquid sulphur dioxide is conducting, Ph_3C^+ Cl$^-$.

2,3,5-triphenyltetrazolium chloride. Colourless solid (decomp. 243°C), turning yellow on exposure to light. It is reduced by aldoses and ketoses to the water-insoluble red pigment, triphenylformazan, and is used as a reagent for reducing sugars, for the detection of dehydrogenases, and for staining plant and animal tissues. See also TETRAZOLIUM BLUE.

2,3,5-triphenyltetrazolium chloride

triple bond.   See COVALENT BOND.

triple point.   Invariant point (temperature and pressure fixed) at which three phases of a ONE-COMPONENT SYSTEM coexist at equilibrium. The CHEMICAL POTENTIAL of the component is the same in all phases. The total number of triple points in a system is given by

$$[p\,(p-1)\,(p-2)]/3!$$

where p is the total number of possible phases.

triplet. *See* MULTIPLET.

triplet state. *See* ELECTRONIC SPECTRO-SCOPY, FLUORESCENCE, MULTIPLICITY, SELECTION RULES.

triptycene (tribenzobicyclo[2.2.2]octatriene, $C_{20}H_{14}$). Colourless solid (mp 253°C), obtained by a DIELS–ALDER REACTION between BENZYNE and ANTHRACENE.

tristearin. *See* STEARIN.

tritane. *See* TRIPHENYLMETHANE.

triterpenes. *See* TERPENES.

tritiation. Replacement of hydrogen in a molecule by TRITIUM.

tritide ($^3H^-$). Anion of TRITIUM.

tritium (^3H, T). Radioactive ISOTOPE of HYDROGEN, produced by bombarding DEUTERIUM compounds with DEUTERONS ($^2H + {}^2H \rightarrow {}^3H + {}^1H$) or lithium-6 with neutrons ($^6Li + {}^1n \rightarrow {}^3H + {}^4He$). Its major use is in thermonuclear bombs ($^3H + {}^2H \rightarrow {}^4He + {}^1n + 18$ MeV), also as a TRACER for hydrogen (*see* ISOTOPE EFFECTS). It is particularly useful in biochemistry where it has been used to study, among other systems, the formation and action of DNA and RNA (*see* NUCLEIC ACIDS). Tritium is a weak beta-emitter and is followed by mixing with a suitable phosphor and measuring the SCINTILLATION.

triton ($^3H^+$). Cation of TRITIUM.

triton. Trade name for surfactants based on alkylaryl polyether. alcohols, sulphonates, sulphates; non-ionic, anionic and cationic types. *See also* SURFACE-ACTIVE AGENTS.

tritonation. Transfer of a TRITON to a substrate.

trityl chloride. *See* TRIPHENYLMETHYL CHLORIDE.

trityl group. The group Ph_3C- (*see* TRIPHENYLMETHYL). Trityl chloride reacts with sugar derivatives to give trityl ethers.

tRNA (transfer ribonucleic acid). *See* NUCLEIC ACIDS.

trona. *See* SODIUM.

3-tropanol (tropine). *See* ALKALOIDS.

tropilidene. *See* CYCLOALKENES.

tropine. *See* ALKALOIDS.

tropolone. *See* 2-HYDROXY-2,4,6-CYCLO-HEPTATRIEN-1-ONE.

tropylium ion (cycloheptatrienyl cation, $^+C_7H_7$). Very stable CARBONIUM ION, exhibiting aromatic character with six π-electrons (*see* HÜCKEL'S RULE) distributed evenly over seven carbon atoms (*see* Figure). It is prepared by abstraction of the hydride ion from 1,3,5-cycloheptatriene with trityl salts (e.g., $Ph_3C.BF_4$) or by removal of hydrogen bromide (on heating) from tropylium bromide with behaves as an ionized salt in solution. Alkaline hydrolysis gives tropone (2,4,6-cycloheptatriene-1-one, C_7H_6O). It forms complexes with metal carbonyls (e.g., h^5–$C_7H_7Mo(CO)_3$.-BF_4) (*see* SANDWICH COMPOUNDS).

tropylium ion

Trouton's rule. Molar ENTHALPY of vaporization divided by the normal BOILING POINT (i.e. the molar ENTROPY of vaporization) is approximately constant

$$\Delta S_{vapn} = L_e/T \sim 88 \text{ J K}^{-1} \text{ mol}^{-1}$$

The law is valid for a large number of non-polar liquids with a range of boiling points. Associated liquids (e.g., water, ethanol) with hydrogen bonding (*see* HYDROGEN BONDS) in the liquid state have higher values, whereas non-associated liquids (e.g., methanoic acid, ethanoic acid) that form dimers in the vapour phase have lower values.

Trp (tryptophan). *See* AMINO ACIDS.

trypsin. *See* ENZYMES.

tryptophan (Trp). *See* AMINO ACIDS.

$$R_2CHCH_2OH \xrightarrow[\text{NaOH}]{CS_2} \overset{R_2C}{\underset{H_2C}{\diagdown}} \overset{H}{\underset{O}{\diagup}} \overset{S}{\underset{C}{\parallel}} \overset{S^-Na^+}{\underset{\diagdown}{\diagup}} \xrightarrow{\text{MeI}} \overset{R_2C}{\underset{H_2C}{\diagdown}} \overset{H}{\underset{O}{\diagup}} \overset{S}{\underset{CSMe}{\parallel}} \xrightarrow{200^\circ C} R_2C{=}CH_2 + MeSH + COS$$

Tschugaev reaction

Tschugaev reaction. Pyrolysis of xanthates formed from the action of carbon disulphide in sodium hydroxide on alcohols to give an alkene by cyclic elimination.

TSH (thyroid-stimulating hormone). *See* HORMONES.

tubocurarine. *See* CURARE.

tungstates. *See* TUNGSTEN COMPOUNDS.

tungsten (W). Group IVb, third row TRANSITION ELEMENT. It is extracted from wolframite ((Fe,Mn)WO$_4$), ferberite (FeWO$_4$), scheelite (CaWO$_4$) and stolzite (PbWO$_4$) as WO$_3$ after sodium hydroxide fusion and treatment with acid. WO$_3$ is reduced to the metal with hydrogen. Tungsten is alloyed in STEEL to which it imparts strength at high temperatures, and is used in electrical devices (filaments in lights, electron guns). Tungsten carbide has replaced diamond in cutting and drilling applications. Tungsten is a hard, lustrous silvery-white metal which is relatively chemically inert. It does not react with acids, except mixtures of hydrofluoric and nitric acids. A protective film of WO$_3$ is formed in oxygen, further reaction occurring only at elevated temperatures. Many non-metals, with the exception of hydrogen, react with tungsten at high temperatures.
$A_r = 183.85$, $Z = 74$, ES [Xe] $5d^4\,6s^2$, mp 3410°C, bp 5930°C, $\rho = 19.3 \times 10^3$ kg m^{-3}. Isotopes: ^{180}W, 0.14 percent; ^{182}W, 26.4 percent; ^{183}W, 14.4 percent; ^{184}W, 30.6 percent; ^{186}W, 28.4 percent.

tungsten compounds. Oxidation states −2 to +6 are found, the most stable compounds are of tungsten(VI). Many compounds are polymeric, and NON-STOICHIOMETRIC COMPOUNDS exist. The chemistry resembles that of MOLYBDENUM and CHROMIUM. Yellow WO$_3$ is formed on heating the metal in oxygen, successive reduction gives W$_{20}$O$_{58}$, W$_{18}$O$_{49}$ and WO$_2$. BRONZES M$_x$WO$_3$ (M = alkali metal or

hydrogen) are formed by the action of the element on tungsten. WO$_3$ dissolves in sodium hydroxide to give sodium tungstate (Na$_2$WO$_4$). Other tungstates are formed by ion exchange. Tungstic acid is an amorphous yellow powder. Isopolytungstates, containing WO$_6$ units, are produced on acidification of solutions of tungstates. Typical ions are [HW$_6$O$_{21}$]$^{5-}$, [H$_2$W$_{12}$O$_{42}$]$^{10-}$ and [W$_{12}$O$_{39}$]$^{6-}$. The less common oxidation states are seen in halides WF$_5$ (WF$_6$ + W), WF$_4$ (heat on WF$_5$), WCl$_3$ and WCl$_2$ (hydrogen reduction of WCl$_6$). The lower halides exist as complexes containing metal clusters. W(CO)$_6$ is similar to Mo(CO)$_6$. Various compounds with non-metals are formed (e.g., WA$_2$, where A = N, B, P, Si, S).

tunnel effects. Effects arising from the penetrations of an energy barrier in a quantum mechanical process. An example (*see* Figure) is in (a) FIELD EMISSION MICROSCOPY in which an electron tunnels through a barrier made thinner by an electric field, and in (b) FIELD ION MICROSCOPY in which an imaging gas atom becomes ionized by an electron tunnelling back into the metal.

turbidimetry. Method of quantitative analysis in which the TURBIDITY of a colloidal dispersion (*see* COLLOID) is measured spectrophotometrically.

turbidity. Cloudy appearance of a colloidal dispersion (*see* COLLOID) due to scattering of light (*see* TYNDALL EFFECT). The turbidity (τ) is defined by lg $I_0/I = \tau l$, where I_0 and I are the incident and transmitted intensities, respectively, and l is the path length.

turbulent flow. Form of fluid flow in which particles of the fluid move in an irregular path resulting in exchange of momentum from one portion of the fluid to another. It takes over from LAMINAR FLOW at high values of REYNOLD'S NUMBER. *See also* NEWTONIAN FLUID.

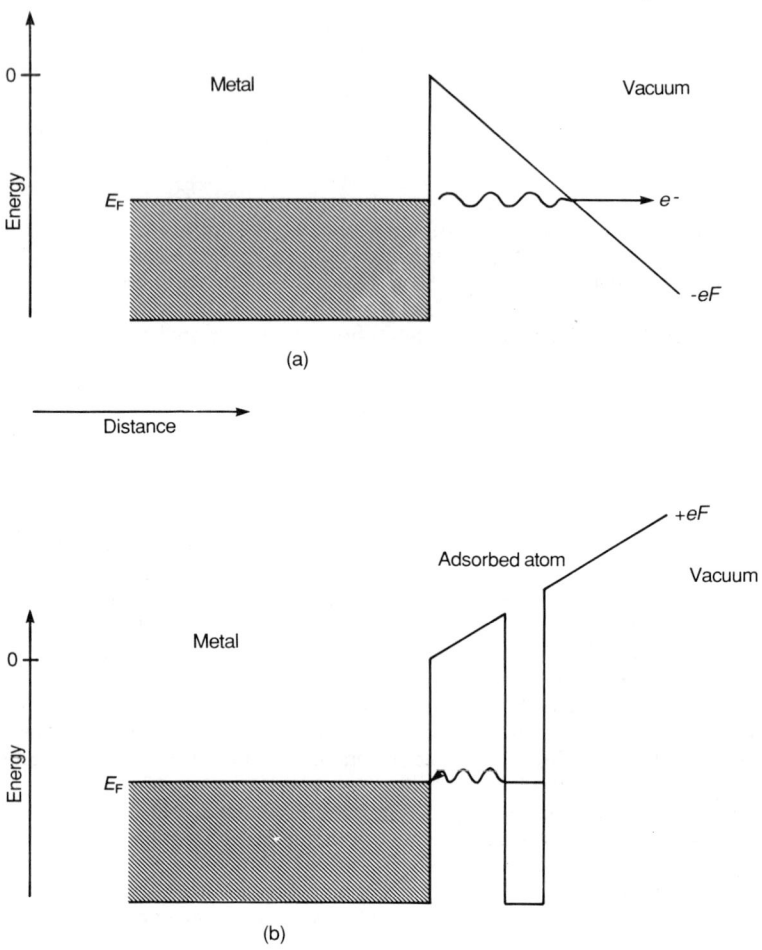

(a)

(b)

tunnel effects
(a) Field emission of electron from a metal Fermi level E_F subjected to an electric field $-F$ V m^{-1}. (b) Field ionization of an adsorbed atom at a metal subjected to an electric field $+F$ V m^{-1}.

Turnbull's blue. *See* CYANOFERRATES.

turnover rate. Number of reactions per second in an heterogeneous catalytic reaction (*see* HETEROGENEOUS CATALYSIS).

turpentine. *See* ESSENTIAL OILS, TERPENES.

twaddle (°Tw). Unit used in reporting specific gravities greater than that of water. A twaddle reading multiplied by 5 and added to 1000 gives a specific gravity with reference to water as 1000. It is used mainly in the acid industry.

two-component condensed system. Solid–liquid equilibrium in which the components are completely miscible in the liquid phase. Since the vapour phase can be ignored and one DEGREE OF FREEDOM removed, the PHASE RULE can be written as

$$p + f' = c + 1$$

The remaining variables—temperature and composition—are plotted on equilibrium diagrams constructed from a series of cooling curves (*see* THERMAL ANALYSIS).
(1) Simple eutectic system (*see* Figure) in which only pure A and B crystallize from the melt or solution. Where the two freez-

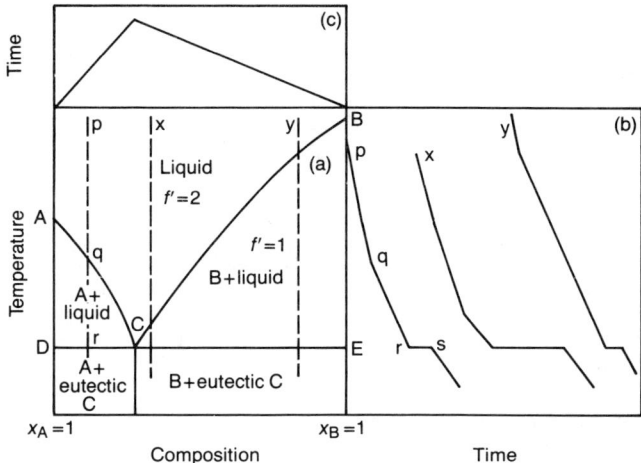

two-component condensed system
No compound formation. (a) Phase diagram; (b) cooling curves; (c) duration of eutectic arrest.

ing point curves (AC and BC) meet at C solid A, solid B and liquid are in equilibrium; this is the EUTECTIC MIXTURE. At temperatures below the eutectic temperature the system is solid (e.g., gold/thallium).

(2) Systems in which a solid compound (AB, AB_2,. . .) with a congruent MELTING POINT is formed (e.g., aluminium/magnesium, Al_3Mg_2; salt hydrates, sulphuric acid/water, three hydrates).

(3) Systems in which the components form a compound with an incongruent melting point. The compound is not stable in the liquid state and decomposes below its melting point (e.g., gold/tin; salt hydrates, $Na_2SO_4.H_2O$).

(4) Systems in which the two components are completely miscible in the solid state and form SOLID SOLUTIONS (*compare* COMPLETELY MISCIBLE LIQUID SYSTEMS). Complete miscibility of solid phases occurs when the atoms of the two components are similar in size and can substitute for one another in the lattice (e.g., copper/nickel; potassium chloride/potassium bromide).

two-dimensional chromatography. Procedure for improved separation between closely related substances by PAPER CHROMATOGRAPHY or THIN-LAYER CHROMATO-GRAPHY achieved by running the chromatograms in two different solvent systems at right angles to one another. In the first solvent system, a single spot in the corner of a square of stationary phase is separated into a row of spots which then form the origins for running the second solvent system at right angles. It is also applicable to electrophoretic techniques.

Tyndall effect. Scattering of light by an heterogeneous fluid which is seen as TURBIDITY. Elastically scattered light is proportional to $1/\lambda^4$ (i.e. blue light is scattered more than red light). This explains why the sky appears blue and a setting sun red. As no fluid is ultimately homogeneous, light is scattered from the molecules and atoms of any liquid or gas (*see* RAYLEIGH SCATTERING, RAMAN SPECTROSCOPY). However, the Tyndall effect has found most use in the study of COLLOIDS and solutions of POLYMERS and natural macromolecules. The measurement of the scattered light as a function of angle and polarization allows an estimate to be made of particle size, shape and interactions.

Tyr (tyrosine). *See* AMINO ACIDS.

tyrosine (Tyr). *See* AMINO ACIDS.

U

U. *See* URANIUM.

U. *See* INTERNAL ENERGY.

u. *See* IONIC MOBILITY.

Ullmann reaction. Synthesis of biaryls (Ar–Ar) by condensation of aromatic halides with themselves or other aromatic halides in the presence of copper powder (e.g., the preparation of diphenyl derivatives). The condensation of an aryl halide with a metal phenolate produces a diaryl ether.

ultracentrifuge. SEDIMENTATION of COLLOIDS under gravity is too slow for particles less than about 1 μm in diameter. In an ultracentrifuge—a high-speed centrifuge—the colloid is spun rapidly, and the high forces thus created (up to 400 000 g) increase the rate of sedimentation. The boundary of solid material is observed using Schlieren or interference optics. At high rotation speeds, the movement of the boundary is studied with time, at lower speeds an equilibrium distribution is attained when sedimentation and diffusion balance each other. *See also* SEDIMENTATION COEFFICIENT.

ultrafiltration. Application of pressure to force solvent and small colloidal particles across a membrane while the larger particles are retained (*see* DIALYSIS). The membrane is supported on a mesh or deposited on sintered glass. *See also* ELECTRODIALYSIS.

ultra-high vacuum. *See* VACUUM.

ultramicroscope. Microscope used to examine colloidal particles (*see* COLLOID) by dark-field illumination in which light scattered by the TYNDALL EFFECT is observed.

ultraviolet photoelectron spectroscopy. *See* ELECTRON SPECTROSCOPY.

ultraviolet spectra. *See* ELECTRONIC SPECTROSCOPY.

uncertainty principle. *See* HEISENBERG'S UNCERTAINTY PRINCIPLE.

Unh. Unnilhexium (*see* TRANSACTINIDE ELEMENTS).

uniaxial crystal. Double-refracting crystal having only one optic axis.

unified mass unit. *See* ATOMIC MASS UNIT.

unimolecular film. *See* SURFACE FILM.

unimolecular reactions. *See* ABSOLUTE RATE THEORY, LINDEMANN THEORY, MOLECULARITY OF REACTION, RRKM THEORY.

unit cell. Characteristic grouping of atoms or ions in a crystal that is repeated in three dimensions in a CRYSTAL LATTICE.

units. *See* CGS UNITS, MKS UNITS, SI UNITS.

universal gas constant. *See* GAS CONSTANT.

unnilhexium (Unh). IUPAC systematic name for element 106 (*see* TRANSACTINIDE ELEMENTS).

unnilpentium (Unp). IUPAC systematic name for element 105 (*see* TRANSACTINIDE ELEMENTS).

unnilqualium (Unq). IUPAC systematic name for element 104 (*see* TRANSACTINIDE ELEMENTS).

unnilseptium (Uns). IUPAC systematic name for element 107 (*see* TRANSACTINIDE ELEMENTS).

Unp. Unnilpentium (*see* TRANSACTINIDE ELEMENTS).

Unq. Unnilqualium (*see* TRANSACTINIDE ELEMENTS).

Uns. Unnilseptium (*see* TRANSACTINIDE ELEMENTS).

unsaturated compound. *See* SATURATED COMPOUND.

UPS (ultraviolet photoelectron spectroscopy). *See* ELECTRON SPECTROSCOPY.

uracil. *See* PYRIMIDINES.

uranium (U). One of the ACTINIDE ELEMENTS widely found in nature as the oxide ore pitchblende (UO_2), also dissolved in the oceans (~ 4 tonnes km^{-3}). The metal is extracted by the reduction of the tetrafluoride by magnesium or calcium at 700°C. The importance of uranium as a nuclear fuel arises from the neutron-induced decay of uranium-235

$$^{235}U + n \rightarrow {}^{140}Ba + {}^{96}Kr + 3n + 200\,MeV$$

As more neutrons are released than consumed, a CHAIN REACTION may be set up, controlled in a power plant, uncontrolled in an atomic bomb. The natural abundance of uranium-235 (0.72 percent) may be enriched by gaseous diffusion of UF_6. Uranium is the starting point for later actinide elements (directly neptunium-237, plutonium-239). Uranium is a dense, electropositive metal existing in three phases below its melting point. Uranium reacts with most elements, water, acids and air, but not with alkalis. The powdered metal spontaneously ignites in air. $A_r = 238.0289$, $Z = 92$, ES [Rn] $5f^3\,6d^1\,7s^2$, mp 1132°C, bp 3818°C, $\rho = 19.07 \times 10^3$ kg m^{-3}.

uranium compounds. Oxidation states +3 to +6 are found, +6 being the most stable. Uranium forms several oxides: UO (gaseous, stable above 1800°C), UO_2, a series of NON-STOICHIOMETRIC COMPOUNDS UO_{2+x} to UO_3. Hydrated UO_4, a peroxide, is precipitated from solutions of uranyl ions by hydrogen peroxide. Known halides are UX_n ($n = 3, 4, 5, 6$ X = F, Cl; $n = 3, 4$ X = Br, I). In aqueous solutions, complex ions of uranium(IV) and uranium(VI) are formed (e.g., $[UNO_3(H_2O)_4]^{3+}$). UO_2^{2+}, the uranyl ion, forms the most common series of uranium salts, crystallizing with six, three or two molecules of water. They are yellow and fluoresce green. Uranyl nitrate ($UO_2(NO_3)_2.6H_2O$), dissolves in many organic solvents, this property being used in the extraction of uranium. Other uranyl salts are formed with halides, sulphate, organic acids, etc. The SANDWICH COMPOUND (h-$^8C_8H_8)_2U$ is known.

uranium decay series. *See* RADIOACTIVE DECAY SERIES.

urea (carbamide, $(NH_2)_2C=O$). White solid (mp 132°C), prepared by the WOHLER'S SYNTHESIS, by the addition of water to cyanamide or ammonia to carbonyl chloride. It is manufactured from ammonia and carbon dioxide (200°C, 400 atm). Urea forms molecular complexes with straight-chain (not branched) hydrocarbons having more than five carbon atoms. The hydrocarbon may be regenerated by treatment with water. Major use is in urea-methanal resins (*see* POLYMERS) and also as a fertilizer.

urea–formaldehyde resins. *See* POLYMERS.

urea–methanal resins. *See* POLYMERS.

urease. *See* ENZYMES.

urethane polymers. *See* POLYMERS, RESINS.

urethanes (NH_2COOR). Esters of carbamic acid (*see* CARBAMATES). The ethyl ester ($NH_3 + ClCOOC_2H_5$), colourless crystals (mp 50°C), is known trivially as urethane and is used as an anaesthetic and hypnotic. It reacts with ammonia to give UREA and ethanol. The reaction is reversible and is a route to urethanes. N-Substituted urethanes are prepared by the CURTIUS REACTION. Polyurethanes are condensation POLYMERS formed from polyhydroxy compounds and polyisocyanates; they are extensively used as FOAMS.

uric acid. *See* PURINES.

uridine.   *See* NUCLEOSIDES.

5'-uridylic acid.   *See* NUCLEOTIDES.

uronic acids.   Formed by the oxidation of the only terminal –CH$_2$OH group in ALDOSES while the aldehyde group remains intact, to yield HOOC(CHOH)$_4$-CHO. They exhibit the reactions of an aldose and an acid. They are easily decarboxylated on boiling with hydrochloric acid; D-glucuronic acid yields D-xylose. The most important are D-glucuronic acid, D-mannuronic acid and D-galacturonic acid. They are widely distributed in plant and animal tissues in complex polysaccharides (*see* AMINO SUGARS). Polymers of the uronic acids are the constituents of GUMS and MUCILAGES; alginic acid (*see* SEAWEED COLLOIDS) and many bacterial capsules are polymers of mannuronic acid. Pectic acid is a polymer of galacturonic acid. *See also* TEICHURONIC ACIDS.

urushiol.   *See* PHENOLS.

UV spectra.   *See* ELECTRONIC SPECTROSCOPY.

V

V. *See* VANADIUM, VOLT.

V. *See* POTENTIAL, VOLUME.

v. *See* VELOCITY.

vacancy. *See* DEFECT STRUCTURE.

vacuum. Many processes require a re-duced pressure, for example ELECTRON SPECTROSCOPY, MASS SPECTROMETRY, VACUUM DISTILLATION and the investiga-tion of adsorption and catalysis on clean metal surfaces. A vacuum is classed as high (pressure$<10^{-3}$ mmHg), very high (pres-sure$<10^{-6}$ mmHg) and ultra high (uhv, pressure$<10^{-9}$ mmHg). At 10^{-6} mmHg, a monolayer of adsorbed gas is formed in about 1 s, the time taken being inversely proportional to the pressure. Therefore a background pressure in the uhv region is required to preserve clean surfaces for times of the order of chemical experi-ments. A vacuum may be created by mechanical means (rotary pump, water pump, turbomolecular pump), by adsorp-tion (MOLECULAR SIEVES, GETTERS, ion pumps) and by diffusion pumps. Pressures are measured by ionization gauges, mer-cury manometers, capsule gauges and by mass spectrometry.

vacuum distillation. DISTILLATION at reduced pressure of high-molecular-mass compounds which would decompose at temperatures needed to effect vaporization at atmospheric pressure. It is used to separate mixtures of liquids, in laboratory synthesis and in the PETROLEUM industry.

Val (valine). *See* AMINO ACIDS.

valence. Former term for OXIDATION STATE. The valence electrons are the outermost electrons responsible for bond-ing. The theory of valency is concerned with the attempt of atoms in molecules to attain a NOBLE GAS ELECTRONIC STRUC-TURE. Simple IONIC BONDS, COVALENT BONDS and COORDINATE BONDS are explained in this way, but transition element complexes (*see* COORDINATION COMPOUNDS) often do not achieve such a structure (*see* EFFECTIVE ATOMIC NUMBER, EIGHTEEN-ELECTRON RULE).

valence band. *See* BAND THEORY OF SOLIDS.

valence bond force field model. *See* FORCE CONSTANT.

valence bond orbital (vbo). Electron-pair bond. For molecular hydrogen, the valence bond orbital (also known as the Heitler–London orbital) is

$$\psi_{vb} = N(\phi_{1sA}(1)\phi_{1sB}(2) + \phi_{1sA}(2)\phi_{1sB}(1))$$

where 1sA, 1sB refer to s-orbitals on each hydrogen atom and (1) and (2) to each electron. This is to be compared to the molecular orbital approach which gives

$$\psi_{mo} = N(\phi_{1sA}(1) + \phi_{1sB}(1)) \times$$
$$(\phi_{1sA}(2) + \phi_{1sB}(2))$$
$$= N[\phi_{1sA}(1)\phi_{1sB}(2) +$$
$$\phi_{1sA}(2)\phi_{1sB}(1) +$$
$$\phi_{1sA}(1)\phi_{1sA}(2) +$$
$$\phi_{1sB}(1)\phi_{1sB}(2)]$$

The terms which are not common to the valence bond and molecular orbital WAVE-FUNCTIONS represent the case when both electrons are on atom A or both on atom B, giving an ion pair (e.g., $H_A^+ H_B^-$). These are given too great an importance in the molecular orbital wavefunction and have been ignored completely in the valence bond wavefunction. Application of valence bond theory to benzene leads to the concept

503

of resonance between the different Kekulé, Dewar and ionic structures (*see* BENZENE).

valence shell electron pair repulsion theory (VSEPR). Theory of the shapes of molecules based on the repulsion between valence electron pairs. Lone pairs, which are spatially more diffuse (bonding pairs have their maximum density between the nuclei) repel each other more than bonding pairs. Thus the order of the strengths of repulsions is lone pair–lone pair ≫ lone pair–bond pair > bond pair–bond pair. VSEPR theory has been used to explain the shapes of, for example, COORDINATION COMPOUNDS, halogen halides (*see* INTERHALOGEN COMPOUNDS), and NITROGEN OXIDES (i.e. why NO_2^- is linear, NO_2 bent and NO_2^+ more so). *See also* SHAPES OF MOLECULES.

valency. *See* VALENCE.

valentinite. *See* ANTIMONY OXIDES AND OXOACIDS.

valeric acids. Pentanoic acids (*see* CARBOXYLIC ACIDS).

valine (Val). *See* AMINO ACIDS.

Valium. Trade name for DIAZEPAM.

vanadium (V). Group Vb, first row TRANSITION ELEMENT. It is extracted, usually with other metals from ores by acid leaching and fusing the resulting oxide with calcium or aluminium. The metal is also produced by reduction of the chloride by magnesium. Most of the vanadium produced is used in alloying in STEEL in which it exists as the carbide.
$A_r = 50.9415$, $Z = 23$, ES [Ar] $3d^3 4s^2$, mp 1890°C, bp 3380°C, $\rho = 5.96 \times 10^3$ kg m^{-3}. Isotopes: ^{50}V, 0.25 percent; ^{51}V, 99.75 percent.

vanadium compounds. Oxidation states −1 to +5 are found (+4 and +5 are the most common). Vanadium(v) is covalent in complexes (e.g., VF_5, V_2O_5, NH_4VO_3). Vanadium(IV) is blue in aqueous solution, VO^{2+} (vanadyl) is stable and found in complexes (e.g., $VOSO_4$, VOF_2). VCl_3 is a violet solid. ALUMS and other double salts are formed from $V_2(SO_4)_3$. Vanadium(II)

(e.g., VSO_4) is strongly reducing. Vanadium forms the SANDWICH COMPOUND vanadocene $(h^5\text{–}C_5H_5)_2V$. Vanadates are anionic derivatives of vanadium, mostly in oxidation state +5, but the parent compound, vanadic acid, is not known. V_2O_5 dissolves in bases to give $VO_3(OH)^{2-}$, $VO_2(OH)_2^-$. Acidification gives complex species (e.g., hexavanadates, $[H_2V_6O_{17}]^{2-}$; decavanadates, $[V_{10}O_{28}]^{6-}$).

van Arkel–de Boer process. *See* CHROMIUM.

van Deemter equation. Equation relating column efficiency, in terms of HEIGHT EQUIVALENT OF A THEORETICAL PLATE (HETP, *H*) to controllable variables such as flow rate, particle size and packing characteristics

$$H = A + B/v + Cv$$

where *A*, *B* and *C* are constants and *v* is the flow rate. The first term, the eddy diffusion term, allows for the different paths followed by solutes due to passing round particles of different sizes. It is related to particle size, geometry and tightness of packing of the stationary phase, but is independent of the solvent, solute and operating conditions of the column, and is the smallest contributor to *H*. The second term, the molecular diffusion term (inversely proportional to *v*) is the contribution due to normal diffusion of molecules passing along the column, *B* is related to the diffusion coefficient of molecules in the mobile phase. At low flow rates, the HETP is mainly determined by this term; at high flow rates its contribution is negligible. The third term, the mass transfer term, allows for the finite rate of transfer between the two phases. *C* depends on the thickness of the immobilized film, the diffusion coefficient of the solute in this liquid and the comparative volumes of the mobile and stationary phases. This term, which increases with flow rate, is the most important at high flow rates.

van der Waals adsorption. *See* ADSORPTION, CHEMICAL ADSORPTION.

van der Waals equation. EQUATION OF STATE obtained by modification of the ideal gas equation (*see* GAS LAW)

$(P + n^2a/V^2)(V - nb) = nRT$

where the term n^2a/V^2 (the internal cohesive force) reflects the decreased pressure on the walls of the container as a result of attractive forces between the molecules, and nb (the covolume) allows for the finite volume occupied by the molecules. The van der Waals constants a and b ($= 4V_m$) are temperature-dependent and can be calculated from measured values of the CRITICAL CONSTANTS.

van der Waals forces. *See* INTERMOLECULAR FORCES.

van der Waals radii. *See* ATOMIC RADIUS.

vanillin (4-hydroxy-3-methoxybenzaldehyde, $CH_3OC_6H_3(OH)CHO$). Odiferous principle of vanilla pod; occurs extensively in nature. It is a cheap byproduct of the paper industry, and is very important in flavouring (foodstuffs) and perfumery (toiletries).

van't Hoff factor (i). Factor that appears in equations for colligative properties (e.g., osmotic pressure, $\Pi V = iRT$), which is equal to the ratio of the actual number of particles present to the number of undissociated particles.

van't Hoff isochore. Equation that expresses the variation of the EQUILIBRIUM CONSTANT with temperature

$(\partial \ln K / \partial T)_P = \Delta H^\ominus / RT^2$

assuming that $\Delta H^\ominus$ is independent of temperature, the above equation on integration becomes

$\ln K = -\Delta H^\ominus/RT + \text{const}$

or more generally (*see* KIRCHHOFF'S EQUATION)

$\ln K = -\Delta H^\ominus/RT + (\Delta a \ln T)/R + \Delta b\,T/2R$
$\qquad + \Delta c T^2/6R - J/R$

where J is a constant, obtained from a knowledge of K at a given temperature T.

van't Hoff isotherm. Equation for calculating the change in the GIBBS FUNCTION for a reaction from a knowledge of the EQUILIBRIUM CONSTANT, or $\Delta G^\ominus$, and the initial and final ACTIVITY values (concentrations or partial pressures) of reactants and products. For the general reaction

$aA + bB + \ldots \longrightarrow lL + mM + \ldots$

$\Delta G = \Delta G^\ominus + RT \ln (a_L^l a_M^m/a_A^a a_B^b)$

$\qquad = -RT \ln K_{\text{therm}} +$

$\qquad\qquad RT \ln (a_L^l a_M^m/a_A^a a_B^b)$

where a_A, a_B, . . . and a_L, a_M, . . ., the initial and final activities of reactants and products, respectively, in the reaction, can be replaced by concentrations or partial pressures assuming ideal behaviour.

vaporization. Endothermic phase change liquid to vapour.

vapour. Substance in the gaseous state at a temperature below its critical temperature; it can be liquefied by pressure alone. *See also* CRITICAL CONSTANTS.

vapour density. Density of a gas or vapour relative to that of hydrogen under identical conditions of temperature and pressure. The vapour density is half the relative molecular mass of the gas.

vapour–liquid equilibrium. *See* ONE-COMPONENT SYSTEM.

vapour pressure (p; dimensions: $m\ l^{-1}\ t^{-2}$; units: mmHg, N m^{-2}). Pressure at which liquid (solid) and vapour can coexist in equilibrium at a given temperature (from the PHASE RULE, $f = 1$ for such a ONE-COMPONENT SYSTEM). Vapour pressure is independent of the relative amounts of the two phases, but depends on the temperature (CLAPEYRON–CLAUSIUS EQUATION). The vapour pressure of a solvent is lowered on addition of a non-volatile solute (*see* RAOULT'S LAW). The vapour pressure of liquids may be measured by static, dynamic or transpiration methods, and that of solids by the EFFUSION method. *See also* COMPLETELY MISCIBLE LIQUID SYSTEMS.

variation principle. Expectation value of the Hamiltonian for a non-exact wavefunction must be greater than the eigenvalue of the Hamiltonian. This principle is used in the variational method in determining coefficients in a LINEAR COMBINATION OF ATOMIC ORBITALS when forming molecular orbitals. Better values of these coefficients, and thus better molecular wavefunctions, give ever-lower energies from solutions of the SCHRÖDINGER EQUATION. *See also* MOLECULAR ORBITAL CALCULATIONS.

Vaska's compound. *See* IRIDIUM COMPOUNDS.

vat dyes. DYES containing an OXO GROUP that are rendered soluble by chemical reduction (usually with sodium dithionite) to a LEUCO COMPOUND. This vat solution, which may be of a different colour from the final dye, is used to treat the material. Air oxidation restores the oxo group and the now water-insoluble dye. *See also* INDIGOTIN.

vbo. *See* VALENCE BOND ORBITAL.

vehicle. Term used in paint technology to indicate the liquid portion of a PAINT, composed of drying oil, solvent and thinner, in which the solid components are dissolved or dispersed.

velocity (v; dimensions: $l\ t^{-1}$; units: m s^{-1}). Vector quantity, defined as the rate of displacement of a body; it is the SPEED of a body in a specified direction.

verbenone. *See* TERPENES.

verdigris. *See* COPPER.

vermiculite. Hydrated SILICATE mineral containing magnesium, iron and aluminium, which crystallizes in the monoclinic system. On heating to 1000°C it exfoliates with a 10-fold increase in bulk due to rapid evolution of water. Expanded vermiculite, because of its low density, is widely used as a thermal insulator and as a rooting medium and soil additive.

vermilion. *See* MERCURY INORGANIC COMPOUNDS.

veronal. *See* BARBITURATES.

vesicle. Small, fluid-filled, membrane-bound sac within the cytoplasm of a living cell.

$\bar{V}_i$.   *See* PARTIAL MOLAR VOLUME.

vibration. *See* INFRARED SPECTROSCOPY, NORMAL MODE.

vibrational partition function. *See* PARTITION FUNCTION.

vibrational selection rules. *See* SELECTION RULES.

vibrational spectra. *See* INFRARED SPECTROSCOPY.

vibration–rotation spectra. Rotational fine structure seen in IR (*see* INFRARED SPECTROSCOPY) or Raman vibration spectra (*see* RAMAN SPECTROSCOPY). For an IR active transition between vibrational level 0, rotational level J and vibrational level 1, the allowed rotational transitions for a heteronuclear diatomic molecule are to $J+1$ (*R*-branch), J–1 (*P*-branch) and for molecules with an odd number of electrons J (*Q*-branch). The frequencies of these transitions are given in terms of the frequency of the vibrational transition ω and the rotational constant for vibrational state v, B_v (*see* ROTATIONAL SPECTRA).

$$P = \omega - (B_1 + B_0)J + (B_1 - B_0)J^2$$
$$Q = \omega + (B_1 - B_0)J(J + 1)$$
$$R = \omega + 2B_1 + (3B_1 - B_0)J + (B_1 - B_0)J^2$$

The inequality of B_0 and B_1 causes the rotational lines to be more closely spaced at higher values of J. Eventually a maximum frequency, the bandhead, is reached. Band heads, in fact, are not usually seen in infrared spectroscopy, but often in the fine structure of ELECTRONIC SPECTROSCOPY. Three branches may also be seen in Raman spectra with $J = -2$ (*O*-branch), J = 0 (see above) and $J = +2$ (*S*-branch) with frequencies

$$O = \omega + 2B_1 - (3B_1 + B_0)J + (B_1 - B_0)J^2$$
$$S = \omega + 6B_1 + (5B_1 - B_0)J + (B_1 - B_0)J^2$$

vicinal (*vic-*). Term referring to neighbouring or adjoining positions on a carbon ring or chain.

vinyl acetate.   *See* ETHENYL ETHANOATE.

vinyl alcohol.   *See* ETHENOL.

vinyl aldehyde.   *See* PROPENAL.

vinylation.   *See* ETHENYLATION.

vinylbenzene.   *See* STYRENE.

vinyl chloride.   *See* CHLOROETHENE.

vinyl chloride polymers.   *See* POLYMERS.

vinyl cyanide.   *See* PROPENENITRILE.

vinyl ether.   *See* DIETHENYL ETHER.

vinyl group.   *See* ETHENYL GROUP.

vinylidene group.   The group $H_2C=C<$.

virial equation.   EQUATION OF STATE that attempts to account for the behaviour of real gases in terms of empirical constants. This takes the form

$$PV_m = RT\,[1 + B(T)/V_m + C(T)/V_m^2 + \ldots]$$

where $B(T)$ is known as the second virial coefficient; $C(T)$, which is usually less important, is the third, etc. The coefficients are temperature-dependent, and there will be a temperature at which $B(T)$ is zero; in which case $PV_m \approx RT$ over an extended pressure range. This temperature is the BOYLE TEMPERATURE.

viscoplastic system.   *See* NEWTONIAN FLUID.

viscose.   *See* FIBRES.

viscosity (dynamic) (η; dimensions; $m\,l^{-1}\,t^{-1}$; units: kg m^{-1} s^{-1}, N s m^{-2}). Measure of the resistance to flow of a fluid when subjected to a shear stress. For a NEWTONIAN FLUID.

$$F = \eta A\,(\mathrm{d}v/\mathrm{d}x)$$

where η is a constant known as the coefficient of viscosity. For an IDEAL GAS η, which is defined as follows

$$\eta = m\bar{c}/3(2)^{1/2}\pi d^2$$

is independent of the density and the pressure and, unlike liquids, increases with increase in temperature (the average speed, $\bar{c}$, is proportional to $T^{1/2}$). When a spherical solid particle, of radius r, moves through a fluid with constant velocity (v) the gravitational force must balance the viscous force given by STOKES' LAW. Thus:

$$\eta = 2\,r^2(\rho_p - \rho_f)\,g/9v$$

where ρ_p and ρ_f are the densities of the particle and fluid respectively. This equation, which forms the basis of the falling sphere method for determining the viscosity of a liquid, is also used in the determination of the size of macromolecules in solution (*see* ULTRACENTRIFUGE, SEDIMENTATION COEFFICIENT).

η for liquids depends on the size, shape and chemical nature of its molecules and on the existence of INTERMOLECULAR FORCES (e.g., HYDROGEN BOND). For a series of similar compounds, η increases with the molar mass. For liquids, η decreases with increase in temperature and increases with increase of applied pressure.

In the study of colloidal solutions, suspensions, etc. viscosity increments are of more significance than absolute values of η. The following functions of viscosity are defined: η_0 is the viscosity of the pure solvent or dispersion medium; η the viscosity of the solution or dispersion. The relative viscosity $\eta_{rel} = \eta/\eta_0$, the specific increase in viscosity $\eta_{sp} = \eta_{rel} - 1$, the reduced viscosity (viscosity number) = η_{sp}/c and the intrinsic viscosity

$$[\eta] = \lim_{c \to 0} \eta_{sp}/c$$

c is the concentration (usually g dm^{-3}).

For polymer solutions the intrinsic viscosity is related to the molecular mass of the polymer by an empirical equation of the form

$$[\eta] = KM_r^{\alpha}$$

where K and α are constants which depend on the solvent and the type of molecule.

viscous.   Describing a fluid with a high VISCOSITY: for example, tar.

visible spectra.   *See* ELECTRONIC SPECTROSCOPY.

vitamins

Table 1. Occurrence, preparation and properties of some vitamins.

Vitamin	Structural formula	Occurrence, synthesis, properties	Comments
Vitamin A (axerophthol, retinol) $C_{20}H_{30}O$ $M_r = 286.44$	Primary alcohol of monocyclic diterpenes. All double bonds are in *trans*-configuration	Occurs in animal fats, butter, yolks of eggs, fish liver oils. Extracted from fish liver oils. Total synthesis from β-ionone and a propargyl halide. Pale yellow prisms; mp 64°C. Sol. in fat; insol. in H_2O. Unstable in soln on heating in air. $E(1\,cm/1\%)$ (324 nm) = 1835. In mammals β-carotene undergoes oxidative cleavage to give 2 mol of the aldehyde retinal; biochemical reduction of –CHO yields vitamin A	Absence in diet leads to loss of weight, failure of growth in young animals, eye diseases, night blindness. Daily dose about 1.5 mg (5000 U)
Vitamin B₁ (aneurin, thiamine hydrochloride) $C_{12}H_{18}Cl_2N_4OS$ $M_r = 327.27$		Occurs in plants, animal tissues, rice husk, yeast, liver, eggs, green leaves. Colourless, bitter-tasting crystals; mp 233–4°C. Sol. in water giving acid soln; slightly sol. in Et_2O, PhH, C_6H_{14}, $CHCl_3$. Destroyed by alkalis and alkaline drugs, oxidizing and reducing agents	Absence in human diet leads to beriberi. Min. daily dose about 2 mg. Most commercial B₁ is synthetic (E101)
Vitamin B₂ (riboflavin) $C_{17}H_{20}N_4O_6$ $M_r = 376.37$		Occurs in milk, eggs, liver, heart, kidney, leafy vegetables; richest natural source is yeast. Synthesized from *o*-xylene, D-ribose and alloxan. Several organisms synthesize large amounts. Yellow–orange needles; decomp. 280°C. Insol. in H_2O, Et_2O, propanone; slightly sol. in EtOH. In alkaline soln. deteriorates rapidly, accelerated by light; stable in acid soln. $[\alpha]_D^{25} = -112$ to $-122°$ in aq. ethanolic alkaline soln. pK = 1.7; 10.2; pI = 6.0	Flavoprotein component of electron transport chain. Daily dose about 1.7 mg

Vitamin	Structural formula	Occurrence, synthesis, properties	Comments
Vitamin B$_6$ (pyridoxine hydrochloride) $C_8H_{12}ClNO_3$ $M_r = 204.64$		Occurs in many foodstuffs especially yeast, liver, cereals. Synthesized from: (1) ethyl 2-oxopropanoate, ethyl glycinate and 1,4-diethoxy-2-butanone; (2) oxazole and diethyl propanedioate; mp (free base) 160°C. V. sol. in H$_2$O; slightly sol. in EtOH, propanone. Acidic aq. soln. is stable. E (1 cm/1%) (291 nm) = 422 at pH 2.1	Coenzyme for general metabolism of amino acids
Vitamin B$_{12}$ (cyanocobalamin) $C_{63}H_{88}CoN_{14}O_{14}P$ $M_r = 1355.42$		Occurs in liver from which it was first extracted. Produced during growth of certain organisms. Hygroscopic dark red crystals. Sol. in H$_2$O, EtOH; insol. in Et$_2$O, CHCl$_3$, propanone. Aq. soln neutral; max. stability in soln at pH 4.5–5.0. $[\alpha]_D^{23} = -59 \pm 9$°C	Antipernicious anaemia factor. Given in microgram doses

continued overleaf

Vitamin	Structural formula	Occurrence, synthesis, properties	Comments
Vitamin C (ascorbic acid) $C_6H_8O_6$ $M_r = 176.14$	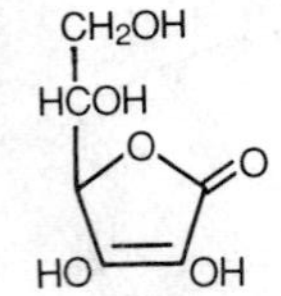	Occurs in vegetables, fruits, especially rosehips, blackcurrants, citrus fruits. Synthesized from glucose or sorbose by oxidation with nitrogen dioxide. Pleasant sharp acid taste, colourless crystals; mp 192°C (some decomp.). $[\alpha]_D^{25} = 22°$. Stable to air when dry. Sol. in H_2O giving acid soln, which is sensitive to oxidation ($pK_1 = 4.17$, $pK_2 = 11.57$); less sol. in EtOH; insol. in Et_2O, trichloromethane, benzene, oils, fats. Strong reducing agent ($E^{\ominus'} = +0.127$ V)	Essential for formation of intercellular material, development of bones, teeth and healing of wounds. Deficiency causes scurvy. Not stored to any extent in animals. Used as antioxidant in foostuffs to prevent rancidity (E300)
Vitamin D_2 (calciferol) $C_{28}H_{46}O$ $M_r = 396.63$	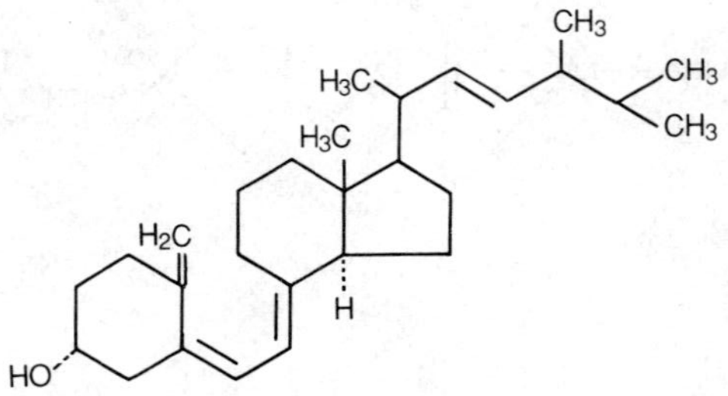	Occurs in animal fats, butter, milk, eggs, fish liver oil. Irradiation of ergosterol with UV gives vitamin D_2. Irradiation of 7-dehydrochol-esterol yields vitamin D_3, which only differs in side chain CH_3 $CH(CH_2)_3CH(CH_3)_2$ Insol. in H_2O; sol. in fats, organic solvents. Oxidized and inactivated by moist air in few days. D_2: mp 115–17°C; $[\alpha]_D^{25} = +82.6°$ in propanone; E (1 cm/1%) (264.5 nm) = 458 ± 7.5 in hexane. D_3: mp 82–3°C; $[\alpha]_D^{20} = +84.8°$ in propanone; E (1 cm/1%) (264.5 nm) = 450–90 in ethanol or hexane	Antirachitic vitamin, essential for development of teeth. Deficiency leads to the development of rickets; not uncommon among Asians consuming unleavened bread

Vitamin	Structural formula	Occurrence, synthesis, properties	Comments
Vitamin E (α-tocopherol, 5,7,8-trimethyl-tocol) $C_{29}H_{50}O_2$ $M_r = 430.69$		Occurs in seed embryos and green leaves. β- and γ-tocopherols (5,8-dimethyltocol and 7,8-dimethyltocol) also show vitamin E activity. Yellow viscous oil; mp 2.5–3.5°C. Insol. in H_2O; freely sol. in oils, fats, organic solvents. Stable to heat and alkalis in the absence of oxygen. E (1 cm/1%) (294 nm) = 71	Deficiency in rats leads to abortion in females and infertility in males. Interrelationship with selenium; essential for tissue respiration and for protection against dietary liver necrosis and muscle degeneration (E307)
Vitamin K_1 (phylloquinone, 2-methyl-3-phytyl-1,4-naphtho-quinone) $C_{31}H_{46}O_2$ $M_r = 450.71$		First isolated from lucerne, but shows widespread distribution in higher green plants. Commercial preparations contain up to 20% *cis*-isomer which shows little vitamin K activity. *trans*-Isomer: yellow viscous oil, mp −20°C, decomp. > 120°C. Insol. in H_2O; sol. in fats, vegetable oils, EtOH, propanone, $CHCl_3$, PhH. Stable to air, but decomp. in sunlight. Unaffected by dil. acids, but is destroyed by soln. of alkalis and reducing agents. $[\alpha]_D^{20} = -0.4°$ (in benzene)	Necessary for the production of blood clotting factors and proteins necessary for normal calcification of bone. In neonates, deficiency may occur because the gut is sterile and there is no synthesis by *Escherichia coli*

continued overleaf

vitamins

Table 2. Factors that are generally regarded as members of the vitamin B complex.

Vitamin	Structural formula	Occurrence, synthesis, properties	Comments
Vitamin B_x (p-aminobenzoic acid) $C_7H_7NO_2$ $M_r = 137.14$	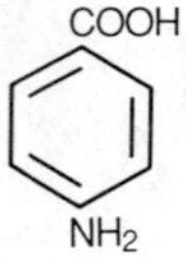	Widely distributed in nature as B complex factor; yeast is good source. Manufactured by oxidation of 4-nitrotoluene followed by reduction of nitro group. Yellow–red crystals; mp 186–7°C. Sol. in H_2O, EtOH, Et_2O. $pK_a = 4.65$	Essential metabolite for certain bacterial cells. Animal feed supplement. Used in manufacture of sunburn preventives and as a sulphonamide antagonist
Choline (β-hydroxyethyltri-methylammonium hydroxide) $C_5H_{15}NO_2$ $M_r = 121.18$	$(CH_3)_3NCH_2CH_2OH$ $\overline{\ }$ OH	Basic constituent of lecithin; occurs in all animal and vegetable tissues, less common as free base. Synthesized by reaction of 1,2-epoxyethane with trimethylamine. Strongly alkaline, very hygroscopic syrup. Sol. in H_2O, EtOH; insol. in Et_2O	Apart from methionine, only substance known to take part in methylation reactions in the body. Animal feed supplement
Biotin (vitamin H) $C_{10}H_{16}N_2O_3S$ $M_r = 244.32$	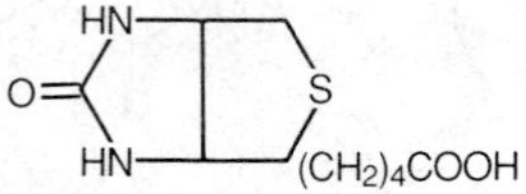	Present in living cells (mainly bound to proteins and polypeptides), yeast, egg yolk, liver. Content high in cancerous cells. Fine needles; mp 232–3°C. Slightly sol. in H_2O, EtOH. $pI = 3.5$; $[\alpha]_D^{25} = +92°$	Coenzyme for a series of enzymes, each of which catalyzes the fixation of CO_2
Folic acid (vitamin B_c, pteroyl-L-glutamic acid) $C_{19}H_{19}N_7O_6$ $M_r = 441.41$ 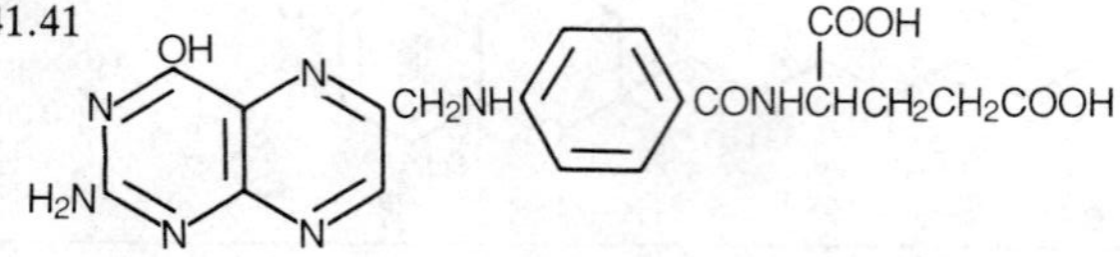		Widespread in nature as tri- and hepta-glutamyl peptides. Green leaves are a rich source. Yellow needles; decomp. 250°C. Sol. in hot H_2O, EtOH. $[\alpha]_D^{25} = +23°$	Specific growth factor for certain organisms; provided in mammals by intestinal bacteria. Deficiency causes poor growth and nutritional anaemia. Active form (tetrahydro– derivative) is coenzyme involved in purine, serine and glycine biosynthesis

Vitamin	Structural formula	Occurrence, synthesis, properties	Comments
meso-Inositol (hexahydroxycyclohexane) $C_6H_{12}O_6$ $M_r = 180.16$	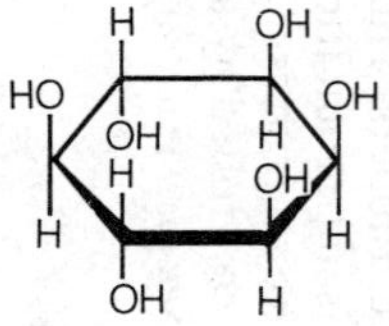	Only *meso*- (of 9 isomers) is widely distributed in yeast, plants (phytic acid). Obtained from corn steep liquor by precipitation as calcium phytate. Sweet-tasting solid; mp 225°C. V. sol. in water; slightly sol. in EtOH, insol. in Et_2O	Essential growth factor for certain microorganisms and animals; absence stops growth of hair in rats and mice
Lipoic acid (thioctic acid) $C_8H_{14}O_2S_2$ $M_r = 206.33$	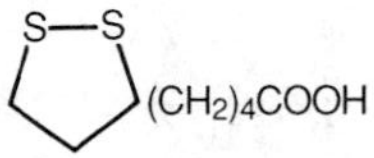	Occurs in liver and yeast. mp 47.5°C. Sparingly sol. in H_2O. $[\alpha]_D^{25} = +96.7°$	Coenzyme for the enzyme pyruvate dehydrogenase
Nicotinamide (vitamin PP) $C_6H_6N_2O$ $M_r = 122.13$	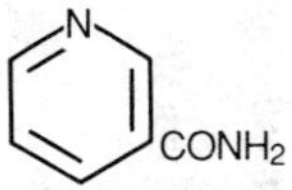	Occurs in plants and animals, usually in the enzyme systems. Synthesized from 3-nicotinic acid (oxidn of quinoline) via methyl ester. White solid; mp 129–31°C. Sol. in H_2O, EtOH	Pellagra-preventive vitamin. Part of the NAD, NADP coenzymes
Pantothenic acid $C_9H_{17}NO_5$ $M_r = 219.24$	$HOCH_2C(CH_3)_2CH(OH)CONH-$ CH_2CH_2COOH	Present in plant and animal (liver) tissues. Yellow, viscous unstable oil, v. hygroscopic. V. sol. in H_2O, Et_2O, $[\alpha]_D^{25} = +37.5°$	Chick antidermatitis factor. Required by higher organisms and some bacteria. Constituent of coenzyme A

vitamins. Substances, other than proteins, carbohydrates, fats and mineral salts, that are essential constituents in the diet of animals and man. In their absence certain deficiency diseases or other abnormal conditions develop. They are substances that play an essential part in metabolic processes, but which animals cannot synthesize. Their precise mechanism of action is poorly understood. They may be classified as water-soluble (vitamin B complex and C) or fat-soluble (vitamins A, D, E, and K) (*see* Table 1). There are several water-soluble 'factors', generally regarded as members of the vitamin B complex (*see* Table 2) which appear to be quite important in animal and plant metabolism. Their importance in human nutrition is questionable, but they serve as normal constituents of cells and play important roles in cellular metabolism. In humans, they are synthesized by the bacterial flora in the intestine; they are also present in many food sources.

vitreous state. *See* GLASS.

V_m. *See* MOLAR VOLUME.

void volume (interstitial volume). Total volume of mobile phase within the length of a chromatography column. In GAS–LIQUID CHROMATOGRAPHY, it corresponds to the retention time for air and hence to the dead volume of the column. *See also* ELUTION VOLUME.

volatile. Describing a substance (solid or liquid) that is capable of vaporizing easily. At the specified temperature, volatile substances have a high VAPOUR PRESSURE.

Volhard's method. TITRATION of silver nitrate solution with potassium thiocyanate, using iron(III) ammonium sulphate as indicator. A red coloration indicates the end-point.

volt (V; dimensions: $\varepsilon^{-1/2} \, m^{1/2} \, l^{1/2} \, t^{-1}$; units: $V = kg \, m^2 \, s^{-3} \, A^{-1} = J \, A^{-1} \, s^{-1}$). SI unit of ELECTROMOTIVE FORCE; defined as the difference of potential required to make a current of 1 AMPERE flow through a resistance of 1 ohm.

voltaic cell. *See* CELLS.

voltammetric titration. *See* TITRATIONS.

voltammetry. Electroanalytical technique in which the voltage is controlled. *See also* LINEAR SWEEP VOLTAMMETRY, POLAROGRAPHY, STRIPPING VOLTAMMETRY.

volta potential. *See* OUTER POTENTIAL.

volume (V; dimensions: l^3; units: m^3). Space occupied by a body or mass of fluid. *See also* GAS LAW, PARTIAL MOLAR VOLUME.

volume of unit cell. *See* CRYSTAL STRUCTURE.

volumetric analysis. *See* TITRATIONS.

VSEPR. *See* VALENCE SHELL ELECTRON PAIR REPULSION THEORY.

vulcanization. Use of sulphur to form cross-links between carbon–carbon double bonds in rubbers. This creates the elasticity which is important for the restoration of the structure after stress. The rate of vulcanization is controlled by accelerators and additives. *See also* POLYMERS.

W

W. *See* TUNGSTEN, WATT.

w. *See* WORK.

Wacker process. Aqueous oxidation of an ALKENE to an ALDEHYDE in the presence of a palladium(II)/copper(II) chloride catalyst, for example

$$C_2H_4 + 1/2O_2 \rightarrow CH_3CHO$$

In the reaction, palladium(II) is reduced by ethene to palladium(0) and is then reoxidized by oxygen and copper(II).

Walden inversion. Optical inversion occurring at an asymmetric carbon atom when the entry of a NUCLEOPHILE and the departure of the leaving group are synchronous, the so-called S_N2 MECHANISM. It was first observed in the conversion of (+)-hydroxybutanedioic acid to the (–)-isomer by treatment with phosphorus pentachloride to (–)-chlorobutanedioic acid and reaction of this with silver oxide.

Walden's rule. Product of the MOLAR IONIC CONDUCTIVITY at INFINITE DILUTION (Λ^0) and the viscosity (η) of the electrolyte should be a constant. As $\Lambda^0 \propto u$ (ionic mobility and $u \propto 1/\eta$ the rule should hold. For solvents in which the degree of solvation changes (thus changing the effective radius of the ion) or those in which a Grotthus mechanism (*see* GROTTHUS THEORY) operates, great deviations from the rule are found.

Walsh diagram. Graph showing the variations of the energies of MOLECULAR ORBITALS of a particular type of molecule with bond angle. The general term 'correlation diagram' is used to show how energies vary with any internal coordinate of the molecule. Inspection of such diagrams gives qualitative information about the shapes of molecules and their excited states.

warfarin (3-(α-acetonylbenzyl)-4-hydroxycoumarin). White solid (mp 161°C), prepared by the MICHAEL CONDENSATION of benzilidene propanone (propanone + PhCHO + OH⁻) with 4-hydroxycoumarin. The acidic enol forms metallic salts and an ethanoate, and the ketone forms (*see* Figure) an oxime and hydrazone. Warfarin is highly toxic, depressing the formation of prothrombin leading to haemorrhages. It is used as a rodenticide; sodium salt is used therapeutically as an anticoagulant.

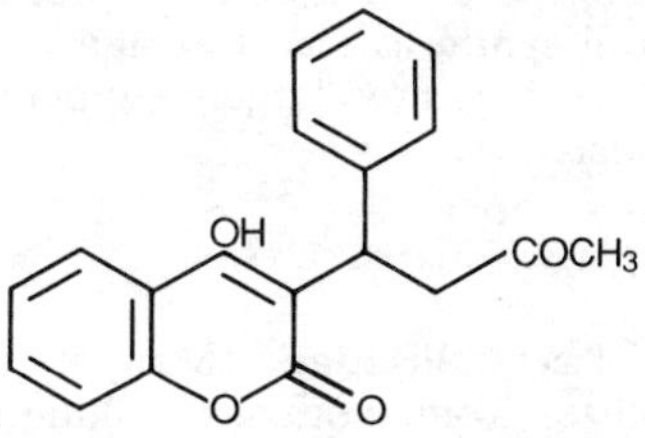

warfarin

water (H_2O). Liquid (mp 0°C, bp 100°C, $\rho = 0.999\ 87 \times 10^3$ kg m⁻³ at 0°C). Most simple oxide of hydrogen (*see also* HYDROGEN PEROXIDE). The melting and boiling points are anomolously high for such a small molecule. In the solid state (*see* ICE) and also in liquid, water forms HYDROGEN BONDS which results in long-range structure. The electrical conductivity is high for a non-metallic liquid because of the dissociation into ions

$$2H_2O \rightleftharpoons H_3O^+ + OH^-$$

$K_w = 1.00 \times 10^{-14}$ at 25°C. Water acts both as a Brønsted base and acid in the above reaction (*see* ACIDS AND BASES, AUTOPROTOLYSIS). Water has a high relative permittivity ($\varepsilon_r = 78$) and is an excellent solvent for ions or polar molecules and those to which it may form hydrogen bonds. Ionic salts may crystallize from aqueous solution with WATER OF CRYSTALLIZATION. Water is a weak oxidizing agent, reacting with electropositive metals to give basic hydroxides: for example

$$Na + H_2O \rightarrow NaOH + 1/2H_2$$

With non-metallic oxides, acidic solutions are formed (*see* HYDROLYSIS). Water is essential to life, particularly important is the supply of pure water to human communities. *See also* ELECTRODIALYSIS, HARDNESS OF WATER.

water equivalent. Mass of water that would have the same HEAT CAPACITY as a given object. It is used for the calibration of calorimeters.

water gas (blue water gas). Mixture of carbon monoxide and hydrogen in roughly equal proportions formed by passing steam through a bed of heated coke. It is used as a fuel to enrich coal gas and as a precursor in chemical synthesis (*see* FISCHER–TROPSCH PROCESS). It is now largely supplanted by SYNTHESIS GAS.

water glass. *See* SILICATES.

water of crystallization. Many metal ions crystallize from aqueous solution with molecules of water. Six molecules octahedrally coordinated or four in a square planar or tetrahedral coordination are common. Copper sulphate has five molecules of water of crystallization, four coordinated and one hydrogen bonded to the sulphate ion.

water softener. *See* HARDNESS OF WATER.

watt (W; units: kg m^2 s^{-3} = J s^{-1}). SI unit of POWER. In electrical terms the power is the product of the current and voltage (1 W = 1 V × 1 A).

wave equation. *See* SCHRÖDINGER EQUATION.

wavefunction. Function, the square of which, gives the electron density at a given point. If this function is an exact solution of the SCHRÖDINGER EQUATION, it is known as an eigenfunction (*see* EIGENVALUE). A normalized wavefunction has the property

$$\int \psi^*\psi \, d\tau = 1$$

where $\psi^*\psi \, d\tau$ is the probability of finding the electron in a region of space x, y, z to $x+dx$, $y+dy$, $z+dz$. *See also* ATOMIC ORBITAL, MOLECULAR ORBITAL.

wavelength (λ; units: m). Distance between the maxima in a wave. For ELECTROMAGNETIC RADIATION it is related to the frequency (ν) and speed of light (c) by $\lambda = c/\nu$.

wave mechanics. *See* QUANTUM MECHANICS.

wavenumber ($\bar{\nu}$; units: m^{-1}). Reciprocal of the WAVELENGTH. It is used as an energy unit ($E = h\nu = hc\bar{\nu}$) (*see* ELECTROMAGNETIC RADIATION, Table).

wax. Ester of a long-chain CARBOXYLIC ACID that is a plastic, hydrophobic and spreadable solid. Any organic compound that has these properties is also known as a wax; for example, some polyethers (*see* POLYMERS). Natural waxes are often esters of hexadecanoic acid (e.g., hexadecanyl hexadecanoate in spermaceti).

weed killer. *See* HERBICIDE.

weight (dimensions: $m\ l\ t^{-2}$; units: N). Force exerted on a body by gravitational attraction (i.e. the weight of a MASS m is m g). The term weight is often loosely used for mass.

weight average molecular mass. *See* NUMBER AVERAGE MOLECULAR MASS.

Weston cell. *See* STANDARD CELL.

wetting agent. Alternative name for a SURFACE-ACTIVE AGENT.

white gold. *See* PALLADIUM.

Wien displacement law. *See* BLACK BODY RADIATION.

Wilkinson's catalyst ($Rh(Ph_3P)_3Cl$). HYDROFORMYLATION, homogeneous catalyst (*see* HOMOGENEOUS CATALYSIS) that converts

SYNTHESIS GAS at 100°C and 30 atm. It also catalyzes oxidation by molecular oxygen and the abstraction of carbon monoxide.

willemite.   *See* SILICATES, ZINC.

window.   Band of electromagnetic wavelengths that is able to pass through a particular medium with little absorption or reflection. Wavelengths typical for window materials are: LiF, 150–3800 nm; fused silica 180–3500 nm; silicate glass 380–2000 nm; NaCl 200–15 000 nm; KBr 200–30 000 nm.

Wiswesser line notation.   Notation for expressing the structures of organic compounds particularly suited to computer processing. The set of symbols used are A–Z, 0–9, minus, slash, ampersand and space. Single atomic symbols are retained and common groups are given individual symbols (e.g., R = benzene ring, V = oxo group, Z = amine). Numerals represent the number of carbon atoms in a saturated alkyl chain.

Wittig reaction.   Reaction between a phosphorus YLIDE and an aldehyde or ketone to give an ALKENE. The exact position of the double bond is known, but the reaction is not completely stereospecific; *cis–trans* mixtures can be formed and the *trans* isomer is favoured in non-polar solvents.

Wohl degradation.   Method for descending the sugar series in which the aldose oxime is treated with ethanoic anhydride and then with silver oxide. A more recent method consists of treating the oxime with 1-fluoro-2,4-dinitrobenzene in aqueous sodium hydrogen carbonate. The yield of the lower sugar in either case is about 50 percent.

Wohler's synthesis.   Intramolecular rearrangement, on heating, of the inorganic compound ammonium cyanate (NH_4OCN) to urea (($NH_2)_2CO$).

wolframite.   *See* TUNGSTEN.

wood alcohol.   *See* METHANOL.

Wood's metal.   Low-melting (71°C) ALLOY

of bismuth (50 percent), lead (25 percent), tin (12.5 percent) and cadmium.

wood sugar (xylose).   *See* PENTOSES.

Woodward-Hoffmann rule.   Conservation of orbital symmetry in pericyclic reactions of organic molecules. Combination of p_z-orbitals occurs by overlap of the lobes of like sign (*see* ATOMIC ORBITAL). In thermal reactions, the highest occupied molecular orbitals (HOMO) (*see* LOWEST UNOCCUPIED MOLECULAR ORBITALS, LUMO) of the reactants are considered. In photochemical reactions (*see* PHOTOCHEMISTRY), because an excited state is formed, the HOMO of one reactant and the LUMO of the other are important. Reactions fall into three categories.
1) Electrocyclic reactions involving a cyclic molecule containing (n–2) π-electrons and a linear molecule of n π-electrons. If n = 4, 8, 12, etc., the thermal reaction is conrotatory (*see* CONROTATION) and the photochemical reaction disrotatory. For n = 2, 6, 10, etc., the symmetry of the reactions is vice versa.
2) Cycloaddition reactions in which two unsaturated molecules give a cyclic product with (n–4) π-electrons (e.g., DIELS–ALDER REACTION). Thermal reactions involving ($4n$+2) π-electrons and photochemical reactions of $4n$ π-electrons are generally more facile because of the SUPRAFACIAL–suprafacial interaction.
3) Sigmatropic reactions in which there is no change in the total number of electrons but a shift in a σ-bond. They are classed in terms of the number of atoms in each fragment. The stereochemistry of the reaction is given by the Woodward–Hoffmann rule. *See also* FRONTIER MOLECULAR ORBITAL THEORY.

work (w or dw; dimensions: m l^2 t^{-2}; units: J).   Work is the product of a force and a displacement:

$$dw = -F\,dl$$

where dw represents the work done on the system by the surroundings (i.e. a positive (negative) numerical value signifies that the surroundings (system) has done work on the system (surroundings)). Work is not a STATE FUNCTION, its value depends on how the process is performed (*see* Table).

work

Process	w
Isothermal irreversible expansion at constant pressure (P_2)	$-P_2\Delta V = -P_2(V_2 - V_1)$
Isothermal reversible expansion	$-nRT \ln (V_2/V_1) =$ $-nRT \ln (P_1/P_2)$
Adiabatic irreversible expansion	$-C_V\Delta T$
Adiabatic reversible expansion	$-(P_2V_2 - P_1V_1)/(1 - \gamma)$ $= -nR(T_2 - T_1)/(1 - \gamma)$
Mechanical work	$-mgh$
Electrical work	$w' = -nFE$

Maximum work (w_{max}) is obtained when the process is carried out reversibly. For an isothermal REVERSIBLE PROCESS, $w_{max} = \Delta A$ (the increase in the HELMHOLTZ FUNCTION). Net work (w') is the useful work (e.g., electrical work) that a system can provide over and above that of expansion

$$w' = w + P\Delta V = \Delta G \text{ or } dw' = dG$$

The decrease in the GIBBS FUNCTION (G) is the useful work that can be obtained from that process. For the reversible expansion of an ideal gas

$$w = -\int_{V_2}^{V_1} P\,dV$$

(i.e. the area under the P–V curve).

work function. *See* HELMHOLTZ FUNCTION.

wurtzite lattice. *See* CRYSTAL FORMS.

Wurtz–Fittig   reaction. Asymmetrical condensation of alkyl and aryl halides in the presence of sodium to give aryl–alkane hydrocarbons, for example, bromobenzene with bromoethane gives mainly ethylbenzene, but with some diphenyl and n-butane as byproducts.

Wurtz reaction. Formation of symmetrical hydrocarbons (R–R) by the condensation of alkyl halides (RX) in the presence of sodium or other metals.

$$RX + 2Na + RX \rightarrow R\text{–}R + 2NaX$$

X

χ. *See* REACTION COORDINATE.

xanthates (dithiocarbonates). Salts of the unstable acids ROC(S)SH, where R may be an alkyl or aryl group. Sodium salts are formed by treating an alkoxide with carbon disulphide and the esters by reaction of the salt with an alkyl halide. Pyrolysis of a methyl xanthate gives an alkene. The free acids are very unstable; other metallic salts are formed by double decomposition with the sodium salt, copper(I) xanthates are yellow. Cellulose xanthate is formed in the viscose process for making artifical silk. Xanthates are used in ore flotation processes.

xanthine. *See* PURINES.

xanthophylls. *See* CAROTENOIDS.

Xe. *See* XENON.

xenobiotics. Compounds that are foreign to the human body, such as drugs, pollutants, pesticides.

xenon (Xe). Most reactive noble gas (GROUP O ELEMENT) obtained by the fractionation of liquid air (8.7×10^{-6} percent of air); nine natural isotopes and 22 unstable nuclides exist. Xenon is slightly soluble in water and reacts only with fluorine. Compounds with oxidation states from +2 to +8 are known, some are exceedingly stable. The linear molecule xenon difluoride (Xe + F_2 at high pressure) is hydrolyzed to xenon and oxygen, and it forms adducts (e.g., $XeF_2.2SbF_5$). Square planar xenon tetrafluoride (Xe + F_2 at 400°C + pressure) is a powerful fluorinating agent, specifically fluorinating the ring in substituted arenes. Xenon hexafluoride (Xe + F_2 at high temperature and pressure) reacts with quartz giving the oxofluoride ($XeOF_4$) and silicon tetrafluoride

and forms complexes such as $[XeF_5]^+$ $[PtF_6]^-$ in which there is interaction between the anions and cations by fluorine bridges. The tetra- and hexafluorides are violently hydrolyzed by water to give the explosive, crystalline xenon trioxide (XeO_3). In alkali solution, XeO_3 gives $HXeO_4^-$ which slowly disproportionates to perxenate (XeO_6^{4-}) (xenon(VIII)) and xenon; yellow XeO_6^{4-} are powerful and rapid oxidizing agents. Xenon is used in fluorescent tubes and bubble chambers. $A_r = 131.29$, $Z = 54$, ES [Kr] $4d^{10} 5s^2 5p^6$, mp −111.9°C, bp −107.1°C, $\rho = 5.877$ kg m^{-3}. Isotopes: ^{124}Xe, 0.013 percent, ^{126}Xe, 0.09 percent; ^{128}Xe, 1.92 percent; ^{129}Xe, 26.44 percent; ^{130}Xe, 4.08 percent; ^{131}Xe, 21.18 percent; ^{132}Xe, 26.89 percent; ^{134}Xe, 10.4 percent; ^{136}Xe, 8.87 percent.

x_i. *See* MOLE FRACTION.

XPS (X-ray photoelectron spectroscopy). *See* ELECTRON SPECTROSCOPY.

X-ray crystallography. Use of X-ray diffraction patterns to determine crystal or molecular structures.

X-ray diffraction. Technique for the study of crystal structures. The wavelengths of X-rays are comparable in size to the distances between the atoms in a crystal lattice and the repeated pattern of the lattice acts like a diffraction grating for X-rays. Strong scattering of the rays occurs in various directions according to BRAGG'S LAW. The technique consists of directing a monochromatic beam of X-rays at the sample and recording the diffracted rays on a photographic plate or electronically. The diffraction pattern consists of a pattern of spots on the plate, and the crystal structure can be established from the positions and intensities of the spots. Various techniques are available for the application to the study of

519

single crystals, powders, fibres, etc., and by computational analysis three-dimensional electron density maps of the solid lattice can be built up.

X-ray fluorescence. *See* X-RAY SPECTRO-METRY.

X-ray photoelectron spectroscopy. *See* ELECTRON SPECTROSCOPY.

X-rays. *See* ELECTROMAGNETIC SPECTRUM.

X-ray spectrometry (X-ray fluorescence analysis, XRF). Spectrophotometric method of analysis in which a sample is irradiated with X-rays (*see* ELECTRO-MAGNETIC SPECTRUM). Core electrons are ejected (*see* ELECTRON SPECTROSCOPY) and replaced as outer electrons fall back with the emission of a characteristic X-ray photon. The process is most efficient for elements of high atomic number ($Z > 24$). The practical lower limit is $Z = 12$, when absorption by air requires a vacuum or helium atmosphere and when the AUGER EFFECT becomes important. XRF is used widely in the identification of TRANSITION ELEMENTS in geological samples.

XRF analysis (X-ray fluorescence analysis). *See* X-RAY SPECTROMETRY.

xylan. Strongly laevorotatory (*see* OPTI-CAL ACTIVITY) HEMICELLULOSE which occurs with cellulose in the wood of deciduous trees, bran and straw. Purified xylan is composed of 18–20 xylopyranose units joined by 1,4-β-linkages with a terminal arabinofuranosyl residue.

xylenes (dimethylbenzenes, $C_6H_4(CH_3)_2$). Three isomers, 1,2, 1,3 and 1,4-dimethyl-benzenes (isomeric with ETHYLBENZENE), present in the light oil fraction of coal tar, are prepared commercially by the catalytic HYDROFORMING of n-octane. Since the isomers have similar boiling points (144, 139.3 and 137–8°C) separation by distillation is difficult; separation is best achieved by gas chromatography. Oxidation (dil. HNO_3 or $KMnO_4$) gives first a toluic acid ($CH_3C_6H_4COOH$) and finally a BENZENE-DICARBOXYLIC ACID ($C_6H_4(COOH)_2$). The 2-isomer is used in the manufacture of PHTHALIC ANHYDRIDE and the 3- and 4-isomers in the manufacture of isophthalic acid (1,3-benzenedicarboxylic acid) and terephthalic acid (1,4-benzenedicarboxylic acid), respectively.

xylenols. *See* HYDROXYDIMETHYLBENZ-ENES.

xylitol. *See* SWEETENING AGENTS.

xylose. *See* PENTOSES.

Y

Y. *See* YTTRIUM.

Y. *See* PLANCK FUNCTION.

Yb. Ytterbium (*see* LANTHANIDE ELEMENTS).

yield value. *See* NEWTONIAN FLUID.

ylides. Internal salts in which the hetero atom (nitrogen, phosphorus, sulphur, arsenic) is formally in the cationic form. Phosphorus ylides are generated by the reaction of an appropriate salt with a strong base (NaH + DMSO)

$$CH_3I + Ph_3P \rightarrow Ph_3P^+ \; CH_3I^- \rightarrow$$

$$Ph_3P=CH_2 \rightleftharpoons Ph_3P^+ \; {}^-CH_2 \text{ (ylide)}$$

They readily condense with aldehydes and ketones (*see* WITTIG REACTION) giving an alkene

$$Ph_3P=CH_2 + RR'C=O \rightarrow$$

$$RR'C=CH_2 + Ph_3P=O$$

yohimbine. *See* ALKALOIDS.

ytterbium (Yb). *See* LANTHANIDE ELEMENTS.

yttrium (Y). Group IIIb, second row TRANSITION ELEMENT. It is extracted from the silicate ore gadolinite. Chemistry is of the +3 oxidation state. It forms an oxide (Y_2O_3), which with europium as activator is the red PHOSPHOR in cathode ray tubes. Yttrium iron garnets ($Y_3Fe_5O_{12}$) are used in radar and other short-wave devices. Yttrium aluminium garnets (YAG) may be substituted for diamonds. A_r = 88.9059, Z = 39, ES [Kr] $4d^1 \, 5s^2$, mp 1523°C, bp 3337°C, ρ = 4.47 × 10^3 kg m^{-3}. Isotope: ^{89}Y, 100 percent.

Z

Z. *See* ATOMIC NUMBER, COLLISION FREQUENCY, COMPRESSION FACTOR.

z. *See* CHARGE NUMBER, VALENCE.

ζ-potential. *See* ZETA-POTENTIAL.

Z-. Stereochemical descriptor (*see* ISOMERISM).

zeaxanthin. *See* CAROTENOIDS.

Zeeman effect. Splitting of rotational lines in a magnetic field. The shift in energy for the rotational transition $J=0$ to $J=1$ in a magnetic field of B teslas is $\Delta E = \beta m_l B$, where m_l is the azimuthal quantum number and β the BOHR MAGNETON. The anomalous Zeeman effect arises from the off-axis alignment of the total angular momentum and total magnetic moment when $\Delta E = g\beta M_J B$. g is the LANDÉ g-FACTOR and M_J the quantum number of the $(2J+1)$ states of a given electronic configuration (*see* TERM SYMBOLS).

zeolite. Crystalline metal aluminosilicate (*see* SILICATES), characterized by a structure having channels and cavities of molecular dimensions. The general formula for a zeolite is $M_{2/n}O:Al_2O_3:xSiO_2:yH_2O$ where M^{n+} is a metal ion (ALKALI METAL or ALKALINE EARTH METAL), and $x=2.0$ for type A zeolites and $x = 2.5$ for type X. The structure is a three-dimensional framework of SiO_2 and AlO_4 tetrahedra cross-linked by oxygen atoms such that the ratio Si/Al = 2 (*see* Figure). Electrical neutrality is maintained by the metal cations. The pore size, typically 0.8–1.1 nm, is then determined by the size and position of the cation, which may be exchanged for other ions (*see* CHROMATOGRAPHY, HARDNESS OF WATER, ION EXCHANGE CHROMATOGRAPHY). Zeolites

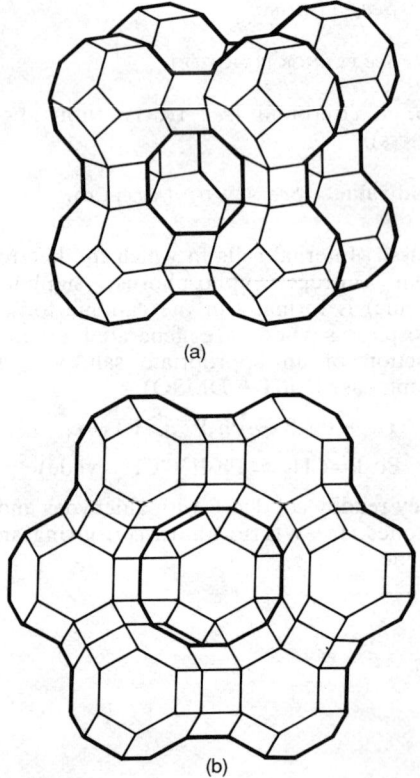

(a)

(b)

zeolite
Structures of two zerolites showing connection of silicon and aluminium: (a) zeolite A, (b) Faiyasite.

are also known as molecular sieves after their ability to selectively adsorb molecules. For example, linear hydrocarbons may be allowed into a channel, whereas branched hydrocarbons are excluded. They have a high coulomb attraction for molecules and thus high heats of adsorption. In particular, water is strongly adsorbed by some zeolites (e.g., 5A), making them useful as drying agents.

Zeolites are also used as catalysts or catalyst supports. Type Y zeolites are petroleum CRACKING catalysts.

zeotropic mixture. COMPLETELY MISCIBLE LIQUID SYSTEM that can be separated into the two pure components by distillation (e.g., benzene/toluene). *See also* IDEAL SOLUTION.

zero-order reaction. *See* ENZYME KINETICS, ORDER OF REACTION.

zero-point energy (ε_0). Residual vibrational energy of a molecule in its GROUND STATE. For a HARMONIC OSCILLATOR $\varepsilon_0 = h\nu/2$.

zeroth law of thermodynamics. Any two bodies or systems in thermal equilibrium with a third body or system are in thermal equilibrium with each other. There is some property, the TEMPERATURE, whose value is the same for all systems in thermal equilibrium. A thermometer inserted in a system indicates its own temperature, which is presumed to be that of the system with which it is in equilibrium.

zeta-potential (ζ). Potential difference, measured in the liquid, between the shear plane (i.e. the outer limit of that part of the double layer which is fixed to the phase moving relative to the liquid) and the bulk of the liquid beyond the limits of the ELECTRICAL DOUBLE LAYER. The potential cannot be measured directly, but can be calculated from measurements of the electrophoretic mobility (*see* ELECTROPHORESIS) of particles using the Smoluchowski equation

$$u = \varepsilon_0 \varepsilon_r \zeta / \eta$$

The SURFACE CHARGE may be calculated from the zeta-potential. In a 1:1 electrolyte

$$\sigma_D = (N_A \varepsilon k T / 500\pi)^{1/2} c^{1/2} \sinh ze\zeta/kT$$

Ziegler–Natta catalyst. Heterogeneous catalyst formed by the reaction of a transition metal halide (e.g., $TiCl_4$, VCl_3) with organometallic compounds (e.g., R_3Al, R_2AlCl, R_2Zn) in a hydrocarbon solvent. These catalysts give syntactic polymers (*see* TACTICITY) from unsaturated monomers (e.g., polypropene (*see* POLYMERIZATION)).

zinc (Zn). Group IIb metal (*see* GROUP II ELEMENTS). It is found as zinc blende or sphalerite (ZnS), calamine ($ZnCO_3$ in Europe, $ZnSiO_4$ in USA), smithsonite ($ZnCO_3$) and willemite (Zn_2SiO_4). It is extracted from the oxide by roasting with carbon. Zinc is a bluish-white metal with good thermal and electrical conductivity. It reacts with the air, steam, halogens and sulphur at high temperatures, and dissolves in dilute acids and hot alkalis. Zinc is used in alloys (e.g., BRASS) and as a protective coating of other metals (*see* GALVANIZING). The reducing properties of zinc are used in some chemical syntheses. Some zinc compounds (e.g., willemite) are phosphorescent and so find use in TV tubes, X-ray screens, etc. Zinc is a necessary trace element in biological systems. It is used in pharmaceutical preparations of insulin to extend the duration of action.
A_r = 65.39, Z = 30, ES [Ar] $3d^{10} 4s^2$, mp 420°C, bp 907°C, ρ = 7.14 × 10^3 kg m^{-3}. Isotopes: ^{64}Zn, 48.9 percent; ^{66}Zn, 27.8 percent; ^{67}Zn, 4.1 percent; ^{68}Zn, 18.6 percent; ^{70}Zn, 0.6 percent.

zinc blende. *See* ZINC.

zinc blende lattice. *See* CRYSTAL FORMS.

zinc compounds. Zinc is an ELECTROPOSITIVE metal that forms ionic salts containing Zn^{2+} ($E^{\ominus}_{Zn^{2+},Zn}$ = −0.76 V). Hydrated Zn^{2+} is generally six-coordinate and in general zinc forms six- or four-coordinate complexes with oxygen or nitrogen ligands.

Zinc oxide has a variety of uses: as a medical ointment, in rubber technology, in semiconductors and as a mould growth inhibitor and disinfectant. Zinc hydroxide is precipitated by alkalis from solutions of Zn^{2+} and then redissolves to zincates which are predominantly $[Zn(OH)_4]^{2-}$ and $[Zn(OH)_3.H_2O]^-$. Zinc sulphate (white vitriol, $ZnSO_4$) is used in the textile industry and in agricultural arsenical sprays. Zinc halides (ZnX_2) are known and find uses in medicines, glues, cements and fluxes. Zinc sulphide (ZnS) is found in two crystal phases, the zinc blende structure and the wurtzite structure (*see* CRYSTAL FORMS). It is also used as a pigment (*see* LITHOPONE).

Organozinc compounds were historically among the first organometallics studied by

Franklin. Grignard-like alkyls (RZnX) (RX + Zn) and dialkyls (R$_2$Zn) (heat on RZnX) show relatively mild reactivity toward some organic functional groups. R$_2$Zn reacts with other GROUP II ELEMENTS (M) to give MZnR$_4$. Organozinc compounds are used with titanium alkyls (*see* ZIEGLER–NATTA CATALYST) to catalyze the polymerization of alkenes. *See also* REFORMATSKY REACTION.

zingiberene. *See* TERPENES.

zircon. *See* ZIRCONIUM.

zirconium (Zr). Group IVb, second row TRANSITION ELEMENT. It is found as the ores zircon (ZrO$_2$.SiO$_2$) and baddeleyite (ZrO$_2$), always in conjunction with HAFNIUM. Uses of zirconium compounds are in the ceramic industry to improve corrosion resistance. The metal has a low neutron cross-section (0.18 barn) and is used as cladding in nuclear reactors. Zirconium is extracted by the KROLL PROCESS and purified by the van Arkel–de Boer process (Zr + I$_2$, the volatile ZrI$_4$ decomposing on a heated wire). Zirconium is a lustrous, silvery metal which is unstable with respect to all non-metals. In air, the metal is protected by a thin layer of oxide. Compounds formed by the reaction with nitrogen, boron and carbon are used as high-temperature refractories in non-oxidizing environments.
A_r = 91.224, Z = 40, ES [Kr] $4d^2 5s^2$, mp 1852°C, bp 4377°C, ρ = 6.51 × 10^3 kg m^{-3}. Isotopes: ^{90}Zr, 51.46 percent; ^{91}Zr, 11.23 percent; ^{92}Zr, 17.11 percent; ^{94}Zr, 17.40 percent; ^{96}Zr, 2.80 percent.

zirconium compounds. Compounds are mostly in the +4 oxidation state. In aqueous solution, polymeric species are formed

(e.g., [Zr$_3$(OH)$_4$]$^{8+}$). Monatomic Zr^{4+} is not known, the ion (4ZrO$_2$.16H$_2$O.nH$^+$)$^{n+}$ exists in some halides and [ZrO(SO$_4$)$_2$]$^{2-}$ in sulphates. Lower oxidation states are found in the halides (e.g., ZrCl$_3$, ZrCl$_2$ and ZrCl), which are stable only in the solid state. Complexes are readily formed showing high coordination numbers (e.g., [ZrF$_8$]$^{4-}$ is a square antiprism, [ZrF$_7$]$^{3-}$ is pentagonal bipyramidal). Organometallics are known: alkoxides (ZrCl$_4$ + alcohol in ammonia), SANDWICH COMPOUNDS (zircocene dichloride), σ-bonded compounds (tetrabenzene zirconium) (*see* CARBON-TO-METAL BONDS).

Zn. *See* ZINC.

zone chromatography. *See* CHROMATOGRAPHY.

zone electrophoresis. *See* ELECTROPHORESIS.

zone refining. Method of purifying solids by moving a small melted region through a column of the material. This is done by drawing the solid through a furnace or by differential heating within a furnace. Impurities are concentrated in the liquid phase and so are eventually removed from the material. It is used for high-melting metals such as tungsten.

Zr. *See* ZIRCONIUM.

zwitterion. Molecule with both positive and negatively charged groups that is overall electrically neutral. Many AMINO ACIDS have zwitterionic forms with ammonium –NH$_3^+$ and carboxylate –COO$^-$ ions (e.g., glycine $^+$H$_3$N–CH$_2$–COO$^-$). *See also* ISOELECTRIC POINT.

Appendices

Appendix Table 1
Fundamental physical constants in SI units.

Physical quantity	Value	SI unit
Atomic mass unit	$1.660\,565\,5 \times 10^{-27}$	kg
Avogadro constant	$6.022\,045 \times 10^{23}$	mol^{-1}
Bohr magneton	$9.274\,078 \times 10^{-24}$	$J\,T^{-1}$
Bohr radius	$5.291\,770\,6 \times 10^{-11}$	m
Boltzmann constant	$1.380\,662 \times 10^{-23}$	$J\,K^{-1}$
Charge to mass ratio for electron	$1.758\,804\,7 \times 10^{11}$	$C\,kg^{-1}$
Curie	3.7×10^{10}	disintegrations s^{-1}
Einstein constant	$8.987\,551\,79 \times 10^{16}$	$J\,kg^{-1}$
Electron rest mass	$9.109\,534 \times 10^{-31}$	kg
Electron volt	$1.602\,189\,2 \times 10^{-19}$	J
Electronic charge	$1.602\,189\,2 \times 10^{-19}$	C
Faraday constant	$9.648\,456 \times 10^4$	$C\,mol^{-1}$
Gas constant	$8.314\,41$	$J\,mol^{-1}\,K^{-1}$
Gravitational acceleration	$9.806\,65$	$m\,s^{-2}$
Ice-point temperature	273.15	K
Landé g-factor for free electron	$2.002\,319\,313\,4$	—
Loschmidt constant	$2.686\,754 \times 10^{25}$	m^{-3}
Molar volume of ideal gas at stp	$0.022\,413\,83$	$m^3\,mol^{-1}$
Normal atmosphere	$1.013\,25 \times 10^5$	$N\,m^{-2}$
Nuclear magneton	$5.050\,824 \times 10^{-27}$	$J\,T^{-1}$
Permeability of vacuum	$4\pi \times 10^{-7}$	$J\,s^2\,C^{-2}\,m^{-1}$ $(H\,m^{-1})$
Permittivity of vacuum	$8.854\,187\,82 \times 10^{-12}$	$J^{-1}\,C^2\,m^{-1}\,(F\,m^{-1})$
Planck constant	$6.626\,176 \times 10^{-34}$	$J\,s$
Proton rest mass	$1.672\,648\,5 \times 10^{-27}$	kg
Rydberg constant	$1.097\,373\,177 \times 10^7$	m^{-1}
Speed of light in vacuum	$2.997\,924\,58 \times 10^8$	$m\,s^{-1}$
Standard pressure	$1.013\,25 \times 10^5$	$N\,m^{-2}$
Standard temperature	273.15	K
Stefan-Boltzmann constant	$5.670\,32 \times 10^{-8}$	$W\,m^{-2}\,K^{-4}$
Triple point of water	273.15	K

Appendix Table 2
Special names and symbols for certain derived SI units.

Physical quantity	Name of SI unit	Symbol for SI unit
Capacitance	farad	F
Electric charge	coulomb	C
Electrical conductance	siemens	S
Electric potential	volt	V
Electric resistance	ohm	Ω
Energy	joule	J
Force	newton	N
Frequency	hertz	Hz
Illumination	lux	lx
Inductance	henry	H
Luminous flux	lumen	lm
Magnetic flux	weber	Wb
Magnetic flux density	tesla	T
Power	watt	W
Pressure	pascal	Pa

Appendix Table 3

SI conversion tables. Factor relating a physical quantity in SI units to the value in cgs units (or other commonly used units).
Value in SI units = value in cgs units × conversion factor.

Physical quantity (symbol)	SI unit	Definition of SI unit	cgs unit	Conversion factor	Dimensions
CONCENTRATION UNITS					
Concentration (c)	mol m^{-3}		mol l^{-1} (M)	1.0×10^3	l^{-3}
	mol dm^{-3}		mol l^{-1} (M)	1.0	l^{-3}
Ionic strength (I)	mol m^{-3}		mol l^{-1} (M)	1.0×10^3	l^{-3}
	mol dm^{-3}		mol l^{-1} (M)	1.0	l^{-3}
Molality (m)	mol kg^{-1}		mol kg^{-1}	1.0	m^{-1}
Mole fraction (x_i)			Dimensionless qty	1.0	1
ELECTRICITY AND MAGNETISM					
Capacitance (C)	F	$A^2\,s^4\,kg^{-1}\,m^{-2} = A\,s\,V^{-1} = C\,V^{-1}$	esu^2 erg^{-1}	1.112×10^{-12}	$\varepsilon\, l$
Conductance (G)	Ω^{-1}, S	$kg^{-1}\,m^{-2}\,s^3\,A^2 = A\,V^{-1}$	ohm^{-1}, mho	1.0	$\varepsilon\, l\, t^{-1}$
Conductivity (κ)	Ω^{-1} m^{-1}, S m^{-1}	$A\,V^{-1}\,m^{-1}$	ohm^{-1} cm^{-1}	1.0×10^2	$\varepsilon\, t^{-1}$
Electric charge, quantity of electricity (Q)	C	$A\,s$	esu	3.335×10^{-10}	$\varepsilon^{1/2}\, m^{1/2}\, l^{3/2}\, t^{-1}$
			emu	10.0	$\mu^{-1/2}\, m^{1/2}\, l^{1/2}$
Electric current (I)	A		esu sec^{-1}	3.335×10^{-10}	$\varepsilon^{1/2}\, m^{1/2}\, l^{3/2}\, t^{-2}$
			emu sec^{-1}	10.0	$\mu^{-1/2}\, m^{1/2}\, l^{1/2}\, t^{-1}$
Electric dipole moment (p_e)	C m	$A\,s\,m$	esu cm	3.335×10^{-12}	$\varepsilon^{1/2}\, m^{1/2}\, l^{5/2}\, t^{-1}$
Electric displacement (D)	C m^{-2}		¼π esu cm^{-2}	2.654×10^{-7}	$\varepsilon^{1/2}\, m^{1/2}\, l^{-1/2}\, t^{-1}$
Electric field strength (E)	V m^{-1}		dyne esu^{-1}	2.998×10^4	$\varepsilon^{-1/2}\, m^{1/2}\, l^{-1/2}\, t^{-1}$
Electric mobility of ion i (u_i)	m^2 V^{-1} s^{-1}		cm^2 V^{-1} sec^{-1}	1.0×10^{-4}	$\varepsilon^{1/2}\, m^{-1/2}\, l^{3/2}$
Electric polarization (P)	C m^{-2}		esu cm^{-2}	3.335×10^{-6}	$\varepsilon^{1/2}\, m^{1/2}\, l^{-1/2}\, t^{-1}$
Electric potential (V)	V	$kg\,m^2\,s^{-3}\,A^{-1} = J\,A^{-1}\,s^{-1}$	erg esu^{-1}	2.998×10^2	$\varepsilon^{-1/2}\, m^{1/2}\, l^{1/2}\, t^{-1}$
			erg emu^{-1}	1.0×10^{-8}	$\mu^{1/2}\, m^{1/2}\, l^{3/2}\, t^{-2}$
Inductance (L)	H	$kg\,m^2\,s^{-2}\,A^{-2} = V\,A^{-1}\,s$	emu of inductance	1.0×10^{-9}	$\mu\, l$
Magnetic field strength (H)	A m^{-1}		oersted (Oe)	79.578	$\mu^{-1/2}\, m^{1/2}\, l^{-1/2}\, t^{-1}$
Magnetic flux (ϕ)	Wb	$kg\,m^2\,s^{-2}\,A^{-1} = V\,s$	maxwell (Mx)	1.0×10^{-8}	$\varepsilon^{-1/2}\, m^{1/2}\, l^{1/2}$
Magnetic flux density (B)	T	$Wb\,m^{-2} = kg\,s^{-2}\,A^{-1} = V\,s\,m^{-2}$	gauss (G)	1.0×10^{-4}	$\varepsilon^{-1/2}\, m^{1/2}\, l^{-3/2}$

continued overleaf

Physical quantity (symbol)	SI unit	Definition of SI unit	cgs unit	Conversion factor	Dimensions
Magnetic susceptibility (χ_m)			Dimensionless qty	4π	1
Molar conductivity (Λ)	$\Omega^{-1}\,m^2\,mol^{-1}$		ohm^{-1} cm^2 mol^{-1}	1.0×10^{-4}	$\varepsilon\,l^2\,t^{-1}$
Permeability (μ)	$H\,m^{-1}$	$kg\,m\,s^{-2}\,A^{-2}$	Dimensionless qty	$4\pi \times 10^{-7} = 1.257 \times 10^{-6}$	μ
Permittivity (ε)	$F\,m^{-1}$	$C\,V^{-1}\,m^{-1}$	Dimensionless qty	8.854×10^{-12}	ε
Poynting vector (S)	$W\,m^{-2}$		erg cm^{-2} sec^{-1}	1.0×10^{-3}	$\varepsilon^{-1/2}\,m^{1/2}\,l^{-3/2}$
Relative permeability (μ_r)			Dimensionless qty	1.0	1
Relative permittivity (ε_r)			Dimensionless qty	1.0	1
Resistance (R)	Ω	$kg\,m^2\,s^{-3}\,A^{-2} = V\,A^{-1}$	ohm	1.0	$\varepsilon^{-1}\,l^{-1}\,t$
Resistivity (ρ)	$\Omega\,m$	$kg\,m^3\,s^{-3}\,A^{-2} = V\,A^{-1}\,m$	ohm cm	1.0×10^{-2}	$\varepsilon^{-1}\,t$
Surface charge density (σ)	$C\,m^{-2}$		esu cm^{-2}	3.335×10^{-6}	$\varepsilon^{1/2}\,m^{1/2}\,l^{-1/2}\,t^{-1}$
Velocity of ion i (v_i)	$m\,s^{-1}$		cm sec^{-1}	1.0×10^{-2}	$l\,t^{-1}$
MECHANICS					
Density (ρ)	$kg\,m^{-3}$	$kg\,m^{-3}$	$g\,ml^{-1}$, $g\,cm^{-3}$	1.0×10^{3}	$m\,l^{-3}$
Dynamic viscosity (η)	$N\,s\,m^{-2}$	$kg\,m^{-1}\,s^{-1} = J\,m^{-3}\,s$	dyne sec cm^{-2}	0.1	$m\,l^{-1}\,t^{-1}$
			poise (P)	0.1	$m\,l^{-1}\,t^{-1}$
Energy (E)	J	$kg\,m^2\,s^{-2}$	erg	1.0×10^{-7}	$m\,l^2\,t^{-2}$
			cal	4.184	$m\,l^2\,t^{-2}$
Force (F)	N	$kg\,m\,s^{-2} = J\,m^{-1}$	dyne	1.0×10^{-5}	$m\,l\,t^{-2}$
Kinematic viscosity (ν)	$m^2\,s^{-1}$		Stokes (St)	1.0×10^{-4}	$l^2\,t^{-1}$
Mass (m)	kg		g	1.0×10^{-3}	m
Power (P)	W	$kg\,m^2\,s^{-3} = J\,s^{-1}$	erg sec^{-1}	1.0×10^{-7}	$m\,l^2\,t^{-3}$
Pressure (p, P)	$N\,m^{-2}$	$kg\,m^{-1}\,s^{-2} = J\,m^{-3}$	dyne cm^{-2}	0.1	$m\,l^{-1}\,t^{-2}$
Specific volume (V)	$m^3\,kg^{-1}$		cm^3 g^{-1}	1.0×10^{-3}	$l^3\,m^{-1}$
Surface tension (γ)	$N\,m^{-1}$	$kg\,s^{-2} = J\,m^{-2}$	dyne cm^{-1}	1.0×10^{-3}	$m\,t^{-2}$
Work (w)	J	$kg\,m^2\,s^{-2}$	erg	1.0×10^{-7}	$m\,l^2\,t^{-2}$
			cal	4.184	$m\,l^2\,t^{-2}$

RADIATION, SPECTROPHOTOMETRIC AND OTHER OPTICAL QUANTITIES

Physical quantity (symbol)	SI unit	Definition of SI unit	cgs unit	Conversion factor	Dimensions
Angle of rotation (α)	rad		degree	$\pi/180 =$ 1.745×10^{-2}	
Decadic absorbance, extinction (A)			Dimensionless qty	1.0	
Illumination (E)	lx	$\text{lm m}^{-2} = \text{cd sr m}^{-2}$	lm cm^{-2}	1.0×10^{4}	
Luminance (L)	cd m^{-2}	cd sr	cd cm^{-2} (stilb)	1.0×10^{4}	
Luminous flux (ϕ)	lm		lm	1.0	
Molar extinction coefficient (ε)	$\text{m}^2\,\text{mol}^{-1}$		$\text{l mol}^{-1}\,\text{cm}^{-1}$	0.1	l^2
Molar optical rotatory power (α_n)	$\text{rad m}^2\,\text{mol}^{-1}$		$\text{deg dm}^{-1}\,\text{l mol}^{-1}$	1.745×10^{-2}	l^2
Molar refraction (R_m)	$\text{m}^3\,\text{mol}^{-1}$		$\text{cm}^3\,\text{mol}^{-1}$	1.0×10^{-6}	l^3
Refraction (R)	m^3		cm^3	1.0×10^{-6}	l^3
Refractive index (n)			Dimensionless qty	1.0	
Specific optical rotatory power (α_m)	$\text{rad m}^2\,\text{kg}^{-1}$		$\text{deg dm}^{-1}\,\text{cm}^3\,\text{g}^{-1}$	1.745×10^{-4}	$l^2\,m^{-1}$

SPACE AND TIME

Physical quantity (symbol)	SI unit	Definition of SI unit	cgs unit	Conversion factor	Dimensions
Acceleration (a)	m s^{-2}		cm sec^{-2}	1.0×10^{-2}	$l\,t^{-2}$
Angular frequency (ω)	Hz	s^{-1}	sec^{-1}	1.0	t^{-1}
Area (A, S)	m^2		cm^2	1.0×10^{-4}	l^2
Frequency (v, f)	Hz	s^{-1}	sec^{-1}	1.0	t^{-1}
Length (l)	m		cm	1.0×10^{-2}	l
			Ångström (Å)	1.0×10^{-10}	
Molar volume (V)	m^3		$\text{cm}^3\,\text{mol}^{-1}$	1.0×10^{-6}	l^3
Period (T)	s		sec	1.0	t
Relaxation time (τ)	s		sec	1.0	t
Rotational frequency	s^{-1}		sec^{-1}	1.0	t^{-1}
Velocity (v)	m s^{-1}		cm sec^{-1}	1.0×10^{-2}	$l\,t^{-1}$
Volume (V)	m^3		cm^3 (cc, ml)	1.0×10^{-6}	l^3
Wavelength (λ)	m		cm	1.0×10^{-2}	l
Wavenumber ($\bar{v}$)	m^{-1}		cm^{-1}	1.0×10^{2}	l^{-1}

continued overleaf

Physical quantity (symbol)	SI unit	Definition of SI unit	cgs unit	Conversion factor	Dimensions
THERMODYNAMICS					
Enthalpy[a] (H)	J		cal	4.184	$m\,l^2\,t^{-2}$
Entropy[a] (S)	$J\,K^{-1}$	$kg\,m^2\,s^{-2}\,K^{-1}$	cal deg^{-1}	4.184	$m\,l^2\,t^{-2}\,\theta^{-1}$
Gibbs function[a] (G)	J		cal	4.184	$m\,l^2\,t^{-2}$
Heat capacity (C_P, C_V)	$J\,K^{-1}$	$kg\,m^2\,s^{-2}\,K^{-1}$	cal deg^{-1}	4.184	$m\,l^2\,t^{-2}\,\theta^{-1}$
Helmholtz function[a] (A)	J		cal	4.184	$m\,l^2\,t^{-2}$
Internal energy[a] (U)	J	$kg\,m^2\,s^{-2}$	cal	4.184	$m\,l^2\,t^{-2}$
Joule–Thompson coefficient (μ)	$N^{-1}\,m^2\,K$	$kg^{-1}\,m\,s^2\,K = J^{-1}\,m^3\,K$	deg atm^{-1}	1/101 325	$m^{-1}\,l\,t^2\,\theta$
Molar entropy (S_m)	$J\,K^{-1}\,mol^{-1}$	$kg\,m^2\,s^{-2}\,mol^{-1}\,K^{-1}$	cal mol^{-1} deg^{-1}	4.184	$m\,l^2\,t^{-2}\,\theta^{-1}$
Quantity of heat (q)	J	$kg\,m^2\,s^{-2}$	cal	4.184	$m\,l^2\,t^{-2}$
Specific heat capacity (c_P, c_V)	$J\,kg^{-1}\,K^{-1}$	$m^2\,s^{-2}\,K^{-1}$	cal g^{-1} deg^{-1}	4.184×10^3	$l^2\,t^{-2}\,\theta^{-1}$
Thermal conductivity (λ)	$W\,m^{-1}\,K^{-1}$	$J\,s^{-1}\,m^{-1}\,K^{-1}$	cal sec^{-1} cm^{-1} deg^{-1}	4.184×10^2	$m\,l\,t^{-3}\,\theta^{-1}$

[a] The corresponding specific quantity is denoted by corresponding lower case letter; units: add kg^{-1}; conversion factor: 4.184×10^3. The corresponding molar quantity is denoted by subscript m (e.g., S_m, G_m); units: add mol^{-1} (see S_m); conversion factor: 4.184.

Appendix Table 4

Atomic masses of the elements

Name	Symbol	Atomic number	Atomic mass	Name	Symbol	Atomic number	Atomic mass
Actinium	Ac	89	227.028	Molybdenum	Mo	42	95.94
Aluminium	Al	13	26.981 54	Neodymium	Nd	60	144.24
Americium	Am	95	243.0	Neon	Ne	10	20.179
Antimony (stibium)	Sb	51	121.75	Neptunium	Np	93	237.0482
Argon	Ar	18	39.948	Nickel	Ni	28	58.69
Arsenic	As	33	74.9216	Niobium	Nb	41	92.9064
Astatine	At	85	210.0	Nitrogen	N	7	14.0067
Barium	Ba	56	137.33	Nobelium	No	102	259.0
Berkelium	Bk	97	247.0	Osmium	Os	76	190.2
Beryllium	Be	4	9.012 18	Oxygen	O	8	15.9994
Bismuth	Bi	83	208.9804	Palladium	Pd	46	106.42
Boron	B	5	10.811	Phosphorus	P	15	30.973 76
Bromine	Br	35	79.904	Platinum	Pt	78	195.08
Cadmium	Cd	48	112.41	Plutonium	Pu	94	239.13
Caesium	Cs	55	132.9054	Polonium	Po	84	209.0
Calcium	Ca	20	40.078	Potassium (kalium)	K	19	39.0983
Californium	Cf	98	251.0	Praseodymium	Pr	59	140.9077
Carbon	C	6	12.011	Promethium	Pm	61	145.0
Cerium	Ce	58	140.12	Protactinium	Pa	91	231.036
Chlorine	Cl	17	35.453	Radium	Ra	88	226.025
Chromium	Cr	24	51.9961	Radon	Rn	86	222.0
Cobalt	Co	27	58.9332	Rhenium	Re	75	186.207
Copper	Cu	29	63.546	Rhodium	Rh	45	102.9055
Curium	Cm	96	247.0	Rubidium	Rb	37	85.4678
Dysprosium	Dy	66	162.50	Ruthenium	Ru	44	101.07
Einsteinium	Es	99	252.0	Samarium	Sm	62	150.36
Element 108	—	108	—	Scandium	Sc	21	44.955 91
Erbium	Er	68	167.26	Selenium	Se	34	78.96
Europium	Eu	63	151.96	Silicon	Si	14	28.0855
Fermium	Fm	100	257.0	Silver (argentum)	Ag	47	107.8682
Fluorine	F	9	18.998 403	Sodium (natrium)	Na	11	22.989 77
Francium	Fr	87	223.0	Strontium	Sr	38	87.62
Gadolinium	Gd	64	157.25	Sulphur	S	16	32.066
Gallium	Ga	31	69.723	Tantalum	Ta	73	180.9479
Germanium	Ge	32	72.59	Technetium	Tc	43	98.9062
Gold (aurum)	Au	79	196.9665	Tellurium	Te	52	127.60
Hafnium	Hf	72	178.49	Terbium	Tb	65	158.9254
Helium	He	2	4.002 602	Thallium	Tl	81	204.383
Holmium	Ho	67	164.9304	Thorium	Th	90	232.0381
Hydrogen	H	1	1.007 94	Thulium	Tm	69	168.9342
Indium	In	49	114.82	Tin (stannum)	Sn	50	118.710
Iodine	I	53	126.9045	Titanium	Ti	22	47.88
Iridium	Ir	77	192.22	Tungsten (wolfram)	W	74	183.85
Iron (ferrum)	Fe	26	55.847	Unnilqualium	Unq[a]	104	261.0
Krypton	Kr	36	83.80	Unnilpentium	Unp[a]	105	262.0
Lanthanum	La	57	138.9055	Unnilhexium	Unh[a]	106	263.0
Lawrencium	Lr	103	260.0	Unnilseptium	Uns[a]	107	262.0
Lead (plumbum)	Pb	82	207.2	Uranium	U	92	238.0289
Lithium	Li	3	6.941	Vanadium	V	23	50.9415
Lutetium	Lu	71	174.967	Xenon	Xe	54	131.29
Magnesium	Mg	12	24.305	Ytterbium	Yb	70	173.04
Manganese	Mn	25	54.9380	Yttrium	Y	39	88.9059
Mendelevium	Md	101	258.0	Zinc	Zn	30	65.39
Mercury (hygragyrum)	Hg	80	200.59	Zirconium	Zr	40	91.224

[a] Symbols based on IUPAC systematic names.

New notation
Previous IUPAC form
CAS version

1 Group IA	2 IIA	3 IIIA IIIB	4 IVB IVB	5 VA VB	6 VIA VIB	7 VIIA VIIB	8	9 VIIIA VIII	10	11 IB IB	12 IIB IIB	13 IIIB IIIA	14 IVB IVA	15 VB VA	16 VIB VIA	17 VIIB VIIA	18 VIIIA
1 H																	2 He
3 Li	4 Be											5 B	6 C	7 N	8 O	9 F	10 Ne
11 Na	12 Mg											13 Al	14 Si	15 P	16 S	17 Cl	18 Ar
19 K	20 Ca	21 Sc	22 Ti	23 V	24 Cr	25 Mn	26 Fe	27 Co	28 Ni	29 Cu	30 Zn	31 Ga	32 Ge	33 As	34 Se	35 Br	36 Kr
37 Rb	38 Sr	39 Y	40 Zr	41 Nb	42 Mo	43 Tc	44 Ru	45 Rh	46 Pd	47 Ag	48 Cd	49 In	50 Sn	51 Sb	52 Te	53 I	54 Xe
55 Cs	56 Ba	57 La ★	72 Hf	73 Ta	74 W	75 Re	76 Os	77 Ir	78 Pt	79 Au	80 Hg	81 Tl	82 Pb	83 Bi	84 Po	85 At	86 Rn
87 Fr	88 Ra	89 Ac ▲	104 Unq[a]	105 Unp[a]	106 Unh[a]	107 Uns[a]											

★ Lanthanide series

58 Ce	59 Pr	60 Nd	61 Pm	62 Sm	63 Eu	64 Gd	65 Tb	66 Dy	67 Ho	68 Er	69 Tm	70 Yb	71 Lu

▲ Actinide series

90 Th	91 Pa	92 U	93 Np	94 Pu	95 Am	96 Cm	97 Bk	98 Cf	99 Es	100 Fm	101 Md	102 No	103 Lr

[a] Symbols based on IUPAC systematic names.